American Institute of Chemical Engineers

9th Topical Conference on Refinery Processing 2006

Held at the 2006 AIChE Spring National Meeting

April 23-27, 2006
Orlando, Florida, USA

Printed from e-media with permission by:

Curran Associates, Inc.
57 Morehouse Lane
Red Hook, NY 12571
www.proceedings.com

ISBN: 978-1-60423-528-9

Some format issues inherent in the e-media version may also appear in this print version.

Welcome to the 9th Topical Conference on Refinery Processing.

This Topical has over 90 papers scheduled for presentation. The Conference runs from Monday morning through Wednesday afternoon. The sessions on Control, Hydroprocessing and Opportunity Crudes are especially well represented accounting for over 60% of the papers.

The level of business in refining has provided a great demand for skilled engineers in licensing, design, construction and operations. This conference will provide us with an opportunity to learn from many of the industry's most knowledgeable engineers.

Tutorials on desalting, sulfur technology and analysis and characterization are scheduled for Monday morning. Sessions on Desalting, FCC, Fouling Mitigation, Energy Conservation and Aromatics are also planned.

We would like to thank the session chairs, the speakers and paper co-authors for their efforts. In addition the staff of AIChE, Dr. Peter Pujado and Dr. Gavin Towler have been very supportive of our programming efforts.

If you have suggestions for improvements or are interested in getting involved with future programming in this area, please let me know at Dennis.Obrien@Jacobs.com.

Thank you,

Dennis O'Brien
Jacobs Consulting

Lori McDowell
Praxair

Kirtan Trevidi
ExxonMobil

9th Topical Conference on Refinery Processing

Session 19 - Analysis/Characterization Tutorial

This session provides an introduction to hydrocarbon mixture analysis and characterization.

CoChair: Isabelle Henaut

Chair: James Speight

1 Interactions of Synthetic Additives with Petroleum Aggregates Probed by Small-Angle Neutron Scattering
Keith L. Gawrys, Paul M. Lindemuth

2 Near-Infrared Modeling of Conjugated Diolefins in Selective Hydrogenation Units
Clementina Lopez-Garcia, Hélène Biguerd, Nathalie Marchal-George, François Wahl

12 Measurements of the Association and Stability of Asphaltenes - a Multidiciplinary Approach
Simon Ivar Andersen

Session 20 - Bottom of the Barrel Processing

This session will include a Keynote speech for the first paper.

CoChair: Jinwen Chen

Chair: Syamal K. Poddar

13 Changes in Asphaltene Structure and Stability during Hydrotreating
Simon Ivar Andersen, Diep Duong, Jesper Bartholdy

14 Coke Quenching in the Delayed Coking Process
Keith Wisecarver, Michael Volk

15 Stability of Water-in-Crude Oil Emulsions in the Burgan Oilfield
Adel Elsharkawy, Harvey Yarranton, Taher Al Sahhaf, Mohamed Fahim

Session 21 - Desalting Tutorial

Desalter operation has a direct impact on crude column overhead corrosion control, metals content in conversion unit feed, and refinery waste water treatment. Heavier, higher metals and sulfur content crude oils are now frequently processed in many refineries. Finding ways to improve desalter performance has become a high priority in most refineries. At the same time, the processing of opportunity crude oils has made desalter operation more challenging. This session will concentrate on papers describing aspects of basic desalter operation. Papers are invited which cover best practices in current desalter vessel design, equipment, and operating strategies.

CoChair: Andrew B. Woodside

Chair: Larry Kremer

16 Developing Crude Pre-Treating Programs to Improve Downstream Operations
Sandra N. Garcia-Swofford, Tony Potter

17 Solids Handling in Crude Oil
Alan Goliaszewski, David Engel, Cato McDaniel, Harold Eggert

18 Controlling Quality Variations in the Feed to Desalters
Larry Kremer

Session 55 - Advances in Desalting

New Developments in crude treatment

CoChair: Andrew B. Woodside

Chair: Larry Kremer

25 Parallel Comparison of Two Different Desalting Technologies—a Case Study
Shaya Movafaghian, Tom Collins, James Chen

37 Bimodal Modulation for Enhanced Desalting
Gary W. Sams, Kenneth W. Warren

44 New Control Technology for Desalter Optimisation
Paul Hewitt, **Scott Vidrine**

45 Desalter Interface Control and Diagnostics Utilizing Sonar Transducers
R. Paul Clewis

46 Removal of Calcium and Other Metal Species Form Crude Oil in the Desalting Process, Part 2
Larry Kremer, Joe Nguyen, Jerry Weers

Session 56 - Advances in FCC

This session discusses recent trends in FCC operations, catalysts, reliability and any other area of interest. Papers are desired from FCC Operators, FCC designers, FCC catalyst vendors and anyone doing active research on FCC improvements.

CoChair: Marty Poole

Chair: Lori T. McDowell

52 Heat of Cracking for Naphtha in Risers of Fcc Units
Carlos Alberto Dantas Moura, Andrea de Rezende Pinho, Jose Mozart Fusco

60 How Fccu Trickle Valves Affect Catalyst Losses
Donald F. Shaw, Richard E. Walter

Session 57 - Aromatics

The session on Aromatics will focus on new developments in production of aromatics (BTX) and aromatic derivatives, such as, but not limited to, ethylbenzene, cumene, linear alkylbenzene(LAB)styrene, phenol, bis-phenol, teraphthalic acid, and polyethyleneteraphthalate. Papers covering best practices in operation, safety, maintenance,controls and optimization are also invited.

Cochairs: B. Erik Henry
Robert J. Schmidt

Chair: Bipin V. Vora

Session 89 - Advances in Hydroprocessing I

Environmental regulations mandating the production of ultra low sulfur diesel (ULSD, < 10 wppm S) and ultra low sulfur gasoline (ULSG, < 30 wppm S and ultimately < 10 wppm S) have challenged the refining industry to develop and implement new catalyst and process technology. Papers covering advances in catalyst and process development in hydroprocessing technology designed to help the refiner meet these challenging specifications are solicited for this symposium. Hydroprocessing applications in distillate hydrotreating, FCC feed pre-treating, desulfurization of FCC gasoline are welcome as well as novel technologies that address the production of clean fuels. Papers describing commercial experience in the application of new hydroprocessing catalysts and processes are also encouraged.

CoChair: Stuart Shih

Chair: Peter Kokoyef

136 Impact of Future Fuel and Crude Demand on Refinery Hydprocessing
William Keesom

137 Limiting Factors in Deep Hydrodesulfurization of Diesel
Teh C. Ho

138 Effect of the Oxide Precursors on the Sulfurability of Hydrotreating Como(P) Catalysts
Anne D. Gandubert, Christelle Legens, Denis Guillaume, Christophe Pichon, Edmond Payen

149 Study about Formation Al2o3-Tio2 Nanofibers as Supports for Hydroprocessing Catalysts
Jose Antonio Muñoz Lopez, Esteban Lopez Salinas, Jose Antonio Toledo Antonio, Carlos Angeles Chavez, Jose Escobar Aguilar

150 Unsupported Sulphide and Nitride Catalysts for Hydrotreating: New Preparation Approaches
Pavel Afanasiev

158 Catalyst Selection for Ultra Low Sulfur Diesel Operations: Activity and H_2 Consumption Considerations
Lawrence S. Kraus, Katsuhisa Fujita, Yuji Noguchi

Session 90 - Advances in light hydrocarbon processing

This session is for papers on catalytic reforming improvements in process flow, operations and catalysts.

CoChair: Russ Anderson

Chair: Ken Peters

178 Comparision of Existing and Emerging Alkylation Technology
Cal Hodge

179 The AlkyClean® Process - Demonstrated New Standard for Alkylation Technology
Vincent J. D'Amico, John C. Gieseman, Emiel H. Van Broekhoven

189 Platinum Level Effects in C5/C6 Isomerization
Richard Rosin, Mark Lapinski, James Vassilakis, Veronica Godfrey

Session 100 - Petrochemicals from Heavy Oils

Papers in this session will highlight the use of heavy feedstocks to make petrochemicals

CoChair: Stuart Shih

Chair: Michael C. Oballa

Session 103 - Upgrading and Processing of Opportunity Crudes I

"Opportunity" crudes are generally characterized by a variety of properties undesirable to a refiner, such as high total acid number (TAN), high sulfur, nitrogen and aromatics content, high viscosity, etc. Although these crudes can cause corrosion and fouling problems in a downstream facility, they offer attractive discounts in crude prices. Effective methods and technologies to upgrade and process opportunity crudes can resolve these problems and provide attractive margins to the refiner. This session is geared towards exploring the current R&D and technology status on upgrading and processing of opportunity crudes. Papers are solicited on this topic in the areas of research studies to understand the fundamental science, current industrial strategies and technologies, and the development of new processes and technologies.

CoChair: Priya Rangarajan

Chair: Jinwen Chen

Session 130 - Advances in Hydroprocessing II

2nd session of same subject

CoChair: Stuart Shih

Chair: Peter Kokayeff

Session 133 - Energy Conservation

Energy conservation is becoming increasingly important in the refining and petrochemical industries. Economic pressures and rising fuel costs encourage more efficient use of energy within these industries so that operating costs can be reduced. Regulatory pressures are also forcing these plants to become more energy efficient so that CO2 emissions can be reduced. This session will address new developments related to energy conservation in refineries and petrochemical plants, which may include topics such as new equipment and processes, cogeneration, integration of refineries and power plants, and design techniques and methodologies that identify energy saving opportunities. Papers are encouraged from refiners, petrochemical producers, vendors, and academics, as well as anyone else with an interest in this area.

CoChair: Kirtan K. Trivedi

Chair: Michael A. Schultz

Session 145 - Upgrading and Processing of Opportunity Crudes II

2nd session of same subject

CoChair: Priya Rangarajan

Chair: Jinwen Chen

Session 175 - Control and Optimization in Refining II

2nd session of same subject

CoChair: Tim Olsen

Session 177 - Advances in Hydroprocessing III

3rd session of same subject

CoChair: Stuart Shih

Chair: Peter Kokayeff

419 Meeting the Challenge of Delayed Coker Naphtha Hydroprocessing
Terry A. Reid

420 Developing Knowledge from Spent Hydroprocessing Catalyst Chemical Analysis
Terry A. Reid

421 Recombination: A Complicating Issue in FCC Naphtha Desulfurization
Laura E. (Jones) Leonard, Peter Kokayeff

422 A Novel Oxidative Desulfurization (Oxyds) Process for Diesel and Vgo
Tzong-Bin Lin, Hsun-Yi Huang, **Jyh-Haur Hwang**, Hung-Chung Shen, Karl T. Chuang

432 Mathematical Modeling of a Trickle Bed Bio-Desulfurizer of Hydro-Treated Diesel with Recycle for the Production of Ulsd (Ultra-Low Sulfur Diesel)
Manasi Mukhopadhyay, **Prof. Ranjana Chowdhury**, Prof. Pinaki Bhattacharya

Session 188 - Upgrading and Processing of Opportunity Crudes III

3rd session of same subject

CoChair: Priya Rangarajan

Chair: Jinwen Chen

442 Opportunity Crudes - Addressing the Challenges of Mechanical Reliability
Jerold I. Danis

452 Crude Oil Management: Reduce Operating Problems While Processing Opportunity Crudes
Larry Kremer

458 Metallurgy for Opportunity Crudes
Tim Ruggles

463 Impact of H2s Scavengers and Other Tramp Amines on Refinery Operations and Equipment Integrity
Lawrence R. White, Craig Winslow, James Rue

476 A Comprehensive Approach to Assessing Opportunity Crudes
Doug Milton, James Feather, Kenneth Bagnoli, James McLaughlin

477 Refinery Optimum Crude Blending and Operation
Kiyomi Meason

Session 215 - Control and Optimization in Refining I

This session offers an opportunity to hear about process control success stories within the refining and petrochemical industry. The preferred subject matter includes innovative solutions to process control issues along with the benefits of optimization programs. If you have a process control application that would be interesting for others to learn about, please submit an abstract for approval.

CoChair: James Harris

Chair: Tim Olsen

487 Apriori Alarm Management during Project Feed Stage Using the Results from the Hazop
E. Michael Stafford, **Harry West**

488 User Approval of Safety Instrumented System Devices
Angela Summers

496 Advances in Abnormal Situation Prevention in Refineries and Petrochemical Plants
Ahmad A. Hamad, **Ravi Kant**

507 Strategic Proof Testing
Dr. William M. Goble

512 Fired Heater Safeguarding Survey
Edward M. Marszal

Session 216 - Control and Optimization in Refining III

3rd session of same subject

CoChair: Tim Olsen

526 A Self-Tuning Regulator for Fluid Catalytic Cracking Units
Felix Severino Farias Junior, Rubens Maciel Filho

527 Robust Dynamic Principal Component Analysis for Process Performance Monitoring
Daniel Castro-Rodriguez, Gary Montague, Elaine Martin

543 Wireless Instrumentation Enables New Best Practices in Monitoring and Automation
Clifford Lewis

544 Using Industrial Wireless Sensors to Monitor Safety Showers and Eye-Wash Stations
Doug Eberhart

545 Advanced Control Strategy to a Fermentation Process to Obtain Ethanol
E. R. Duarte, Laércio Ender, **R. Maciel Filho**

Session 220 - Fouling Mitigation

Today fouling mitigation in refineries is particularly critical. Since there is a shortage of refinery capacity, there are strong incentives for each refinery to maximize the days on stream. In addition, keeping insulating foulants off heat exchanger surfaces greatly reduces energy consumption and carbon dioxide emissions. Therefore, this session on fouling mitigation will review the causes of refinery fouling and case studies of fouling mitigation. Causes will include asphaltene fouling, corrosion induced fouling, and polymerization of conjugated olefins.

Mitigation methods include additives, oil compatibility prediction, heat exchanger design, heat exchanger inserts, and optimum cleaning scheduling.

CoChair: Kirtan K. Trivedi

Chair: Irv Wiehe

19a Interactions of Synthetic Additives with Petroleum Aggregates Probed by Small-Angle Neutron Scattering

Keith L. Gawrys and Paul M. Lindemuth. Downstream Research, Nalco Company, Energy Services Division, 7705 Highway 90-A, Sugar Land, TX 77478
The mechanisms, interactions, and association behaviors of synthetic additives necessary for crude oil production, refining, and flow assurance are poorly understood. For example, refinery desalter operations involve the intimate mixing of wash water and crude oil to remove water-soluble contaminants and oil-insoluble particulates. The desalting process involves the formation of water-in-crude oil emulsions that are treated with synthetic demulsifying agents to promote water droplet coalescence. Furthermore, synthetic dispersants are used to mitigate asphaltene deposition in various refinery process units and flow assurance applications. The mechanisms of interaction between these chemicals and petroleum asphaltenes are not well characterized. Small-angle neutron scattering (SANS) is a useful tool for studying the aggregation behavior in complex petroleum fluids, whereas other characterization methods employing visible light are often precluded due to the high refractive index and visual opacity of petroleum mixtures. SANS experiments were performed on mixtures of raw crude oil with various amounts of perdeuterated heptane (0, 0.25, 0.5, and 0.75 w/w) to induce asphaltene flocculation. SANS scattering curves for untreated crude oils at each solvent condition were compared to those in which the crude was treated with 1000 ppm of dispersant and/or demulsifier. Two heavy crude oil blends (i.e., Crude A and Crude B) were initially studied. Guinier analyses indicated that neither of the synthetic additives affected the aggregation state of asphaltenes in Crude A. Dispersant addition to Crude B appeared to decrease the nano-particle volume fraction (i.e., with respect to the untreated crude) at some solvent strengths, but it did not modify the aggregate size. This observation seems to contradict the widely accepted belief that dispersants work by disrupting micro-particle flocs, thus increasing the concentration of nano-particles. The addition of demulsifier disrupted micro-particle flocs, as manifested by a decrease in low Q scattering intensity. It also modified the size of nano-particle aggregates with respect to the untreated crude. Observations concerning the role of chemical additives on asphaltene aggregation behavior were directly related to the chemical composition of the crude blends, SARA fractionation, and abilities to form stable water-in-oil emulsions.

NEAR-INFRARED MODELING OF CONJUGATED DIOLEFINS IN SELECTIVE HYDROGENATION UNITS

Clementina López-García, Hélène Biguerd, Nathalie Marchal-George, François Wahl
Institut Français du Pétrole, IFP Lyon, Vernaison, France

Abstract

Fluid Catalytic Cracking (FCC) gasolines contain unsaturated aliphatic compounds such as olefins and diolefins. Olefins are valuable compounds since they have high octane numbers. However, conjugated diolefins are undesired compounds because they can polymerize easily, provoking plugging problems and deteriorating gasoline stability. To remove these diolefinic compounds, FCC gasolines are treated in Selective Hydrogenation Units (SHU). The main objective of this process is to selectively hydrogenate the conjugated diolefins while minimizing the hydrogenation of olefins. The conjugated diolefins are usually measured via indirect methods such as the Maleic Anhydride Value (MAV) method. This method is based on the Diels-Alder reaction between the conjugated diolefins and maleic anhydride. For daily monitoring, however, the MAV analysis is impractical since it is very time consuming (over 5 hours per sample) and not suitable for automation. For these reasons, Near InfraRed (NIR) spectroscopy was used to model the MAV content in SHU effluents. The NIR technique has the advantages of being fast and repeatable, but is generally not adapted for trace analysis. Since the diolefins contents in the SHU products are very low (<1 wt%), a specific spectra preprocessing methodology was developed to model the differences of MAV (Delta_MAV) in SHU products. The resulting Partial Least Squares (PLS) model does not depend on the feed hydrocarbon matrix. Excellent results were obtained since 98% of the calibration samples are predicted in the confidence interval of the chemical MAV method. A very good agreement between experimental and predicted MAV values was obtained for the external validation data. The NIR model was also tested for SHU pilot plant monitoring; the model can correctly predict the MAV variations when the process operating conditions are modified. Hence, the NIR prediction of the conjugated diolefins content in selectively hydrogenated FCC gasolines via the delta MAV is a very attractive tool to determine the performances of the selective hydrogenation units in real time.

Introduction

The rising worldwide demand for transportation fuels calls for additional conversion of heavy petroleum cuts into valuable products such as gasolines or diesels. With over 350 units in operation, Fluid Catalytic Cracking (FCC) is currently the dominant refinery conversion process that allows the upgrading of heavy cuts to produce gasolines. However, the FCC gasoline quality does not meet the current commercial specifications, particularly the total sulfur content. Therefore, these cracked naphthas must undergo further refining prior to their integration into the gasoline pool. FCC gasolines also contain unsaturated hydrocarbons such as olefins and conjugated diolefins. The conjugated diolefins (or dienes) are particularly detrimental since they easily polymerize, inducing plugging problems in other refining units and promoting gasoline instability.

The objective of the Selective Hydrogenation Units (SHU) is to remove the conjugated diolefins from FCC gasolines while minimizing the hydrogenation of olefins, which have high octane numbers. The total conjugated diolefins content in these gasolines is commonly measured via the Diels-Alder reaction between the conjugated diolefins and maleic anhydride. These methods are known as the Maleic Anhydride Value (MAV) or the Diene Value[1] (DV). The main disadvantage of both procedures is that they are extremely time consuming (over 5 hours per sample) and not suitable for automation. This makes daily process monitoring tedious and difficult.

Several techniques have been reported in literature to determine the total diolefins content in gasolines[2,3,4,5]. Differential pulsed polarography [2,3] was used to reduce the conjugated diolefins in presence of a proton donor. This method is fast (15 min) and was applied to pyrolysis gasolines and their SHU products[2]. However, some hindered molecules like 2,5-dimethyl-2,4 hexadiene may not be measured[3]. An HPLC-UV method[4] was also tested but the diolefins content was systematically underestimated. Albuquerque et al.[5] also worked on the quantification of conjugated dienes in gasolines by supercritical fluid chromatography (SFC) with UV detection. This method is selective to conjugated diolefins but the SFC-UV response is gasoline feed dependent.

Near InfraRed Spectroscopy (NIR) is a wide-spread industrially used technology. The main NIR advantages are: rapidity (~10 min), excellent repeatability, possibility of multi-properties prediction with one spectrum and no sample preparation. The most important NIR refining application is on-line analysis for gasoline and diesel blending operations to minimize the fuel cost while meeting the current legal specifications. The main NIR drawbacks are that the calibration requires a complex procedure that must be carried out on a diversified database. Furthermore, NIR is not adapted for trace analysis.

The industrial MAV specification in SHU products is ≤ 2 mg/g, which corresponds to about 0.3 wt% of conjugated diolefins. The NIR modeling of the diolefins content in steam cracking gasolines has already been reported [6,7]. However, these gasolines have very high MAV contents (> 100 mg/g). No applications to FCC gasolines and their SHU effluents were found.

This work presents the NIR modeling of the MAV content in SHU products. A specific spectrum treatment was elaborated to model the low MAV contents in these samples. Subsequently, the model was successfully applied to the daily process monitoring.

Mathematical Tools for NIR modeling

Principal component analysis [8] (PCA) was used to analyze the variability of the spectra database. PCA transforms a high number of correlated variables into a fewer number of new uncorrelated variables called principal components (factors). The first principal component accounts for as much of the variability in the data as possible and each succeeding orthogonal component successively takes into account the residual variance of the data. PCA is applied to problems having a high number of variables compared to the number of samples and is commonly used for NIR applications.

The NIR modeling was carried out with a Partial Least Squares regression (PLS) algorithm [9]. The PLS multivariate linear regression technique correlates the NIR absorbances from the spectra calibration database with the property to be predicted using, as PCA, new uncorrelated variables. The PLS algorithm maximizes the covariance between the X-scores (X: spectral absorbances of the samples) and the corresponding Y-scores (Y: vector of the predicted property of all the calibration samples). The number of PLS factors was chosen according to a leave-one-out cross validation method. This approach consists in removing a calibration sample from the calibration set, calculating the corresponding PLS model based on the remaining samples in the calibration database, predicting the left out sample with the resulting model, and evaluating its prediction error as the difference between predicted ($\hat{y}_i$) and actual values (y_i). This operation is repeated for every sample in the calibration set and an average error, called Root Mean Square Error (RMSE), is calculated according to Equation 1 where n is the total number of samples. The number of PLS factors is chosen through the minimum RMSE calculated by cross validation.

$$RMSE = \sqrt{\frac{1}{n}\sum_{i=1}^{n}(\hat{y}_i - y_i)^2} \qquad \text{Equation 1}$$

Experimental Section

Analyses

Maleic Anhydride Value analyses (MAV): The MAV analyses were carried out according to the following procedure. A weighed sample of gasoline is added to 50 mL of a saturated maleic anhydride solution using toluene as solvent. The mixture is heated to boiling point and refluxed one hour. The non-reacted maleic anhydride is hydrolyzed to maleic acid, extracted from the reaction mixture and then titrated with a strong base (NaOH 0.5 M). The MAV is reported as the number of milligrams of maleic anhydride that react with 1 g of sample. The quantification limit of the method is 0.5 mg/g.

Other global analyses: The density at 15°C was determined according to the ASTM D4052 standard. The distribution of hydrocarbons by chemical family was carried out with an in-house capillary gas chromatography (GC) method derived from the ASTM D6733-01 method.

NIR Spectroscopic Instrument and spectra acquisitions: The spectrometer used for spectra acquisitions is a Fourier Transform BOMEM MB160 that operates in transmission mode. The spectrometer detector is deuterated tri glycide sulfide (DTGS). A 2 mm (±0.02) QX quality cell was used for all the analyses. The cell temperature was set to 27.5 ± 0.5°C controlled by Peltier effect. Each spectrum was obtained with 100 scans per sample in the 4000 – 12000 cm^{-1} region at a 4 cm^{-1} resolution. The software used for spectra acquisition is GRAMS32 from Bomem. The reference spectrum was recorded every hour with the empty cell filled with dried nitrogen.

Global characteristics of the samples

All the gasolines analyzed in this work are SHU products of several FCC gasolines. All the samples were obtained from IFP pilot plants. The database is constituted of 87 samples: 58 were selected as calibration database and 29 for external calibration purposes. The calibration database is only used for estimation of the NIR model parameters. The calibration samples come from 5 different gasoline feeds referred to as Feed 1 to Feed 5. The 29 external validation samples are only used to evaluate the quality of the NIR model predictions. These samples were obtained after SHU treatment of 3 different feeds: Feed 5, 6 and 7. Although one of the feeds (Feed 5) was the same as for some of the calibration samples, the SHU effluents were obtained at different operating conditions.

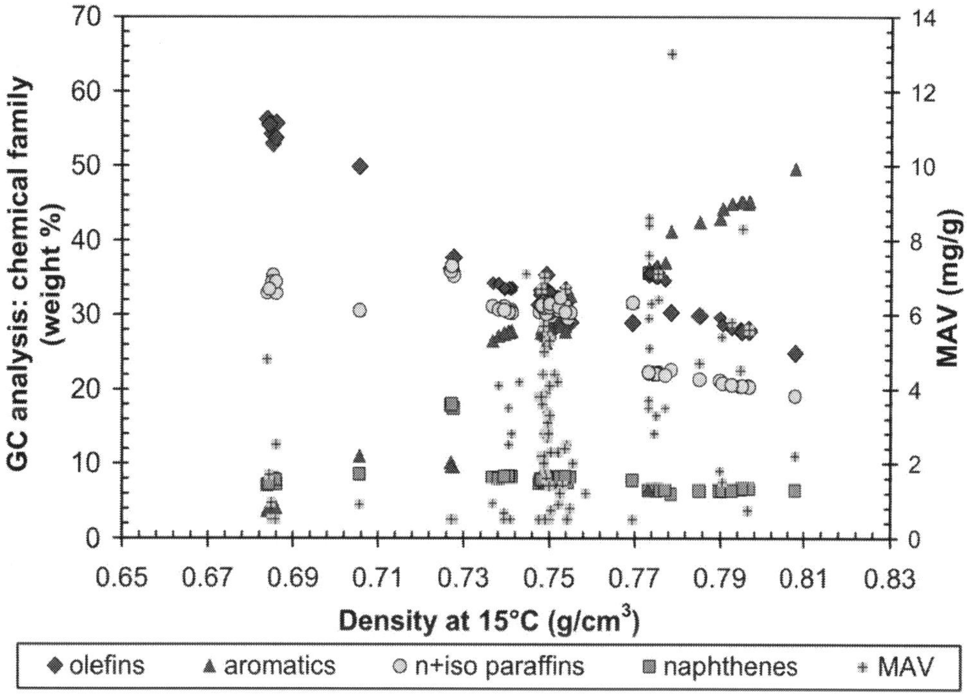

Figure 1: Global characteristics of the samples density at 15°C vs. detailed hydrocarbon analysis and MAV

Figure 1 represents the distribution of the major hydrocarbon chemical families (olefins, aromatics, saturates) and MAV as a function of the density at 15°C. Only the points for which all the properties were available are shown. As expected, the aromatics content is higher for heavier samples; the olefins content is higher for gasolines of low density. Most of the samples are included in the 0.74-0.76 g/cm^3 density range. Concerning the MAV values, it should be noted that the conjugated diolefins content is not correlated to the major hydrocarbon composition. The MAV values in the calibration and external validation database vary from 0.5 to 12.5 mg/g and are distributed as indicated in Table 1. These MAV values represent approximately 0.06 to 2 weight percent (wt%) of conjugated diolefins. Thirty-three percent of the calibration samples are inside the 0.5-2 mg/g range which is in agreement with the industrial specifications for SHU effluents (MAV < 2 mg/g).

Table 1: MAV distribution of the database

MAV range (mg/g)	Number of samples	
	Calibration	External Validation
0.5 - 1	1	0
1 - 3	18	2
3 - 5	11	6
5 - 7	6	8
7 - 9	5	8
9 - 11	13	5
11 - 13	4	0
Total	58	29

NIR spectra

Typical NIR spectra are shown in Figure 2. An NIR spectrum is commonly depicted as the chemical fingerprint of the sample. All the spectra of the calibration database superimpose well since the samples have the same chemical families (paraffins, olefins, aromatics, naphthenes) but at different concentrations. It must be pointed out that the conjugated diolefins are chemically a sub-family of olefins. Hence, the NIR absorption[12] of the olefins will hamper the signal of the conjugated diolefins and make the prediction of their concentration difficult. Moreover, olefins are present in the samples in a much higher content (20-60 wt%) than diolefins (< 2 wt%).

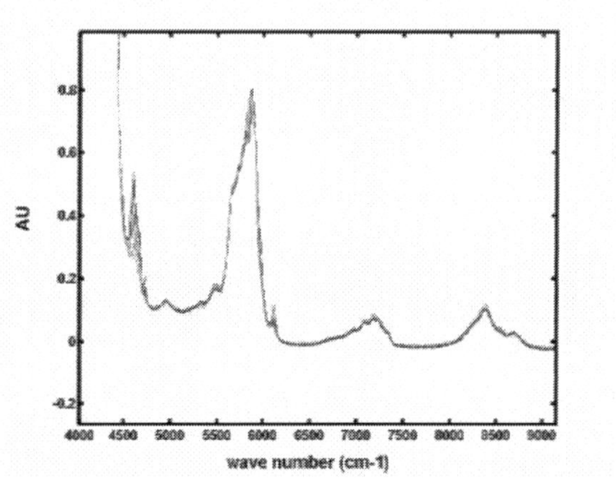

Figure 2: NIR spectra of calibration database

NIR multivariate data analysis

Classic spectrum pretreatment

Since the raw spectra are sometimes baseline shifted, a baseline correction and a mean centering were applied to all spectra. This procedure helps to extract real variations between the samples before PCA and PLS in the NIR region used for multivariate analysis.

Principal component analysis

The principal component analysis (PCA) of the calibration data was carried out in the 4450 to 9000 cm^{-1} NIR region that contains most of the spectral information of SHU products. The 1st and 2nd PCA scores are presented in Figure 3 and account for 99% of the database variance. The SHU effluents obtained from the feeds that have a similar content of aromatics or olefins are grouped together (ex. feeds 2, 3 and 4). Hence, the different effluents are gathered according to their global chemical composition. Since the conjugated diolefins content is extremely low (< 2 wt%), their spectral information is overshadowed by the spectral absorbances of the principal hydrocarbons, and more particularly of the olefins. Therefore, it will be difficult to model the MAV directly from the pretreated raw spectra.

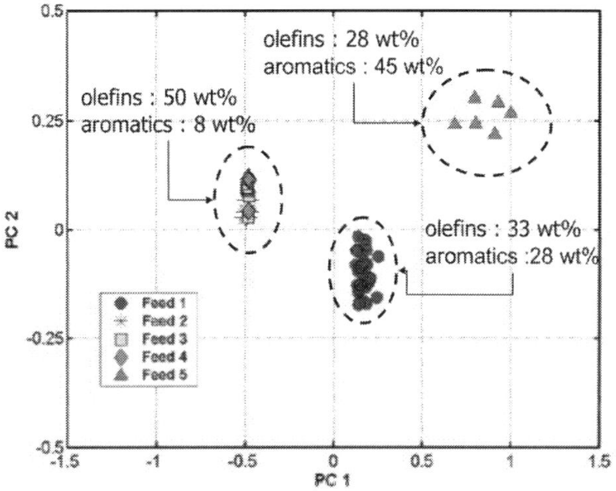

Figure 3: Calibration database PCA scores. SHU effluents obtained from Feeds 1 to 5.

Delta_MAV Modeling: In order to extract the spectral information of the conjugated diolefins from the complete spectrum and to correlate it to the MAV content, this work tackles the NIR modeling in a particular way, which consists in modeling the MAV differences. As indicated before, the objective of the SHU process is to selectively hydrogenate the conjugated diolefins while minimizing the olefins hydrogenation (< 5%). Since the other chemical families (paraffins, naphthenes and aromatics) do not react in the SHU, the hydrocarbon matrix is almost identical between the reactor inlet and outlet (feed and product). Hence, their spectra are also very similar. For a set of SHU products derived from the same feed, one of the sample is chosen as reference. Subtracting the "reference" spectrum from the spectra of the other products should eliminate the absorption bands of the hydrocarbon matrix and will therefore mainly represent the absorption differences that are produced by the differences in the amount of conjugated diolefins of these samples. Hence, the spectra differences were used to model the MAV variations, which are referred to as Delta_MAV and defined in Equation 2:

$$Delta\_MAV_i = MAV_{reference} - MAV_i \qquad \text{Equation 2}$$

Figure 4 illustrates the PCA scores of the spectra differences. Five sets of SHU products are represented. For each set, a reference sample was chosen. It is observed that the different sets of samples are no longer lumped according to their main hydrocarbon composition as was the case in Figure 3. Therefore, modeling the Delta_MAV instead of the MAV has two major combined advantages: it strongly decreases the "feed effect" and boosts the conjugated diolefins signal.

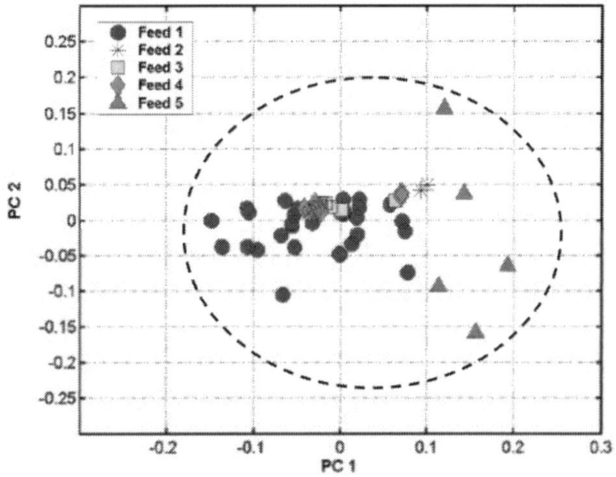

Figure 4: Calibration database PCA scores of spectra differences:SHU effluents obtained from Feeds 1 to 5.

MAV modeling results

Calibration and external validation results

A Partial Least Squares model was developed to predict the Delta_MAV values using the spectra differences. The root mean square error of the calibration calculated by cross validation or RMSECV (see Equation 1) did not decrease significantly after 6 factors were kept in the PLS model. Since the NIR model predicts the Delta_MAV, for each sample "i", the MAV can be calculated as the difference between the actual MAV from the reference sample ($MAV_{reference}$) and the Delta_MAV_i calculated by NIR (Equation 2). The actual and predicted MAV obtained on the calibration and external validation sets are illustrated in the parity plot of Figure 5. The dotted lines delimit the confidence interval of intra-laboratory reproducibility (CI'_R) of the chemical MAV method. It can be noticed that all the calibration samples are very well predicted: 98% of the effluents are predicted within the CI'_R. There is no bias with respect to the MAV range or the feed type. The external validation points are also well predicted: 22 from 26 points are predicted within the CI'_R. The other four samples are very close to the confidence interval limits.

The NIR modeling results for the calibration and external validation sets are summarized in Table 2. The experimental standard deviation σ_{exp} was calculated as a function of the measured MAV for each dataset since the reproducibility is a function of the MAV content. The root mean square error (RMSE) is commonly compared to the experimental standard deviation. The cross validation model uncertainty on the calibration samples (RMSECV) is the same as the experimental standard deviation. For the external validation set, the prediction error (RMSEP) has the same order of magnitude as the experimental uncertainty. The average relative errors of the calibration and validation datasets are also very close, 10% and 11% respectively. These results confirm that the model can accurately predict the very low MAV contents in SHU products.

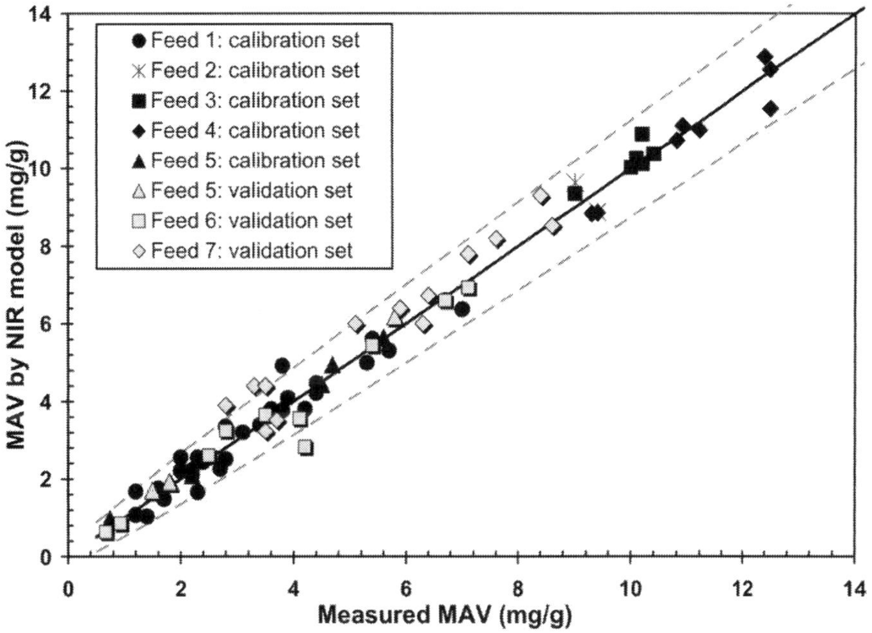

Figure 5: Delta_MAV NIR Model. Parity plot of calibration and external validation sets.

Table 2 : NIR modeling results of calibration and external validation datasets

	Number of samples	MAV range (mg/g)	RMSE (mg/g)	σ_{exp} (mg/g)	average relative error
Calibration	53	0.7 – 12.5	0.49*	0.49	10%
External validation	26	0.7 – 8.6	0.59**	0.45	11%

* RMSECV: Root mean square error of cross validation
** RMSEP: Root mean square error of prediction

Application of the NIR model to continuous monitoring of SHU units

The NIR model was applied for the daily monitoring of SHU pilot plants at IFP. The real time MAV monitoring is critical for the industrial SHU operation because it allows to determine whether the product satisfies its specifications and to avoid plugging in the downstream refining units. NIR is perfectly adapted to monitoring since the NIR results are available in about 10 minutes, compared to over 5 hours for the chemical MAV measurement.

Since our NIR model is applied to R&D process optimization, it must be able to cover the treatment of a large variety of industrial FCC feeds under a wide range of operating conditions. The NIR model sensitivity was tested for daily SHU monitoring. The objective is to verify whether the model is able to predict in real time the changes in MAV when the process operating conditions are modified.

Figure 6 illustrates an example of SHU monitoring of a pilot plant test carried out with FCC Feed 6. The x-axis represents the monitoring day of the test. The left y-axis is the MAV content and the right y-axis the reactor temperature. The variation of the SHU operating conditions other than temperature (pressure or hydrogen flowrate) is indicated by the black arrows on the temperature series. During the pilot plant monitoring, the MAV predicted with the NIR model is determined daily (1 to 2 measures) and the MAV by the chemical reference method (chemical MAV) is carried out once the unit reaches steady state for a given set of operating conditions. For this reason, there are more NIR MAV data than chemical MAV data. As illustrated, the model is able to predict precisely and in real time the MAV variations. This continuous monitoring is useful to assess the effect of the operating conditions on the MAV content or to determine if the unit has reached the steady state. For example, on the 16th monitoring day, the reactor temperature was increased by 20°C. The diolefins hydrogenation was therefore favored and the predicted MAV decreased from 6.6 to 3.7 mg/g. This value is in agreement with the chemical MAV (3.5 mg/g). These results proved that the Delta_MAV NIR model can be used to accurately follow the performances of SHU units.

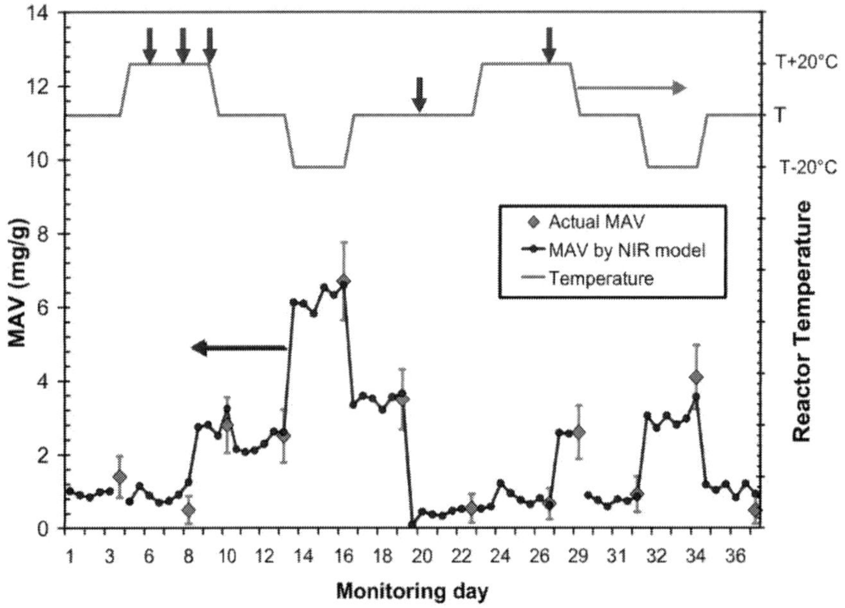

Figure 6: Example of SHU daily monitoring results (selective hydrogenation of Feed 6)

Conclusions

FCC gasolines contain conjugated diolefins that cause plugging and stability problems, and need to be further refined. The diolefinic compounds are converted in selective hydrogenation units (SHU) where the diolefins content is reduced to less than 0.3 wt%. The MAV method is used to indirectly quantify the conjugated diolefins content in the SHU feeds and products. This method is extremely time-consuming (over 5 hours per sample) and difficult to carry out for continuous SHU monitoring. For this reason, a NIR MAV model was developed in this work to replace the chemical MAV method. The main challenges of such a model are: the extremely low concentrations of conjugated diolefins, the presence of high contents of olefins (20-60 wt%) that interfere with the diolefins and the strong influence of the other major hydrocarbons (aromatics, saturates). To overcome these difficulties, a novel spectra treatment has been applied prior to multivariate calibration. This pretreatment consists in subtracting spectra of samples that have very similar hydrocarbon matrix. This treatment can be applied because in the SHU process the hydrocarbon matrix remains almost constant between the reactor inlet and outlet. The only changes in composition are mainly related to the variation in conjugated diolefins and a slight olefins hydrogenation. The property predicted by the NIR model is the MAV variation called Delta_MAV. The application of this spectrum pretreatment strongly diminishes the hydrocarbon matrix influence and enhances the conjugated diolefins NIR signal. Taking into account chemical MAV precision, our results show that the NIR model could accurately predict the MAV values for the calibration data and a large external validation set. This model was also used to continuously monitor the performances of SHU units. The MAV variations could be instantaneously determined when the SHU operating conditions were changed. The very good agreement between experimental and NIR predicted values was also verified. Hence, predicting the MAV in selectively hydrogenated FCC gasolines via this NIR model appears to be an attractive tool to determine the SHU performances in real time.

References

1. "Laboratory Test Methods for Petroleum and its Products", Universal Oil Products, Method N° 326-82, Universal Oil Products, Des Plaines, Illinois, USA.
2. Polák J., Janáĉek L., Jiři V. (1986) "Polarographic determination of conjugated dienes in hydrogenation products of pyrolysed gasoline", *Analyst*, 111, pp. 1207-1210.
3. Swarin S. J., Perry K. L. (1989) "Determination of conjugated dienes in gasoline by differential pulse polarographie", *Anal. Chem.*, 61, pp. 1502-1504.
4. Ghazvini V.M., Jäckle H.W., Steinhauer E., Determination of conjugated doilefins in gasolines by high performance liquid chromatography, *Analytik*, 1996, 112, 5, pp. 210-212.
5. Albuquerque F. C. (18-22/08/2002) "Determination of conjugated dienes in petroleum products by supercritical fluid chromatography and UV detection", *224th ACS National Meeting. Fuel Chemistry Division Preprints*, 47(1).
6. Descales B., Lambert D., Llinas J.R., Martens A., Osta S., Sanchez M., Bages S., (2000) United States Patent US006070128A.
7. Descales B., Lambert D., Llinas J.R., Martens A, Osta S., Sanchez M. (1988) United States Patent US005712797A.
8. Burns D. A., Ciurczak E. W. and Mark H. (1992) Data analysis: Multilinear regression and Principal Component Analysis in Handbook of Near Infra-red Analysis, Marcel Dekker, New York.
9. Aji S., Zanier-Szydlowski N., Faraj A. (2004) "Partial least square modeling for the control of refining processes on mid-distillates by near infrared spectroscopy", *Oil Gas Sci. Technol.*, 59, 3, pp. 303-321.
10. American Society for Testing and Materials. ASTM D4052-96(2002)e1 Standard Test Method for Density and Relative Density of Liquids by Digital Density Meter. Annual Book of ASTM Standards, 2005, 05.02.
11. American Society for Testing and Materials. ASTM D6733-01 Standard Test Method for Determination of Individual Components in Spark Ignition Engine Fuels by 50-Meter Capillary High Resolution Gas Chromatography. Annual Book of ASTM Standards, 2005, 05.04.
12. Yalvac E.D., Seasholtz M.B., Crouch S.R. (1997) "Evaluation of Fourier Transform Near-Infrared for the simultaneous analysis of light alkene mixtures", *Appl. Spectrosc.*, 51, 9, pp. 1303-1310.

19c Measurements of the Association and Stability of Asphaltenes - a Multidiciplinary Approach

Simon Ivar Andersen, HaldorTopsoe A/S, Ny Møllevej 55, Kgs. Lyngby, DK-2800, Denmark

The phase separation of asphaltenic constituents of petroleum during production, handling and refinery processing continues to be a challenge. This presentation gives an overview of the application of microcalorimetry, SANS, X-ray diffraction and other techniques combined with the chemical alteration of asphaltenes in order to better understand the stability mechanisms. It is known that association and aggregation are important mechanisms in the formation of asphaltenic sludges. Two types of overall mechanisms, the pi-pi aromatic stacking and the hydrogen bonding, are possible association steps. However, the application of methylation and silylation to asphaltenes, which removes the hydrogen binding sites, clearly indicates that these specific interaction sites may indeed participate in a positive way to prevent sludge formation. For native asphaltenes we have demonstrated that especially SH and NH are important interactions. Hence, during HDS and HDN processes, these stabilizing bonding sites are removed, which may lead to additional instability. The mechanism is currently interpreted as smaller asphaltenes (higher polarity, low MW) interacting with large asphaltenes (lower polarity, high MW) preventing these from associating with large molecules. The presentation discusses the pi-pi stacking and the type of possible ordering of such stacks by comparison with crystal formation in pure polyaromatic components. The presentation also gives an overview of the balance between the various forces involved.

20a Changes in Asphaltene Structure and Stability during Hydrotreating

Simon Ivar Andersen, Diep Duong, and Jesper Bartholdy. HaldorTopsoe Research Laboratories, Ny Møllevej 55, Lyngby, DK-2800, Denmark

This tutorial covers our experience within the area of asphaltene alteration during hydrotreating of residues to remove sulfur (HDS), nitrogen (HDN) and metals (HDM) either as a final upgrading, a coker pretreatment or as an FCC pretreatment process. Aspects of various reactor schemes will be described. The aim is to give an understanding of the relationship between structure, function, phase stability and formation of organic heavy sludge and coke that may impair other processes as well as storage facilities. It was previously shown that the sludge is intimately linked to the remaining asphaltenes in the process. Data in support of this will be reviewed, and the balance between hydrogenation and cracking documented. This balance will eventually be dominated by cracking as the process temperature is increased to meet HDS specifications. Cracking reactions will increase the molecular incompatibility between asphaltic material and the product maltene fraction, which will become more and more aliphatic. Stability tests have been carried out combined with thorough chemical analyses. It is well known that fractionation of asphaltenes may give a much better idea of the variations within the material. However, this method is used only infrequently. The present work includes both heptane-toluene extraction with increasing heptane content and extraction with n-methyl-pyrrolidone (NMP), and a correlation is made with the development of stability. NMP has been shown to generate a non-fluorescent insoluble fraction. The content of this change with process severity. The classical instability approach only takes hydrocarbon structure changes into account. It will be shown that molecular interactions through functional groups may be an important stabilizing mechanism inhibiting the asphaltene destabilization.

20b Coke Quenching in the Delayed Coking Process

Keith Wisecarver and Michael Volk. Department of Chemical Engineering, University of Tulsa, 600 S. College Ave., Tulsa, OK 74104
Proper quenching of the coke bed in the delayed coking process is vital to safe operation. Numerous refinery accidents and near accidents have resulted from "hot spots" in the coke bed due to improper quenching. We present here experimental results on steam stripping and water quenching in a delayed coking pilot unit system with a three inch diameter coke drum. Results show the dependence of coke cooling profiles on the coke density and water feed rates. A model is presented that predicts the coke bed temperature profile over time and thus the optimum quenching conditions. The effect of coke morphology (shot vs. sponge) and the quenching of beds with mixed/varying coke morphology are shown and discussed.

20c Stability of Water-in-Crude Oil Emulsions in the Burgan Oilfield

Adel Elsharkawy, Petroleum Engineering Department, Kuwait University, B.O. Box 5969, Safat 13060, Kuwait, Harvey Yarranton, University of Calgary, Calgary, T2N 1N4, Canada, Taher Al Sahhaf, Chemical Engineering, Kuwait University, B.O. Box 5969, Safat 13060, Kuwait, and Mohamed Fahim, Kuwait University, Kuwait City, Kuwait.
Asphaltenes and resins separated from emulsion samples collected from Burgan oil field were used with heptane-toluene mixtures as model oil to study the effect of oil aromaticity, resin content, and pH of the aqueous phase on the stability of water in model emulsions. It was confirmed that, as long as the asphaltenes are completely solubilized, increasing aromaticity leads to less stable emulsions. A consistent correlation between emulsion stability and relative resin mass content (R/(R+A)) was observed for all three of the field samples. There was a sharp decrease in stability when the R/(R+A) value exceeded 0.75. Emulsion stability was enhanced at high pH and possibly at very low pH (< 2).

21a Developing Crude Pre-Treating Programs to Improve Downstream Operations

Sandra N. Garcia-Swofford and Tony Potter. Downstream, Nalco Company, 7705 Hwy 90A, Sugar Land, TX 77478
Crude oil is amongst the largest expense in a refinery. Some refineries will run more than 100 different crudes a year and various blends a day. This adds a level of complexity in the operation and can make meeting refining goals a major challenge. Crude oil contains various contaminants, which include water, and salts that can cause problems with corrosion and fouling in downstream units. Removal of these contaminants is essential to maximize refining profitability. The first step towards doing this is crude pre-treatment. Evaluating the use of pre-treatment programs before desalting can help improve desalter operation and performance. With the growing trend to process heavier crude oils having the flexibility to pre-treat certain crude blends can help avoid downstream upsets. For example, some crudes that are received via tanker ships can have problems with high water content that can cause water slugs downstream in the desalters. A proactive approach to receiving these type of crudes can avoid such problems.

This paper will cover guidelines that should be evaluated when looking at developing an optimum crude handling program. It will focus on the basic operational, mechanical and chemical elements that should be evaluated or implemented to improve crude quality before the desalter. Aspects such as testing and monitoring will also be reviewed. Examples of existing applications will be referenced.

21b Solids Handling in Crude Oil

Alan Goliaszewski, David Engel, Cato McDaniel, and Harold Eggert. GE Water & Process Technologies, 9669 Grogans Mill Road, The Woodlands, TX 77380
Solid particulates contained in crude oil can have a significant impact on refinery operation. Solids can contribute to fouling of atmospheric distillation pre-heat trains or furnaces. They can drive refinery products off spec, particularly the residual streams and coke product. Difficulty with desalter operation can result from solids stabilizing emulsions or accumulating in the bottom of the desalter. Thus, the refiner needs to effectively deal with crude oil solids.

Crude oil has been found to contain four major types of solids. They may be found individually or in combination. Characterization of these solids will be presented through chemical analysis, optical microscopy and scanning electron microscopy.

Strategies for effectively dealing with the general types of solids will be reviewed. Chemical treatments can be utilized in the desalter to assist in the delivery of the solids to the desired oil or water phase, or used to diminish the stabilization of the interfacial layer in a desalter. Mudwash strategies can be used to improve desalter performance.

Results from a refinery application will be presented that demonstrates the effective control of solids. Routine high rates of solids removal resulted in meeting the refiner's objectives for product quality and desalter operation

Controlling Quality Variations in the Feed to Desalters

By Larry Kremer
Baker Petrolite

The quality of crude oil charged to the desalter can vary day to day or even with in minutes to produce desalter upsets. Several tools and tests can be used to monitor crude quality and help in the development of mitigation strategies. One of the most useful mitigation strategies to control quality variations is crude oil pretreatment. Crude pretreatment resolves complex emulsions in the crude oil that could otherwise lead to desalter upsets. The pretreatment also conditions solids so that they may more easily be handled in the desalter.

The feed to the desalter typically contains three streams: crude oil (90-97%), wash water (3-10%), and slop oil (0-5%). In some cases refineries will occasionally feed higher levels of slop oil, and use higher levels of wash water. All three of these streams can vary in quality. Since crude oil is the most important component, it will be covered in greatest detail. However slop oil and wash water will also be discussed for completeness.

I. CRUDE OIL QUALITY VARIATIONS AND MITIGATION STRATEGIES

In many refineries the charge to the desalter is often a blend of various crude oils. As the component crude oils change and the blend ratios change, the feed quality can change causing variation in performance of the desalter. Even refineries running the same crude oil can experience quality variations that can lead to desalter upsets. There are many sources of this quality variation prior to receipt by the refinery.

Crude Oils Are Blends

Purchased crude oil is almost always a blend (Ref. 1) of oil produced by multiple wells and in some cases different formations. Producers will blend production from various sources and quality to obtain a more consistent average quality. In some cases diluents such as condensate are added to meet pipeline specifications for viscosity. In some cases refineries accept crude oil of unknown quality to be transferred into their crude storage tanks from collector trucks.

Crude Oil Degradation During Transportation

Crude oil can experience mixing with other shipments during transportation. For example, when crude oil is transferred into tankage, either at marine terminals or at intermediate storage on pipelines, it is mixed with the heel from prior shipments in the bottom of the tank or with trail back from prior shipments through a pipeline. Shippers try to minimize this but some mixing is inevitable.

Crude Oil Is Altered By Additives

Many materials are added to crude oil either to enhance production or to improve the properties of the crude oil during transportation. For example, hydrogen sulfide scavengers are added to many crude oils for safety reasons, to protect workers handling the crude oil. Some off shore platforms must add hydrate inhibitors, such as methanol, to prevent the wells from being shut in. There have been documented cases where the methanol slugs have reached the desalter causing desalter upsets. The methanol is mostly water soluble, so it ends up increasing the COD/BOD in the desalter effluent going to the WWTP and has caused environmental permit excursions.

Refinery Tank Farm Management

As crude oil is transferred into a storage tank it tends to stir up material (sludge) on the bottom of the tank. After the tank has been filled the BS&W in the incoming crude oil and any stirred up sludge begin to settle. When the tank is put on line the top of the tank tends to have 'good dry oil' and the bottom of the tank wetter oil. Even in cases where the operators can adequately drain the tank prior to charging to the crude unit, there will be a variation in BS&W going to the crude unit during the day. If not drained well, a water slug can be sent to the crude unit, with a possible desalter upset as a consequence.

Mitigation Strategies

Operational and mechanical improvements

Some refineries have installed swing arm or floating suction draws on their crude storage tanks. This enables the good dry oil from the top of the tank to be charged first to the crude unit. When the wet oil is reached in the tank, a fresh tank can then be switched on for the crude charge.

Most crude tanks have draws near the bottom of the tank. Where possible, it is best to drain as much settled water as possible from the tank before it is charged to the crude unit. There are many limitations to successfully draining water from the crude tank. In many refineries the potential for environmental releases to the atmosphere or to a proper water treatment facility has eliminated the possibility of draining water from the tanks. Where draining is possible it is often difficult for the operators to visually distinguish oil from black water. To overcome this difficulty refineries have used chemicals to produce a clean break (discussed in the next section) or in line water detectors (such as those supplied by Agar Corporation) to enable the operators to drain clean water and stop the process when oil is reached.

Even when the water has properly been drained from the tanks water slugs can still hit the crude unit. Sludge accumulates in most crude storage tanks. The sludge is generally unevenly distributed. Lakes of brine can accumulate in the mountains of sludge which can sometimes break and send slugs of water to the crude unit. There are many cases where the migrant sludge has plugged the transfer line to the crude unit. Chemical pretreatment programs have been successfully used to control the sludge in crude storage tanks.

Whether water is drained or not, it is generally good practice to homogenize the crude oil as much as possible by using mixers on the side of the tank, prior to transferring crude to the crude unit. There are many limitations to good mixing including broken mixers and mixers that are under designed for the gravity and viscosity of crude oil currently purchased by the refinery. Often the exit line is very near if not identical to the inlet to the tank. This allows for short circuiting and poor mixing between the incoming and outgoing crude oil. Inhomogeneity in the crude tank is a recurring problem for refinery operations that is difficult to fix.

Chemical treatment of crude oil transferred into storage solves many problems

Baker Petrolite has developed numerous successful crude tank treatment programs around the world over the last three decades. The type of chemistry used and application are dependent on the specific objectives of the refinery. While the benefits to oil movements varies, the general benefit to the crude unit is consistently good quality crude oil, resulting in few desalter upsets. Improved desalter performance includes reduced hydrocarbon contamination of the effluent water, reduced rag layer build up, and better dehydration.

Case 1. Baker Petrolite manufactures chemicals designed to dehydrate the crude oil in storage so that the oil charged to the crude unit is low in BS&W. This procedure requires that the refinery is capable of draining brine from the bottom of the tank prior to charging the tank to the crude unit. There are several good reasons why a refinery would want to dehydrate crude in tankage. Some refineries do not have desalters, so tank dehydration is the only method available for removing corrosive salts from the crude charge. Even in refineries with desalters, tank dehydration can reduce the salt going to the crude unit enough to enable desalters to reach desired salt levels in the desalted crude oil.

In some cases custody transfer costs are established by gauging shipments in storage. Any emulsion layer in the bottom of the tank is paid for as crude oil. By dehydrating this material, the refinery no longer pays for brine water at crude oil prices.

The water draining process is often difficult for operators. Some crude oils hold up water so it takes excessively long times to settle the tanks. When draining the water it is often oily and black making it difficult for the operator to judge when to stop draining water. In line water monitors can be very helpful in determining when the transition from water to oil occurs. Properly formulated dehydration chemistries will also result in a clean oil/water break and reduce the oil under carry in the drained water. This makes it much easier for the operators to determine when to stop draining tanks.

The data in Table I shows the results of tank dehydration that illustrates several of these points. A cargo of difficult to dehydrate crude was received by ship and transferred into two tanks in the refinery. The crude in tank 1 was not chemically treated. The same crude oil in tank 2 was treated with BPR 23040 tank pretreatment additive. Samples were taken from the top, middle and bottom of both tanks from two different locations and analyzed for quality parameters. Table I is the salt content of the crude oil samples in pounds per thousand barrels (PTB).

<u>**Table I**</u>: Crude Oil Salt Content Reduction with Chemical Pretreatment

Salt Content (PTB)

Settling	Without Chemical Tank1						Baker Petrolite Treatment Tank 2					
Time	"A" Point			"B" Point			"A" Point			"B" Point		
(hrs.)	TOP	Middle	Bottom	TOP	Middle	Bottom	TOP	Middle	Bottom	TOP	Middle	Bottom
0	>100	>100	>100	>100	>100	>100	95	>100	>100	>100	>100	>100
6							21	>100	>100	22	>100	>100
24	>100	>100	>100	36	>100	>100	12	12	>100	13	16	>100
30	25	>100	>100	30	>100	>100	13	14	88	12	14	>100
48	19	>100	>100	24	>100	>100	12	12	70	13	11	20
72	22	>100	>100	26	>100	>100	8.5	8		8	9	14
96	19	20	>100	30	60	>100	8	8	8	8.2	8.2	8.2

It is clear that after 24 hours of settling the tank is not uniform in the salt distribution with the 'good oil' in the top samples. As time progresses the salt content decreases as brine droplets fall to the bottom of the tank. This process is greatly accelerated by the addition of the Baker Petrolite BPR 23040 tank pretreatment additive. It can also be seen that the salt level is ultimately lower in the treated tank at all levels than it is in the untreated tank.

Chemical treatment produces a more uniform distribution of salt and BS&W in the tanks well as better quality, thus the desalter can expect better, more uniform performance.

Case 2. Baker Petrolite developed chemistries that would reduce the volume of sludge in storage tanks and first applied it over 20 years ago in Australia. Since then the technology has been successfully applied throughout the world. Since the sludge reduction process has been described in detail elsewhere (Ref 4), there will not be a lengthy discussion here.

There is often a concern that the sludge reduction process will suspend solids in the crude causing upset problems in the desalters as the solids are just being moved from the tank bottoms to the desalter. In actual practice this is not the case. The sludge generally is a complex emulsion containing over 50% hydrocarbon, considerable brine and small amounts of inorganic solids (<10%). The sludge reduction chemistry breaks up this sludge emulsion so that the hydrocarbon is recovered into the crude oil. The water is released to be drained out of the tank or be easily removed in the desalter. Most of the inorganic solids are large enough to settle out in the storage tank. Any fine solids that are sent to the desalter have been cleaned up by the surfactants in the desludging compounds, so that they are more easily transferred to the water phase in the desalter and removed without problem.

Once again the result of the tank treatment program is to produce a better quality crude oil which is more easily processed in the desalter.

Case 3. In many cases refineries are not able to drain any water from the storage tanks, so the tanks are generally continuously mixed and no settling is allowed. We have found that specialized pretreatment programs can still provide benefits to the refinery in terms of producing a crude charge of uniformly better quality. There are several reasons why this improvement in desalter performance is observed. First of all the chemicals are not effective until they diffuse through the crude oil to a location where they can do their job. Adding the chemical treatment to the crude oil prior to tankage gives maximum time for the molecules to diffuse to the necessary locations.

Second, we have found that asphaltenes can stabilize emulsions in desalters (Ref. 2 & 3). When crudes are blended there is the possibility of some asphaltene destabilization. This can manifest itself at the oil water interface as increased emulsion stability. Over 15 years ago, we discovered that by mixing additives with asphaltene control properties into the crude oil, **prior** to blending of crude oils in the refinery, the asphaltene could better be controlled at the oil water interface, thus leading to better emulsion resolution in the desalter. Since this phenomenon was described in detail at the AIChE 2002 Spring Meeting, it will not be discussed further here.

Third, inorganic solids in the crude oil are often partially coated with organic materials, like asphaltenes and high melting waxes that cause them to accumulate at the oil water interface. They often are associated with oil under carry in the desalter effluent water. By adding solids control chemistries to the pretreatment programs we have been able to water wet the solids so they are more easily removed in the desalter without causing upset conditions.

Finally, as mentioned before, selection of the proper chemistry can break up sludge in the storage tank and prevent finely suspended sludge from being charged to the crude unit. Since sludge is a complex emulsion, it is not easily resolved in the electrical fields of the desalter.

All of these factors of a good pretreatment program result in better quality crude oil being sent to the crude unit, thus reducing desalter upsets, oil under carry, emulsion band build up and high BS&W in the desalted crude oil.

As an example, one of our customers had a crude storage tank on continuous mixing.. The crude oil types processed by the refinery remained constant for years. Crude oil was constantly being charged to the tank and crude was constantly removed from the tank to be charged to the crude unit. There was no possibility of draining brine in the tank farm. Over the years we had determined that when filterable solids increased above 60 PTB, it was very likely they would have a desalter upset, measured in terms of increased recovered oil production. An evaluation was begun with BPR 23025 pretreatment additive. Table II shows the reduction in desalter upsets once the pretreatment was begun.

Table II: Desalter Upsets On Days With High Filterable Solids In Crude Oil

No Pre treatment			BPR 23025 Pretreatment		
No of Days TFS>70 PTB	No of Upset Days	Chance of Upset	No of Days TFS>70 PTB	No of Upsets	Chance of Upset
5	5	100%	6	0	0%
No of Days TFS> 60 PTB	No of Upset Days	Chance of Upset	No of Days TFS> 60 PTB	No of Upset Days	Chance of Upset
12	8	66%	11	1	9%

With no treatment there was a 66% chance of having an upset when the filterable solids were greater than 60 PTB and 100% chance when they were greater than 70 PTB. Once pretreatment was started, these oil under carry upsets were virtually eliminated. At the same time the BS&W in the desalted crude oil dropped by 20%.

II. SLOP OIL QUALITY VARIATIONS AND MITIGATION STRATEGIES

It is common practice to inject slop oil into the crude as a way of reprocessing the recovered oil. In some cases the slop oil is transferred into the crude storage tanks and in other cases it is injected into the crude charge to the desalter. The quality variations of the slop oil and the method in which it is injected can both cause operational problems and upsets.

Slop oil can be recovered from many sources in the refinery (Ref 5). Quality parameters such as water content, solids content, and solvency (paraffin/aromatic ratio) can vary widely depending on slop oil source. A few % water typically does not cause a problem, but some slops have 20% or more water which can lead to desalter upsets. It is well known that solids can stabilize emulsions. Oils containing catalyst fines or coke fines can cause desalter upsets, so slop oil from those units are generally not fed directly to the crude unit. These fine particles can be injected indirectly into the crude charge through recovered oil from the waste water treatment plant.

The quality of slop oil can often be improved by a combination of mechanical and chemical treatment. Mechanical treatment can range from tank settling to more complex operations involving centrifugation. Addition of the proper chemical can produce cleaner separations by the mechanical devices and can control whether the solids partition into the oil phase or the water phase. It is generally best to remove as many solids as possible into the water where it can be more easily handled in the waste water treatment plant. Such treatments can result in slop oils with BS&W consistently below 1% and very low in filterable solids.

Paraffinic slop oils such as off spec naphtha can destabilize asphaltenes in the crude oil and lead to difficult to resolve emulsions in the desalter. Since off spec fuels and many other slops are low in salt and water, there is no reason other than convenience to run them through the desalter. It is often better to inject them downstream of the desalter to minimize problems.

How the slop oil is injected before the desalter will also affect the occurrence of upsets. Best practice is to continuously inject slop oil at a low dosage (< 2%) controlled by the crude unit. More common is to inject unknown or variable amounts of slop oil to the unit. A poor strategy is to begin injecting slop oil at a low dosage and keep raising the rate of injection until a major upset occurs. The slop oil is then shut off completely until the slop oil inventory reaches a critical level. It is easier for the operators to control the desalter with a uniform quality of feed than to respond to spikes of bad quality feed.

III. WASH WATER QUALITY VARIATIONS AND MITIGATION STRATEGIES

The importance of good wash water quality has been discussed elsewhere (Ref 6,7), so will not discussed in detail here. Some key quality parameters and their typical acceptable operating limits are summarized in Table III.

Table III. Acceptable Wash Water Quality Parameters

pH	5-8
Hardness	<150 ppm as $CaCO_3$
Suspended solids	<30 ppm
Ammonia	<35 ppm

Waters that exceed these limits have been used in desalting, but they should be investigated carefully to make sure there will be no problems.

Atmospheric column overhead sour water, vacuum column overhead condensate and stripped sour water are commonly used with good desalting results. Refineries can reduce water flow to the waste water treatment plant by reusing these waters in the desalter. The primary concern with stripped sour water is that if not carefully attended, the pH can exceed 8 and result in stable emulsions forming in the desalter vessel. If working improperly, there can also be high levels of ammonia in the stripped sour water. More expensive to use is well water and clarified surface water, which can be high in hardness and suspended solids. While boiler feed water is of excellent quality for desalting it is also relatively expensive so only used in special circumstances.

When the desalter is in upset condition, switching to higher quality wash water can sometimes improve performance. For example at one refinery the operators have found that when the pH of the stripped sour water is too high they can switch to surface water and improve desalter performance.

SUMMARY

The quality of the three material feeds to the desalter can affect performance of the unit. Proper screening of potential sources of crude oil, slop oil and wash water can identify potential problems. There are several mitigation strategies available to the refinery including blending, operational changes, and chemical treatment programs. One of the most successful methods of improving crude oil quality is to pre-treat the crude oil with specially formulated products as the crude is transferred into storage tanks.

REFERENCES

1. Lawrence N. Kremer, "Crude Oil and Quality Variations," Petroleum Technology Quarterly, Autumn 2004, pp 87-97
2. Joseph L. Stark, Joe Nguyen and Lawrence N. Kremer, "Crude Stability as Related to Desalter Upsets," 5th International Conference on Refinery Processing, AIChE Spring National Meeting 2002, New Orleans, LA March 11-14
3. Joseph L Stark, Joe Nguyen, and Lawrence N. Kremer, ""New Method Prevents Desalter Upsets From Blending Of Incompatible Crudes," *Oil & Gas Journal*, Mar. 18, 2002, pp 89-91
4. Lawrence N. Kremer and Joe Nguyen, "Emulsified Sludge Accumulation in Crude Oil Storage Tanks," presented at the 1999 AIChE Spring Meeting, Houston, TX, March 14-18
5. Lawrence N. Kremer, "Using Emulsion Breaking Chemicals to Recycle Waste Oil," presented at 1st International Congress on Liquid Waste Recycling, San Francisco, CA, May 23-27, 1994

6. Lawrence N. Kremer, "Desalter Tutorial: Best Practices for Best Results" 5th International Conference on Refinery Processing, AIChE Spring National Meeting 2002, New Orleans, LA , March 11-14
7. Robert P. Adamski, "Improve Desalter Performance by Optimizing Wash Water Selection and Use," 5th International Conference on Refinery Processing, 2002 AIChE Spring National Meeting, New Orleans, LA, March 11-14

AIChE Spring National Meeting 2006 Orlando, Florida
9th Topical Conference on Refinery Processing
TA105 Advances in Desalting

Authors:
Shaya Movafaghian (AIChE)
James Chen
Tom Collins
Petreco div. of Cooper Cameron Corp.

Parallel Comparison of Two Desalting Technologies—A Case Study

Abstract:

This paper details the performance of the Bilectric™ desalter and its operational and economic value-added. Paper presents historical operating data and operating data particular to this study, collected by an independent source. Data includes a comparative study for an existing desalting train at U.S. Gulf Coast region refinery, were consumption of excessive chemicals and frequent operation and maintenance problems along with high oil-in-water concentration from the desalters were chronic to the process.

After careful evaluation and analysis of the three existing desalting trains, one vessel was revamped with the new technology. The operating conditions of this vessel were compared with vessels on the parallel trains.

Comparative studies of operating data indicated improvements in operational flexibility, quality of the treated crude and discharged effluent water. Economic value-added was derived through monitoring and measurement of the process variables for several months. Important saving sources, as reported by the refiner, were: lower chemicals usage, lower power consumption, lower oil concentration in the effluent water and lower operating and maintenance costs during upset conditions. The payback period for the invested capital was reported to be less than one year.

Within 12 months after initial trials of the first successful revamp, two other vessels on the parallel trains were replaced with this technology.

Introduction

The importance of the desalting process is well understood by the refining industry. Crude oil as received at the refinery may carry large quantities of inorganic salts, dissolved in the residual water droplets. The dispersed water droplets are typically small and highly stabilized by man-made emulsifying agents—left over from production and pipeline operations—or naturally present emulsifying agents.

The inorganic salts dissolved in the water phase, if not removed, may cause severe corrosion problems across the refinery. For example, under operating conditions of the distillation tower, some salts decompose through hydrolysis reaction which results in release of chlorine. The chlorine gas in combination with water vapor forms corrosive hydrochloric acid, which will condense in the overhead lines and the condenser. In addition, these salts can cause fouling in downstream heaters and exchangers as well as contamination of residuum products. The secondary benefits of a properly designed and operated crude unit desalter in a refinery are known to extend beyond the immediate downstream units[1].

Bilectric™ Design (BIL)

The concept of electrostatic coalescence is well established and proven time and again over the historical course of this technology in the oil industry. In theory, two major mechanisms have been identified as the sources of separation for a homogeneous dispersion in a low conductivity continuous phase[2]:

- Induced dipole, resulting from electrical forces between particles in an AC electric field.
- Electro-fining, resulting from the electrophoresis in a uniform electrical field and in the area immediately adjacent to the electrodes.

Dipole coalescence operates on a macro scale and throughout the bulk fluid as a function of AC field gradient. This phenomenon is the most prominent factor in electrostatic desalting.

The first BIL unit was introduced in 1980s and today hundreds of these units are in operation across the globe. This design is intended to maximize utilization efficiency of the vessel's internal space, by special emulsion distribution mechanism. Unlike other available configurations, in BIL arrangement feed is distributed evenly between the energized electrodes in the horizontal plane and normal to the direction of falling water droplets. This will reduce the hydraulic resistance to falling water droplets, resulting in an increase in the treatment capacity per unit horizontal area of the vessel.

Feed in the form of crude-water emulsion, enters the distribution header at the bottom of the desalter and branches into the vertical risers as illustrated in Figure 1. The riser assembly comprised of bottom and top distributors, which are designed specifically for uniform distribution of the emulsion between the lower-middle and middle-upper electrode pairs.

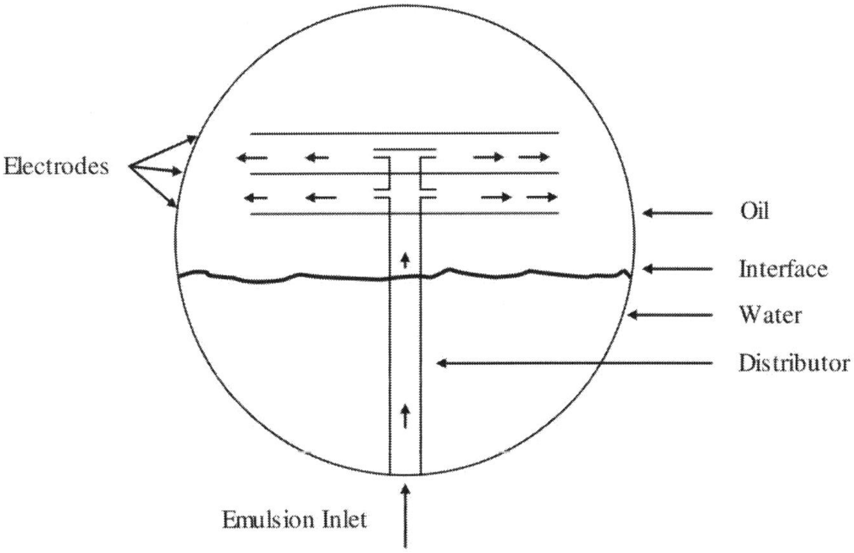

Figure 1, Internal Arrangement

Each electrode is typically energized by a single transformer and the electrodes are 120° out of phase to create a balanced load. Emulsion leaves the tip of the distributor in a radial pattern. As it travels in the horizontal plane between the grids, the induced forces between the adjacent droplets initiate the coalescence process. The droplets grow in size and fall vertically towards the oil water interface. This vertical motion promotes a non-turbulent zone between the oil and water sections in the vessel, preventing unwanted disturbance at the interface. The hydraulic pattern of the emulsion jet and the controlled turbulence of the flow in the space between the electrodes increase the coalescence probability of the stray droplets which pass within the coalescing distance of one another. Optimum separation therefore, is achieved by proper operation to arrive at a commercially viable hydraulic and electric field balance. The small size water droplets, escaped from the lower electrode space, will flow vertically to the upper electrode zone, benefiting from a second chance for coalescence. The unique internal hydraulics of this design is also known to promote better filterable solids removal.

Historical Data

Over two decades, the performance of BIL technology has been proven in variety of applications and locals. Followings are a few examples. Similar improving trends have been seen in almost all installations with various objectives, ranging from de-bottlenecking to improve in treated crude or water quality. With this bench mark, similar trends in increased performance were expected for this new installation.

Chart 1
US Gulf Coast 30-35 API Crude

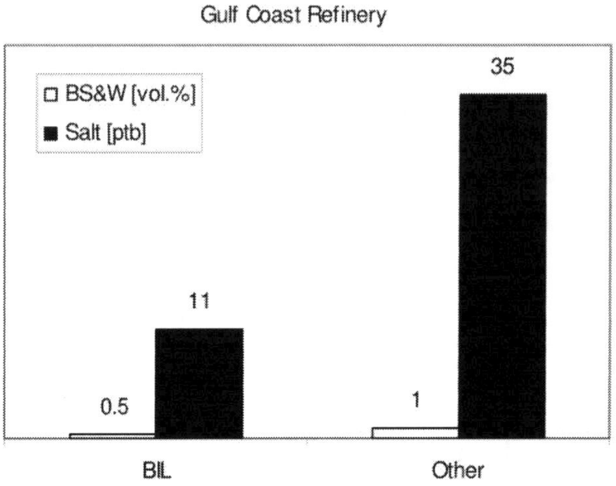

Chart 2,
US West Coast 15-17 API SJV Crude

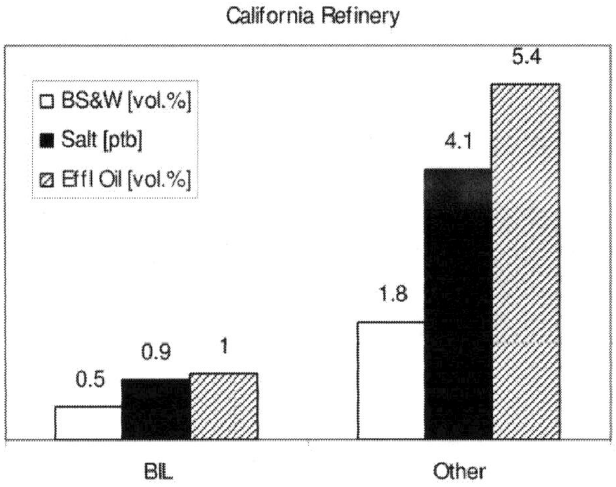

Chart 3
US Rockies 38-40 API Sulfonated Crude

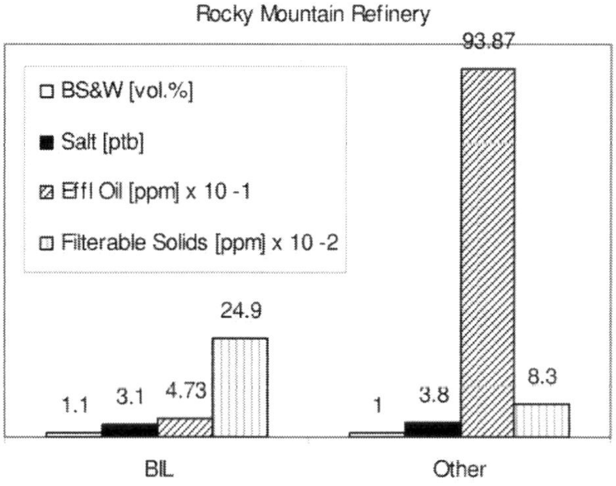

Chart 4
Canada 20-22 API Lloydminster Blend

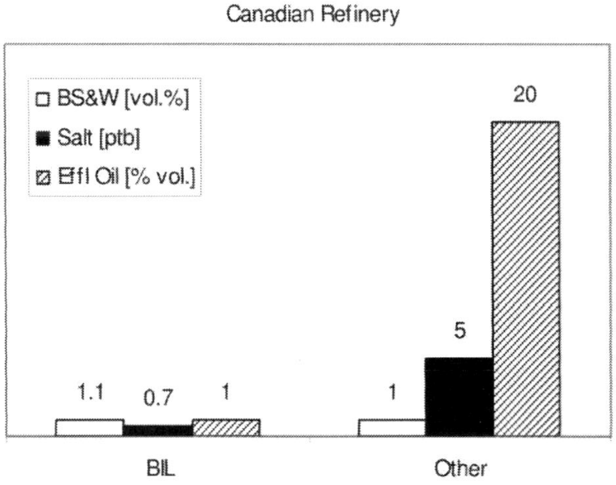

Case Study

At the U.S. Gulf Coast region, a refiner was encountering persistent operational and performance problems with the desalter train. Desalter trains were comprised of three parallel vessels as first stage, to treat a slate of heavy South American crude from Petrozuata region in Venezuela. Within the first two years of operation, all efforts by the refiner to optimize the performance of the desalter train did not produce an acceptable result. Decision was made to experiment with an alternative technology by converting one of the vessels to BIL.

A thorough operational and economic evaluation by the refiner revealed significant improvements in the newly revamped train. The operating data was collected for one-year period, through continuous testing and monitoring of chief process variables.

Operating Data

A parallel analysis of the operating data was performed over the same periods, prior and post refurbishment. During these periods, all three desalters were operating under similar conditions to obtain a common and fair comparison basis.

The following data has been extracted from operation data logs. The comparison is made between the newly revamped BIL desalter, located at the train A, and existing desalters with similar vessel size and flow rate, located at the train B. The desalter train schematics Figure 2, below, shows this arrangement. Operating data for the third desalter is omitted from some of the graphs presented in this section, due to similarity of the data.

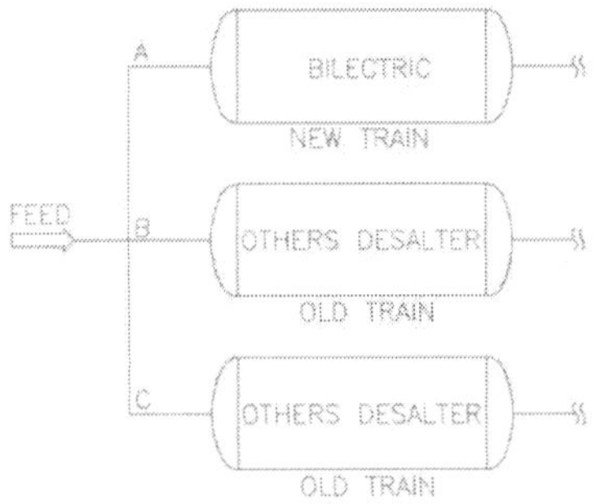

Figure 2, Desalter Train Schematics

Figure 3 illustrates a comparison between the capacity of desalters on the new and old trains. The graph demonstrates similar throughput in the pre-revamp period. It is also apparent that the new desalter was subject to larger throughput fluctuations due to operational limitations.

However, it will be shown that the BIL design was less responsive to these anomalies and provided more consistent result. Near the end of the test period, the new design showed a 10,000 bopd lead over the existing desalter while maintaining a better performance.

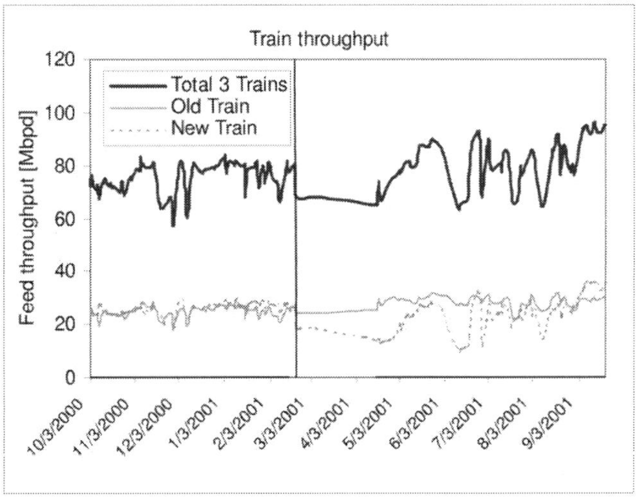

Figure 3, Train Throughput

As one of the major performance metrics, basic sediment and water (BS&W) content of the treated crude was monitored at the outlet of each desalter as shown in Figure 4. Again, the comparison is a composite data from the pre and post revamp periods. The average values of the BS&W in the post-revamp period points out a significant reduction in BS&W of the treated crude from the new desalter, while the existing desalter maintains same levels of BS&W, similar to the pre-revamp period.

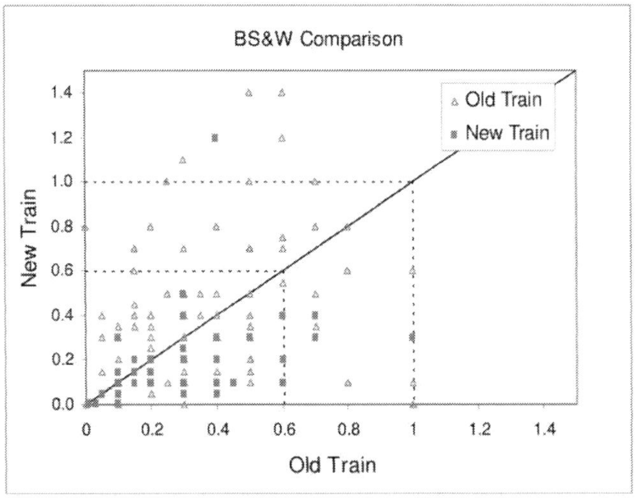

Figure 4, Old and New Trains BS&W

This significant improvement is an be seen that the old train BS&W data is inconsistently distributed in a range up to 1.0% v/v, while the new desalter shows a consistent concentration of values within 0.6% boundary limits.

When compared to the previously presented capacity data, the consistent and reliable operation of the new design under inconsistent throughput conditions is clearly verified.

Perhaps the most important function of the desalter is its salt removal efficiency under realistic operating condition. Figure 5, is a comparison of the salt content of the treated crude. During the pre-revamp period, both desalters are identical and demonstrating same levels. After the revamp, data from the new train reveals significant improvements in salt removal, while subject to similar feed fluctuations, compared to the old unit. Significant to point out, is the reduced magnitude of inconsistency in the salt content for the BIL, while other technology maintains the same level of salt removal with the magnitude of inconsistency similar to the pre-revamp period.

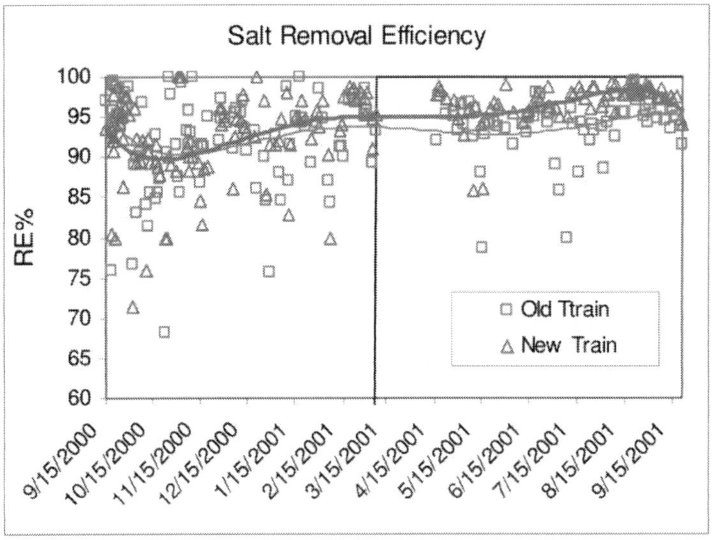

Figure 5, Salt Removal Efficiency

As described earlier, BIL unique internal arrangement, combined with the controlled hydraulics, provides a quite interface zone between the oil and water phases within the vessel. Stability of the interface is a major contributing factor to unnecessary current draw from the grids, reducing the transformer load. Figure 6, is a demonstration of this advantage, accomplished by BIL's unique internal arrangement. The composite graph shows the variations of transformer current, drawn from both desalters in the periods prior to revamp and thereafter. High and inconsistent current loading and uncontrolled modulation of the transformer load, is recorded in the graphs for desalters prior to revamp. A major reduction in current is recorded for after installation of the BIL internals in the desalter located on the new train (lower graph).

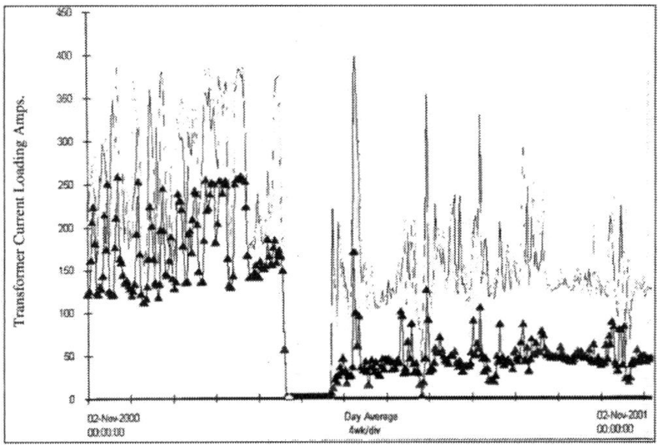

Figure 6, Transformer Load Comparison

The quality of the effluent water or brine from any desalter is a direct function of oil-water interface characteristics. The contribution of the internal design and the influences of electrical and hydraulic parameters are among the major variables for the control and resolution of the rag layer.

When these variables are not properly optimized, persistent build up of a rag layer compromises the performance of the desalter. To overcome this problem, operators tend to lower the oil-water interface, which in turn contributes to additional oil and grease carry-under in the desalter's effluent water.

Another area of improvement in this experiment was the amount of oil and grease in the effluent water. Figure 7 illustrates the amount of oil and grease carry-under in pre revamp period with chemical dosage at 30-40 ppm.

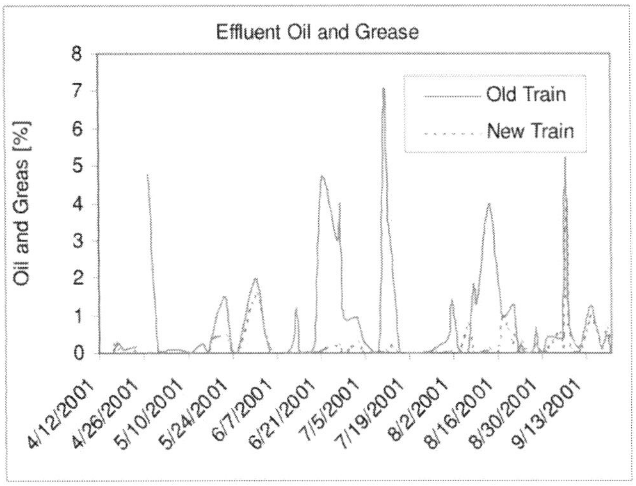

Figure 7, Effluent Quality

Figure 8, compares post revamp oil and grease carry-under data between the old and the new trains. Variations and inconsistencies continued in the results from competitor's design, while maintaining 30-40 ppm chemical dosage. On the other hand, the revamped desalter's oil and grease under-carry was negligible compare to the old train, with the chemical dosage reduced to 7-12 ppm.

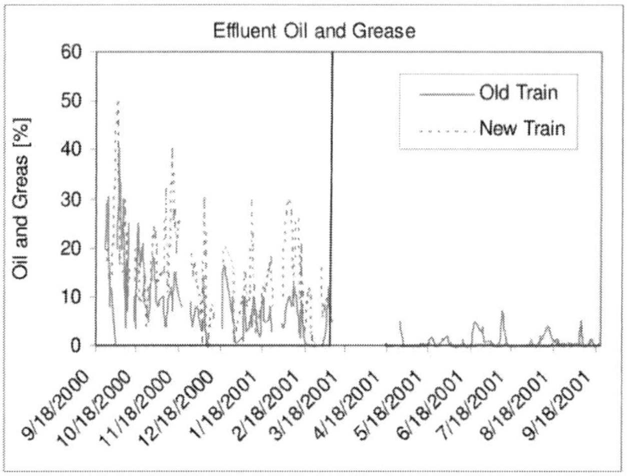

Figure 8, Effluent Water Quality Comparison

Relative response of a desalter to operational upset is extremely important to the operators. In the real world, the feed to the desalter varies in rate and composition, depending upon the refiner's operation preference or limitations. Figure 9, captures the response of the desalter train to a tank-switch operation. The two desalters located at the old trains, have experienced a major short-out in the transformer current loading, while the BIL desalter (bottom curve) have dampened the effect successfully, and maintained a consistent current draw from the grids.

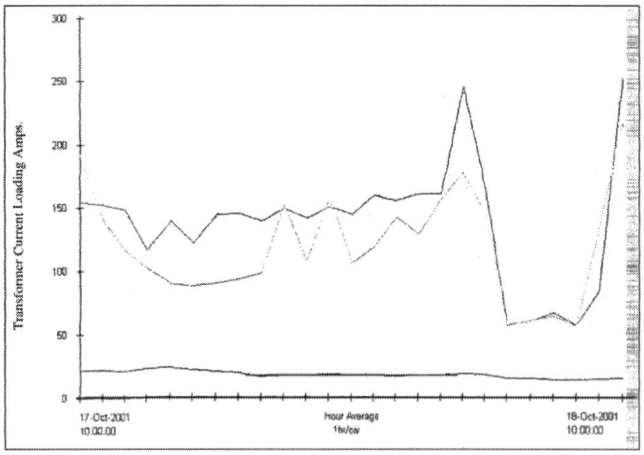

Figure 9, Upset Condition

Formation of a persistent emulsion at the oil-water interface is an undesirable phenomenon, which the refiner must be prepared to encounter. Common approach to control this rag layer is usually chemical treatment of the feed. Figure 10, is demonstrating the refiner's attempt to control the rag layer by dosing demulsifying agent to the feed. The upper curve captures the transformer current loading for the old train desalters. The dosing rate of the demulsifier is 27 ppm. The refiner then attempts to reduce the dosage of the demulsifier agent to 20 ppm. A major spike in the current drawn from the transformer on the old train desalter, is indicative of presence of a growing rag layer, which results in short-out of the grids. Secondly, it reveals total reliance of the old desalters on chemical treatment to control the rag layer and to maintain operability. The operator, then increases the demulsifier dosage back to its original levels. In response the old desalter current levels drops to its original values. The experiment was repeated again to show repeatability. The BIL desalter, on the other hand, maintains its original current level through out the experiment down to zero ppm dosage (lower curve). The refiner concludes that the BIL design has an inherent rag layer resolution effect and, therefore, performs satisfactorily, at lower demulsifier dosage.

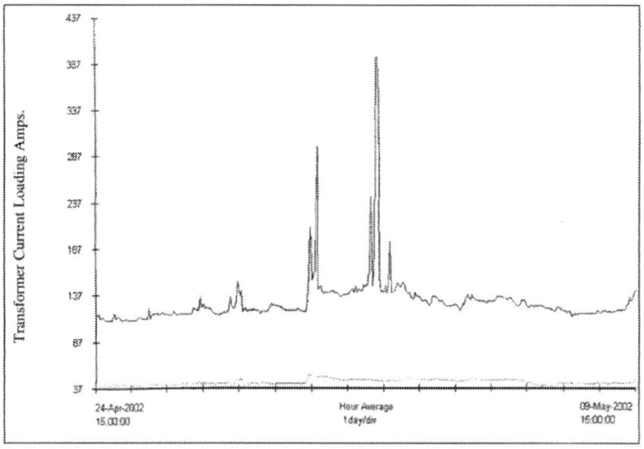

Figure 10, Chemical Treatment

Economics

This study, as performed by an independent source, has proven superiority of the BIL technology by demonstrating its performance and associated cost-savings. Refiner's economic analysis during the course of this study reveals cost-savings in four major areas:

• Energy savings: The BIL current loading was averaged at 44 amps as compared to alternative technology of 155 amps average.

• Operational stability: The desalter train was subject to throughput fluctuations due to unstable operation. It was determined that BIL design was less susceptible to these fluctuations and treated crude from this unit showed less variations to the specifications..

• Oil carry-under: BIL technology has less oil-in-water in the effluent, as a direct result of the unit's operational stability.

• Reduction in chemical additives: To maintain an acceptable performance, alternate technology required 2-3 times more demulsifier agent than BIL.

The refiner reported major cost-savings per year by converting the existing units to BIL.

Based on these results, the refiner adopted the BIL technology for the two remaining trains within short period.

Most recent reports from the refiner indicate similar results to date. Current total train capacity is reported at about 100k bopd for a 16 API slate with reported 0.5% vol. BS&W.

Conclusion

BIL technology continues to outperform alternative technologies. This design is a field-proven-technology over the past two decades and continues to perform exceptionally in the downstream industries.

This design employs a unique emulsion distribution mechanism which optimizes the effective space utilization of the vessel. The balance between the dynamics of the fluid phases within the vessel and the electric field provides a quite zone at the interface, which in turn, contributes to the stability of operation and performance consistency. These operational advantages, translated in economic benefits, have provided significant cost-saving opportunities for the refiners in the past two decades.

References

1. Jack W. Barnett: "Desalters can remove more then salt and sediment", Oil and Gas Journal, 1988.

2. L.C. Waterman: "Electrical Coalescers", Chemical Engineering Progress, Vol. 61, No. 10, p. 51-57, 1965.

FREQUENCY MODULATION FOR ENHANCED DESALTING

Gary Sams, NATCO Group Inc., Tulsa, Oklahoma
Kenneth Warren, NATCO Group Inc., Houston, Texas

2006 AIChE Spring National Meeting,
Topical A: 9th Topical Conference on Refinery Processing (TA)
#55 - Advances in Desalting (TA015)
Orlando, Florida

Abstract

The process of crude oil desalting can be considered as two major sub-processes: (1) ideally, homogeneous dispersion of the wash water into the oil followed by (2) deep dehydration of the subsequent mixture. An increase of the overall desalting efficiency results from improvements to one or the other of these sub-processes. The application of dual frequency technology to the desalter power supply allows the operator to adjust the energy supplied to the drop interface as necessary for overcoming interfacial tension while also adjusting the time available for electrostatic field decay as required to minimize the effects of oil conductivity. It has not previously been possible to directly address either of these parameters. It has been shown in both laboratory and field tests that significant improvement in dehydration results from application of these techniques. NATCO has been granted a U.S. patent[1] for a dual frequency transformer and is marketing it under the trade name Dual Frequency®. Similar improvement in desalting efficiency is also seen.

Introduction

The process of desalting crude oil requires the mixing of dilution water of lower salinity than the entrained brine into the wet crude oil with sufficient intensity to disperse the dilution water into drops of approximately the same size range as the entrained brine drops. This assures optimum contact efficiency between the dilution water and entrained brine as the drops begin to coalesce. Additionally, the energy imparted to the system during mixing assists in bringing about drop-to-drop interaction.

However, this mixing energy can have both positive and negative consequences. It has long been recognized that excessive mixing intensity can result in such finely divided dispersions that coalescence in a reasonable time frame becomes impractical. Therefore, it was often necessary in actual practice to compromise between the desired brine/dilution water contact efficiency and the ability to dehydrate the mixed oil/water system in the conventional electrostatic dehydrator.

Modulation of the electrostatic field in order to utilize electrostatic energy for both mixing and coalescence has been successfully applied to minimize the necessity of process compromise.[2] The application of voltage modulation to electrostatic mixing has been commercially applied for about eighteen years. In this application, a microprocessor-based controller in the desalter power supply modulates the voltage applied to the desalter electrodes between a low, coalescing voltage and a point in excess of the critical voltage gradient. A portion of the dilution water is injected above the electrodes and flows counter-current to the oil as the water drops are repeatedly broken up and coalesced. This process results in a highly efficient brine/dilution water contact and also minimizes dilution water usage.

Another approach to utilization of the electrostatic field for both mixing and dehydration lies in the use of frequency modulation of the field to energize the water drops sufficiently to cause drop shattering and coalescence.

Dual Frequency Modulation

In combination AC/DC electrostatic systems,[3] as shown in Figure 1, both types of electrostatic fields are applied within the dehydrator/desalter to maximize dehydration through the bulk dehydration characteristics of the AC field and the superior polishing or haze removal of the DC field. The DC component is produced by two half-wave rectifiers which alternately charge adjacent electrodes. The charge is then maintained by the capacitance of the electrode system.

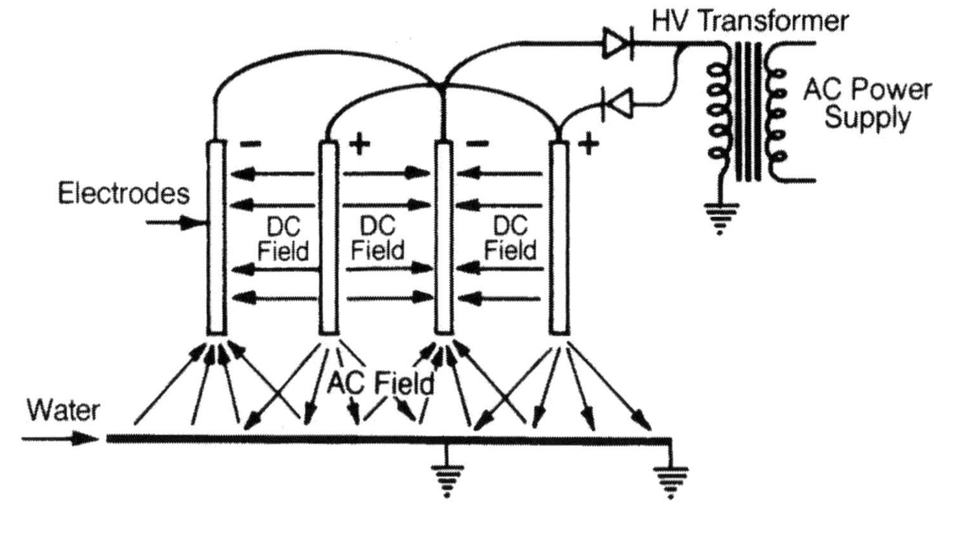

Figure 1 – Combination AC/DC Electrostatic Fields

As Figure 2 shows, due to conductivity of the oil phase between the electrodes, this charge decays with the depth of discharge depending upon the conductance of the oil.[4] This

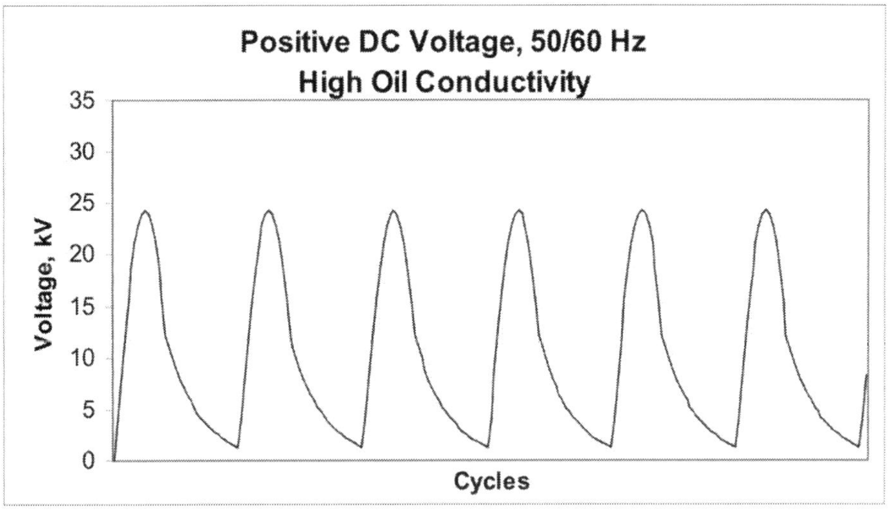

Figure 2 – Detrimental Voltage Decay

decay reduces the average intensity of the DC field, and if the oil is sufficiently conductive, the loss of field strength may negatively impact performance.

When used in conjunction with AC/DC electrostatic field, it has been shown that high frequency modulation, in the range of 1 kHz, can dramatically increase coalescence.[5] As Figure 3 shows this higher frequency maximizes the energy applied to the electrodes and strengthens both the AC and DC fields resulting in a subsequent increase in droplet interactions. Increasing the frequency of the AC current prior to rectification results in shorter times between recharge of the electrodes and increases the average DC field intensity.

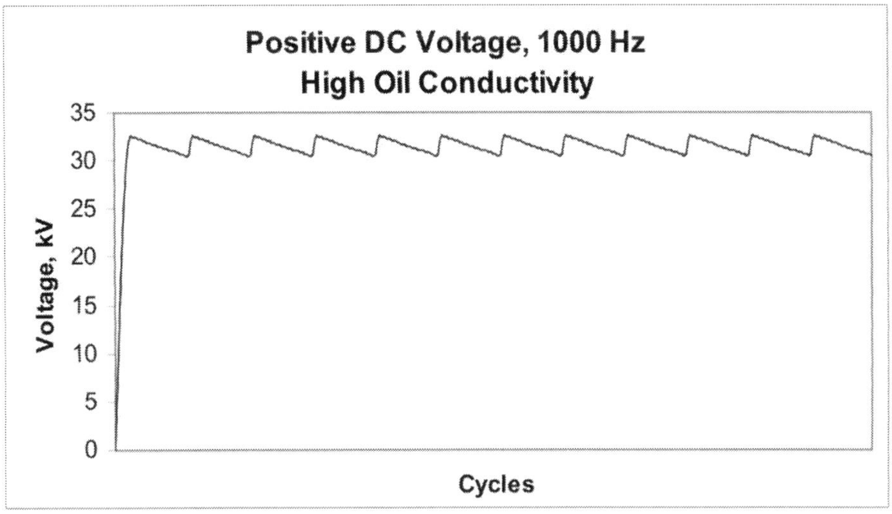

Figure 3 – Limited Voltage Decay

Adjustment of the AC power supply frequency can thus be used to counteract the effects of conductive oils. The AC power supply frequency will be referred to as the base frequency.

Modulation of the electrostatic field in dehydrators and desalters has been practiced for many years, either as a means of increasing dehydration effectiveness or for utilizing the electrostatic forces for both mixing (drop shattering) and coalescence. These processes use low frequency modulation (0.2 to 2 Hz). This effect appears to result from energizing the dispersed drops in the range of their resonant frequency. Under these conditions the drops absorb energy from the electrostatic field, resulting in the perturbed drops becoming distorted as the interfacial tension holding them in a spherical shape is overcome. These energized drops are then able to interact more readily with each other in the coalescence process. The modulation frequency producing this effect will be referred to as the pulse frequency.

Dual frequency dehydration results when a base frequency adjusted to minimize the effects of oil conductivity is then amplitude-modulated with a pulse frequency adjusted to maximize energy absorption by the dispersed phase drops. The shape of the pulse modulation may be varied over a wide variety of waveforms of various periods. It may be helpful to visualize this process in terms of AM radio transmission in which a carrier wave (base frequency) is modulated in amplitude by the sound wave (pulse frequency) to be transmitted.

Application of Modulated Dual Frequency Desalting

Previous use of the electrostatic field for both mixing and coalescence relies upon modulating the voltage so it exceeds the critical voltage gradient at some point, producing drop shattering, and then is reduced gradually through the range of coalescing voltage. However, drops may also be energized by pulsing them at their resonant frequency, and if this is done with sufficient energy input, drops may also be shattered in this manner. For example, the resonant frequency of a 1500 micron droplet is about 1.5 Hz while the resonant frequency of a 50 micron droplet is 200 Hz.

The ability to produce drop shattering electrostatically enables the use of counter-flow dilution water for desalting with the attendant increase in extraction efficiency seen in such a process as opposed to a single-stage, mixer-settler process as seen in conventional desalters using mixing valves prior to dehydration. The use of frequency-based dispersion extends this process to more difficult oils. Sweeping the pulse frequency from 1.5 Hz to 200 Hz will coalesce droplets up to 1500 microns and then shatter them to 50 microns.

Dual Frequency Dehydration Results

Dual frequency dehydration is achieved by using a constant base frequency to maximize the electrostatic field strength and modulating the voltage amplitude at the resonant frequency required to achieve optimum dehydration.[6]

Numerous pilot tests have documented the benefits of dual frequency dehydration on oils ranging in gravity from 16° to 40° API. In comparison with AC/DC results, Table 1 shows the residual BS&W (basic sediment & water) can be maintained while increasing oil flux using dual frequency dehydration. The optimum frequency and voltage parameters were determined for each of these tests and the temperature and chemical treatment were identical. There are five parameters for each test including base frequency, pulse frequency, maximum voltage, minimum voltage and waveform. In these three tests the oil flux was increased from as little as 25% to as much as 144%. The residual BS&W was maintained in nearly every test.

Table 1 – Increased Oil Flux			
Oil Gravity °API	**Flux (BOPD/ft^2) @ BS&W**		**Comments**
	AC/DC	**Bimodal**	
16	100 @ 0.63%	125 @ 0.56%	Diluted SAGD, Canada
20.2	90 @ 0.38%	220 @ 0.4%	West Sak, Alaska
20.7	90 @ 0.35%	175 @ 0.4%	Bohai, China

Table 2 shows the reduction in residual BS&W that could be achieved at a constant oil flux. In these four tests the oil flux was maintained at a pre-determined rate and the electrostatic technology was changed from AC/DC to dual frequency. The BS&W reduction ranged fro 38% to nearly 100%.

Table 2 – Constant Oil Flux			
Oil Gravity °API	**BS&W @ Flux (BOPD/ft^2)**		**Comments**
	AC/DC	**Bimodal**	
20.8	0.9% @ 150	0.55% @ 150	North Sea at 50% over original design
30	0.7% @ 175	0.22% @ 175	Okemah, Oklahoma
34	1.2% @ 100	0.8% @ 100	Gulf of Mexico
40	0.1% @ 225	Trace % @ 225	Bristol, Oklahoma

Dual Frequency Results

Laboratory tests were conducted to demonstrate the mixing benefits achievable using dual frequency modulation of the electrostatic field. These mixing experiments were conducted on two medium gravity crude oils using an electrostatic susceptibility tester capable of producing a dual frequency modulated electrostatic field. The final salinity results were compared with those from an AC/DC electrostatic susceptibility tester.

These tests were structured to show the level of mixing that could be achieved utilizing a modulated pulse frequency. These tests were conducted by injecting 3% of dilution water into a 200 ml oil volume. The mixture was blended mechanically using a commercial blender at a slow speed and short time duration. The sample was then coalesced electrostatically using three electrostatic technologies. For each technology the electrostatic field was applied for 3 minutes.

For each oil, three tests were conducted for comparison. Test 1 was conducted in a traditional AC/DC electrostatic field. Test 2 was conducted in a dual frequency electrostatic field with the base frequency and the pulse frequency held constant. Test 3 was conducted in a dual frequency field accompanied by modulation of the pulse frequency. The base frequency was held constant and the pulse frequency swept slowly from about 1 Hz to a maximum of 200 Hz.

After each test the residual salt was extracted by hot water and measured by titration of the separated water. The test results for both oils are presented in Tables 3 and 4.

Table 3

API Gravity	27°		
Initial Salinity, ppm	175,000		
	Water Salinity, ppm	% Salinity Reduction	% Mix Efficiency
AC/DC	86,605	50.5	23.8
Dual frequency	99,844	42.9	17.6
Mod. dual frequency	49,326	71.8	59.4

As Table 3 shows Test 1, using AC/DC dehydration, the initial salinity decreased from 175,000 ppm to 86,605 ppm for a 50.5% salinity reduction and a modest 23.8% mix efficiency. Test 2, using dual frequency electrostatic, reduced the salinity to 99,844 ppm for a 42.9% salinity reduction and a lower 17.6% mix efficiency. Test 3, using modulated dual frequency electrostatics, reduced the salinity to 49,326 ppm for a 71.8% salinity reduction and a high 59.4% mix efficiency. The mix efficiency increased a significant 250% when a modulated dual frequency electrostatic field was applied.

Table 4

API Gravity	20.7°		
Initial Salinity, ppm	3,492		
	Water Salinity, ppm	% Salinity Reduction	% Mix Efficiency
AC/DC	2,762	20.9	7.9
Dual frequency	2,124	39.2	21.5
Mod. dual frequency	1,771	49.3	27.5

As Table 4 shows Test 1, using AC/DC dehydration, the initial salinity decreased from 3,492 ppm to 2,762 ppm for a 20.9% salinity reduction and a low 7.9% mix efficiency. Test 1 results are related to the mechanical mixing accomplished by the commercial mixer. Test 2, using dual frequency, reduced the salinity to 2,124 ppm for a 39.2% salinity reduction and an improved 21.5% mix efficiency. Test 3, using modulated dual frequency electrostatic, reduced the salinity further to 1,771 ppm for a 49.3% salinity reduction and 27.5% mix efficiency. The mix efficiency increased a significant 343% when a modulated dual frequency electrostatic field was applied.

Conclusions

Both pilot and field tests have demonstrated the dehydration and capacity benefits that can be achieved by utilizing a dual frequency electrostatic field. NATCO has been granted a U.S. patent for a dual frequency transformer and is marketing it under the trade name Dual Frequency®. The results have consistently shown that a dual frequency transformer can be used to improve the performance of existing dehydration processes by either reducing the effluent BS&W or increasing the dehydrator capacity. Improvements ranging from 38% to 144% have been achieved in pilot tests.

Laboratory bench tests on a few oils have demonstrated that a significant desalting benefit can be achieved by sweeping the pulse frequency over a wide range. Mixing between the dilution water and conate water were improved by resonant frequency shattering.[7] This improved mixing reduced the residual salt by an addition 20 to 30% over conventional AC/DC dehydration.

References

1. U.S. Patent No. 6,860,979, "Dual-Frequency Electrostatic Coalescence", issued March 1, 2005.
2. Warren, Kenneth W., Sams, Gary W. and Nakayama, Toshio, 1998. Electrostatic Fields: Essential Tools For Desalting, *AIChE 1998 Spring Meeting*, 9-11 March.
3. Burris, Donald R., 1977. Dual Polarity oil dehydration, *Petroleum Engineer*, 30 – 41.
4. Byers, Charles H. and Amarnath, Ammi, 1995. Understand the potential of electro-separations, *Chemical Engineering Progress*, 63-69.
5. Eow, John S. and Ghadiri, Mojtaba, 2003. Motion, deformation and break-up of aqueous drops in oils under high electric field strengths, *Chemical Engineering and Processing*, 42, 259 – 272.
6. Sams, Gary W. and Wallace, Harry G., 2003. Field Trials Scheduled for New Dehydration Technology, *Offshore Technology Conference*, 15353.
7. CIP Application, USPTO, "Multiple Frequency Electrostatic Coalescence", filed February 15, 2005.

55c New Control Technology for Desalter Optimisation

Paul Hewitt and **Scott Vidrine**. Tracerco, Suite 200, 1100 Hercules, Houston, TX 77058
NEW CONTROL TECHNOLOGY FOR DESALTER OPTIMISATION

Within many refineries there is a growing tendency to process heavier crude or blend these into lighter feedstocks. During crude oil refining it is advantageous to optimise the blend to reduce overall feedstock costs whilst at the same time ensuring the final mix can be adequately processed with minimal upset.

One of the fundamental needs is the efficient separation of wastewater from the crude using a desalter vessel after water washing the feedstock to reduce chloride levels.

Upon a request from a major oil refiner, Tracerco has redesigned its' Tracerco ProfilerTM instrument that is commonly used within three phase separation devices within the oil and gas industry to withstand the elevated operating temperatures typically experienced within a desalter vessel.

This paper offers information on the development of an instrument that offers phase density measurement vertically throughout a desalter vessel to 1" resolution allowing the position and extent of oil, emulsion, water and solids to be measured within the control room in real time.

55d Desalter Interface Control and Diagnostics Utilizing Sonar Transducers

R. Paul Clewis, Transducer USA, Inc., 2890 W. Sam Houston Parkway S., Suite A013, Houston, TX 77042

Improving interface control in the desalting process remains one the salient issues in optimizing refining operations; all the more important because of the desalting systems's role as "gatekeeper" for primary feeds to the overall refining process. The constant tug-of-war of improvements in equipment design and emulsion breaking chemistries versus the stress of opportunity crude mix and feed rates above rated capacities requires that the control of all system operating parameters be as stable as possible. Interface control technologies developed to date have ranged from simple averaging devices to complex combinations of analytical devices: the former stressing functional simplicity at the expense of control stability, and the latter providing impressive separation profiling at the cost of a burdensome complication of the operator's involvement. A recent entry into the field of desalter interface control has proven to provide an impressive compromise between control stability and complexity. This technology; an industrialized sonar transducer, shows promise in providing both the basic control functions required and the diagnostic input required to evaluate critical matters such as emulsion growth and solids accumulation. This paper presents both laboratory and field evaluation of this technology, with emphasis on the quality of interface control and process diagnostics. The observed performance is further compared to existing control technologies, both in terms of reliability and end-user complexity.

Removal of Calcium and other Metal Species from Crude Oil in the Desalting Process, Part 2

By Lawrence Kremer
Joe Nguyen
Jerry Weers
Baker Petrolite Corporation

Summary

This article is an update of the 2004 AIChE paper entitled "Removal of Calcium and other Metal Species from Crude Oil in the Desalting Process" (Ref 1). Some crude oils from West Africa, the North Sea, China and other locations have sufficiently high levels of calcium naphthenate to cause product quality and operational concerns at some refineries. Use of specialized chemistry converts calcium into a water soluble species that can be removed in the desalter. Several successful field applications will be discussed. There are several application details involving metals removal programs that need to be considered before beginning an application in the refinery.

Background

A number of crude oils from Africa, the North Sea, China, and other locations are being produced with high levels of calcium naphthenate. This form of calcium is relatively difficult to remove by normal desalting techniques and can cause downstream problems such as catalyst poisoning, downgrading of bottoms products, and fouling and emulsion issues (Ref 1). In order to take advantage of favorable pricing of these opportunity crude oils, many refineries have implemented special treatment programs to enhance the removal of metal species from the crude oil. It should be pointed out that while it is possible to get the calcium levels in the desalted crude oil down to a few parts per million, it is not always necessary to reach such low levels in order to mitigate negative downstream impacts.

Recently Baker Petrolite developed EXCALIBUR Metals Removal Technology for calcium and other metals. The EXCALIBUR program includes not only various complexing agents to remove metals but also specially developed emulsion breakers, corrosion inhibitors and scale inhibitors to optimize metals removal and control secondary concerns. In the last three years we have had several successful evaluations in Europe, Asia, and North America. Since these successes have been detailed elsewhere (Ref 3, 4) they will only be mentioned as supporting material in this paper. The discussion here will center around several key questions such as: Why does calcium removal work? What is the impact on the waste water treatment plant? What is the impact on scale formation? What is the impact on the desalter?

Why does calcium removal work?

Calcium naphthenate is not significantly soluble in water in its salt form. It has been found that adding metals removal agents into the water contacting the crude oil in the desalter can improve the water solubility of calcium and protonate the naphthenic acid, forcing it to partition more into the oil phase. The general equation is:

$$2\ H^+(aq) + Ca\ (OOCR)_2 \rightarrow Ca^{++}\ (aq) + 2RCOOH$$

where $Ca\ (OOCR)_2$ is a generalized formula for calcium naphthenate and RCOOH is a general form for naphthenic acid in the protonated form. Adding metals removal chemistry increases the $H^+(aq)$ and drives the reaction to the right. Without the $H^+(aq)$ from the additive the calcium naphthenate is stabilized and remains in the oil. Several refineries have found that increasing amounts of calcium can be removed in the desalter effluent water by increasing the amount of the additive that is injected. This can clearly be seen in the data in Figure 1., which is a plot of the amount of calcium removed as a function of the amount a of additive injected.

Figure 1. Calcium removal as a function of the dosage of calcium removal additive

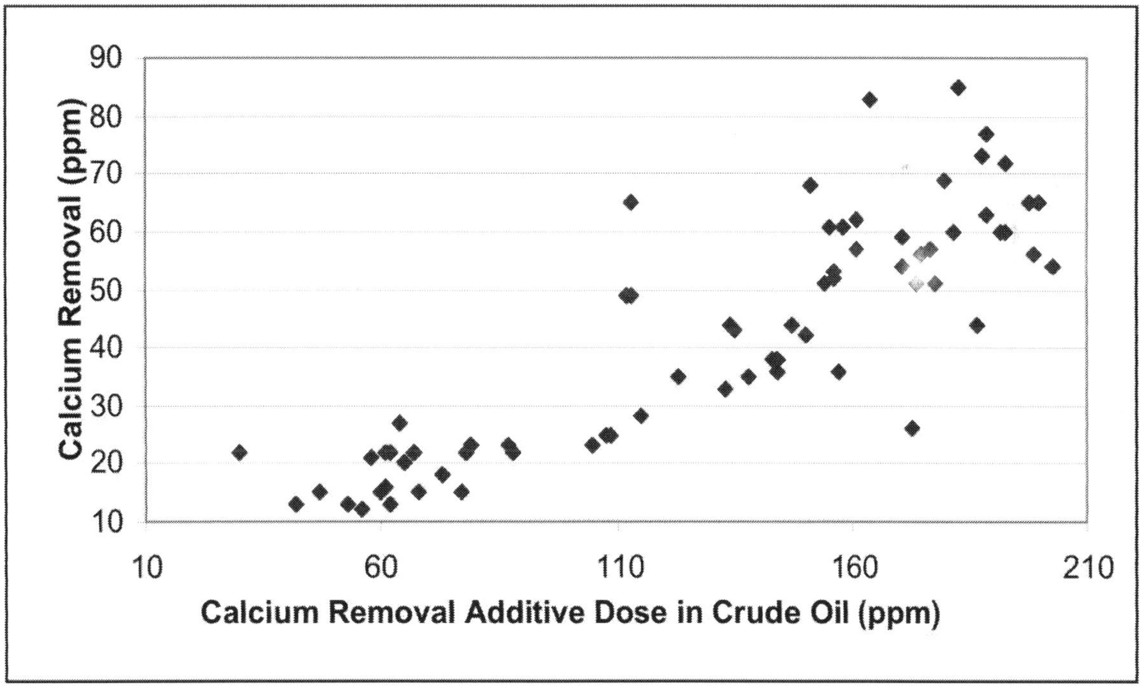

While it is possible to obtain 95+% calcium removal, it is often not required to meet the calcium targets for downstream operations. Refineries set calcium targets in the FCC or coker feed or in the residual fuel oil specifications and then adjust the EXCALIBUR metals removal program to consistently meet the desired targets. At times, no treatment may be required to meet the established targets.

What are the components of a successful metals removal program?

Since effective metals removal technology depends on converting the metal containing species in the oil phase into water soluble species, the logical place to implement this in the refinery is the desalter. Such a metals removal program can be implemented with minimal modification to the desalter equipment and operations. The metals removal program includes a properly selected emulsion breaker program, a metals removal agent, scale inhibitors to prevent fouling of the desalter and downstream equipment, and corrosion inhibitors to protect the desalter wash water injection system. A diagram of a typical EXCALIBUR metals removal program is shown in Figure 2.

Figure 2. Typical EXCALIBUR metals removal program

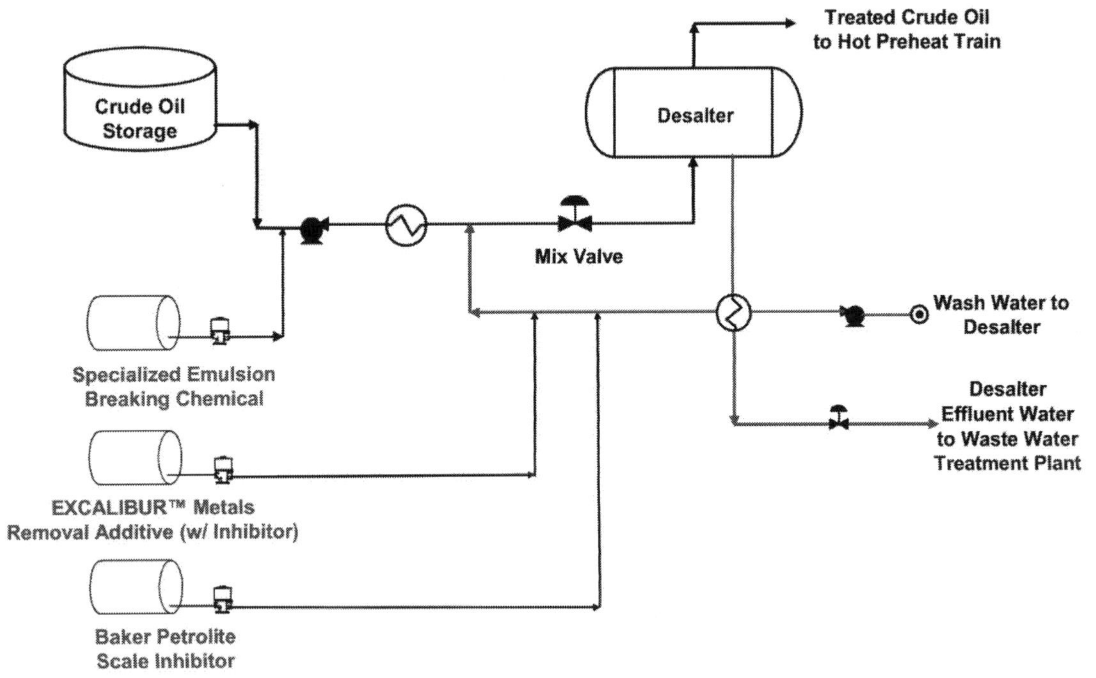

Properly selected emulsion breaker is injected into the crude charged to the unit. The metals removal additive and corrosion inhibitor are injected into the wash water before the mix valve. An appropriate amount of scale inhibitor is also added to the wash water to prevent fouling in the desalter. While the arrangement of this program is not greatly different that a standard desalter treatment program, it is necessary to closely monitor desalter performance so that changes can be made to the injection scheme as results demand. At a minimum the pH of the desalter effluent water and calcium levels in the desalted crude oil must be monitored. The chemical treatment program can then be adjusted to meet the targets of the program.

What is the impact on waste water treatment?

It is sometimes anticipated that adding metals removal chemicals at the desalter will increase the amount of organic materials sent to the waste water treatment plant. However, proper selection of emulsion breaking chemicals and metals complexing agents will actually decrease the desalter impact on the waste water treatment plant. This is because the additives force the transfer of oil and grease and water soluble organics (such as phenols and naphthenates) from the effluent water into the desalted crude oil. Table I lists some of the desalter effluent quality parameters measured in the field during a recent EXCALIBUR metals removal application. Before running the high calcium Doba crude oil, the refinery had their standard desalter treatment program in place. While running the high calcium naphthenate Doba crude oil they utilized the EXCALIBUR metals removal technology. Use of EXCALIBUR technology reduced the oil content of the desalter effluent water from an average of 243 mg/L down to an average of 34 mg/L. There was a similar dramatic decrease in effluent water phenol content. In this case the average effluent water COD increased marginally, but at other refineries the COD and BOD levels have declined.

Table I. Desalter effluent water quality with and without EXCALIBUR

	Oil in Effluent Water, mg/L	Chemical Oxygen Demand, mg/L	Phenol, mg/L
Before Doba Processing	243	2,002	12.3
Average Values During Doba Run	34	2,870	2.3

What is the impact on scale formation?

Another consideration is the potential for formation of scale in the desalter and downstream of the desalter. Several customers felt that the increased amounts of calcium in the desalter effluent water would lead to scale formation in the exchangers, transfer lines and water treatment plant equipment. However, EXCALIBUR technology includes application of scale control agents to prevent this type of deposition. In fact at one application the customer found that the scale began to diminish once the EXCALIBUR program was initiated.

Selection of the proper complexing agent can play a significant role in the potential for scaling. Understanding of salt solubilities can be complex, as the solubility is dependent on other ions in the water, in addition to water temperature and pH. In general as the pH is lowered (excess acid added) the salts become more soluble. Some salts such as calcium formate show regular solubility, *i.e.* they are more water soluble at higher temperatures. Others, such as calcium sulfate show inverse solubility, so they tend to scale up on hot surfaces. Low solubility calcium salts have the potential of scaling up desalters and downstream equipment.

Table II lists solubilities of various calcium salts as listed in vendor literature or the *CRC Handbook of Chemistry and Physics*. Refineries are well aware of scaling problems associated with calcium sulfate and calcium carbonate. Use of calcium complexing agents such as citric acid and tartaric acid would be expected to pose a scaling problem.

Many desalters have limited wash water volumes available. When crude oils with high calcium are being treated, the solubility of the calcium salt becomes important. For some high calcium crude slates the limited wash water rate can thus limit the ability to remove calcium when the salts are of low solubility. Thus selection of a complexing agent that forms a very soluble salt becomes important.

<u>Table II</u>. Solubility of calcium salts in g/100cc of water solution at low and high temperatures

Calcium Salt	Cold Solubility	Low T (°C)	Hot Solubility	High T (°C)
Acetate [V]	27	0	23	100
Carbonate	0.00153	36	0.0019	76
Citrate	0.096	23	--	--
Formate	16.2	0	18.4	100
Lactate	3.1	0	7.9	80
Oxalate	0.00067	12	0.0014	96
Sulfate	0.209	30	0.1619	100
d-Tartrate	0.0266	0	0.0689	87.6

All data taken from *CRC Handbook of Chemistry and Physics* 65th Edition B-81-84 except slats marked by [V], indicating information from vendor data.

<u>What is the impact on wash water corrosion?</u>

EXCALIBUR metals removal technology is typically most effective at a desalter effluent water pH less of than 7.0, but these conditions make the water more corrosive to carbon steels commonly used in construction of desalter vessels, brine exchangers and associated wash and effluent water piping. For this reason EXCALIBUR™ metals removal additives contain corrosion inhibitors designed to maximize metal protection while minimizing any impact on the waste water treatment plant. In refinery applications of EXCALIBUR technology, after an initial period of passivation the corrosion rate on mild steel drops to less than 5 mils per year. In several cases electrical resistance corrosion probe readings have shown desalter wash water system corrosion rates to be lower when using EXCALIBUR technology than before the program was started.

What are the impacts on emulsion resolution in the desalter?

Many of the high calcium naphthenate crude oils are sufficiently different than the normal crude slates processed that a different emulsion breaking formulation must be used. Since calcium, salts and other undesirable constituents are transferred to the water phase, it is critical to remove as much of the water as possible from the desalted crude oil to minimize downstream impacts. Baker Petrolite has developed new emulsion breakers specifically to work in conjunction with metals removal programs. In some cases the existing emulsion breaker package could be used, but in many of our applications it was necessary to introduce a product that could provide the level of crude oil dehydration required.

Conclusions

Application of EXCALIBUR metals removal technology in refinery desalting operations enables refineries to process crude oils high in calcium naphthenate content. These crude oils are often available at a discount, providing higher net profit margins for the refinery. Use of this technology can lower calcium content in the desalted crude to levels which prevent the downstream problems that can be caused by high calcium crude oils. This technology has also been proven to effectively address the increased potential for system scaling and corrosion. In some cases the switch to high calcium naphthenate crude blends will require that a more effective emulsion breaking chemical be used.

EXCALIBUR is a trademark of Baker Hughes Incorporated

References

1. Lawrence N. Kremer, Joe Nguyen, Jerry Weers, “Removal of Calcium and other Metal Species from Crude Oil in the Desalting Process,” Seventh Topical Conference on Refinery Processing, Recent Advances in Desalting, AIChE Spring National Meeting 2004, New Orleans, LA, April 25-29, 2004
2. Pitter, P. and Chudoba, J., “Biodegradability of Organic Substances in the Aquatic Environment,” 1990, CRC Press, Boca Raton, FL
3. Jerry Weers, Joe Nguyen, “A Metals Removal Process for Doba Crude Oil,” ERTC 9th Annual Meeting, 2004, Prague
4. Jerry Weers, Scott Bieber “Calcium removal from high TAN crudes,” Petroleum Technology Quarterly, Q3 2005, pp 21 – 25

Heat Of Cracking For Naphtha In FCC Units Risers

Carlos Alberto Dantas Moura - Petrobras Research Center – Rio de Janeiro - Brazil
Andrea de Rezende Pinho - Petrobras Research Center – Rio de Janeiro - Brazil
José Mozart Fusco - Petrobras Research Center - – Rio de Janeiro - Brazil

Abstract –
Experience in Fluidized Catalytic Cracking Units has shown that it is advantageous to introduce a naphtha stream in the FCC riser to produce light olefins like ethylene and propylene, high valuable petrochemical feedstocks. The reactions are mainly endothermic and the products are fuel gas, LPG, cracked naphtha, medium to heavy oil fractions, and coke deposited on the catalyst.

To model this type of operation one needs to know the heat of reaction for the naphtha. One possible path is to combine operating data and analyses of the naphtha stream and products either from a pilot plant or from an industrial unit. The analyses include ASTM D-2887 distillation, density, PIANIO chromatography and heats of combustion of the naphtha feed, cracked naphtha, and other products. These data are combined with heats of reaction for the pure compounds and operating data in a computer routine which performs the heat balance.

This work presents results of heats of reaction for the cracking of naphtha obtained in a FCC pilot plant of our Research Center.

1- Introduction

Petrobras has developed a computer program to simulate many types of the Converter of Fluidized Catalytic Cracking (FCC) Units, called ***SimCraq***^{OT}. The typical feed to the FCC risers is Gasoil or Residue. But if one desires to increase the production of olefins, like ethylene and propylene one of the possibilities that has been commercially practiced is to introduce unconventional feedstock like naphtha in addition to the gasoil.

Petrobras has a good industrial experience in processing naphthas in FCC Units. In the beginning of the 1990s two "dual internal risers" Orthoflow C Units were converted to external risers, one of them dedicated exclusively to the cracking of Coke Naphtha. The purpose was twofold: a)to improve the quality of the coking naphtha (increasing the octane number and minimizing gum formation) and to generate light olefins. The naphtha riser was designed to operate in a wide range of the reaction temperature and the Cat/Oil ratio. See Ramos *et alii* (9).

Corma *et alii* (3) simulated in the laboratory different options for processing naphtha streams in the FCC unit and evaluated the best alternatives to process light straight run and fluid catalytic cracking naphthas to produce light olefins and/or to reduce olefins content in commercial gasoline.

To model this type of operation using ***SimCraq***^{OT} we need to know the variation of enthalpy in the cracking reaction for the naphtha. For gasoil there is the very important work of Dart and Oblad (4), which has remained, to the best of our knowledge, a milestone used in FCC simulators and even as a classical industrial example in the book by Edmister (7).

. In a first attempt the values of the heat of reaction for naphtha used in the simulator ***SimCraq***^{OT} were estimated from results of industrial tests. Because of their inherent imprecisions they needed confirmation from experimental data of a pilot plant.

There was the need of trying a similar approach for the heat of cracking of naphtha and the ideal place to measure it is in a pilot plant where you can use only naphtha and no gasoil as the feedstock and peform a grater number of measurements and analyses.

In Petrobras' Research Center the FCC Pilot Unit has been recently used to study the yields obtained from naphtha cracking, as a function of operating conditions. By making only two additional measurements, the heats of combustion of the naphtha feed and of a liquid effluent comprising cracked naphtha, medium to heavy oil fractions, the problem of estimating the heat of the cracking reaction was greatly simplified.

2- The FCC Pilot Plant

A Davison Circulating Riser (DCR), licensed by Grace Division, was used in this work and it is represented in Figure 1 by a schematic diagram. See Young (12). The parts of interest to this study are the reactor and the separation column. In the reactor the naphtha feed and the regenerated catalyst enter at the bottom of the riser reactor, with the help of streams of Nitrogen and dispersion steam. It is convenient to group the reaction products in three, two of which go to the separation column and the third one, coke deposited on the catalyst, goes to the stripper.

The separation column has a condenser with Freon but no internals and no reflux. So, the separation is not sharp and some pentanes and a very small amount of C_5^+ go to the gaseous top product. It also contains some 75-79 % of Nitrogen and steam. The bottom product, which will be called liquid effluent contains very little propane, some butanes, cracked naphtha, and medium to heavy oil fractions.

We measured :

- the gas chromatography
- the distillation D-2887 and PIANIO of the liquid effluent

Then it is possible to correct the yields by :

- incorporating the butanes to the LPG yield
- incorporating pentanes and C_5^+ into the cracked naphtha
- separating the Light Cycle Oil and the Decanted Oil from the liquid effluent

The Simulated Distillations D-2887 of the naphtha feed and the liquid effluent allow us to estimate the amount of the medium to heavy oil fractions. This is important to measure the yields. The LCO is defined as the fraction 221-343 °C and the Decanted Oil the fraction 343-End Boiling Point (°C) in the D-2887, weight %. On the other hand, having the heat of combustion of the entire liquid effluent simplifies the calculation of the heat of reaction by diminishing the number of products.

3- Methodology

Any reaction can be written in the symbolic form :

$$aA + b\,B + ... \rightarrow m\,M + n\,N + ... \quad (1)$$

The heat of reaction for (1) , ΔH_R is calculated by equation (2) :

$$\Delta H_R = (\, m\ \Delta H_{F,M} + n\ \Delta H_{F,N} + ...\,) - (\, a\ \Delta H_{F,A} + b\ \Delta H_{F,B} + ...\,) \qquad (2)$$

where $\Delta H_{F,A}$ is the enthalpy of formation for compound A and a, b, m, n, represent either the mass of the substances or the number of moles.

If we write the reactions of combustion for the reactants and the products and subtract the second group from the first the moles of CO_2 and H_2O will cancel out and ΔH_R may be calculated by equation (3) :

$$\Delta H_R = (\, a\ \Delta H_{C,A} + b\ \Delta H_{C,B} + ...\,) - (\, m\ \Delta H_{C,M} + n\ \Delta H_{C,N} + ...\,) \qquad (3)$$

In equation (3) $\Delta H_{C,X}$ represents the heat of combustion for component X.

The difference between (3) and (2) is that the enthalpies of formation were replaced by enthalpies of combustion and the order of the subtraction (Products – Reactants) reversed.

When ΔH_R is positive the reaction is endothermic, meaning that we have to supply heat to the reaction to keep the temperature constant. So, all ΔH_C are negative.

The net heats of combustion are reported at 25 °C and we calculate the gas heat combustion using the heats of combustion of the pure components at the same temperature, taken from DIPPR (5) or TRC(11).

For the heat of combustion of the coke we used the same approach of Dart and Oblad, shown in Figure 4 of their paper. We slightly improved their graph using data from DIPPR for pure components with the lowest (H/C) weight ratio of 5.25 % and a total of 18 hydrocarbons. The equations we fit to the data were :

$$y = 35245.8 + 462.568\,x \qquad (4.1)$$
$$y = 33449.2 + 850.4007\,x - 18.11667\,x^2 \qquad (4.2)$$

y = heat of combustion of solid hydrocarbons (J/g) ; x = 100 (H/C). Equation 4.2 has an average error of 0.27 % and maximum error of 1.16 % while for equation 4.1 the errors are 0.62 % and 1.48 %. The data points used and the linear fit are shown in Figure 2.

We also fit an equation to the variation of the specific heat of graphite (J/(kgK) with temperature from 250 to 1700 K, using data from the JANAF Tables (7) :

$$C_P = -503.7104 + 5.239696\,T - 4.40335D\text{-}3\,T^2 + 1.72241D\text{-}6\,T^3 - 2.516962D\text{-}10\,T^4 \quad (5)$$

The average error for C_P in equation (5) was 0.36 %

Equation (5) is necessary to calculate the value of ΔH_R at different temperatures. For the variation of the enthalpy of pure components we used the model of Aly and Lee (1) for the ideal gas with the constants given in DIPPR (2001). Because the pressure is about 35 psia, the isothermal correction of the enthalpy is negligible. The enthalpy basis temperature is 25°C.

For the naphtha and the liquid effluent the variation of enthalpy with temperature is estimated by the work of Kesler and Lee (8) with the equations given in the API Technical Data Book (2).

Our reaction of cracking may be simplified to :

naphtha $\rightarrow$ gas + liquid effluent + coke (6)

In the example below equations (3) and (4) were used for one of our pilot runs:

	Naphtha Feed	Liquid Effluent	
d 20/4	0.74330	0.77490	
d 60/60	0.74791	0.77954	(*)
K UOP	11.669	11.222	(*)
Qnet (MJ/kg)	40.500	41.600	
Sulfur (ppm)	794.0	562.0	

(*) - Calculated
Qnet = Net Heat of Combustion = - (Enthalpy of Combustion)
Simulated distillations D-2887 (% weight) - Measured
ASTM distillation (% volume) - Calculated by Riazi-Daubert (10)

Naphtha feed			Liquid Effluent		
%	D-2887 (C)	D-86 (C)	%	D-2887 (C)	D-86 (C)
0.0	0.50	37.76	0.0	-14.00	28.67
10.0	32.50	54.00	10.0	32.50	53.41
30.0	70.00	74.14	30.0	72.00	75.20
50.0	105.00	100.14	50.0	112.00	107.14
70.0	141.50	132.77	70.0	144.50	134.78
90.0	185.00	169.94	90.0	194.50	178.92
100.0	223.50	193.62	100.0	392.50	248.66

Gas Composition (mol fraction)

H2	0.0932	CH4	0.1186	C2H6	0.0333	C_2H_4	0.1002
C3H8	0.0229	C3H6	0.3224	i-C4H10	0.0412	n-C_4H_{10}	0.0125
1-C4H8	0.0349	i-C4H8	0.0599	t-2-C4H8	0.0420	c-2-C4H8	0.0304
i-C5H12	0.0562	n-C5H12	0.0050	1,3-C4H6	0.0017	(C5)+	0.0258

Gas Molecular weight = 39.703
Coke Heat of Combustion = -37675290 J/kg = -16198 BTU/lb (H/C = 0.05)
Gas Heat of Combustion = -46142392 J/kg = -19838 BTU/lb
Flow rates (g/h) :
Naphtha Feed = 598.5000 Gas + LPG = 148.6420
Liquid Effluent = 439.9229 Coke = 9.9351
Naphtha Enthalpy of Cracking at 25 C = 2162987 J/kg = 929.92 BTU/lb
Naphtha Enthalpy of Cracking at 579.90 °C = 2117380 J/kg = 910.31 BTU/lb

4- Influence of the precision of the Heats of Combustion

Heat of Combustion measurements presented an error of 0.6 MJ/kg, or about 1.5 %. Although this error appears to be small, it is greater than the one that can be obtained by the Precision Method for determining the heat of Combustion of Liquid Hydrocarbons Fuels by Bomb Calorimeter, ASTM D 4809-00 which has a repeatability of 0.091 MJ/kg or about 0.2 %. This error has a strong influence on ΔH_R which is calculated as a difference of large similar numbers. In the example given if we use Qnet = 41.0 MJ/kg instead of 41.6 we will obtain ΔH_R = 740.31 BTU/lb, 20.4 % less ! This problem is well known in the estimation of heats of formation ΔH_F from heats of combustion for pure substances.

5- Relation between ΔH_R and reaction severity

Figure 3 shows the influence of the reaction severity on the value of ΔH_R for the same feed and fifteen different tests. In these tests the reaction temperature was varied between 520 and 620 °C for each different catalyst. The catalyst/oil ratio was varied to achieve different temperatures of reaction. Higher severities are related to higher temperatures of reaction and higher yields of Fuel Gas and LPG. Because of the error in the Heats of Combustion for the naphtha feed and the liquid effluent, there is some dispersion of the values of ΔH_R. Tests were made with a straight-run naphtha and a naphtha from a Delayed Coking Unit but some laboratory analyses are still in course. For the straight run naphtha the pilot plant results are in accordance with the findings of Corma *et alii* (2005) that high severities are required to start the cracking reactions. Of course the use of more precise heats of combustion will improve the quality of our results. This will be undertaken in our future tests.

References

1. Aly,F.A. & L.L. Lee - Self consistent equations for calculating ideal gas heat capacity, enthalpy and entropy. Fluid Phase Equilibria, 6(3-4), 169-179, 1981
2. API Technical Data Book-Petroleum Refining. Washington,D.C. - Fourth Edition, Tenth Revision, 1987 , Chapter 7
3. Corma,A. *et alii* – Different process schemes for converting light straight run and fluid catalytic cracking naphthas in a FCC unit for maximum propylene production. Applied Catalysis A:General 265(2), 195-206, 8 July 2004
4. Dart, J.C. & A.G. Oblad – Heat of Cracking and Regeneration in Catalytic Cracking. Chemical and Engineering Progress, 45(2), 110-118, Feb 1949
5. DIPPR - Design Institute for Physical Properties - American Institute of Chemical Engineers / Brigham Young University (Provo, Utah) , Project 801 - Evaluated Process Design Data 2001
6. Edmister,W.C. - Applied Hydrocarbon Thermodynamics. Volume 2, 2nd Edition, Gulf Publishing Company , Houston, 1988, 240 pp.
7. JANAF Thermochemical Tables, Third Edition, Chase,M,W, *et alii* - Journal of Physical and Chemical Reference Data, Volume 14, 1985 , Supplement No. 1 1856 pp, 2 Volumes, American Chemical Society
8. Kesler,M.G. & Lee, B.I. - Improve prediction of enthalpy of fractions. Hydrocarbon Processing, 55(3), 153-158, March 1976
9. Ramos, J.G.F; Pinho, A.R.; Castillero, J.A.M.; Pedrosa Neto, P., Double Riser FCC - An Opportunity for the Petrochemical Industry, NPRA Annual Meeting, AM-06-13, Salt Lake City, 2006.
10. Riazi,M.R. & Daubert,T.E. -Analytical correlations interconvert distillation-curve types. Oil & Gas Journal 84(34) 50-54,57 Aug 25, 1986
11. TRC Thermodynamic Tables. Hydrocarbons in 12 Volumes ; Non-Hydrocarbons in 9 Volumes. Thermodynamics Research Center, The Texas A&M University System, College Station, Texas , 1992
12. Young,G.W. – Realistic Assessment of FCC Catalyst Performance in the Laboratory, Fluid Catalytic Cracking : Science and Technology, Studies in Surface Science and Catalysis, Vol. 76, Elsevier Science Publishers, 1993

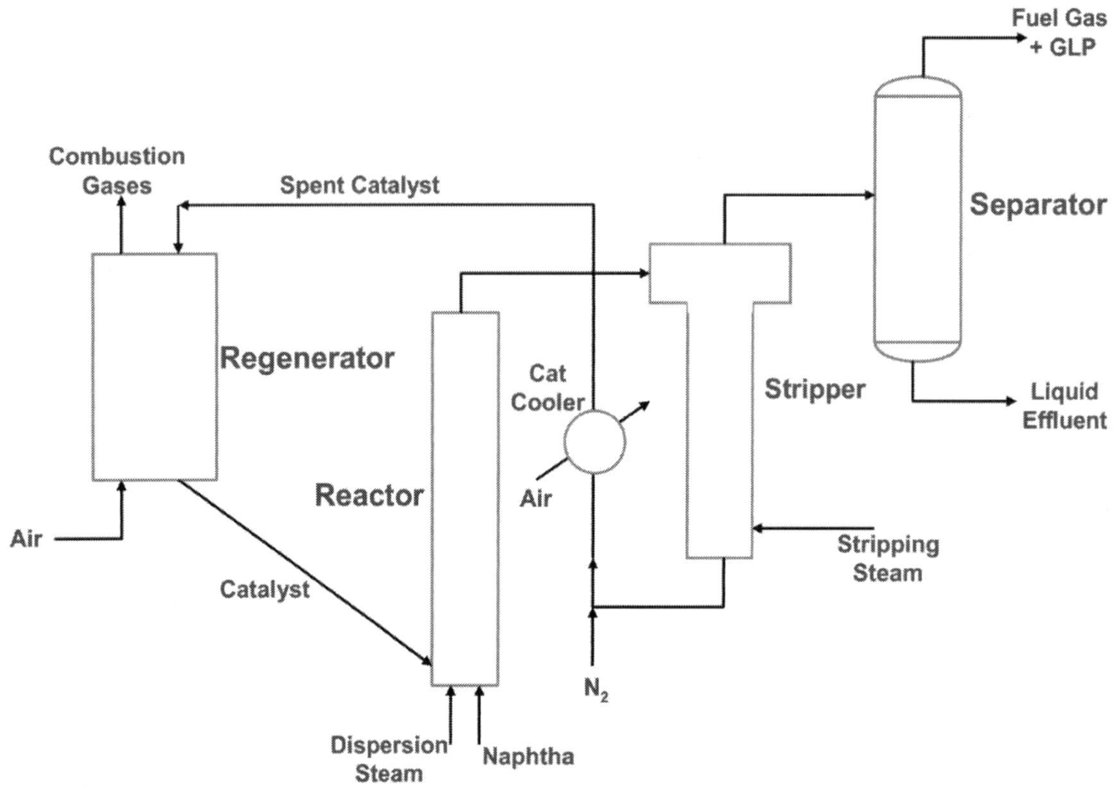

Figure 1 – Pilot Plant

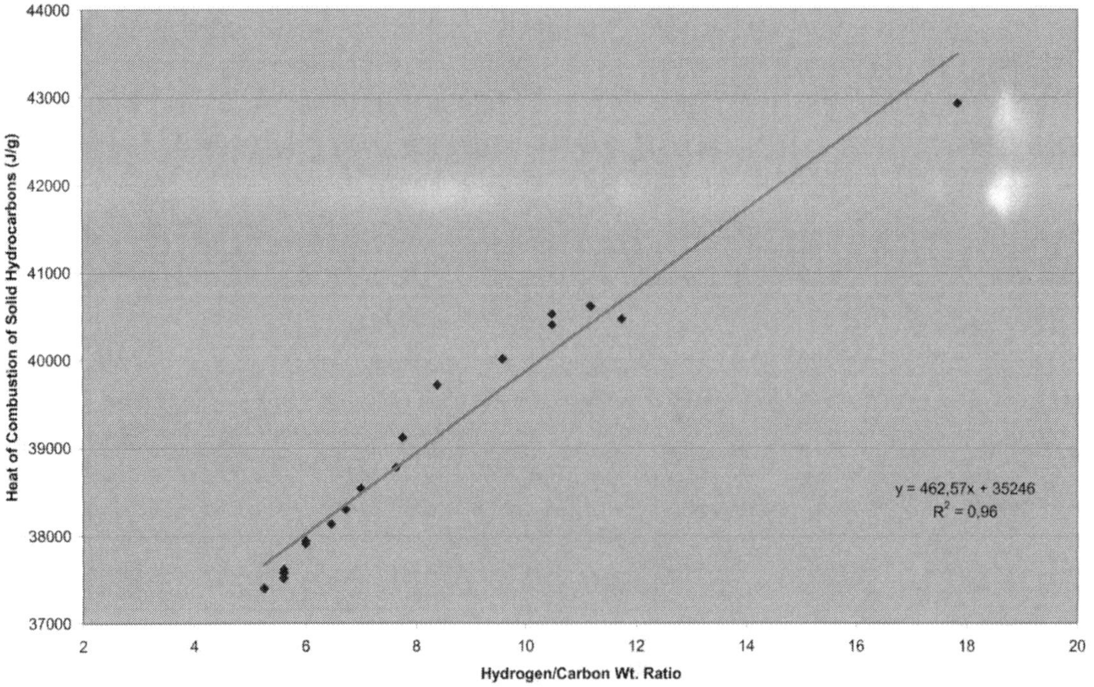

Figure 2 – Heat of Combustion of Hydrocarbons versus (H/C) weight %

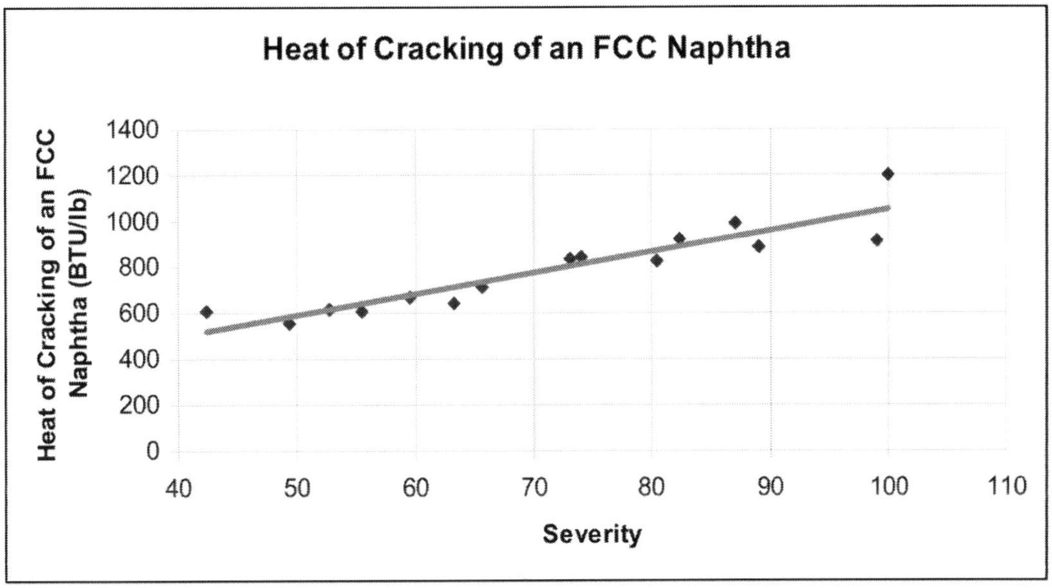

Figure 3 – Heat of Cracking of an FCC naphtha versus Severity

HOW FCCU TRICKLE VALVES AFFECT CATALYST LOSSES (Paper # 41770)

Donald F. Shaw, Senior Engineering Advisor, Carmagen Engineering, Inc.
Richard E. Walter, FCCU Process Engineer, Carmagen Engineering, Inc.

Abstract

High or abnormal catalyst losses from Fluid Catalytic Cracking Unit (FCCU) cyclone systems are sometimes attributed to plugged diplegs, often believed to be caused by one or more malfunctioning trickle valves. Sometimes, units are shut down and catalyst unloaded to "clear" a plugged dipleg. This paper will look at the fundamental forces that affect the opening and closing of a trickle valve for both positive and negative pressure cyclones. These forces are caused by the trickle valve design, weight of the flapper or counterweight and catalyst level in the dipleg. The opening and closing forces will be compared for the various types of trickle valves that are widely used on FCCU's. The features of various trickle valve configurations will also be discussed, including the advantages and disadvantages for various applications as well as measures taken to improve their performance, including their erosion resistance.

Background

Fluid solids processes circulate huge quantities of solids to achieve the conversion of the feed stocks to desired products, e.g. vacuum gas oils to gasoline in a Fluid Catalytic Cracking Unit (FCCU). Catalyst circulation rates can range from 10 to over 100 tons/minute between the reactor and the regenerator.

Cyclone systems typically consist of primary and secondary stage cyclones in both the reactor and regenerator to recover the catalyst solids from the reactor products and the regenerator flue gases. As more catalyst is collected with the cyclone systems, fewer processing problems occur with the product recovery circuit and the lower the potential solids emissions will be in the regenerator flue gas stream.

The role of the trickle valve has become increasingly more important as the FCC technology has advanced from primarily bed cracking to Short Contact Time Riser reactors and awareness of environmental concerns has driven refiners to minimize catalyst losses during both normal operations and startup and upset situations. Initially the FCC's did not utilize trickle valves on the cyclone diplegs. This worked fine once a bed level was established above the level of the dipleg discharges. The bed acts as the seal for the diplegs and low losses can be achieved. However, high catalyst losses were observed during unit startups before a bed was established or during upset situations where the bed levels were reduced below the level of the dipleg.

The addition of the trickle valve to the end of the dipleg was successful in maintaining low catalyst losses during the unit startup phase before bed levels were established and during upset situations. The trickle valves also help maintain the dipleg seals when negative pressure reactor cyclones are utilized in SCT or other operations which discharge the catalyst into a dilute phase.

While overall experience with trickle valves has been good, now there was a mechanical device which adds to the possible list of reasons for increased catalyst losses from either the reactor or regenerator. In addition to the catalyst and other hardware, there is now the concern that the trickle valves could be directly involved in causing the increased catalyst losses. While it is possible for trickle valves to be the cause of high catalyst losses, the following will describe how a properly designed trickle valve should not be designated as the primary cause for high catalyst losses.

Purpose of the Trickle Valve

Simply put, the purpose of the flapper valve on the discharge end of the dipleg is similar to that of a check valve in a process line in that the flapper prevents catalyst or gas flow up the dipleg. A cyclone will not work properly with gas flow up the dipleg. Upflow is prevented either by pressure differential, a mechanical device, or by massive down flow of solids. Figure 1 shows how a typical cyclone dipleg works. Cyclones can operate at a positive or negative pressure relative to the containment vessel. Also the diplegs can discharge into a bed or into a dilute phase. All these factors affect how the flapper valves function and what the flapper valve is expected to do.

The purpose of trickle valves that are inserted into the bed is primarily to minimize cat losses during start up and during other periods when the bed level is below the valve. Typically trickle valves that operate in the bed are provided with a shroud arrangement to prevent a gas bubble from entering the dipleg. Few problems are typically experienced with these trickle valves if they are immersed in a well fluidized bed. Therefore limited discussion will be directed toward this application.

Also positive pressure cyclones are not expected to experience flow up the diplegs except during start up or other low rate periods. Typically the diplegs are inserted in a bed with positive pressure cyclones in an attempt to create a backpressure and to minimize gas flow down the dipleg. Except for certain coupled cyclone systems in reactors, cyclones in FCCU's typically operate at a negative pressure relative to the containment vessel.

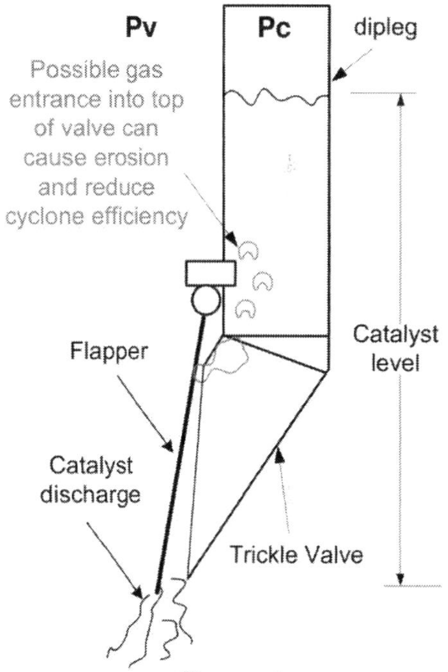

Figure 1

Figure 1 shows a typical negative pressure cyclone system with pressure in the vessel as Pv and the pressure in the cyclone as Pc, which is less than Pv. Typically Pc is less by the pressure drop through the cyclones. This differential pressure tends to push the flapper closed. To balance this closure force, a level of catalyst must build up in the dipleg creating a "hydrostatic" force to open the flapper. The weight of the flapper (or counterweights) creates an additional closure force that must also be overcome by the catalyst level. This paper will provide insight into and compare the forces that cause the opening and closing of several widely used trickle valves. Also some of the advantages and disadvantages of each type trickle valve will be presented.

Types of Trickle Valves

There are at least 5 different types of Trickle Valves that are widely used in FCCU's as listed below.

Typical considerations for selection of valve type:

- Is cyclone in negative or positive pressure service?
- What is the cyclone DP, or required catalyst level in dipleg?
- Will dipleg discharge in dilute phase or in bed?
- What is the mass flux rate down the dipleg?
- How large should the dipleg diameter be?
- What catalyst level is required to operate the valve?

The following summarizes some of the pros and cons for the various trickle valve applications.

Conventional Miter Body Trickle Valve	
• Pros – Relatively simple design – inexpensive – Depends on gravity to close – No tight clearance to seize or bind thus, flapper operation not affected by coke formation or refractory spalls – free draining – Can operate in bed and dilute phase – Support components subject to minimum wear – Blockage of flapper requires a large chunk of refractory or coke • Cons – Large vertical opening (approximately 2D) + when flapper rotates, gas can enter the trickle valve in dilute phase operation + height of catalyst in dipleg may not reach the top of the opening for large valves and/or low cyclone delta P – Relatively small seating forces due to small hanging angle	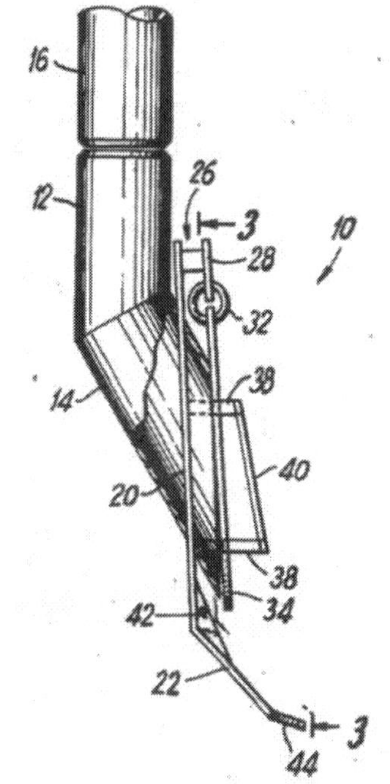 **Figure 2** **U.S. Patent 4,074,691**

Miter Trickle Valve With Refractory Lined Flapper and Blank Outs

- Pros
  - Similar to conventional miter trickle valve
  - Developed to address the vapor leakage by blanking a portion and adding refractory to help control erosion
  - Increased flapper weight increases seating forces and stability
- Cons
  - Similar to conventional miter trickle valve
  - Provides a wide seating surface
  - Excessive blank out can cause operating problems

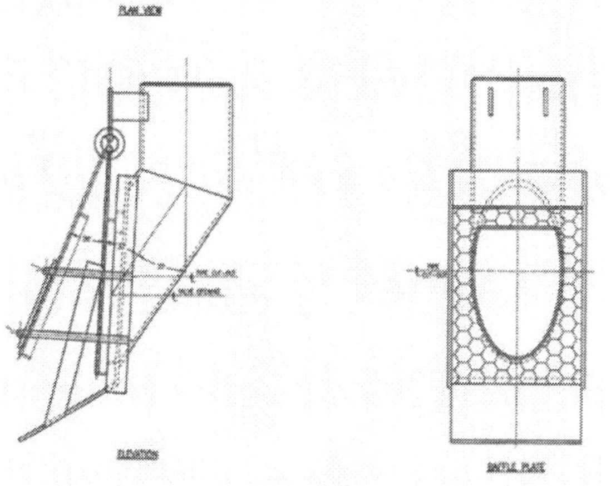

Figure 3

Conventional Trickle Valve With Elbow

- Pros
  - Similar to conventional trickle valve except:
    - + Reduced height of vertical opening better controls gas leakage at top of flapper (1.5 D)
    - + Erosion resistant refractories applied to critical surfaces can control erosion
    - + Increased flapper weight increases seating forces and stability
- Cons
  - Similar to conventional miter trickle valve

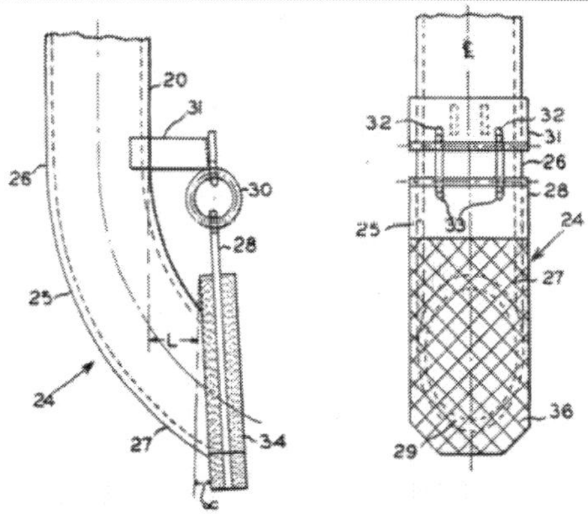

Figure 4
U.S. Patent 4,996,028

Horizontal Counterweighted Trickle Valve

- Pros
  - Addresses miter trickle valves large vertical opening concerns by eliminating and making seat horizontal with a positive seating force
  - Intended primarily for dilute phase service
- Cons
  - More complex design with counterweights, bearings, tight clearances, etc and greater fabrication control required
  - May have limitations in bed service
  - May be vulnerable to catalyst or coke build up on horizontal plate in service
  - Refractory or coke chunks may hang up on the seat resulting in flapper hanging open
  - Coke can form on the counterweight increasing the seating forces or on bearing surfaces binding operation – either open or closed

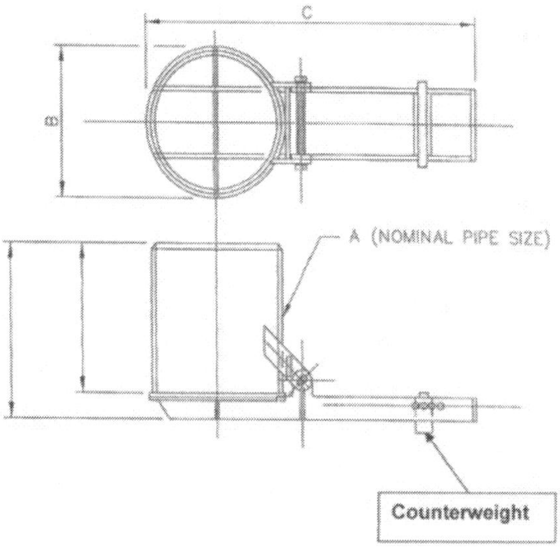

Figure 5

Angled, Counterweighted Trickle Valve

- Pros
  - Developed to address the concerns with a horizontal counterweighted plate holding up debris - angle of flapper and seat tend to clear out any debris
- Cons
  - Similar to horizontal counterweighted trickle valve except catalyst and debris is not expected to build up on flapper plate

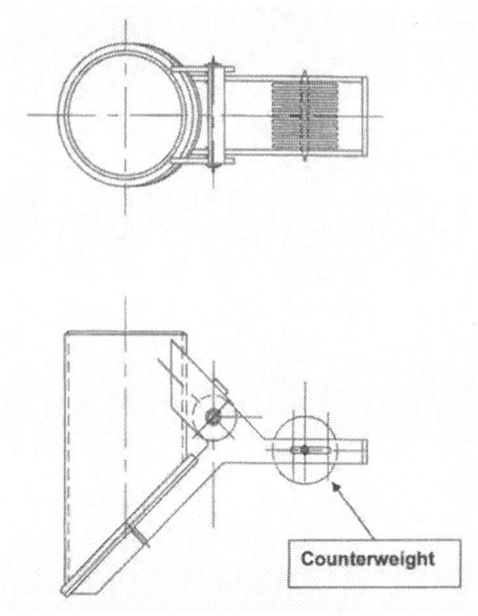

Figure 6

What Forces Act on a Counterweighted Trickle Valve

The forces acting on a counterweighted trickle valve are similar for both the horizontal and angled valve and these will be discussed first since it is relatively straightforward. Looking at Figures 5 and 6 we can see that the closure force that is built into the design of the counterweighted design is established by hanging a temporary weight on the center of the flapper plate and moving the counterweight along a moment arm to balance that weight (including the flapper plate). Based on calculations, it is relatively easy to determine the equivalent hydrostatic pressure that is required to just open the flapper (offset the counterweight force). It should be noted that these calculations ignore any accumulation of catalyst or coke on the horizontal flapper or any dynamic forces associated with the momentum of the catalyst. It is assumed that some allowance based on experience has been included in the settings for these considerations. The pressure due to catalyst height required to open against the counterweight (plus flapper) is simply:

P = W (hanging weight)/dipleg area (based on ID)

The actual pressure may vary depending on the trickle valve manufacturer; however, this is typically in the 0.1 to 0.2 psi range. The catalyst level in the dipleg must balance the counterweight plus the cyclone system Delta P. For a secondary cyclone the typical accumulated Cyclone Delta (Pv-Pc) is on the order of 1.5 psi so that the total closure pressure is on the order of 1.6 to 1.7 psi. For a catalyst density of 29 lb/ft^3 a catalyst dipleg level of almost 8 feet is required to just open the valve against the cyclone DP and the counterweight.

What Forces Act on a Miter Body Trickle Valve

The forces that act on and open a traditional miter bend trickle valve are also quite simple but are more complicated to calculate. The following addresses the forces and moments that control the opening and operation of a typical cyclone trickle valve installed in the dilute phase of a fluidized bed reactor or regenerator using negative pressure primary or secondary cyclones. For diplegs that operate in a bed, there are additional hydrostatic forces due to the turbulent nature of the bed that are not considered in this study. Also for diplegs submerged in a bed, the hydrostatic head of the catalyst bed height above the trickle valve discharge must be accounted for. This calculation basis covers typical trickle valve type as shown in Figure 2.

The calculations are based primarily on the forces acting on the trickle valve flapper plate as it begins to open. The model for the equations is based on the fact that the flapper plate is supported at the top with two hinge rings and that the plate tends to rotate around the hinge rings as it opens. As opening forces act on the flapper plate, the rotation around the support hinges causes the opening at the bottom of the seat to increase more than the opening at the top of the plate. Of course the hinge ring is not an absolute pivot point and there can be some small lateral translation of the plate. This small translation, however, does not have a significant affect on the calculations. Also, these calculations do not include any dynamic forces which will tend to further open the trickle valve once it has opened and catalyst is flowing. See Figure 7 for the terminology used in the equations for miter trickle valve.

Trickle Valve Forces

The following forces act upon a trickle valve to seat and unseat the flapper plate against the opening in the trickle valve body.

An opening force arises from the pressure acting on the flapper plate (area at the valve opening) due to the hydrostatic head of catalyst in the dipleg. This pressure acts horizontally on the surface of the flapper plate. For simplification, we are taking the calculated pressure at the middle of the trickle valve opening and spreading this over the entire area of the elliptical trickle valve opening. This assumption is valid as long as there is a reasonable catalyst level in the dipleg.

A net closing force occurs due to the differential gas pressure that acts across the flapper plate, i.e., the external pressure that exists inside the vessel (Pv) minus the pressure inside the cyclone body (Pc). This net differential pressure only acts on the area of the valve discharge opening, and tends to close the valve in a negative pressure cyclone system since the vessel pressure is higher than the pressure inside the cyclone.

Since the trickle valve flapper plate hangs from its support ring at a small angle (approximately 3 degrees from vertical), the weight of the flapper plate creates a seating force against the valve opening. This seating force is determined by the moment created by the horizontal offset (y) of the flapper plate mass from the vertical centerline through the hinge ring support.

Dynamic forces have not been included. Dynamic forces are caused by the momentum of the falling gas/catalyst stream in the dipleg impinging on the flapper plate. Instead, the calculations determine the condition where the trickle valve just opens and any such dynamic forces will tend to further open the valve. As the valve opens further, the offset (y) increases, thus increasing the closure forces until the opening forces/moments (catalyst level) are in balance with the closing forces/moments.

Calculation of Trickle Valve Moments

To calculate the effect of the trickle valve flapper weight, catalyst head in the dipleg, and the differential pressure between the vessel and inside the cyclone, one takes moments around the flapper plate vertical support point (point C in Figure 7).

For static equilibrium the summation of Moments (Mc) around point C must equal zero.

$$\sum Mc = 0 \qquad \text{Eq. (1)}$$

At the point of the trickle valve opening, the moment due to pressure on the inside of the flapper (Pc + static catalyst head in the dipleg) must equal the closure moment due to the weight of the trickle valve flapper and the pressure in the vessel. Using the moment arm lengths shown in the attached Figure 7, the following are determined:

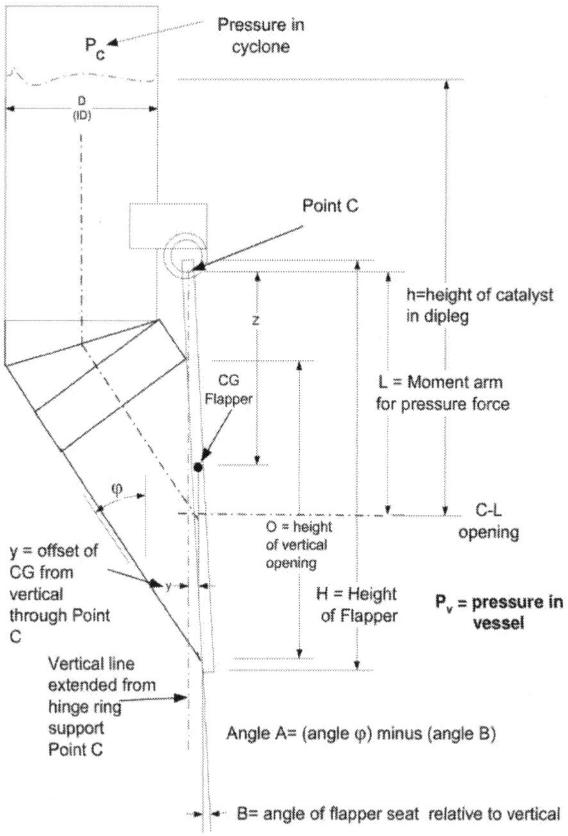

Figure 7

M_1 - Opening Moment - Moment due to catalyst head in the dipleg (counterclockwise moment)

$$M_1 = L * h * \frac{\rho}{144} * A_o = L * h * \frac{\rho}{144} * \frac{\pi}{4} * \frac{D^2}{\sin(A)}$$

Eq. (2)

Where:
- M_1 = moment due to catalyst head, inch-lb
- ρ = density of catalyst in the dipleg, lb/ft^3
- h = height of catalyst in dipleg, feet
- A_o = area of trickle valve opening, inch^2
- L = moment arm length, inch (see Figure 7)
- D = dipleg ID, inch
- φ = angle for the miter bend
- B = angle of the trickle valve flapper seat
- A = angle of flapper seat on miter = (φ - B) (see Figure 7)

M_2 - Moment due to Pressure in cyclone (counterclockwise moment)

$$M_2 = L * Pc * A_o = L * Pc * \frac{\pi}{4} * \frac{D^2}{\sin(A)} \qquad \text{Eq. (3)}$$

Where: M_2 = moment due to pressure in the cyclone, inch-lb
Pc = pressure in cyclone body, psig

M_3 - Moment due to Pressure (Pv) in vessel (clockwise moment)

$$M_3 = L * Pv * A_o = L * Pv * \frac{\pi}{4} * \frac{D^2}{\sin(A)} \qquad \text{Eq. (4)}$$

M_4 - Moment due to trickle valve flapper weight (clockwise moment)

$$M_4 = y * W_f \qquad \text{Eq. (5)}$$

Where: M_4 = moment due to weight of flapper, inch-lb
Wf = weight of flapper, lb
y = horizontal offset for trickle valve flapper center of gravity (CG) from the hinge ring due to angle (B), inch [y = z * tan (B)]
z = distance from the flapper support hinge to flapper center of gravity (CG)

Considering clockwise moments as positive and counter clock wise moments as negative we have:

$$-M_1 - M_2 + M_3 + M_4 = 0 \qquad \text{Therefore,} \qquad \text{Eq. (6)}$$

$$-\left(L * h * \frac{\rho}{144} * \frac{\pi}{4} * \frac{D^2}{\sin(A)}\right) - \left(L * Pc * \frac{\pi}{4} * \frac{D^2}{\sin(A)}\right) + \left(L * Pv * \frac{\pi}{4} * \frac{D^2}{\sin(A)}\right) + (y * W_f) = 0 \quad \text{Eq. (7)}$$

Combining the pressure moments to get a differential pressure moment one obtains the following:

$$-\left(L * h * \frac{\rho}{144} * \frac{\pi}{4} * \frac{D^2}{\sin(A)}\right) + \left(L * \frac{\pi}{4} * \frac{D^2}{\sin(A)}\right) * (Pv - Pc) + (y * W_f) = 0 \qquad \text{Eq. (8)}$$

Simplifying this equation we get:

$$\left(L * \frac{\pi}{4} * \frac{D^2}{\sin(A)}\right) * \left(h * \frac{\rho}{144} + (Pc - Pv)\right) = y * W_f \qquad \text{Eq. (9)}$$

Solving for the head of catalyst in the dipleg required to open the trickle valve one obtains:

$$h = \left[\left(\frac{1}{L} * \frac{4}{\pi} * \frac{\sin(A)}{D^2} * y * W_f\right) + (Pv - Pc)\right] * \frac{144}{\rho} \qquad \text{Eq. (10)}$$

For an elbow type valve shown in Figure 3 the equation would be modified as follows:

$$h = \left[\left(\frac{1}{L} * \frac{4}{\pi} * \frac{0.67}{D^2} * y * W_f\right) + Pv - Pc\right] * \frac{144}{\rho} \qquad \text{Eq. (11)}$$

Example of Calculation of the Moments

The following is an example of the moments calculated for a typical 6 inch miter bend trickle valve.

For a typical 6 inch valve on a secondary dipleg (without refractory lining on the flapper)

φ	=	35 degree
B	=	3 degrees
A	=	$\varphi - B$ = 32 degrees
L	=	9.9 inch
D	=	6 inch
y	=	0.392 inch
W_f	=	17 lb (1/2" thick x 8" x 15", no refractory)
Pv-Pc	=	1.5 psi
ρ	=	29 lb/ft^3

The overall moment balance is determined using equation (2), $-M_1 + (M_3 - M_2) + M_4 = 0$

Calculating the individual moments for the example one determines the moments as follows:

Starting with Eq. (2) for the trickle valve opening moment:

$$M_1 = 9.9 * h * \frac{29}{144} * \frac{\pi}{4} * \frac{6^2}{\sin(32)}$$

$$M_1 = 106.4 * h$$

Combining Eqs. (3) and (4) to get the differential pressure moment:

$$M_3 - M_2 = L * \frac{\pi}{4} * \frac{D^2}{\sin(A)}(Pv - Pc) = 9.9 * \frac{\pi}{4} * \frac{6^2}{\sin(32)} * (1.5)$$

$$M_3 - M_2 = 792.3 \text{ in} - \text{lb}$$ - Due to differential pressure

To calculate the trickle valve flapper weight moment, start with Eq. (5):

$$M_4 = y * W_f$$

$$M_4 = 0.392 * 17$$

M_4 = 6.7 in-lb - due to flapper weight

Solving the equation for M_1, the opening moment, one obtains:

$M_1 = (M_3\text{-}M_2) + M_4$ = 792.3 + 6.7 = 799 in-lb

Since M_1 = 106.4 * h

Therefore, h (height of catalyst in dipleg) = 7.5 ft

It is important to note that for normal operation the height of catalyst in the dipleg required for the pressure balance is not materially affected by the flapper weight for the small 3 degree hanging angle. The weight effect of the flapper is less than 1% of the total opening moment for the trickle valve flapper. If the weight of the flapper is doubled, the opening moment increases to 13.3 inch-lb and the height of catalyst in the dipleg is increased by approximately 0.06 feet to approximately 7.6 ft.

Another impact considered is that during normal operation, the flapper is opened wider than 3 degrees; therefore, the "weight effect" is increased as the "y" distance increases. If we assume that the flapper is operating at a 20 degree angle (large for a secondary cyclone, but possibly typical for a primary) with a double weight flapper, the corresponding "weight moment" increases to 87 in-lb, and the required height of catalyst increases to approximately 8.2 ft. In this case it is seen that the weight effect of the flapper on total opening moment is still only approximately 10% of the pressure moment for a secondary cyclone.

Effect of Flapper Weight Not Significant on Closure Forces

The above calculations demonstrate that flapper weight does not have a significant effect when compared to Delta P on the opening of a trickle valve where the normal hanging angle is 3 degrees from vertical. It should be noted, however, that increasing the weight does have an positive effect on the stability of the flapper and the dipleg level since small pressure fluctuations are dampened out by the weight of the flapper or changes in the level of the catalyst in the dipleg.

The weight stability effect has also been reported in cold flow studies of light plastic flappers where the catalyst level in the dipleg can fluctuate significantly because small pressure fluctuations cause erratic movement of the flapper. Cold flow studies with steel flapper plates simulating an actual trickle valve also report that increasing the weight of the trickle valve flapper can stabilize the catalyst level in the dipleg, especially as the flapper plate weight is increased (due to the addition of refractory).

Elbow Type Trickle Valves

Figure 4 shows a type of trickle valve used on FCCU diplegs in dilute phase service which employs a patented elbow configuration (U.S. Patent 4,996,028) rather than a miter to provide the seating surface. The principal benefit of the elbow is that the vertical opening (O in Figure 7) is reduced by approximately 25%. Also, this design increases the closure force since it adds weight to the flapper with refractory on both sides. Typically the vertical opening for a 35 degree miter valve is approximately 2*D whereas the vertical opening for an elbow valve is typically 1.5*D.

Similarly, it should be noted that cold flow studies that artificially increase flapper weight by adding weights on pegs should be considered as suspect since the weights are generally offset from the center of mass of the flapper and the moment arm effect is exaggerated. Simple mechanics demonstrate that the flapper does not simply rotate around the hinges with the eccentric weight, but rather the top of the flapper tends to open more than the bottom thus allowing more gases to enter through the opening at the top of the valve.

Based on observing erosion patterns on trickle valve flapper plates it has been determined that as the flapper opens, gases from the vessel tend to enter the dipleg at the top of the flapper. A principal advantage of the elbow valve and trickle valve with blank out is that they reduce the quantity of gas that enters the dipleg. Typical erosion patterns on flappers have been observed as horseshoe in shape (inverted), see Figure 8. As these gases enter the top, they tend to cause flow disruption in the dipleg and erratic dipleg levels. It should be noted that the vertical opening for a typical 24 inch dipleg is 48 inches high with a miter valve. This height opening can be significant compared to a typical dipleg level in a primary cyclone.

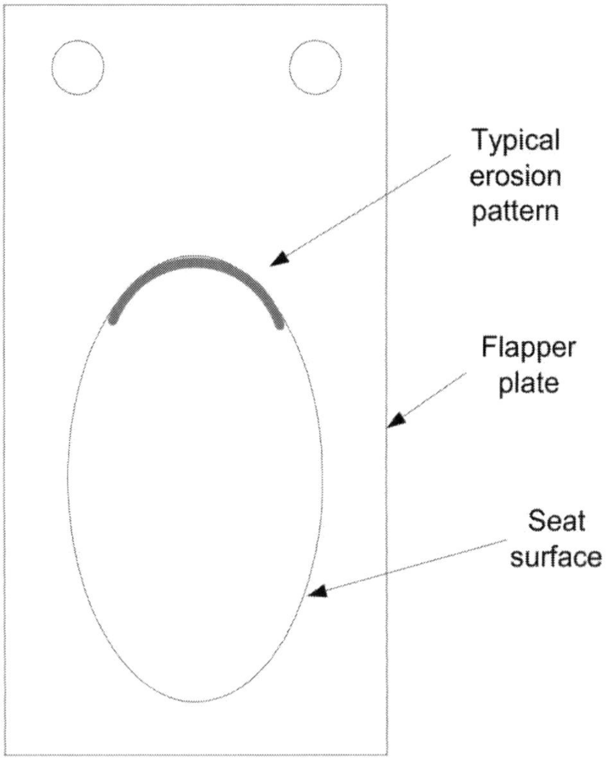

Figure 8

Effect of Flapper Weight

Figure 9 illustrates the effect of trickle valve flapper weight on the dipleg catalyst level. Note that the catalyst level required due to flapper weight is given in inches of catalyst therefore a 1/2" steel plate with 1" refractory on each side requires only about 2" of catalyst to open.

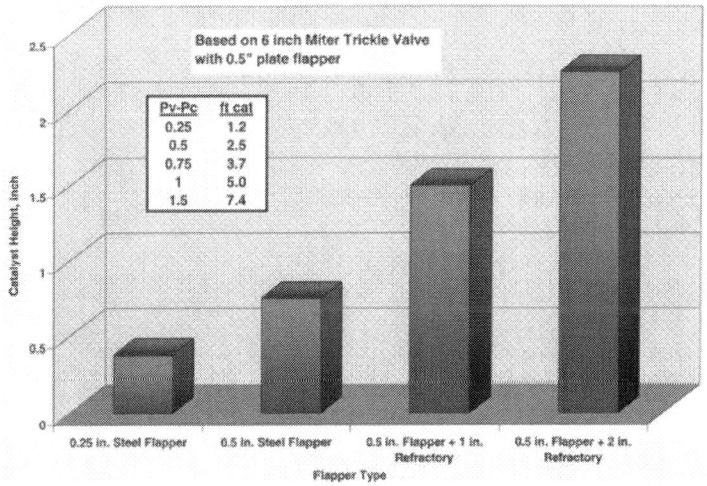

Figure 9

Effect of Increasing Seat Angle

Figure 10 shows how the seating force is increased with increasing the seat angle for trickle valves. Seat angles between 3 and 10 degrees have been investigated. For example increasing the seat angle from 3 to 6 degrees approximately doubles the seating forces. These forces are small relative to the closure force due to the differential pressure as shown in the table. Based on this, it is suggested that consideration be given to increasing the seat angle for trickle valves to control leakage up to the secondary dipleg during startup operation.

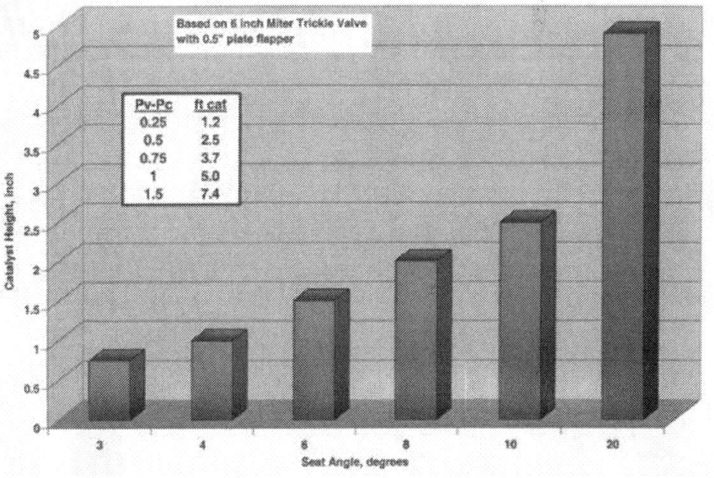

Figure 10

Possible Causes of Trickle Valve Malfunction

- Gas leaking up the diplegs
- Binding of bearings on counterweighted trickle valves
- Chunks of refractory/coke hanging flapper open
- Coke, refractory or debris between flapper and stop bars keeping flapper closed
- Erosion of seating surfaces
- Impingement from discharging solids in close proximity

Summary

A variety of trickle valve designs have been used in FCCU's. Experience is an important consideration in selecting the valve type. Counterweighted valves have primarily been used in dilute phase service. Trickle valves with flappers hanging at a small angle have been used in both bed and dilute phase service. The counterweighted trickle valves typically provide a greater seating force than a conventional miter trickle valve. The seating forces associated with a trickle valve flapper are quite small and suggests that there is little risk in increasing flapper weight and/or increasing the flapper hanging angle.

Each of the trickle valve options offers some advantages and disadvantages as discussed. The horizontal counterweighted trickle valves are more complex and may be more vulnerable to binding at the internal bearings or hanging up due to coking or chunks of refractory. The angled counterweight valve alleviates some of these concerns. The conventional miter trickle valve has the advantage of being much simpler but has the disadvantage for certain services of providing a long vertical opening that can allow gas to enter at the top while discharging catalyst at the bottom. There are several enhancements that improve on this limitation. The perceptions that the weight of the flapper plate is a significant factor in the maloperation of a trickle valve are not supported by the calculations with a negative pressure cyclone. The major forces controlling the opening of the trickle valves are the differential pressure forces across the flapper. These forces are balanced by the catalyst head in the dipleg. A "heavy" flapper is an unlikely cause of high cat losses.

In a negative pressure cyclone, the primary closing force arises from a net pressure differential between the vessel and the cyclone body. In a negative pressure cyclone this tends to push the flapper closed and the opening of the flapper is created by a catalyst level in the dipleg which balances the differential pressure. Because the flapper hangs at a small angle (approximately 3 degrees) the weight effect (at opening) of the flapper is approximately 1% of the moment due to differential pressure.

Submerging the bottom of the dipleg in the bed offers some advantages; however this is not always desirable from the process viewpoint. The secondary trickle valve in dilute phase is often considered by many as the most difficult application since the catalyst flux can be quite small. Also gas leakage into the dipleg during periods with low flux can cause problems if the gap between the flapper and the seat allows too much gas to leak into the dipleg. For these cases the fabrication tolerances can be critical.

As has been discussed, a properly designed trickle valve should not be the first item on the list of reasons for high catalyst losses. Many other items, such as mechanical damage, high attrition rate catalyst, potential coking or coke spalling from the cyclones, unexpected high velocity attrition source, should be heading the list of potential causes for the high catalyst losses.

The authors wish to thank EMTROL, LLC and Fisher-Klosterman, Inc. for providing some sketches on typical valve types.

56c Heat Transfer and Heat Removal in Flowing Dense Phase Fluidized Beds

Peter Van Opdorp and david lomas. development, uop llc, 25 east algonquin road, des plaines, IL 60016

Abstract

As refiners considered the processing of less expensive atmospheric residues from crude oil in the early 1980s, the removal of excess heat from commercial Fluidized Catalytic Cracking (FCC) units became an important issue. In addition to the traditional vacuum gas oil portion of the crude (650–1050°F), these residues contained heavier 1,050+°F material that possess a higher concentration of coke precursors. Cracking of the heavier oil and its contaminant carbon residue posed an operating situation requiring higher than normal heat release in the FCC regenerator. In combination with higher regenerator temperatures, other contaminants in the heavier feedstocks such as nickel and vanadium cause both accelerated catalyst deactivation and poorer yield performance. A heat removal device was needed to remove excess heat from the circulating catalyst and allow the refinery operator a means of controlling the regenerator and the FCC unit's heat balance.

This paper relates the history and work process for the successful development of such a heat removal device and the key engineering and fluidization parameters to quantitatively determine the heat transfer coefficients and heat duties are examined. The pertinent heat transfer data and correlations from the literature and the mathematical treatment of the device are illustrated. The result is an understanding of flowing gas and emulsion phase influences, surprisingly high heat transfer capabilities and simple designs that are still valid, reliable and valuable 25 years later.

Molecular Based Kinetic Modeling of Fcc Process

Ryuzo Tanaka, Craig A. Bennett, Wei Wei, Gang Hou and Michael T. Klein

School of Engineering, Rutgers, The State University of New Jersey, 98 Brett Rd, Piscataway, NJ 08854-8058

Abstract

Computational tools for building and editing kinetic models for paraffin fluid catalytic cracking (FCC) were developed. A special emphasis was placed on the relationship between the mechanistic- and pathways-level models. Mechanistic models incorporate more fundamental kinetic information and hence, in principle, have a wider range of applicability. Their main drawback is the explosion in the number of molecular and intermediate species and the mathematical stiffness associated with the relative time constants of these types of species. These issues are absent in pathways-level models, which leads to a dramatic reduction in the solution time and justifies their use for cases of practical interest. But this often is accompanied by the loss of some chemical significance. The practical resolution of this conflict was to guide the development of the pathways level model by mechanistic insights into the fluid catalytic cracking chemistry, including the incorporation of the role of catalyst sites' number and strength, at the fundamental level, and the Cat/Oil ratio, at the process level. Catalyst deactivation by coking was also incorporated.

Introduction

Fluid catalytic cracking (FCC) is the quintessential process for the conversion of heavy oils into lighter and more-valuable products. FCC can handle a wide range of feeds, from resid-containing mixtures to paraffinic crudes and cycle stocks. The ability to control, manipulate and optimize this process is enhanced by a rigorous representation of the process chemistry. This motivates the present interest in a molecule-based FCC kinetics model.

Traditional hydrocarbon conversion process models have implemented lumped kinetics schemes, where the molecules are aggregated into lumps defined by global properties, such as boiling point or solubility. Molecular information is obscured due to the multi-component nature of each lump. However, increasing environmental concerns and the desire for better control and manipulation of the process chemistry have focused attention on the molecular composition of both the feedstocks and their refined products. Modeling approaches that account for the molecular fundamentals underlying reaction of complex feeds and the subsequent prediction of molecular properties require an unprecedented level of molecular detail. This motivated the development of computer tools not only to solve and optimize but also to formulate the model.

This paper describes these tools in the context of their application to FCC chemistry. The basic approach and overview of the tools are presented first. This is followed by a detailed description of the steps involved in model formulation, optimization and use.

Approach and Overview

The Kinetic Model Editor Overview

The Kinetic Model Editor was originally based on the OdeGen[1,2] parser and was designed as a set of modeling tools to facilitate the conversion of reaction networks into mathematical equations written in the C language. A spreadsheet-based user interface was designed to allow users to carry out a series of modeling procedures automatically, including the reaction network conversion, process model configurations, data entry, compilations, executions and result analysis and visualization. The user interface not only provides a step-by-step guide to the model building, but also ensures the consistency of the internal connections between different parts of the kinetic model.

KME was designed to convert a list of chemical reactions constituting the reaction network into mathematical equations for the mass and energy balances and kinetic rate calculations. Figure 1 shows a typical reaction file as the input in which the components and reactions are listed in the KME syntax that allows KME to recognize the type of reactions, reactants, products and calculate the associated rate expressions.

	REACTIONS
1	[L] + P04 <-> MO02 + P02 + [L]
2	[L] + P05 <-> MO02 + P03+ [L]
3	[L] + P05 <-> MO03 + P02 + [L]
4	[L] + P06 <-> MO02 + P04 + [L]
5	[L] + P06 <-> MO03 + P03 + [L]
6	[L] + P06 <-> MO04 + P02 + [L]
7	[L] + P07 <-> MO02 + P05 + [L]
8	[L] + P07 <-> MO03 + P04 + [L]
9	[L] + P07 <-> MO04 + P03 + [L]
10	[L] + P07 <-> MO05 + P02 + [L]
11	[L] + P08 <-> MO02 + P06 + [L]
12	[L] + P08 <-> MO03 + P05 + [L]
13	[L] + P08 <-> MO04 + P04 + [L]
14	[L] + P08 <-> MO05 + P03 + [L]
15	[L] + P08 <-> MO06 + P02 + [L]
16	[L] + P09 <-> MO02 + P07 + [L]
28	[L] + P10 <-> MO08 + P02 + [L]
29	[L] + P11 <-> MO03 + P08 + [L]
30	[L] + P12 <-> MO03 + P09 + [L]
31	[L] + P13 <-> MO03 + P10 + [L]
32	[L] + P14 <-> MO03 + P11 + [L]
33	[L] + P15 <-> MO03 + P12 + [L]
34	[L] + P16 <-> MO03 + P13 + [L]

Figure 1. Reaction List in KME Syntax

KME uses this input file to create a subroutine program, Ode.c, written in the C language, in which the reaction rate calculations and species mass balance equations are specified. The Ode.c program code is compiled with a set of other supporting model codes and together they become the kinetic model program itself. Among these subroutines, the Driver.c program is designed to handle the kinetic model inputs including initial compositions, initial reactor configurations and operating conditions, kinetic rate parameters, etc. The Model.c program is designed as the process model used to organize the model calculations according to different type of reactors and different type of chemistries, i.e., homogeneous chemistry, heterogeneous chemistry, catalyst deactivation, etc.

To an average model user, these model pre-configuration processes can be quite complicated and disordered. A spreadsheet-based graphical user interface was thus developed to guide the user through these complicated modeling procedures while still being able to focus on fundamental chemistries and kinetics by shifting the burden of programming and bookkeeping to the KME computer software. This interface environment was chosen because of its comprehensive data processing features and the fact that almost all scientists and engineers are comfortable working with it. Moreover, its built-in Visual Basic for Applications (VBA) is a powerful and extremely efficient graphical interface development tool.

Building the Kinetic Model Equations

KME's "Build Model" function converts a list of reactions into ordinary differential equations in the Ode.c file. The Ode.c file contains the reaction rate expressions including kinetic rate constant calculation subroutines. As an example, a reaction network can be as simple as

A + B -> C

C -> D

These reactions are processed in KME (Build Model) to create code such as that shown in Figure 2, which contains the reaction rate expression for the two reactions and the differential equations for each species, organizing their individual balance equations according to the reactions in which they are involved.

```
/* begin reactions */
 rxn[0] =
k(lgA1,E1)*(y[A]*y[B]);
   rxn[1] = k(lgA2,E2)*(y[C]);
 /* end reactions */

   /* ODE for A */
  dydt[A] = 0 - rxn[0];
   /* ODE for B */
  dydt[B] = 0
 - rxn[0];
   /* ODE for C */
  dydt[C] = 0 - rxn[1] + rxn[0];
```

Figure 2. Output of KME

The species appearing in the reaction list can be gas-phase species, catalyst sites or surface species. Gas-phase species can be denoted directly by their chemical formula, i.e. O2, H2O, or by any kind of conventional names, i.e. Water, Butane, etc. The catalyst sites are denoted with the name of the catalyst site enclosed by a pair of square brackets, e.g., [L]. Surface species are simply denoted as the name of gas phase species followed by the specific catalyst site, for example H2O[L1] represents adsorbed water on the catalyst site L1.The special syntax used in the reaction network allows KME to discern not only species types but also reaction types and rate laws.

For example, the reaction "A + [L] <-> A[L]" can be identified as an adsorption reaction where the gas phase species A is adsorbed onto the catalyst site L, denoted by the square bracket, resulting in the surface species A[L]. The "<->" sign indicates that the reaction is a reversible reaction which would be later reflected through the format of the reaction rate parameters. In the case of irreversible reactions, a single direction arrow "->" is used. In the above example, the reaction rate equation will be automatically written as:

$$r^0 = k^* (C_A{}^*\Theta_L - \Theta_{AL}/Ke)$$

where r^0 in the unit of moles per moles of catalyst site per unit time (equivalent to [1/s]); C_A is the concentration of gas phase species A in moles per volume; Θ is the dimensionless fractional coverage of empty catalyst site [L] or that of the surface species A[L] normalized by the total number of catalyst sites $[L]_T$; k is the rate constant and, in this case, has the unit of [mole/mole-site*time*(mole/volume)$^{-1}$] meaning moles per mole site per unit time and per unit concentration of gas phase species; and Ke is the equilibrium constant with the unit of [1/(mole/volume)]. Note that the units of rate constant and equilibrium constants will depend on the type and order of reactions, e.g. adsorption, desorption, surface reaction, etc. The rate constant k and equilibrium constant Ke are calculated using the standard Arrhenius format with $A_{forward}$, E, ΔH and $A_{reverse}$ as input values provided by the user or by parameter tuning.

Still using the reaction A + [L] -> A[L] as an example, in a plug flow reactor, the ordinary differential equations for species A, catalyst site [L] and surface species A[L] can be written as:

$$dF_A/dV = - (\rho L)^*r^0; \qquad (1)$$

$$d\Theta_L/dV = - r^0; \qquad (2)$$

$$d\Theta_{AL}/dV = r^0; \qquad (3)$$

where F_A is the molar flow rate of A, V is the reactor volume and ρL is the concentration of total catalyst sites per unit volume of reactor. Note that the catalyst site [L] and surface species A[L] can be written in the form of ODE or DAE equations.

Automatic Data Entry Templates

The "Build Model" function of KME also generates input templates for the user to enter the values of the input concentration, operating conditions and reaction rate parameters. These templates are later conveniently displayed as a data-entry form through the spreadsheet user-interface with which the user can easily manipulate the data accordingly.

The gas phase and surface species are easily distinguished by their units, mol/s or surface coverage fraction Θ. In this example, L is the vacant catalyst site. All the surface species are renamed by connecting the name of the gas phase species with the name of their attached sites via a "_" symbol. The catalyst loading is specified in operating conditions file.

Other data entry templates include the reaction rate constant file and the operating conditions file. If model optimization is to be performed, another two template files are generated: the observed product composition file and the objective function weight file specifying the level of confidence in of each observation values. Each file is automatically generated by KME according to the type of reactors and other specific model options.

Catalyst Activity Model

Many chemical processes involve a time dependency of catalyst activity. In some cases, catalyst promoters are added along with the feedstock to activate or enhance the catalyst activity. Therefore KME allows for simulation of not only the main reactions but also the reactions leading to time-dependent catalyst activity.

KME simulates these processes by modeling catalyst activity changes as a set of slow reactions, with a "Time On Stream" time characteristic time of, typically, of order days or months, as opposed to seconds for the main reactions. Due to this significant difference in the two time scales, the main and activity reactions can be decoupled as two separate reaction systems to avoid the numerical issues associated with solving PDE equations. This activity transient reaction network is used to calculate the rate of catalyst aging or catalyst activity changes. After an iteration of the main reaction calculation, the secondary reaction network can carry out calculations to update catalyst site availability and consumption rates using the local compositions provided by the main reaction simulation.

This additional reaction network can be entered through the KME user interface using the same syntax as the main reaction. In fact, it can be entered together with the main reactions by using a keyword "SAM" (catalyst selectivity and activity model).

The KME User Interface

The spreadsheet-based KME interface is developed with VBA and embedded in the workbook. Upon opening of the workbook, VBA macros automatically build an extra set of menu buttons designated to KME modeling functions, including reaction network editing, model customization, data entry, program code compilation, execution and result plotting. These menu buttons can be enabled (black) or disabled (grey) depending on the model type or the current status of the model development. For example, the execution function will not be unlocked before program codes are generated and compiled. This feature restricts and, at the same time, guides the user through the model building process.

C16 Paraffin FCC Model Development

Reaction Mechanism

The mechanism of paraffin FCC chemistry is, essentially, the carbenium ion chemistry of acid catalysis coupled with background thermal reactions.[3-8] Herein we focus on the acid cracking component. Figure 3 summarizes the elementary steps used to guide the development of the pathways-level C16 FCC kinetic model.

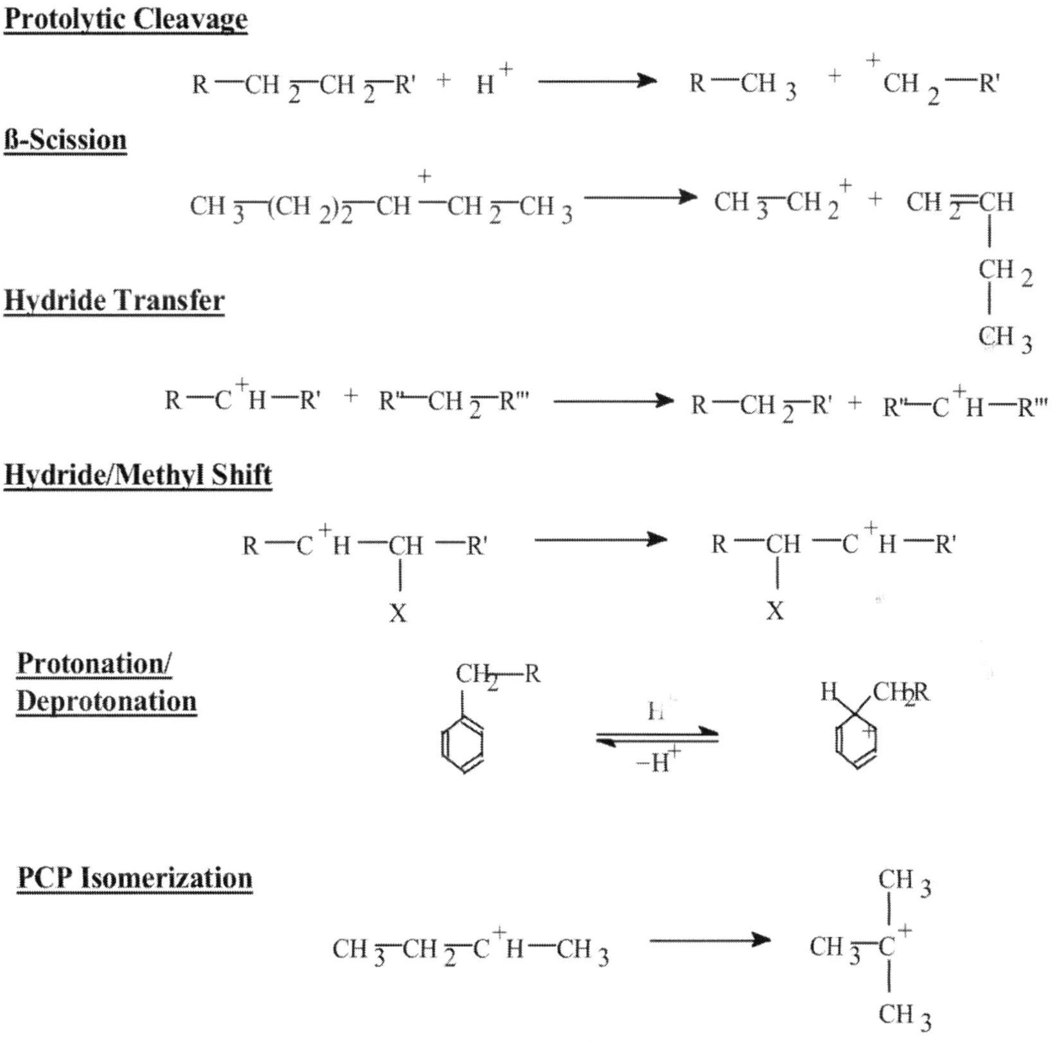

Figure 3. Elementary steps of catalytic cracking reactions.

These mechanistic steps were mapped into a set of pathways-level reaction families that captured the kinetics without the burden of the explicit mathematical description of the catalyst intermediates. The model species were also classified into paraffins, olefins, naphthenes and aromatics (PONA) categories as well as the acid sites [L].

The model building process was guided by a set of reaction approximations that pruned the model from the essentially infinite set of feasible reactions to the kinetically significant subset. These rules were based on two ultimate foundations: kinetics and analytical chemistry. Kinetics were used to "rule out" feasible but kinetically insignificant reactions, such as reactions leading to the formation of primary carbenium ions. A set of organizational rules was deployed to maintain a reasonable correspondence between the level of detail in the model and in available analytical measurements. The reaction rules were summarized in Table 1.

Table 1. Reaction rules for paraffin catalytic cracking reactions.

Reaction Steps	Reactant Rules
β-Scission	Methyl and primary carbenium ions were not allowed to form.
Hydride Shift and Methyl Shift	Primary carbenium ions were not allowed to form Migration to a stable ion or branched ions Number of reactions allowed is a function of branch number.
Protonation / Deprotonation	Primary carbenium ions were not allowed to form
PCP Isomerization	Primary and methyl carbenium ions were not allowed to form PCP-isomerization increases either the number of branches or the length of the side chains. PCP-isomerization to form vicinal branches was not allowed. Only methyl and ethyl branches were allowed; A maximum of three branches was allowed. Number of reactions allowed is a function of branch and carbon number.

Automated Model Building

Mechanistic insights were used to develop and refine the pathway level reaction network. The catalytic cracking components and reactions were organized into a small number of compound classes and reaction families, respectively. The catalytic cracking reaction mixture consists of a few observable compound classes (paraffins, iso-paraffins, cyclo-paraffins, olefins, iso-olefins, cyclo-olefins, diolefins and aromatics.) which react through a limited number of reaction families (cracking, dealkylation, isomerization, ring expansion/contraction, ring closure/opening, hydrogenation/Dehydrogenation, amortization and coking. Table 2 shows the various reaction families for C16 paraffin catalytic cracking at the pathways level.

Table 2. Reaction families for C16 paraffin catalytic cracking at the pathways level.

Reaction	Applies to	Example
1 Cracking/Addition	Paraffins, Olefins, Diolefins, Isoparaffins, Isoolefins	
2 Dealkylation	Cycloparaffins, Cycloolefins, Atomatics	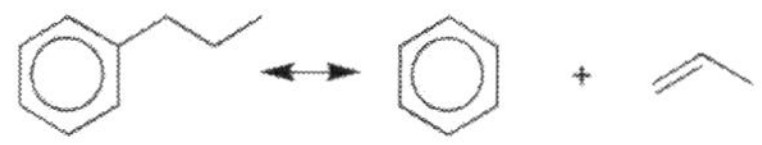
3 Isomerisation	Paraffins, Olefins	
4 Ring Expansion/Contraction	Cycloparaffins, Cycloolefins	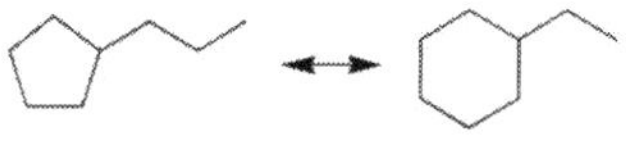
5 Ring Closure/Opening	Olefins, Diolefins	
6 Hydrogenation/Dehydrogenation	Paraffins, Isoparaffins, Cycloparaffins, Olefins	
7 Aromatization	Cycloolefins (6 membered)	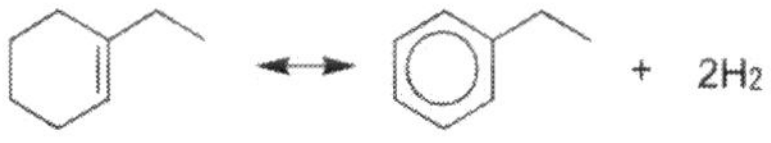
8 Coking	Cycloolefins (5 membered)	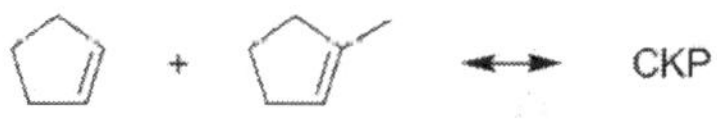

The reactions were entered into KME interface based on the reaction families. The rate laws for the reactions were made automatically according KME syntax. The feed composition and reaction conditions were entered through the input and conditions sheets of KME. In cases of parameter tuning, the rkt sheet was also used.

The C16 Paraffin FCC Model Diagnostics

A C16 FCC model with 87 species and 342 reactions was built automatically in only 3 CPU seconds on a Pentium 4 PC. The corresponding PFR model with molar expansion was then generated automatically and solved in 2 CPU seconds in once-through mode. Table 3 summarizes the characteristics of the C16 model.

Table 3. Characteristics of the C16 paraffin catalytic cracking model.

Species	#	Reactions	#
Hydrogen	1	Cracking	120
Paraffins	16	Dealkylation	27
Iso-paraffins	8	Isomerization	16
Cyclo-paraffins	15	Ring expansion/contraction	10
Olefins	10	Ring closure/opening	22
Iso-olefins	8	Hydrogenation/Dehydrogenation	34
Cyclo-olefins	11	Amortization	5
Diolefins	7	Coking	21
Aromatics	11	(adsoption/desorption on catalyst site)	87
Total Number of Species	87	Total Number of Reactions	342

Model Results and Validation

The C16 model was tuned with laboratory data[4-8] using KME in tuning mode. The objective function was the square of the difference between predicted and experimental yields weighted by the experimental standard deviation, as shown in Eq. 4,

$$F = \sum_{i=1}^{M} \sum_{j=1}^{N} \left(\frac{y_{ij}^{model} - y_{ij}^{exp}}{\varpi_j}\right)^2 \qquad (4)$$

where i is the experiment number and j is the species or lump number, and ω_j is the experimental measurement deviation.The A factors for the acid chemistry were inferred from the literature (Watson et al., 1996, 1997) and held constant during optimization. Further, the activation energies were constrained by the relation between activation energies and heat of reaction shown in Eq. 5.

$$E_{backward} - E_{forward} = \Delta H_{rxn} \qquad (5)$$

In sum, the activation energies of the prevailing reaction families were tuned to fit the data.

Figures 4a and 4b show the parity plots for the C16 paraffin cracking. The agreement between the predicted and experimental data sets for all the high and low yield compounds is good.

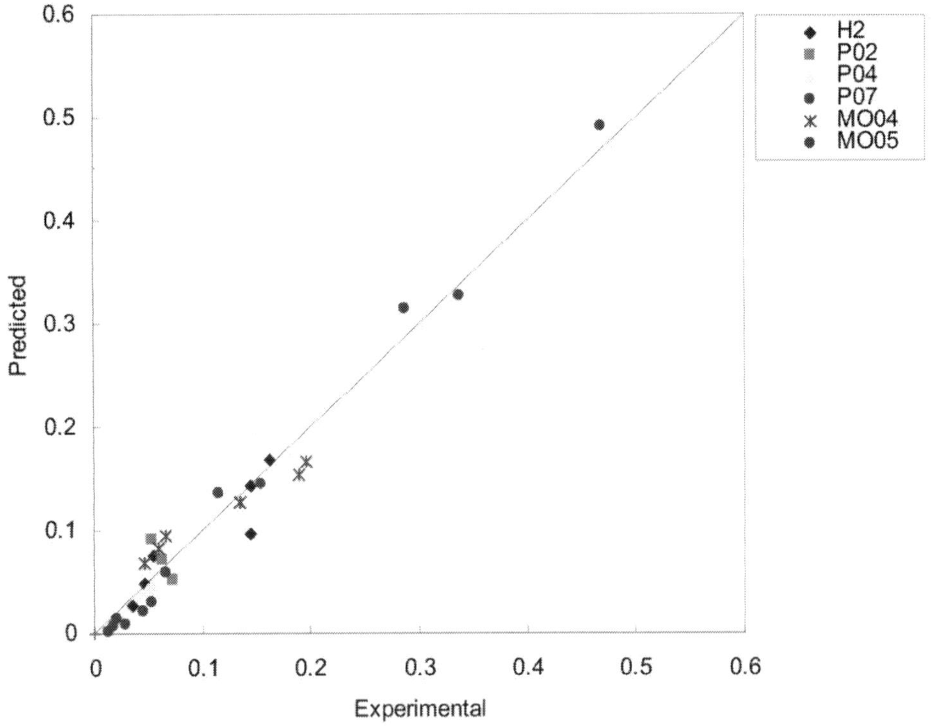

Figure 4. Parity plots for C16 paraffin catalytic cracking model at various operation conditions (T, P, and LHSV).

Summary and Conclusions

The KME modeling tools were used to construct a paraffin cracking model. The automated approach was successfully demonstrated for a C16 paraffin cracking model with 87 species and 342 reactions at the pathways level. The once-through and tuning solve times were 3 and 2 CPU second, respectively, on a Pentium 4 machine.

References

1. Broadbelt, L. J., Stark, S.M., and Klein, M. T., "Computer Generated Pyrolysis Modeling: On-the-Fly Generation of Species, Reactions and Rates", Ind. and Eng. Chem. Res., 33, pp. 790-799 (1994)

2. Hou, G., "Integrated Chemical Engineering Tools for the Building, Solution, and Delivery of Detailed Kinetic Models and Their Industrial Applications", Ph.D. Dissertation, University of Delaware (2001).

3. Gates, B. C., "Catalytic Chemistry", John Wiley & Sons, New York, (1992).

4. Watson, B. A., M. T. Klein and R. H. Harding, "Mechanistic Modeling of n-Heptane Cracking on HZSM-5," I&EC Research, 35, 1506-1516, 1996.

5. Watson, B. A, M. T. Klein and R. H. Harding, "Mechanistic Modeling of n-Hexadecane Cracking on REY,"Energy & Fuels, 11, 354-363, 1997.

6. Watson, B. A., M. T. Klein and R. H. Harding, "Catalytic Cracking of Alkylbenzenes: Modeling the Reaction Pathways and Mechanisms," Applied Catalysis, 13-39, 1997.

7. Watson, B. A., M. T. Klein and R. H. Harding, "Mechanistic Modeling of a 1-Phenyloctane/n-Hexadecane Mixture on REY," I&EC Research, 36, 2954-2963, 1997.

8. Watson, B. A., M. T. Klein and R. H. Harding, "Catalytic Cracking of Alkylcyclohexanes: Modeling the Reaction Pathways and Mechanisms," International Journal of Chemical Kinetics, 29, 545-560, 1997.

Evaluation of the PID Performance for FCC Units

[1]Trajano P. R.; Maciel Filho R. and Wolf – Maciel M. R.
Laboratory of Optimization, Design and Advanced Control (LOPCA)
Faculty of Chemical Engineering – State University of Campinas (UNICAMP)
PO BOX 6066, ZIP 13081-970, FAX +55193788395, Campinas, SP, Brazil
[1]*e-mail:* pleyci@feq.unicamp.br

ABSTRACT

Catalytic cracking reactions are high mass carbon chain molecules breaking of heavy oils such as gasoil and petroleum refining residues. This process is extremely important because derived products of petroleum of high molecular weight are transformed in products of higher commercial values. This work objective is the implementation of a deterministic mathematical model using the mass and energy balances that describe a FCC conversor, and the development of a FORTRAN language simulator. It is intended, also, to study cases in order to explore the potential of the model to simulate the process control strategy. It is considered an evaluation of PID controllers under disturbances in some controlled variables important for operation. Good performance results have been achieved through simulations carried out in two distinct cases: at first, perturbations on the air temperature for the regenerator and, at second, on the feedstock, considering the PID action in some meshes of the process catalytic cracking.

Keywords: Fluid Catalytic Cracking, PID Control, Model Evaluation

1. CATALYTIC CRACKING PROCESS

Catalytic cracking reactions promote chain breaking, yielding lighter hydrocarbons, with higher commercial value as gasoline and nafta. Particularly, in the Brazilian case, the development importance of such process is reinforced by the increasing heavier oil reservoir discoveries recently. Any conversion efficiency upgrade for heavier feedstock, allied to a large production, might represent a significant gain.

The catalytic cracking process occurs in equipments called converters. A FCC (Fluid Catalytic Cracking) converter is composed basically by a riser, a separating vessels/stripper and a regenerator. Because of the interaction of the process variables, and because of the complexity of the chemical kinetics through the process of the catalytic reactions, it has several dynamical problems and, consequently, difficulty to operate and control. (VIEIRA, 2002).

A way of controlling commonly used, for its simplicity and industrial performance, is the PID control. The PID (Proportional Integral Derivative) is a conventional way of control, used in many industrial processes. For this specific catalytic cracking process, the PID controllers are used for maintaining the process in operational conditions avoiding the violation of restrictions that could cause any operational risks or loss of catalytic mass. In this case, those controllers act in a lower level of the controlling process.

2. PROCESS DESCRIPTION

The feeding stoke enters in the FCC conversor riser, where the catalytic cracking reaction occurs. The final products that contain the less weigh hydrocarbon fractions are taken to the recovery step, this step occurs in the distillation columns or " fractioners". The non reactive catalyst is taken to the regenerator and then, after the regeneration, is fed continuously back to riser.

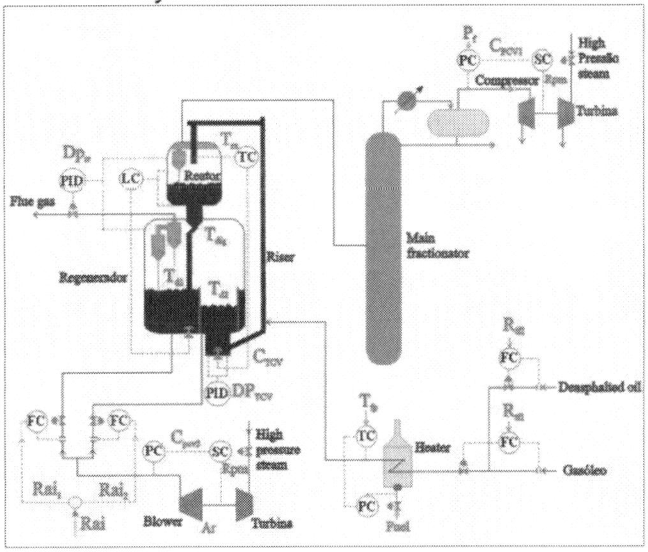

Figure 1 – Catalytic Cracking "FCC" Unit

The conversor, as seen in figure 1, is divided in three parts: riser, reactor and regenerator. The riser is a large diameter tube where the mix of vapors of hydrocarbons and catalyst goes up. The vaporized molecules penetrate in the pores of the catalyst where the cracking reactions effectively occurs as, progressively, o coke is deposited in the surface of the solids, and, at the same time, is the responsible for the deactivation of the catalyst.

The reactor is the place where the gaseous products are separated from the catalyst by gravity. The product (gases) that came from the reaction, are taken to a main distillation columns and are separated by their ebullition point, as the coked catalyst (deactivated) goes through a "stripper", a place where the inertial separation occurs between the catalyst and the gases from the cracking process, combined with a water vapor shower used to take out the excess of hydrocarbons in the solid. This process culminates in the third part of the FCC unit, the regenerator.

The used catalyst (with coke) is sent to the regenerator, which has the function of burning the lodged carbon deposits in the surface of the catalyst, turning them into combustion gases while, due to the elimination, the catalyst particles are reactivated. That combustion occurs because of the high temperature when the catalyst reaches the regenerator (490°C – 550°C), the presence of the deposited coke and the outflow of injected air by the inferior part of the regenerator.

The airflow in the catalytic mass inside the regenerator causes the formation of bubbles, producing an effect similar to a liquid in boiling process. The contact with the air, progressively transformed into combustion gases, and the solids makes possible the formation of a fluidized bed so that particles tends to behave as they are in liquid state. The region, where the solid mass predominates, is known as dense phase. Above the bed, there is a region where the combustion gases predominate and, however, there are many particles, known as diluted phase. Those dragged fines of catalyst are almost totally recovered by the set of cyclones of the regenerator, normally of double stage (two cyclones in series). (ABADIE, 1997)

The catalyst, with low coke text, and, because of that, with its recouped activity, is returned to the riser by a transport line called "stand-pipe", passing through a temperature control valve, TCV. Its flow controls the temperature of the catalyst in the top of the riser and it is a function of the opening and of the differential of pressure between the regenerator and the riser. Arriving at the riser, the catalyst finally is mixed to the feeding stoke, completing the cycle. (ABADIE, 1997)

In this work, the catalytic cracking unit control was realized with the implementation of PID controllers in some of the phases of the process. The loops can be observed in figure 2:

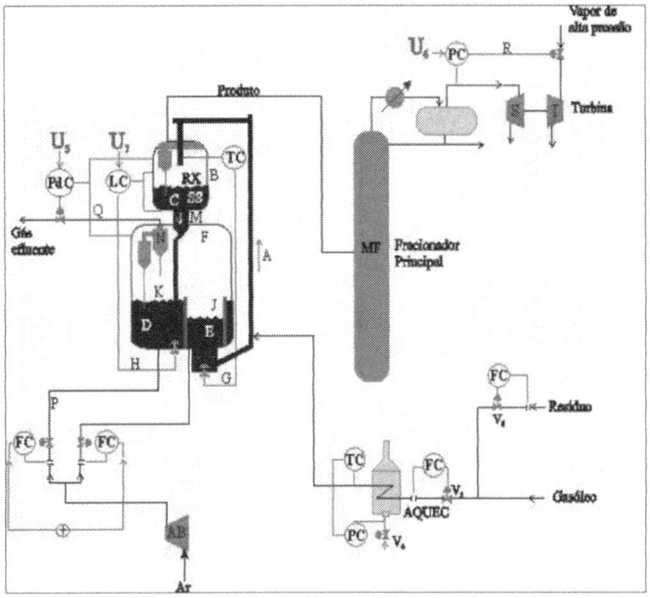

Figure 2 – Control Loops in the FCC

- Mesh u_5: The pressure of the regenerator (P_{reg}) is controlled by manipulating the outflow of the effluent gas. This control is very important so that the reaction mixture in riser does not return to the regenerator, provoking an explosion.
- Mesh u_6: The control of the pressure of the compressor (P_{suc}) is made through the manipulation of the vapor outflow in the turbine.
- Mesh u_7: The level of the stream bed catalyst ($H_{\_ra}$) deactivated in the regenerator is controlled by the valves of outflow of deactivated catalyst that comes from the reactor. The control of this level controls the level of catalyst in the phases rg_1 and rg_2.

From the variables of the process, some can be manipulated, as: RTF – Total feed flow rate to riser; TFP – Feed temperature at riser entrance; RAI – Air flow rate to regenerator; CTCV – Regenerated catalyst valve operation. The variables controllers in process are: TRX – Temperature of cracking mixture in riser; TRG1 (or TRG2) – Temperature of dense phase at regenerator first and second stages and conversion of the reactor (Severity of cracking reaction).

3. MATHEMATICAL MODELLING

The model considered here basically follows the work of Moro (1992), where is considered the riser cracking and two regenerator stages with partial CO burning. The reactor vessel and the dense dilute phases of the two regenerator stages are considered systems of lumps parameters. The mass and energy balances lead to set of nonlinear ordinary differential equations that can be integrated by any available mathematical package.

To following equations are used to describe the main phenomena taking place in the system:

- Coke Balance in the Reactor

$$Hra\frac{dCcat}{dt} = -Rrc.Ccat + 100.Rcf \tag{1}$$

- Energy Balance in the Riser

$$Hris.Sc.\frac{dTrx}{dt} = Sc.Rrc.(Trg_2 - Trx) + \frac{Sf.Dtf.Rtf.(Tfp - Trx)}{1440} - \frac{\Delta Hfv.Dtf}{1440 - \Delta Hcr.Roc} \tag{2}$$

- Catalyst Inventory in the Reactor

$$\frac{dHra}{dt} = Rcr - Rsc \tag{3}$$

- Energy Balance in Reactor Vessel

$$Hra.\frac{dTra}{dt} = Rrc.Trx - Rsc.Tra \tag{4}$$

- Heat Balance on the Dense Phase of the Regenerator First Stage

$$Hrg\,1.Sc.\frac{dTrg\,1}{dt} = Sc.Rsc.Tra - Sc.Rrc\,1.Trg\,1 + 0{,}0215\,.Sa.Rai\,1.(Tai - Far\,12.Trg\,1) - \frac{Fgk\,1.Sa.Trg\,1}{100} - \frac{\Delta Hrg\,1.Carb\,1.12}{1000} \tag{5}$$

- Catalyst Mass Balance on the Dense Phase of the Regenerator First Stage

$$\frac{dHrg1}{dt} = Rsc - K_w(h - hv)^{0,5} \tag{6}$$

- Balance of Coke in the Regenerator First Stage

$$\frac{dCrc1}{dt} = \left(\frac{Rsc.Csc - Rrc1.Crc1}{Hrg1.Rcb1}\right) - R_{cb1} \tag{7}$$

- Oxygen Balance in the Dense Phase of the Regenerator First Stage

In the dense phase of the regenerator the oxygen mass balance corresponds to:

$$\frac{Vde1.\rho^m}{100}.\frac{DOfg1}{dt} = 0{,}21.R_{mai1} - \frac{F_{gm1}.Ofg1}{100} - \frac{10}{12}.Rcb1.Hrg1.Fat1 \tag{8}$$

The mass balance of O_2 in the dilute phase of the regenerator first stage:

$$\frac{Vdi1.\rho^m}{100}.\frac{d[O_2]}{dt} = F_{gm}.\frac{(Ofg1 - [O_2])}{100} - \frac{Rco}{2.Vdi1.60} \tag{9}$$

- Heat Balance in the Dilute Phase

$$\frac{dTdi1}{dt} = \frac{F_g}{\rho.Vdi1.(Trg1 - Tdi1)} + 67636\frac{Rco.60}{\rho.Sa} \tag{10}$$

- Balance of Regenerator Pressure

$$Vdi1.\frac{d\rho}{dt} = Fi - Fo \tag{11}$$

4. RESULTS

The computational program was developed in language FORTAN based on the model and the work of Moro and Odloak (1995), which represents a unit of catalytic cracking.

Extensive simulations were carried out to verify the prediction and compared with real values in the refining process. Perturbations were done in the system to verify the response of the variables controlled by two distinct cases.

In the first test it was applied a step change into a variable that influences directly the conversor regenerator, the air temperature to the regenerator – TAI, modifying its value in 10%. In the second, the 10% step was applied in reference to the initial value of RTF – feeding flow rate, that directly influence all the conversor behavior, more specifically in the catalytic cracking reaction and in the reaction severity.

In all tests, the analyzed variables were:

- Temperature of cracking mixture in riser - TRX;
- Temperature of dense phase at regenerator first stage – TRG1;
- Temperature of dense phase at regenerator second stage – TRG2;
- Temperature of dilute phase at regenerator second stage;
- Temperature of general phase exit at regenerator;
- Severity of the cracking reaction (Conversion of the reactor).

i) First case: Change in the air temperature of the regenerator - TAI:

For this first case it was implemented a step disturbance of 10% in the initial value of TAI – Regenerator air temperature, the initial value was 190°C for 209°C, a difference of +19°C. In Figures 3 and 4 can be observed that the line indicating the start of the disturbance degree in the system.

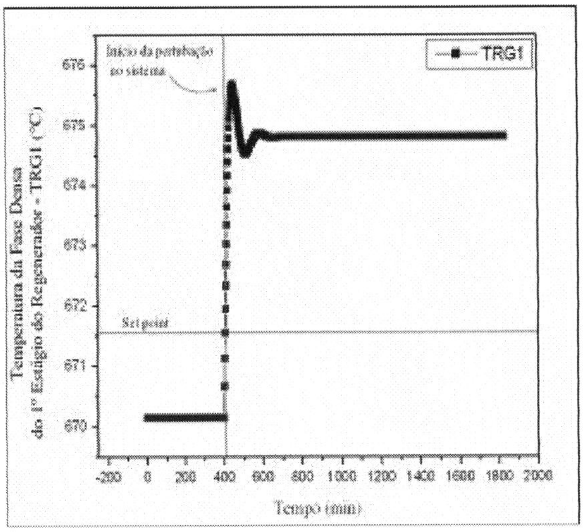

Figures 3 and 4 – Response after distur

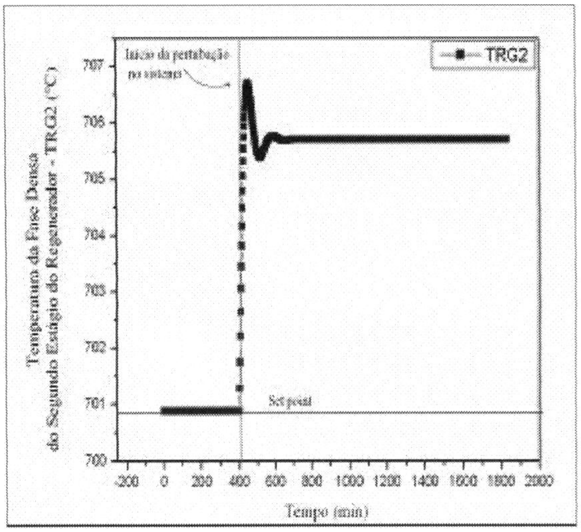

bance in TAI for TRG1 and TRG2.

Figures 3 and 4 correspond to the response of the variables TGR1 (temperature of the dense phase of the first period of the regenerator) and TRG2 (temperature of the dense phase of the second period of the regenerator). As waited, as an addition in the temperature of air in the regenerator was given, it was observed an increase in the temperatures of the dense phases of the regenerator.

As the temperature of the regenerated catalyst increases, the reactor temperature TRX also increases with consequences in the severity of the reaction, as can be seen in figures 5 and 6. The temperature of riser must be monitored and controlled to allow the attainment of an adequate conversion of the cracking reaction.

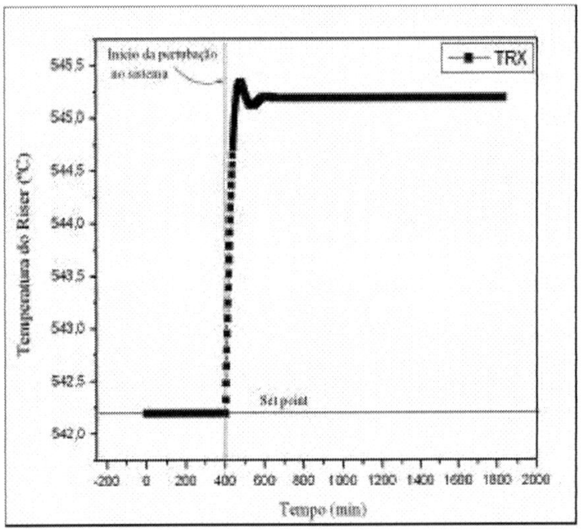

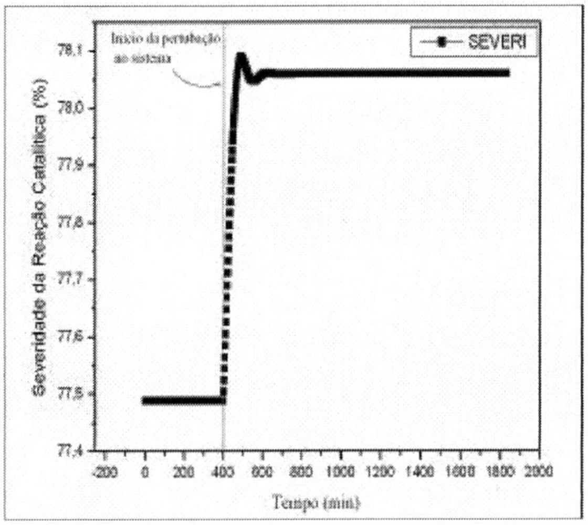

Figures 5 and 6 – Response after disturbance in TAI for TRX and Sever

Figures 7 and 8 present the responses in the variables: exit of the diluted phase temperature of the regenerator - TD2 and the Temperature of exit of the general diluted phase - TDG. As observed in the dense phase, Figures 3 and 4, also in the diluted phase it can be observed the increase of the temperature.

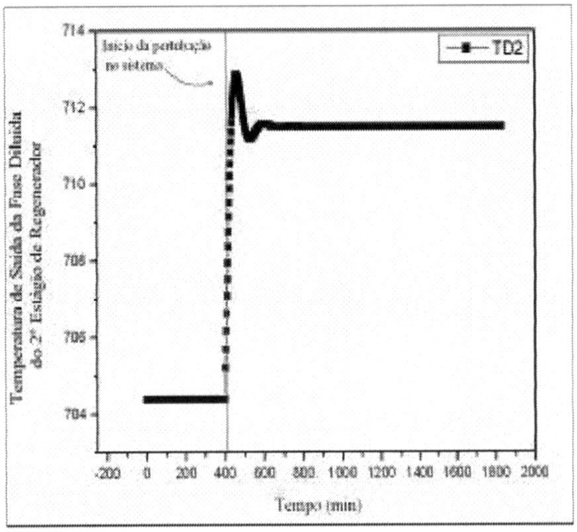

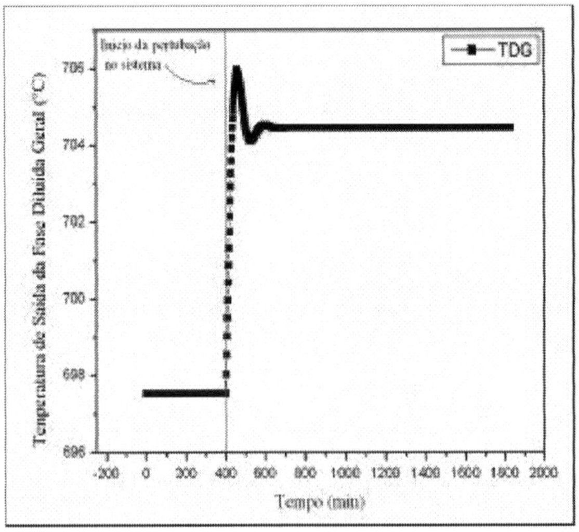

FFigures 7 and 8 – Response after disturbance in TAI for TD2 and TDG

ii) Second case: Change in the riser Total feed flowrate - RTF:

For the second case it was implemented a step change of 10% in the initial RT. The initial value in stationary state was of 9700 m3/d and an addition of 970 m3/d was carried out.

For all variables, it was observed the reduction in the initial variable conditions, due to the larger amount of load entering in the converter, without modifying other variable as, for example, temperature of air for burning of the coke in the regenerator. This disturbance or increase in the entrance reflects directly in the necessity of increasing the temperature of the regenerator. This in results in an alteration of the amount of catalyst in circulation, what, in its turn, also modifies the temperature of the reactor and the conversion. Thus, it led to a decrease in the temperature of riser as well as in the temperatures of the dense and diluted phases of the regenerator, as can be observed in Figures 9 (TRG1), 10 (TRG2), 11 (TRX), 12 (Sever), 13 (TD2), 14 (TDG).

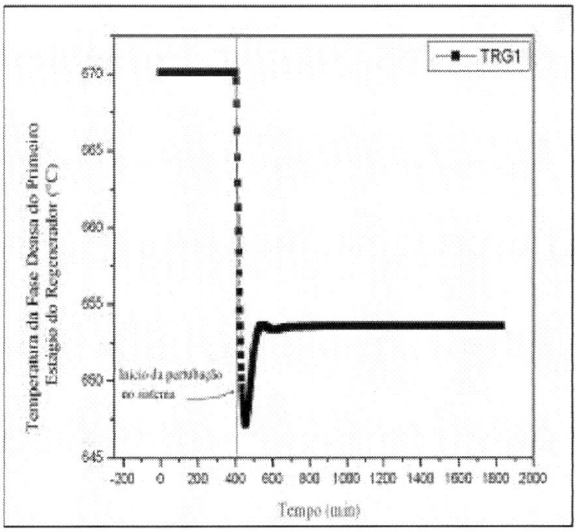

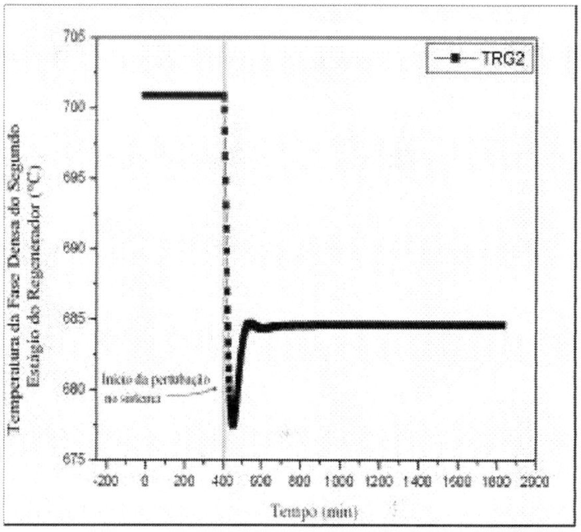

Figures 9 and 10 – Response after disturbance in RTF for TRG1 and TRG2

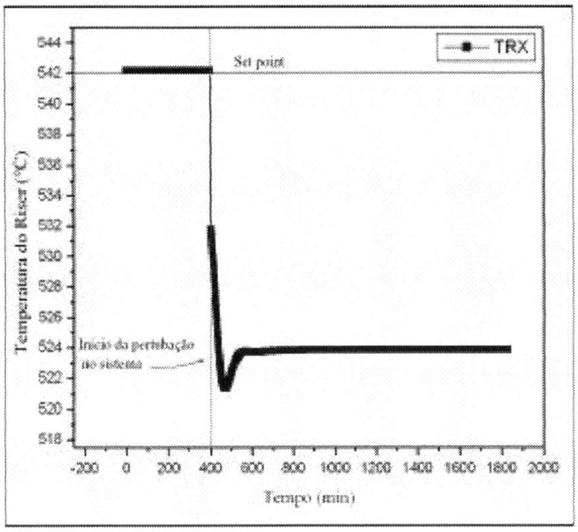

Figures 11 and 12 – Response after

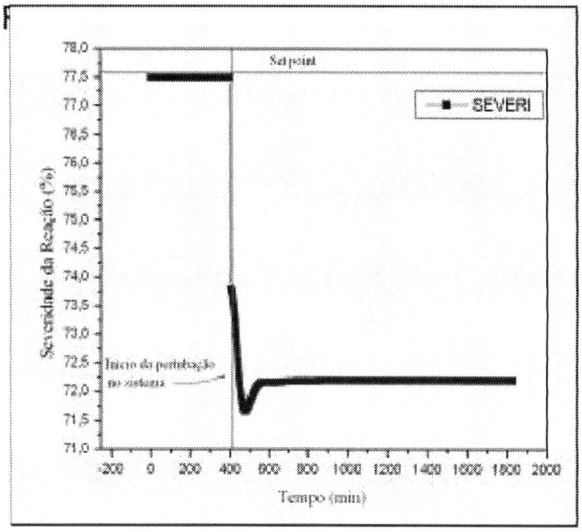

disturbance in RTF for TRX and Sever

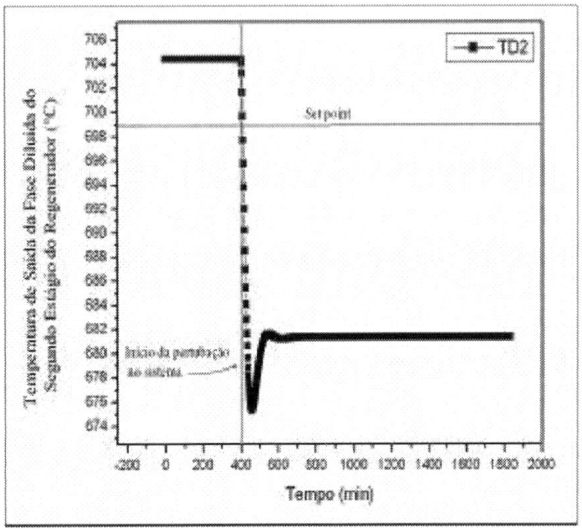

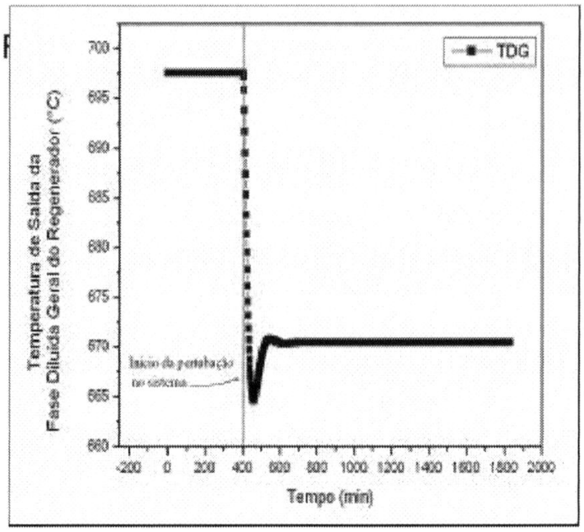

Figures 13 and 14 – Response after disturbance in RTF for TD2 and TDG

5. CONCLUSIONS

The results show that PID controllers used in the control loops allowed a safe operation even when significant disturbances were been introduced in the plant. This allows strategies of hierarchic control with predictive controllers type to be developed so that it is possible to generate set-points to operate the FCC in high performance with possible fluctuations of load and operational conditions.

6. REFERENCE

ABADIE, E. *Craqueamento Catalítico. Relatório SEREC/CEN-SUD,* Rio de Janeiro, 1997.

MORO, L. F. and ODLOAK, D. *Constrained multivariable control of fluid catalytic cracking converters.* J. Proc. Control, vol. 5, no. 1, pp. 29-39, 1995.

MORO, L. F. *Desenvolvimento de um Controlador Preditivo Multivariável para um Conversor Industrial de Craqueamento Catalítico.* Dados. Dissertação (Tese de Mestrado) USP-SP, 77 p, 1992.

SPANDRI R. *Sintonia de Controladores Regulatórios.* Boletim Técnico Petrobras, RJ, 2003.

VIEIRA, W. G. *FCC: Controle Preditivo e Identificação Via Redes Neurais.* Dados. Dissertação (Tese de Doutorado), UNICAMP-SP, 2002.

56f Optimizing FCC: LPG rate and Cat Gasoline Octane Constraints

Marty Poole, Valero, Houston, TX 77032
Draft submitted for company approval

Adsorption of 2, 5-dimethyl benzene sulfonyl chloride to a palladium (100) surface

Bradley R. Atkinson and Rayford G. Anthony
Kinetics, Catalysis and Reaction Engineering Laboratory, Artie McFerrin Department of Chemical Engineering, Texas A&M University, College Station, TX 77843.

Aromatic thiols are of industrial importance due to their use in the manufacture of a wide variety of specialty chemicals. This list includes, but is not limited to, agrochemicals, polymers, pigments, dyes and pharmaceuticals [Roy, 2003; Roberts, 1997]. The thiol production reaction that involves the hydrogenation of 2, 5-dimethyl benzene sulfonyl chloride to 2, 5-dimethyl benzene thiol is focused on here. A new process has been patented for the hydrogenation reaction [Jacobsen, 2001] that eliminates the typical heavy metal salt by-product, which is an environmental, ecological and biological contaminant [Poland, et al., 2001; Tsai, et al., 2003; Piao, et al., 2003; Zerahn, et al., 1999; Bruce, et al., 2003].

As the first step in an investigation to find the transition state of the rate limiting step in the reaction system, adsorption of the 2, 5-dimethyl benzene sulfonyl chloride to a series of 4 (100) palladium matrices is studied by quantum mechanical simulation. These investigations utilized the Gaussian03 software simulation package. Density functional theory (DFT) was used for all calculations at the B3LYP level with a mixed basis set. H, C, and O utilized the 6-31G(d') basis set while the S, Cl and Pd molecules all were based with LANL2DZ sets. S and Cl molecules each took advantage of additional d-shell mathematical representations while Pd used the basic LANL2DZ basis set. Additionally, S, Cl and Pd all were restricted with effective core potentials (ECP) to ease the computational load on the computing cluster. The calculations were spin restricted to a neutral charge and singlet spins.

The creation of the palladium matrices for Gaussian03 simulation was the first step. How can the delicate balance of realistic accuracy and computational ease be reconciled on a molecular level? In testing their Pd-based reaction system for the effects of particle size and charge distribution throughout the catalyst surface, Duca and Barone simulated adsorption of various amino-nitro-benzene molecules over a layer of six Pd molecules arranged in a simple, rectangular plane [Barone and Duca, 2002]. Their further work shows a scheme for simulating a three-layered Pd matrix with only 22 atoms [Duca, et al., 2001]. The matrices used and designed in this study are shown in figure 1.

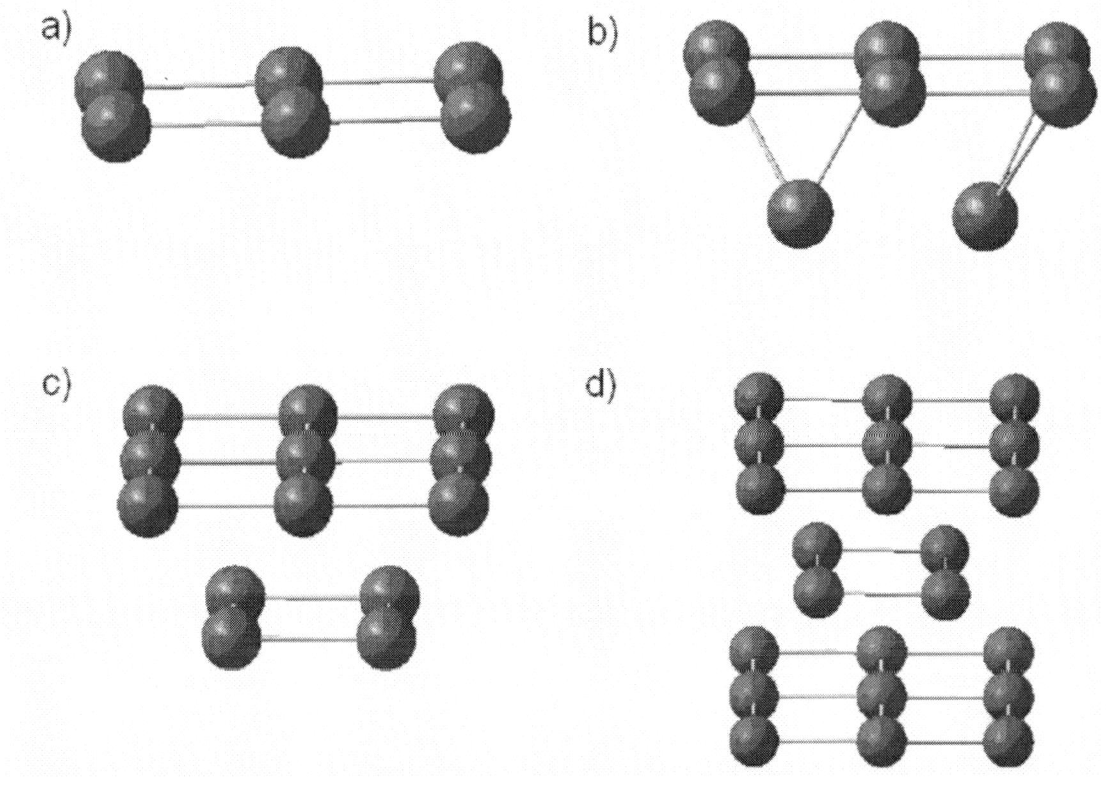

Figure 1 – A diagram of each of the four palladium structures. Each are described briefly below by letter, the enthalpies of formation shown are calculated single point energies of the locked-in structures. a) Pd6 – a single layered six atom structure, $\Delta H_f = -4.772 \times 10^5$ kcal/mol. b) Pd8 – a two layered, 8 atom structure in a dual inverted pyramid scheme 6 atoms over 2 atoms, $\Delta H_f = -6.363 \times 10^5$ kcal/mol. c) Pd13 – a two layered, 13 atom structure in an inverted pyramid scheme 9 atoms over 4 atoms, like two Pd8s side by side, $\Delta H_f = -1.034 \times 10^6$ kcal/mol. d) Pd22 – a three layered, 22 atom structure in an hourglass scheme 9 atoms over 4 atoms over 9 more atoms, like Pd13 with a extra layer on the bottom, $\Delta H_f = -1.75 \times 10^6$ kcal/mol.

Possible adsorptive modes were investigated over the simplest palladium matrix for ease of calculation. All of the aromatic molecules showed approximately the same behavior: a backing away from the palladium matrix and a rotation into a "sulfonyl chloride down" position. These rotations from the original adsorption modes are shown in figure 2. These studies showed that the adsorption modes for 2b) and 2c) (explained in the figure caption) are the main adsorption modes.

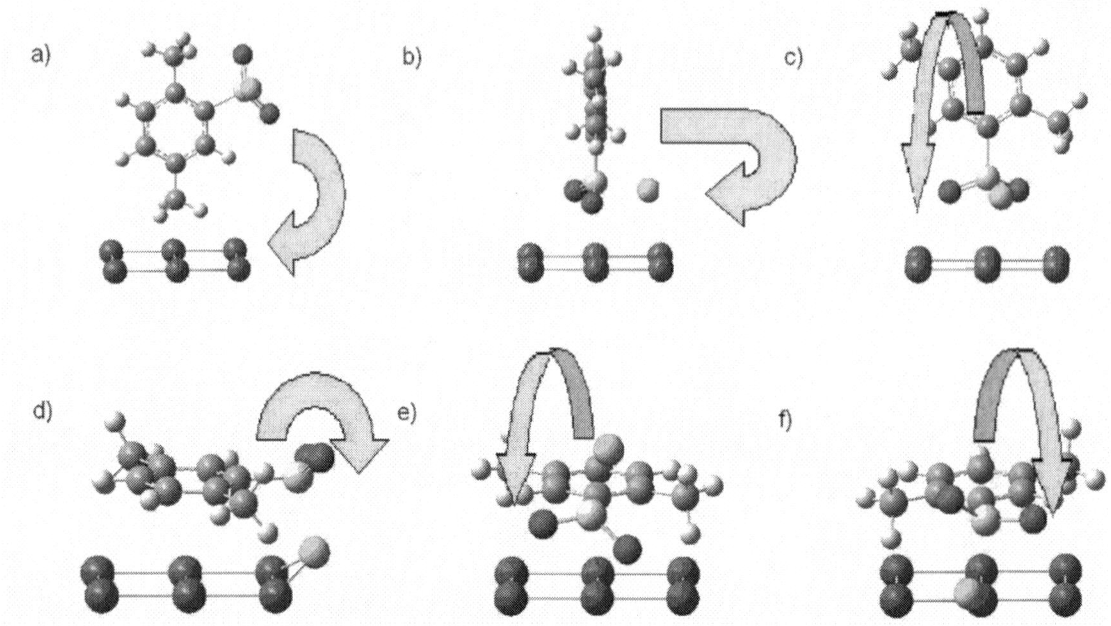

Figure 2. The adsorption modes considered during the course of the investigation. Sclads is an abbreviated filename used for the sulfonyl chloride adsorption calculations. a) a representative structure for the adsorption of the molecule via one of the methyl groups in the two orientations possible on the non-symmetrical Pd6 matrix (others not shown). All of the structures represented in this single example showed the same behavior. b) a vertical adsorption mode via the sulfonyl chloride functional group. c) same as 2b) but in the opposite Pd6 orientation. d) a horizontal adsorption mode with the chlorine molecule interacting with the surface immediately, similar to another adsorption mode by flipping the molecule so that the chlorine is farthest from the Pd 6 matrix (not shown) and exhibiting similar behavior. e) a horizontal adsorption that is simply 2d) flipped and in the opposite Pd6 orientation. f) a horizontal adsorption that is simply 2d) in the opposite Pd6 orientation. The movements from the first few steps of the DFT minimizations are represented by the arrows. In every case, the driving force was the need for one or both of the oxygen atoms bonded to the sulfur to interact directly with one or more of the palladium atoms.

The orientations from 2b) and 2c) were setup on the other, more complicated palladium matrices as well. Each step in the energy minimization calculation brings a lower energy and thus a more stable state. How does this simulation represent what is happening in the reactor? The turning of the molecules into the "sulfonyl chloride down" position shows that this is the most energetically stable of the postulated adsorption conditions. . The further minimizations are a picture of what the aromatic molecules would be doing on the surface of the catalyst without the presence of hydrogen. The results of the full minimizations over the 6 and 8 palladium matrices are shown below. The 22 and 13 atom matrices have proven beyond the current hardware/software configuration. Calculations with newly implemented convergence criterion and step processes are being run currently for these matrices.

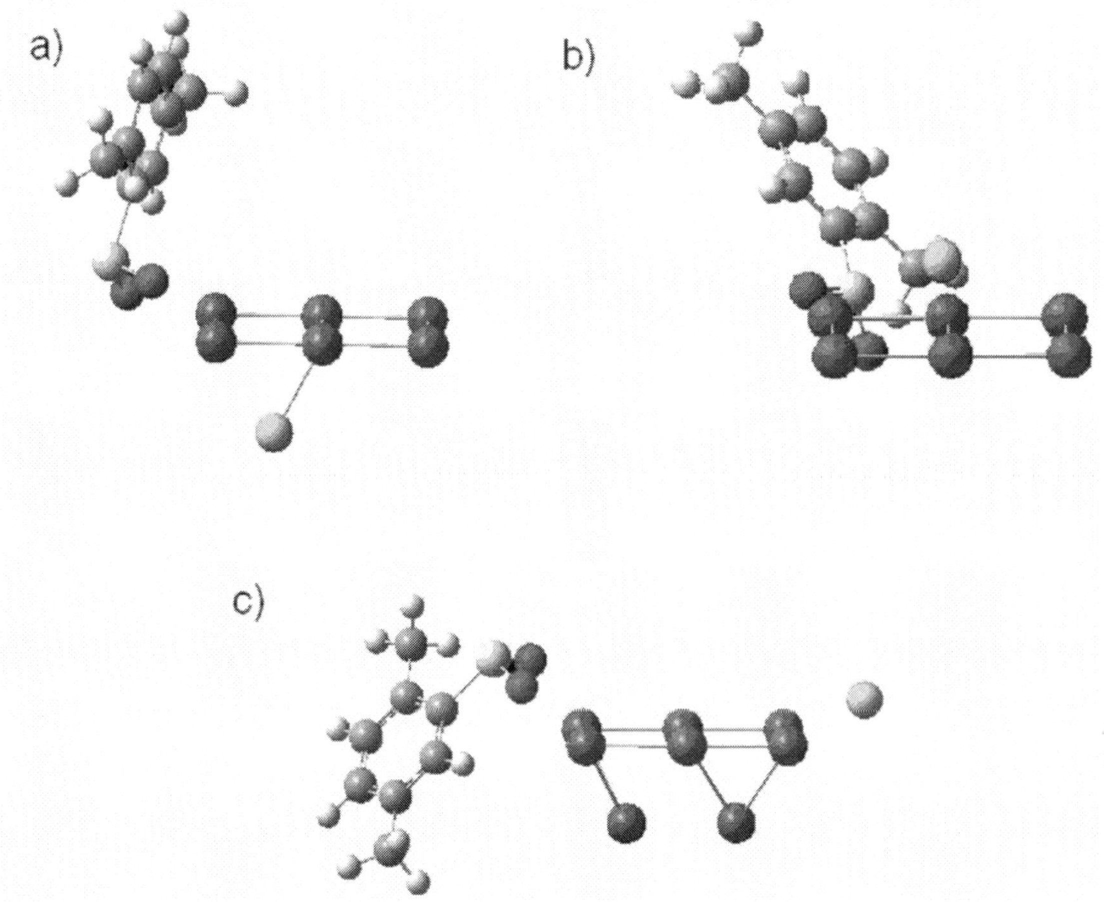

Figure 3. The results of the DFT optimization calculations on 2b), 2c) and the equivalent 2b) adsorption mode over Pd8 are shown here. These represent the minimums found by the Gaussian03 package. a) The endpoint structure of the 2b) calculation, ΔH_{ads} = -51.965 kcal/mol. b) The endpoint of the 2c) calculation, ΔH_{ads} = -12.995 kcal/mol c) The endpoint of the 2b) mode calculation over the Pd8 palladium matrix, ΔH_{ads} = -56.2816 kcal/mol.

These simulated adsorption energies for a) and c) are reasonable when compared to the individual adsorption energies of the atoms of the sulfonyl chloride group, but b) has a marked difference in energy from the other two. This has to do with the energetic differences between certain palladium atoms as compared to others. The convergence of the larger matrices will shed more light on the phenomena that are involved in this reaction.

References

[1] Kathrin-Maria Roy, Ullman's, Encyclopedia of Industrial Chemistry, Wiley-VCH, New York, 2003, pp.625-652.

[2] John S. Roberts, Encyclopedia of Chemical Technology, Vol. 24, John Wiley and Sons, New York, 1997, pp. 19-34.

[3] Stephen E. Jacobsen. (13 September 2001) Hydrogen Reduction of Sulfonyl Chlorides to Thiols. International Patent Classification: C07C 319/02, 309/86. International application number: PCT/SU01/06129. International Publication Number: WO 01/66517 A1. Patent Assigned to E.I. Du Pont De Nemours and Company.

[4] John S. Poland, Scott Mitchell, Allison Rutter, Cold Regions Science and Technology 32 (2001) 93-105.

[5] W. T. Tsai, H. P. Chen, W. Y. Hsien, C. W. Lai, M. S. Lee, Resources, Conservation and Recycling 39 (2003) 65-77.

[6] Fengyuan Piao, Kazuhito Yokoyama, Ning Ma, Toru Yamauchi, Toxicology Letters 145 (2003) 28-35.

[7] B. Zerahn, A. Kofoed-Enevoldsen, B. V. Jensen, J. Mølvig, N. Ebbehøj, J. S. Johansen, I. –L. Kanstrup, Respiratory Medicine 93 (1999) 885-890.

[8] S. L. Bruce, B. N. Noller, A. H. Grigg, B. F. Mullen, D. R. Mulligan, P. J. Ritchie, N. Currey, J. C. Ng, Toxicology Letters 137 (2003) 23-34.

[9] Giampaolo Barone, Dario Duca, Journal of Molecular Structure (Theochem) 584 (2002) 211-220.

[10] D. Duca, G. Barone, Z. Varga, G. La Manna, Journal of Molecular Structure (Theochem) 542 (2001) 207-214.

[11] M. J. Frisch, G. W. Trucks, H. B. Schlegel, G. E. Scuseria, M. A. Robb, J. R. Cheeseman, J. A. Montgomery, Jr., T. Vreven, K. N. Kudin, J. C. Burant, J. M. Millam, S. S. Iyengar, J. Tomasi, V. Barone, B. Mennucci, M. Cossi, G. Scalmani, N. Rega, G. A. Petersson, H. Nakatsuji, M. Hada, M. Ehara, K. Toyota, R. Fukuda, J. Hasegawa, M. Ishida, T. Nakajima, Y. Honda, O. Kitao, H. Nakai, M. Klene, X. Li, J. E. Knox, H. P. Hratchian, J. B. Cross, C. Adamo, J. Jaramillo, R. Gomperts, R. E. Stratmann, O. Yazyev, A. J. Austin, R. Cammi, C. Pomelli, J. W. Ochterski, P. Y. Ayala, K. Morokuma, G. A. Voth, P. Salvador, J. J. Dannenberg, V. G. Zakrzewski, S. Dapprich, A. D. Daniels, M. C. Strain, O. Farkas, D. K. Malick, A. D. Rabuck, K. Raghavachari, J. B. Foresman, J. V. Ortiz, Q. Cui, A. G. Baboul, S. Clifford, J. Cioslowski, B. B. Stefanov, G. Liu, A. Liashenko, P. Piskorz, I. Komaromi, R. L. Martin, D. J. Fox, T. Keith, M. A. Al-Laham, C. Y. Peng, A. Nanayakkara, M. Challacombe, P. M. W. Gill, B. Johnson, W. Chen, M. W. Wong, C. Gonzalez, and J. A. Pople, Gaussian 03, Revision C.02, Gaussian, Inc., Wallingford CT, 2004.

Commercialization of a New Ultra High Purity/ High Yield Phenol Process

R. J. Schmidt
UOP LLC
25 East Algonquin Rd
Des Plaines, IL 60017

S.R. Keenen
Sunoco Chemicals
Frankford Plant - Margaret and Bermuda Streets
Philadelphia, PA 19137

Prepared for presentation at the AIChE Spring Meeting, Orlando, Florida, April 23-27, 2006

Unpublished

Abstract

Using the latest phenol technology from Sunoco and UOP, producers are now able to manufacture the highest quality phenol on the market today. With total GC impurities at <10 wt.ppm and total cresols at <10 wt.ppm, overall phenol purity is approaching 100%.

These results have now been observed in several recent commercial start-ups. Thus a new benchmark has been established for producing ultra high purity phenol in support of growing markets such as bis-phenol A (BPA) and high performance polycarbonates (PC). This outcome is mainly due to improvements in the phenol purification section of the process coupled with very high purity cumene feedstocks coming from advanced zeolitic cumene technology that replaces older SPA cumene technology.

Yields have also been improved dramatically with cumene/phenol consumption ratios now approaching 102% of theoretical consumption with the new process. This has been the result of several recent advances in the low pressure oxidation section of the process as well as introduction of an advanced CHP decomposer design.

Detailed results from the recent start-ups along with the implications for the future of the phenol market are reviewed in this paper.

Keywords: Alkylation, benzene, propylene, cumene, zeolite, phenol, cumene hydroperoxide, acetone, alpha-methylstyrene,

Introduction

The Sunoco/UOP Phenol process produces high-purity phenol and acetone by the cumene peroxidation route, using oxygen from air. This process features low-pressure oxidation for improved yield and safety, advanced CHP cleavage for high product selectivity, an innovative direct product neutralization process that minimizes product waste, and an improved, low cost product recovery scheme. The result is a very low cumene feed consumption ratio of <1.31 wt cumene/wt phenol that is achieved without acetone recycle and without tar cracking. The process also produces an ultra high product quality at relatively low capital and operating costs. Extensive commercial experience validates these claims. These improvements are discussed for each of the key major sections of the process.

The Sunoco/UOP Phenol Process is jointly licensed by Sunoco and UOP. The photograph shown in Figure 1 is of the Sunoco phenol plant in northeast Philadelphia. This plant has been producing phenol and acetone from cumene since 1954. Sunoco is the North Americas largest producer of phenol. Having Sunoco as a partner in phenol is a great asset to UOP in that Sunoco provides world class R&D in addition to providing access to operating units and the unique ability to train licensees in an operating plant. Note that the location of the phenol plant is in the heart of a major metropolitan area adding great confidence to the safety aspects of the design of the Sunoco-UOP phenol technology.

The first UOP designed Cumox Process started up in 1965. Three units were subsequently licensed by UOP prior to the merger with Allied (now Sunoco) having capacity of 205 KMTA. UOP/Allied technologies were merged in 1986 to form what is now the Sunoco/UOP phenol process. Since the merger, eight units representing 1,050 KMTA of licensed capacity have been added. In 2005, UOP became a wholly owned subsidiary of Honeywell. In addition to phenol production, about 70% of worldwide acetone production comes from cumene-based phenol units with 30% coming from isopropyl alcohol (IPA).

Refinement of the process over the past decade has been driven by several market forces namely:

1. the growing bis-phenol A market, driven by the electronics and automotive industries demand for epoxy and polycarbonate resins and
2. the growing phenolic resin market, driven by housing and furniture industries for composite wood resins.

Also, because competition in these areas for product differentiation and price has become so keen, phenol technology suppliers have been driven to improve production yield, process safety and phenol/acetone product quality to remain competitive. The Sunoco/UOP phenol process represents the most advanced technology for phenol production via the cumene peroxidation route.

Current Challenges Facing Phenol Producers

Many challenges face current phenol producers as companies strive to become more competitive in a variety of geographic areas and under varying market forces including:

1. Lower energy and utility costs
2. Feedstock cost and tight/limited feedstock availability
3. Downstream users demanding higher purity products
4. Tightening environmental regulations

A major short term issue for producers is the high cost of benzene due to demands from other petrochemical producers and limited on-purpose production of benzene in the current market. Also, depending on location, the availability of propylene has been tight as competition from polypropylene producers for this feed has been keen. This has created a short term situation of very low margins for phenol product pricing over the last 12 months especially in the Far East. As more propylene capacity comes on stream over the next several years from planned fluid catalyst cracking (FCC) and stream cracking of naphtha projects, the propylene supply issue should be relaxed. Details regarding the production of cumene as the primary feedstock for phenol manufacturing are covered by UOP using a zeolite catalyst in a previous AIChE publication [(1)]. In the future, the current low margins on phenol should also be improved as more downstream planned capacity is added including new BPA and polycarbonate (PC) projects.

Table 1 shows the typical phenol product quality expected from the process for two different grades of phenol including: (1) resin-grade purity suitable for most merchant market applications and many bisphenol-A markets and (2) PC-grade purity required by the most demanding bisphenol-A and polycarbonates markets. Note that both products can be accommodated in the Sunoco/UOP phenol process by making appropriate adjustments to process operating and fractionation conditions. Because of the high growth expected for the PC market, demand for the PC-grade purity product is expected to grow.

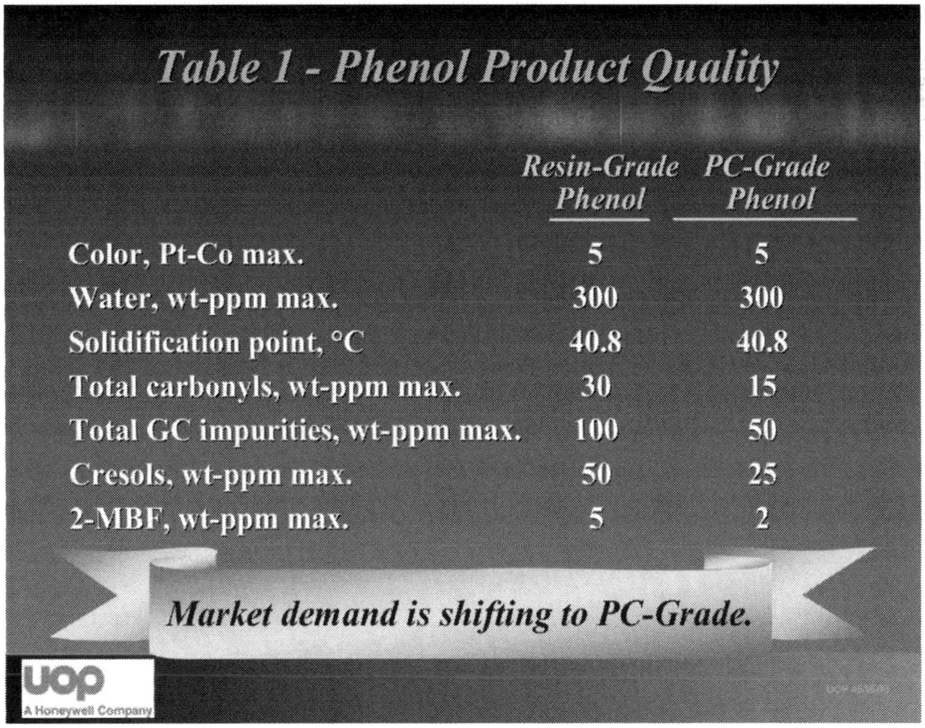
Table 1 - Phenol Product Quality

	Resin-Grade Phenol	PC-Grade Phenol
Color, Pt-Co max.	5	5
Water, wt-ppm max.	300	300
Solidification point, °C	40.8	40.8
Total carbonyls, wt-ppm max.	30	15
Total GC impurities, wt-ppm max.	100	50
Cresols, wt-ppm max.	50	25
2-MBF, wt-ppm max.	5	2

Market demand is shifting to PC-Grade.

uop
A Honeywell Company

Table 2 shows the typical high quality acetone product that is achieved in the process. Note that this specification is sufficient to meet or exceed that typically required for bis-phenol A manufacture. Even in this market however, demand is expected to grow for a higher purity product as consumption for the acetone byproduct becomes more out of balance with the high rate of growth of phenol consumption for PC and producers look to further differentiate their product quality to more fewer downstream users.

Other drivers include the need to improve the overall yield performance of the process to compete in a tight feedstock supply environment. Figure 2 shows the improvement in overall yield as measured by the cumene/phenol consumption ratio and the AMS yield on a weight and mole basis respectively over time for the Sunoco/UOP phenol process. Results show a fairly dramatic improvement over the past decade to the current state where a very close approach to the theoretical minimum cumene/phenol consumption of 1.277 is achieved. Thus the yield efficiency of the process is very high, helping to keep production costs low as the majority of the production cost of phenol is due to the fresh cumene feed consumption.

Table 2 - Acetone Product Quality

	Typical Spec
Acetone Purity, wt% min.	99.7
Color, Pt-Co max.	5
Specific Gravity, 20/20 °C	0.790-0.792
Water Content, wt% max.	0.3
Reducing Substances, hrs. min.	>2
Distillation Range, °C (including 56.1 °C)	0.5
Non-Volatile Matter, g/100ml	0.001
Acidity (as acetic acid) wt%	0.001
Alkalinity (as ammonia), wt-ppm	10
Total GC Impurities, wt-ppm max.	250

Competitive market requires tighter specs.

UOP
A Honeywell Company

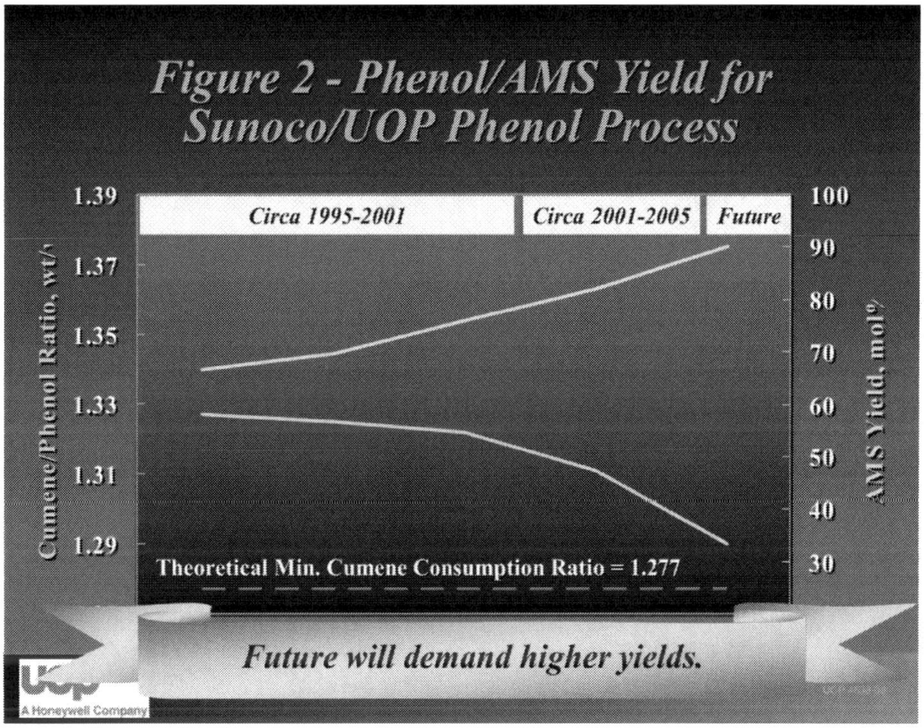

Sunoco/UOP: Meeting the Challenges Facing Phenol Producers

Meeting the challenges facing phenol producers requires using the best available technology to achieve the lowest possible cost of production in a highly competitive environment. Sunoco and UOP have a long history of introducing phenol technology innovations since the technology was first introduced the 1950's. Some of the very latest improvements are highlighted below:

1. Improved low pressure oxidizer design
2. Heat-integrated concentration section
3. Higher yield decomposer system
4. Simplified direct neutralization design
5. Optimized acetone purification design
6. Phenol purification utilizing chemical / resin treatment
7. Improved phenol recovery from waste streams

The impact of each of these innovations will be discussed in detail to demonstrate how recent producers are able to improve the efficiency of existing and new unit operations.

The Sunoco/UOP Cumene Peroxidation Route to Phenol Production

Figure 3 is a block flow diagram showing the major processing steps required to manufacture phenol. The major processing steps include: (1) liquid phase oxidation of cumene to cumene hydroperoxide (CHP), (2) concentration of CHP, (3) acid-catalyzed decomposition of concentrated CHP to phenol and acetone, (4) neutralization of acidic decomposition product, (5) fractionation of the neutralized decomposition product for recovery of acetone, phenol, AMS, and residue, (6) recovery of phenol and the effluent waste water via an extraction process to prepare it for further downstream treatment required to meet effluent quality specifications, and (7) hydrogenation of AMS back to cumene for recycling to synthesis; or, optionally, refining of AMS for sale as a product.

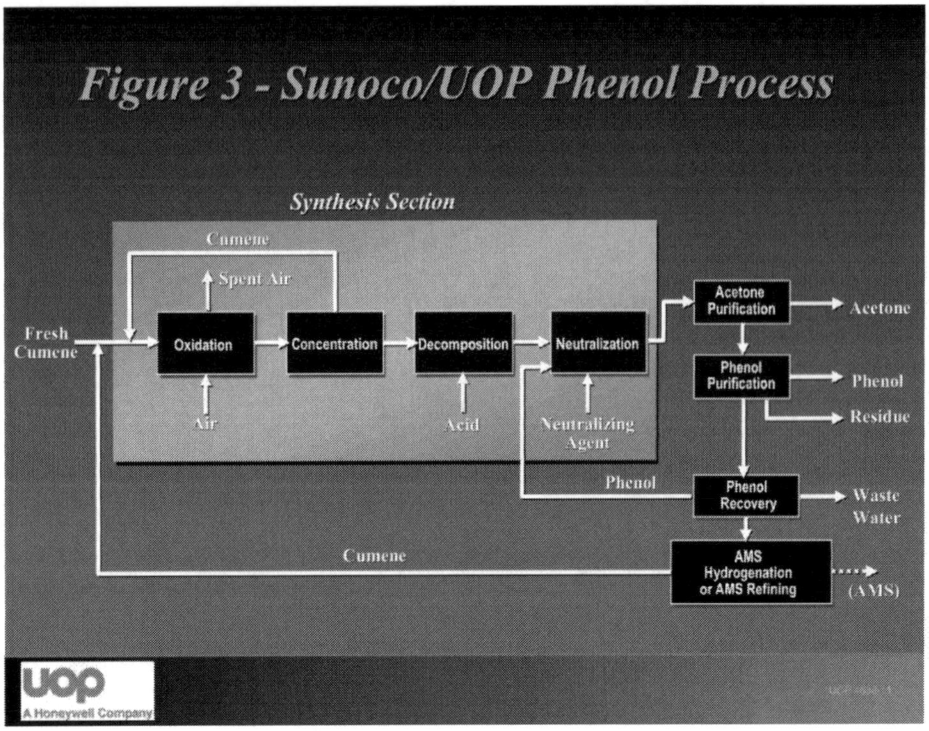

The capital cost breakdown for the process gives an indication of the mechanical complexity and equipment demand for each section of the process. The main oxidizer reactor section is by far the majority of the capital cost followed by the phenol purification and CHP concentration section with a total of roughly 3/4ths of the cost of the plant. The detailed breakdown is given below:

Oxidation	40%
Concentration	11%
Decomposition	5%
Neutralization	<1%
Acetone Purification	9%
Phenol Purification	27%
Phenol Recovery	5%
AMS	3%

In the oxidation section, cumene is oxidized to a target of 25 wt% terminal CHP concentration. This is achieved utilizing a two vessel reactor system to control the very exothermic autocatalytic oxidation reaction. The CHP is then concentrated and then decomposed to phenol and acetone using sulfuric acid, water, and temperature to control exothermic cleavage reaction in a two step process using a decomposer and a dehydrator (produces AMS). The products are then treated with an amine to remove any acids. This is a achieved by utilizing a new direct neutralization procedure which replaces the ion exchange resin process used in the past.

For product recovery, acetone is purified using caustic to react out aldehydes which are difficult to remove by simple distillation due to similar boiling points. After separating the acetone, the phenol is purified to remove trace contaminants such as 2-methylbenzofuran (2-MBF) and mesityl oxide (MO) in a new high temperature ion exchange resin process. This replaces the capital and utility intensive phenol stripper column used in the past. Plant aqueous effluent streams are processed to recover phenol and to prepare the effluent water for any additional downstream treatment required to meet effluent quality specifications.

In the final step AMS can either be hydrogenated back to cumene and recycled to the synthesis section or refined for sale as a net product.

Overall Process Description/ Chemistry

The main reactions for phenol and acetone production via cumene peroxidation are shown in Figure 4. Both reactions are highly exothermic. Oxidation of cumene to cumene hydroperoxide (CHP) proceeds via a free-radical mechanism that is essentially auto-catalyzed by CHP. The decomposition reaction is catalyzed by strong mineral acid and is highly selective to phenol and acetone. In practice, the many side reactions which take place simultaneously with the reactions are minimized by optimization of process conditions.

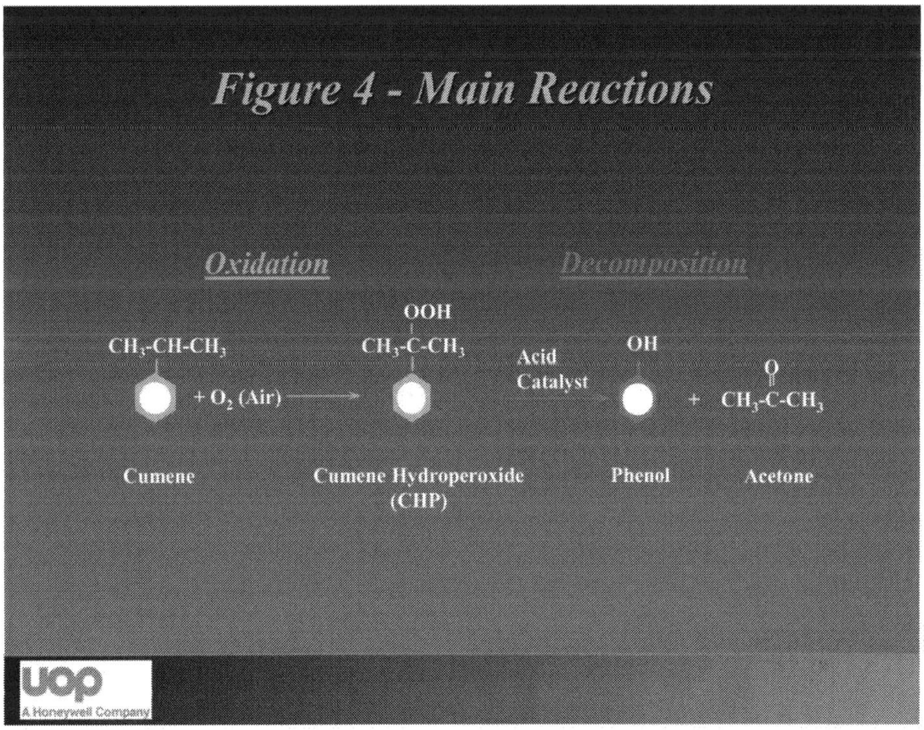

DMPC or dimethylphenylcarbinol (2-phenyl-2-hydroxypropane) is the main oxidation by-product, and the DMPC/AMS reactions play a significant role in the plant as shown in Figure 5. The DMPC can be dehydrated to produce AMS and water. If desired, AMS can be hydrogenated back to cumene. More details regarding the process chemistry for the Sunoco/UOP phenol process can be found in previous publications on this subject [(2)].

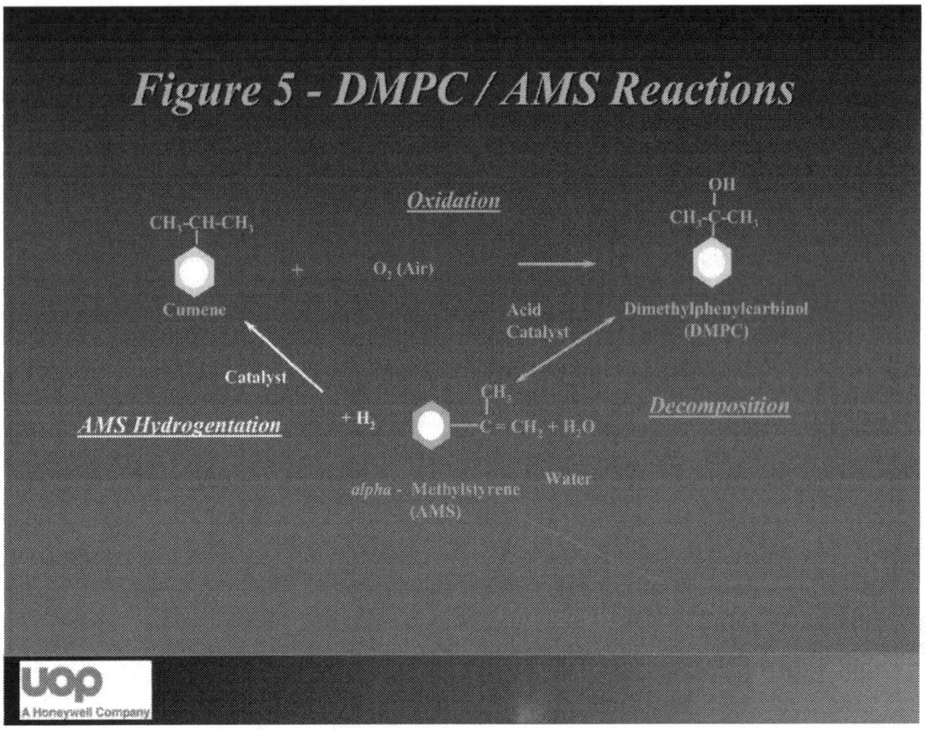

Process Flow and Recent Technology Improvements

Oxidation Section

Figure 6 shows a typical series flow for a two oxidation reactor configuration for the low-pressure technology. Additional reactors can be used in multiple reaction trains depending on the capacity of the unit, location, and processing objectives. This flexibility allows incremental investment staging to meet growing capacity demands.

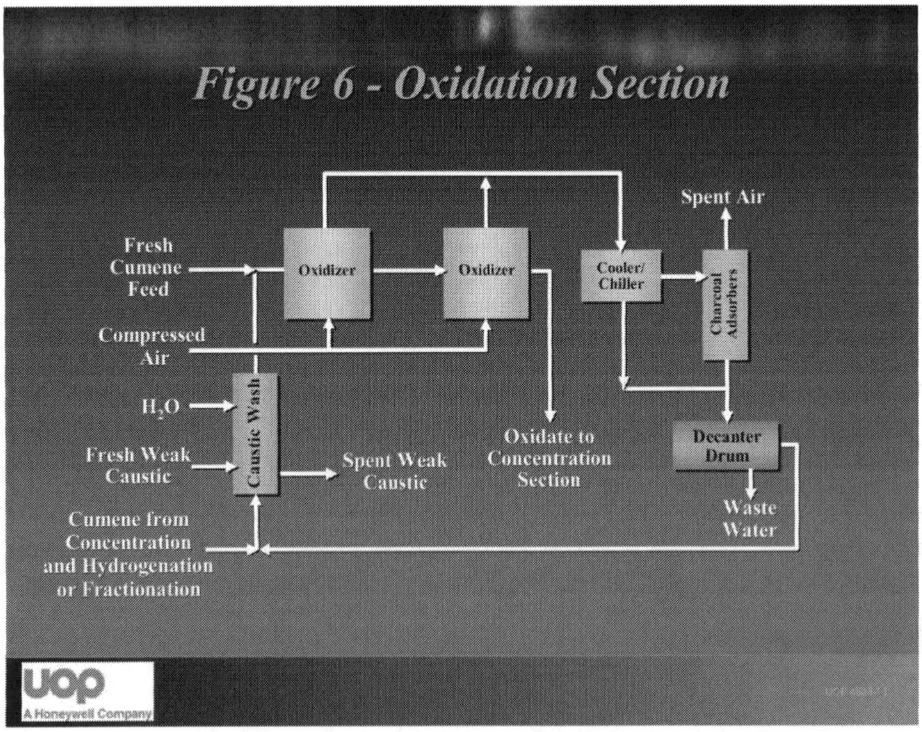

For large phenol units, it is economical to recover the heat of reaction from the second oxidizer by heat integration with the concentration section. The hot oxidizer circulating liquid stream is used to supply heating to the pre-flash column upper vaporizer. The net oxidate from oxidizer No. 2 (effluent from oxidation section) flows directly to the concentration section.

Figure 6 also shows the typical cumene recovery system from spent air. Spent air streams from both oxidizers are combined and routed through a water-cooled condenser, a chilled condenser and an entrainment separator for the maximum removal of hydrocarbon and cumene. From the entrainment separator, the air flows to the regenerable charcoal adsorber system. Two of the adsorbers are always on line in series flow, while the third one is being regenerated with steam.

The cleaned air from the charcoal adsorbers is vented to safe atmospheric disposal. A catalytic incinerator is typically not required to meet emission standards, but one can be provided if regulations stipulate incineration as the emission control method. The cumene collected by the charcoal adsorber is recovered by desorption with low-pressure steam followed by condensing the steam and decanting the cumene and water phases. The cumene is then recycled to the feed pre-wash column.

Only two oxidizers are required for units up to 300 KMTA. The low operating pressure used in the Sunoco/UOP process minimizes the cost of the oxidizers and air compressor (centrifugal). There are 15 to 20 temperature sensing points in each oxidizer along with two independent cooling loops for safety. The first oxidizer runs at slightly higher temperature than the lag in order to get better conversion rate and to maximize selectivity with a target of about 25% terminal CHP concentration. Competing high pressure oxidizer technologies typically operate at a much higher temperature and achieve only about 20% terminal CHP which increases the recycle rate. The low operating pressure used in the Sunoco/UOP process results in less heavies formation so that tar cracking and acetone recycle in the decomposition section are not required to achieve high yield.

Caustic wash is used to remove acids and other organic salts in the recycle cumene. With Sunoco/UOP low-pressure oxidation, the air stripping combined with partitioning of the acids to the condensate in the spent air cumene recovery system and the weak-caustic scrubbing of the recycle cumene is so effective that no other method of acid removal is required. Acidic conditions in the oxidizer terminate the cumene-to-CHP reaction and can lead to CHP decomposition which is very exothermic and dangerous.

Recent improvements to the oxidation section include:

1. Improved oxygen distribution and internal circulation systems resulting in the requirement of only two oxidizers at world scale capacities up to 300 KMTA.
2. Reduced O_2 in the vent gas resulting in smaller air compressors and smaller charcoal beds.
3. Use of a high efficiency charcoal adsorption as a result of optimal temperature and humidity selection. This results in longer cycle times with fewer regenerations and a steam consumption reduced by up to 40% with less waste water production.

Concentration Section

Figure 7 shows an example of the typical flow for the concentration section of the process. The main objective is to concentrate CHP in the oxidizer reactor effluent to a level of about 75 to 85 wt% prior to decomposition to phenol and acetone.

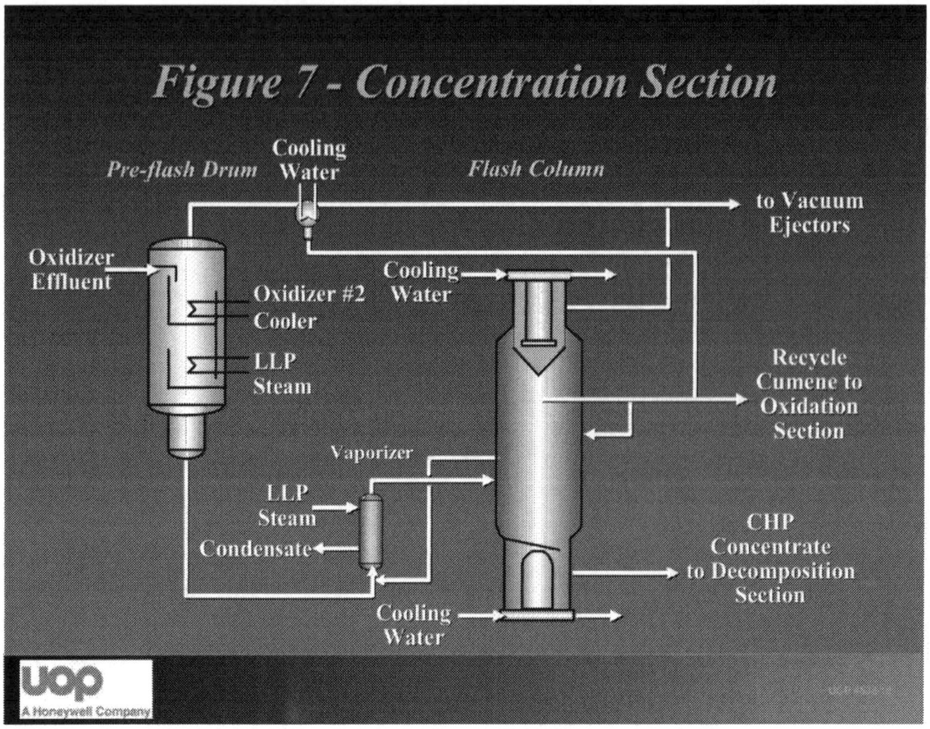

For large phenol units, it is economical to use a two-column concentration system, in which the heat of reaction from oxidizer No.2 and very low pressure steam are used to vaporize cumene in the first (pre-flash) column, reducing the size of the main flash column. The pre-flash drum and flash column operate under vacuum to minimize the temperature necessary to concentrate the CHP. The vacuum is typically generated by an ejector system. Final CHP concentration is achieved in the flash column vaporizer and flash column, both of which operate under deeper vacuum than the pre-flash drum.

A cumene quench tank is also provided in this section for automatic emergency quenching of various strategic sections of the concentration section if necessary to maintain safe operating temperatures in the event of an incipient CHP decomposition excursion.

The concentration section takes CHP from the Oxidizer effluent at about 25 wt% and concentrates this stream to about 82-84 wt% CHP. The concentrated CHP collects in the integral receiver/cooler at the bottom of the flash column, where it is cooled to a safe temperature. The flash is done under a deep vacuum to keep temperatures low in order to prevent thermal decomposition. The reflux on the flash column is used to control the CHP in the recycle cumene in the range of 0.5 to 4 wt%. An internal flash condenser minimizes the pressure drop in the overhead. The internal bottoms condenser reduces the amount of concentrated CHP in the column for safety. No reboiler is used.

Recent improvements in the concentration section technology include:

1. A new simple, two-stage design resulting in reduced capital cost and improved ease of operation.
2. Heat integration with the oxidation section resulting in low steam consumption.
3. An improved low holdup design and elimination of intermediate storage of hot CHP that increases process safety.

Decomposition Section

The decomposition or cleavage section of the process involves the catalytic decomposition of concentrated CHP in the presence of ppm levels of acid to crude phenol and acetone. The most effective technology for this section is a unique two-step process described by Sifniades [3] in 1982. The process involves the use of a backmixed reactor section at low temperature/ higher contact time for the main CHP decomposition step followed by a plug flow dehydration section at higher temperature/ short contact time for conversion of dicumylperoxide (DCP) and residual DMPC to AMS. The process represents a breakthrough in AMS yield improvement and, is currently being used by all licensors as the process of choice for modern high yield phenol technology.

Figure 8 shows an example of the most advanced decomposer technology offered by Sunoco/UOP. It consists of a very simple but elegant drum and loop reactor design where concentrated CHP from the concentration section flows into the decomposer drum, along with a metered amount of water to maintain optimal reaction conditions in the decomposer recycle loop. Sulfuric acid is injected via injection pumps into the loop to provide the catalyst required for the decomposition of CHP to phenol and acetone.

A circulation pump is provided to circulate the contents of the decomposer. Sulfuric acid is injected into the circulating stream to such an extent that the decomposition of CHP and dehydration of dimethylphenylcarbinol (DMPC), a key by-product of oxidation reaction, are precisely controlled. The level of unreacted CHP and DCP is monitored via calorimeters, to which part of the acid catalyst flow is routed. Only a very small slip stream is required from the decomposer to measure the CHP and DCP concentrations. A two stage delta T reading with the first measuring CHP concentration and second measuring DCP concentration is typically obtained. Acid is injected to ensure complete conversion of the CHP and DCP as an indirect measure of the concentrations of these species that are present in the decomposer. The typical acid concentration in decomposer is very low at about 20-30 wt-ppm.

The effluent from the decomposer is pumped to the dehydrator in which the effluent is heated to a temperature where remaining DMPC is dehydrated and DCP converted to AMS at very high yield. This is a unique advantage of the Sunoco/UOP decomposition technology. The Sunoco/UOP process produces approximately 90% AMS yield from DMPC. This also results in higher phenol yield, thus lower cumene consumption and less residue (e.g., tar) formation.

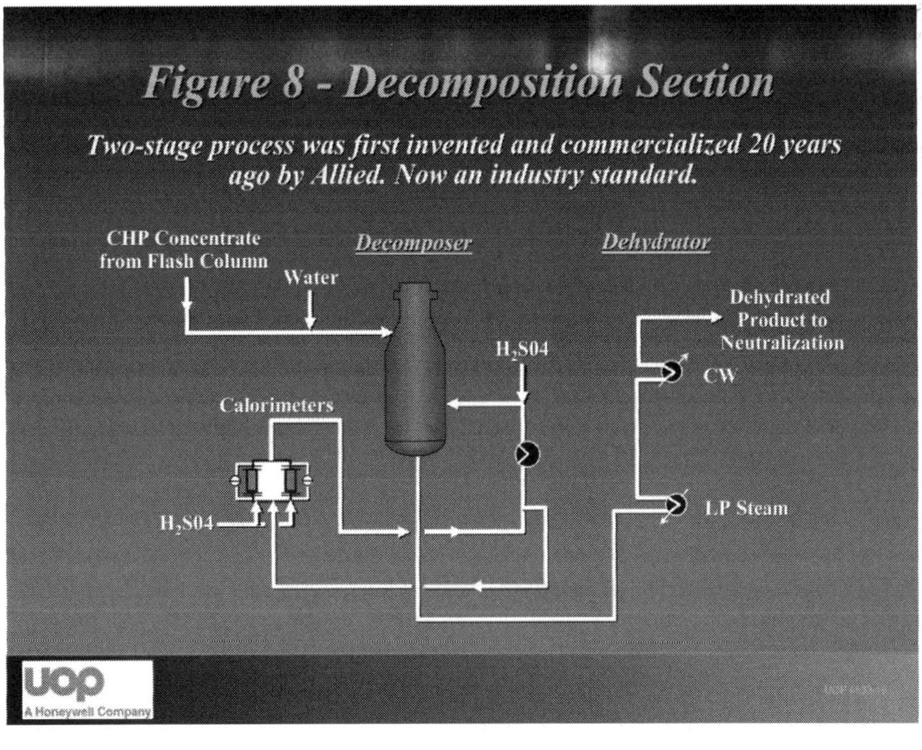

Most accidents have occurred in the decomposition section because acid injection was stopped and then was restarted at a time when the CHP level was allowed to increase. This risk is minimized in the UOP design with the use of three independent acid injection points.

Recent improvements in the decomposition section technology include:

1. Recycle ratio lowered to 25:1 resulting in reduced utility requirements and lower capital.
2. Use of multiple independent acid injection points for increased process safety.
3. Optimization of operating conditions resulting in higher process yields. Yields as high or higher than 99.5 mol % Phenol/CHP, 99.7 mol % Acetone/CHP, and 90 mol % AMS/DMPC can be expected from the process.

Neutralization Section

Limitations of traditional neutralization systems using ion exchange resins or high density salt solutions include:

1. High caustic usage
2. Salt disposal
3. Fouling in purification section requiring frequent plant shutdown for cleaning
4. High capital
5. High utilities for distillation
6. High wastewater generation

The acid catalyst that is added in the decomposition section must be neutralized to prevent yield loss due to side reactions and protect against corrosion in the fractionation section. The Sunoco/UOP Phenol process uses a novel approach for neutralization described by Keenan [(4)] in 2001. The process neutralizes the acid catalyst used in the decomposer by injecting a stoichiometric amount of a diamine which does not need to be removed from the process as shown in Figure 9.

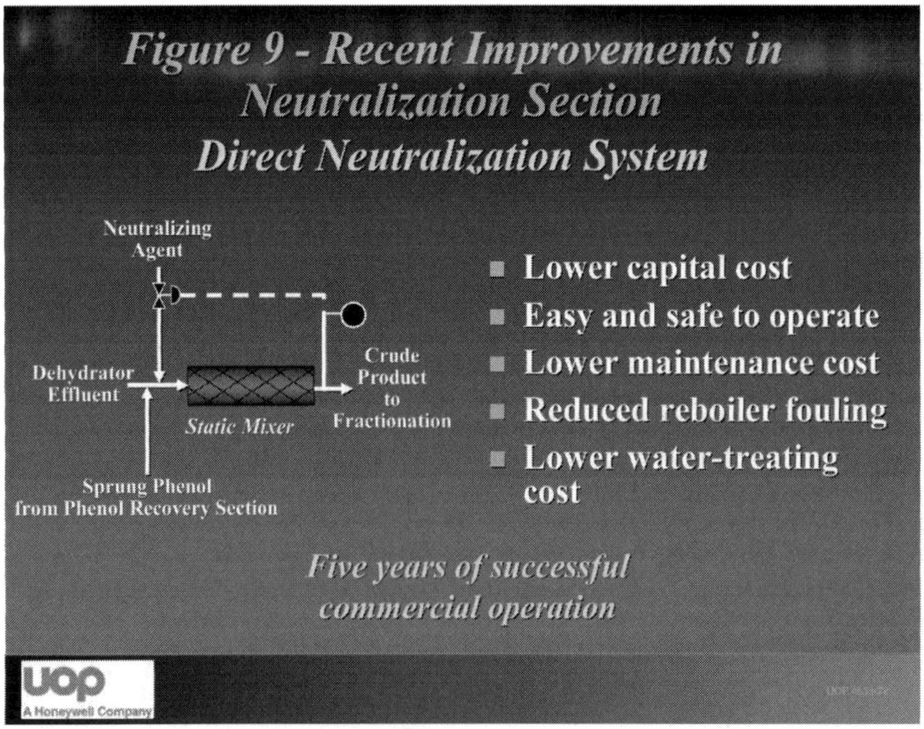

Direct diamine neutralization is better than the IX resin previously used resulting in reduced capital costs, waste water, complexity/easier operation. The process also provides a more constant feed upstream to the distillation section as measured by pH. Good control of crude product pH is critical for minimizing undesirable side reactions during purification. By replacing the older ion exchange resin technology with the new direct neutralization process, the phenolic waste water production can be reduced by 45% or more in this section at lower capital cost.

In addition, the diamine salts formed in the direct neutralization process are more soluble than those formed in competing processes (e.g., using ammonia) and results in less downstream fouling of exchangers. Sunoco has been using this design in their Frankford, PA plant for over five years and have observed a significant reduction in crude acetone column exchanger fouling. The fouling in this exchanger was previously attributed to the IX resin system and was virtually eliminated after the introduction of the new technology. In addition, the diamine salt does not contribute significantly to SO_x or NO_x when the residue is used as boiler fuel.

Acetone Refining Section Process Flow Optimization

Acetone is a major by-product of the cumene oxidation process for the production of phenol. The overall economics of the process are highly dependent on production of high quality acetone (e.g., 99.7 to 99.9% purity) for sales in the solvents and bis-phenol A markets. A new reactive distillation process scheme was described by Bhinde [5] in 2001. Figure 10 shows the typical acetone refining flowscheme for the Sunoco/UOP process and utilizes a unique reactive distillation approach to control product quality and reduce costs.

The phenol process produces some impurities such as aldehydes that boil with the acetone. These aldehyde impurities (which are monitored in part by the Permanganate Fading test) are removed in the finished acetone column (FAC) along with water. Caustic is injected into this column to catalyze the condensation of aldehydes with acetone to form an aldehyde-alcohol product. The heavier condensation products are less volatile, the bulk of which leave with the FAC bottoms in the oil phase as a heavy by-product.

Continued optimization of the reactive distillation section has resulted in lower wastewater generation, lower utilities requirements, and improved product quality.

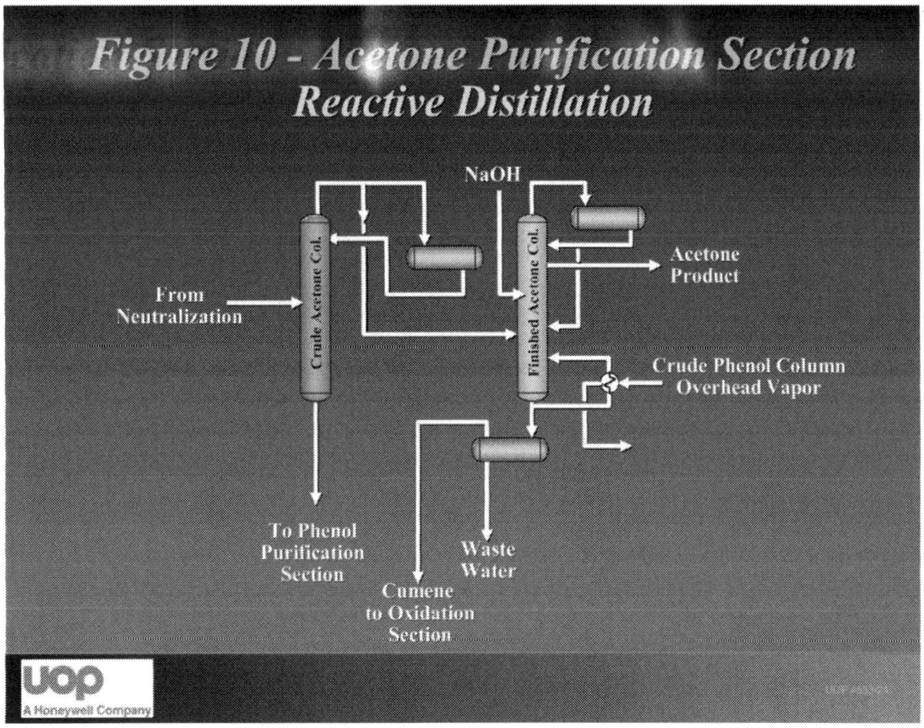

Phenol Fractionation and Purification Section

Once the crude phenol has been produced, it must be further fractionated to prepare a finished product that is of sufficient purity to meet downstream user specs. An example of the phenol purification is shown in Figure 11 using AMS hydrogenation as a means of recycling the by-product AMS to maximize phenol production. Note the separate residue stripper column section allows the final stripping of phenol from the residue to be conducted at higher vacuum, which allows both the crude phenol column and the residue stripper section to be reboiled with medium-pressure steam. As a result of this design, no high-pressure steam is required anywhere in the phenol plant.

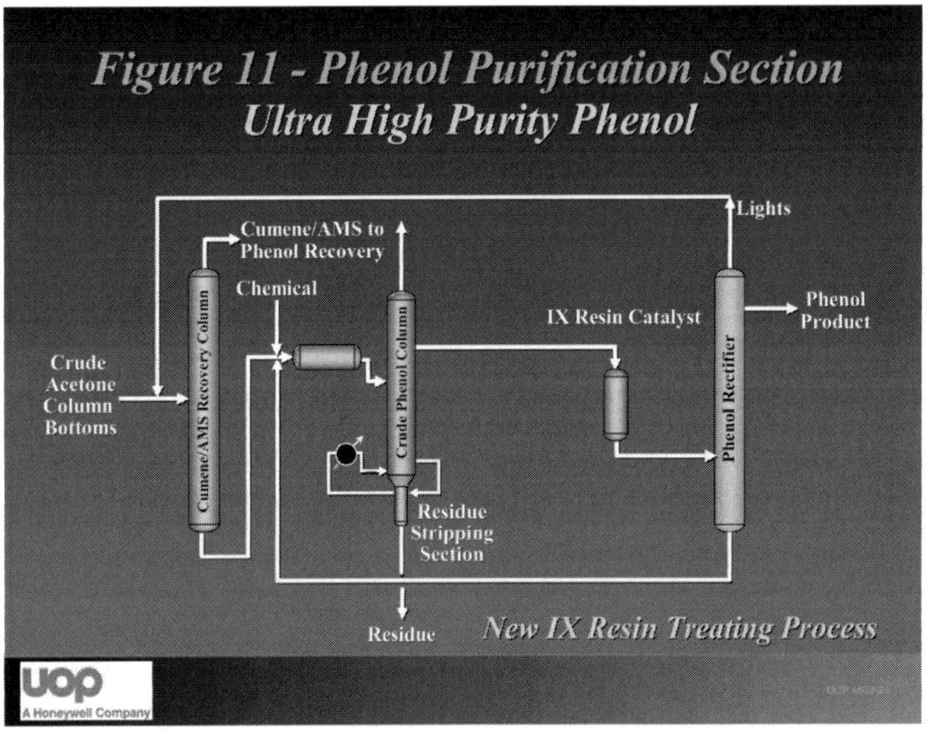

Figure 11 - Phenol Purification Section
Ultra High Purity Phenol

The cumene/ AMS recovery column (CRC) separates cumene from heavier phenol and AMS tar products while minimizing AMS in the cumene and minimizing cumene in the bottoms. The CRC operates under a heavy vacuum to keep temperatures low thus minimizing undesirable side products. A chemical agent is injected into the bottoms stream from the CRC to remove carbonyl impurities such as acetol (*alpha*-hydroxyacetone) and mesityl oxide from the phenol. Immediately after chemical injection, the bottoms from the column is routed through a chemical treatment reactor to provide residence time for the chemical treatment reactions. The crude phenol column has a top pasteurizing section to remove the small amount of light by-products generated during distillation.

The main product from the column is taken off as a side-cut and flows to ion exchange (IX) resin treaters, in which the IX resin catalyzes conversion of methylbenzofurans (MBF) and residual AMS to high-boiling components. MBF and AMS are otherwise difficult to remove by distillation. The IX resin bed replaces the previously used phenol stripper which was the largest utility consumer in the plant. The IX resin process was further described by Jenczewski [6] in 1995 and Wijesekera [7] in 2002.

The effluent from the IX resin treaters goes to the phenol rectifier, where the heavy components along with some phenol are distilled to the bottoms and recycled back to the crude phenol column. The phenol rectifier is designed to remove the remaining 2-MBF and AMS from the phenol. The phenol rectifier also has a top pasteurizing section for removal of small amounts of light by-products generated during distillation. Phenol product flows by gravity from the rectifier side cut to storage.

Use of an IX resin catalyst is much more effective in controlling total carbonyls and 2-MBF in the phenol and results in a very high purity phenol product in the range of <10ppm total carbonyls and <1ppm 2-MBF. This is compared to claims of <20ppm total carbonyls and <5ppm 2-MBF in some competing cumene hydroperoxide processes.

Recent improvements in the phenol fractionation and purification section technology include:

1. Incorporation of stab-in condensers resulting in the elimination of overhead receivers and pumps (lower equipment count) and a reduced pressure drop compared to a conventional condenser system.
2. Introduction of IX resin catalyst to reduce MBF, AMS, MO, and Hydroxyacetone resulting in the elimination of the energy intensive phenol azeo-stripping column and removal of the acid neutralizer system
3. A more versatile chemical treatment system

Phenol Recovery Section

Figure 12 shows the typical fractionation scheme needed to recover high quality AMS. The phenol recovery process utilizing a unique sodium phenate solution to recover phenol was described by Emblidge [8] in 2003. To achieve a high yield of AMS (typically >93% recovery), the crude AMS must first be caustic washed to remove phenol prior to fractionation. The washed product is then fractionated to remove cumene followed by fractionation in the AMS finishing column to remove the heavy by-products.

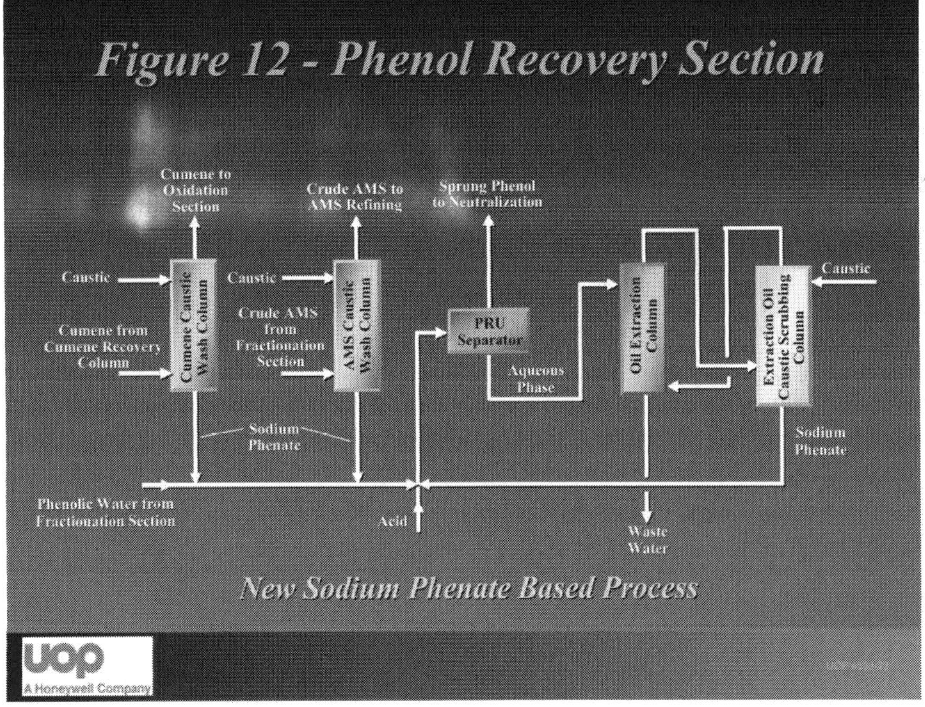

The phenol recovery unit (PRU) takes the highly phenolic streams that come from several sodium phenate sources and acidifies them in order to regenerate phenol from the salt. The somewhat limited solubility of the phenol "sprung" from the salt into the water phase results in a separate phenol phase. The phenol phase is then sent to neutralization. The temperature and sodium sulfate concentration must be carefully controlled to minimize the loss of acetone and phenol to the waste water. The aqueous phase from the PRU separator may contain up to 1-2 wt% phenol.

In the oil extraction column (OEC) and OEC caustic scrubbing column, cumene is used to extract phenol from the aqueous phase. Caustic is added into OEC caustic scrubber to facilitate extraction of phenol from cumene. The caustic wash columns are liquid full and have a caustic utilization rate of >85%.

Recent improvements include:

1. Introduction of the sodium phenate solution resulting in: (a) higher solubility for phenol and improved recovery of phenol from aqueous waste streams, (b) reduced caustic consumption as validated by the Sunoco-Frankford plant with a >40% reduction using the new process, (c) an increased phenol product yield, and (d) an improved effluent product quality.
2. Lower salt content of recovered phenol resulting in a significant reduction of fouling in the fractionation section.

Commercial Validation of Recent Improvements

All of the improvements described above have now been commercially demonstrated in two recent plant start-ups in 2005 and are designated as Units "A" and "B" for discussion purposes. Construction of these new units was mechanically completed in early 2005 and demonstration of the process improvements at full design throughput was completed by mid 2005. Both units have achieved ultra high purity phenol product quality, high phenol yields, and reduced utility consumption compared to previous unit operations. All new designs from Sunoco/UOP currently incorporate these technology advancements.

Results in Tables 3 and 4 show the typical high purity phenol and acetone product purities observed in these new units. It is now possible to produce phenol at a quality that well exceeds the normal requirement for BPA/PC. As the market for BPA/PC has grown in recent years, the demand for higher phenol product purity has increased. This trend is expected to continue in the future as more applications for PC are developed.

The definition of phenol quality shown as "ultra high purity" is based mainly on the total carbonyl content of the phenol product. "Total carbonyls" is a common phenol specification. Carbonyl compounds such as mesityl oxide (MO) and alpha-hydroxyacetone (acetol) can cause discoloration in some downstream products such as polycarbonates. Some applications have very specific limits on other impurities. A 2-methylbenzofuran (2-MBF) content of 1-2 ppm or less is readily achieved, but is not a requirement for all applications. In addition, the typical cresol content is less than 20 wt-ppm . In the case of the recent Unit "A" start-up, the phenol product contained approximately 8 ppm cresols, well exceeding the 25 ppm maximum spec for BPA/PC as shown in Table 3. Competitive technologies generally require intensive distillation to achieve these low cresol levels. Further, Sunoco/UOP can customize the unit design to produce lower grades of phenol if required, such as resin grade phenol, resulting in lower capital and operating costs.

A very high quality acetone product meeting or exceeding 99.85 wt% purity was also demonstrated in the new units as shown in Table 4. Of particular note is the very low water content in the finished acetone product that is well below the 0.3 wt% max. spec typically used as a design basis for the FAC.

Table 3 - Commercial Phenol Product Quality (Utilizing Recent Improvements)

	PC-Grade Phenol	*Ultra High Purity Phenol Unit "A"*	*Ultra High Purity Phenol Unit "B"*
Color, Pt-Co max.	5	5	5
Water, wt-ppm max.	300	91	105
Solidification point, °C	40.8	40.8	40.8
Total carbonyls, wt-ppm max.	15	7	0
Total GC impurities, wt-ppm max.	50	18	8
Cresols, wt-ppm max.	25	7.6	NA
2-MBF, wt-ppm max.	2	0.9	1.7

A new standard in phenol purity

UOP A Honeywell Company

Table 4 - Commercial Acetone Product Quality (Utilizing Recent Improvements)

	Typical Spec	*Unit "A"*	*Unit "B"*
Acetone Purity, wt% min.	99.7	99.85	99.87
Color, Pt-Co max.	5	5	5
Water Content, wt% max.	0.3	0.13	0.13
Reducing Substances, hrs. min.	2	>2	>3
Non-Volatile Matter, g/100ml	0.001	0.000	0.000
Acidity (as acetic acid) wt%	0.001	0.000	0.000
Alkalinity (as ammonia), wt-ppm	10	0	0
Total GC Impurities, wt-ppm max.	250	<250	160

Exceeding market expectations

UOP A Honeywell Company

Figures 13 and 14 highlight some of the economic benefits that can now be achieved using the new technology. This is demonstrated in terms of improved yield, reduced steam consumption, and reduced cost of production. Of particular note are results for Unit "A" shown in Figure 13 in which an overall yield that exceeded the plant design basis was observed. This result is now within about 2% of the theoretical limit as measured by the cumene/phenol consumption ratio. Further improvements are under development to bring the consumption ratio to within about 1% of theoretical. Thus the Sunoco/UOP process is highly efficient from a feed utilization standpoint. This is extremely important as feedstock consumption is the major component of the cost of production. Figure 14 highlights the trend toward lower cost of production and utilities as new improvements have been commercialized by Sunoco/UOP over the past 20 years. The latest technology offers a performance that reduces both the cost of production by about 20% and the total steam consumption by about 50% compared to technology offered in the 1980's.

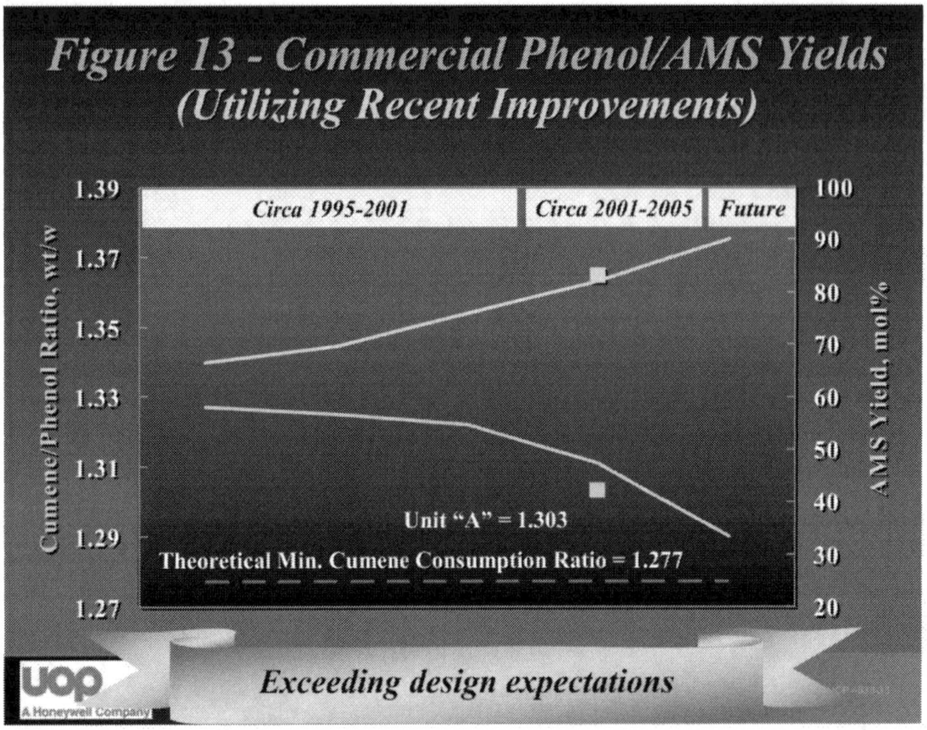

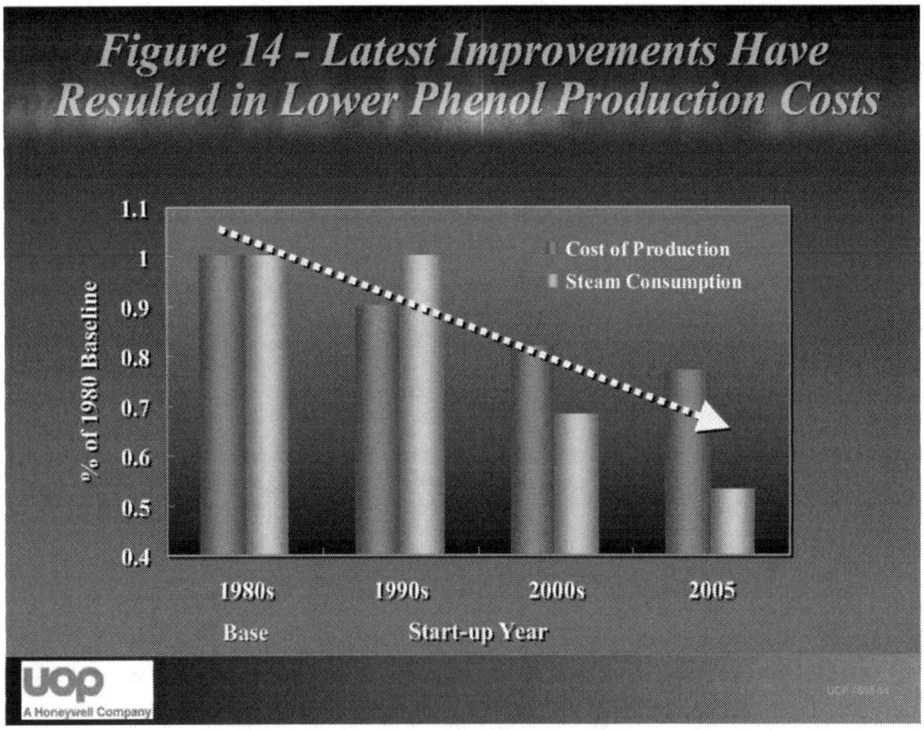

Conclusion

Phenol producers continue to face a challenging marketplace. Recent process improvements by Sunoco/UOP enable licensees to gain the upper hand in this environment by achieving an ultra high purity phenol product, at high yield, and at a significantly reduced cost of production. Development programs continue at Sunoco/UOP to further improve product yields and lower production costs.

Sunoco/UOP is the world leader in cumene-to-phenol process technology. Sunoco is the largest phenol producer in North America with a production capacity of 590,000 MTA. A total of 14 Sunoco/UOP Phenol Process plants have been licensed with twelve in operation and two currently in design and construction with over 2 million MTA in combined phenol capacity.

REFERENCES

1. R. Schmidt, A. Zarchy, and G. Petersen, *New Developments in Cumene and Ethylbenzene Alkylation - Paper 124b*, 1st Annual Aromatic Producers Conference – AIChE Spring Meeting, New Orleans, LA., March 10-14, 2002.
2. R. Schmidt, *Industrial Catalytic Processes – Phenol Production*, Applied Catalysis A: General, Vol 280, Issue 1, pp.89-103, February 25, 2005.
3. U.S.Patent 4,358,618 (November 9, 1982), S. Sifniades, A.A.Turnick, F.W. Koff (to Allied Corporation).
4. U.S.Patent 6,201,157 (March, 13, 2001), S.R. Keenan (to Suncoco, Inc. – R&M).
5. U.S.Patent 6,303,826 (October 16, 2001), M.V. Bhinde, S.R. Keenan (to Suncoco, Inc. – R&M).
6. U.S.Patent 5,414,154 (May 9, 1995), T.J. Jenczewski, L. Cresentinin, J.A. Kweeder (to Allied Signal, Inc.).
7 U.S.Patent 6,388,144 (May 14, 2002), T.P. Wijesekera, S.R. Keenan (to Suncoco, Inc. – R&M).
8. U.S.Patent 6,667,178 (December 23, 2003), R.W.Emblidge (to Suncoco, Inc. – R&M).

57c Production of Metaxylene within the Aromatic Complex

Philibert Leflaive, Luc Wolff, and Karin Barthelet. Institut Francais du Pétrole, BP 3, Vernaison, 69390, France

Although the metaxylene (MX) market is much smaller than the paraxylene one, there is however an interest in the production of pure metaxylene within the aromatic complex.

Different options for MX production within the C8-aromatics loop were developed and analyzed. More particularly, two interesting options can be further detailed :

- The first option is the adsorption based ELUXYL-MX™ metaxylene production process, new process available at Axens in the ParamaX portfolio. This process relies on both a high performance adsorbent-desorbent system and the industrially proven ELUXYL™ simulated moving bed (SMB) technology for PX world-scale capacities. When coproduction of pure PX and pure MX is wanted, an excellent synergy can be found putting the two SMB processes in series, the ELUXYL™ raffinate is sent to the ELUXYL-MX™ SMB unit dedicated to the production of MX with high purity (99.5+%) and high recovery.

- The second option relies on the ELUXYL™ Twin Raffinate technology [1], a modified version of the ELUXYL™ PX production process. In ELUXYL™ Twin Raffinate process, in addition to the conventional inlets and outlets (desorbent and feed injections, extract and raffinate withdrawals) of the simulated moving bed, a second raffinate is used allowing the recovery of a pure MX/OX mixture. This raffinate 2 is sent to an MX/OX splitter, which allows to get an MX enriched stream at the column head and an OX concentrated stream at the bottom. Then, either the head or the bottom stream may be sent to a crystallizer for production of MX or OX respectively. An interesting point of this process scheme is the flexibility offered for either producing MX or OX.

ELUXYL™ Twin Raffinate and ELUXYL-MX™ technologies can also be associated.

We propose to detail how the ELUXYL-MX™ was developed in terms of adsorbent intrinsic performances and process design and how benefits can be taken from particular arrangements involving ELUXYL™ Twin Raffinate technology for pure PX and MX coproduction.

[1] L. Wolff, Ph. Leflaive, "ELUXYL™ Twin Raffinate Technology", AIChE National Meeting 2004

INTEGRATING PETROCHEMICALS INDUSTRY WITH A FUEL ORIENTED REFINERY

Ahmed Sirag Khogeer(ahmed.khogeer@aramco.com), Research & Development Center, Saudi Aramco and Mohamed Balamesh (balamema@aramco.com), Process & Controls System Dept. Saudi Aramco, PO Box 5836, Dhahran, 31311, Saudi Arabia and

Abstract

Refining and petrochemicals businesses had shifted the operations dramatically between fuels, olefins and aromatics production. The main objective of integration the petrochemicals in a fuel oriented refinery is to increase the profit and take advantage of the high rise on some petrochemicals prices such as propylene and aromatics compared to LPG and gasoline. For long time, petrochemicals including Aromatics and olefins were produced in chemical or petrochemical plants that may be a part of a refining corporation but not integrated in a refinery.

Both Refining and petrochemical businesses go on cycles between low and high margins. However, the duration of these peaks are different. Many refining corporation are looking to optimize the resources to produce valuable petrochemical feedstock from the cheapest refinery streams. One of the best places to do this is within the refinery itself. Some of this optimization can be done with a minimum process condition change while other requires major revamp. However, even these revamps can payback very quickly. In this paper, we will show real cases in which the refining and petrochemical businesses had shifted operations dramatically between fuels, olefins and aromatics production. In addition, it will cover some consideration taken into account in designing some grass root refineries and a refinery upgrade for the maximum benefit from fuels-petrochemicals integration.

Introduction

This business suffered during the last two decades of declining margins and more stringent environmental regulations which shrunk the number of US refineries by 40 refineries in only eight years. No grass root refinery was built in the United States since 1980. For those who remain in the business, optimizing the performance and adding value means the difference between life and death for their refineries. Although the refining margins and the oil prices soared during the last two yeas, refining corporation do not trust current prices for long term heavy capital investment. One of the promising areas for profits is petrochemical integration within a refinery that was originally built to supply fuels to transportation markets. For long time, petrochemicals including Aromatics and olefins were produced in chemical and petrochemical plants that may be a part of a refining corporation.

Both refining and petrochemicals industries go into cycles between different peaks. These cycles start when the demand increases more than the industry can supply, thus from basic demand and supply relations, prices will increase. With new higher prices, more new entrant are attracted to enter the business and current manufacturers are encouraged to increase capacity to utilize higher margins. With the time required for designing, approving, financing and constructing a new refinery or a petrochemical plant can take years, the result is a surplus capacity after commissioning of new plants and expansions. This leads to a prices and margins drop and a lower plant utilization rates. The situation continues until a natural demand increase consumes most of the excess capacity or some older plants exit the business due to failure to meet the regulations or due to declining margins. The cycle for the refining business is about 15 years while for petrochemical plants it is different for each product but is in the range of 7-10 years.

Petrochemicals industry Investments usually requires less capital than in the refining industry. However, the marketing channels of petrochemical products are more complex, thus usually big stakeholders of final or intermediate petrochemicals enter as partners in the in petrochemicals plants. Petrochemicals showed much higher returns and growth as compared to the more mature refining business as shown in table 1. Returns on incremental investment to achieve integration are significantly higher, 28% on average. The low refinery margins can be offset by the stronger petrochemicals returns.

Table 1: Refining Versus Petrochemicals statistics

	Refining	**Petrochemical**
Return on replacement capital in (USA)	2.4 %	15.6 %
Demand growth	2 %	4-7%
Returns	5-12 %	7-23%

Petrochemicals growth rates estimates defer in the literature, however, it is believed that ethylene and propylene will have the highest growth rates which were sited in the literature at 3.5-6 % annually as shown in table 2.

Table 2: Petrochemicals growth estimate [9]

Product	**% Growth 2007-2012**	**Additional Required Capacity by 2012 (MMTPA)**
Para Xylene	4.6	7.3
Mixed Xylene	4.1	9.8
Propylene	3.5	15.7
Ethylene	3.2	13.4
Benzene	2.8	5.8
Cumene	2.6	3.4

Today the refining and petrochemicals businesses had shifted the operations dramatically between fuels, olefins and aromatics production. The main objective of integration the petrochemicals in a fuel oriented refinery is to increase the profit and take advantage of the high rise on some petrochemicals prices such as propylene and aromatics compared to LPG and gasoline. For example the FCC in the refinery can be utilized to produce Cumene which is high value product. The incremental benefit for installing the Cumene operation is an additional $20 million per year margin at a capital investment of $ 23 million, yielding a 39% incremental internal rate of return (IRR). On the other side Toluene and Xylene can be used for petrochemicals instead of gasoline blending by using the toluene disproportionation and Xylene isomerization technologies to produce more benzene and o-Xylene. The incremental benefit for this case was $37 million per year margin at a capital investment of $ 68 million, for an IRR of 27%.

Economics are not the only driving force for refinery petrochemicals integration. Environmental regulations are another factor. This force the refineries to look into extraction of benzene, toluene or Xylene from reformate. The Traditional production methods for mixed Xylenes are primarily the catalytic reforming of naphtha; and to a lesser extent produced as a by-product of the steam cracking of naphtha. The two production methods yield a mixed Xylenes stream. With the technology improvement, toluene became superior feed stock for making Para- Xylene and benzene through a new process called Selective Toluene Disproportionation.

Besides the conventional technology approaches, new avenues are being pursued to extend benefits for integrated producers. The first example of this strategy is improving feedstock properties to increase petrochemical yields. This approach derives from an understanding of the kinetic properties of each process operation and seizing on the ability to improve efficiency by tailoring the feeds. UOP's newly introduced MaxEne process is one example. By altering the composition of naphtha feed to ethylene steam crackers, this process gives the ability to increase steam cracker ethylene yield by 30% further integration is possible by sending the rejected $C7^+$ hydrocarbons that are non-normal to the catalytic reformer. This enrichment also has the benefit to increase octane barrel production.

The second example of advanced process technology hinges on changing the refinery process to maximize the conversion of crude oil fractions to petrochemicals. UOP has developed a modification of the conventional FCC process to shift the product slate heavily towards petrochemicals. The Petro FCC process will yield 56% product into petrochemicals while traditional FCC will yield 17% into petrochemicals.

Benefits and areas for integration:

Among the benefits of integrating petrochemicals into existing fuel oriented refinery operations are [3]:

- Reduced capital investment or working capital due to efficient infrastructure utilization.
- Reduced fixed costs as a result of shared services.
- Optimization of overall refinery product yields.
- Higher value of transfer streams and products.
- Petrochemical downturns tempered by more stable refining cycles.
- Promotes new business & creates employment opportunities.
- Generates capital & operating cost synergies.
- Diversifies business portfolio.

- Maximize Recovery and Utilization of Valuable Raw Materials.
- Reduce Environmental Emissions and meet fuels product specs more economically.

There are many areas for petrochemicals integration in a fuel oriented refinery in order to have an efficient system such as:

- Utilities, Waste Treatment,
- Intermediate Product Storage,
- Safety and Loss Prevention,
- Waste heat and Fuel Streams,
- Olefins Purification and Usage,
- Aromatics Recovery and Usage, and
- Hydrogen Utilization.

Table 3 and figure 1 show refinery streams that provide feedstock for petrochemical plants. Fluid catalytic cracking units (FCC) provide ethylene, propylene, and butylenes. These products can also be produced from Delayed cokers but in smaller quantities. Catalytic reformers produce aromatic, such as Benzene, Toluene, and mixed Xylenes known in industry as BTX.

A typical aromatics complex produces benzene and paraxylenes. It consists of a BTX separation and extraction plant and a mixed Xylenes separation and isomerization plant. Toluene and $C9^{+}$ aromatics, byproducts of the aromatics extraction facility are used as high-octane gasoline blendstock. Raffinate, also a byproduct, is used for the refinery light-naphtha product.

Table 3: Petrochemicals from Refinery Streams [3]

Refinery stream	**Petrochemical stream**	**Refinery use**
FCC off gas	Ethylene	Fuel gas
Refinery propylene (FCC product)	Propylene	Alkylation/polygasoline
Reformate	Benzene, toluene, Xylenes	Gasoline blending
Naphtha and LPG	Ethylene	Fuel gas
Dilute ethylene (FCC and delayed Coker off gases)	Ethylbenzene	Fuel gas
Refinery propylene (FCC product)	Polypropylene, Cumene, Isopropanol, Oligomers	Alkylation
Butylenes (FCC and delayed Coker)	MEK (methyl ethyl ketone)	Alkylation, MTBE
Butylenes (FCC and delayed Coker)	MTBE	Alkylation, MTBE
Refinery benzene and hydrogen	Cyclohexane	Gasoline blending
Reformate	Orthoxylene	Gasoline blending
Reformate	Paraxylene	Gasoline blending
Kerosene	n-Paraffin	Refinery product
FCC light cycle oil	Naphthalene	Diesel blendstock

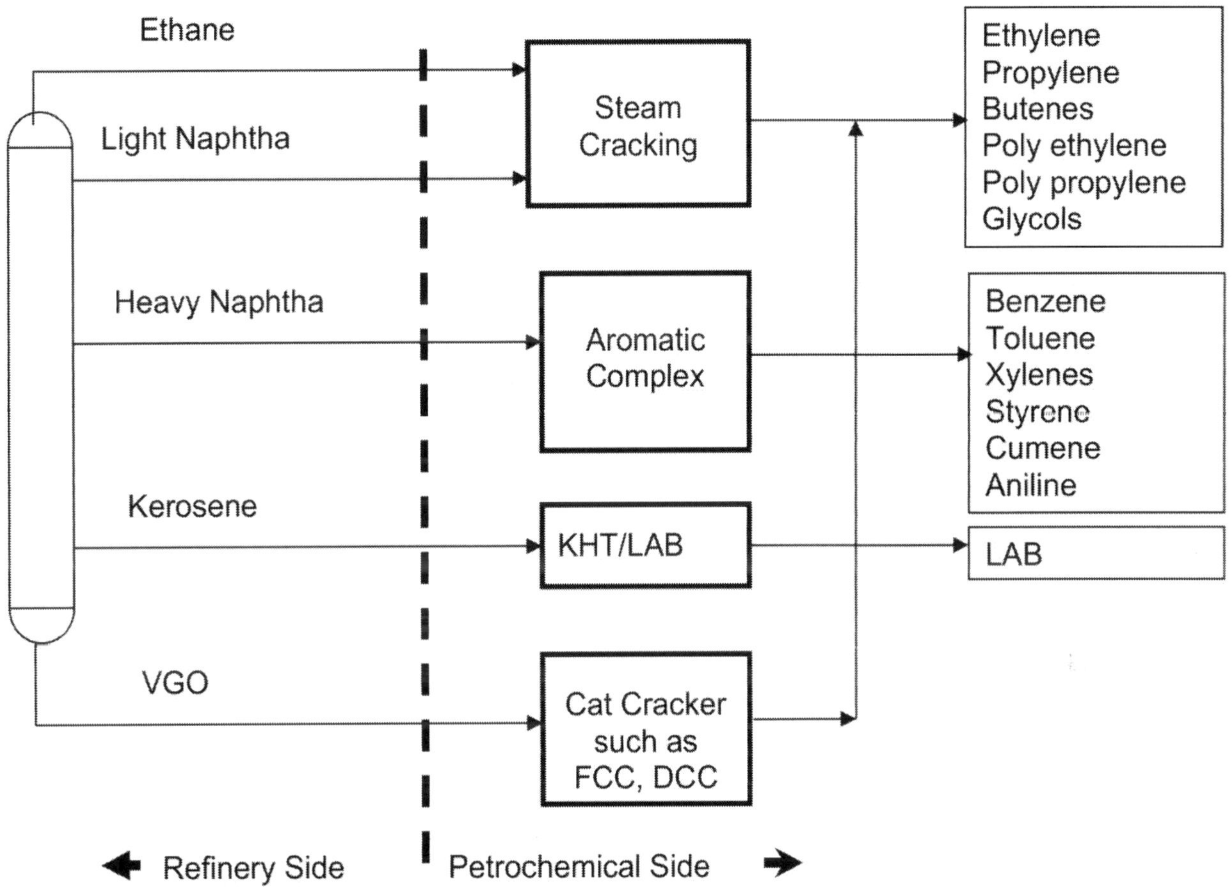

Figure 1: Integration Scheme

Opportunities for certain streams in Integration

Olefins can be recovered from the FCC unit. Currently, propylene is produced at small quantity from ethane crackers at petrochemical plants. These small quantities may not make propylene recovery economical. When combined with the propylene form an FCC unit which can range from 2-4 % wt for a conventional FCC unit, recovery of propylene to a PP plant becomes more economically attractive. The same way, recovery of FCC produced ethylene cannot be justified and is usually sent to fuel gas system. In an integrated plant, this stream can be recovered in the ethane cracker for added value. Same thing is for butylenes.

Aromatics Opportunities: A big concern for refiners is the gasoline benzene spec which limits the gasoline benzene content to 1%. As can bee seen in table 4, many refinery gasoline blending components such as reformate and FCC naphtha are rich in benzene. At some cases, it may be required to hydrogenate benzene which leads to octane loss. On the other hand, ethylene naphtha crackers produce a significant amount of Py-Gas rich in benzene, toluene and mixed Xylenes. Thus, the recovery of benzene from both sources in an integrated plant will be a Win/Win solution and solve several problems including refinery environmental problem and Increase amount of benzene available for chemical use while returning toluene to the refinery side for gasoline octane boosting. Xylene can be recovered from reformate and also from py-gas and valuable stream is sent for polyester manufacturing while other less chemically desired Xylenes are valuable gasoline blending component.

Table 4: Aromatics from refinery streams

Component	CCR Reformate	FCC Gasoline	Petro-FCC Gasoline
Benzene	1-10	0.5-1.5	3
Toluene	15-25	5-10	15
Xylenes	30	2-12	26
C9+ Aromatics	9	12-18	35
Total Aromatics	60-75	20-40	Up to 80

Hydrogen Opportunities: Steam crackers are net Hydrogen producers which are not fully utilized in petrochemical plants and can be sent to fuel gas system. On the other hand, refineries are net Hydrogen consumers for hydrotreating and hydrocracking purposes. A Win/Win situation for both parties is to transfer hydrogen from steam cracker to the refinery which adds value to the hydrogen and substantially lowers refinery investment.

Fuel Systems Integration: Py-Gas and oil from ethane crackers can be sent to the refinery for further processing instead of using it as a fuel. On the other hand, excess refinery methane and ethane which are normally used as fuel gas and excess LPG can be sent to the steam cracker to produce valuable ethylene product.

Case study:

From the literature background, it is concluded that the best candidates for refinery petrochemicals integrations are olefins and BTX especially P-Xylene. It is also known that the best thing is to have this integration in the early design phases in order to integrate other systems such as heat recovery, fuel gas, waste treatment, etc... However, this is not always the case. In this case study, we will see how we can apply the integration concept to a real life case for a system of seven refineries and see the best candidates for this integration and how specific products may apply.

In this section we will take a system of refineries is in the Middle East as a case study. Middle East refineries are in general simple refineries geared mainly towards supplying the local markets with some state of art export refineries. Since the countries in this study are net oil exporters and thus gain more money by exporting crude oil than from refining. Another constraint on integration is that the main outlet for petrochemical product is export to international markets. The secondary and tertiary petrochemical industries are much less than the supply of basic petrochemicals thus logistics and marketing can play a major rule in the decision to expand into petrochemicals.

In this work, we started with a model consisted of seven refineries; in which five are domestic refineries with variable capability of exporting excess products to other refineries or international markets. These refineries where built with the vision of supplying local markets so there are limitations on the type and quality of products they can produce. The other two are export refineries and are state of the art refineries with capabilities of meeting stringent products specifications. For a feasible refinery –petrochemicals integration, the minimum refinery capacity is usually 200 KBPD. Thus most of these refineries can be theoretically integrated in to a petrochemicals facilities.

Table 5: Summary of Refinery Processes

	Crude Capacity, KBPD	Reformer	FCC	HC
Refinery # 1	560	Y	N	Y
Refinery # 2	120	Y	N	Y
Refinery # 3	80	Y	20	N
Refinery # 4	450	N	N	N
Refinery # 5	210	Y	N	N
Refinery # 6	400	Y	94	N
Refinery # 7	400	Y	N	Y
New export refinery #1	400	?	?	?
New export refinery #2	400	?	?	?

The most two attractive integration from refineries are the olefins production and the BTX. The first one refers to ethylene and propylene. The main source for these olefins currently is from cracking of ethane and naphtha. A concern usually is to produce a petrochemicals grade propylene that can be sent to a poly propylene plant. This usually requires high purity of 99.5 % +.

From the table above, the best candidate for olefins recovery especially propylene is refinery # 6 with its huge cat cracker. As mentioned, a conventional FCC can produce upto 4% wt propylene.

The decision of which refinery to upgrade and integrate is governed by the supply commitments and the complexity of the refinery. Thus it may appear that refinery # 1 is the best candidate for a new cat cracking unit. However, as it appears, the Hydrocracker unit does not leave as much feed for the new FCC unit thus the maximum capacity is limited. The same applies for refineries # 2 & # 7. These refineries were built with different objectives.

Refinery # 4 is the best candidate for a new catalytic cracking unit which can supply both fuels and petrochemicals market. For a new grass root unit, it is very important to decide the orientation of the refinery. For a fuel oriented refinery, the refinery may select a conventional FCC operating either at a gasoline or distillate modes. If the refinery is fuels oriented, and then the attention will be to what kind of FCC to build, is it a VGO feed or a direct Residue? RON will be among the important targets. Olefins will be considered as relatively secondary products that are less valuable and to some degree undesired even of the refinery has an Alkylation or catalytic condensation units. The design of the whole gas concentration system will be for a propylene content of about 4 % which will effect the decision of installing a polypropylene unit and the necessary to produce petrochemical grade propylene requiring about 60 theoretical stages.

Technology selection for olefins max process:

For refinery # 4, the intention was to enter the petrochemical markets. Even thought the refinery is a fuel oriented one, propylene prices are now about the double for gasoline. There are three major derivation of catalytic cracking unit for olefins max production. These are the high severity HS-FCC process, petro FCC and deep catalytic cracking unit. The first two are not commercially proven at the time of the study. The DCC technology was commercialized successfully and several units are operating with propylene yield of about 18% of the feed. This successful application of the DCC technology made it is considered for refinery # 4 expansions. Note that the substantial increase in the propylene content makes it unsuitable for revamps but only for new installation since gas recovery and concentration section equipments such as towers and compressors should be designed to accommodate the increased gas and olefins make. That is why it was not considered as a revamp for any of the FCC in the study.

In this refinery, even thought, the refinery is looking into exploring the petrochemicals markets; it did not to expand into the BTX business at this time. As can be seen from the table below for general refinery economics, the most attractive investment is in the olefins from ethane crackers. BTX is just marginally attractive.

Table 6: economic evaluation for some refineries integration

	%IRR
Benzene	0
Cumene	>15
Aromatics	<10
Styrene	>15
Steam Crackers	
Ethane	>25
Ethane/Propane	20
Naphtha	>10
FCC Propylene	> 20
Mixed Xylene	>15

Ethane crackers are popular around the world and it may be one of the best ways to produce high quality ethylene. However, for some countries, ethane supplies can be limited which requires the use of other means to produce olefins in which more emphasis is given to the DCC technology.

For the two new export refineries, the objective is high quality fuels for the international markets thus again; the technology chosen may not be a petrochemical oriented FCC but a conventional one. The decision considered not only the FCC unit but the gas recovery section, the gasoline pool will be affected seriously and thus requires additional octane boosters the olefins will require different storage and shipping means which needs to be planed in advance fore a successful integration... can

Conclusion

1. Integrating refining and petrochemical business can substantially boost a refinery profit margin.
2. Integration into an existing refinery is not attractive as if it is carried out on a grass rot refinery.
3. Revamps to an existing FCC units into an olefins oriented unit is governed by the downstream treating and concentration capacities and is usually not attractive.
4. Geographic location and presence of secondary and tertiary petrochemical industries can effect the decision of petrochemicals integration particularly for BTX.
5. Even thought that BTX recovery from both sides may not be as attractive as olefins production, it can critically help meet gasoline specification for both Benzene content and RON.

Nomenclature

ABBREVIATION	Description
RON	Research Octane Number
KBPD	Thousand barrels per day
FCC	Fluidized Catalytic Cracking unit
CCR	Continuous Catalyst Regeneration (Type of Catalytic reforming)
HC	Hydrocracker

References

1. Anon; Petrochem complex shields refining profits; Oil & Gas Journal; v 96 n 31; Aug, 1998; ISSN:00301388
2. Khogeer, A. Multi Objective Multi Refinery Optimization with Environmental and Catastrophic Failure Effects Objectives. PhD Dissertation, Colorado State University. 2005.
3. Liu, David. Strategic Integration of Refinery and Petrochemical Plants. Hydrocarbon Engineering. September 2001.
4. Zukauskas, Vladas. Keys to profitably producing aromatics. UOP LLC 25 E. Algonquin Road. Des Plaines, Illinois 60017
5. Partha Maitra John Folkers Peter Coles. Integrating a refinery and petrochemical complex. PTQ AUTUMN 2000.
6. SRI, In Search of Advantage-Emergence of a Dipolar investment Pattern. 1999.

Factors That Affect the Quality of the Essential Lemon Oil (Citrus Aurantifolia) during the Distillation

Gamarra, F.M.C.[1], Sakanaka, L.S.[2], Tambourgi, E. B.[3] , Cabral, F.A.[4]

1-DESQ-FEQ/State University of Campinas, CP 6121, CEP 13081-970-Campinas/SP,Brazil
gamarra@feq.unicamp.br

2-DTAM-CCA/ State University of Londrina, CP 6001, CEP 86051-970-Londrina /PR, Brazil
lyssa@uel.br

3-DESQ-FEQ/ State University of Campinas, CP 6121, CEP 13081-970-ampinas/SP,Brazil
eliastam@feq.unicamp.br

4-DEA-FEA/ State University of Campinas, CP 6121, CEP 13081-970-Campinas/SP,Brazil
cabral@fea.unicamp.br

Abstract

The essential oil of key lime (*Citrus aurantifolia*) was obtained by steam distillation at normal conditions (1.0 bar/25°C) with steam at 110°C, during 10h. The GC analysis identified about 10 principle substances, being limonene, p-cymene, myrcene and β-bisabolene the most significative compounds. Aldehyde content increased significantly during distillation time. After 10h of process, oil has presented more than 3% of aldehydes content due to oxidative reactions.

Key words: essential key lime oil, steam distillation, capillary GC, aldehyde, C*itrus aurantifolia.*

1 Introduction

In addition to juice production, essential oil is one of the main byproducts of citrus processing [1]. Citric essential oil consists on one of the most important essential oils, like the oil extracted from lemon that is an important flavor agent, normally obtained by distillation process. Some processes require two steps, being the second one also called desterpenization. In this process, the terpenes are removed so oil becomes more soluble in water. This component thus can be utilized as flavoring agent of non alcoholic carbonated beverages.

Oil of lemon is one of the most important flavoring oils, used widely in all kind of beverages, soft drinks, soft drinks powders and tablets, and in baked goods, such as cakes, pastries, pie fillings, confectionery, soft and hard center candies, gelatin desserts, ice creams, etc. The oil is also employed in perfumes, toilet waters, eaux de Cologne, and in cosmetics to which it imparts a refreshing top note [2].

Lemon essential oil are complex mixtures of chemical compounds like limonene, γ-terpinene, citral, linalool and β-caryophyllene among others, which can be represented by three main classes: terpenes, oxygenates and sesquiterpenes. The most significant flavor compound is citral, while linalool possesses highly distinctive organoleptic characteristics. In addition, limonene, myrcene, octanal, and γ-terpene among others contribute with high aroma flavor of lemon oil [3, 4].

The quality of essential oil depends on different factors. Among them are the chemotype and biotype of the plant, the climatic conditions as well as extraction process [1,5]. There is more than 130 volatiles identified in lemon oil, being most of them unsaturated, like the terpenes, which can be easily modified, causing looses on the flavor and aroma of the essential oil [4]. Essential lemon oil basically consists of 75% of terpenes, 12% of oxygenated compounds and 3% of sesquiterpenes.

Actually, there is an increasing industrial demand for limonene, the main component of the essential lemon oil. The sensory and physical-chemical properties of this oil are directly related to the aldehydes content [1]. Quantities above 3% of these compounds are undesirable and prohibited when food application is requested, what make the acquisition of accurate compositional data highly desirable.

Considering the great interest on the production of these essential oils, the aim of the present work was to study the influence of distillation time extraction on the sensory quality and physical-chemical properties of the essential oil obtained of key lime (*Citrus aurantifolia*) by cold pressing.

2 Experimental

2.1 Material

The fruits (*Citrus aurantifolia*) used in this study were collected from December 1998 to April 1999, on three different plantations belonged to Backus industry and located at northern Peru (Lambayeque), at mature stage. These plantations were chosen so the industry could standardized their raw material. Extraction was performed at the same day it was collected.

2.2 Extraction of the Essential Oil by Distillation Process

For the isolation of the essential oil distillation process was conducted at Backus industry, using industrial equipment. In the distillator, the oil glands in the flavedo are ruptured by pressure in an equipment (6 m high and 1.8m of diameter) especially constructed to the industry. The emulsion, composed of lemon juice and oil, is collected quickly and the solids are separated from the liquids by a screening device (screw-type finisher), which removes the particles of the fruit peels. Than, the emulsion is transferred to the distillation tank and steam, at 110°C, is injected.

This process undergoes in normal conditions (1.0 bar, 25°C). The condensed volatiles follows to a condenser fed with cold water (3°C). The condensed volatiles are recovered at temperature between 35 and 40°C, and separated in oil and water by a decanting flask of 200ml. Than, the oil is sent to a homogenizer tank, where is homogenized at 10 rpm before it is stored in tambours at 25°C. Samples of 10 ml were taken during distillation process and collected every hour, in triplicate, during the 10h of process. Samples were immediately analyzed.

2.3 Essential Oil Analysis

2.3.1 Sensorial

Sensory quality of key lime essential oil was performed by eight trained people. They evaluate the following attributes: presence of particles in suspension, color, flavor and taste. Each appraiser was oriented to give a note for each of the attribute from 0 to 10.

2.3.2 Gas Chromatography

Volatile oil samples were analyzed on an Agilent – 6850, Serie GC system equipped with a split-splitless injector and capillary non-polar column (10 m long, 0.100 mm ID, 0.20 μm film thickness) type Agilent DBWAX (J&W Scientific). Helium (AGA, 99.995%) was used as the carrier gas. The analysis procedure involved injecting 0.1μL of essential oil submitted to 9 hours of steam distillation. The samples were diluted in ethanol at a ratio of 0.5 mg of sample in 1 ml ethanol and temperature program of 318K. To recognize elution times for each component, pure compounds injections in the GC were performed and the peak assignation was carried out. The composition of lemon oil was then obtained by assuming a direct proportion between the response of the integrator of the GC and the mass fraction of the mixture.

2.3.4 Physical Determination

Optical rotation [6], refractive index [6] and specific gravity [6] were determined in triplicate.

2.3.5 Aldehyde Determination

The determination of aldehydes by hydroxylamine method consists on the use of a solution of hydroxylamine hydrocloride and the subsequent neutralization with standardized alkali of the dydrochloric acid liberated by the reaction. After the reaction with the aldehyde, the mixture is titrated with standardized acid. The quantity of aldehyde is expressed in percentage [6].

2.4 Statistical Analysis

An analysis of variance (ANOVA) was used. Duncan's multiple-range test was applied for the calculation of the significant differences of the samples among the distillation time process.

3. Results and Discussion

The degree of ripeness is an important factor affecting the concentration and composition of the volatile compounds in the fruit [7]. These researchers found that limonene is most abundant in oils from greenish-yellow peel coloration fruits (intermediate maturation). The highest yield of volatiles secondary metabolites was also observed at this stage of maturation, what indicates that limonene could be used as a functional index of ripeness. Nevertheless, in this work, essential oil was extracted of key lime at mature stage, once is in this stage where most juice is extracted and is economically viable for the industry.

Essential key lime oil presented crystalline, without particles in suspension, with a yellow color, a fresh and piquant odor and a lightly bitter taste.

Table 1 shows the mean composition of the oils used in the experimental runs after 9 hours of hydrodistillation. Only the more significant compounds are listed. Terpenes (around 60%) are represented basically by the monoterpene hydrocarbon limonene (which alone accounts for 50%) followed by both α-thujene and p-cymene, other important terpene hydrocarbons components that account for about 5%. The class of oxygenates includes different components such as aldehydes, alcohols and acetates.

Table 1 - Principal Key lime essential oil components (milliliters per kilogram of essential oil) obtained within 9 h of steam distillation process. Values represent the mean values of oils from the 3 different regions.

Composition	ml/Kg
α-thujene	1.14323
α-pinene	0.86805
Sabinene	0.28189
β-pinene	0.87092
Myrcene	1.26942
p-cymene	1.56630
d-limonene	49.65704
β-caryophyllene	0.47698
trans-α-bergamottene	0.79584
β-bisabolene	2.29847

The most important oxygenate is aldehyde citral, present in the form of its stereoisomers neral and geranial. Citral contributes significantly to the quality of lemon flavor and aroma [8]. Finally, as regards sesquiterpenes (condensed terpenes characterized by very high molecular weight and very low volatility) the highest weight fraction was registered for bysabolene (2.50%).

These results are in accordance to the ones obtained by Benvenuti et al. [3], although the concentrations of their components are higher. In their work, they reported a concentration of d-limonene around 70% and 0.84% of bysabolene in essential oil of peel from Sicilian lemon variety.

Combariza et al. [7] also found that monoterpenes represent the main (86-88.79%) family of components in *Citrus volkameriana* peel oil. After limonene (78.2-79.4% depending on the state of ripeness of the fruit), the most abundant monoterpene encountered was γ-terpinene, followed by myrcene, α-pinene, sabinene and p-cymene. Moreover, they found that steam distillation yielded a higher fraction of oxygenated compounds than cold pressing. The amount of oxygenated compounds in key lime essential oil can also be associated with the extending distillation process time, once other researches have reported steam distillation process of maximum 3 h at laboratory scale [1, 5, 7].

No significant differences ($p \leq 0.05$) were encountered between the essential oils provided by the different regions of Peru, during steam distillation process and related to optical rotation (Figure 1). As distillation time increases from 1 to 10h, oils optical rotation diminished from a range of 42 to 49 to values in the range of -4.7 and 0.8.

This result can be explained by the fact that during the first hours of distillation, terpenes and some of the oxygenated compounds are distilled. These compounds are responsible for the levogene behavior due to its chemical composition structure. Lately, during the process, optical rotation values becomes low and levogene by the increase of these oxygenated compounds, like aldehydes, alcohols and esters.

On the other hand, refractive index showed a different behavior during distillation time, for the oils extracted of lemons from the three different regions (Figure 2). Refractive index showed a slight increase on the values from 1.475 (first hour) to values in the range of 1.479 and 1.485, for the samples of M1 and M2 regions. It was observed a gradual increase and oscillating behavior for samples of M3 region that could be explained by the presence of more unstable oxidative compounds in this sample. As for optical rotation, the low values for refractive index encountered in the first hours of the distillation process are due to the higher amount of terpenes, whereas at the final distillation stage, the oxygenated compounds presence prevails.

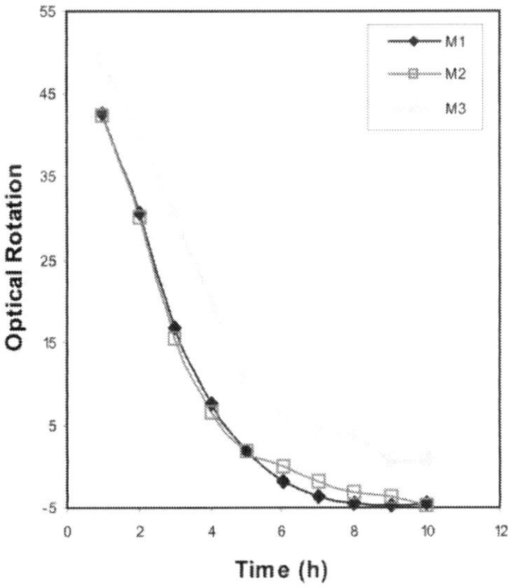

Figure 1. Optical Rotation of essential Key Lime oil from 3 regions of northern Peru (M1, M2, M3) obtained by steam distillation (1bar/25°C).

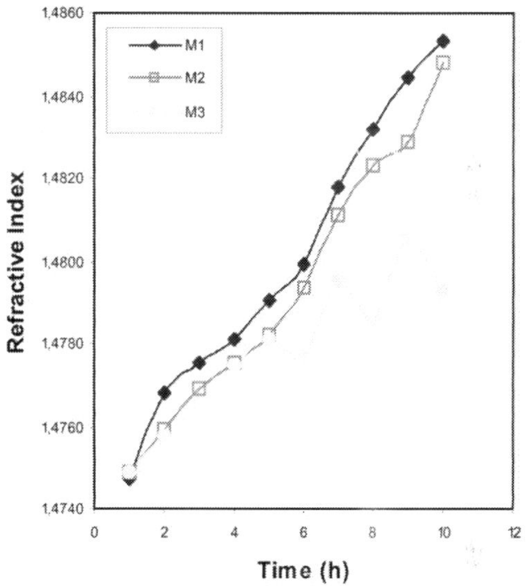

Figure 2. Refractive Index of essential Key Lime oil from 3 regions of northern Peru (M1, M2, M3) obtained by steam distillation.

The values of the essential key lime oil specific gravity maintained practically constant during the distillation process, increasing slightly from 0.847 to 0.882 g/l (Figure 3). This result can be explained by that fact that since the distillation process is slow, it causes gradual changes in the oil, so the volatiles compounds, as terpenes, volatizes slowly and gradually. At the same time, carbonylic compounds are also gradually formed. Since these oxygenated compounds have a higher molecular weight, the oils specific gravity increases.

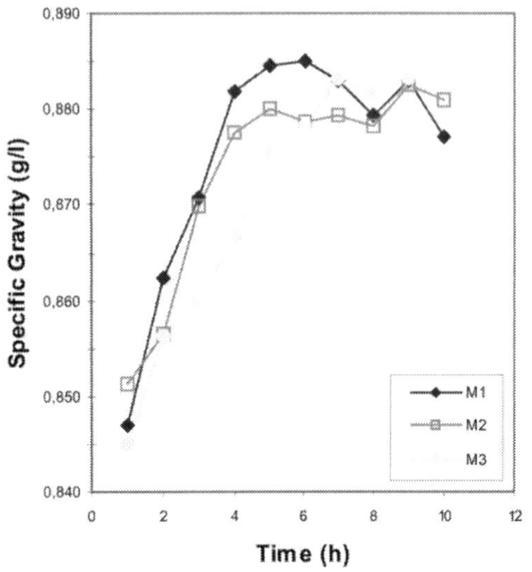

Figure 3. Specific Gravity of essential Key Lime oil from 3 regions of northern Peru (M1, M2, M3) obtained by steam distillation (1bar/25°C).

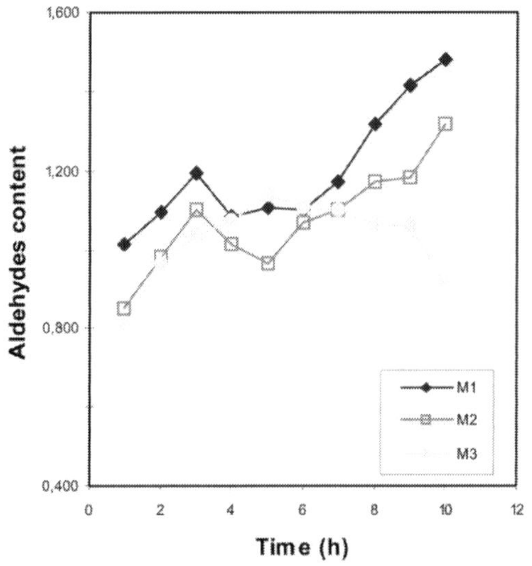

Figure 4. Aldehyde content of essential Key Lime oil from 3 regions of northern Peru (M1, M2, M3) obtained by steam distillation (1bar/25°C).

In relation to the aldehydes content, the samples from the different regions showed different behaviors during distillation process (Figure 4). For the samples of M3 region, the fact that the emulsion juice-oil remained in contact for longer time than for the other regions, allowed a higher oxidation of its compounds, thus altering its aldehyde content. Longer distillation time resulted in essential oils with higher amounts of aldehydes (samples from M1 and M2 regions).

The aldehyde content is the most important indication of the quality of the essential oils, and it is related to the presence of the oxygenated compounds. These compounds, especially the carbonylics, have an important influence on the aroma of the essential oil.

Essential key lime oil with good quality can be obtained within 10 h of steam distillation process at normal conditions. The recovery of essential oil from the condenser at a temperature of 35°C is also important to maintain a high oil quality, once it avoids more oxidative reactions and looses of the terpenes at this stage. Above 10 h of process, the presence of carbonylic compounds in the oil over 3% compromises its applications in foods. This information could be of great interest for flavor industries to better utilize the raw material.

4 Conclusions

Steam distillation time process influences the quality of essential key lime oil. Under the conditions performed in this study, essential oil with 1.10% of carbonylic compounds was obtained within 10 h of process. This essential oil can be used as flavoring agent of non alcoholic carbonated beverages.

References

[1] Vekiari, S.A.; Protopapadakis, E.E.; Papadopoulou, P.; Papanicolaou, D.; Panou, C.; Vamvakias, M. Composition and seasonal variation of the essential oil from leaves and peel of a cretan lemon variety. Journal of Agricultural and Food Chemistry, 50, 147-153, 2002.

[2] Guenther, E. "The Essential Oils", vol. III, Ed. Fritzsche Brothers INC, New York, USA, 777p, 1955.

[3] Benvenuti, F; Gironi, F.; Lamberti, L. Supercritical deterpenation of lemon essential oil, experimental data and simulation of the semicontinuous extraction process. Journal of Supercritical Fluids, 20, 29-44, 2001.

[4] Van Straten, S.; Maarse, H. Volatile compounds in food, Division for Nutrition and Food Research, TNO: Zeist, 1983.

[5] Blanco Tirado, C; Stashenko, E.E.; Combariza, M.Y., Martinez, J.R. Comparative study of colombian citrus oils by high-resolution gas chromatography and gas chromatography-mass spectrometry. Journal of Chromatography A, 697, 501-513, 1995.

[6] Guenther, E. "The Essential Oils", vol. I, Ed. Fritzsche Brothers INC, New York, USA, 427p, 1955.

[7] Combariza, M.Y.; Blanco Tirado, C.; Stashenko, E.; Shibamoto, T. Limonene concentration in lemon (*Citrus volkameriana*) peel oil as a function of ripeness. Journal of High Resolution Chromatography, 17, 643-646, 1994.

[8] Gramshaw, J.W.; Sharpe, K. Estimation of citral in lemon oil by gas-liquid chromatography using a capillary column. Journal of Science and Food Agricultural, 31, 93-98, 1980.

89a Impact of Future Fuel and Crude Demand on Refinery Hydprocessing

William Keesom, Solutions and Services, UOP, 25 East Algonquin Road, Des Plaines, IL 60017

Crudes are becoming more difficult to process while at the same time, refined product quality is going up and there is a growing shift from gasoline to diesel. Hydrogen costs are going up as natural gas is becoming more expensive. Understanding future crude, natural gas, and product trends is key to developing cost effective new hydroprocessing technologies and implementing them in refineries.

89b Limiting Factors in Deep Hydrodesulfurization of Diesel

Teh C. Ho, Corporate Strategic Research Laboratories, ExxonMobil Research and Engineering Co., 1545 Rt 22 East, Clinton Township, Annandale, NJ 08807
Many countries are mandating substantial sulfur reduction in transportation fuels. For example, the U. S. Environmental Protection Agency ruled that on-road diesel sulfur level should be reduced to 15 ppm by June 2006, from the current average of 350-500 ppm. This is why there has been a massive, worldwide effort in developing new or much improved hydrodesulfurization technologies in recent years. The problem of central concern is how to remove sulfur from 4-substituted and 4,6-disubstituted dibenzothiophenes (DBTs). These species are the most refractory sulfur species due to steric hindrance around the sulfur atom. The present paper aims to identify limiting factors bearing on the chemistry and catalysis of removing sulfur from DBTs with steric hindrance. Among the topics discussed are: reactivity-property relationships, catalyst functionalities, inhibiting effects, poisoning dynamics, nature of active sites, and new catalytic materials.

Effect of the oxidic precursors on the sulfurability of hydrotreating CoMo(P) catalysts

A. D. Gandubert,[1] C. Legens,[1] D. Guillaume,[1] C. Pichon,[1] E. Payen[2]

[1] *IFP-Lyon, BP3, 69390 Vernaison, France*
[2] *Laboratoire de Catalyse de Lille, Bat C3, USTL, 59650 Villeneuve d'Ascq, France*

INTRODUCTION

One of the most important step in the petroleum refining is the hydrodesulfurization that is used to remove sulfur compounds of the petroleum fractions. This catalytic process involves CoMo(P)- or NiMo(P)-type catalysts. Their active phases are consisting of molybdenum disulfide slabs promoted by cobalt or nickel atoms that are well dispersed on a high specific surface area oxide support such as alumina. As the highest sulfur content allowed in gasoline and diesel is continually decreasing, more and more efficient hydrotreatment catalysts are required. Improvement of those catalysts imposes to have a better understanding of the structure and genesis of the CoMoS active phase. Thus, a procedure to characterize this mixed phase has been established. In this context, a multi-techniques approach correlating X-ray photoelectron spectroscopy (XPS) and transmission electron microscopy (TEM) with activity measurements has been conducted. This methodology of characterization was used to study the effect of the nature of the oxide phase precursor on the genesis of the active phase.

EXPERIMENTAL SECTION

Catalysts preparation. The CoMo and CoMoP oxidic precursors were prepared by incipient wetness impregnation of γ-alumina extrudates with the impregnating solution described below. The impregnated extrudates were placed in water-saturated atmosphere for at least three hours. This maturation step permits the diffusion of the deposited species into the alumina porous framework. The solids were then dried at 120°C overnight to eliminate water. Finally the solids were calcined under air at 500°C for two hours.

The CoMoP impregnating solution was prepared by dissolving molybdenum oxide MoO_3 and cobalt hydroxide $Co(OH)_2$ in phosphoric acid aqueous solution.[1] The CoMo impregnating solution was obtained by dissolution of the decamolybdocobaltate cobalt salt $Co_3Co_2Mo_{10}O_{38}H_4$, a new starting material.[2] Both CoMoP and CoMo catalysts had a Mo loading of 20 wt % as MoO_3 and a Co loading of 5.6 wt % as CoO, corresponding to a Co/Mo atomic ratio of 0.5. Those catalysts will be hereafter referred to as *CoMoP* and *CoMo*.

Sulfidation conditions. The oxidic precursors were pre-sulfided with a $15\%H_2S/85\%H_2$ gas mixture at atmospheric pressure with a gas flow of 2L/h/gram of catalyst. This sulfidation was performed at various temperatures from room temperature (20°C) to 600°C. The samples were heated under the sulfiding mixture at a rate of 5°C/min up to the desired temperature, and maintained at this temperature for two hours. They were then cooled down to room temperature at a rate of 20°C/min under the reactive mixture. The pre-sulfided samples were finally transferred into glas vials under vacuum in order to avoid any contact with air. The nomenclature of the pre-sulfided samples will be the same as the oxidic one to which -

Sx will be added, x corresponding to the sulfidation temperature. For example, *CoMoP-S400* will refer to the CoMoP catalyst sulfided at 400℃.

XPS analysis. The XPS sampling of the sulfided catalysts was performed in a glove box under argon atmosphere, with controlled oxygen and water level (less than 20 ppm) in order to avoid their partial reoxidation. The samples were crushed and pressed onto an indium foil that was attached to the sample holder via a double side carbon tape. The sample holder was then moved directly to the introduction chamber of the XPS spectrometer, thanks to the special connection of the glove box to the XPS spectrometer. The XPS spectra were recorded on a KRATOS AXIS ULTRA instrument with Al monochromator source (1486.6 eV) and an hemi-spherical analyzer operating at fixed pass energy of 40 eV. The measurements were made at 293 K in steps of 0.05 eV for cobalt and 0.1 eV for molybdenum, and at a pressure lower than 1×10^{-9} Pa in the sample analysis chamber. Binding energies (BE) of the various elements have been referenced to the C1s level of the contamination carbon at 284.6 eV.

The decomposition of Mo3d and Co2p XPS spectra were performed using the appropriate oxide and sulfided references as supported monometallic catalysts and with an unsupported MoS_2 compound.[3] Analysis of the Mo3d spectrum gives the sulfidation rate into MoS_2 whereas the Co2p spectrum allows us to quantify the amount of CoMoS phase. In the case of cobalt species it is necessary to take into account the shake-up signals. The shake-up are resulting of multi-electronic excitations that take place during the irradiation of the sample by the X-rays.

A relative quantification of metallic atoms was obtained from the measurement of the corresponding peak intensity according to the methodology described in a previous work.[3] To approach the effective atomic surface concentrations, we have taken into account all the atoms except carbon as it is considered mainly as surface contaminant. For one atom, the signal obtained can be decomposed into different contributions relative to various chemical species. The calculation of the respective contribution of each peak intensity will lead to the knowledge of the relative proportion of each chemical form for a given atom. Two samples have been analyzed for each catalyst and the data presented correspond to the average values of the two analysis. The relative error is estimated to ± 10%.

TEM analysis. The pre-sulfided samples were analyzed by TEM. The extrudates were crushed into a fine powder and a small amount of powder was diluted with ethanol. Two drops of this solution were deposited on a carbon-coated Cu grid and the solvent was evaporated under infrared light. TEM images were taken using a 200 kV JEOL-2010 transmission electron microscope equipped with digital camera.

Catalytic activity measurements. The pre-sulfided samples were tested in toluene-hydrogenation in hydrotreatment conditions. The catalytic tests were carried out in a fixed bed high pressure flow microreactor under the following conditions: 2 cm^3 of catalyst in crushed shape (300-1000 μm) diluted in SiC, total pressure of 60 MPa, reaction temperature of 350℃, liquid hourly space velocity (LHSV) of $2h^{-1}$, hydrogen-to-feed ratio of 450 L/L. The feed was consisting of dimethyldisulfur (5.9%) and toluene (20%) mixture in cyclohexane. After the unit was pressure tested at room temperature, the temperature was increased at 3℃/min to 350℃. Steady-state conversion was measured after 2 hours on stream. During the increase of temperature and the plateau at 350℃, dimethyldisul fur is decomposed into methane CH_4 and hydrogen sulfide H_2S and by consequence the pre-sulfided sample is sulfided once again. So the catalysts were pre-sulfided *ex situ* at various temperatures (between 20°C and 600°C) a nd then sulfided again under the feed at 350°C. Reacti on products were analyzed by on line gas chromatography. The activity in toluene-hydrogenation was expressed considering a first order of reaction.

RESULTS AND DISCUSSION

XPS characterization. The oxidic *CoMo* and *CoMoP* catalysts were pre-sulfided with a 15%H_2S/85%H_2 gas mixture, at atmospheric pressure, at various temperatures, from room temperature (20°C) to 600°C, and characterized afte r sulfidation. The XPS characterization allowed us to determine which species are present on the surface of the catalysts. By using the aforementioned methodology of quantification, we have been able to follow the evolution of the amount of the various species with the sulfidation temperature.[3]

Molybdenum species. The Mo3d spectrum is composed of molybdenum with three different oxidation degrees. First there is Mo(6+), with a binding energy of the main peak $Mo3d_{5/2}$ measured at 232 ± 0.1 eV, that corresponds to residual molybdenum oxide. Then there is a contribution of Mo(5+), with a binding energy of the main peak measured at 229.8 ± 0.1 eV, attributed to molybdenum oxysulfide according to literature.[4,5] The last contribution detected corresponds to Mo(4+), with a binding energy of the main peak measured at 228.7 ± 0.1 eV, attributed to molybdenum disulfide. The two peaks at lower binding energy are sulfur 2s contributions and are not taken into account for quantification as S2p signal is preferable because more sensitive. Figure 1 shows a typical decomposition of the XPS Mo spectrum obtained for *CoMo* sulfided at 300°C.

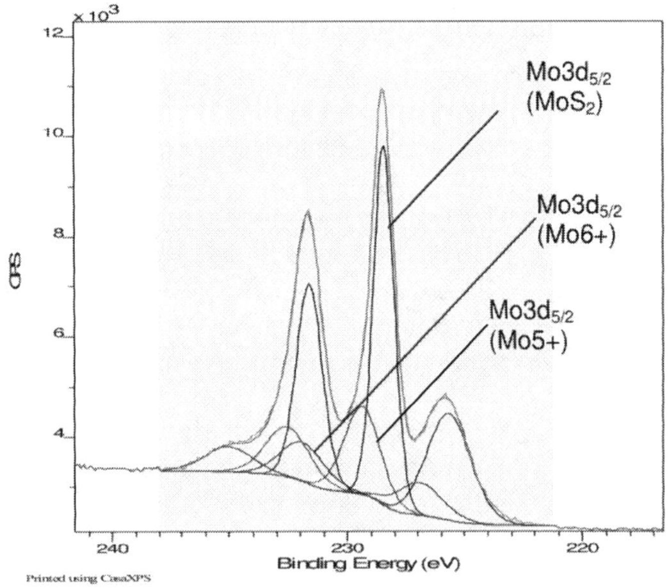

Figure 1: XPS molybdenum spectrum for *CoMo* catalyst, pre-sulfided at 300°C.

Figure 2 shows the evolution of the relative amount of each aforementioned species with the sulfidation temperature for *CoMo*. The proportion of molybdenum oxide Mo(6+) is decreasing while the proportion of sulfide MoS_2 is increasing progressively with the temperature increase. The largest amount of MoS_2 phase is reached for temperature equal or higher than 350℃. For those catalysts about 70% of the molybdenum oxide are transformed into MoS_2. On the contrary the XPS quantification shows that the amount of Mo(5+) is constant: whatever the sulfidation temperature there is about 20-25 % of oxysulfide detected at the surface of the catalysts. It means that a part of the molybdenum oxide impregnated on the catalysts is resistant to the treatment by H_2S/H_2 and can not be sulfided into MoS_2.

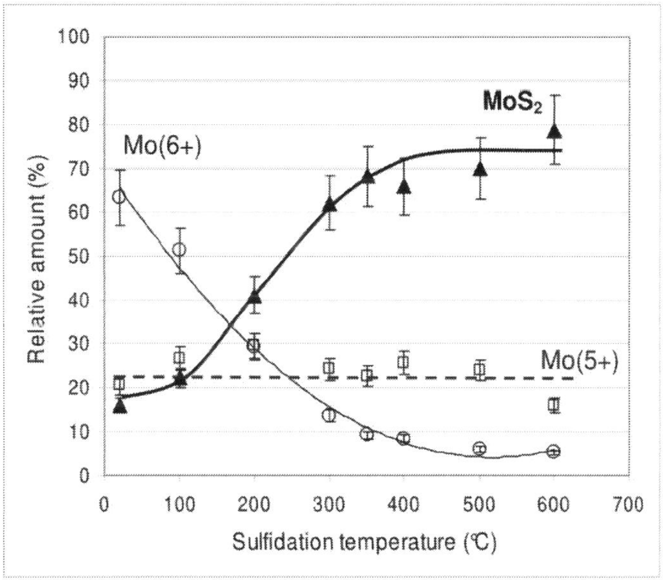

Figure 2: Evolution of molybdenum species detected by XPS for *CoMo* catalyst, pre-sulfided at various temperatures.

The quantification of the molybdenum species was also applied to *CoMoP* and is presented in Figure 3. For sulfidation temperature increasing from 20℃ to 350℃, the amount of Mo(6+) is decreasing and the amount of MoS_2 is increasing. For temperature higher than 350℃, the amounts of oxyde and sulfide are stable upon increasing sulfidation temperature. According to these results, it is clear that a sulfidation temperature equal or higher than 350℃ is required to well sulfide the *CoMoP* and obtain the largest amount of MoS_2, about 80% of the molybdenum oxide being transformed into MoS_2. Regarding the species Mo(5+), its evolution with the sulfidation temperature is particular. Indeed the amount of oxysulfide increases with the temperature between 20℃ and 200℃ to reach a maxim um amount of about 40%. For temperature higher than 200℃ the amount decreases with the temperature increase. These results show that the sulfidation of *CoMoP* proceeds in two steps, under 15%H_2S/85%H_2 gas mixture: first the molybdenum oxide is transformed into both oxysulfide and sulfide species, for sulfidation temperature below 200℃. Then from 200° C to 600℃, the formation of sulfide MoS_2 results at the same time from the sulfidation of the oxysulfide Mo(5+) and the non transformed oxyde Mo(6+).

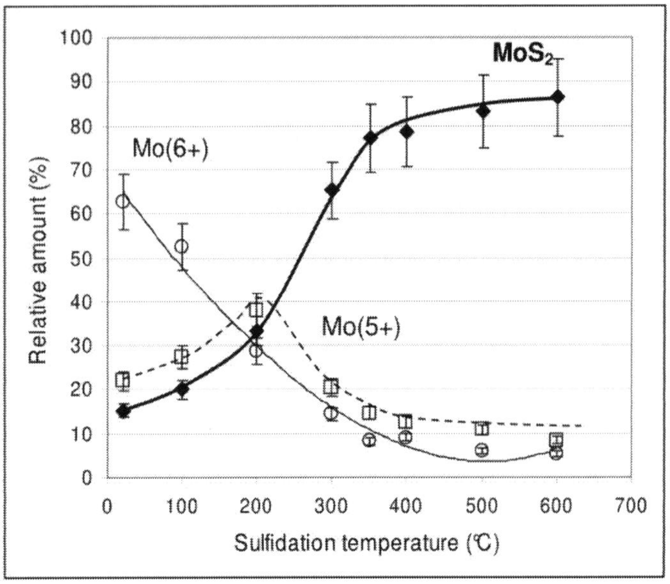

Figure 3: Evolution of molybdenum species detected by XPS for *CoMoP* catalyst, pre-sulfided at various temperatures.

Cobalt species. The Co2p spectra were also decomposed and three different species of cobalt were detected at the surface of the catalysts and quantified by XPS: cobalt oxide Co(2+), characterized by a main peak $Co2p_{3/2}$ at 781.5 ± 0.1 eV, CoMoS mixed phase, with a binding energy of the $Co2p_{3/2}$ measured at 778.6 ± 0.1 eV, and cobalt sulfide Co_9S_8, with the main peak at 778.1 ± 0.1 eV. Figure 4 shows a typical decomposition of the XPS Co spectrum obtained for *CoMo-S300*.

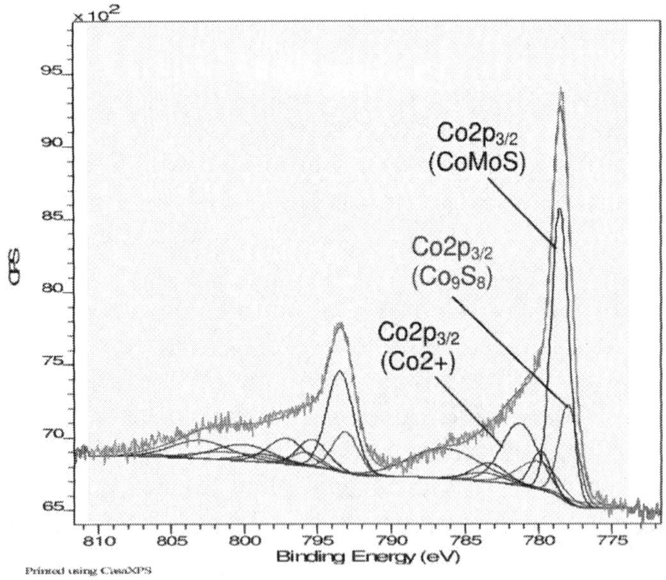

Figure 4: XPS cobalt spectrum for *CoMo* catalyst, pre-sulfided at 300℃.

Figure 5 presents the evolution of the relative amounts of cobalt species with the sulfidation temperature for *CoMo*. For sulfidation temperature between 20℃ and 350° C, the amount of residual oxide Co(2+) decreases while the amount of CoMoS mixed phase increases with the temperature increase. For higher temperature, the amount of residual oxide is stabilized (about 30%), and the amount of MoS_2 decreases, drawing a volcano curve. The maximal amount of CoMoS phase corresponds to a proportion of 50% of cobalt atoms engaged in the mixed phase. Concerning Co_9S_8, the amount increases between 20℃ and 100℃ and i s stablilized at about 20% until 350℃. For higher temperature the a mount increases again to reach 40% at 600℃. It seems that the cobalt engaged in the mixe d phase migrates to form a single sulfur species.

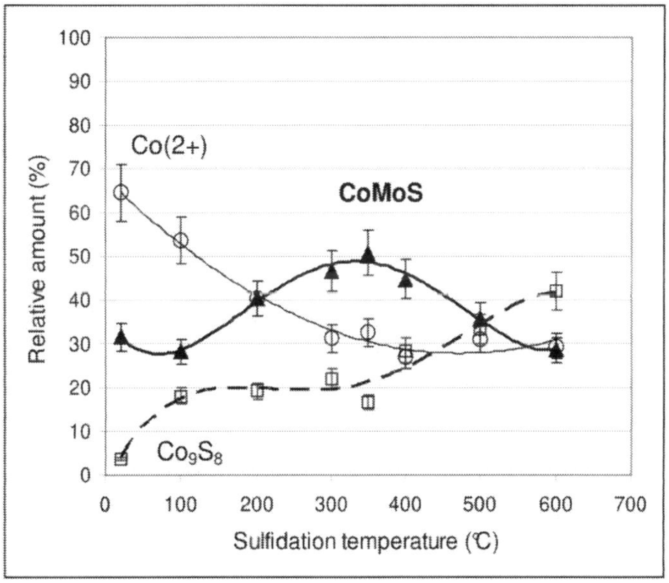

Figure 5: Evolution of cobalt species detected by XPS for *CoMo* catalyst, pre-sulfided at various temperatures.

Figure 6 shows the quantification of the cobalt species, for *CoMoP* sulfided at various temperatures. The amount of active CoMoS phase increases with the sulfidation temperature between 100℃ and 350℃ while the amount of oxide C o(2+) decreases. For temperature higher than 350℃, the amount of residual oxide is stabilized and the amount of mixed phase decreases, the maximum of CoMoS phase being reached with about 55% of cobalt atoms in decoration position of the MoS_2 slabs. On the contrary, the proportion of Co_9S_8 sulfur increases continually with the sulfidation temperature.

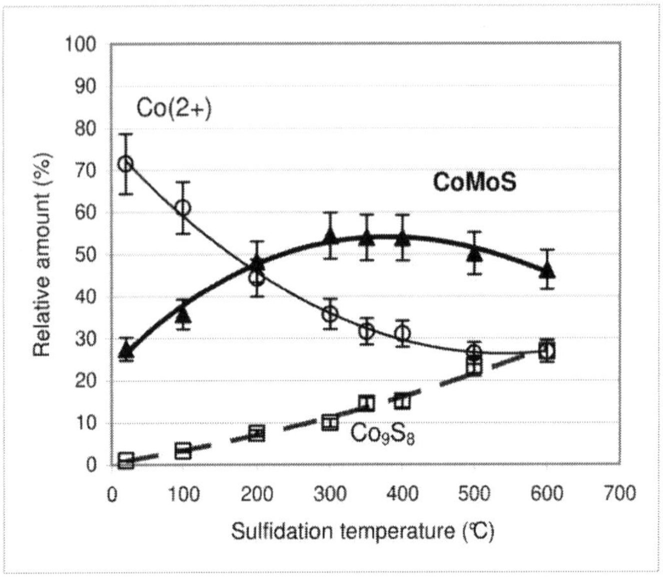

Figure 6: Evolution of cobalt species detected by XPS for *CoMoP* catalyst, pre-sulfided at various temperatures.

Comparison between CoMo and CoMoP. Concerning the Mo species, the main difference between *CoMo* and *CoMoP* is the evolution of oxysulfide Mo(5+) with the sulfidation temperature. For *CoMo* the amount of Mo(5+) is not greatly affected by the sulfidation temperature. On the contrary this amount reaches a maximum at 200℃ for *CoMoP* (Figure 3). So it seems that Mo(5+) is not involved in the process of formation of MoS_2 for *CoMo* whereas it is an intermediate species in the sulfidation of *CoMoP* precursor. Moreover for both catalysts, the Mo spectra showed that a sulfidation temperature higher than 350℃ is required to obtain the best sulfidation of the Mo, which means to obtain the maximal amount of sulfide MoS_2 and minimal amounts of residual oxide and oxysulfide at the surface of the catalysts. The highest MoS_2 proportion is about 70% for *CoMo* and 80% for *CoMoP*. The lower sulfurability of *CoMo* is assigned to the presence of oxysulfide species that are resistant to sulfidation even at high temperature. In case of *CoMoP* both oxide and oxysulfide species are sulfided into MoS_2 leading to a higher sulfidation rate.

The Co spectra of the *CoMo* and *CoMoP* are similar and for both catalysts the amount of *CoMoS* draws a volcano curve, with a maximum of about 50-55% of cobalt in decoration position of MoS_2 slabs reached for sulfidation temperature between 300℃ and 400℃. We can also notice that at high temperature (500℃ and 600 ℃) the amount of Co_9S_8 increases in case of *CoMo* up to 40% whereas it remains constant for *CoMoP*.

TEM analysis. For both *CoMo* and *CoMoP* catalysts, the TEM micrographs showed that below 200℃ the MoS_2 phase is not structured into slabs but consists of non organized particles. On the contrary the micrographs of the samples pre-sulfided at temperatures higher than 200℃ showed the presence of more or less stac ked MoS_2 slabs. Analyzing more than 400 crystallites for each samples allowed us to estimate the average length and stacking of the slabs. Figure 7 presents the histogram obtained for the length of the slabs detected by TEM on *CoMo-S500*.

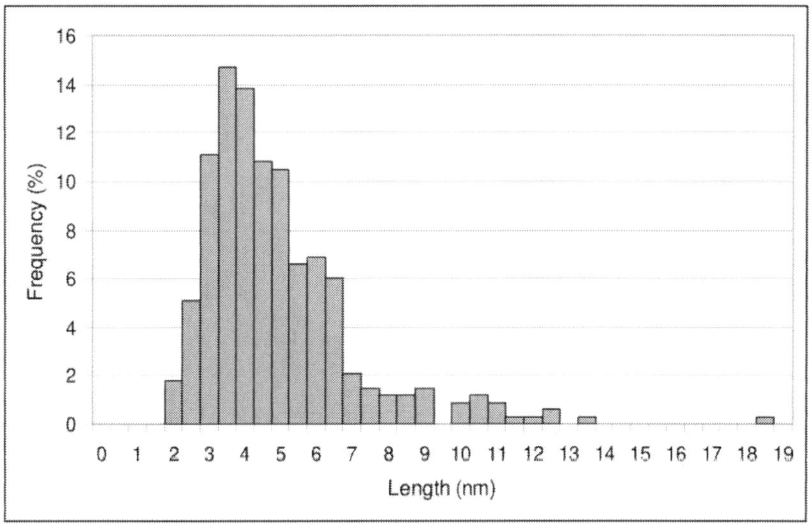

Figure 7: Histogram of length of MoS_2 slabs realized from the TEM analysis of *CoMo* pre-sulfided at 500℃.

Figures 8 presents the average length of MoS_2 slabs for *CoMo* pre-sulfided from 350℃ to 600℃. For lower temperature of pre-sulfidation, th e micrograph did not show enough slabs for a correct estimation of their mean length. These results show that the pre-sulfidation temperature did not greatly affect the morphology of the slabs: the samples sulfided between 350℃ and 600℃ all contain slabs of about 4.3 ± 0. 2 nm long, the average stacking being 1.9.

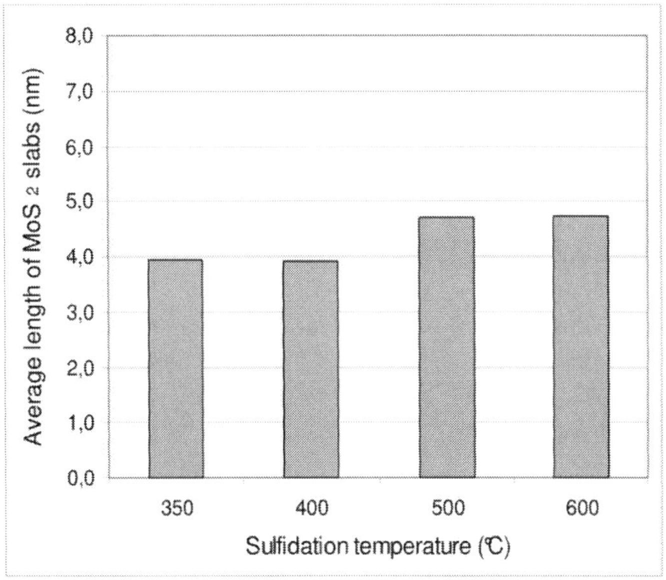

Figure 8: Average length of MoS_2 slabs determined by TEM, for *CoMo* catalyst pre-sulfided at various temperatures.

Figure 9 shows the average length of the slabs for the *CoMoP* pre-sulfided from 300℃ to 600℃. For this catalyst the sulfidation temperatur e had a large effect on the morphology of the MoS_2 slabs. For the samples sulfided at 300℃, 350℃ an d 400℃, the slabs are about 3.3 ± 0.2 nm long. For higher temperatures, the slabs are twice as long (6.6 ± 0.2 nm), which suggests a sintering of the MoS_2 particles. On the other hand, no effect of the sulfidation temperature was noticed on the stacking of those slabs: whatever the temperature, the average stacking is 1.7.

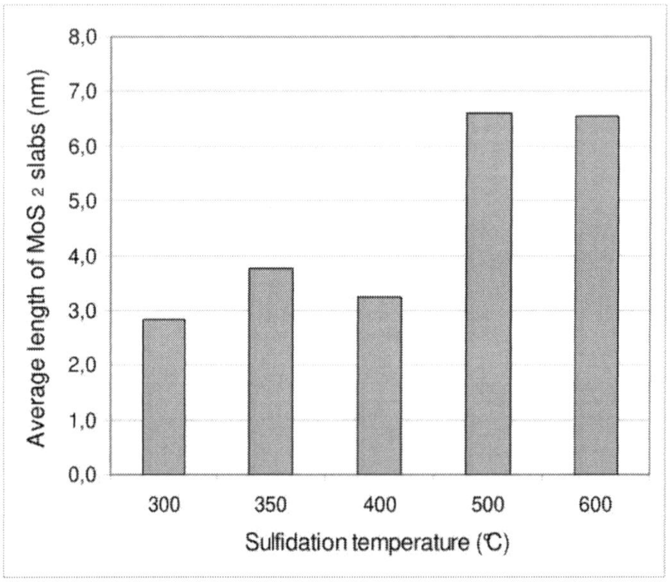

Figure 9: Average length of MoS_2 slabs determined by TEM, for *CoMoP* catalyst pre-sulfided at various temperatures.

Activity measurements. Figure 10 presents the toluene-hydrogenation activity versus the sulfidation temperature, for *CoMo* and *CoMoP* catalysts. For both catalysts pre-sulfided from 20℃ to 500℃, the sulfidation temperature had no effect on the activity in toluene-hydrogenation. On the contrary the activities decrease for the samples pre-sulfided at 600℃. This is probably due to the sintering of the MoS_2 slabs observed by TEM in the case of *CoMoP*. For the *CoMo* it may be due to a coking of the catalyst as no variation of the active phase morphology was observed by TEM. Moreover the CoMo has a slightly higher activity than the CoMoP. Nevertheless both catalysts present the same activity in cyclohexane-isomerisation, about 3.7% of conversion. This result implies that *CoMo* and *CoMoP* catalysts have similar acidic properties. Thus the small gain of activity of *CoMo* compared to *CoMoP* does not originate from a gain of acidity.

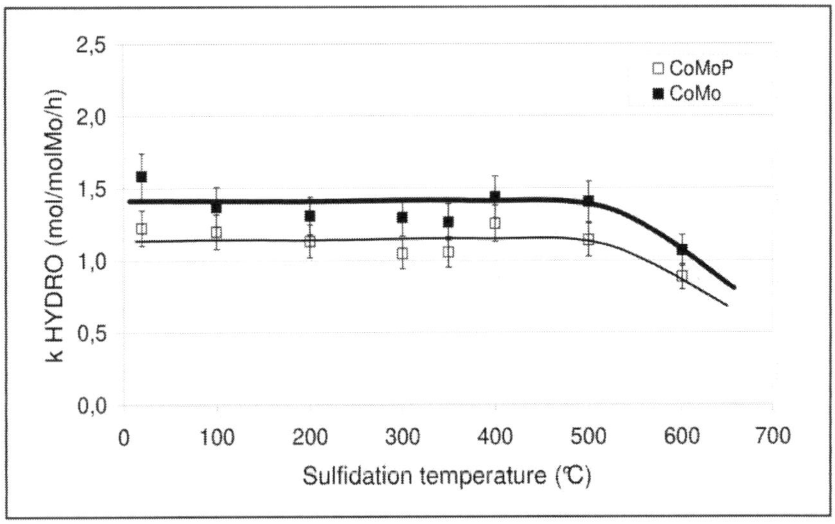

Figure 10: Catalytic activity in toluene-hydrogenation of the *CoMo* and *CoMoP* catalysts, pre-sulfided at various temperatures.

Influence of phosphorus. The role of phosphorus in standard catalysts has been widely studied and various interpretations were proposed in the literature. Regarding the effect of phosphorus on the morphology of molybdenum slabs, Kemp et al. have compared the morphology of MoS_2 slabs obtained after sulfidation of Mo/Al_2O_3, MoP/Al_2O_3, CoMo/Al_2O_3 and CoMoP/Al_2O_3 catalysts prepared by conventional one-step impregnations, with 2wt% or 3wt% as phosphorus loading.[6] They reported that phosphorus addition leaded to an increase of the stacking of the slabs from 1 to 3, and a decrease of their length. The increase of the stacking was confirmed by the studies performed by Ramirez et al. and Poulet et al. on MoP/Al_2O_3 catalysts prepared by coimpregnation of ammonium heptamolybdate and orthophosphoric acid, containing 14wt% Mo and 2.8wt% P.[7,8] According to those authors, the origin of more stacked layers of MoS_2 when phosphorus was added may be due to weaker interaction of MoS_2 with the support surface. The effect of phosphorus addition on the HDS activity of CoMo catalysts was also investigated by many authors and they gave various explanations. According to Iwamoto and Grimblot, Bouwens et al., Eijsbouts et al., phosphorus has no detectable promotion effect on thiophene-hydrodesulfurization.[9,10,11] Studies on the effects of phosphorus addition observed on CoMo catalysts are numerous in the literature. To summarize we can conclude that the "doping" effect of phosphorus depends on various parameters including the way of preparation of the catalysts or the phosphorus loading.

Our work on *CoMo*/Al_2O_3 and *CoMoP*/Al_2O_3 catalysts, prepared by using cobalt salts of $Co_2Mo_{10}O_{38}H_4^{6-}$ and $P_2Mo_5O_{23}^{6-}$ respectively, allowed us to study the effect of phosphorus on such catalysts. Under the sulfidation conditions used for this study (15%H_2S/85%H_2 gas mixture, atmospheric pressure, temperature between 20℃ and 600℃) we have observed that phosphorus has no significant effect on the sulfurability of those two catalysts. Indeed both catalysts require a sulfidation temperature between 300℃ and 400℃ to obtain at once the maximal amounts of molybdenum sulfide MoS_2 and of cobalt engaged into CoMoS active phase. For those optimal samples, about 75% of oxide Mo(6+) are transformed into MoS_2 slabs and 55% of cobalt atoms are in decoration position of the slabs. Moreover TEM analysis showed that the length and stacking of MoS_2 slabs are similar for *CoMo* and *CoMoP* sulfided between 300°C and 400°C. This result proves that, i n this range of temperature, phosphorus does not affect the MoS_2 slabs morphology of these catalysts prepared with heteropolycompounds as starting material. The main difference is observed at 200°C where the Mo(5+) species reaches a maximum in the presence of phosphorus. XPS is not able to distinguish oxysulfide species from MoS_3 as the main Mo3d peak is at the same binding energy.[5] EXAFS experimentations are therefore in progress in order to see if phosphorus may induce the formation of different intermediates of sulfidation. Moreover at higher sulfidation temperature (above 500°C), phosphorus leads to an i ncrease of the length of MoS_2 slabs whereas the CoMoS phase seems to be more stable with a lower transformation of cobalt into sulfur monometallic species. Regarding the toluene-hydrogenation activity, once again phosphorus addition does not improve the activity of the catalyst that we studied.

CONCLUSION

The sulfurability of CoMo and CoMoP oxidic precursors prepared with heteropolyanions based impregnation solution was investigated. XPS characterization allowed us to quantify the MoS_2 and CoMoS active phases present on the catalysts after sulfidation at various temperatures. Combining XPS, TEM, and activity measurements allowed us to determine the optimal sulfidation temperature range: for both catalysts, the maximal amounts of MoS_2 and CoMoS phases are obtained for the samples sulfided between 300°C and 400°C. Moreover, whatever the oxidic precursor (CoMo or CoMoP), the MoS_2 slabs present the same morphology (length and stacking). This work allowed us to conclude that the two different systems studied here, CoMo and CoMoP prepared by using respectively $Co_2Mo_{10}O_{38}H_4^{6-}$ and $P_2Mo_5O_{23}^{6-}$ cobalt salts solutions, present similar properties under the used experimental conditions (sulfidation between 20°C and 600°C, by 15%H_2S/85%H_2, at atmospheric pressure). Indeed the CoMo and CoMoP catalysts studied here were calcined at 500°C before being sulfided at various temperatures, a calcination that can act as a levelling step of the preparation of the active phases. Studies are currently in progress to investigate the behaviour and the sulfurability of catalysts directly pre-sulfided after drying step of their preparation.

REFERENCES

1. Griboval A., Blanchard P., Payen E., Fournier M., and Dubois J.L., *Catal.Today*, **45** (2001) 173
2. Martin C., Lamonier C., Fournier M., Mentré O., Harlé V., Guillaume D., and Payen E., *Chem.Mater.*, **17** (2005) 4438
3. Gandubert A.D., Legens C., Guillaume D., and Payen E., *submitted to Oil and Gas Science and Technology*
4. Da Silva P., PhD thesis, Paris VI (1998)
5. Echard M., PhD thesis, Caen (2001)
6. Kemp R.A., Ryan R.C., and Smegal J.A., *Proc.9th Int.Cong.Catal.*, **1** (1988) 128
7. Poulet O., Hubaut R., Kasztelan S., and Grimblot J., *Bull.Soc.Chim.Belg.*, **100** (1991) 857
8. Ramírez J., Castaño V.M., Leclercq C., and López Agudo A., *Appl.Catal.*, **83** (1992) 251
9. Bouwens S.M.A.M., Vissers J.P.R., de Beer V.H.J., and Prins R., *J.Catal.*, **112** (1988) 401
10. Eijsbouts S., Van Gestel J.N.M., Van Veen J.A.R., de Beer V.H.J., and Prins R., *J.Catal.*, **131** (1991) 412
11. Iwamoto R. and Grimblot J., *Stud.Surf.Sci.Catal.*, **106** (1997) 195

89e Study about Formation Al2o3-Tio2 Nanofibers as Supports for Hydroprocessing Catalysts

Jose Antonio Muñoz Lopez, Esteban Lopez Salinas, Jose Antonio Toledo Antonio, Carlos Angeles Chavez, and Jose Escobar Aguilar. Instituto Mexicano del Petroleo, Eje Central Lázaro Cárdenas No. 152 Col. San Bartolo Atepehuacan, Mexico, Mexico
Introduction: Refineries have undergone restructuring and revamping units due to changes of petroleum feeds properties and the demand of more environment friendly products. The increasing demand of lighter products and middle distillates will be continuing in near future and the demand of heavy crudes, at the same time, will be decreased. To meet this demand, a substantial quantify of heavy crudes has to be upgraded and hydroprocessing is one of the mostly used processes in this regard [1]. The need for better hydroprocessing catalysts capable of complying with the present and the future environmental regulations has promote research on the use of new supports and active phases. Among the new supports, alumina-titania, have shown interesting properties [2]. In this regard, different studies concerning various aspects of hydrotreating catalysts on Al_2O_3-TiO_2 have been undertaken by several groups [3]. They have reported that modification of alumina by TiO_2 improves not only the hydrodesulfurization but also the hydrogenation reaction [4]. In this work, the textural and structural property variations in alumina with varying of incorporation titania method and titania source were studied and discussed. Experimental: Aluminas were prepared by a pH-swing method, which is based on the exponential U-shaped solubility curve of hydroxylated aluminas, and described elsewhere [5]. The effect of TiO_2 incorporation (15 % wt) into the Al_2O_3 using three different methods such as precipitation, co-precipitation and multi-deposition of titania, and the effect of titania source ($K_2[TiO(C_2O_4)_2]2H_2O$ and $TiOSO_4xH_2SO_4xH_2O$) are reported. All supports were characterized by nitrogen physisorption, XRD, TEM, UV-vis and Raman spectroscopies. Results and conclusions: Nano-fibrillar morphology Mixed oxides Al-Ti were prepared and BET surface area of the Al_2O_3-TiO_2 was in the range 306-343 m^2/g and after calcination at 500 °C in the range 311-342 m^2/g (due to fibrillar morphology, this materials exhibits strong resistance to sintering), average pore size of 13-15 nm and total pore volume in the range of 1.0-1.15 cm^3/g. The results from XRD, texture, TEM and Raman indicate that better titania dispersions are obtained by deposition method and the titania source have great impact in the titania dispersion in this fibrillar aluminas.
References: 1. Ramirez J., Macías G., Cedeño L., Gutierrez-Alejandre A., Cuevas R., Castillo P., "The Role of Titania in Supported Mo, CoMo, NiMo, and NiW Hydrodesulfurisation Catalysts: Analysis of Past and New Evidences", Catalysis Today, 98, 19-30, 2004. 2. Rana M. S., Ancheyta J., Rayo P., Maity S. K., "Effect of Alumina Preparation on Hydrodemetallisation and Hydrodesulfuration of Maya Crude", Catalysis Today, 98, 151-160, 2004. 3. Maity S. K., Ancheyta J., Alonso F., Rana M. S., "Preparation, Characterization and Evaluation of Maya Crude Hydroprocessing catalysts", Catalysis Today, 98, 193-199, 2004. 4. Santes V., Herbert J., Cortez M. T., Zarate R., Díaz L., " Catalitic Hydrotreating of Heavy Gasoil FCC Feed on Alumina-titania supported NiMo Catalysts", Appl. Cat. A, 281, 121-128, 2005. 5. Ono T., Ohguchi Y., Togari O., Preparation of Catalysts III, edited by Poncelet G., Grange P., and Jacobs P., Elsevier Science, Amsterdam-Netherlands ,631-641, 1983.

Unsupported sulphides and nitrides for hydrotreating: new preparation approaches and catalytic properties.

P. Afanasiev.

Institut de Recherches sur la Catalyse, 2, Avenue Albert Einstein, 69626 Villeurbanne Cedex, France.

pavel.afanasiev@catalyse.cnrs.fr

Due to societal needs to minimize the negative environmental effects of exhaust emissions, the authorized sulfur level in motor and non transportation fuels has been decreasing during the last years. Many different approaches were proposed to eliminate sulfur-containing molecules from the petroleum feedstocks (1). Because of relatively low investment costs, the utilization of more active catalysts is the most attractive solution for petroleum refiners. Such a demand for more active HDT catalysts incited a strong increase of research activity. These studies have shown that one of the ways to improve the catalysts could be the use of new bulk catalytic materials with increased density and/or performance of the active sites. The conventional HDS is usually carried out over sulfided $CoMo/Al_2O_3$ and $NiMo/Al_2O_3$ catalysts, known since the beginning of the last century (1-3). However the emergence of highly loaded sulfide catalysts such as NEBULA, STARS and by Criterion and Akzo Nobel, demonstrates that sulfide based systems still have a great potential of improvement.

Beside sulfides, other bulk phases are extensively studied, including nitrides, carbides and phosphides (2). Much controversy exists on the literature on the properties of these systems. While sometimes demonstrating high activity, these compounds have lower stability, being easily oxidized in air. Moreover their transformation to the corresponding sulfides is possible, making more difficult the analysis of experimental data.

In our work we targeted on the alternative ways of thermally stable macro – and mesoporous transition metal sulfides, carbides and nitrides. Various solution preparations, including aqueous, non aqueous and soft hydrothermal techniques, as well as decomposition of molecular precursors were investigated. Through control of precursors, solvents and reaction conditions, bulk sulphide and nitride catalytic materials were prepared, with structural and morphological tuning over the micro- and mesoscopic length scales. The examples of synthesis approaches are presented below.

1. Soft chemistry routes to highy dispersed sulphide catalysts.

The most important part of work concerned the synthesis of Co(Ni)-Mo(W) sulfides. The possible ways of association between cobalt and molybdenum are versatile. They include either direct reaction in the solution leading to the formation of mixed Ni(Co)-Mo-S phases or the preparation of high surface area bulk sulphide followed by its promotion with appropriate cobalt precursor. All the syntheses include three main variables, which are: the solvent used, the molybdenum precursor and the precipitating agent (the last might be represented by cobalt or nickel salt).

Table 1 provides several examples of the solution reactions from our recent work, in which we obtained either dispersed molybdenum sulphide alone , or the one pot synthesis of promoted sulfides.

Table 1. Solution reactions used for the synthesis of the unsupported sulphide catalysts.

Solvent, admixture	Mo source	Precipitating (reducing/sulfiding) agent	[a]Ssp , m^2/g	Conditions, comments, reference
H_2O, eventually CTBA surfactant	$(NH_4)_2MoS_4$	N_2H_4 NH_2OH	210	Soft aqueous preparation (353 K), high surface area bulk or supported materials (5,8, 11).
H_2O	$(NH_4)_2MoS_4$	No one	90	573 K hydrothermal, elongated MoS_2 stacks (6).
H_2O-acetone	$(NH_4)_2Mo_2S_{12}$	Ambient oxygen	70	298-330 K, microtubular, hollow sphere or vesicle like nanostructured objects (7,9, 10).
H_2O	$(NH_4)_2MoOS_2$	Diluted acid (HCl)	60	298 K, microtubular $MoOS_2$ sulfide (12)
H_2O-ethyleneglycol, eventually Triton surfactant	$(NH_4)_2MoS_4$ $(NH_4)_2Mo_2S_{12}$	$Ni(NO_3)_2$	150	298 K One-step preparation of Ni(Co)MoS sulfides (13)
Ethyleneglycol	$(NH_4)_2MoS_4$ $(NH_4)_6Mo_7O_{24}$	$Ni(NO_3)_2$	80	463 K glycothermal. Onion-like MoS2 shells.

[a] maximal value of specific surface area after sulfidation at standard conditions (673 K, H_2S/H_2)

The syntheses reported in table 1 provide wide a variety of particular morphology types of molybdenum sulfide. Depending on the reaction used, microspherical, open-pore fractal, or hollow tubular morphologies were obtained (Fig.1). High surface areas of dispersions ascertained high catalytic activity of the corresponding promoted catalysts.

In most cases the morphology of precipitates obtained in the solution was preserved after sulfidation of the materials at 673 K. Therefore open porous system advantageous for the reactants diffusion was created.

Fig.1 SEM images of sulfide materials: A - hydrothermally prepared dispersion of MoS_2; B – microtubules of $MoOS_2$ solid from the acid precipitation; C – amorphous Co-Mo-S precursor from the reaction between $Co(NO_3)_2$ and $(NH_4)_2Mo_2S_{12}$; D – microtubules of amorphous MoSx sulfide obtained by water –acetone reflux of $(NH_4)_2Mo_2S_{12}$.

The preparation methods employed include soft hydrothermal syntheses, surfactant – aided, or ultrasound-induced precipitations in the mixed non aqueous solutions, as well as solventothermal (polyol) reactions allow controlling the length and stacking of the MoS_2 fringes. Thus, hydrothermal (and glycothermal) methods led to the elongated slabs with considerable stacking, whereas the mixed solvent surfactant assisted precipitations yield very short and weakly stacked fringes (Fig. 2).

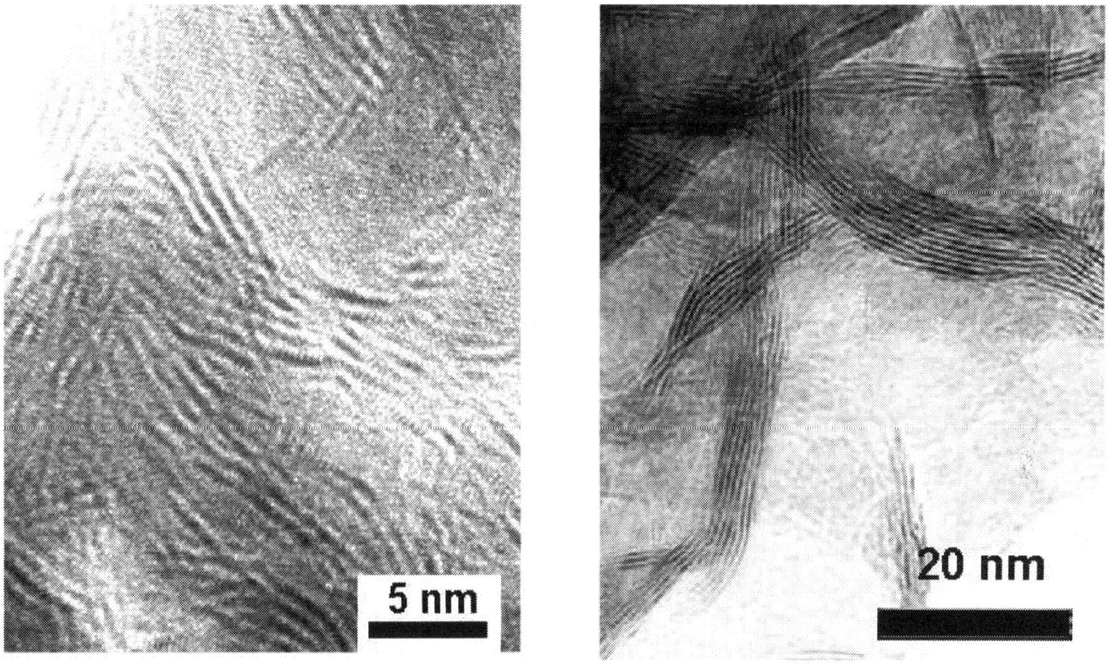

A **B**

Fig. 2 Transmission microscopy images of the Co-Mo-S catalysts: A- prepared by surfactant assisted precipitation; B – hydrothermal preparation (573 K).

Overall, it has been shown that reactions of thiomolybdate MoS_4^{2-}, thiodimolybdate $Mo_2S_{12}^{2-}$ and oxothiomolybdate $MoOS_2^{2-}$ cores with nickel or cobalt salts in mixed solutions produce amorphous solids (pre-catalysts) with Ni(Co)/Mo ratio depending on the conditions and particularly on the nature of molybdenum thioanion applied. The morphology of Co(Ni)-Mo-S solids obtained after sulfidation of the pre-catalysts depends mostly on the preparation technique used. Homogeneity of the sulfided catalysts strongly depends on the reaction conditions in the solution and on the nature of precursors used. Worth noting that the residual carbon in these solids plays an important role in stabilizing their morphology and catalytic activity. Its content varies between 3 and 4 % wt from a catalyst to another. The absence of this residual carbon corresponds to the collapse of porous system and low catalytic activity.

Numerous characterizations including XRD, TPR, scanning and transmission electron microscopy were used to understand the reactions occurring under the hydrothermal or in mixed solvents. EXAFS at the Co, Ni and Mo edges reveals evolution of transition metals coordination from the amorphous pre- catalysts to the Ni(Co)MoS sulfided phases.

The goal of our preparations was to obtain highly active catalysts of deep HDS. The heaviest fractions of gasoline and diesel pools such as bottom FCC naphtha and coker naphtha, contain mainly alkylated benzothiophenes, dibenzothiophenes (DBT) and alkyldibenzothiophenes. The HDS of these least reactive sulfur compounds was therefore estimated in a micro pilot reactor working at 15-45 bar, using 4,6 DMDBT dissolved in dodecane. Optimization of the promoted NiMo and CoMo unsupported systems, issued from the ADTM reactions (fourth raw in Table 1) allows to obtain catalysts with very high activities in the thiophene and 4,6-DMDBT hydrodesulfurization, several times greater than those of commercial references available (Fig. 3). It is well known that HDS of thiophenic compounds proceeds via two reaction pathways. In the hydrogenolysis pathway the sulfur is directly removed from the molecule by C-S bond cleavage (hydrogenolysis pathway). In the hydrogenation pathway the aromatic ring is first hydrogenated and then sulfur is removed. Which pathway will be dominating depends on the structure of sulphur-containing molecule and the type of catalyst used. Thus, nickel-containing unsupported catalysts were more selective to partially hydrogenated products; therefore the second pathway was favoured in them compared to Co-Mo-S counterparts. At the same time we observed that at the same conversion and temperature, thiophene reacted preferably via the hydrogenolysis pathway, whereas for 4,6-DMDBT both the hydrogenation and hydrogenolysis routes seemed to be important. By this reason comparative activity of our catalysts in the thiophene and 4,6-DMDBT reactions were very different.

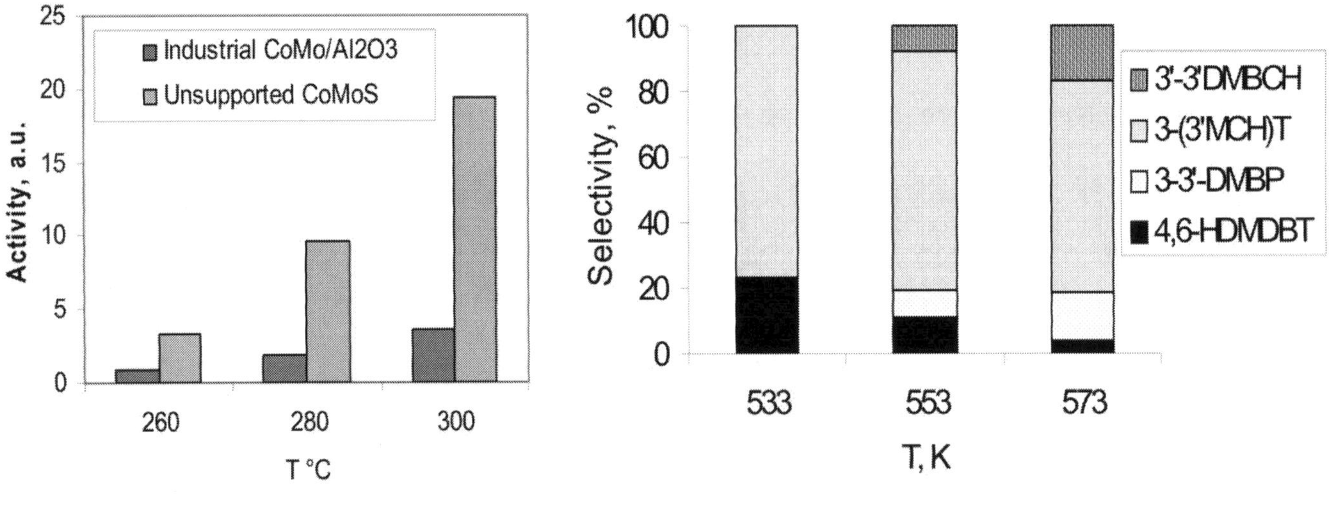

Fig. 3. A - comparison of (4, 6)- DMDBT HDS activity for the unsupported CoMoS system and an industrial alumina -supported reference; B – selectivity of (4, 6)- DMDBT transformation as a function of temperature (20-30 % conversion). HDMDBT – non – desulfurized partially hydrogenated DMDBT; white areas -DMBP direct HDS product; dashed areas – partially hydrogenated HDS products.

High catalytic activity of the bulk mixed phases does not warrant the possibility of their direct industrial application. The problems of mechanical resistance and regeneration of such systems remain important. The presence of some binder is necessary which would provide mechanical cohesion of the catalyst and avoid the reactor plugging due to accumulation of dust. Preliminary results of catalytic activity tests show that the introduction of moderate amounts of alumina binder (up to 20 % wt) allowed not only preserve but even increase the specific activity of our unsupported catalysts.

2. Unsupported molybdenum and bimetallic nitrides

Another class of extensively studied bulk catalytic systems are binary or ternary nitrides. When dealing with hydrogenation reactions, noble metal- like behavior of the TM nitrides and carbides seems well proven (14). By analogy with promoted Co-Mo-S type systems, ternary nitrides were syhthesized and tested in HDS and HDN catalysis, but the data on their activity are contradictory. While some promising results have been reported, the following questions still have no definite response:

-Which is the nature of active phase in the nitride catalysts under HDS or HDN conditions? Is it intrinsic nitride activity, or all these the catalysts are more or less sulfided, and real catalytic centers are the sulfide species on the surface of nitrides?

-Is there any real promotion in ternary nitrides, or their enhanced activity is again just a consequence of sulfidation and creation of some Co-Mo-S species?

In our recent work, we tried to shed some additional light on these questions. To do that, we prepared several dispersed binary and ternary nitrides and compared their HDS and HDN activities, paying particular attention to the state of these catalysts after reaction. Unsupported systems were chosen since it is much easier in this case to control the state of the metals, compared to the supported counterparts.

To obtain the unsupported nitrides, decomposition of molecular precursors was applied. Indeed, thermodynamics of nitrides formation is much less favourable compared to sulfides, therefore their direct preparation in the solutions seems impossible. However, decomposition of molecular complexes allows to decrease considerably the preparation temperature. As opposed to the traditional 'heat and beat' solid-solid and gas-solid procedures, the preparation technique based on the use of monometallic and bimetallic complexes tried to make advantage of the decrease of the diffusion distances detween the metallic atoms and those of the reducing agent which is homogeneously distributed within the precursor volume.

New molecular complexes of hexamethylenetetramine (HMTA) with transition metals including Co, Ni, Mo, V and their combinations have been prepared by simple aqueous reactions between HMTA and salts of the corresponding metals such as nitrates or acetates. This way, monometallic or bimetallic complexes were obtained. Some of them were characterised structurally (such as Co and Ni monometallic complexes (Fig. 4)), for others only the stoichiometry and unit cell parameters are established for the time being.

Decomposition of these HMTA complexes produced highly dispersed materials, appropriate for catalytic use. Decomposition of Mo, V, Co-Mo and Ni-Mo compounds, containing binary nitrides such as Mo_2N (15), Mo_3N_2 or VN, as well as ternary nitrides

Co_3Mo_3N and Ni_2Mo_3N (Fig. 5, 6). These materials demonstrate specific surface areas of 100-150 m^2/g. Decomposition pathways, as studied by mass spectrometry shown to be the auto reduction of the transition metals species by HMTA, occurring with formation of water and carbon oxides. It appears that slight modification of the gas flow composition allows qualitative control of the nature of the products formed. Thus, slight changing of the treatment atmosphere might very strongly modify the nature of phases obtained in the case of molubdenum HMTA complex, as sketched in Fig. 7. The reasons of such versatile behaviour are in the comparable values of the equilibrium constants of the corresponding reactions leading to nitride, carbide or oxide. Therefore addition of only 1% of hydrogen to argon led to switching from nitride to carbide, whereas the presence of several mbar of water vapour yielded the oxide product

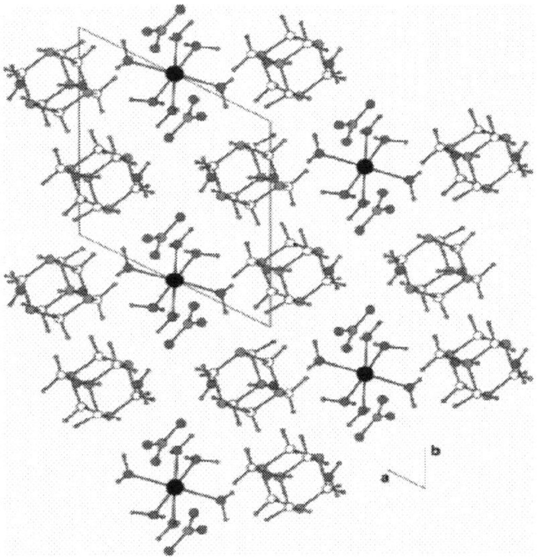

Fig. 4. Structure of the molecular Ni(Co)-HMTA nitrate complex Black circles – Ni(Co); grey circles – N; white circles – C.

Fig. 5 Lamellar Mo_2N crystallites obtained from the decomposition of Mo-HMTA complex

The catalytic activities of as synthesised binary and ternary nitrides have been evaluated in thiophene HDS and quinoline HDN reactions. If the molybdenum nitride with surface area 150 m^2/g had thiophene HDS activity twice higher than unsupported MoS2 reference, we failed to observe any promotion effects in the nitride based systems. Neither the catalysts prepared from the monometallic Mo_2N by further addition of Co of Ni, nor the bimetallic nitrides showed any promising HDS activity. Oxidation with further re-sulfidation might generate some non negligible HDS performance, which remains however much lower compared to bulk sulphide systems described in section 1. As thorough analysis of literature shows, the earlier reports on high intrinsic activity of ternary nitrides should be related to such re-assembling of Co-Mo-S type species.

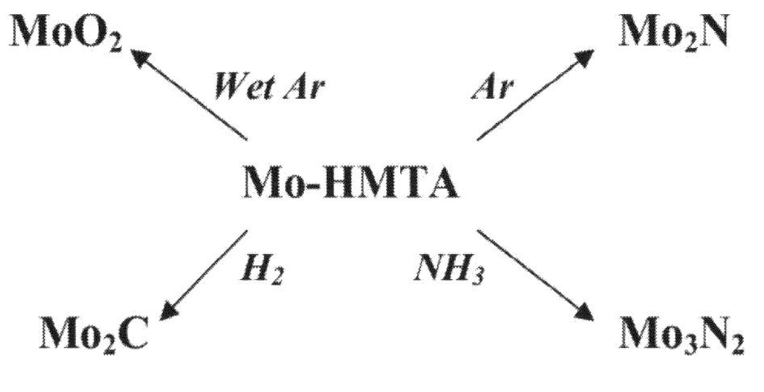

Fig. 6 Variation of the decomposition products for the Mo-HMTA complex as a fucntion of treating atmosphere

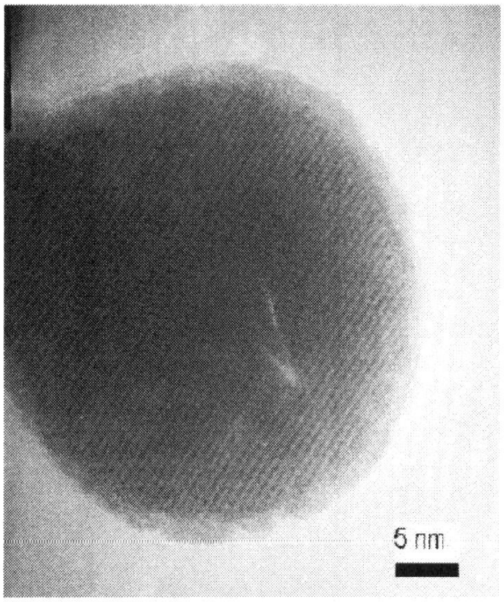

Fig. 5 Particle of Co_3Mo_3N nitride obtained from the decomposition of Co-Mo HMTA complex

Our preliminary data suggest that the systems synthesised have considerable HDN activity. Again, in the presence of sulphur, surface sulfidation seems to be the main reason of the activity of these bimetallic systems. Overall, dispersed nitrides were synthesized via original way and tested in model HDS and HDN reactions. However, their HDS activities were much lower compared to the corresponding sulphide systems. No synergy between Co(Ni) and Mo in the ternary nitrides could be inferred from our results.

References

1. D.D. Whitehurst, T. Isoda and I. Mochida, Adv. Catal. 42 (1998) 345.
2. R. Prins In: G. Ertl, H. Knözinger and H.J. Weitkamp, Editors, Handbook of Heterogeneous Catalysis, VCH, Weinheim (1997), p. 1908.
3. S. Brunet, D. Mey, G.Pérot, C. Bouchy, F. Diehl, Appl. Catal. 278 (2005) 43.
4. L.A. Gerritsen, S. Eijsbouts, Y. Inoue, P.H. Desai, Catal Courier (Akzo Nobel) N34.
5. P. Afanasiev, G.F.Xia, G. Berhault, B. Jouguet, M. Lacroix, Chem. Mater. 11 (1999) 3216.
6. E. Devers, P. Afanasiev, B. Jouguet, M. Vrinat. Catal. Lett., 82 (2002) 13.
7. P. Afanasiev, C. Geantet, C. Thomazeau, B. Jouguet, Chem. Comm. 12 (2000) 1001.
8. I. Bezverkhyy, P. Afanasiev, M. Lacroix, Inorg. Chem. 39 (2000) 5416.
9. P. Afanasiev, I. Bezverkhyy, Chem Mater, 14 (2002) 2826
10. P. Afanasiev, I. Bezverkhy, J. Phys. Chem. B. 107 (2003) 678.
11. I. Bezverkhyy, P. Afanasiev, M. Lacroix, J. Catal. 230 (2005) 133.
12. D. Genuit, I. Bezverkhyy, P. Afanasiev, J. Solid State Chem. 178 (2005) 2759.
13. D. Genuit, P. Afanasiev, M. Vrinat, J. Catal. 235 (2005) 302.
14. D. J. Sajkowski, S. T. Oyama, Appl. Catal. 134 (1996) 339.
15. P. Afanasiev, Inorg. Chem. 41 (2002) 5317.

CATALYST SELECTION FOR ULTRA LOW SULFUR DIESEL OPERATIONS: ACTIVITY AND H_2 CONSUMPTION CONSIDERATIONS

Lawrence S. Kraus
Albemarle Catalysts Company LP
Houston, Texas USA

Katsuhisa Fujita and Yuji Noguchi
Nippon Ketjen Co., LTD.
Tokyo, Japan

Abstract

Ultra Low Sulfur Diesel (ULSD) operations target significantly lower product sulfur levels, typically ≤ 10 wppm, than current diesel hydrotreating operations. State-of-the-art catalysts are required to achieve reasonable cycle lengths in ULSD operations. ULSD production results in significantly higher H_2 consumption than is observed in LSD production (product S < 500 wppm). Maximizing catalyst activity and balancing H_2 consumption with unit H_2 supply capabilities are critical considerations in ULSD catalyst selection. This paper discusses catalyst selection and application in ULSD production. Pilot plant data are presented comparing high activity STARS™ CoMo and STARS™ NiMo catalysts to elucidate the impact of catalyst selection, process conditions, and process severity on performance (hydrodesulfurization, hydrodenitrogenation, aromatics saturation, color improvement, H_2 consumption). A ULSD catalyst system optimization study of STARS™ CoMo and STARS™ CoMo/Nebula™ demonstrates that proper catalyst system design can achieve improved catalyst activity while managing H_2 consumption and maintaining catalyst stability. Commercial data illustrating the impact of catalyst system design on H_2 consumption/unit H_2 supply balance and catalyst deactivation/cycle length are discussed. Catalyst selection guidelines for ULSD operations are presented.

Introduction

ULSD operations will target significantly lower product sulfur levels (≤ 10 wppm) than currently required (≤ 500 wppm). Reduction of the product S to this low level will require greater catalyst activity to achieve acceptable cycle lengths. The demands of ULSD production have made catalyst activity a premium in diesel hydrotreating. Catalyst suppliers have developed high activity catalysts for use in ULSD production and other high performance applications, such as hydrocracker feed pretreat. Albemarle Catalyst Co. LP and Nippon Ketjen Co., LTD. have developed high activity STARS™ CoMo, STARS™ NiMo, and Nebula™ catalysts for these applications. STARS™ (Super Type II Active Reactive Sites) catalysts are supported metal catalysts that utilize a manufacturing process that prevents strong interaction of the active metals (CoMo or NiMo) with the alumina support. This results in 100% Type II active metals sites. Type II sites have greater intrinsic activity than Type I sites where the active metals interact more strongly with the alumina support. A thorough discussion of Type I and Type II sites can be found elsewhere.[1] Nebula™ is a catalyst formulation that has a very high concentration of base metal active sites without a conventional support. This formulation results in maximum active site density in the reactor. Table I lists Albemarle/Nippon Ketjen catalyst options for ULSD production. This table shows the hydrodesulfurization (HDS) and hydrodenitrogenation (HDN) RVA (Relative Volume Activity) values of new-generation STARS™ CoMo, STARS™ NiMo, and Nebula™ catalysts in ULSD (product S ≤ 10 wppm) production. Table I also shows the relative H_2 consumption of these catalysts during ULSD operations. There are several grades of STARS™ CoMo catalysts. Therefore, the performance properties of STARS™ CoMo catalysts are presented as a range. From Table I it is seen that as catalyst activity for ULSD production increases, the corresponding H_2 consumption observed increases.

Table I
ULSD Catalyst Options: Activity and H_2 Consumption

CATALYST:	HDS RVA (Product S < 10 wppm)	HDN RVA (Product S < 10 wppm)	RELATIVE H_2 CONSUMPTION (Product S < 10 wppm)
High Activity CoMo (STARS™ CoMo)	100 - 150	100 - 130	100 - 105
High Activity NiMo (STARS™ NiMo)	190	150	135
Bulk Active Metal (NEBULA™)	> 250	> 200	> 175

Table II shows typical organic sulfur species found in diesel streams. This table also contains information on the relative reactivity of these organic sulfur species and the D-86 distillation end point (EP) where these species appear in diesel hydrotreater feeds. The dibenzothiophene-derived species with substitution in the 4 position or the 4 and 6 positions are the most difficult to convert.(2) These species are referred to as sterically-hindered dibenzothiophenes (SH-DBT's).

When processing diesel feeds with typical endpoints, producing ULSD involves conversion of sulfur species having the form of 4,6-dimethyl-dibenzothiophene (4,6 DM-DBT) and its derivatives. Conversion of 4,6-DM-DBT can occur via a direct route or an aromatic saturation route, as illustrated in Figure 1. The aromatic saturation route is 3+ times faster than the direct route for conversion of 4,6-DM-DBT and its derivatives.(3) Consequently, catalysts with high aromatic saturation activities will have high activities for 4,6-DM-DBT conversion and will be effective for ULSD production. Catalysts with high aromatic saturation activities also have high activities for saturation of sulfur-free aromatics species. This results in high H_2 consumption from aromatics saturation accompanying high ULSD activity. High aromatic saturation activity is the source of the intrinsic high ULSD HDS activity of STARS™ NiMo and Nebula™ catalysts.

Table II
Diesel Organic Sulfur Species

Sulfur Compound Structure	Name	Relative Reaction Rate	D-86 EP When Observed
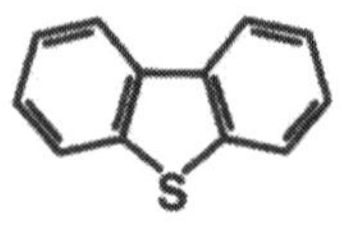	dibenzothiophene	6 - 12	590°F
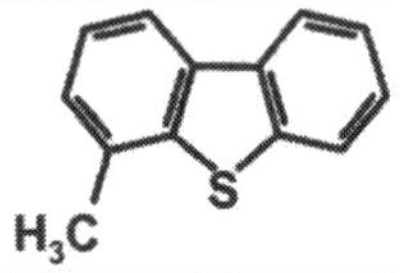	4-methyl-dibenzothiophene	2 - 4	600-620°F
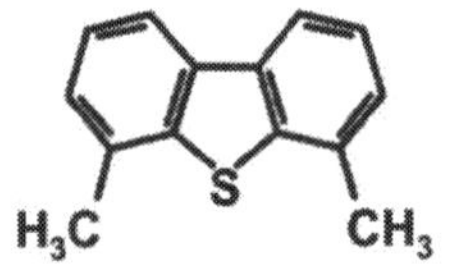	4,6-dimethyl-dibenzothiophene	1	630-650°F
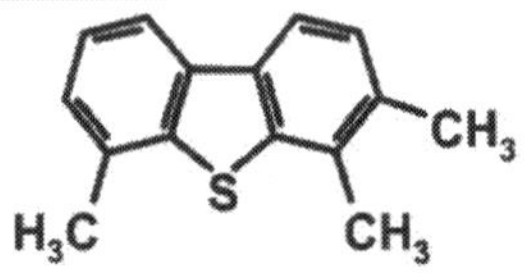	3,4,6-trimethyl-dibenzothiophene	1	660-680°F

STERICALLY HINDERED DI-BENZO-THIOPHENE (SH-DBT) SPECIES

Figure 1
Reaction Paths for Refractory S Species

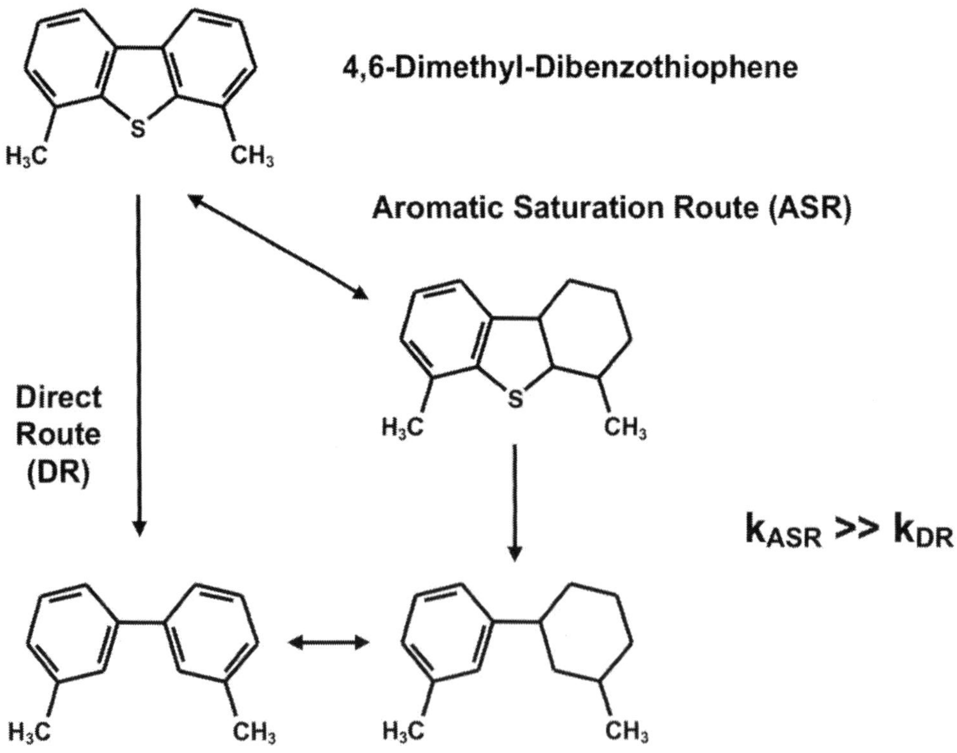

Commercial hydroprocessing units have fixed capabilities for H_2 recycle and make up H_2 supply. The H_2 consumption in a hydroprocessing unit must be compatible with the unit H_2 supply capabilities. The makeup H_2 supply must exceed the H_2 consumption in the unit and H_2 circulation must be sufficient to ensure reasonable hydrogen partial pressure (ppH_2) at the reactor outlet. A typical measure for the ability of a unit to maintain ppH_2 at the reactor outlet is H_2 Availability:

$$\mathbf{H_2\ Availability} = \frac{\mathbf{H_2/Oil_{Feed}\ (SCFB)}}{\mathbf{H_2\ Consumed\ (SCFB)}}$$

The H_2 Availability is the ratio of the total H_2 fed to the unit (make up + recycle + quench) to the H_2 consumed in the reactor. H_2 Availability is an indicator of the ability of a unit to maintain ppH_2 through the reactor. Low H_2 Availability indicates low reactor outlet ppH_2. Since reactor temperatures are highest at the outlet, maintaining reactor outlet ppH_2 is especially important for controlling fouling/catalyst deactivation. Low H_2 Availability results in high catalyst deactivation rates and short cycle length. As shown in Table I, H_2 consumption during ULSD production can vary significantly depending on the catalyst used. The high activity of new-generation catalysts cannot be fully utilized if insufficient H_2 is available to maintain catalyst activity. Albemarle/Nippon Ketjen recommend a minimum H_2 Availability of 3 when processing straight run feeds and a minimum H_2 Availability of 4 when processing cracked feedstocks. ULSD catalyst selection must be made to optimize catalyst activity/H_2 consumption and meet product property requirements without product quality (e.g. cetane) giveaway. Catalyst selection cannot be made simply to maximize catalyst activity in a reactor without considering these other factors.

This paper contains data comparing the performance (activity, H_2 consumption, product properties) of STARS™ CoMo and STARS™ NiMo catalysts in diesel hydrotreating. Data are also presented illustrating catalyst system design optimization using combined beds of STARS™ CoMo and Nebula™ catalyst to provide significant catalyst activity and performance improvements over STARS™ CoMo while managing H_2 consumption and catalyst stability. A commercial example illustrating the critical importance of catalyst selection and H_2 consumption management is presented. Finally, catalyst application guidelines in ULSD operations are discussed.

STARS™ CoMo and STARS™ NiMo Comparison

Results from two experimental series comparing STARS™ CoMo and STARS™ NiMo catalysts are reported. The first had the objectives of demonstrating the impact of process conditions on the performance of STARS™ CoMo and STARS™ NiMo catalysts and catalyst selection. The second series had the objectives of demonstrating the impact of STARS™ CoMo and STARS™ NiMo catalysts and severity (product S level) on H_2 consumption and product properties during diesel hydrotreating.

Process Condition Study

A series of experiments was conducted to demonstrate the relationships of process conditions and STARS™ CoMo and STARS™ NiMo performance. These experiments were conducted in isothermal pilot plant equipment at Albemarle facilities in Amsterdam, The Netherlands. The feed used was a blend of straight run diesel (SRD) and FCC light cycle oil (LCO). Feed properties are shown in Table III and process conditions are found in Table IV. The relationships of product S, product N, and H_2 consumption with hydrogen partial pressure (ppH_2) are shown in Figure 1 and with weight average bed temperature (WABT) in Figure 2.

Table III
Process Condition Study: Feed Properties

Feed Source:	**30/70 Vol/Vol LCO/SRD**
Sulfur (wt%)	**1.38**
Nitrogen (wppm)	**392**
API Gravity (°)	**31.0**
Br No. (g/100 g)	**5**
D-86 Distillation (°F)	
IBP	**336**
T–50	**527**
FBP	**734**
Aromatics (wt%)	
Mono-	**22.3**
Di-	**19.7**
Tri+ -	**4.7**
Total	***46.7***

Table IV
Process Condition Study: Process Conditions

LHSV (hr^{-1})	**0.5 – 3.0**
ppH_2 (psia)	**290 - 1160**
WABT (°F)	**605 - 685**
H_2/Oil (SCFB)	**1200**
Catalysts	**STARS™ CoMo STARS™ NiMo**

LHSV = Liquid Hourly Space Velocity
ppH_2 = Hydrogen Partial Pressure
WABT = Weighted Average Bed Temperature
H_2/Oil = Hydrogen-To-Oil Ratio
SCFB = Standard Cubic Feet/Barrel

The data in Figure 1 were obtained at fixed WABT = 644°F, LHSV = 1.0 hr^{-1}, and H_2/Oil = 1200 SCFB. From this figure it is seen that STARS™ NiMo is the superior performer at higher pressure (ppH_2 > 500 psia) and lower product S levels. STARS™ NiMo catalyst is advantageous at high pressure and low product S (product S < ~ 200 wppm) where hydrogenation activity is beneficial for converting 4,6-dimethyl-dibenzothiophene and related species. STARS™ CoMo is the superior performer at lower pressure (ppH_2 < 500 psia) and low HDS levels where the sulfur species being removed are converted via the direct desulfurization route and do not require high aromatic saturation for efficient removal. STARS™ NiMo has superior HDN performance at all pressure levels. The HDS of STARS™ NiMo becomes greater than that of the STARS™ CoMo catalyst at conditions where product N is very low (< 10 wppm). At very low pressure levels the H_2 consumption of the STARS™ NiMo catalyst is only slightly higher than that of the STARS™ CoMo catalyst and the HDS performance of the STARS™ NiMo is less than that of the STARS™ CoMo. The increase of H_2 consumption with increasing ppH_2 is greater with the STARS™ NiMo.

Figure 1
Process Condition Study
Relationship of Product S, Product N, H_2 Consumption and Hydrogen Partial Pressure
STARS™ CoMo & STARS™ NiMo

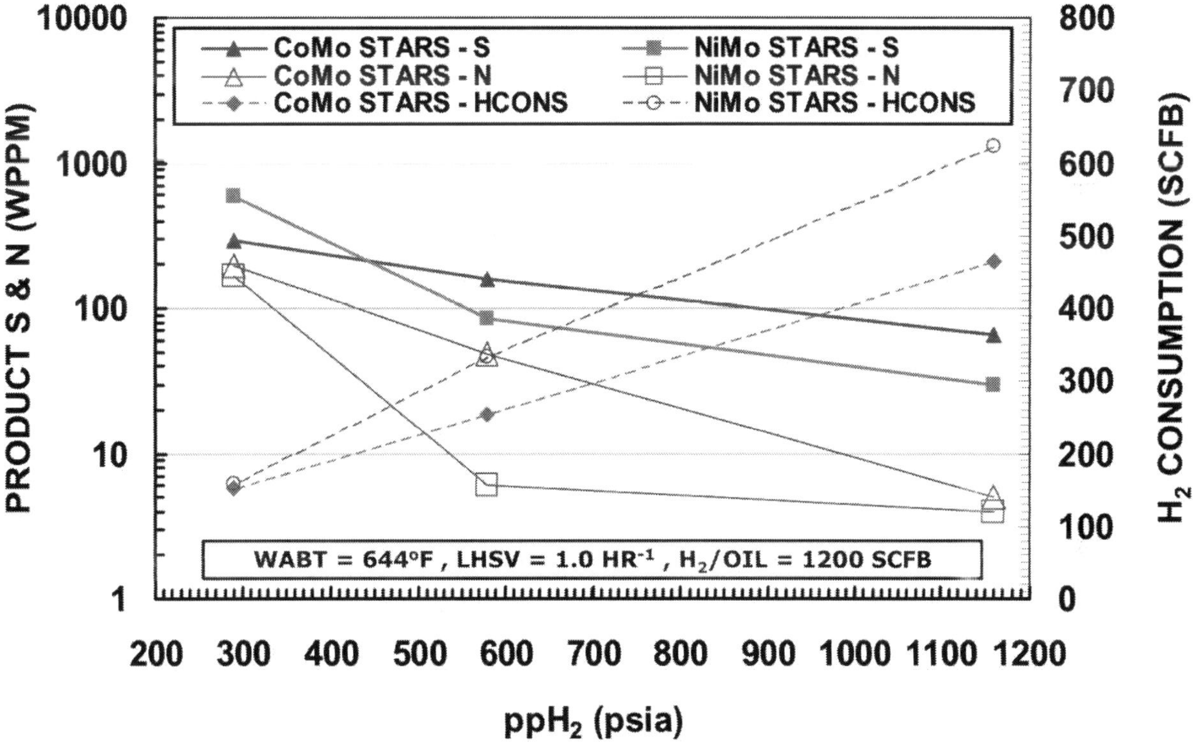

Figure 2 shows the relationships of product S, product N, and H_2 consumption with WABT for STARS™ CoMo and STARS™ NiMo catalysts. These data were obtained at fixed ppH_2 = 580 psia, LHSV = 1.0 hr^{-1}, and H_2/Oil = 1200 SCFB. At product S < ~ 200 wppm the STARS™ NiMo catalyst had superior HDS activity. STARS™ NiMo was the superior HDN catalyst at all temperature conditions and shows superior HDS activity at product N levels < ~ 20 wppm. H_2 consumption was higher with the STARS™ NiMo at all temperatures, even when its HDS performance was lower than that of the STARS™ CoMo. As the temperatures increased, the H_2 consumption of the STARS™ CoMo catalyst approached that of the STARS™ NiMo catalyst. This is expected because catalyst H_2 consumptions become similar as temperatures increase and aromatic saturation equilibrium is approached.

Figure 2
Process Condition Study
Relationship of Product S, Product N, H_2 Consumption and WABT
STARS™ CoMo & STARS™ NiMo

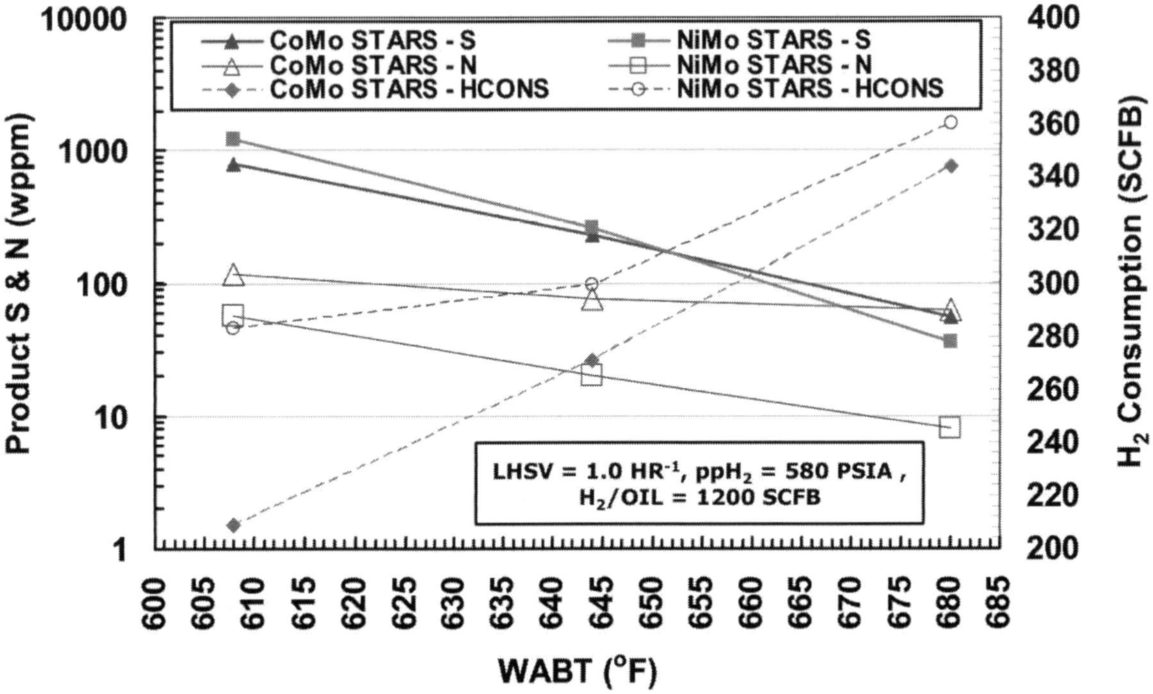

Figure 3 shows the H_2 consumption of STARS™ CoMo and STARS™ NiMo catalysts as a function of LHSV. At LHSV of 3.0, 1.0, and 0.5 the H_2 consumption of STARS™ NiMo is 17%, 31%, and 65% higher, respectively. With both catalysts there is a strong, non-linear space velocity effect at low LHSV. In the case of LHSV = 0.5 hr^{-1}, the H_2 Availability with STARS™ NiMo is approximately 2. Operating under this condition would result in rapid catalyst deactivation and short cycle length. For large volume units, application of STARS™ CoMo could be useful as a means to manage H_2 consumption.

In general, STARS™ NiMo is more active for HDS than STARS™ CoMo catalysts for high severity (low product S) diesel HDS. H_2 consumption is significantly higher with STARS™ NiMo and this must be accounted for when considering this catalyst for diesel hydrotreating operations.

When selecting catalyst for a commercial ULSD operation it is important to realize that the unit will be targeting a specific product S specification, not a specific temperature. Catalyst comparisons should be made at constant product S level to ensure a direct evaluation of relative catalyst performance and H_2 consumption at the required product S specification.

Figure 3
Process Condition Study
Relationship of H_2 Consumption and Liquid Hourly Space Velocity
STARS™ CoMo & STARS™ NiMo

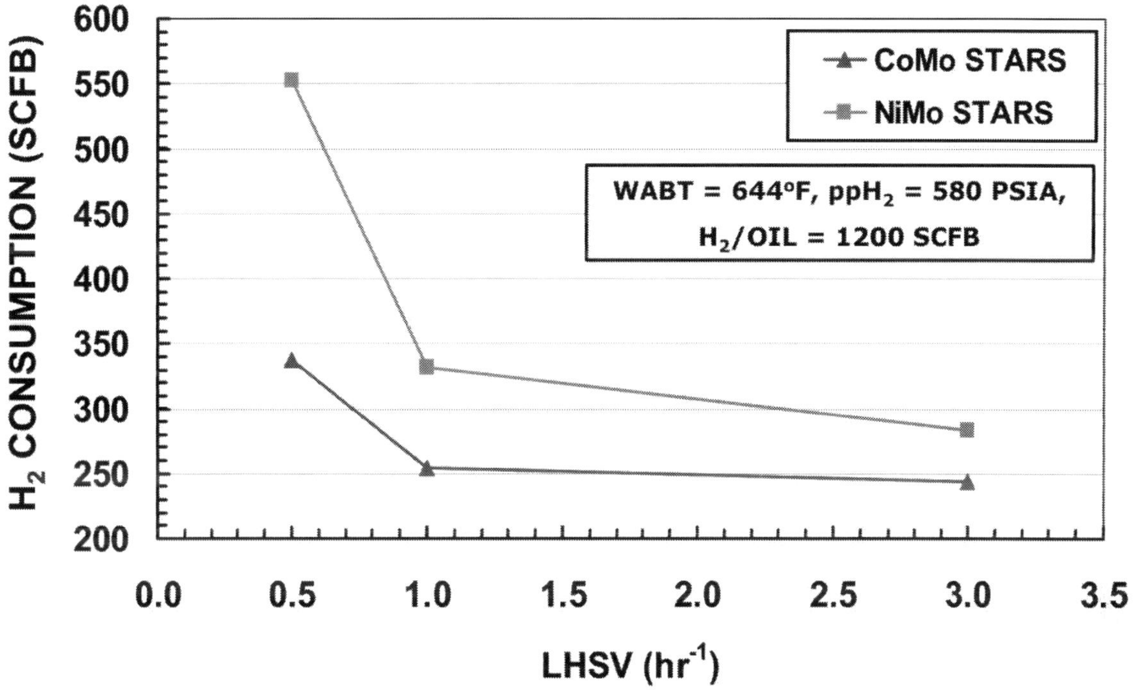

Product Property Study

A second series of experiments demonstrates the impact of catalyst selection and product S level on other product properties. These experiments were conducted using STARS™ CoMo and STARS™ NiMo catalysts in isothermal pilot plant equipment at Nippon Ketjen facilities in Niihama, Shikoku, Japan. The feed used was a SRD. The properties of the feed are shown in Table V and the process conditions employed are found in Table VI.

Figure 4 shows the relationships of product S and N and WABT. Values for WABT are represented as the difference in the Actual WABT and a Base WABT. As the WABT is increased to levels that result in product S < 400 - 500 wppm, STARS™ NiMo becomes more active for HDS than STARS™ CoMo, as observed in the process condition study. With the feed and process conditions used ULSD product S levels (≤ 10 wppm) are achieved with STARS™ NiMo at 9°F lower WABT than that required for STARS™ CoMo. Also, as observed in the process condition study, STARS™ NiMo becomes the more active catalyst when product nitrogen levels are very low, in this case < 10 wppm. Organic nitrogen inhibits HDS, HDN, and aromatics saturation reactions. Therefore, deep HDN results in low levels of nitrogen inhibitors being present on the catalyst surface and full utilization of catalyst intrinsic activity.[4]

ULSD can be achieved at lower temperatures with STARS™ NiMo, but activity is not the only consideration. Using STARS™ NiMo will have an impact on H_2 consumption and product properties. Figures 5 - 7 show H_2 consumption, product aromatics, and product color as a function of product S level, respectively.

Table V
Product Property Study: Feed Properties

Feed Source:	SRD
Sulfur (wt%)	1.46
Nitrogen (wppm)	182
API Gravity (°)	32.7
Br No. (g/100 g)	< 1
Saybolt Color	7
D-86 Distillation (°F)	
IBP	426
T–50	581
FBP	694
Aromatics (wt%)	
Mono-	21.0
Di-	15.5
Tri+ -	1.0
Total	*37.5*

Table VI
Product Property Study: Process Conditions

LHSV (hr^{-1})	1.0
ppH_2 (psia)	725
WABT (°F)	550 - 660
H_2/Oil (SCFB)	1800
Catalysts	STARS™ CoMo STARS™ NiMo

LHSV = Liquid Hourly Space Velocity
ppH_2 = Hydrogen Partial Pressure
WABT = Weighted Average Bed Temperature
H_2/Oil = Hydrogen-To-Oil Ratio
SCFB = Standard Cubic Feet/Barrel

Figure 5 shows that the H_2 consumption with STARS™ NiMo is higher at all product S levels, as expected. At the same level of HDS this additional H_2 consumption is due to higher aromatics saturation. With the feed processed and process conditions used, the H_2 consumption is ~ 10% higher with STARS™ NiMo than with STARS™ CoMo during ULSD production. This is lower than the differences usually observed with these two catalysts, which is typically around 25 - 30%. Based on the data in Figure 5, at product S = 8 wppm, the H_2 consumptions with both the STARS™ CoMo and STARS™ NiMo catalysts would be about 60 - 75 % higher than that observed with these catalysts at LSD conditions (product S = 400 - 500

Figure 4
Product Property Study
Relationship of Product S & N and WABT
STARS™ CoMo & STARS™ NiMo

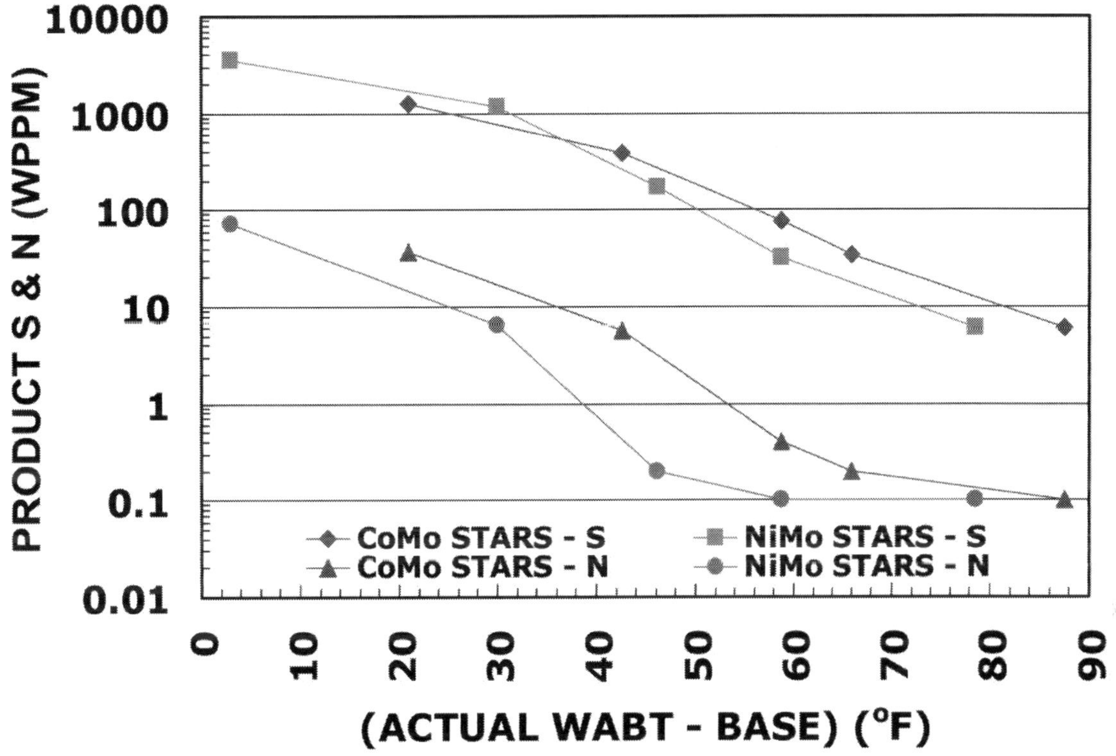

wppm). This illustrates the greatly increased H_2 consumption that is a critical fundamental difference between current diesel hydrotreating operations and future ULSD operations.

At product S = 8 wppm, the estimated H_2 Availabilities for the STARS™ CoMo and STARS™ NiMo catalysts were 3.7 and 3.3, respectively. At product S = 400 - 500 wppm, the H_2 Availabilities for the STARS™ CoMo and STARS™ NiMo catalysts were > 6 and > 5, respectively.

The additional H_2 consumption with STARS™ NiMo catalysts will be an extra cost, but does have potentially beneficial impacts on product properties. Figures 6 and 7 show the impact of STARS™ NiMo catalyst and the extra H_2 consumption associated with it on product aromatics and color, respectively, as functions of product S levels.

Figure 5
Product Property Study
Relationship H_2 Consumption and Product S
STARS™ CoMo & STARS™ NiMo

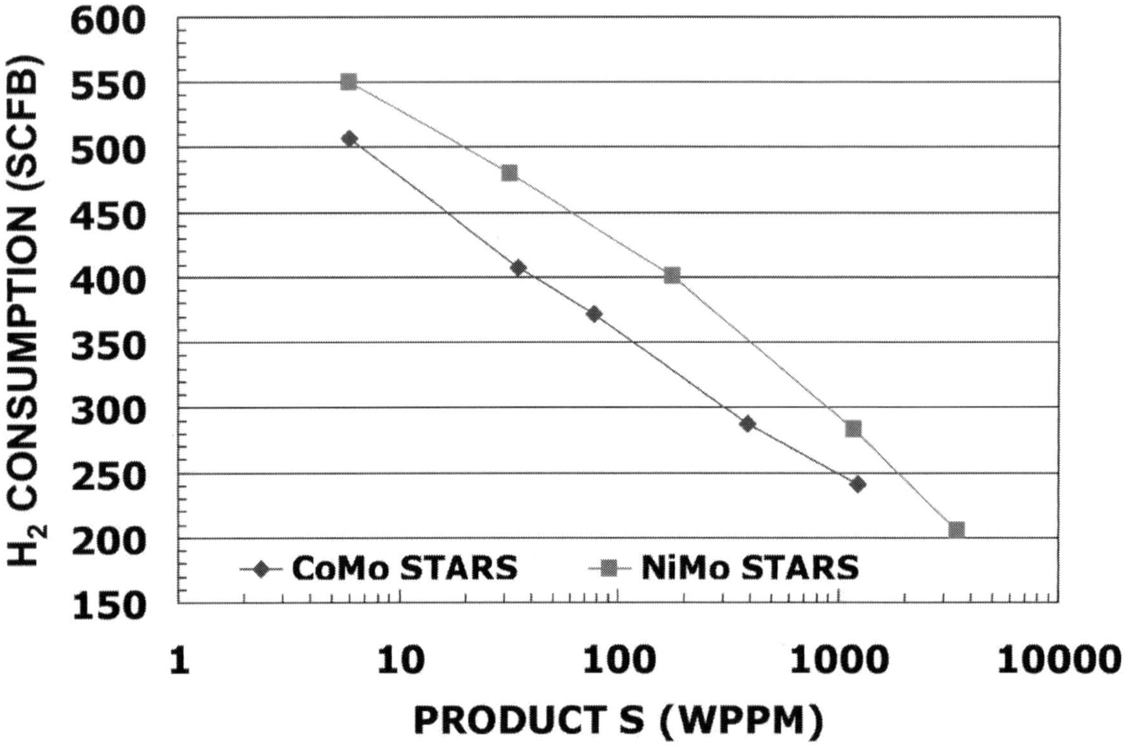

Figure 6 shows significantly lower products aromatics when using STARS™ NiMo catalyst. With both catalysts significant net reduction of total aromatics occurs at ULSD conditions: STARS™ CoMo - 36%, STARS™ NiMo - 44%. At process conditions where product S = 400 - 500 wppm no net reduction of total aromatics is observed with STARS™ CoMo catalyst and only a small net reduction of total aromatics is observed with STARS™ NiMo.

Product color can limit catalyst cycle length. Figure 7 shows the impact of catalyst selection on product color as product S is reduced. The degradation of color quality with increasing process severity (temperature) is typical.[5] At product S levels below current LSD requirements (400 - 500 wppm), STARS™ NiMo provides significant product color benefits. At ULSD conditions (product S < 10 wppm), STARS™ NiMo provides an improvement of 12 - 18 color numbers. STARS™ NiMo could be used as a means to help control color problems. Figure 7 also shows the extreme sensitivity of product color to product S level (operating temperature). Based on these data, increasing the operating severity to reduce product S from 10 to 6 wppm (a WABT increase of 5 - 6°F) is estimated to result in a ~ 4 color number reduction with STARS™ NiMo and an ~ 8 color number reduction with STARS™ CoMo. In cases where product color is an important property, overtreating will have significant negative consequences for product color properties in addition to increasing catalyst deactivation.

Figure 6
Product Property Study
Product Aromatics and Product S
STARS™ CoMo & STARS™ NiMo

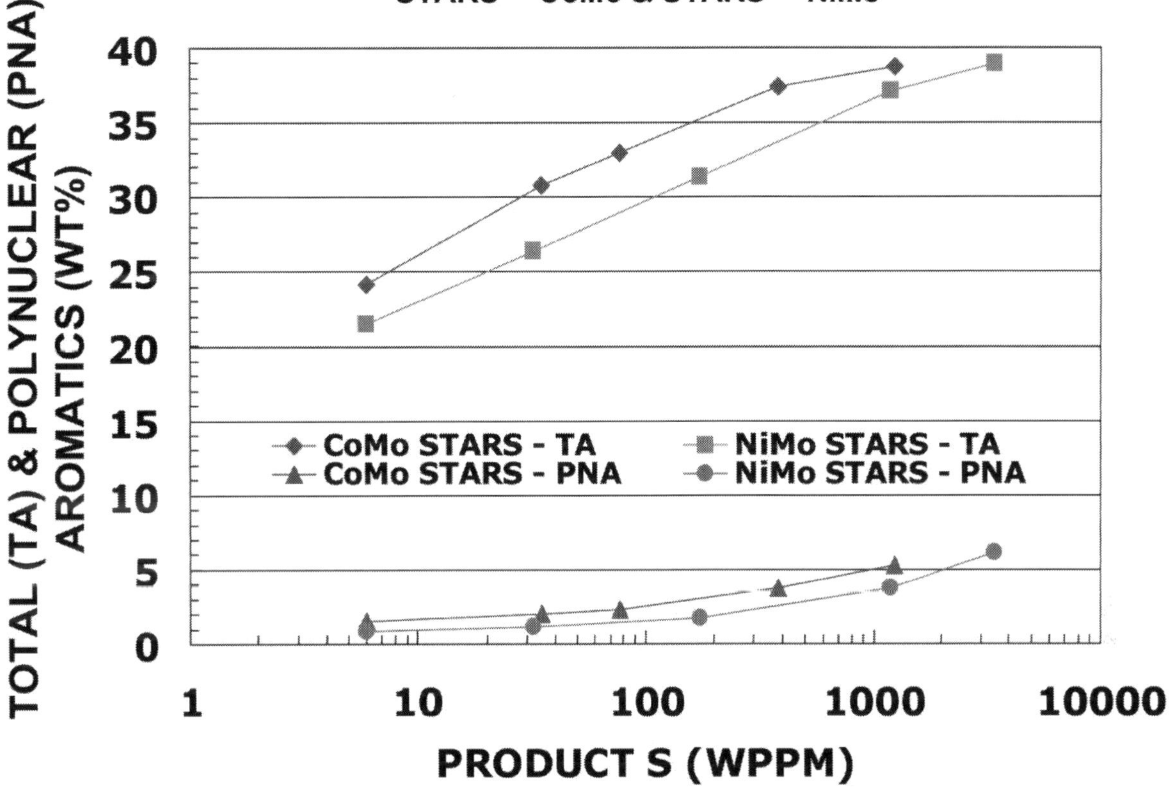

Figures 4 through 7 demonstrate that STARS™ NiMo can provide not only additional catalyst activity, but also product property improvement. The critical aspect of applying STARS™ NiMo catalyst is the need to ensure that the increased H_2 consumption occurring with it is compatible with the H_2 supply of the unit where it will be applied.

Application of high activity, high H_2 consumption catalysts like STARS™ NiMo in ULSD service can be managed by combining them with STARS™ CoMo catalysts to capture the activity benefits of STARS™ NiMo without exceeding reasonable H_2 consumption levels. In the following section a comparison of STARS™ CoMo and STARS™ CoMo/Nebula™ is made to demonstrate the benefits from using combined catalyst loads.

Figure 7
Product Property Study
Product Color and Product S
STARS™ CoMo & STARS™ NiMo

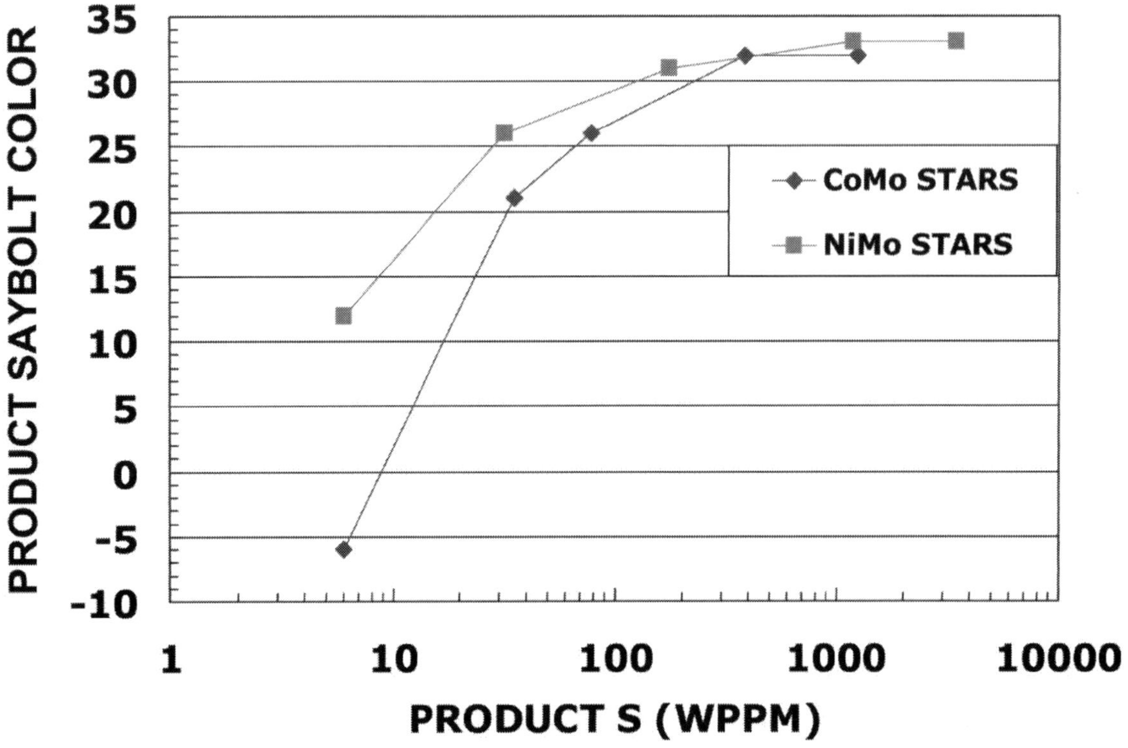

Catalyst System Optimization: STARS™ CoMo and STARS™ CoMo/Nebula™

Application of high activity catalysts in ULSD operations is an opportunity to increase cycle length and improve product properties. The high H_2 consumption associated with these catalysts makes it critical to ensure that H_2 consumption is balanced with H_2 supply to avoid excessive catalyst deactivation and stability problems. One approach to utilizing high-activity, high-H_2-consumption catalysts, like Nebula™, is to use them in combination with other catalysts. A STARS™ CoMo/Nebula™ combined catalyst system can significantly boost catalyst system activity while keeping H_2 consumption at a manageable level.

These experiments were conducted using STARS™ CoMo and STARS™ CoMo/Nebula™ catalyst systems in isothermal pilot plant equipment at Nippon Ketjen facilities in Niihama, Shikoku, Japan. The STARS™ CoMo/Nebula™ system was ~ 25% Nebula™. The feed used was a SRD. The properties of the feed are shown in Table VII and the process conditions employed are found in Table VIII. The feed in these experiments is the same as that used in the tests demonstrating the impact of STARS™ CoMo and STARS™ NiMo and product S level on product properties. Experiments were conducted by adjusting pilot plant WABT to maintain product S = 8 wppm.

Figure 8 shows the normalized WABT required to maintain product S = 8 wppm for the STARS™ CoMo and STARS™ CoMo/Nebula™ systems as a function of time on stream

(TOS). The normalized WABT data are presented as the difference in Normalized WABT required and a Base WABT, rather than as an absolute temperature. During the first 90 days of operation, the deactivation of the STARS™ CoMo and STARS™ CoMo/Nebula™ catalyst systems were both in the 4.5 - 5.0 °F/month range. At TOS > 90 days, both catalyst systems show a greatly reduced rate of deactivation, around 0.7 °F/month. In the stable operation regime (TOS > 90 days), the STARS™ CoMo/Nebula™ system was on average 27°F more

Table VII
Catalyst System Optimization: Feed Properties

Feed Source:	**SRD**
Sulfur (wt%)	**1.46**
Nitrogen (wppm)	**182**
API Gravity (°)	**32.7**
Br No. (g/100 g)	**< 1**
Saybolt Color	**7**
D-86 Distillation (°F)	
IBP	**426**
T–50	**581**
FBP	**694**
Aromatics (wt%)	
Mono-	**21.0**
Di-	**15.5**
Tri+ -	**1.0**
Total	***37.5***

Table VIII
Catalyst System Optimization: Process Conditions

LHSV (hr^{-1})	**0.9**
ppH_2 (psia)	**725**
WABT (°F)	**600 - 680**
H_2/Oil (SCFB)	**1800**
Product S (wppm)	**8**
Catalysts	**STARS™ CoMo** **STARS™ CoMo/Nebula™**

LHSV = Liquid Hourly Space Velocity
ppH_2 = Hydrogen Partial Pressure
WABT = Weighted Average Bed Temperature
H_2/Oil = Hydrogen-To-Oil Ratio
SCFB = Standard Cubic Feet/Barrel

active than STARS™ CoMo. This is a significant improvement in HDS activity from incorporation of only 25% Nebula into the catalyst load. At an average catalyst deactivation rate of 1.0°F/month, this is equivalent to 2+ years of additional cycle length. The STARS™ CoMo/Nebula™ catalyst system has equivalent stability to the STARS™ CoMo system. This demonstrates that with proper catalyst system design, significant activity improvements can be achieved while maintaining catalyst stability.

Figure 8
Catalyst System Optimization
Normalized ΔWABT VS. Time On Stream
STARS™ CoMo & STARS™ CoMo/Nebula™

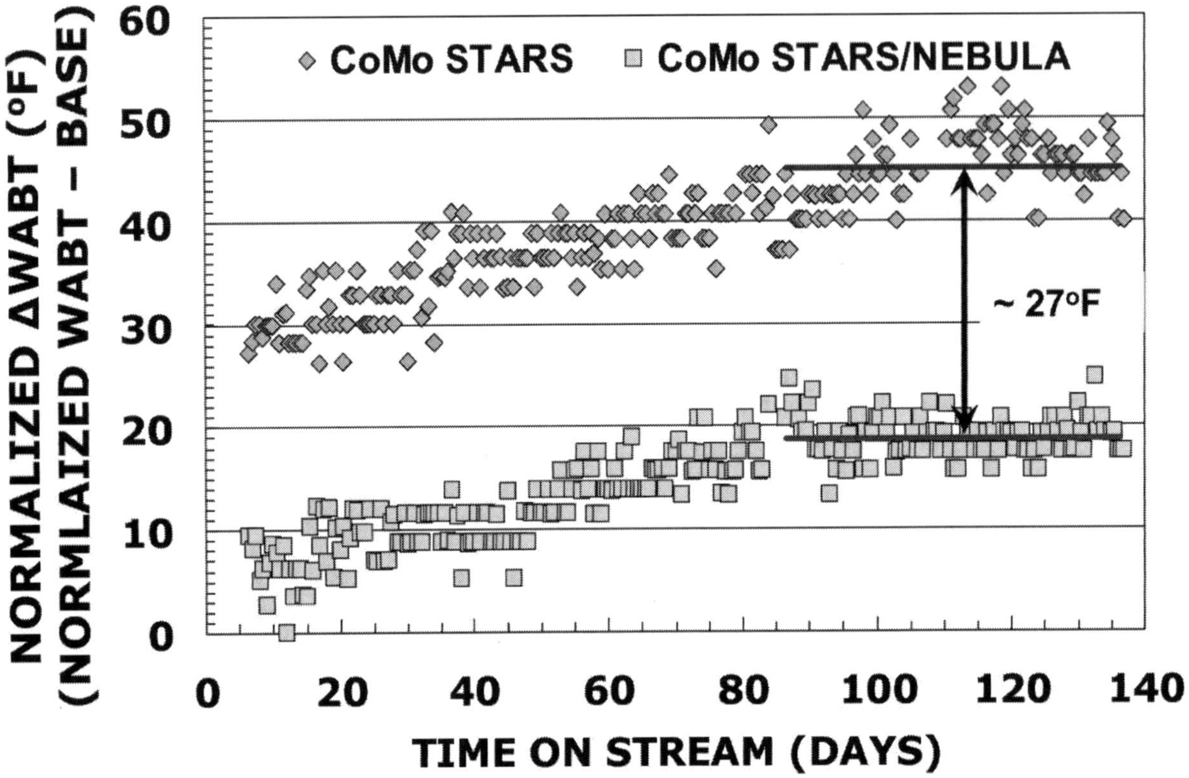

Figure 10 shows the H_2 consumption with TOS for the catalyst systems tested. At start of run (SOR) the H_2 consumption with the STARS™ CoMo/Nebula™ catalyst system was 14 - 16% higher than with the STARS™ CoMo. At TOS > 90 days, the STARS™ CoMo/Nebula™ catalyst system H_2 consumption was 9 - 10% higher. During stable operation (TOS > 90 days), the H_2 Availability with the STARS™ CoMo system was in the 3.8 - 3.9 range. For the STARS™ CoMo/Nebula™ catalyst system, H_2 Availability ranged from 3.5 - 3.6 during this period.

Table IX shows product aromatics, density, and color properties obtained with the two catalyst systems during the stable operation period (TOS > 90 days). During this period the product total aromatics and product polynuclear aromatics were approximately 10% lower and 30+% lower, respectively, with the STARS™ CoMo/Nebula™ catalyst system.

Figure 9
Catalyst System Optimization
H_2 Consumption VS. Time On Stream
STARS™ CoMo & STARS™ CoMo/Nebula™

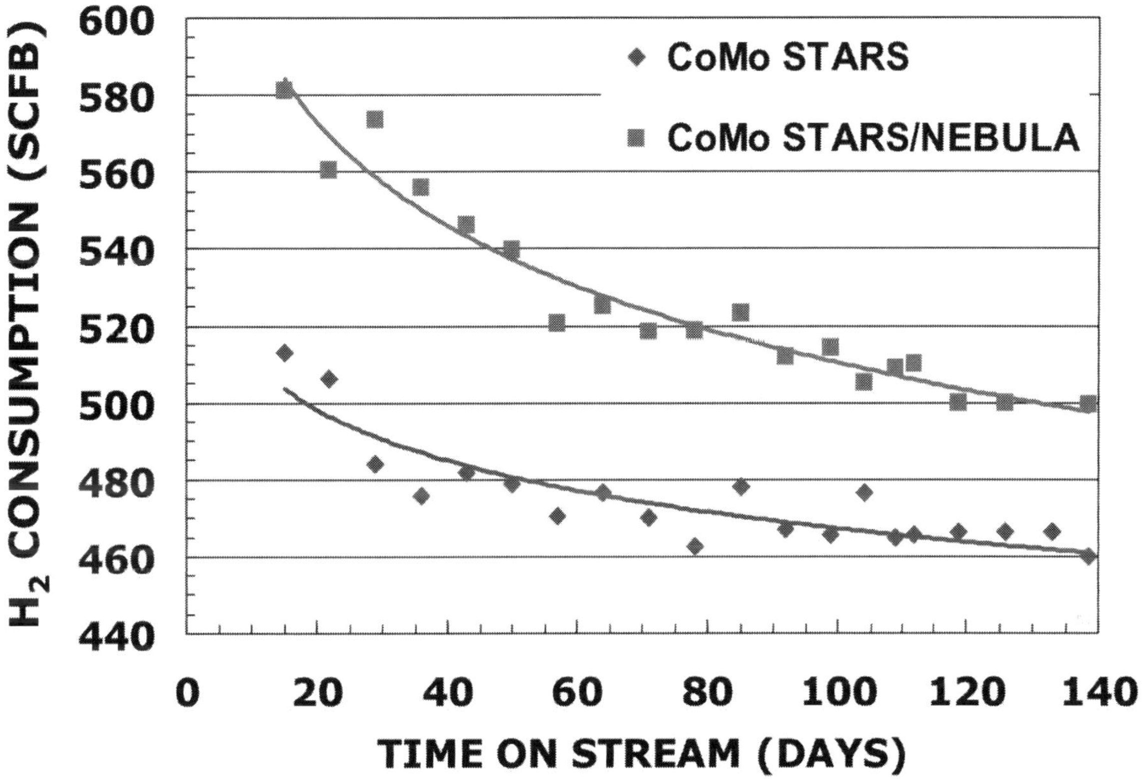

Table IX
Catalyst System Optimization: Stable Operation Product Properties

	Total Aromatics (wt%)	Polynuclear Aromatics (wt%)	Density (API Gravity °)	Saybolt Color (Color Number)
Feed	37.5	16.5	32.7	7
STARS™ CoMo	27.4	2.0	37.4	4
STARS™ CoMo/Nebula™	24.9	1.4	37.6	17

Also, the product from the STARS™ CoMo/Nebula™ system had a lower density and significantly better color properties than the STARS™ CoMo product.

The STARS™ CoMo/Nebula™ catalyst system was able to provide higher activity, improved product properties, and equivalent catalyst stability. This demonstrates that with proper catalyst system design significant catalyst activity improvements can be achieved with no adverse impact on catalyst stability.

Commercial Example: Impact of Catalyst Selection/H_2 Consumption on Catalyst Stability

Commercial diesel hydrotreating operations have fixed H_2 supply capabilities. This makes it critical to properly select catalyst systems to ensure that the H_2 consumption in the unit is compatible with the H_2 supply. A commercial diesel hydrotreating unit illustrating the importance of this is discussed in this section. This unit originally used a catalyst system consisting of STARS™ NiMo/Nebula™. In the subsequent catalyst cycle a STARS™ CoMo/Nebula™ catalyst system was used to better manage H_2 consumption. In both systems the ratio of STARS™ Catalyst/Nebula™ was the same. The feed properties and process conditions are shown in Tables X and XI, respectively.

Table X
Commercial Example: Feed Properties

Feed Source:	**95/5 Vol/Vol SRD/LCO**
Sulfur (wt%)	**0.70**
Nitrogen (wppm)	**150**
API Gravity (°)	**34.1**
Br No. (g/100 g)	**< 1**
D-86 Distillation (°F)	
IBP	**297**
T–50	**558**
FBP	**694**

Table XI
Commercial Example: Process Conditions

LHSV (hr^{-1})	**2.0**
ppH_2 (psia)	**750**
WABT (°F)	**640 - 765**
H_2/Oil (SCFB)	**1200**
Product S (wppm)	**10**
Catalysts	**STARS™ CoMo/ Nebula™** **STARS™ NiMo/ Nebula™**

LHSV = Liquid Hourly Space Velocity
ppH_2 = Hydrogen Partial Pressure
WABT = Weighted Average Bed Temperature
H_2/Oil = Hydrogen-To-Oil Ratio
SCFB = Standard Cubic Feet/Barrel

In Figure 10, the Normalized WABT as a function of TOS data from the two catalyst cycles are shown. Here it is seen that the rate of deactivation of the STARS™ NiMo/Nebula™ catalyst system is ~ 3.5 times higher than that of the STARS™ CoMo/Nebula™ catalyst system. This can be attributed to the 20 - 25% higher H_2 consumption with the STARS™ NiMo/Nebula™. This is expected given the higher aromatic saturation activity of STARS™ NiMo as compared with STARS™ CoMo and the identical STARS™/Nebula™ ratios used in both catalyst systems. Also, in Figure 10, it is seen that the initial activity of the STARS™ NiMo/Nebula™ catalyst system is significantly higher (lower SOR WABT required) than the STARS™ CoMo/Nebula™. However, this higher activity cannot be utilized because of the rapid rate of deactivation with the STARS™ NiMo/Nebula™ system. These data are also an illustration of the potential danger in basing catalyst system comparisons and catalyst selections on initial or SOR activity/performance without accounting for the impact of H_2 consumption on catalyst stability. Clearly the STARS™ NiMo/Nebula™ system had SOR higher activity (lower temperature requirement). However, a longer catalyst cycle could be achieved with the STARS™ CoMo/Nebula™ system because of its greater stability.

Figure 10
Commercial Example
Impact of Catalyst System on Catalyst Stability

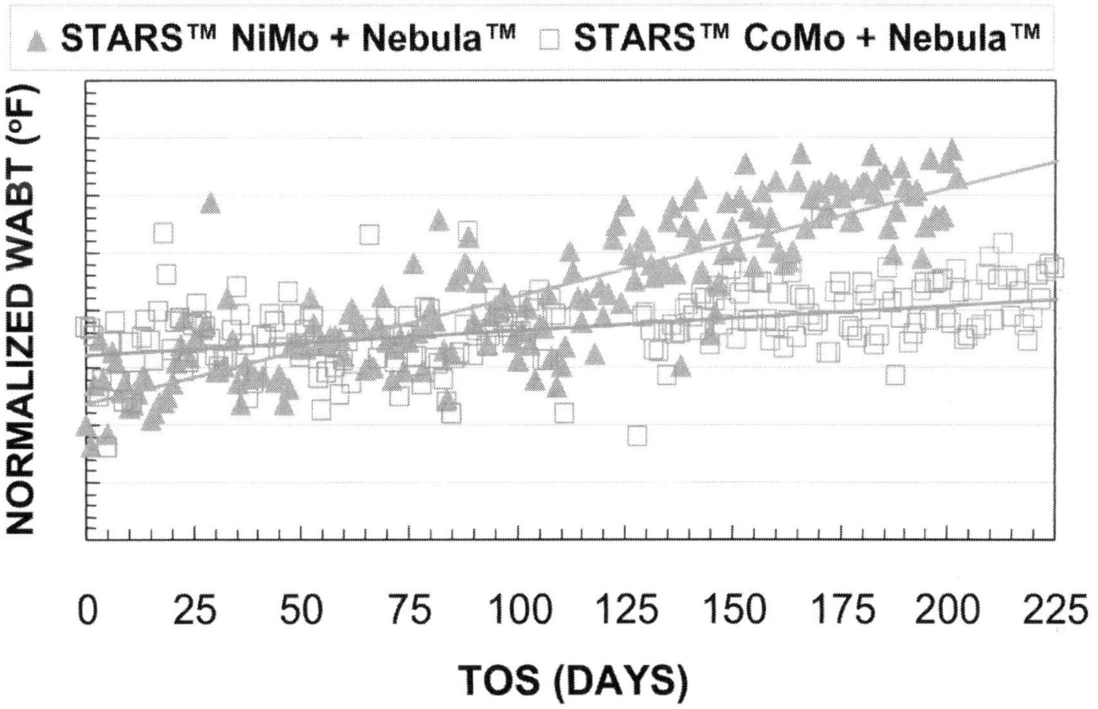

This commercial example clearly shows the critical impact of catalyst system design and selection on H_2 consumption/H_2 supply balance and catalyst cycle length. Also, this example illustrates that simply applying higher catalyst activity does not ensure an increase in catalyst cycle length.

ULSD Catalyst Selection Guidelines

The following section discusses general guidelines for ULSD catalyst selection/application. Every catalyst application should be evaluated individually in order to account for unique feed properties, unit capabilities, constraints, etc.

CoMo: CoMo catalysts are preferred when reduced H_2 consumption is an objective. This makes CoMo the best catalyst choice for units constrained by limited H_2 supply and requiring closely managed H_2 consumption for good catalyst stability/cycle length. CoMo is the preferred application for low ppH_2 units that cannot utilize the high aromatics saturation activity of NiMo and Nebula™ catalysts. For large volume, low space velocity reactors CoMo catalysts can be applied to manage/control H_2 consumption. When processing feeds with low SH-DBT's and organic nitrogen CoMo catalysts are the preferred catalyst choice because of reduced aromatics saturation and HDN requirements. This makes CoMo generally the preferred application for straight run feeds. The reduced H_2 consumption with CoMo also can be used as a means to avoid excessive H_2 costs and manage/avoid product quality giveaway (e.g. cetane giveaway) in high pressure units.

NiMo: The best applications for NiMo catalysts are units able to utilize and benefit from high aromatics saturation and HDN activity. These include units with moderate to high pressure (reactor outlet $ppH_2 > 600$ psia), high H_2 Availability, and feeds containing high levels of SH-DBT's and organic nitrogen. This makes NiMo a good application with cracked stocks, if the high H_2 consumption with these feeds can be managed. NiMo is also the preferred catalyst when maximizing aromatics saturation, density reduction, and cetane improvement are economically favorable. NiMo can be used to help control product color problems. NiMo catalysts show high H_2 consumption in ULSD service. It is critical when applying NiMo catalysts that the H_2 consumption in the unit is in balance with the H_2 supply available. This will be especially important with cracked stocks.

NiMo/CoMo: Combined NiMo/CoMo catalyst loads are a means to optimize/manage H_2 consumption. This type of catalyst system can be used to increase catalyst activity above that of a CoMo system while keeping H_2 consumption at an acceptable level. The relative amounts of CoMo and NiMo can be adjusted to balance the H_2 consumption and the H_2 supply available and ensure good catalyst stability. These combined loads should be used in moderate to high pressure (reactor outlet $ppH_2 > 600$ psia) units in order to capture the aromatic saturation and HDN activity advantages of the NiMo catalyst in the load.

Nebula™: Nebula™ is typically used in combination with STARS™ catalysts to manage H_2 consumption. Combined STARS™/Nebula™ systems offer higher activity than STARS™ catalysts alone with manageable increases in H_2 consumption. The application guidelines for Nebula™ are similar to those for NiMo. Nebula™ should be applied in units with moderate to high pressure (reactor outlet $ppH_2 > 600$ psia) and with sufficient H_2 supply to balance H_2 consumption. Nebula™ is effective for processing feeds with high levels of SH-DBT's and organic nitrogen such as cracked feedstocks. Processing cracked feedstock with Nebula™ will result in high H_2 consumption, which must be accounted for when designing catalyst systems with it. Nebula™ is effective for aromatics reduction, density reduction, cetane improvement, and color improvement. As with NiMo catalysts, the high potential H_2 consumption with Nebula™ makes it critical to carefully balance unit H_2 supply capabilities and H_2 consumption when applying it in ULSD operations.

Summary

In ULSD operations, H_2 consumption will be much higher than in current LSD operations. Conversion of 4,6-dimethyl-dibenzothiophene and related species is necessary to produce ULSD and is favored by high aromatics saturation activity. Increasing catalyst activity for ULSD production is accompanied by increasing H_2 consumption. Proper catalyst selection/application is critical for ULSD units. Catalyst systems must be designed to balance H_2 consumption and H_2 supply to ensure good catalyst stability. Combined catalyst systems like STARS™ CoMo/Nebula™ can significantly increase catalyst system activity and improve product properties while maintaining H_2 consumption at acceptable levels.

References

(1) Topsøe, H., B.S. Clausen, and F.E. Massoth, *Hydrotreating Catalysis, Science and Technology*, Springer-Verlag, Berlin (1996)

(2) a) Ma, X., K. Sakanishi, and I. Mochida, *Ind. Eng. Chem. Res.* **33** (1994) 218

b) Shafi, R. and G. J. Hutchings, *Catalysis Today* **59** (2000) 423

c) Steiner, P. and E.A. Blekkan, *Fuel Processing Technology* **79** (2002) 1

d) Ring, Z., J. Chen, H. Yang, H. Du, and Y. Briker, Proceeding 2004 AIChE National Spring Meeting, Paper 92a, (2004)

(3) Whitehurst, D.D., T. Isoda, and I. Mochida, *Advances in Catalysis* **42** (1998) 345

(4) a) Kwak, C., J.J. Lee, J.S. Bae, and S.H. Moon, *Applied Catalysis B: Environmental* **35** (2001) 59

b) Laredo,G.C., J.A. De Los Reyes, J.L. Cano, J.J. Castillo, *Applied Catalysis A: General* **207** (2001) 103

c) Koltai, T., M. Macaud, A. Guevara, E. Schulz, M. Lemaire, R. Bacaud, and M. Vrinat, *Applied Catalyisis A: General* **231** (2002) 253

d) Egorova, M. and R. Prins, *Journal of Catalysis* **224** (2004) 278

e) Yang, H., J. Chen, C. Fairbridge, Y. Broker, Y.J. Zhu, and Z. Ring, *Fuel Processing Technology* **85** (2004) 1415

f) Rabarihoela,-Rakotovao, V., S. Brunet, G. Berhault, G. Perot, and F. Diehl, *Applied Catalysis A: General* **267** (2004) 17

(5) a) Takatsuka, T., Y. Wada, H. Suzuki, S. Komatsu, and Y. Morimura, *Sekiyu Gakkaishi* **35(2)** (1992) 179

b) Ushio, M., M. Hatayama, T. Waku, M. Akiyama, Y. Okamoto, and A. Inoue, *Proceedings ACS Division of Petroleum Chemistry, 210th National Meeting* (1995) 537

c) Sakanishi, K., M. Ando, S. Abe, and I. Mochida, *Sekiyu Gakkaishi* **34(6)** (1991) 553

90a Comparision of Existing and Emerging Alkylation Technology

Cal Hodge, A 2nd Opinion, Inc., 19 Serenade Pines Place, The Woodlands, TX 77382
This presentation is an update of a similar presentation (AM0413 that was presented at the 2004 NPRA Annual Meeting. Alkylation technology vendors have been sent a copy of that paper with a request for comments on and updates of their technologies as reported in 2004.

The presentation explains why more alkylate is needed and then compares alkylation technologies.

The AlkyClean® Process –
Demonstrated New Standard for Alkylation Technology

Vincent J. D'Amico and John C. Gieseman, ABB Lummus Global, USA
Emiel H. van Broekhoven, Albemarle Catalysts, The Netherlands

Introduction

A reliable and cost competitive alternative to existing liquid acid alkylation technologies has been demonstrated. Based on over two years of successful operation of a prototype, refinery slipstream unit, the new AlkyClean process, employing an environmentally friendly zeolite catalyst, provides refiners with a cleaner and inherently safer alkylation technology.

The new solid acid catalyst (SAC) process produces high quality alkylate without all the drawbacks of the existing hydrofluoric (HF) and sulfuric (H_2SO_4) acid based technologies. Neither acid soluble oils (ASO) nor spent acids are produced, and there is no need for product post-treatment of any kind. Besides these processing advantages, eliminating the use of these toxic and corrosive liquid acids greatly reduces maintenance and monitoring requirements while reducing environmental and personnel safety concerns. Benchmarking efforts have confirmed the overall competitiveness of the new technology.

Alkylate, the product of the reaction of isobutane with light olefins (C_3-C_5), is highly valued as an ideal "clean fuels" blending component in the gasoline pool because it has no olefins or aromatic compounds, a low sulfur content, a limited heavy end, a low vapor pressure and both high research and motor octane numbers. As MTBE use further diminishes, alkylate will continue its growing importance as a means to compensate for lost octane-barrels and to recapture isobutylene back into the gasoline pool. Relative to catalytic polymerization (cat-poly), economic evaluation indicates that replacing an existing cat-poly unit with the AlkyClean process will result in a very attractive payout, which can be less than one year.

Process flow scheme

The process flow scheme for the AlkyClean process is similar to that employed for current liquid acid technologies. As illustrated in Figure 1, the process consists of four main sections: feedstock pretreatment, reactor system, catalyst regeneration, and product distillation.

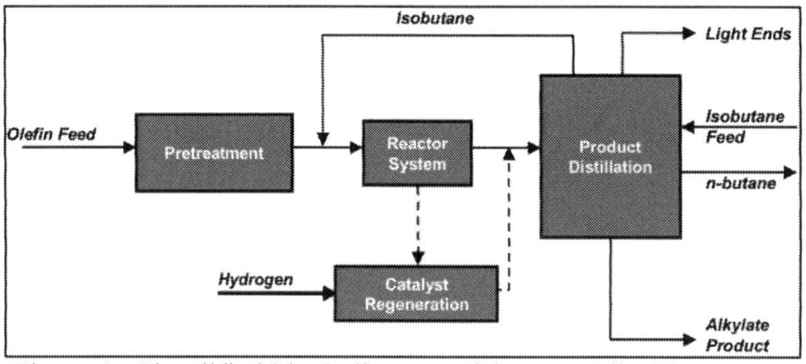

Figure 1. Simplified block diagram of the new AlkyClean process

Feedstock pretreatment requirements will depend on contaminant levels, but are generally expected to be no greater than those prudent for the liquid acids. The zeolite catalyst utilized in the process is rugged, with a good tolerance to typical impurities. It is easily regenerated to its full activity in the event of an accidental exposure to high levels of contaminants.

Table 1 compares the process operating conditions for the AlkyClean process and the liquid acid technologies. The AlkyClean reactor system operates in the liquid phase at higher temperatures, in the range of 50-90°C (122-194°F), thereby, eliminating the costly refrigeration requirements associated with H_2SO_4 processes. To achieve a high octane alkylate and limit by-product production, H_2SO_4 units typically utilize a total reaction section (external) isobutene-to-olefin feed ratio (I/O) of between 8/1 and 10/1, while HF units optimally run at an I/O of between 12/1 and 15/1. With similar quality alkylate product, the I/O for the new SAC process is in the same range of that utilized for these liquid acid processes. The ability to operate the AlkyClean process at comparable I/O is important in two ways. First, it allows for a cost competitive process, as the fractionation requirements associated with isobutane recycle are major capital investment and operating cost components. Second, it facilitates incorporation of the technology in the revamp/de-bottlenecking of an existing liquid acid unit, potentially without major modification to the 'back-end' fractionation/recycle facilities.

Table 1
Operating conditions comparison

	AlkyClean Process	**H_2SO_4**	**HF**
Operating temperature	50-90°C (122-194°F)	4-10°C (39-50°F)	32-38°C (90-100°F)
External I/O	8-15/1	8-10/1	12-15/1

Catalyst

Central to the new technology is the utilization of a novel, "true" SAC. In this situation, "true" SAC means that the catalytic acid function is intrinsic to the solid itself rather than being a separate component, such as an immobilized liquid deposited on a solid substrate. The proprietary formulation is zeolite based. It contains no halogens, has acid sites with sufficient strength for alkylation, yields high quality alkylate with good selectivity, and exhibits the required activity, stability and regenerability characteristics necessary for a successful process. Most recently, this catalyst has been further optimized and successfully manufactured in commercial-scale trials. This optimization, a result of novel discoveries, has significantly increased catalyst activity.

Reactor system and cyclic operation with integrated catalyst regeneration

The AlkyClean process achieves superior performance by coupling its optimized SAC with an alkylation reactor system that efficiently minimizes the peak olefin concentration in the reaction zone (i.e., maximizes the internal I/O). This is accomplished by utilizing serial reaction stages and a unique, easily operable reactor design that achieves the operating conditions essential to attain high product quality and prevent rapid catalyst deactivation.

In the AlkyClean process, multiple reactors are used to allow for continuous alkylate production as individual reactors cycle back and forth between on-line alkylation and rejuvenation following an inventive procedure established during the process development effort.[1] During rejuvenation, olefin addition is stopped and hydrogen is added to achieve a low reactor concentration of dissolved hydrogen, while maintaining liquid phase alkylation reaction conditions. This allows for a seamless switchover between operations and minimizes energy consumption. The rejuvenation step serves to remove accumulating heavier molecular weight species that lead to catalyst pore plugging and deactivation. Over time, however, there is a gradual loss of catalyst activity. To fully recover this activity – with a frequency depending on the operating severity – a reactor is taken off-stream for a moderate temperature (250°C/482°F), vapor phase regeneration with hydrogen. To allow for this hydrogen strip/moderate temperature regeneration (MTR) operation while maintaining full continuous alkylate production, an additional swing reactor is provided.

This cyclic operation of the reactor section is depicted in the Figure 2. The process does not require any transfer of catalyst, either between reactor stages or to a separate regeneration vessel, thereby enhancing operability. The use of a swing reactor provides for additional maintenance flexibility and allows the unit to stay on-stream when catalyst replacement eventually becomes necessary after years of operation. At the end of its useful life, the catalyst is returnable to its manufacturer, eliminating any potential catalyst disposal problem for the refiner.

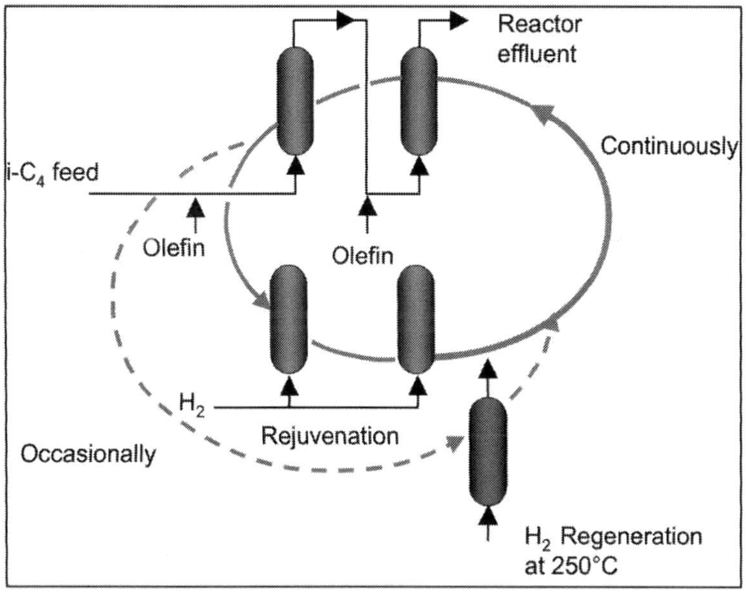

Figure 2: AlkyClean process reactor operating scheme

Demonstration Unit

Construction of a 10 BPSD AlkyClean process demonstration unit in Porvoo, Finland, was completed in 2002. Figure 3 shows the process flow schematic of the demonstration unit, which contains all of the key elements of a commercial plant. Three reactors are included, two under cyclic operation (i.e., alternating between alkylation and rejuvenation) allow for continuous production of alkylate, and the third allows for swing reactor MTR.

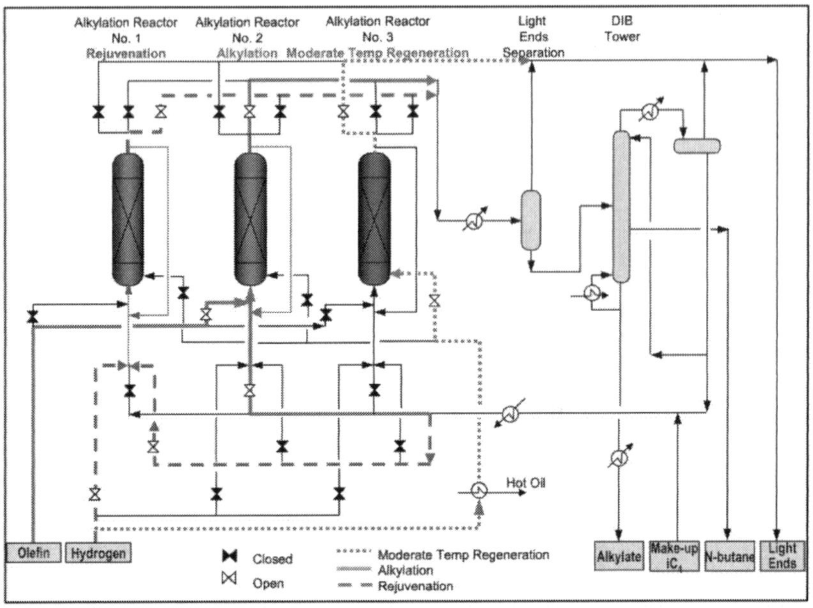

Figure 3: Process flow schematic of the demo unit

Figure 4 is a photograph of the installed reactor section. The reactors are sufficiently large and proportioned to allow for reliable scale-up. As such, each reactor represents a "core" of a much larger reactor and provides for the necessary hydrodynamic similarity to a full commercial-scale reactor system. Equally important to note, these reactors utilize SAC produced in commercial manufacturing trials, ensuring the exact catalyst characteristics of commercial-scale production conditions. Figure 5 provides an outside view of the facility, including the back-end deisobutanizer and alkylate product off-take trailer.

Figure 4: Demo reactor section

Figure 5: Demo unit outside view

Demonstration Unit Operation

After mechanical completion, the unit went through a shakeout and start-up period of about one month. During this period, procedures were refined and proven for the *in situ* activation of the catalyst and the reliable start-up of the reactor section. Subsequently, the unit operated reliably on a near continuous basis over several months utilizing piped slipstreams of the actual feeds processed in the nearby Porvoo Refinery HF alkylation unit, and produced alkylate of comparable high quality.

In addition to proving the operability of the process, key aspects of the technology were demonstrated. First and foremost was continuous cyclic operation – alternating reactors between periods of alkylation and rejuvenation – for periods of up to four weeks before taking a reactor off-line for MTR. In doing so, the durability of the catalyst was demonstrated over hundreds of cycles of rejuvenation and multiple MTRs. Regenerated catalyst samples from the unit were tested in a bench scale unit under benchmark conditions, further confirming the ability to repeatedly regenerate the catalyst and re-establish fresh catalyst activity and performance. After obtaining performance data over a wide range of conditions, which supported the identified objectives, the unit operation was temporarily suspended, allowing for unit modifications to incorporate identified operational improvements. During this hiatus, a newly improved catalyst was manufactured in a commercial scale trial and loaded into the reactors. Following these changes, the demonstration unit was restarted and ran smoothly and continuously for about six months under full cyclic operation, with periodic rotational MTR of the reactors. During this period, the benefits of the operational improvements were verified and the improved activity of the optimized AlkyClean catalyst was confirmed, along with stability and full activity recovery over multiple MTRs. Representative performance data from this latter operating period while processing refinery C_4 olefins are presented in Figures 6-8 for alkylate RON, RVP and C_5+ yield, respectively. This data, based on automated sampling from the on-line analytical system, show both the high product quality achieved and the stability of the cyclic process over time.

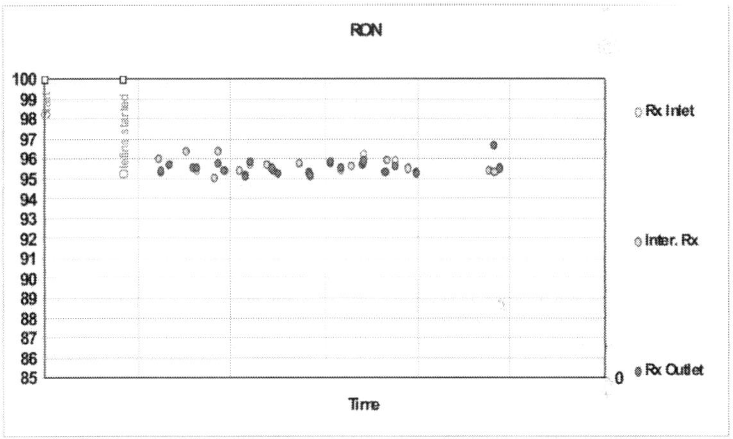

Figure 6: Demo unit performance processing refinery C_4 olefins, RON

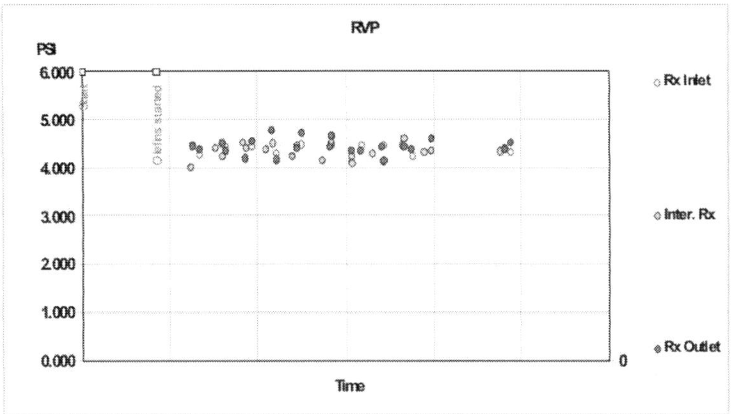

Figure 7: Demo unit performance processing refinery C_4 olefins, RVP

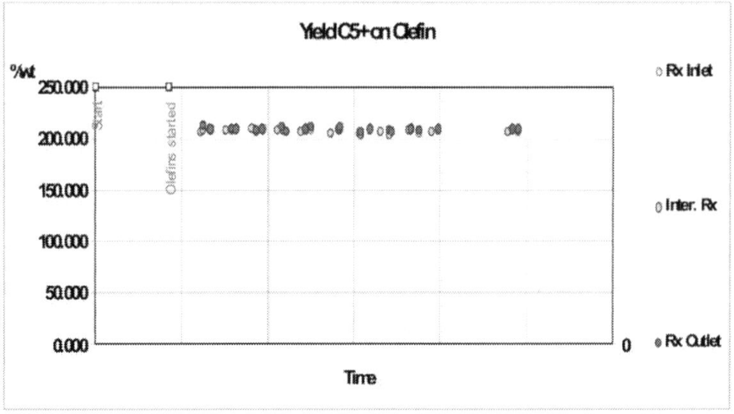

Figure 8: Demo unit performance processing refinery C_4 olefins, C_5+ yield on olefin

Competitiveness vs. liquid acid technologies

Table 2 provides an overview summary of the competitiveness of the AlkyClean process versus the established liquid acid technologies. The performance and economics of the AlkyClean process are fully competitive with current liquid acid technologies. High quality product has been reliably produced in an operation that has proven to be reliable and robust. Sensitivity to feedstock variation is low and tolerance to impurities is high. The economic competitiveness of the new SAC process is enhanced by its low mechanical complexity and the use of common (i.e. non-proprietary) refinery process equipment. The low pressure and mild temperatures employed, along with the absence of either a corrosive or erosive environment, allow for the use of carbon steel construction.

Based on an in-house benchmarking effort, we estimate that the total installed cost (TIC) for an AlkyClean process unit is about 10-15% lower than that for the equivalent H_2SO_4 unit. This comparison excludes off-site costs. When acid regeneration facilities for the H_2SO_4 unit are included, the SAC unit's cost becomes substantially less. We project that this level of investment for the SAC process is about on par with the cost for a HF unit. With respect to total cost of production, the results indicate that the requirements for the new AlkyClean process are comparable to that of the H_2SO_4 process. For both technologies, catalyst consumption cost is a significant component of the variable cost. In the case of the AlkyClean process, this cost was conservatively based on a minimum ultimate catalyst life. Thus, there is considerable upside potential, which would result in the reduction of the AlkyClean process's total production cost. In comparison to the AlkyClean and H_2SO_4 processes, production costs for HF units may be judged, on the surface, to still be somewhat lower; however, the increased costs for maintenance, mitigation, and monitoring, among others, that the HF technology requires offset any perceived advantage.

Table 2
Comparison of AlkyClean process with liquid acid technologies

Parameter	Modern Sulfuric Acid	Modern HF Acid	AlkyClean Process
Base Conditions:			
C_4= feedstock			
Product RON	95	95	95
Product MON	91.5	92.5	92.5
Alkylate Yield	Base	Base	Base or better
Total Installed Cost, ISBL	Base	85% of Base	85% of Base
Total Installed Cost Including OSBL (regeneration facilities, and/or safety installations)	Base	70% of Base	< 50% of Base
Feed Treatment	Base	Higher	Base
Product Treatment	Yes	Yes	No
ASO Yield	Base	Less	None
Equipment Maintenance	High	High	Very Low
Corrosion Problems	Yes	Yes	None
Reliability and On Stream Factor	Average	Average	Expected Above Average / High
Turnarounds Frequency / Duration	Varies / Longer	Varies / Longer	Match FCC or better / Shorter
OPEX	Base	Site Specific Typically Lower	Slightly Higher
Safety	Unit specific safety precautions as well as transport precautions	Very specific safety precautions required that extend throughout refinery	No special precautions other than those for any refinery process unit. Inert catalyst.
Environmental	-	-	No emissions to air, water, or ground.

Economic incentive for employing the AlkyClean process to replace cat-poly

Evaluation of the economics for replacing a cat-poly (or oligimerization) unit (and by extension other similar "indirect alkylation" units) with the AlkyClean process indicates a clear incentive for the installation of the SAC alkylation technology. The basis for this analysis was a typical C_4 FCC feed cut supporting the production of 10,000 BPSD of alkylate, with the corresponding production of cat-poly gasoline (at the same unit feed rate) falling out at 4,623 BPSD. Overall unit flow rates for cat-poly and the AlkyClean process are provided in the block flow diagrams of Figures 9 and 10, respectively. Two economic cases were developed based on historical U.S. Gulf Coast product pricing information derived from CMAI and Platt's. The first was a four-year average (spanning 2001-2005) and the second was for the single month at the end of this four year period, July 2005. For reference, this pricing information is summarized in Table 3.

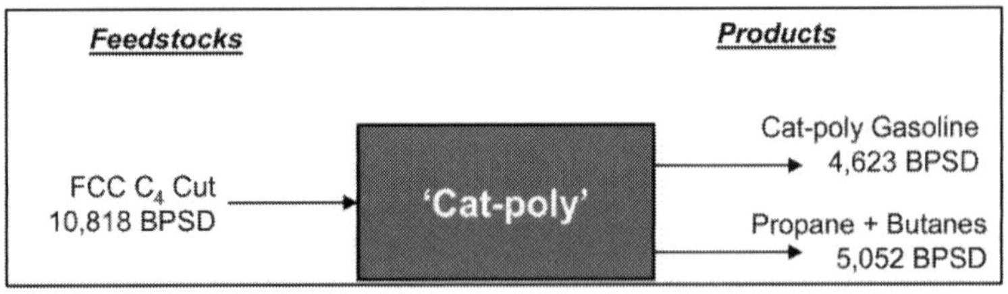

Figure 9: Cat-poly block flow diagram

Figure 10: AlkyClean process block flow diagram

Table 3
Historical pricing information

	4 Year Average Case	Single Month Case (7/05)
	U.S. cents/gallon	
Propane	59.38	83.78
Iso-Butane	73.25	117.51
N-Butane	70.94	101.77
'Cat-poly'	99.49	166.02
Alkylate	111.99	186.24

The results of the evaluation for these two cases (i.e. four-year average and single month) are then summarized in Tables 4 and 5, respectively. In both cases, product slate values for the two processes are calculated net of operating costs and, additionally for the SAC alkylation process, net of supplemental isobutane make-up requirements. The net product slate value differential for the AlkyClean process relative to cat-poly was then calculated on a daily basis and converted to a yearly differential utilizing an on-stream factor of 350 operating days per annum. Finally, the simple payout period for the AlkyClean process investment was calculated based on the unit's in-house estimated total installed cost (TIC) of US$42 million ($4^{th}$ qtr. 2005, USGC location). Based on these calculations, the payouts in both cases are attractive. For the averaged four-year period, the payout is 1.7 years, and based on the single month period of July 2005, the payout is extremely low at only 0.9 years. The reduced payout time for the July 2005 period is not surprising, as it reflects last year's run-up in crude oil prices, which approached US$60 per barrel in July, relative to an average price of about US$35 per barrel for the four year period spanning 2001-2005. Finally, it should be noted that this analysis does not account for the blending advantage that alkylate offers relative to the highly olefinic cat-poly blendstock in terms of ability to produce additional volumes of RFG and CARB gasoline.

Table 4
AlkyClean vs. cat-poly – Four-year-average case

	Cat-poly	AlkyClean
	U.S. $/day	
Gasoline Blendstock	193,176	470,405
$C_3 + C_4$ By-products	152,232	51,561
Isobutane Make-up	-	(88,726)
Operating Cost	(16,002)	(34,213)
Net Product Slate Value	329,406	399,027
AlkyClean Differential	-	69,621
Annual AlkyClean Differential, MM$/yr.		24.4
AlkyClean TIC, MM$		42.0
AlkyClean Payout, yrs.		1.7

Table 5
AlkyClean vs. cat-poly – Single-month (7/05) case

	Cat-poly	AlkyClean
	U.S. $/day	
Gasoline Blendstock	322,354	782,286
$C_3 + C_4$ By-products	231,151	73,884
Isobutane Make-up	-	(142,338)
Operating Cost	(18,288)	(39,100)
Net Product Slate Value	535,217	674,732
AlkyClean Differential	-	139,515
Annual AlkyClean Differential, MM$/yr.		48.8
AlkyClean TIC, MM$		42.0
AlkyClean Payout, yrs.		0.9

Conclusions

The new AlkyClean solid acid catalyst alkylation process is now available to refiners and provides the following benefits:

- Elimination of corrosive / toxic liquid acid usage and associated safety concerns
- Elimination of heavy hydrocarbon by-product (ASO) production
- Elimination of reactor refrigeration and alloy construction material requirements
- High operating reliability and low maintenance costs
- Good tolerance to feedstock impurities
- Reduced sensitivity to changes in feedstock olefin composition
- Production of quality, high octane alkylate product
- Competitive plant investment and production costs
- Equipment reuse potential for more economical retrofitting/de-bottlenecking of existing liquid acid units
- Attractive payout potential for both new applications and as a replacement to catalytic polymerization units

This new "clean fuels" technology has successfully completed its demonstration phase. Spanning a period of more than two years, the many months of successful operation of the demonstration unit has confirmed all key aspects of the AlkyClean process technology, while allowing for both parametric optimization studies and the proofing of operational procedures to ensure the reliability of the new process. The unit operated on slipstreams from a nearby refinery and ultimately utilized an optimized, increased activity catalyst – which was successfully manufactured in a commercial scale trial – to fully prove the 'real-world' performance of the new SAC process.

Based on this effort, the AlkyClean process has proven to be robust, requiring minimal maintenance, while producing high quality product. The zeolite based solid acid catalyst has exhibited tolerance to both upsets and exposure to high levels of contaminants.The catalyst's ability to be easily and repeatedly regenerated, with full activity recovery, has been demonstrated over long periods of operation. With all of its very positive benefits, this breakthrough technology provides a profitable addition to the processing portfolio of refiners, as they strive to meet regulatory-driven demand for both cleaner fuels and greener refining processes.

REFERENCES

1. Broekhoven, E.H. van, Mas Cabré, F.R., Bogaard, P., Klaver, G., Vonhof, M., US Patent No. 5.986.158 (1999).

Platinum Level and Catalyst Density Effects in C_5/C_6 Isomerization

V. M. Godfrey, M. P. Lapinski, R. R. Rosin, J. G. Vassilakis
UOP LLC
Des Plaines, Illinois, USA

Abstract

The effect of Pt level and catalyst density on chlorided-alumina catalyst performance in light-naphtha isomerization has been studied. The approach of this study is to examine these effects on paraffin isomerization independent of the effects on benzene saturation and naphthene reactions. The effects have been also studied on a feed typical to commercial light-naphtha isomerization unit.

Introduction

Isomerate produced from light-naphtha streams has been a significant contributor to the world's octane pool for over forty years. Demand for isomerization capacity is expected to grow. Octane loss from lead phase-out, reduction of gasoline aromatics, or desulfurization of FCC naphtha can be offset by light-naphtha isomerization. Recent EPA rule changes regarding gasoline benzene levels and gasoline oxygenates will further increase the demand for octane. Light-naphtha isomerization also can provide a cost effective solution to benzene management in motor fuels by saturation of benzene or by processing benzene precursors to paraffins.

Many catalysts and technologies currently exist for isomerization of light-naphtha with the catalyst being a critical component (1). All paraffin isomerization catalysts are dual functional, employing both metal function and acid function chemistries on the same catalyst. Most commercial paraffin isomerization catalysts employ platinum as the metal function due to platinum's superior resistance to sulfur poisoning. Commercial paraffin isomerization catalysts differ in their acid function chemistry. Currently three broad classes of acid chemistries are used commercially: chlorided-alumina, sulfated zirconia, and zeolites. Catalysts from these classes of acid chemistries have different activities, selectivities, and stabilities (2). Chlorided-alumina catalysts have the highest activity and selectivity, but are not stable in feeds containing even part per million levels of water. This water sensitivity is managed commercially very successfully by the use of feed dryers, making chlorided-alumina catalysts the most widely used type of catalyst for light-naphtha isomerization.

Recent years have seen a dramatic increase in the price of platinum. The price of platinum has risen 25% per year for the last four years, Figure 1. The high cost of platinum has been the focus of many refiners looking to control the cost of catalyst reloads. Another focus is the density of the catalyst. Catalysts with lower Pt levels and lower densities result in a lower total fill cost for a load of catalyst. Therefore, there is great interest in understanding the effects of catalyst Pt level and catalyst density on the performance in light-naphtha isomerization reactions.

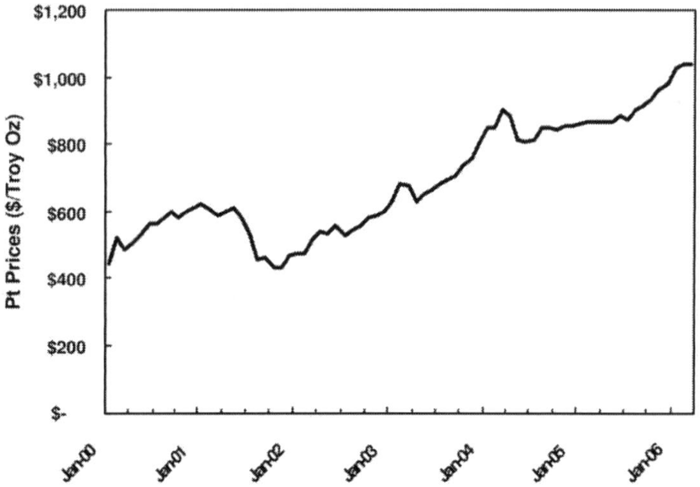

Figure 1

Platinum Prices 2000-2006

Source: Johnson-Matthey Base Prices

This paper details recent studies that were conducted to examine the effects of Pt level and catalyst density on the performance of chlorided-alumina, light-naphtha isomerization catalysts. The approach of this work is to examine the effects of a wide range of Pt levels and catalyst densities inclusive of the levels in currently available commercial catalysts.

EXPERIMENTAL

The catalysts used in this study are described in Table 1. The approximate loaded density is the density that would be expected from a typical dense loading of a commercial reactor. The loaded densities and the loaded Pt on a volume basis of these catalysts are compared in Figure 2.

Table 1

Catalyst Descriptions

Catalyst	Description	Pt Level (wt%)	Approx. Loaded Density
A	Base Pt Level, Base Density	0.24%	55 lb/cu ft (880 kg/m3)
B	Base Pt Level, Low Density	0.24%	48 lb/cu ft (775 kg/m3)
C	Low Pt Level, Base Density	0.12%	55 lb/cu ft (880 kg/m3)
D	Low Pt Level, Low Density	0.12%	48 lb/cu ft (775 kg/m3)

The catalysts were prepared at the semi-works scale in a manner consistent with commercial scale catalyst production. The chloride level per gram for all catalysts is similar. Therefore, varying catalyst density will vary total acid sites on a per volume basis.

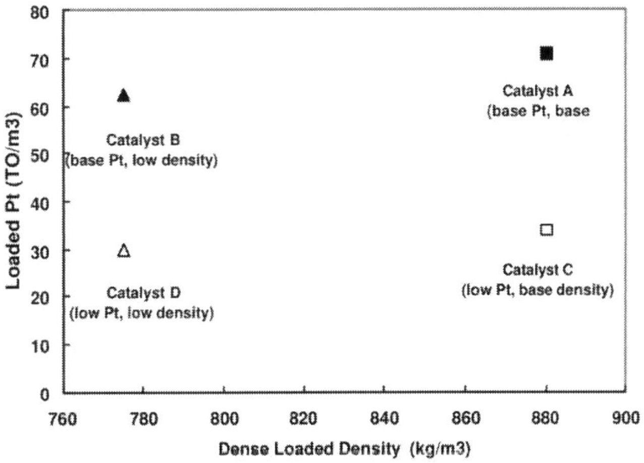

Figure 2

Loaded Densities and Loaded Pt of Experimental Catalysts

One of the benefits of light-naphtha isomerization to the refiner is the ability to control benzene in the gasoline pool. This means that feed to a light-naphtha isomerization process unit will contain not only light paraffins, but also benzene and naphthenes. The reaction network in a commercial light-paraffin isomerization unit is more complicated than just paraffin isomerization. Benzene saturation, naphthene ring opening, and naphthene isomerization are reactions that also occur on the catalyst. These reactions are not simply side reactions independent of the main paraffin isomerization reactions. The presence of benzene and naphthenes will significantly suppress the paraffin isomerization activity.

The experimental approach of this study is to examine this complex reaction network in two parts. First, the effects of catalyst Pt level and catalyst density on paraffin isomerization independent of benzene and naphthene reactions will be studied. Second, the reactions of benzene and naphthenes will be studied in such a way that the selectivities of the naphthene reactions can be determined. Lastly, the performance of these catalysts will be evaluated in feed typical of commercial light-naphtha isomerization feeds.

CATALYST PERFORMANCE – MODEL FEEDS

The catalysts were first tested with a feed containing only paraffins: 30 wt% n-C_5 and 70 wt% n-C_6. All catalysts were tested on an equal volume basis. Variation in catalyst density will result in different amounts of acid sites and metal sites in the reactor. Figure 3 shows the results plotted as the PIN or Paraffin Isomerization Number versus average bed temperature. The PIN number is the summation of the formation ratio of i-C_5, 2,2 di-methyl butane, 2,3 di-methyl butane.

$$\textbf{PIN = I-C}_5\textbf{ ratio + 22DMB ratio + 23DMB ratio}$$

The more active catalyst produces a higher PIN number at given temperature. Paraffin isomerization reactions are equilibrium limited under commercial operating conditions. The equilibrium PIN is also shown in Figure 3.

Catalysts A (base Pt, base density) and Catalyst C (low Pt, base density) have similar activity as do Catalysts B (base Pt, low density) and Catalyst D (low Pt, low density). This result indicates that Pt level has no effect on paraffin isomerization activity over the range of Pt level of this study. However, Catalysts A and C are considerably more active than Catalyst B and D. These two sets of catalysts differ in catalyst density. These results indicate that catalyst density, not loaded platinum, affects directly paraffin isomerization.

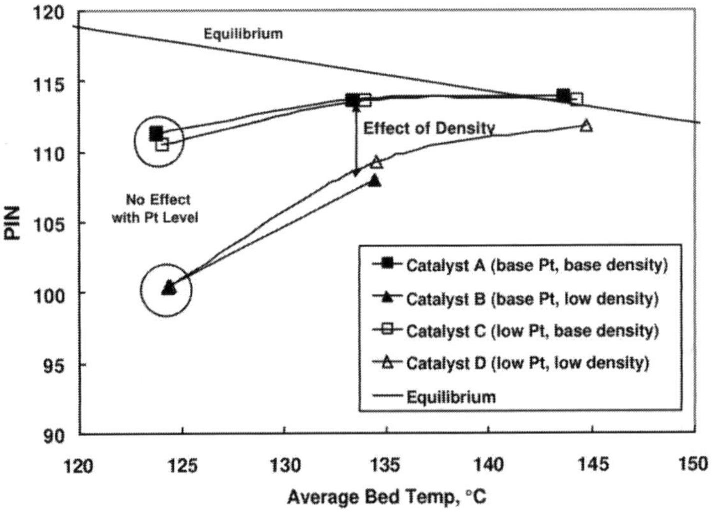

Figure 3

Pt Level and Density Comparison

Paraffin Isomerization Only – 30 wt% n-C_5, 70 wt% n-C_6

Lower catalyst density results in lower paraffin isomerization activity. This can be understood by reasoning that lower catalyst density on an equal volume basis will result in fewer acid sites in the volume or, in the case of a commercial light-naphtha isomerization unit, fewer acid sites in the reactor.

These catalysts were next evaluated for benzene saturation and naphthene reactions. Of particular interest is the selectivity of C_6 naphthene disappearance reactions: ring opening, cracking, or alkylation to heavier naphthenes. In a feed containing both C_5 and C_6 paraffins, it is impossible to determine the extent of ring opening to paraffin. Therefore, the feed of this portion of the study contains C_5 paraffin but no C_6 paraffins. The feed composition is 5% benzene, 20% cyclo-hexane, and 75% n-C_5. All catalysts saturated benzene completely at all temperatures. However, the catalysts differed in their ability to convert C_6 naphthenes. C_6 naphthene conversion activity is critical for a commercial light-naphtha isomerization catalyst because C_6 naphthenes will suppress paraffin isomerization activity. The C_6 naphthene conversion of these catalysts as a function of temperature is shown in Figure 4. Naphthene conversion is defined as:

$$C_6 \text{ Naphthene Conversion} = \{C6N_{in} - C6N_{out}\} / C6N_{in}$$

As benzene is saturated completely by all the catalysts tested, the molar quantity of benzene in the feed is added to the total amount of C_6 naphthenes fed.

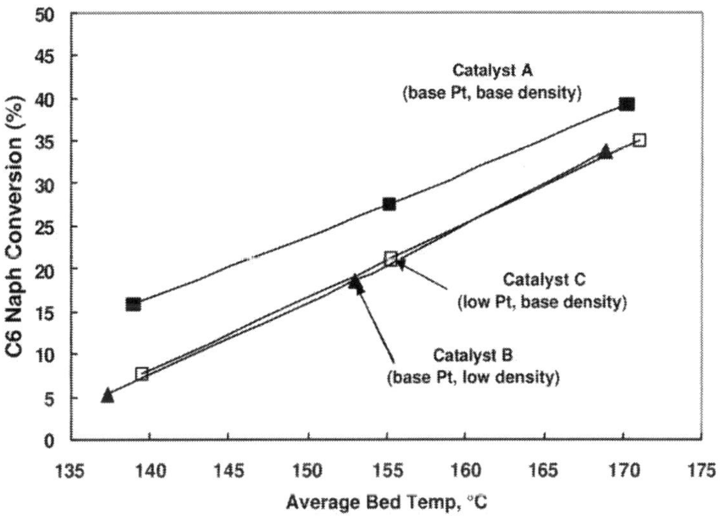

Figure 4

C_6-Naphthene Conversion Activity

5% benzene, 20% cyclo-hexane, 75% n-C_5

Catalyst A (base Pt level and base density) was the most active for C_6 naphthene conversion. Catalyst B (base Pt and low density) and Catalyst C (low Pt and base density) were less active. Lower Pt or lower catalyst density results in lower C_6 naphthene conversion. This result indicates that C_6 naphthene conversion is dependent on both the metal and acid function of the catalyst.

The selectivity of C_6 naphthene disappearance reactions was also investigated. C_6 naphthenes can ring open to form C_6 paraffin, can crack to light ends, or can alkylate with paraffins to form heavy naphthenes. The formation of heavy naphthenes is a particular problem in commercial light-naphtha isomerization units that utilize separation and recycle of unreacted hydrocarbons. These heavy naphthenes can concentrate in a recycle stream and further suppress the paraffin isomerization activity of the catalyst. Figure 5 shows the selectivity of the catalysts for the formation of heavy naphthenes as a function of the C_6 naphthene disappearance. The preferred performance is for high C_6 naphthene conversion and low heavy naphthene selectivity. Figure 5 shows that as the Pt level or the catalyst density is reduced, the tendency to form heavy naphthenes is increased. This result indicates that C_6 naphthene selectivity is dependent on both the metal and acid function of the catalyst. Catalyst A (base Pt, base density) performed the best with highest conversion and lowest selectivity to heavy naphthenes.

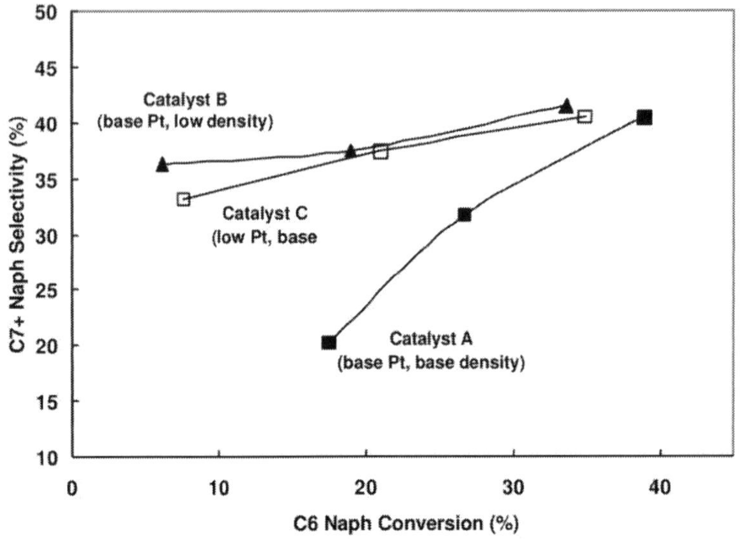

Figure 5

C_6-Naphthene Conversion Selectivity to Heavy Naphthenes

5% benzene, 20% cyclo-hexane, 75% n-C_5

From these experiments with models feeds the picture that emerges is that paraffin isomerization is controlled primarily by acid site density. On an equal volume basis, lower density catalysts have fewer acid sites, and so, lower paraffin isomerization activity. Benzene appears to saturate even at low Pt levels. Naphthene conversion and selectivity to heavy naphthenes is dependent on Pt level and acid site density. Lower Pt levels and lower catalyst density will result in greater formation of heavy naphthenes. These observations are summarized in Figure 6.

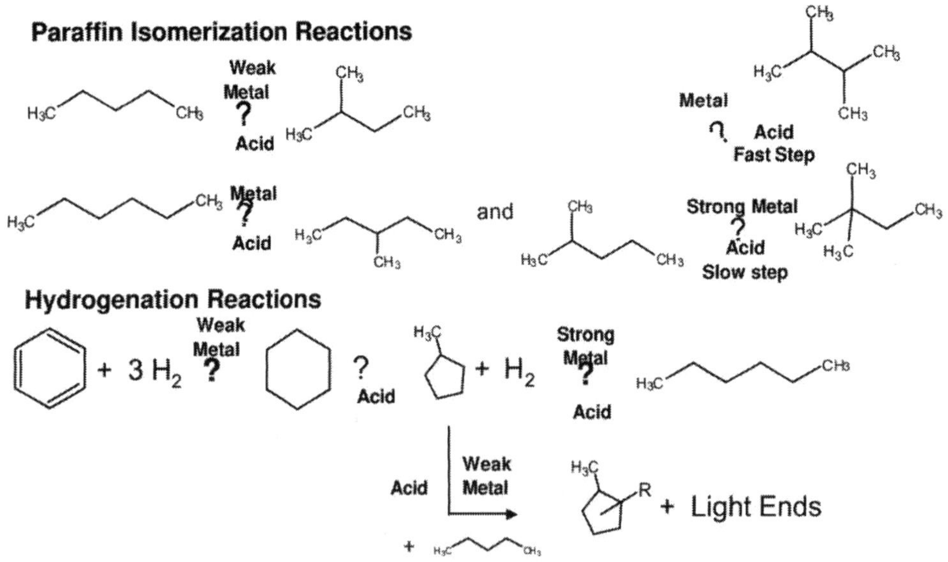

Figure 6

Reaction Network of Light-Naphtha Isomerization Unit

Catalyst Performance - Commercial Feed

Catalysts A, B, and C were tested in a feed typical of commercial light-naphtha feeds. The feed has a C_5 paraffin to C_6 paraffin ratio of 0.9, 1.7 wt% benzene, 6.0 wt% cyclo-hexane, and 1.25 wt% n-C_7. The test conditions were 450 psig (31 barg) total pressure with an H_2 to hydrocarbon mole ratio of 0.05 in excess of saturation.

Figure 7 compares the activity of the catalysts on the basis of PIN versus average bed temperature. Catalyst C (low Pt, base density) is the least active. Catalyst B (base Pt, low density) is the next most active. Catalyst A (base Pt, base density) is the most active. At an average bed temperature of 135C, Catalyst A achieves maximum PIN. Paraffin isomerization reactions can be equilibrium limited over active catalysts. At temperature greater than 135C, the activity of Catalyst A becomes equilibrium limited. At 150C the PIN achieved by Catalyst A is actually less than at 135C because of equilibrium limited behavior. Catalyst B (base Pt, low density) achieves similar PIN performance to Catalyst A only at significantly higher temperatures.

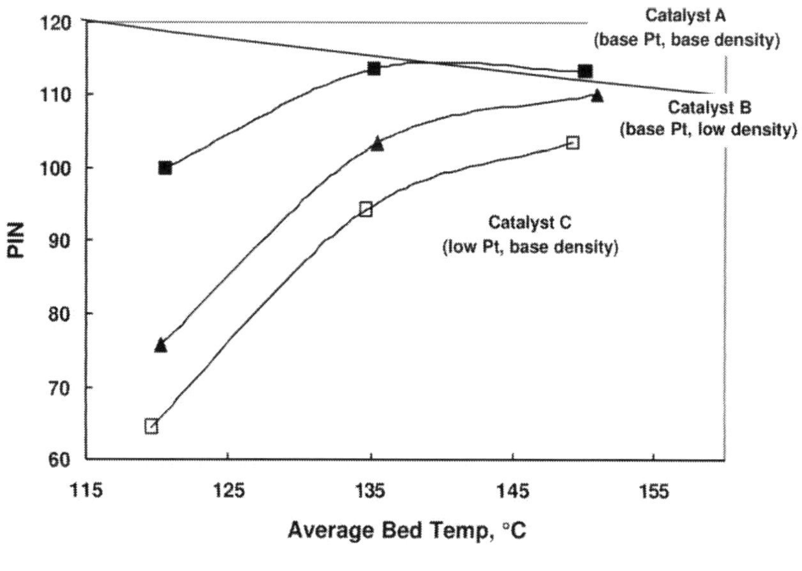

Figure 7

Catalysts' Activity in Commercial Feed

1.7 wt% benzene, 6% cyclo-C_6, 1.25% n-C_7

However, PIN is only one measure of catalyst performance. The product yield of the catalyst must also be considered. In Figure 8, the total yield of components with a molecular formula C_5 or greater, C_5+ Yield, is compared to a calculated research octane number. The C_5+ yield is an inverse measure of a catalysts tendency to crack feed hydrocarbons to C_4 or less. The research octane number or RON is a measure of the isomerate value as a blending component in the gasoline pool.

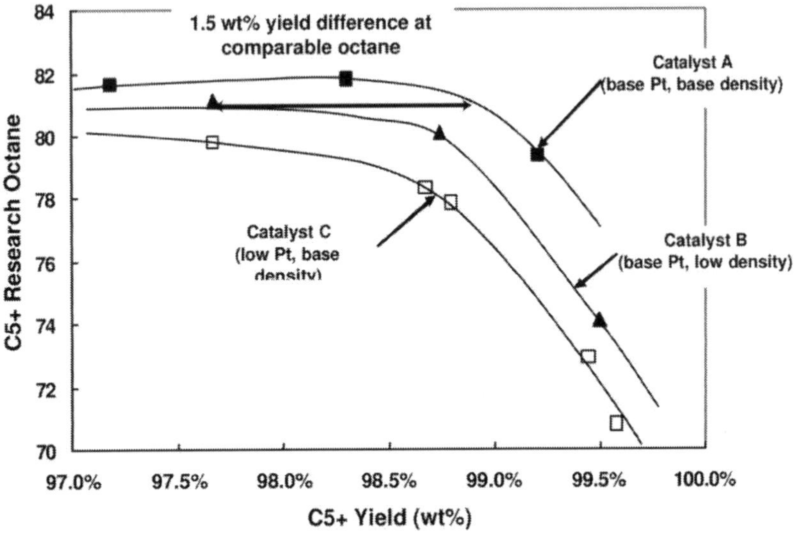

Figure 8

Yield-Octane in Commercial Feed

1.7 wt% benzene, 6% cyclo-C_6, 1.25% n-C_7

Figure 8 shows that Catalyst A (base Pt, base density) achieves the highest octane for a given yield. Catalyst C (low Pt, base density) does not achieve the same octane as Catalyst A over an acceptable range of C_5+ yield. Catalyst B (base Pt, low density) can achieve a similar octane as Catalyst A but only at higher temperatures and significantly lower yields.

These results indicate that for the feed evaluated, lowering Pt significantly lowers catalyst activity. Lowering catalyst density lowers catalyst activity as well, but this lower activity can be compensated for by raising temperatures. However, raising temperatures to compensate for a lower activity catalyst will result in yield loss due to cracking to light ends.

SUMMARY

Model feed testing has indicated which reactions in a light-naphtha isomerization are affected by catalyst platinum level or by catalyst density variations for chlorided-alumina catalysts. Paraffin isomerization is not affected by platinum level, but is affected by catalyst density changes. Lower catalyst density results in less acid sites on a volume basis and lower isomerization activity. Benzene saturation is unaffected by platinum level over the range of this study. Naphthene conversion is strongly affected by both platinum level and catalyst density. Lower platinum levels or lower catalyst density resulted in lower C6 naphthene conversion and higher selectivity to heavy naphthenes.

Testing with a commercial feed blend showed that lowering platinum level or catalyst density results in lower product octane for a given yield. The lower activity of low density catalysts can be compensated for by raising temperatures because of the equilibrium limited nature of paraffin isomerization reactions. However, operation at higher temperatures will result in significantly lower product yields. Any cost savings of a low density catalyst needs to be weighed against the isomerate yield loss.

This work has shown that low platinum catalysts can be used with feeds that contain no or low levels of benzene or naphthenes. Matching the platinum level of the catalyst to the benzene and naphthene content of the feed may be the best way to lower the overall catalyst loading cost to the refiner.

REFERENCES

1. D.J. Shecterle, R.R. Rosin , "Upgrading Light Naphtha", Proceedings of the 7th Topical Conference on Refinery Processing, AIChE Conference, Mar. 21-23, 2004

2. N.A.Cusher, R.D.Gillespie, G.A.Marr, W. Martin, R.R.Rosin, and P.Palmer, "High Activity C5-C6 Isomerization Catalyst for Lower Capital Solutions", Proceedings of the 4th European Catalyst Technology Conference, February 18-19, 2003

Welded Plate Exchangers: More Than Ever Needed in Catalytic Reforming

Paper # 42609

AICHE – Spring 2006 National Meeting

Orlando, Florida

Francois Reverdy
Alfa Laval Packinox
Houston, Texas USA

Cat Reforming: A Process with Two Energy Levels

Operates at 950°F

Cp = 0.8 Btu/lb.°F
ΔT = 800°F

Charge Rate, 1000 bpd	**25**	**50**
Energy Change, MM Btu/hr	**225**	**450**

Operates at 150°F

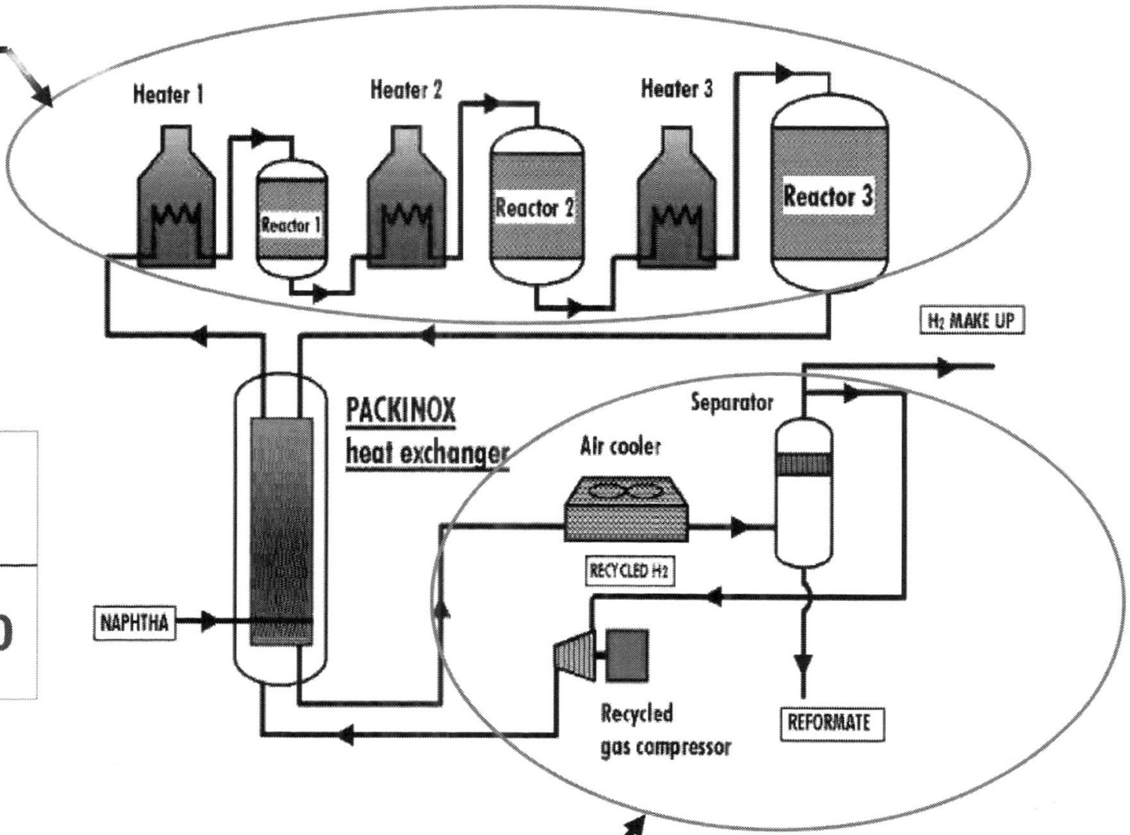

AICHE paper # 42609 - 2006

A Process with Atmospheric Emissions

Fired Heaters = Major Sources of Emissions

Charge Rate, 1000 bpd	**25**	**50**
Overall Firing Duty, MM Btu/hr	**150**	**300**

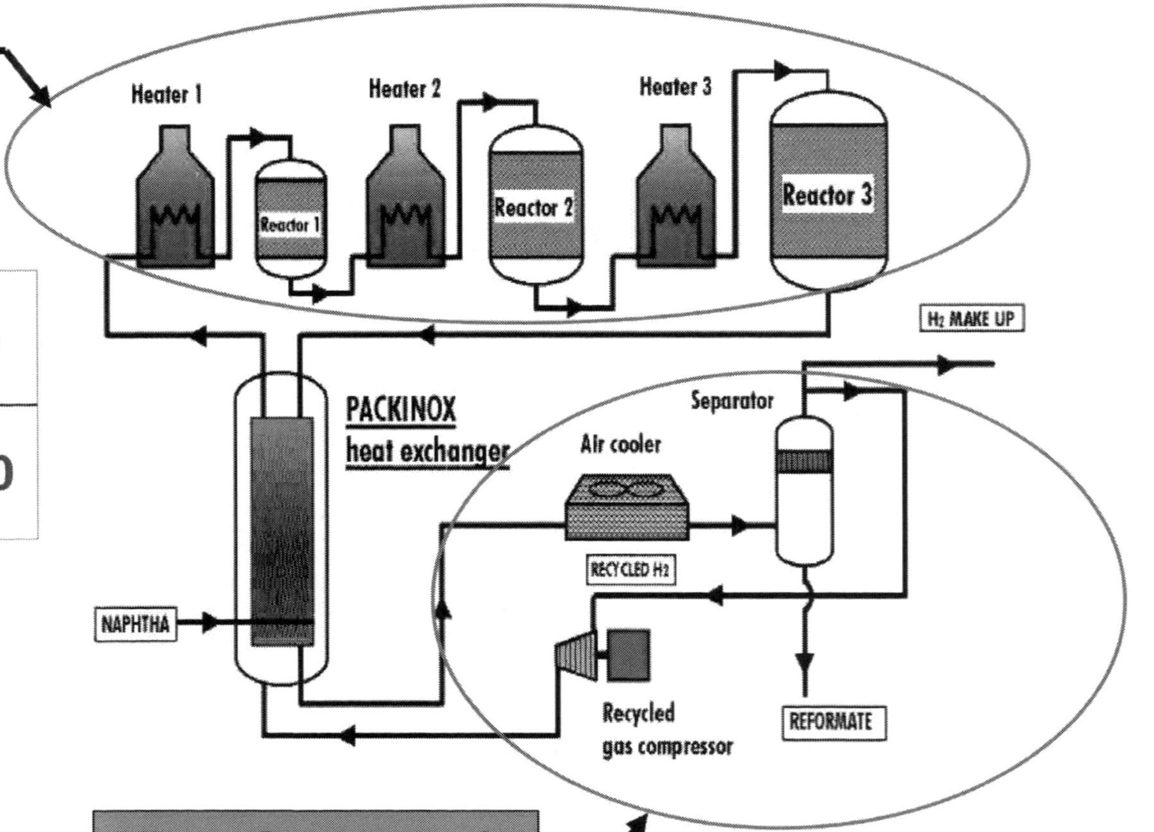

Minor Sources of Emissions

AICHE paper # 42609 - 2006

Heat Recovery with Feed-Effluent HE

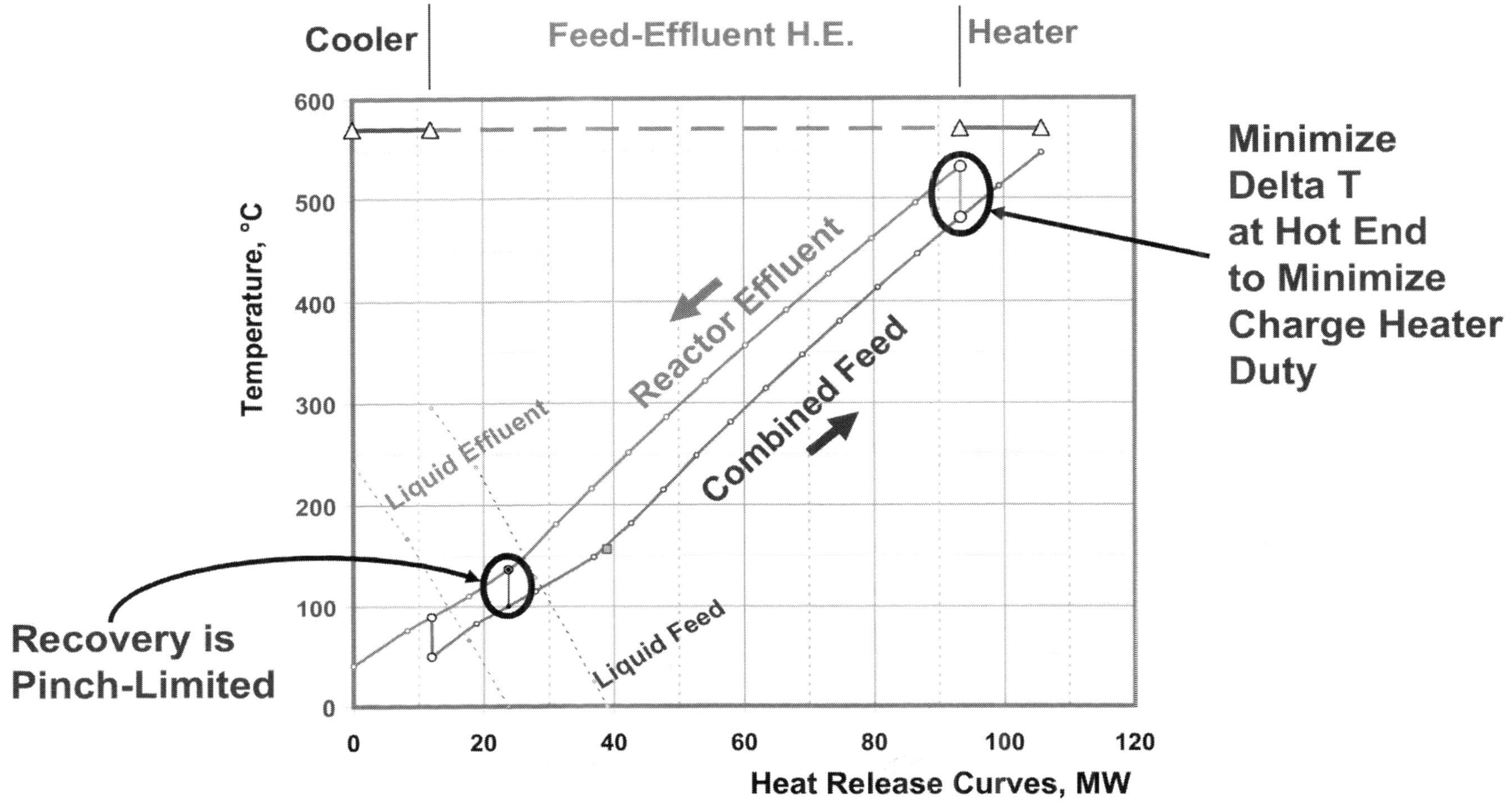

AICHE paper # 42609 - 2006

Value of Incremental Heat Recovery

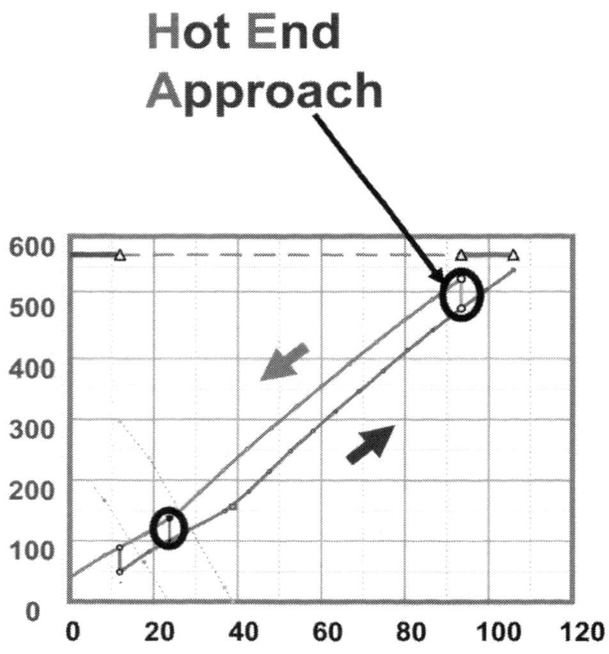

Annual Energy Savings at Fired Heater per Degree F of H.E.A., $1,000		Charge Rate, 1,000 bpd	
		25	50
Energy Cost / MMBTU	$5	$14	$28
	$7.5	$21	$42
	$10	$28	$56

Assumptions:
- **85% overall efficiency for the Fired Heater**
- **8400 hours / year**

AICHE paper # 42609 - 2006

Value of Incremental Heat Recovery (2)

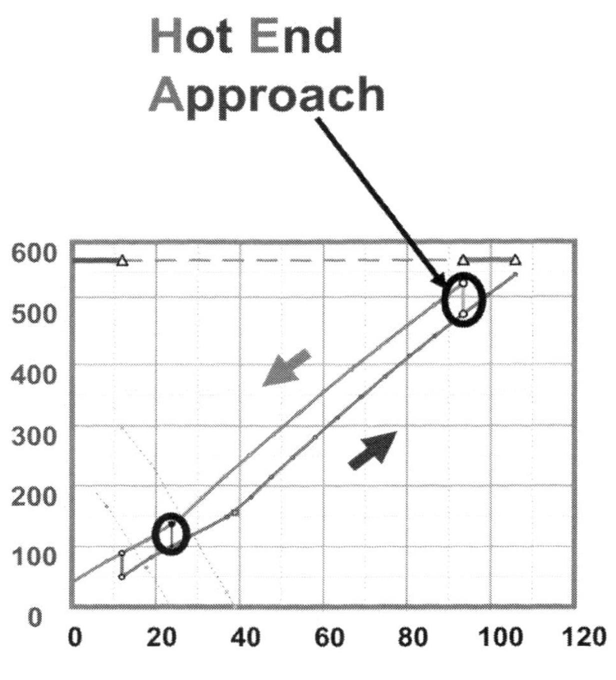

Annual Emissions Reduction at Fired Heater per Degree F of H.E.A., tons	Charge Rate, 1,000 bpd	
	25	50
NOx	1.0	2.0
SOx	1.3	2.6
CO_2	200	400

Assumptions:

- 0.25% N in Heater fuel
- 0.6% S in Heater fuel
- Low to average emissions controls
- 58% Heater process efficiency

AICHE paper # 42609 - 2006

Response to the Challenge

Maximize Thermal Efficiency → Maximize Tube Length

- **Special, Explosion Formed Plates → 150 ft Tube Length Equiv.**
- **True Countercurrent Flow**
- **Fully Welded Construction**

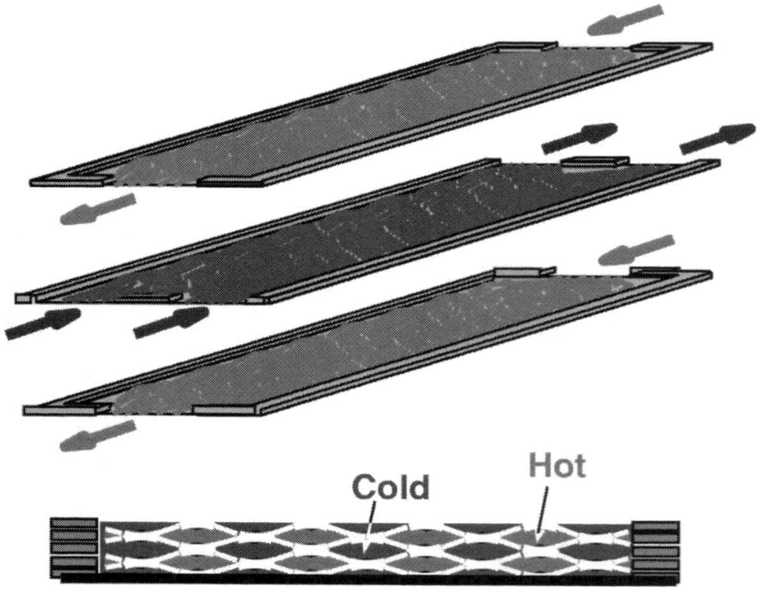

AICHE paper # 42609 - 2006

Response to the Challenge (2)

High Temperature Hydrogen Service

↓

Pressure Vessel For Operator Safety

↓

"SHELL & PLATE" Heat Exchanger

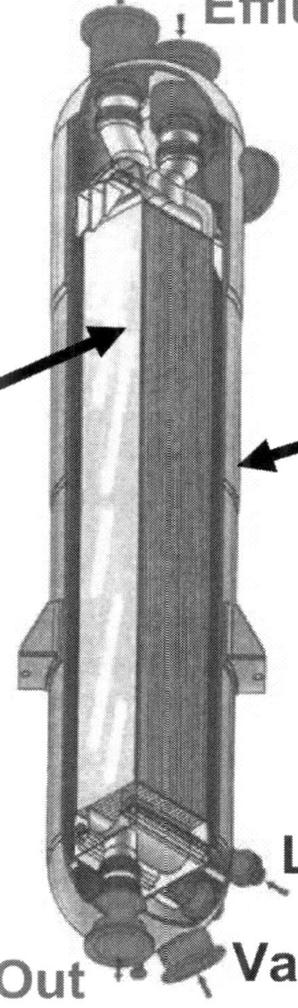

AICHE paper # 42609 - 2006

Response to the Challenge (3)

Dual-Phase Feed → "OPTIMIZE" Flow Distribution

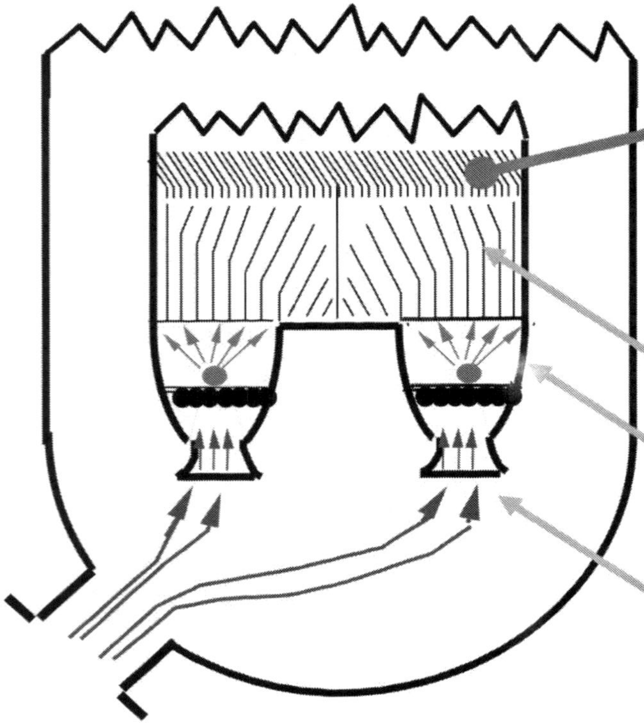

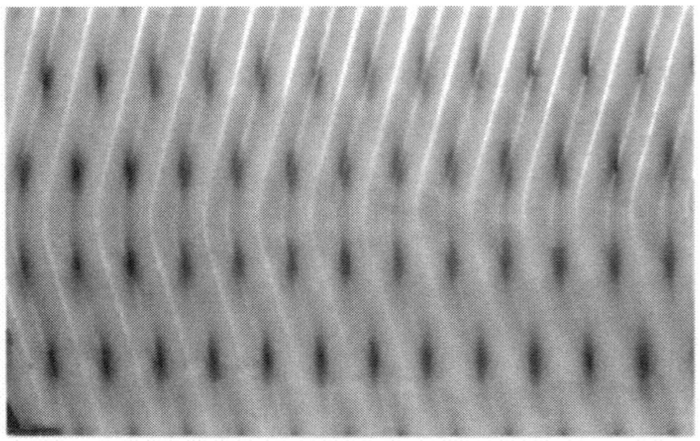

- **High Turbulence by Design**
- **"Full-Wetting" Zone**
- **Liquid Feed Injected for Distribution**
- **Recycle Gas Funneled for Lift**

AICHE paper # 42609 - 2006

Response to the Challenge (4)

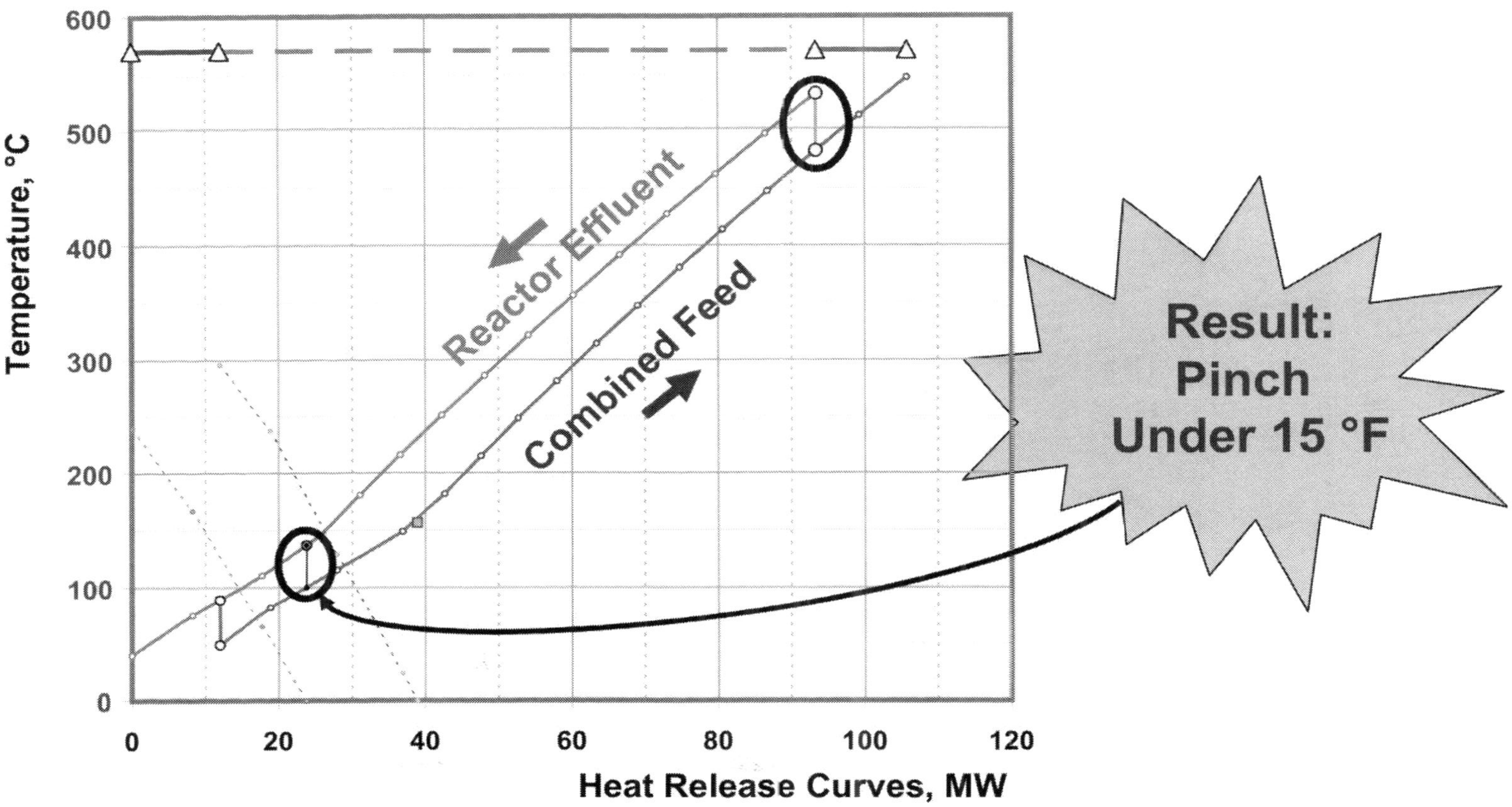

AICHE paper # 42609 - 2006

Value Added

Representative Sample Value Added at 50,000 bpd			
Thermal Efficiency Gain (H.E.A Reduction)			40°F
Energy Savings at Fired Heater	Fuel $ / MMBTU	$7.5	$1,680,000
Emissions Reduction at Fired Heater	NOx, $ / ton	$1000	$80,000
	SOx, $ / ton	$1000	$104,000
	CO_2, $ / ton	$20	$320,000

Total Value Added > $2,000,000

AICHE paper # 42609 - 2006

Compact and Safer for the Operator

- **Single Shell Design**
  - **Up to 85,000 bpd (Cat Reforming)**
- **Fully Welded Pressure Vessel**

AICHE paper # 42609 - 2006

Lower Total Project Cost

- **Lower Installed Cost**
  - **Package "Heater + Cooler + Feed/Effluent"**
- **Lower Operating Cost**
  - **Mainly: Heater Fuel Savings**
  - **Secondary:**
    - **Compressor and Cooler Energy Savings**
    - **Environmental Costs**

Reforming: A Relatively Clean Service

	2001	2002	2003	Rule Of Thumb
No. of PKX In Service on 01/01	**87**	**102**	**110**	
Approx. No. of PKX Cleaned that year	7	8	6	Every 10 to 12 Years
Percentage of Total	8%	8%	6%	
Approx. No. of PKX Decontaminated for Entry	12	19	6	Once In Between
Percentage of Total	14%	19%	6%	

AICHE paper # 42609 - 2006

Faster & Cheaper Cleaning

Fouling Type	Salts	Iron Compounds	Solids Migration	Wet gums	Hardened Coke
Preferred IN SITU Cleaning Method	Chemical	Chemical	Mechanical	Cold Oxidation	Steam-Air Decoking
Cleaning Agent	Water	KMnO4 Mild Acid	Vacuum / Rupture Disk	H_2O_2 at 6% Conc.	Steam / Air Nitro / Air
Typical Duration Of Cleaning Work	24 to 48hr	24 to 48hr	24 hr	24 hr	24 to 72hr
Typical Direct Cleaning Cost	$25k to $50k	$35k to $70k	$25k to $50k	$35k to $70k	$100k to $250K
Typical Effectiveness	100%	100%	90%	100%	100%

AICHE paper # 42609 - 2006

Two Chemical Cleaning Methods

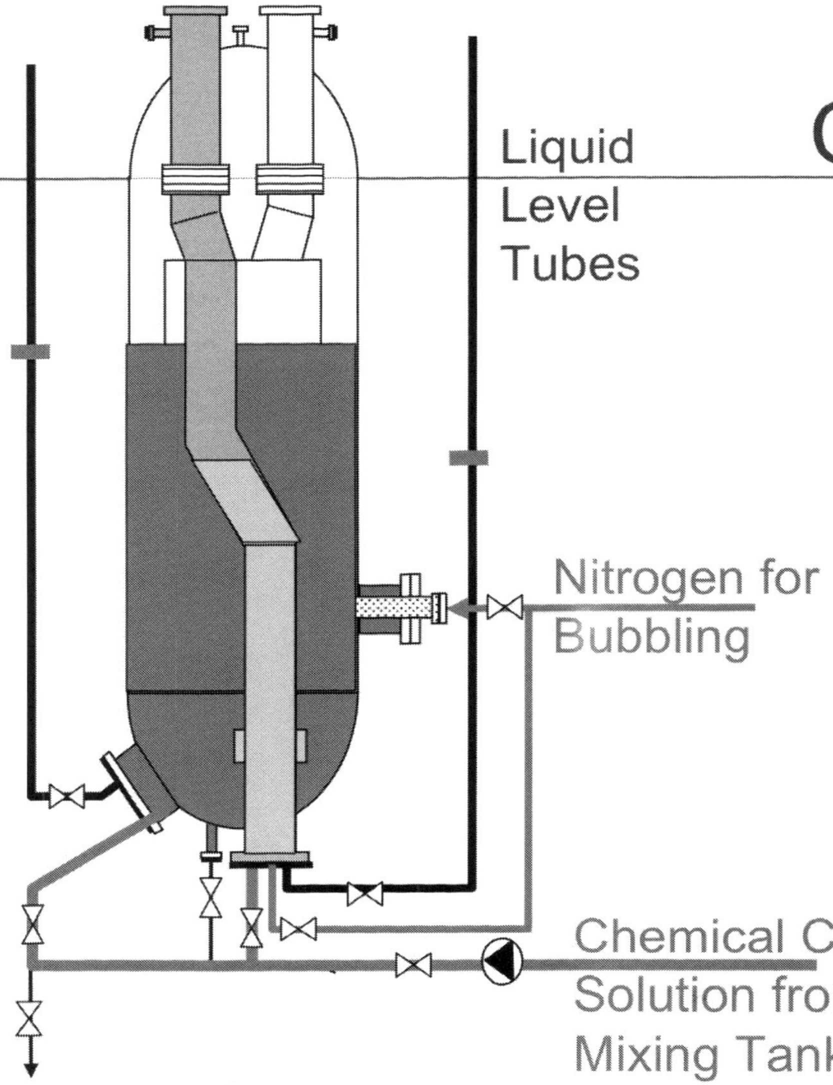

"Fill & Drain"

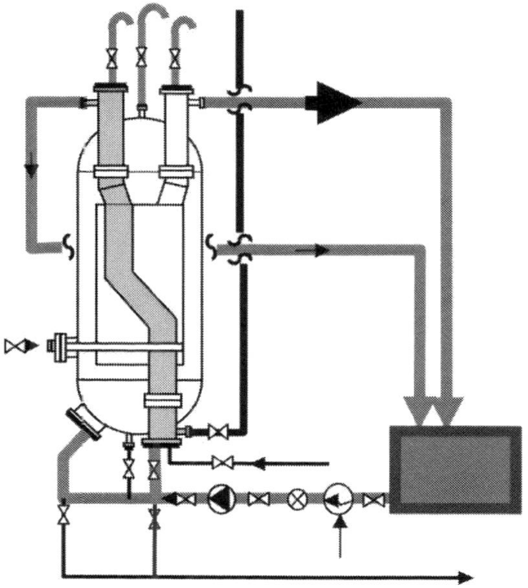

"Recirculation"

AICHE paper # 42609 - 2006

Cleaning by Steam-Air Decoking

Example of Steam-Air Decoking on CCR

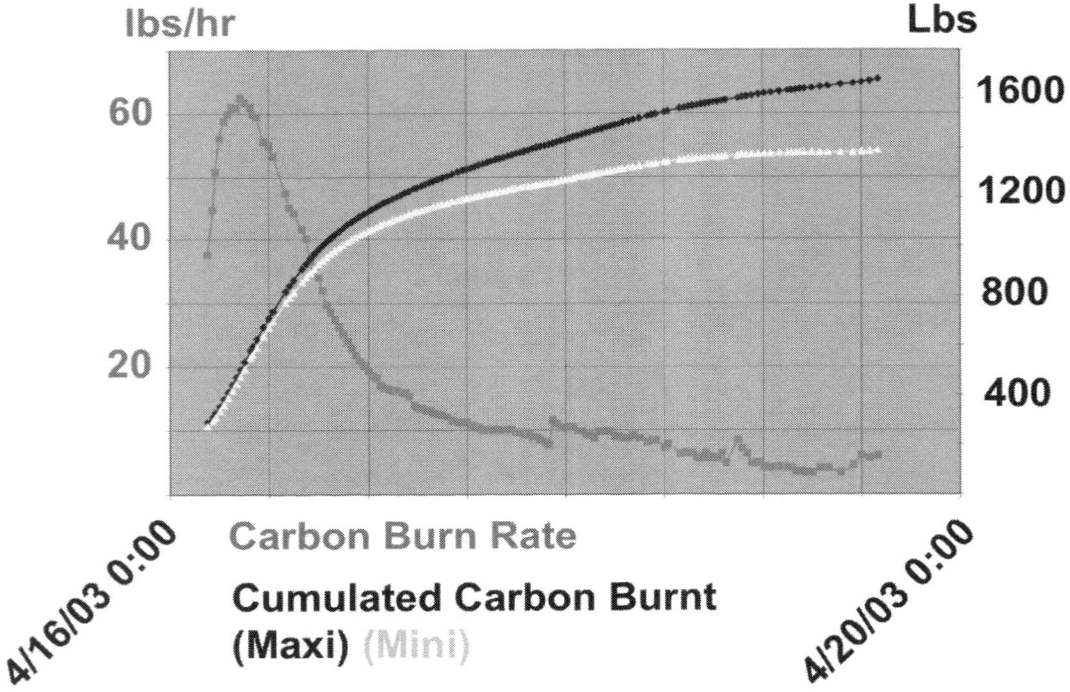

Refer to HTRI Presentation for Details

AICHE paper # 42609 - 2006

90e Alkylene™ - Improved Solid Catalyst Alkylation Technology for Clean Fuels

Douglas A. Nafis, PTE Development, UOP LLC., 25 East Algonquin Rd., Des Plaines, IL 60056, Margaret A. Stine, PTE Refining - Marketing, UOP LLC., 25 East Algonquin Rd., Des Plaines, IL 60056, Dale J. Shields, Light Fuels & Olefin Derivatives T/C, UOP LLC., 25 East Algonquin Rd., Des Plaines, IL 60056, and Gary A. Dziabis, PTE Refining - Development, UOP LLC., 25 East Algonquin Rd., Des Plaines, IL 60056.
Legislated changes in gasoline specifications throughout the world indicate the importance of alkylate as a blending component in today's gasoline pool. Reductions in total aromatics, benzene, vapor pressure, oxygenates, olefins, sulfur, and distillation endpoint collectively affect every major gasoline pool blending component except alkylate. Because of its high octane and paraffinic nature, alkylate is an ideal blendstock for any gasoline pool requirement.

Although liquid acid alkylation technologies have historically been safe, reliable and economical process operations, they are coming under increasing scrutiny due to environmental concerns. UOP's Alkylene™ process overcomes the environmental concerns involved with handling liquid acids by using truly solid alkylation catalyst. The performance characteristics of this solid catalyst are combined with a unique liquid riser reactor design to yield a technology that is both reliable and economically competitive with traditional liquid acid alkylation processes.

Development of the Alkylene™ process began in 1995, and proceeded through a variety of reactor systems (fixed bed, recycled bed, swing bed) before evolving to the current liquid riser process, which provides advantages in catalyst utilization and regeneration over competing reactor systems. UOP is currently working with a commercial customer on the design of the first commercial Alkylene™ unit. This paper will trace the development of the Alkylene™ process in the context of the many technical challenges involved in the development of the catalyst and reactor system.

Transformation of Heavy Gas Oils Derived from Oil Sands to Petrochemical Feedstock

Duke du Plessis, Alberta Energy Research Institute, Suite 2540, AMEC Place, 801 - 6th Avenue S.W., Calgary, AB T2P 3W2, Canada and

Catherine Laureshen, Alberta Energy Research Institute, Suite 2540, Amec Place, 801 - 6th Avenue S.W., Calgary, AB T2P 3W2, Canada

Abstract

Rapid expansion of the oil sands industry in Alberta has guaranteed a large production of heavy oil fractions in the near future. Alberta's petrochemical industry is presently ethane-based. The supply of ethane in Alberta is declining and that places some doubt on potential growth of the petrochemical industry. The growing production of oil from oil sands and accompanying heavy oil fraction breathes hope for growth in the petrochemical industry, provided that these heavy oils can be transformed into feed to petrochemical plants. It will however entail having large complexes where oil sand/upgrading, refining and petrochemical units are integrated. Three industry studies have evaluated different technologies and process schemes for integrating oil sands, refining and petrochemical operations. The common feature and result of the studies is that the higher the degree of processing, the higher the capital costs but also the higher the return on investment. This paper presents some process configuration and studies undertaken, which prove that it is feasible to produce petrochemical feedstock or petrochemicals from vacuum gas oils at competitive prices.

Introduction

With an estimated reserve of approximately 2.5 trillion barrels of bitumen in the ground, the Alberta oil sands are one of Canada's major energy resources. Oil sands are deposits of bitumen, which is heavy black oil that, when treated, can be converted into synthetic crude oil and used by refineries to produce gasoline and diesel fuels. Alberta's oil sand deposits are mostly found in the Athabasca region, mainly in the nearby Fort McMurray area. Oil sand production is now at around 1million bbl/day (approximately 50% of Canada's total crude oil production) and is expected to grow to 3million bbl/day by 2015.

The current process for upgrading the initial product from the oil sands into higher value petroleum products is complex and highly dependent on many factors, including access to refineries that are able to produce the higher value gasoline components. Currently, most Canadian refineries are designed to process only light and medium crude oil; some of the production of oil sands synthetic crude oil refining is being shipped to the United States. Many of the oil sands projects planned for the near future will be coming on-stream within the next ten years. This will significantly increase the total throughput of synthetic crude oil from Alberta, and with these increases, strains may be developed on the current processing capacity of refineries and also on the pipelines shipping the product. New refineries will be required to handle those new synthetic crude oil capacities.
The Alberta Government represented by Alberta Economic Development and Alberta Science and Technology Departments have over the past three years carried out various studies, in

collaboration with industrial partners, to explore the economic viability of integration of refineries and petrochemical plants with bitumen upgraders. All studies to date have indicated that the higher the level of integration the higher the capital costs but also the higher the net present value (NPV) or return on investment. One of the arms of Alberta Science and Technology Department is Alberta Energy Research Institute (AERI).

AERI and NOVA Chemicals have over the past three years collaborated on a project to convert heavy oils derived from oil sands into paraffins, olefins, gasoline and aromatic compounds (BTX) – petrochemical feedstock. The first part of the study sort out technologies, which are under development at international research centers and can be applied to Alberta's intermediate process streams. The process streams from oil/tar sands operations as well as heavy bitumen operations contain abundant amounts of gas oil fractions, lots of heavy aromatics and light gases which could be converted to petrochemical feedstock. Two key technologies that can be used to transform the heavy gas oils into olefins, gasoline and an aromatics-rich fraction – cycle oils were identified. The first part of the processing will convert vacuum gas oils derived from oil sands into olefins, gasoline, aromatics-rich cycle oils, slurry and coke. In the second stage, the aromatic compounds in the cycle oils are then saturated and cleaved to obtain mainly paraffins (steam cracker feed) and BTX.

At the time that this study was started, none of the technologies was in place, because there were no commercially available catalysts to realize those dreams. Working together with the University of Petroleum in Beijing, some catalysts were developed, tested and proven to give high yields of olefins, gasoline and aromatics-rich cycle oils. We also collaborated with the University of Stuttgart, where another catalyst was developed, which carries out the aromatic ring cleavage step to produce paraffins and BTX. All above are now proven at a laboratory stage. This project also developed a process flow scheme, which astutely combines the two technologies with an ethylene plant to produce mainly olefins, gasoline, and BTX. The above were achieved in the first and second phases of the project, which were the proof of concept stages. This paper highlights results obtained from the proof of concept stages of this research project. The results will be shown as NOVA-Heavy Oil Laboratory Catalyst (NHC) Process and Aromatic Ring Cleavage (ARORINCLE) Process. A final integration stage is also discussed.

NHC Process:

This technology starts with vacuum gas oil as feed. The process entails fluidized-bed catalytic cracking (FCC) technology, with a new riser and a specially designed catalyst. It is based on a proven technology platform – the FCC Unit platform. Similar technologies are in existence like the Deep Catalytic Cracking (DCC) and the Catalytic Pyrolysis Process (CPP) Technologies by SINOPEC, and the PetroFCC process by UOP. All these give high yields of olefins from vacuum gas oils or atmospheric residue. However, neither CPP nor PetroFCC processes have handled vacuum gas oils derived from Alberta oil sands that are highly aromatic in nature. They have however been proven to handle residues emanating from paraffinic crude e.g. Daqing atmospheric residue.

A new catalyst was invented in this project, which handles vacuum gas oils derived from oil sands operations, which are highly naphthenic/aromatic. A combination of the new catalyst and hardware, produce mainly ethylene, propylene, butadiene, gasoline and cycle oils (rich in aromatics) and coke.

Equipment:

The reaction unit for the experiments is a confined fluidized bed reactor system which has five sections – the oil and steam feed systems, the reaction zone, the temperature control system, and the product separation and collection system. For each experiment, a set amount of catalyst was loaded into the reactor with an effective volume of about 580 ml. A set amount of distilled water was pumped into the furnace to produce steam for fluidizing the catalyst. The steam was mixed with the feedstock, which was kept in an oven and fed simultaneously by a feed pump. The mixture was heated to approximately 500°C in a preheater, and then entered the reactor. In the reactor, the feed was atomized and contacted with fluidized catalysts, and catalytic pyrolysis occurred. The reaction product from the reactor was cooled and separated into the liquid and gas products, which were collected respectively. After the reaction, the spent catalyst was removed from the reactor by a vacuum pump.

The catalytic reaction product consisted of pyrolysis gas, liquid and coke. An Agilent 6980 gas chromatograph with a Chemstation software was used to determine the volume percentages of pyrolysis gas components, which were converted to mass percentages using the ideal gas equation of state. The pyrolysis liquid was analyzed with a simulated distillation gas chromatograph to obtain the mass percentages of gasoline, diesel and heavy oil. Coke content on the catalyst was determined using a coke analyzer.
Results to date however, are limited to laboratory runs and the catalysts have not been demonstrated in a pilot plant. A few outstanding issues have yet to be resolved at this particular stage before a pilot plant can be built.

Experimental Results:

Some of the results from the NHC process are shown in Tables 1 and 2 below.

Table 1. Effect of Feedstock and Catalyst

	Base Catalyst	NHC-9	Base Catalyst	NHC-12
Feed Type	HAGO	HAGO	HVGO	HVGO
Temp (°C)	660	660	660	660
Ethylene	12.3	11.7	6.9	9.2
Propylene	19.4	22.3	10.7	16.1
Butylenes	9.0	12.0	5.9	9.5
Total Light Olefins (wt.%)	40.7	46.0	23.5	34.8

HAGO = Heavy Atmospheric Gas Oil and HVGO = Heavy Vacuum Gas Oil

Table 1 shows that the new catalyst outperforms a standard base catalyst by over 5 percentage points as far as production of olefins is concerned. This is valid when using either HAGO or HVGO as feedstock. With the heavier feedstock HVGO, the advantage of a new catalyst that is tailored for this type of feed becomes more pronounced.

Table 2. NHC process yields versus steam cracking.

	Steam Cracking	NHC-17	Steam Cracking	NHC-17
Feed Type	HAGO	HAGO	HVGO	HVGO
Temp (°C)	-	660	-	660
Ethylene	18.8	11.9	15.6	8.9
Propylene	11.6	24.2	11.9	17.4
Butylenes	6.0	11.7	6.0	8.5
Total Light Olefins (wt.%)	40.7	46.0	23.5	34.8

Table 2 shows the advantage of this technology over steam cracking for the production of olefins. The heavier the feedstock, the higher is the advantage of this technology. One would argue that the technology makes less ethylene than steam cracking, but this is based on the nature of the cracking mechanism for both reactions. Steam cracking is based on free radical reaction mechanism because in the absence of a catalyst, hydrocarbons crack according to the free radical mechanism [1]. The primary free radical decomposes at the beta position to form ethylene. Therefore, the higher the reaction temperature, the higher is the yield of ethylene. The catalytic cracking process is based on carbenium-ion mechanism, which preferentially produces more propylene than ethylene [2].

A few things remain to complete these studies, which include:

- Optimization to select the best of the developed catalysts
- Completion of reaction runs at a technical scale unit (2kg/hr) located at The University of Petroleum Beijing, to select one out of three excellent catalysts for the NHC process.
- Design of a 50 kg/hr pilot plant based on scale up of results from the 2kg/hr unit, to enable delivery of a simple process design package

ARORINCLE Technology

Introduction

Today the refining industry has to process heavier and heavier feedstocks and refiners have to look for new alternatives to improve economic margins with the increasing share of middle distillates, such as gas oils (GO) or light cycle oils (LCO). LCO is one of the streams from the FCC unit and is normally disposed of as low-value heating oil. It is characterized by a high content of aromatics (> 70 wt.%) and a relatively high amount of heavy aromatic sulphur and nitrogen compounds (~3 and 0.08 wt.%, respectively) [3]. GO may contain more than 50 wt.% aromatics and also relatively high amounts of sulfur and nitrogen (more than 2.5 and more than 0.08 wt.% respectively). LCO and GO upgrading may be achieved by a two-step catalytic process involving hydrotreatment followed by ring cleavage of naphthenic structures resulting in a product containing light paraffins ($C_1 - C_4$) and a BTX rich stream.

The hydrotreating step is there to remove the heteroatoms such as sulfur and nitrogen and selectively hydrogenate poly-ring aromatics with little or no saturation of single ring aromatic compounds. This can be achieved using a combination of commercial NiMo/Alumina and NiW/Alumina based catalysts on staged beds. The selectivity of the saturation of poly-ring aromatics is governed by two major factors:

a) The kinetic factor - the hydrogenation reactions of single ring aromatics are very slow compared to hydrogenation of poly-ring aromatics [4].
b) Adsorptive hindrance factor – poly-ring aromatics have higher adsorption strengths on the catalyst active centers compared to single ring aromatics, which makes active catalytic centers less accessible for single ring aromatics [5-9].

The second step is the naphthenic ring cleavage step, which is carried out on a noble metal modified zeolite catalyst. This kind of catalyst was shown to be effective for naphthenic and aromatic ring cleavage by some [10-13]. The main reaction occurring at this step is hydrocracking of naphthenic and paraffinic compounds in the feed, resulting in the formation of light paraffins. The side chains of the single ring aromatics and condensed naphthenic rings on single aromatic rings are also hydrocracked during this step, resulting in the formation of light paraffins and $C_6 - C_9$ aromatics, which is the BTX rich stream.
Results from research carried out for both steps of the technology as described above are shown below. The feed however is derived from Alberta's Athabasca Oil Sands Upgrader (VGO).

Experimental
Equipment and apparatus

All the experimental runs were carried out in a laboratory scale fixed bed-reactor in the up flow mode. Because the unit contains only one reactor, all the runs were carried out in such a way that the ring saturation step of the technology is carried out first. Thereafter, zeolite based noble metal catalyst was reloaded for the second step reaction to take place. The catalyst used for the first step is a stacked catalyst bed of NiW/Al_2O_3 and NiMo/ Al_2O_3. The volume of the total catalyst bed is 10 ml. The total volume of the catalyst for the ring cleavage step was 8 ml. The runs for both the hydrogenation and the ring cleavage steps showed very good repeatability before the final runs were carried out. To minimize the effect of deactivation of the catalyst, the experimental runs were carried out during the day and the catalyst was regenerated during the night. This allowed catalyst activity to be constant during the runs.

Effect of process parameters on the hydrogenation step

Thirty runs were carried out in order to study the influence of process parameters: pressure, temperature and liquid hourly space velocity (LHSV) – on the hydrogenation of GO. Figure 1 shows obtained results:

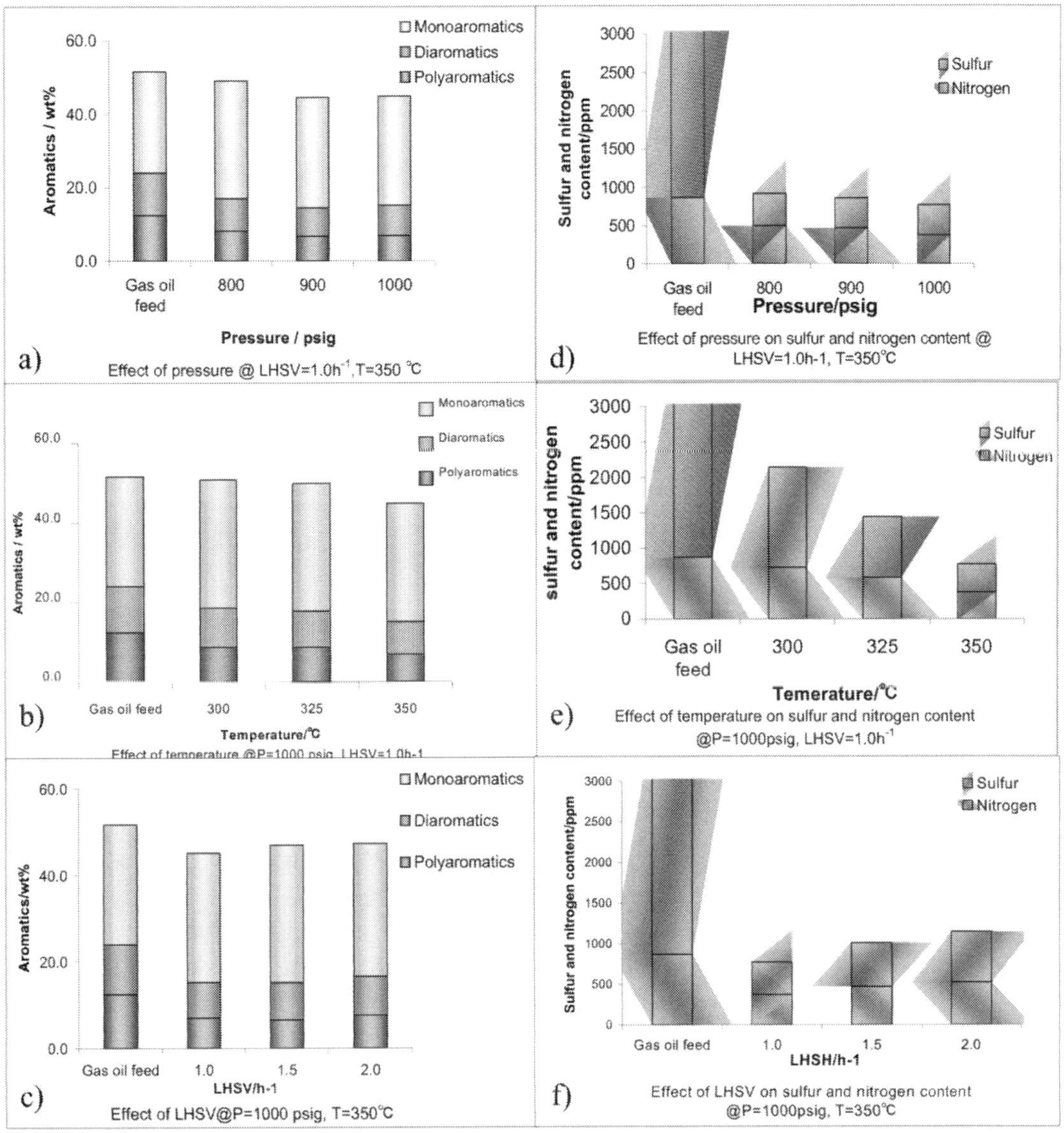

Figure 1: Effect of pressure a)&d), temperature b)&e) and pressure c)&f) on the conversion of aromatic compounds and hetero-atom removal during GO hydrotreating.

It can be seen from Figure 1 that higher pressure, higher temperature and lower LHSV lead to more extensive aromatics saturation and heteroatom removal.

Effect of process parameters on the ring cleavage step

Twenty seven runs were carried our to study the influence of process parameters on the ring cleavage step of hydrotreated GO (HTGO). Figure 2 shows some examples of results obtained.

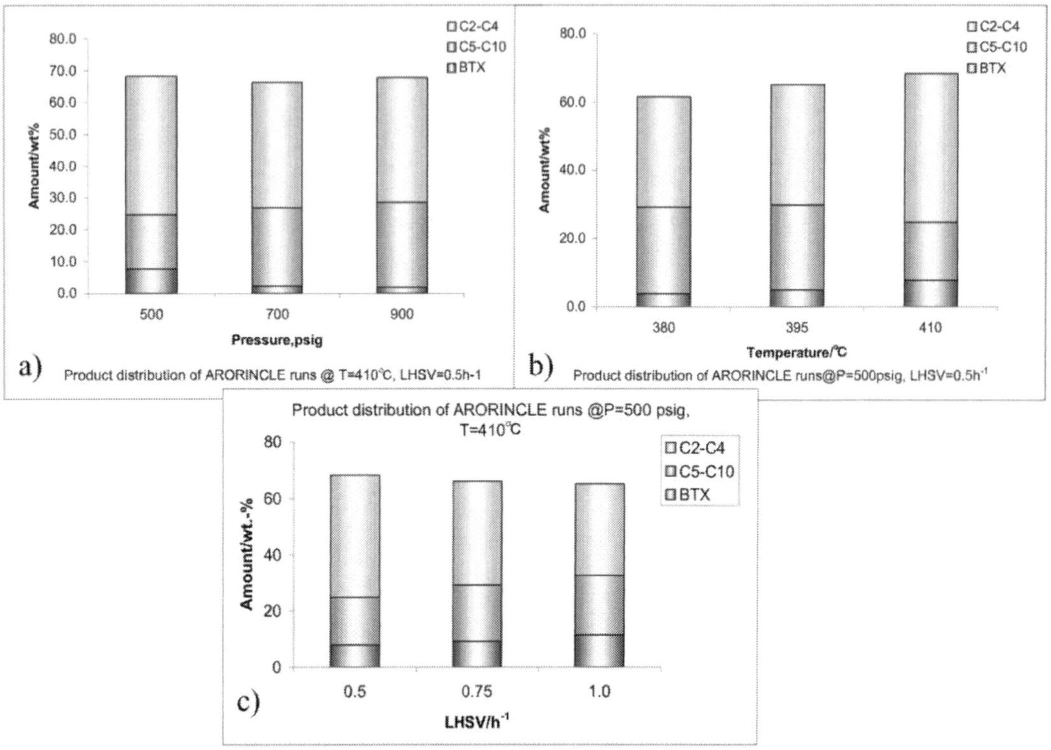

Figure 2: Effect of pressure a), temperature b) and pressure c) on product composition for the ring cleavage step

It is obvious from Figure 2 that lower pressure, higher temperature and lower LHSV promote higher production of light paraffins. One can see from the Figures that one can easily adjust the desired product composition by changing the process parameters. For example, in case, higher yield of BTX is required, the change of LHSV from 0.5 h^{-1} to 1 h^{-1} results in obtaining double the amount of BTX yield (see. Figure 2, c)).

Four additional runs, at the same experimental conditions, but with different feedstocks were carried out for a period of 100h each in order to study the effect of sulfur and nitrogen content of the feedstock on product yields of the ring cleavage step. The conditions for these runs were: Temperature = 410°C, Pressure = 900 psig and LHSV = 0.5 h^{-1}. These runs were necessary, since it is known that heteroatoms can poison noble metal containing catalysts, which is used in the second step of the technology. Figure 3 represents the change in the amount of C_{11+} fraction which remained unconverted during the 100 hours runs versus time-on- stream for four different feedstocks.

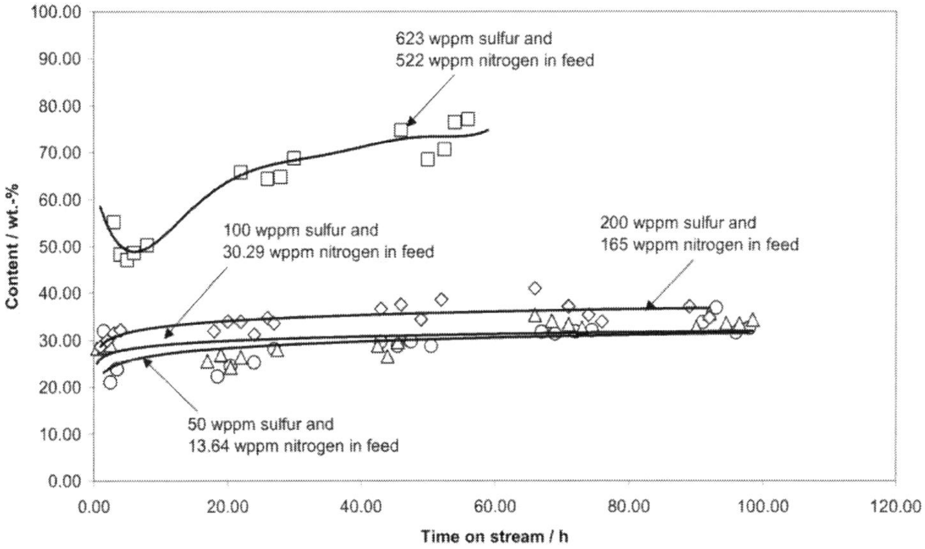

Figure 3: Heteroatoms effect on retaining of C_{11+} fraction during the ring cleavage step of the ARORINCLE technology

The results presented in Figure 3 represent, first of all, the level of tolerance of the catalyst to the presence of sulfur and nitrogen in the feed. From Figure 3, one can also see that the catalyst (Pd / H-ZSM-5), exhibits some level of heteroatoms tolerance over the period tested. The feedstock with the S content of 200 wppm and N content of 165 wppm can be successfully converted on this catalyst within the interval studied. However, the ring cleavage of the feedstock, containing 1145 wppm of total heteroatoms, had to be stopped after 56 hours on-stream due to very strong deactivation. It is obvious, from Figure 3, that the catalyst contacted with the feedstock with the highest content of heteroatoms was deactivated, already after 20 hours on-stream, indicating that the level of heteroatoms is unacceptable. No significant deactivation was observed for the feedstocks containing 130.29 wppm and 63.64 wppm respectively of total heteroatoms. The interval studied is however very short and more rigorous catalyst deactivation study is the object of another forthcoming study.

Summary of the ARORINCLE process

- The feasibility of producing sufficient product with the desired quality, which can be used as feedstock for the ring cleavage step was proved.
- The feasibility of opening and cracking aromatic rings of gas oil derived from Alberta's Athabasca oil sands was demonstrated
- The sulfur and nitrogen tolerance of the Pd / H-ZSM-5 was shown. It was shown that feedstocks, containing less than 150 wppm of total heteroatoms, do not cause any immediate deactivation of the catalyst during 100 hours on-stream.
- The feasibility of simultaneous production of light paraffins and BTX from this gas oil was also demonstrated.
- The study of the ring cleavage step revealed that high amounts of BTX can be produced at low pressure (500 psig) and high LHSV ($1\ h^{-1}$).

Integration of the two processes

The next step was the integration of the two processes as shown in Figure 3. The figure has two sections, the upgrader on the left hand side and the petrochemical complex on the right hand side. Bitumen as feed goes into the Crude Distillation Unit (CDU) where the lighter ends are taken out and the residue goes to the Vacuum Distillation Unit (VDU). The HVGO from the vacuum unit undergoes hydrotreating before being transferred to the NHC Unit. The gases from the NHC Unit go to the ethylene plant for separation. The cycle oils from the NHC unit are transferred to the Aromatic Saturation Unit followed by cracking in the Aromatic ring cleavage Unit for the production of paraffins and crude BTX. Business cases carried out indicate that this combination gives the best return on investment.

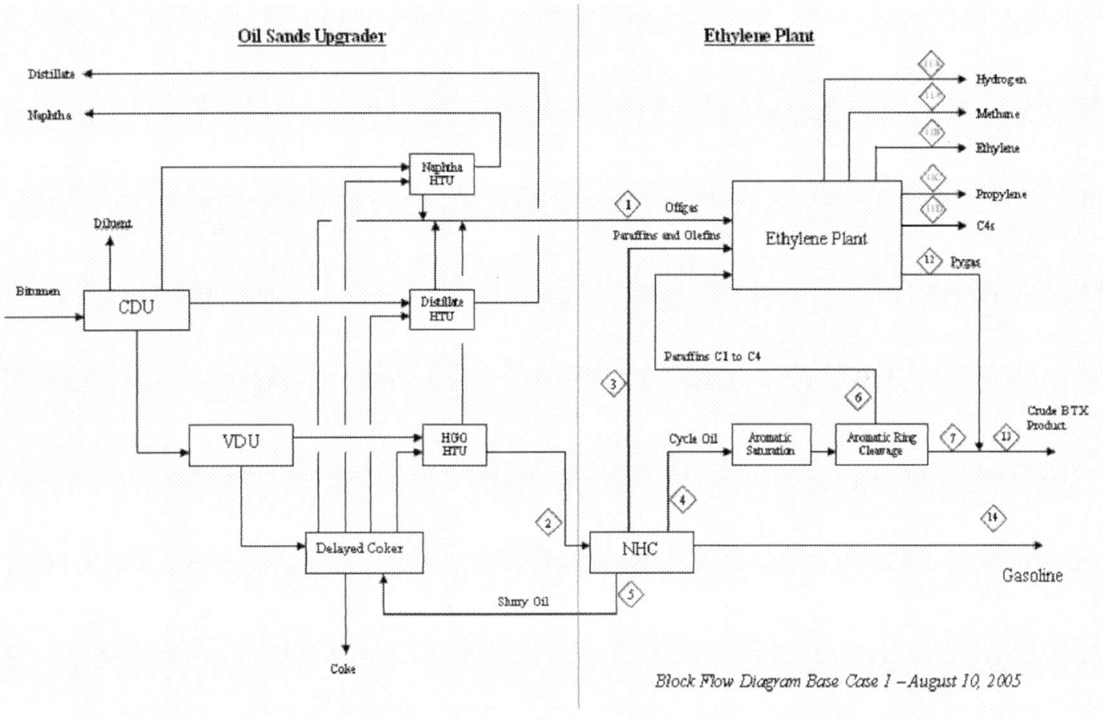

Figure 3: Integration of Oil Sands Upgrader with a Petrochemicals Plant and the NHC and ARORINCLE Units.

Advantages of the processes integrated with an Upgrader and Petrochemicals Complex in Alberta

This project when realized is expected to have the following advantages:

- Co-production of fuels and petrochemicals from bitumen creates higher value products and new market options for Alberta's oil sands products. This increases revenues and reduces the risk in fluctuating markets.
- Reduced natural gas impact in Alberta, as the Petrochemical manufacturer now has the option of utilizing the heavy liquid feedstock or extracting ethane/propane from gas.

- Increased attraction of capital investment in Alberta.
- Maintenance of the "Alberta Advantage" for the petrochemical industries and stimulation of R&D innovation within the province.
- Improved efficiency in the refining process for synthetic crude oil.
- Increased revenues for the Alberta oil sands producer.
- Higher employment; sustained or higher economic growth

References

[1] Hou Dianguo et al., China and Petroleum Processing and Petrochemicals Technology No.4 (2002) 51

[2] Shi Wenyuan et al.,ianguo et al., China and Petroleum processing and Petrochemicals Technology No.2 (2001)15

[3] G.C. Laredo, S. Leyva, R. Alvarez, M.T. Mares, J. Castillo, J.L. Cano, Fuel 81(2002) 1341.

[4] S.M. Yui, AOSTRA J. Res., 5(3) Chem. Abstr. ABSTR. NO. 101854 V112 N.12 (1989) 211.

[5] S.C. Korre, M.T. Klein, R.J. Quann, Ind. Eng. Res. 34 (1995) 101.

[6] W. Chen, Z. Cao, J. of Fushun Petroleum Institute, 16(1) (1996) 1.

[7] S. Jongpatiwut, M. Santikunaporn, J. E. Herrera, D. E. Resasco, W. E. Alvarez, E. L. Sughrue, G. W. Dodwellb, 13 ICC, Paris, 11-16 July (2004) paper P4-001

[8] S. Jongpatiwut, M. Santikunaporn, J. E. Herrera, D. E. Resasco, W. E. Alvarez, E. L. Sughrue, G. W. Dodwellb, App. Cat. A: General 262 (2004) 241.

[9] P. Rautanen, PhD Thesis: Liquid phase hydrogenation of aromatic compounds on nickel catalyst, Helsinki University of Technology (2002)

[10] M. Lylykangas, PhD Thesis: Kinetic modeling of liquid phase hydrogenation reactions, Helsinki University of Technology (2004)

[11] M.S. Lylykangas, P.A. Rautanen, A.O.I. Krause, Ind. Eng. Chem. Res. 41 (2002) 5632.

[12] C. Ringelhan, G Burgfels,. J. G. Neumayr, W. Seuffert, J. Klose, V. Kurth; Catalysis Today, 97(4), (2004) 277-282

[13] B.G. McVicker et al., EP0882001B1 Patent, ExxonMobil Chemical Europe Inc. (2002)

[14 L.B. Galperin et al., US5463155 Patent, UOP (1995)

[15] Y. Mitsuyoshi et al., JP1998292180A Patent, Hokkaido Government Industrial Research Institute (1998)

Selective Ring Opening of Diesel Fuels

R.G. Egeberg, S.I. Andersen, D.D. Whitehurst, G. Hytoft, and K.G. Knudsen
Haldor Topsøe A/S, Nymøllevej 55, DK-2800 Lyngby, Denmark

Introduction

Growth in the demand for high quality, ultra low sulphur transportation fuels means that new processes for upgrading of low quality feedstocks such as light cycle oil (LCO) must be introduced. In order to increase the cetane number and lower the density of diesel oils significantly with a low hydrogen consumption, the concept of selective ring opening (RIO) appears very attractive. However, although some RIO activity has been demonstrated for catalysts tested on model feeds, the selectivity and yields are generally low. We have investigated the potential of treating each hydrocarbon class separately. From studies of RIO mechanisms on model feeds, we have tried to predict the influence of hydrocarbon type transformations on the overall properties of real diesel feeds.

Ring Opening Mechanisms

The conversion of aromatic and naphthenic compounds under typical hydrotreating conditions has been the issue of a number of studies [1-4]. Prior to the opening of a 6-membered napthenic ring, a ring contraction to a methyl-cyclopentane structure occurs. These isomerized intermediates may then be ring opened by noble metals and/or an acid function. The yields reported of RIO products from two-ring structures are quite low - below 30%. The observed products are alkylcyclohexanes or alkylbenzenes, paraffins are not observed. Furthermore, the yield of RIO products is always accompanied by a yield of cracking products. In other words, the process is not selective, leading to loss of valuable diesel molecules and formation of naphtha and gas. Additionally, at most conditions isomerization is a significant side reaction, meaning that high-cetane n-paraffins will be transformed into low-cetane branched paraffins. Only selective metal-catalyzed hydrogenolysis cleaving substituted C-C bonds gives a substantial cetane increase [5].

We have studied the relation between such fundamental studies on model feeds and tests on real feed. From screening experiments with a large number of catalysts on model compound feeds, the best materials were selected for a real feed test on a Swedish Class I diesel. The observed yield structure (boiling point distribution) was very similar for all catalysts, whether zeolite or oxide based, noble metal based or sulphide. However, zeolite based materials required lower temperatures to reach the same shift in boiling point distribution. The results obtained were very similar to those obtained for a typical mild hydrocracking catalyst. So, whilst catalysts with a reasonable RIO activity indeed exist, it seems difficult to maintain a good selectivity when processing a real diesel with several different hydrocarbon types.

RIO Processes

Currently, two types of commercial process options exist for upgrading of low quality diesel. In a one-stage mild hydrocracking process, a slightly acidic sulphur-tolerant catalyst is employed. Hydrocracking reactions give good cetane improvement, however, at the expense of diesel yield. Another option would be to implement a two-stage HDS/HDA process, where the sulphur and nitrogen species are removed in the first stage. A noble metal catalyst is then used in the second stage to saturate aromatics. This process consumes relatively high amounts of hydrogen and will not give a high enough cetane increase in the case of naphthenic feeds. The RIO processes proposed in the literature are one of these types depending on the catalyst type, but no real feed data have been published yet, which demonstrate a better performance than that of the presently available mild hydrocracking or HDS/HDA catalysts. Therefore, it may be worthwhile to consider the potential of alternative process solutions.

In order to improve the selectivity of a RIO process, we have considered how to treat each of the hydrocarbon classes present in a diesel feed. Based on the low RIO selectivity of the catalysts tested so far, it was attempted to evaluate which hydrocarbon types had the highest potential for upgrading. For instance, normal paraffins are best left untreated to avoid the formation of branched low-cetane products. For isoparaffins, naphthenes and aromatics, the best treatment depends very much on the catalyst used and the value assigned to cetane and density improvements. The task is complicated by the scarcity of cetane data for model compounds. In general, the best route for higher aromatics appears to be hydrogenation to monoaromatics followed by ring opening to alkylbenzenes.

Naturally, this approach requires that a reasonably selective separation process exists, so that the hydrocarbon classes can be isolated. By selective adsorption on a number of adsorbents, we have investigated the feasibility of separating a real diesel feed. We find that the most challenging task is the separation of naphthenes from isoparaffins.

References

[1] G.B. McVicker, M. Daage, M.S. Touvelle, C.W. Hudson, D.P. Klein, W.C. Baird Jr., B.R. Cook, J.G. Chen, S. Hantzer, D.E.W. Vaughan, E.S. Ellis, O.C. Feeley, J. Catal. 210 (2002) 137.
[2] M Santikunaporn, J.E. Herrera, S. Jongpatiwut, D.E. Resasco, W.E. Alvarez, E.L. Sughrue, J. Catal 228 (2004) 100.
[3] D. Kubička, N. Kumar, P. Mäki-Arvela, M. Tiitta, V. Niemi, H. Karhu, T. Salmi, D.Y. Murzin, J. Catalk 227 (2004) 313.
[4] M.A. Arribas, A. Corma, M.J. Díaz-Cabañas, A. Martínez, Appl. Catal. A 273 (2004) 277.
[5] R.C. Santana, P.T. Do, M. Santikunaporn, W.E. Alvarez, J.D. Taylor, E.L. Sughrue, D.E. Resasco, Fuel 85 (2006) 643

100c Advanced Technology for Aromatics Saturation

Frank Mey[1], Rainer Schoedel[1], Reinhard Geyer[1], Dirk Lose[1], **Kristi Ann Morris**[2], and Paul Himelfarb[3]. (1) CRI KataLeuna, Leuna, Germany, (2) CRI Catalyst Company, Two Greenspoint Plaza, Suite 1110, 16825 Northchase Drive, Houston, TX 77060, (3) Westhollow Technical Center, Shell Chemicals, LTD, Houston, TX 77251
Nickel based catalyst are widely used for aromatics saturation. A deciding criterion for the catalyst selection is the sulfur content of the feed. The main parameters that influence the performance of nickel catalysts are discussed. A sulfur tolerance study shows that Nickel catalyst deactivation is a function of process conditions and sulfur concentration in the feed. Catalyst tolerance to sulfur poisoning can be increased under certain conditions by well over a factor of threefold thereby greatly increasing cycle length. A new concept for enhanced performance in aromatics saturation is presented. The key of this concept is a dual catalyst/process combining guard bed and aromatics saturation technology minimizing the sulfur poisoning impact by driving the sulfur into the bulk of catalyst. Test results and commercial experience in aromatics saturation will be presented.

Title: Effect of Chlorine, Fluorine and Titania on Surface Structure and Hydroprocessing Activity of $NiMo/Al_2O_3$ Catalysts Supported on Alumina

Deena Ferdous, Narendra N. Bakhshi, Ajay K. Dalai and John Adjaye***

Department of Chemical Engineering
Catalysis and Chemical Reaction Engineering Laboratories
57 Campus Drive, S7N 5A9, Saskatchewan, Saskatoon, Canada

*** Syncrude Canada Ltd. Edmonton Research Center, 9421- 17th Ave*
Edmonton, Alberta Canada T6N 1H4 Canada

Abstract

In this work, three catalysts namely $NiMo/Al_2O_3$-12 wt% Cl, $NiMo/Al_2O_3$-4 wt% F and $NiMo/Al_2O_3$-6 wt% Ti were prepared. All modified supports and fresh catalysts were characterized by BET surface area, pore volume and pore diameter, XRD, TPR, TPD, XPS, and XANES techniques. Addition of chlorine and fluorine formed weak-intermediate acid-centers on the catalyst surface whereas titania formed weak-intermediate-strong acid-centers. Also, these elements increased the formation of polymolydenum oxide on the catalyst as evident from TPR and XPS analysis. XANES analysis showed that incorporation of chlorine, fluorine and titania increased the formation of molybdenum oxide on the catalyst. The initial activity of these catalysts were tested in a trickle-bed reactor using three different gas oils such as light gas oil (LGO), blended gas oil (Blended: 50% LGO and 50% HGO) and heavy gas oil (HGO), all derived from Athabasca bitumen. This research showed that the optimum catalysts (containing 12 wt% chlorine or 6 wt% titania on Al_2O_3) removed over 86 wt% nitrogen compounds from blended feed which was higher than that from commercial catalyst. However, addition of additives in alumina had very little effect on sulfur conversion of blended and HGO.

1. Introduction

The increase in use of oil sand-derived gas oil for producing cleaner transportation fuel has spurred research for developing better catalyst for efficient removal of nitrogen and sulfur compounds from these stocks. The most commonly used support for the hydrotreating catalyst is γ-Al_2O_3. The hydroxyl groups of γ-Al_2O_3 play an important role in determining the surface chemistry of the resulting alumina-based catalysts [1]. The metal species of the catalyst interact with the alumina hydroxyl groups to produce dispersed metal oxide phases [2] during its preparation processes. This metal-support interaction plays an important role on catalyst activity by influencing dispersion of active ingredients and by influencing the reducibility and sulfiidity of this phase [3]. The addition of additives change the nature of the hydroxyl groups which in turn affect the structure and dispersion of metal oxide species [4]. Literature indicates that addition of chlorine, fluorine, and titania to Al_2O_3 can improve the HDN and HDS activities of $NiMo/Al_2O_3$ and $CoMo/Al_2O_3$ catalysts [5].

The fluoride ions increase the acidity of the catalyst which in turn increase the HDN [6] and HDS [7] activities of the catalyst. The influence of fluoride ions on the activity of catalysts depend on many parameters such as concentration of fluoride ions, reaction type, reaction temperature and the order of incorporation of fluoride ions upon the catalyst preparation. It was found that small amounts of fluoride (0.3 -0.8 wt %) increases the number of active centers and the dispersion of active metal [5]. Fluoride ions incorporated before molybdenum and cobalt ions increase HDS activity, and when incorporated in the last stage of impregnation decrease the HDS activity of the catalyst [8]. On the other hand, chloride ions also increase the acidity as well as HDS activity of the catalyst [8]. Lewandowski and Sarbak [9] reported that fluoride ions led to a decrease in the specific surface area of $NiMo/Al_2O_3$ catalyst, whereas chloride ions had little effect on this property. Also, modification of these catalysts using fluoride and chloride ions did not change the pore size distribution of the catalysts.

The first attempt was made with unconventional oxide like TiO_2 as support for various reactions such as the treatment of heavy oils, the hydrocracking of bituminous coal [10] and the hydrotreating of solvent refined coal [11]. TiO_2 support has been reported to provide higher dispersion of active phase [12] and increase in activity of CoMoS catalyst for HDS of benzothiophene [13]. However, this type of support has usually low surface area (≤ 50 m^2/g). On the hand, mixture of alumina and TiO_2 can provide large surface area, increase acidity [6], reduce interaction between MoO_3 and Al_2O_3 and increase reducibility [3] and sulfidity [14] of the catalyst, which are desired for hydrotreating of heavy gas oil.

Although a number of researchers used chlorine, fluorine and titania to modify $NiMo/Al_2O_3$ catalysts, most studies have been conducted using model compounds and light gas oils. Information on performance of these catalysts on the hydroprocessing of actual feed is scarce. To our knowledge, information on performance of chlorine, fluorine and titania promoted $NiMo/Al_2O_3$ on the hydroprocessing of actual feed especially from Athabasca bitumen is not available. In the present work, a systematic study has been conducted to examine the physico-chemical properties of a series of NiMo catalyst supported on chlorine, fluorine and titania modified Al_2O_3 and to correlate these with their HDN and HDS activities using three different gas oils derived from Athabasca bitumen.

2. Experimental

2.1. Catalyst Preparation

In this work, three NiMo catalysts were prepared using chlorine, fluorine and titania modified Al_2O_3 support. Chlorine, fluorine and titania concentrations in Al_2O_3 was maintained at 12, 4 and 6 wt%, respectively using appropriate amount of ammonium chloride, ammonium fluoride and titanium isopropoxide in n-propanol. After impregnation, the modified Al_2O_3 supports were dried at 120 °C for 12 hours and then calcined at 550 °C for 6 hours. NiMo based catalyst using these modified Al_2O_3 were prepared using an ammoniacal solution of ammonium heptamolybdate [$(NH_4)_6Mo_7O_{24}.4H_2O$] and nickel nitrate [$Ni(NO_3)_2.6H_2O$] which were dried for 12 hours at 120°C and then calcined at 500°C for 4 hours. In all catalysts, Ni and Mo concentrations were maintained at 3 and 14 wt%, respectively.

2.2. Catalyst Characterization

BET surface area, pore volume and pore size measurements of different modified supports and catalysts (~0.1 g) were performed using Micromeritics adsorption equipment (Model ASAP 2000, manufactured by Micromeritics Instruments Inc., Norcross, GA USA) at 78 K using liquid N_2. Before analysis, catalyst sample was evacuated at 200 °C for 4 h in a vacuum of 5×10^{-4} atm to remove all adsorbed moisture from the catalyst surface and pores.

Temperature programmed reduction (TPR) analysis of different catalysts was performed using Quantachrome equipment (Model ChemBET 3000, Manufactured by Quantachrome Corporation, FL, USA). The sample U tube containing 0.03 g of catalyst was placed in an electric furnace and heated from 22 to 1050 °C at 20 °C/min at one atmosphere in a reducing gas of 3 mol% H_2 in N_2 (obtained from Praxiar, Mississauga, ON, Canada) with a flow rate of 30 mL/min. H_2 consumption during the TPR experiments was measured with a thermal conductivity detector (TCD). The TPR plots were logged using an on-line data acquisition system.

Temperature programmed desorption (TPD) of ammonia was performed using Quantachrome equipment (Model ChemBET 3000, Manufactured by Quantachrome

Corporation, FL, USA). For TPD analysis, 0.1 g catalyst sample was placed in an adsorption vessel (U-shaped), heated with a flow of 0.1 %NH_3 in N_2 at the flow rate of 20 ml/min and at heating rate of 10 °C/min up to 400 °C and kept at this temperature for one hour. Then it was cooled to room temperature using this gas mixture at the same flow rate. Finally, the TPD was run from room temperature to 1050 °C at heating rate of 10 °C/min with He flow at 20 mL/min. The amount of NH_3 desorbed was measured using a TCD.

Powder X-ray diffraction (XRD) studies were performed using Rigaku diffractometer (Rigaku, Tokyo, Japan) using Cu Kα radiation filtered by a graphic monochromator at a setting of 40 kV and 130 mA. The powdered catalyst samples were smeared on glass slide with methanol and dried at room temperature. The X-ray diffraction analysis was carried out in the scanning angle (2θ) range of 4 to 80 ° at scanning speed of 5 °/min.

Fourier transform infrared spectroscopy (FTIR) analysis of different modified supports and catalysts were performed with and without pyridine adsorption using a PerkinElmer (Spectrum GX) spectrometer equipped with a MCT detector. The catalyst power of 10 mg was used in an in-situ FTIR cell (Thermo Spectra-Tech) and spectra was collected in the range of 400-4000 cm^{-1} wave numbers using 32 scans and a resolution of 4 cm^{-1}. FTIR analysis of pyridine adsorbed samples was also performed in the range of 1250-1650 cm^{-1}. In this case sample was dried for two hours and treated with pyridine for three hours. Pyridine treated sample was transferred to in-situ FTIR cell and spectra were collected using the same conditions.

X-ray photo electron spectroscopy (XPS) analysis was performed using Kratos Axis 165 XPS equipment using monochromated Al K α X-ray radiation. The pressure of the analysis chamber was maintained at $4x10^{-8}$ torr. For each sample, two particles were analyzed and the average data for the peak area and binding energy (BE) was taken.

X-ray absorption near edge structure (XANES) of all catalysts listed was performed at the Synchrotron Radiation Centre (CRC), Madison, WI in order to study the oxidation structure of molybdenum oxides of different catalyst. The measurements were carried out in fluorescence yield mode on the Canadian Double Crystal Monochromator (DCM) beam line. For both kinds of measurement, catalyst samples were pressed into self supported wafers and then put into soft X-ray beam.

2.3. Catalyst Performance Studies

Performance of different catalysts were tested in a micro-trickle bed reactor using three different feed materials such as light gas oil (LGO), blended gas oil (Blended: 50% LGO and 50% HGO) and heavy gas oil (HGO) derived from Athabasca bitumen. The properties of different feed materials are given in Table 1. The temperature, pressure, liquid hourly space velocity (LHSV) and hydrogen flow rate were maintained at 350 °C, 8.7 MPa, 2 h^{-1}, and 50 mL/min for light gas oil (LGO), 375 °C, 8.7 MPa, $1h^{-1}$ and 50 mL/min for blended gas oil (Blended: 50% LGO and 50% HGO) and 385 °C, 8.7 MPa, $1h^{-1}$ and 50 mL/min for heavy gas oil (HGO), respectively. The detail experimental procedures are given elsewhere [15]. The liquid product sample was collected after 24 hours interval. The products were stripped with nitrogen for removing the dissolved ammonia and hydrogen sulfide and then were analyzed for their total nitrogen and sulfur contents by combustion/chemiluminence technique (using Antek 9000 analyzer) following ASTM D4629 method.

Table 1: Nitrogen and sulfur content of different feed materials

	LGO	Blend	HGO
Total nitrogen (ppm)	623	2356	3199
Total sulfur (ppm)	27114	35100	39132

3. Results and Discussion

3.1. Characterization of Modified Supports and Catalysts

All fresh catalysts were characterized with respect to BET surface area, pore volume and pore size measurement. TPR, TPD, XRD, FTIR and XANES spectroscopy analyses were performed to study their physicochemical properties. The detailed discussion on these analyses is given below.

3.1.1. BET surface area, Pore Volume and Pore Size Measurement

BET surface area, pore volume and pore size of different titania modified supports and catalysts are given in Table 2. The experimental error on the measurement of surface area was ±2 m^2/g. No significant change in surface area of the support was observed with the incorporation of 12 wt% chlorine or 6 wt% titania in alumina which indicates a good dispersion of chlorine or titania on the alumina surface. However, there was decrease in average pore diameter indicating somewhat structural changes due to the incorporation of these elements. A significant drop in surface area was observed with the incorporation of 4 wt% fluorine in alumina which indicates the blocking of large pore at this fluorine concentration, confirmed by the decrease in pore volume of support. The data in Table 2 shows that the incorporation of Ni and Mo into modified alumina caused further decrease in surface area. Again, it could be due to the plugging of small pores. Highest surface area of ~210 m^2/g was obtained using catalyst containing 12 wt% chlorine which could be due to the breaking of Al-O-Al at higher chlorine concentration thus increasing in pore volume and pore diameter of this catalyst (see Table 2). Similar trends were also observed by Lewandowski and Sarbak [9].

Table 2: BET surface area, pore volume and pore diameter of different alumina supports and Ni(3wt%)Mo(14 wt%) catalysts

	Support			Catalyst		
	BET area, m^2/g	Pore vol. cm^3/g	Pore diameter, nm	BET area, m^2/g	Pore vol. cm^3/g	Pore diameter, nm
0	240	0.700	11.2	193	0.487	10.1
12 wt %Cl	234	0.690	11.8	201	0.537	11.1
4 wt% F	201	0.682	13.6	140	0.497	14.1
6 wt% Ti	238	0.636	10.6	181	0.468	11.0

3.1.2. Temperature Programmed Reduction

Temperature programmed reduction (TPR) analysis was performed to determine the nature of different oxide species present on the modified alumina and catalyst surface as well as to study the reducibility of the catalyst. The TPR profiles of different modified supports as well as NiMo catalysts supported on modified alumina is given in Figure 1. From Figure 1a it is observed that Al_2O_3 peak intensity decreased significantly with the incorporation of additives. It is known that there are six different hydroxyl groups on γ-Al_2O_3 [16]. The decrease in peak intensity is due to the replacement of hydroxyl groups by additives. Ramirez and Gutierrez-Alejande [17] reported that reduction peak at ~800 °C is due to the reduction of titania species interacted with alumina. Therefore, the peak at 811 °C in case 6 wt% titania containing alumina could be attributed to reduction of titania and alumina. New shoulders at 699 and 773 °C were also observed in case of titania containing alumina which could be due to the different degree of interaction of anatase titania phase with alumina [17].

TPR profiles of different $NiMo/Al_2O_3$ catalysts are given in Figure 1b. Mainly two peaks were observed for all catalysts. The peak at ~530 °C is probably the result of the partial reduction of well-dispersed polymolybdate-like species, i. e. $Mo^{6+} \rightarrow Mo^{+4}$ and reduction temperature shift from 530 to 550 ° C with the incorporation of 12 wt% chlorine or 6 wt%

titania whereas it shifts from 530 to 630 in case of 4 wt% fluorine. Peak at higher temperature (930 °C) represents complete reduction of polymolybdate plus the reduction of the tetrahedrally coordinated molybdate group i. e. $Mo^{6+} \rightarrow Mo^{+4}$ and $Mo^{+4} \rightarrow Mo^{+0}$ [18]. Figure 1b shows a significant increase in peak intensity with the incorporation of chlorine, fluorine and titania to Al_2O_3 which suggests that these elements increased the formation of molybdenum oxides in the catalyst. Modification of alumina support by additives leaves few basic hydroxyl

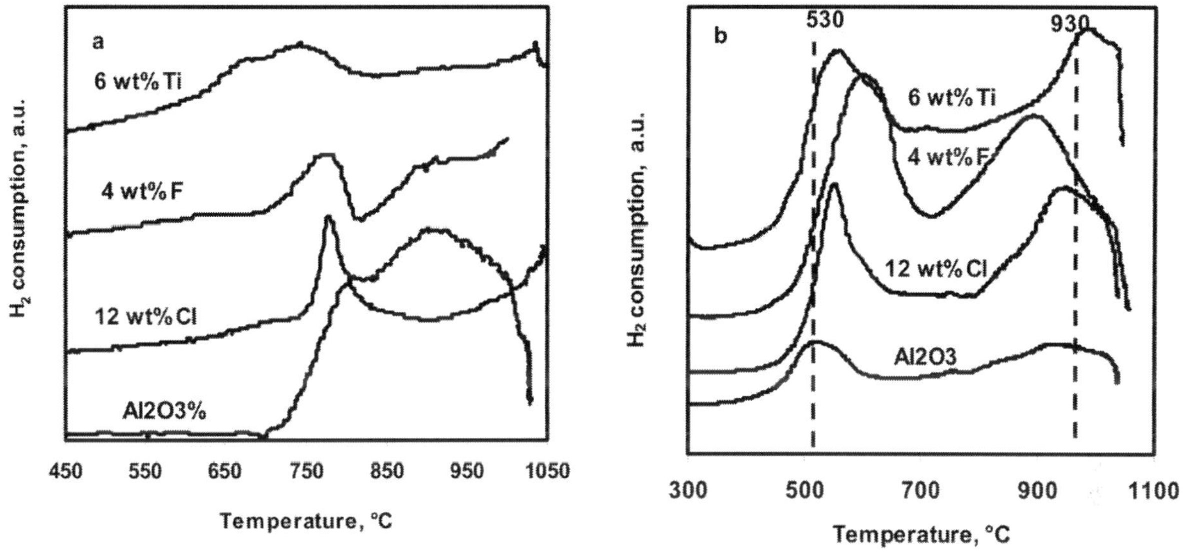

Figure 1. TPR profiles of different modified Al_2O_3 and catalysts (a) modified Al_2O_3 (b) NiMo Catalysts on modified Al_2O_3

groups and only a small number of non-basic hydroxyl groups for the metal metallate anions as evident from the TPR of titania modified alumina. As a result, the metallate anions can only react with the small number of non-basic hydroxyl groups. Therefore, modification of the alumina support may weaken the metal-support interaction which increases the formation of polymolybdate stacks on the catalyst surface. In conclusion, this analysis shows that addition of additives to alumina caused the formation of more molybdenum oxide on the catalyst.

3.1.3. Temperature Programmed Desorption

Temperature programmed desorption (TPD) of adsorbed ammonia was used to measure the acidic properties of different supports and catalysts. NH_3 can be absorbed on the catalyst surface physically and chemically. Ammonia is a strong base and expected to interact with OH species which are responsible for the acidity of Al_2O_3. The physically absorbed NH_3 is desorbed at lower temperature whereas chemically absorbed NH_3 is desorbed at higher temperature.

TPD profiles of different modified supports and NiMo catalysts supported on modified γ-Al_2O_3 are given in Figure 2. There are different ranges of temperature for the desorption of ammonia, namely, 20-200, 200-350 and 350-550 °C, which correspond to weak, intermediate, and strong acid centers, respectively [19]. Pure alumina shows only a broad peak in the temperature range of 105-400 °C which indicates the combination of weak-intermediate-strong acid centers. Incorporation of additives shifts peak to higher temperature which is due to the replacement of some basic hydroxyl groups by chlorine, fluorine or titania. Similar trends were also observed in case of catalyst (see Figure 2b). In summary, TPD analysis showed the formation of weak-intermediate acid centers on chlorine and fluorine containing catalysts and weak- intermediate-strong acid centers on titania containing catalyst.

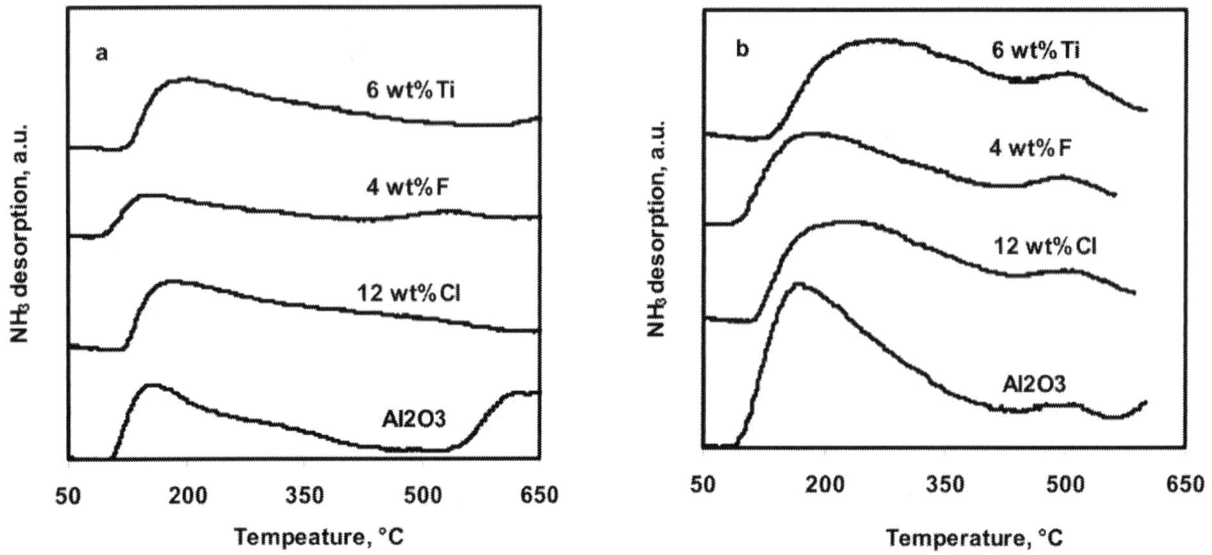

Figure 2. TPD profiles of different modified Al_2O_3 and catalysts (a) modified Al_2O_3 (b) NiMo Catalysts on modified Al_2O_3

3.1.4. X-ray Diffraction

X-ray diffraction (XRD) was used to detect different crystalline phases on the catalyst surface. The analysis of XRD spectra of different modified Al_2O_3 supports and NiMo catalysts supported on modified alumina showed only peaks corresponding to γ-Al_2O_3 which indicates well dispersion of chlorine, fluorine or titania on alumina. XRD analysis of different catalysts showed that addition of chlorine and fluorine in alumina caused the formation of MoO_3 crystals whereas no crystalline MoO_3 was observed in case of titania containing catalyst.

3.1.5. FTIR analysis

Fourier transform infrared (FTIR) spectroscopy is particularly useful to characterize heterogeneous catalysts. Application of this technique to catalysts provides useful micro-structural information that can be correlated with catalytic activity. FTIR analysis was performed in order to study the different functional groups, and Lewis-Bronsted acidity of different modified alumina and NiMo catalysts supported on modified alumina. Each experiment was performed for 2 to 3 times and the results were quite reproducible.

The FTIR spectra of different modified alumina and NiMo catalyst supported on modified alumina shows regions 3000-4000 cm^{-1} and 900-1800 cm^{-1} which are termed as the hydroxyl stretching region and frequency multiplied range (plot not shown). In the present work, a wide peak in between 3751 to 2816 cm^{-1} was observed for all samples which are the combination of different hydroxyl groups. Decrease in peak intensity with the incorporation of additives indicates that some of the OH groups are replaced by chlorine, fluorine or titania. The band at 1670 cm^{-1} was attributed to the presence of the adsorbed water and the band at 1100 cm^{-1} represents Al-O stretching vibration of alumina. Similar trends were also observed for catalyst.

FTIR analysis was also performed on pyridine adsorbed samples to identify Lewis and Bronsted sites. Pyridine is sensitive to both Lewis and Bronsted acid sites. Three bands at 1442, 1487 and 1589 cm^{-1} were observed in case of different supports (Figure 3a). The band at 1442 cm^{-1} is due to the presence of Lewis acid site. The band at 1589 cm^{-1} is attributed to the presence of Lewis acid sites. However, no Bronsted acid sites were observed on the modified alumina. This plot clearly shows that incorporation of addition increases the formation of Lewis acid sites. The trends for different catalysts are very similar to those observed in case of modified alumina (Figure 3b). This plot shows a wide peak at ~1540 cm^{-1} which corresponds to Bronsted acid sites. However, no significant change in Lewis and Bronsted acidity was observed in case of catalyst with the incorporation of chlorine, fluorine and titania.

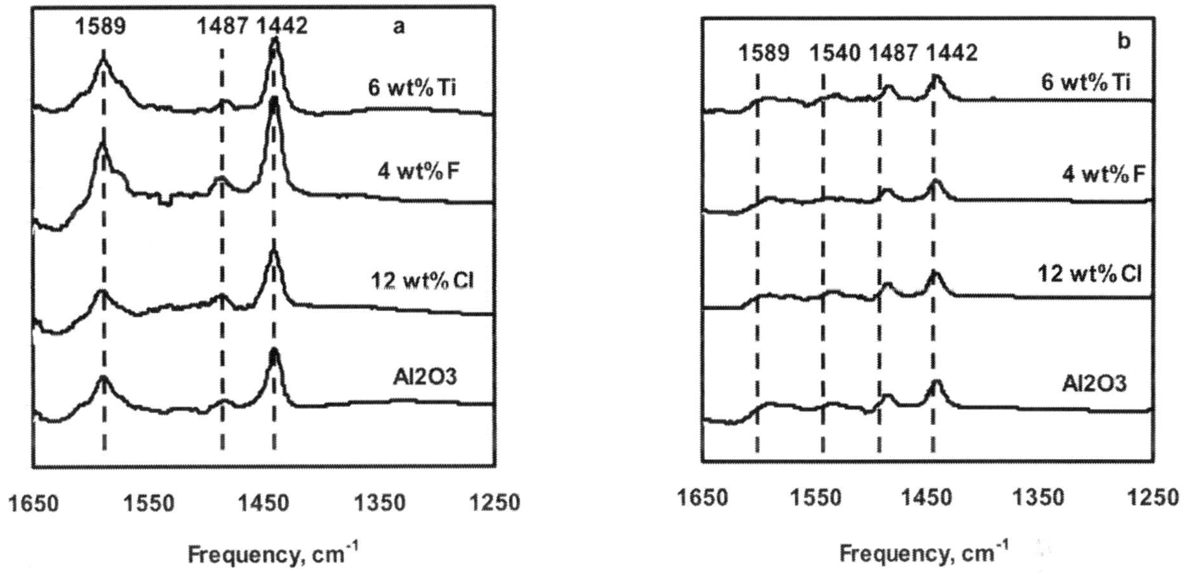

Figure 3. FTIR spectra of different pyridine adsorbed supports and catalysts (a) supports (b) catalysts

3.1.6. X-ray Photo Electron Spectroscopy

X-ray photo electron microscopy (XPS) analysis of different catalysts was carried out in order to study surface structure molybdenum oxides. The Mo 3d spectra of different catalysts are given in Figure 4. Mo 3d XPS shows two peaks (peak at 231.6 eV and peak at 234.4 eV) for all catalysts which could be attributed to the binding energies of electrons in the $3d_{3/2}$ and $3d_{5/2}$ levels of molybdenum in the $Mo+^6$ valance state [20]. From this figure it is clearly observed that 12 wt% chlorine and 6 wt% titania in alumina increases the formation of MoO_3 on the catalyst surface which was also confirmed from TPR

3.1.7. X-ray Absorption Near Edge Structure Spectroscopy

XANES is associated with the excitation process of core electron to unoccupied bounded states. Mo $L_{III,II}$-edges XANES was performed in order to study the nature of different molybdenum oxides of catalysts. This technique also gives information on the electronic states of the X-ray absorption atom and its surrounding local structure. Mo $L_{III,II}$-edges XANES of different catalysts containing 12 wt% chlorine, 4 wt% fluorine and 6 wt% Ti is given in Figure 5 . Mainly two bands at 2527 and 2630 eV were observed for all catalysts which correspond to Mo L_{III} and L_{II} edges, respectively. Both the L_{III} and L_{II} -edges describe the transactions from the 2p core state to the empty states of the d-band. The L_{III} band is originated when the transition proceeds from $p_{3/2}$ core-level to a vacant d final state, while the electronic transition from $p_{1/2}$ level to the same d-band produces the L_{II} band [21]. These two bands were also compared with the reference MoO_3 and MoO_2 compounds to distinguish between poly and tetra molybdenum oxide. Similar two bands were also observed in case of reference compounds. Moreover, it was not possible to identify whether these bands are for polymolybdenum or tetramolybdenum oxide. However, this analysis indicates that incorporation of chlorine, fluorine and titania in alumina increased the formation of

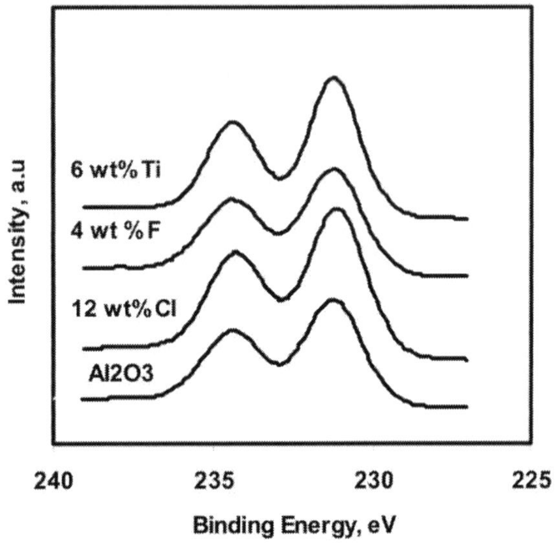

Figure 4. XPS spectra of different catalysts

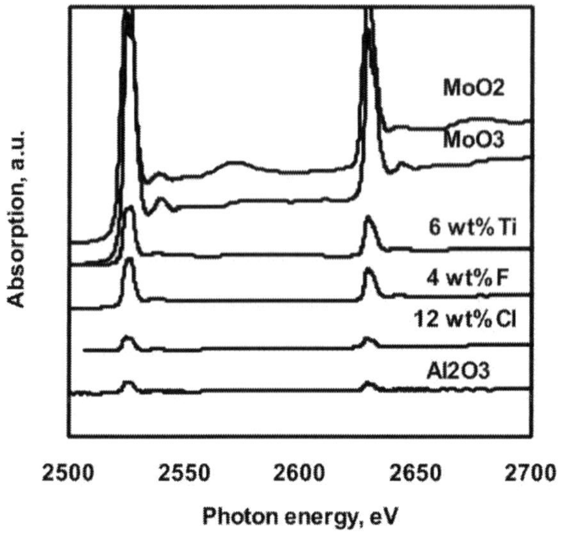

Figure 5. XANES spectra of different catalysts

molybdenum oxide in NiMo catalyst as evident from the increase in peak intensity.

3.2. Studies of HDN and HDS

In this research, the conversions of nitrogen and sulfur were calculated based on equation 1.

$$\% \text{ conversion} = \frac{[\text{Feed amount}] - [\text{Product amount}]}{[\text{Feed amount}]} \times 100 \qquad (1)$$

Where, [Feed amount] and [Product amount] are the concentrations of nitrogen and sulfur in feed and products, respectively. Few experiments were repeated and the reproducibility for the nitrogen and sulfur conversions was in the range of 1-1 .5 wt%. The effect of chlorine, fluorine and titania concentrations on nitrogen conversion for three different feeds are given in Figure 6a. For all feed materials total nitrogen conversions increased significantly with the incorporation of additives in alumina. This increase is significant in case of LGO and blended feed. For example, nitrogen conversion increased from 57 to 71 wt% when 12 wt% chlorine or 6 wt% titania was added in alumina whereas it increased from 57 to 66 wt% when 4 wt% fluorine was added. It is known that during NiMo catalyst preparation, both tetrahedral and octahedral molybdenum oxides are formed on the catalyst surface and only octahedral molybdenum oxides are known to be the precursor of active phase for hydrotreating reaction [14]. The increase in HDN activity with the addition of chlorine, fluorine or titania could be attributed to the increase in octahedral molybdenum oxides formation as evident from TPR, XANES and XPS spectroscopy analyses. Figure 6a shows that rate of increase in nitrogen conversions in case of LGO and blended are much higher than that of HGO. It is due to the presence of less refractory nitrogen and sulfur compounds in these feed materials. Moreover, nitrogen conversion obtained from these catalysts is comparable with that obtained from commercial catalyst.

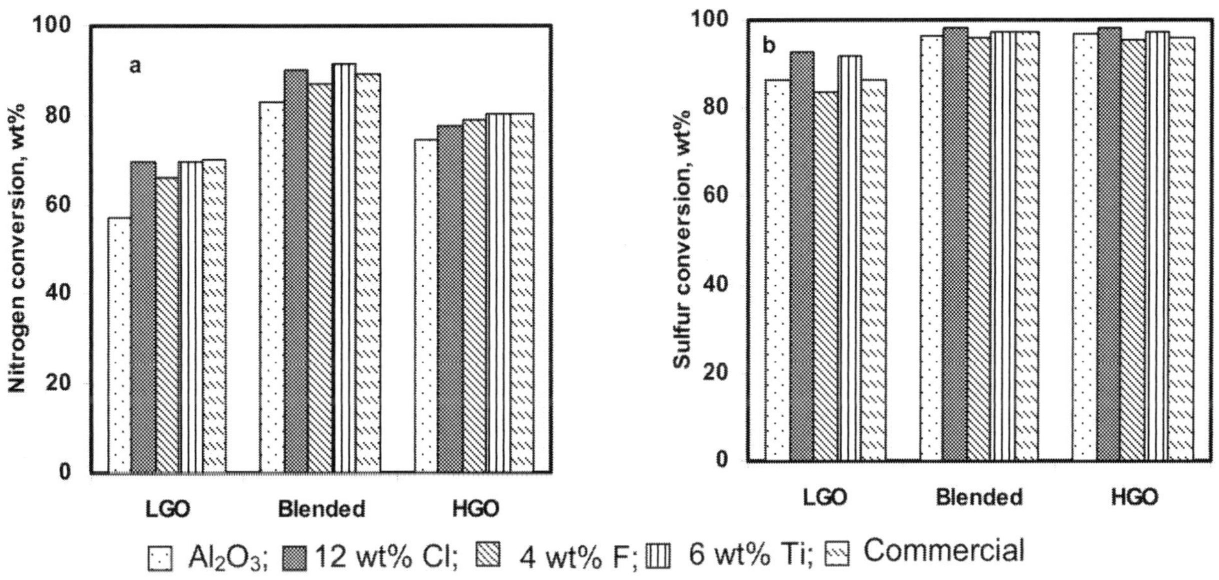

Figure 6 Effect of additives on total nitrogen and sulfur conversion of different gas oils LGO: T=350°C; P=8.7 MPa; LHSV= 2 h^{-1}; Blended: T=375°C; P=8.7 MPa; LHSV= 1 h^{-1}; HGO: T=385°C; P=8.7 MPa; LHSV= 1 h^{-1}; (a) nitrogen; (b) sulfur

The effect of additives on the HDS of different feed materials is given in Figure 6b. In case of LGO, sulfur conversion increased from 86 to ~ 92 wt% with the addition of 12 wt% chlorine or 6 wt% titania in alumina whereas no significant change in sulfur conversion was obtained with the incorporation of 4 wt% fluorine. Increase in HDS activity in case of titania containing catalyst could be attributed to the increase in surface acidity. On the other hand, Ramirez et al. [22] stated that increase in HDS activity in the presence of titania in alumina is due to the synergistic effect. They also proposed that the partial reduction and sulfidation of TiO_2 to Ti^{3+} species can act as promoter to the MoS_2 phase. They indicated that in the case of TiO_2–Al_2O_3 hydrotreating catalysts, after sulfidation and in reaction conditions in the presence of a high hydrogen partial pressure, titania can be reduced and sulfided. Under this condition, the titania sulfide possesses semiconductor properties, and thus has will have metal conductivities. In the sulfided catalyst, the supported molybdenum sulfide active phase also has the semiconductor properties [22]. As calculated by Harris and Chianelli [23], the empty orbital of Mo in MoS_2 lies very near to the d orbital of Ti in TiS_2. At high hydrogen partial pressure, partial reduction of TiO_2 and/or TiS_2 surface layers occurs and electrons from Ti^{3+} species can easily be injected to the Mo3d conduction band. These authors indicated that this is the required conditions for the MoS_2 promotion in HDS catalysts. This explanation indicates that the role of titania in HDS catalysis is not only that of a conventional support but also it acts as an electronic promoter of the molybdenum sulfide phase. Increase in sulfur conversion in case of chlorine containing catalyst could be due to electronic effect of chlorine. Moreover, Figure 6b shows no significant change in sulfur conversion in case of blended and HGO and it was in the range of 93-99 wt%. This is because sulfur conversion reached a point at which further conversion was limited because the remaining sulfur is considerably less reactive than the sulfur which had been converted [24]. No change in sulfur conversion in case of blended and HGO is also due to the complex mixture of nitrogen and sulfur compounds. However, the sulfur conversions obtained for LGO using chlorine and titania containing catalysts were higher than that obtained from commercial catalyst.

4. Conclusions

This work indicates that addition of additives such as chlorine, fluorine and titania to alumina improves HDN and HDS of different gas oils derived from Athabasca bitumen. Formation of molybdenum oxides increased significantly in the presence of additives. Chlorine, fluorine and titania also increased the formation of polymolybdenum oxide in catalyst as evident from TPR and XPS analyses. FTIR analysis showed the formation of more Lewis acid sites on alumina in the presence of additives. However no significant change in Lewis-Bronsted acidity was observed in case of NiMo catalyst supported on modified alumina. Nitrogen conversion increased from 57 to 71 wt% when 12 wt% chlorine or 6 wt% titania was added in alumina whereas it increased from 57 to 66 wt% when 4 wt% fluorine was added. For HGO maximum nitrogen conversion of ~80 wt% was observed in case of 6 wt% titania containing catalyst. Additives had significant effect on LGO sulfur conversion.

References

1. A. Benitez, J. Ramirez, A. Vazquez, D. Acosta, A. A. Lopez, Appl. Catal. A: General. 133 (1) (1995) 103.
2. Y. Okamoto, T. Imanaka, J. Phys. Chem. 92 (1988) 7102.
3. J. R. Grzechowiak, I. Wereszczako-Zielinska, J. Rynkowski, M. Ziolek, App. Catal. A General 250(1) (2003) 95.
4. S. I. Kim, S. I. Woo, J. Catal. 133 (1992) 124.
5. M. Lewandowski, Z. Sarbak, Appl. Catal. A: General. 1997, 156(2), 181.
6. J. A. R. Van Veen, H. A. Colijn, P. A. J. M. Hendriks, A. J. Van Nelsons, Fuel Pro. Tech. 35 (1993) 137.
7. J. F. Cambra, P. L. Arias, M. B. Guemez, J. A. Lagarreta, J. L. G. Fierro, Ind. Eng. Chem. Res. 30 (1991) 2365.
8. H. Matralis, C. Papadopoulou, A. Lycourghiotis, Appl. Catal. A General 116 (1994) 221.
9. M. Lewandowski, Z. Sarbak, Crys. Res. Tech. 31(6) (1996) 773.
10. K. Tanabe, H. Sssasaki, H. Hattori, K. Ouchi, K. Machino, H. Itoh, Fuel Pro. Tech. 2(4) (1979) 253.
11. I. Mochida, T, Oishi, Y. Koral, H. Fujitsu, Ind. Eng. Chem. Process Des. Dev. 23 (1983) 203.
12. E. Lecrenay, K. Sakanishi, I. Mochida, Catal. Today 39(1-2) (1997) 13.
13. H. Tanaka, M. Boulinguiez, M. Vrinat, Catal Today 29 (1996) 209.
14. Z. B. Wei, W. Yan, H. Zhang, T. Ren, Q. Xin, Z. Li, App. Catal. A: General 167(1) (1998) 39.
15. D. Ferdous, A. K. Dalai, J. Adjaye, App. Catal. A General 260 (2004) 153.
16. M. Sun, D. Nicosia, R. Prins, Catal. Today 86(1-4) (2003) 173.
17. J. Ramirez, A. Gutierrez-Alejandre, J. Catal. 170(1) (1997) 108.
18. Y.- C. Park, E.-S. Oh, H.-K. Rhee, Ind. Eng. Chem. Res. 36 (1997) 5083.
19. M. Lewandowski, Z. Sarbak, Fuel 79 (2000) 487.
20. F. P. Rouxinol, B. C. Trasferetti, R. Landers, M. A. B. de Moraes, J. Brazilian Chem. Soc. 15(2) 2004 324.
21. E. J. Lede, F. G. Requejo, B. Pawelec, J. L. G. Fierro *J. Phys. Chem. B.*, 106(32) (2002) 7824.
22. J. Ramirez, L. Cedeno, G. Busca, J. Catal. 184(1) (1999) 59.
23. S. Harris, R. R. Chianelli, J. Catal. 98(1) (1986) 17.
24. J. -W. Kim, F. V. Hanson, Energy (Oxford) 23(3) (1998) 221.

103a Producing Synthetic Crude Oil from Heavy Oils: Process Heat Integration

Jorge Ancheyta[1], Jose A.D. Munoz[1], and Ignacio Elizalde[2]. (1) Instituto Mexicano del Petróleo, Eje Central Lázaro Cárdenas 152, Col. San Bartolo Atepehuacan, Mexico, (2) Facultad de Química, UNAM, Ciudad Universitaria, Col. San Bartolo Atepehuacan, Mexico

Due to the increasing production of heavy crudes and the worldwide demand for cleaner fuels, new alternatives must be found for satisfying those requirements. On the other side, the production of light crudes is decreasing and hence in the future the refineries will have to process higher amounts of heavy crudes. The great challenge for the refineries is to transform heavy crudes and residues into valuable products to maximize the production of light fractions for the elaboration of fuels that satisfy the environmental regulations [1].

The heavy crude hydrotreatment process (HDT), developed at IMP (The Mexican Institute of Petroleum), is an alternative to solve the problematic of heavy crudes. The HDT process has been tested at experimental scale, the details are provided elsewhere [2]. The HDT process employs a system of two reactors in series with selective catalysts, the first for hydrodemetallization (HDM) and the second for hydrodesulfurization (HDS) for processing the heavy fraction. One process option is to hydrotreat the light fraction and the product of both hydrotreating steps is mixed to formulate synthetic crude.

Since the process configuration includes several cold and hot streams, it is then necessary to perform heat integration studies. Additionally, the reactions are exothermic, and they must be quenched with hydrogen recycle streams and with heat integration in order to control the temperature to keep the products quality. The industrial HDT scheme has a lot of potential for process heat integration due to the high temperatures of the catalytic beds effluents and other streams which can be used to preheat cold process streams. Thus, the main objective of this study is to develop process heat integration in a heavy crude oil hydrotreating plant. Experimental data obtained from pilot plant studies, simulations with commercial software, and Pinch analysis are employed to select the best case from an energetic point of view.

References

1. E. Furimsky, Appl. Catal. A 171 (1998) 177-206. 2. J. Ancheyta, G. Betancourt, G. Marroquín, G. Centeno, L.C. Castañeda, F. Alonso, J. A. Muñoz, Ma. T. Gómez, P. Rayo. Appl. Catal. A 233 (2002) 159-170

RESIDUE UPGRADING WITH SYDEC^{SM} DELAYED COKING: BENEFITS & ECONOMICS

John D. Elliott, Foster Wheeler USA Corp., Houston, Texas
David A. Wedlake, Foster Wheeler USA Corp., Houston, Texas

Introduction

Improved refinery margins and profitability are often achieved by taking advantage of lower cost crude oil while increasing the production of high value, high demand distillate products and phasing out low value, low demand products such as high sulfur fuel oil (HSFO). Increasing the use of lower cost crude oils usually involves either long-term deals and/or taking advantage of opportunity crude oils having discount pricing. These crude oils are usually heavy and sour.

Heavy crude oils are characterized by having larger percentages of residues than the lighter benchmark crude oils. Residues from these heavy crudes are frequently higher in sulfur, asphaltenes, metals, nitrogen and often TAN (Total Acid Number) compared to residues from benchmark crude oils. Handling these large volumes of heavy residue without producing excessive volumes of HSFO requires that the refiner have residue, or bottom of the barrel, conversion facilities to turn the residue into more valuable distillate products. Those refiners that produce large amounts of heavy, high sulfur fuel oil generally operate at lower refinery margins and profitability than full conversion refiners. They can also suffer from low production of distillates. High demand for distillates is driven by the clean transportation fuels market. For example, there is increased demand for gasoline in North America and increased demand for diesel in Western Europe.

Private studies have shown that complete bottom of the barrel conversion through delayed coking to minimize or eliminate heavy fuel oil can raise refining margins by US $1.50 to US $4.50 per barrel of crude oil. While this value depends on local situations it provides the incentive for residue conversion. Switching to lower price heavy crude oils and taking advantage of the light-heavy spread in pricing provides further economic incentive for these projects.

Residual oil conversion projects are used to meet these needs by converting low value residual oil. Commonly evaluated primary conversion processes are:

<u>Non-Catalytic Processes</u>
Visbreaking
Solvent Deasphalting
Delayed Coking
Flexicoking or Fluid Coking
Partial Oxidation

<u>Catalytic Processes</u>
Ebullated Bed Hydrocracking
Slurry Phase Hydrocracking
Residual Fluid Catalytic Cracking

All the residue processes listed above have good features and specific uses that the refiner must evaluate and decide which is best for his specific situation. Delayed Coking is the most common choice for these types of projects because it offers the following benefits:

a. Complete conversion of residue feedstocks and elimination of residual fuel oil production. Demetallization is nearly 100 percent.
b. Production of a broad range of liquid distillates that can easily be accommodated in an existing refinery to meet increased demand for clean distillate transportation fuels.
c. Moderate capital investment that is quickly recaptured.
d. The process has been commercialized for over 70 years and is well supported by specialty equipment vendors, chemical suppliers and third party consulting firms.
e. Byproduct fuel grade coke is easily sold at positive netback prices.
f. Modern delayed coker designs are shown to be energy efficient, environmentally friendly and intrinsically safe.
g. The semi-batch nature of the delayed coker makes the operation easier and provides schedule leeway on the rare occasions when equipment fails.
h. Operating and maintenance costs are reasonable.

With respect to utilization of opportunity crude oils, the flexibility of the SYDEC^{SM} Delayed Coking process allows a wide variety of crude oil qualities to be economically processed. Furthermore, the thermal cracking nature of the process is not affected by high TAN residues and does not produce high TAN distillates. Additionally there are a few opportunity, high TAN crude oils that are low enough in sulfur and metals to produce anode grade petroleum coke which has a much higher value than the fuel grade coke produced from the typical high sulfur, high metals crude oil residues.

Recognizing that residue can be upgraded via catalytic or non-catalytic processes, this paper is focused on the most commonly used, commercially proven and reliable technology, namely Delayed Coking.

SYDEC^{SM} Delayed Coking

Foster Wheeler is the leading supplier of delayed coking technology with its SYDEC^{SM} (Selective Yield Delayed Coking) process.

Foster Wheeler's typical SYDEC^{SM} design basis for maximum liquid yield cokers minimizes the coking pressure and recycle ratio within economic and technical constraints while producing clean Heavy Coker Gas Oil (HCGO) distillate product with low contaminant levels of carbon residue and metals. A Foster Wheeler SYDEC^{SM} coker operating at low pressure and ultra-low recycle may produce as much as 25 percent less coke than one operated at traditional pressures and recycle ratios. A coker operation with zero recycle will produce even less petroleum coke and higher yields of liquid products at the expense of poorer quality heavy coker gas oil (HCGO) having higher contaminant levels. If the downstream processing of the HCGO can economically accommodate the higher metals and Conradson carbon, this type of operation can be very attractive for incorporation into the

Sydec^{SM} Delayed Coker design basis. A comparison of theoretical HCGO yields and properties resulting from the three main types of delayed coker design and operation is presented in Table 1.

Table 1

Comparison of Heavy Coker Gas Oil Properties Based on Delayed Coker Operation

HCGO	Conventional	SYDEC^{SM} Ultra-Low Recycle	SYDEC^{SM} Zero Recycle
WT% Yield	24.4	37.9	41.8
End Point, °F	920	1060	1110
Gravity, °API	18.4	15.6	14.7
CCR, wt%	0.25	0.6	1.2
Ni + V, wppm	0.5	0.8	1.8

Delayed Coking products typically require further downstream hydroprocessing because of hydrogen deficiency and sulfur levels Figure 1 shows the typical SYDEC^{sm} process flow diagram.

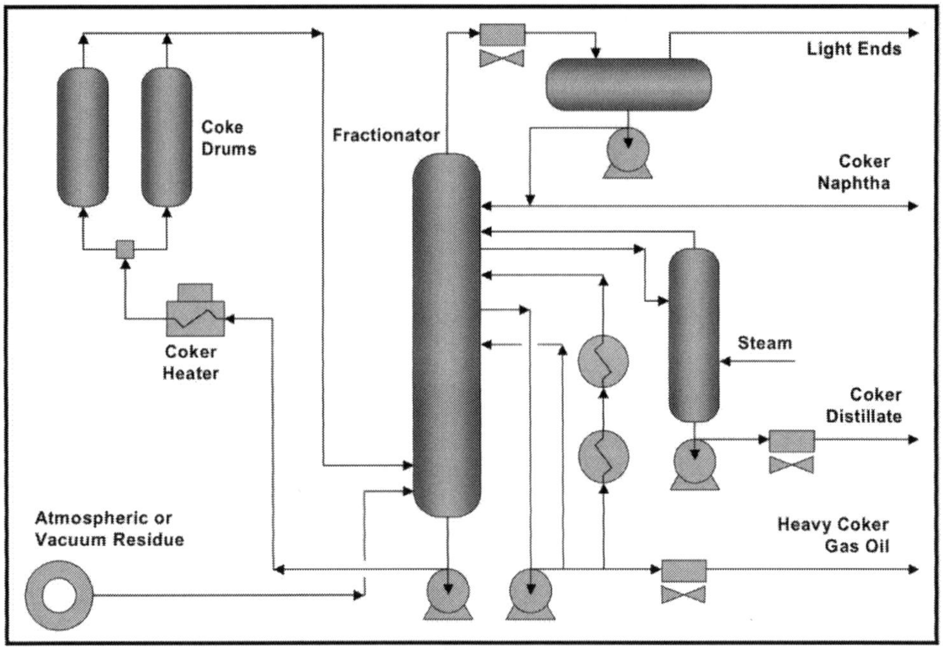

Figure 1

SYDEC^{SM} Delayed Coking Process Flow Diagram

Environmental Features of SYDEC^{SM} Delayed Cokers

- Recovery of unconverted blowdown system hydrocarbons
- Water recovery and reuse, including coke cutting water and process sour water.

- Heater emissions management, such as low NOx burners and Selective Catalytic Reduction.
- Coke handling features for minimizing particulate emissions including enclosed conveyor systems, water sprays and coke pit designs protected from wind.
- Sludge injection facilities to convert refinery sludges.

Petroleum Coke (Petcoke) Disposal

Concerns about coke markets and disposal opportunities for fuel grade coke are unfounded for most geographical locations. Although fuel grade coke is a low value byproduct (compared to distillate transportation fuels), there is significant worldwide trade and demand for fuel coke, because of its high caloric value, low ash and discount pricing relative to coal. This ensures a ready market to power utility stations, cement producers and others. The net price a refiner realizes for fuel grade petcoke is a function of sulfur levels, coke hardness and transportation costs. There are a number of international corporations that will handle and market petcoke for the refiner. Anode coke is a higher value product that can be produced from low sulfur, low metals feedstocks, either straight run residues, hydrotreated residues, specialty tar stocks, or solvent deasphalting routes.

Best-in-Class Features of SYDEC^{SM} Delayed Coking

The SYDEC^{sm} Delayed Coking process features provide economic, safe and robust conversion of residues to lighter products and coke. The following table contains a partial listing of its "benchmark" features.

Table 2
SYDEC^{SM} Features

Feed/Operating Parameters	**Coke Drums**
- Deep vacuum residue feed from heavy crude (1050 °F+ or greater) - Low pressure coking (15 psig or slightly less) - Ultra-low recycle (TPR=1.05) for maximum clean liquid products - Low cycle times • 16 to 18 hours: new unit • 12 to 14 hours: revamped unit - Coke quench time maximized to minimize drum stresses.	- Large Size (28 to 32 ft. dia.) for minimum number of modules - Advanced plate chemistry and finishing specifications - Single wall thickness - Hydraulically powered, remotely operated Advanced drum unheading • Top: Foster Wheeler hinge system • Bottom: Slide valve closed system or Foster Wheeler automated with manual unbolting - Motor driven major isolation valves; interlocked isolation valving - High viscosity antifoam (100,000+ cks) with additive nozzle located away from exit vortex

<table>
<tr><td></td><td>- Fatigue resistant mechanical design optimization</td></tr>
<tr><td>Fractionator
- Spray chamber wash zone
- Flash zone heat shield
- Coke screen & coke fines removal circuit
- Heater charge pump with coke crushing impeller
- On-line water wash capability
- Direct recovery of coke condensate/wax tailings into fractionator</td><td>Coke Dewatering and Handling
- High volume maze for fines recovery and/or secondary fines recovery
- Full drum volume below chute
- Covered conveyors
- Drum drain to pit
- On-site coke storage for draining/cooling</td></tr>
<tr><td>Heater
- Foster Wheeler double fired design
- 1 year plus run length with on-line spalling
- Low NOx burners with SCR if required
- Minimum length transfer line</td><td>Slop/Sludge Handling
- Sludge injection during coke quench
- Slop, wax tailings & coke condensate used as overhead line quench
- Overhead line quench spray nozzle using slop oils
- Slops dewatering using waste heat</td></tr>
<tr><td>Coke Cutting
- Large, high jet pump
- Hydraulically operated winches/rotary joint
- Combination cutting tools
- Free fall arrestor
- Interlocked safety system</td><td>Blowdown System
- Robust Blowdown drum internals
- Continuous operation to handle coke drum PSV discharge
- Direct recovery of vent vapors</td></tr>
<tr><td colspan="2">Safety Interlocks
- Coke drum inlet isolation valves, switch valve, utility valve and bottom unheading
- Coke drum outlet isolation valves and Blowdown valves
- Coke drum vent, PSV isolation and drum top unheading device
- Heater pass flow, fuel pressure and combustion air
- Coke cutting system</td></tr>
<tr><td colspan="2">Safety features incorporated during mechanical design
- Enclosed shelter for coke cutting operator
- Coke drum structure fire protection systems
- Coke drum structure emergency egress pathways</td></tr>
<tr><td colspan="2">Execution & Operability Benchmark Factors
- Fast project delivery: 27.5 months for a benchmarked project from start of basic design to "oil in"
- Fast and safe startups: days not weeks
- High on-stream factors: audited at 0.95 for first six months of operation
- Long periods between turnarounds: 5 years typical</td></tr>
</table>

THE IMPACT OF ADDING SYDEC[SM] DELAYED COKING TO A TYPICAL LOW CONVERSION REFINERY MAKING HIGH SULFUR FUEL OIL

The benefits of incorporating SYDEC[SM] Delayed Coking into a HSFO producing refinery are illustrated in the simplified case study described below. The non-residue conversion refinery consists of the following major process units: Crude & Vacuum Unit, Fluid Catalytic Cracking Unit (FCC) of vacuum gas oil, Hydrotreater for diesel and jet production and Catalytic Reformer for gasoline production

The non-residue conversion refinery can be seen in Figure 2.:

Figure 2

BFD: Base Case Refinery Configuration

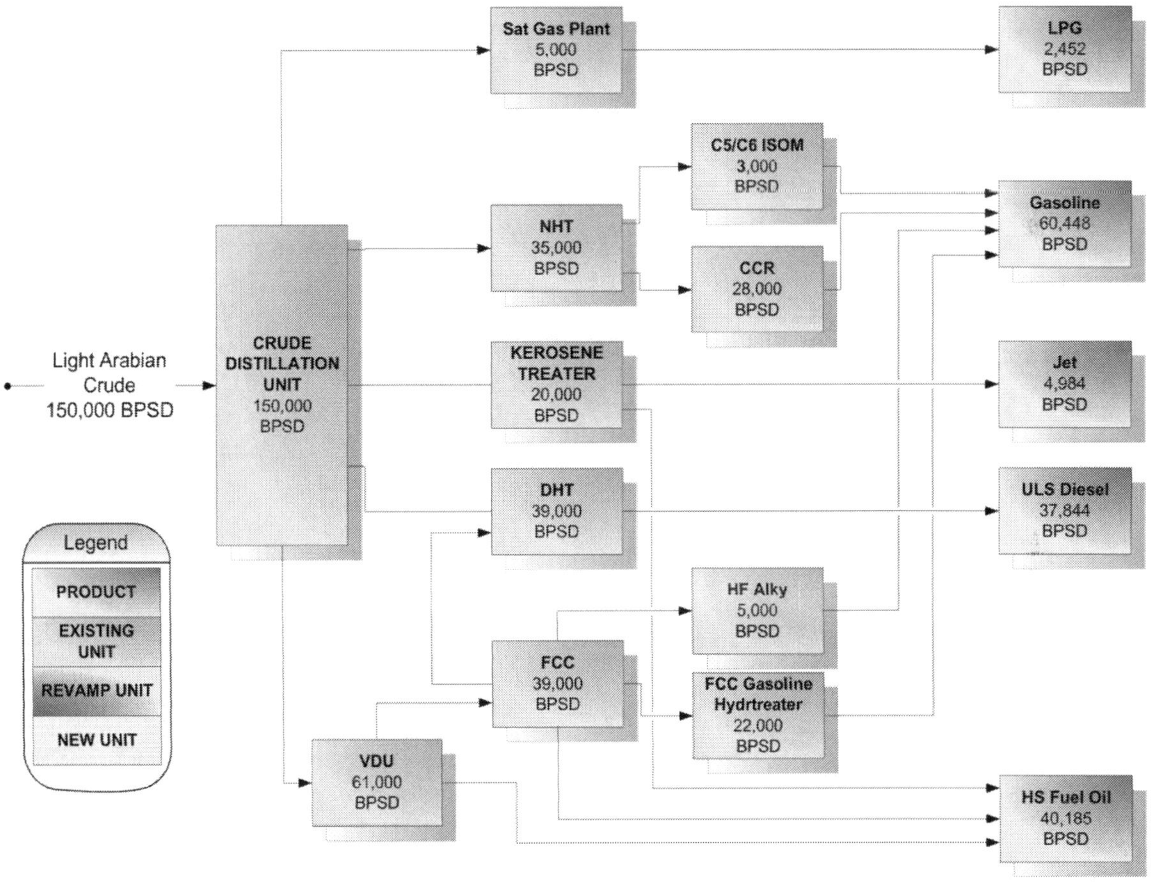

This type of low conversion refinery produces HSFO, which is sold into a declining market. This will in turn eventually require the refiner to reduce the price of HSFO so as not to suffer a loss refinery throughput due to lack of HSFO sales. The demand for the light distillate products will require larger volumes of higher priced light crude oil be processed to meet the demand, thus creating surplus volumes of HSFO relative to the local market. Making the situation worse, future environmental regulations will restrict or eliminate the use of HSFO in the refinery and local markets. This refinery case study will show the benefit of exiting the HSFO market and switching to the use of low priced, heavy crude oils. The basis of this study is the Foster Wheeler SYDEC^{SM} delayed coking process as the bottom of the barrel conversion technology.

A linear program (LP) model of the base refinery processing Light Arabian crude, a typical light crude, was developed. A nominal size of 150,000 BPSD was chosen for the refinery. The comparative benefits were developed through the addition of the SYDEC^{SM} coking project to the base HSFO refinery. To take advantage of the light-heavy crude differential typical of these upgrades, the upgraded refinery uses Maya crude oil.

For modeling purposes, the following key product specifications (Table 3) were used.

Table 3

Refinery Key Product Specifications

Key Product Specifications	
Gasoline	
Specific Gravity	0.72-0.75
RON	92
Sulfur PPM	10
RVP PSI	9
Jet A1	
Specific Gravity	0.80-0.84
Sulfur PPM	3000
Smoke mm	25
Freeze Point C	-47
Diesel	
Specific Gravity	0.82-.845
Sulfur PPM	10
Cetane Index	45
Fuel Oil	
Specific Gravity	0.98 max
Viscosity at 122 F	360
Sulfur wt%	3.5

To calculate differences in net variable refining margin (i.e., product sales minus feedstock costs minus variable costs), crude oil and refined product prices were based on DOE/EIA data for the week of February 10, 2006 (Table 4).

Table 4

Crude & Product Pricing for Evaluation Purposes

Feed and Product Pricing	
Arab Light $/bbl	56.37
Maya $/bbl	47.92
LPG $/bbl	41.52
Gasoline $/bbl	71.10
Jet A1 $/bbl	76.26

Diesel $/bbl	74.76
Fuel Oil $/bbl	46.12
Petroleum Coke $/metric ton	7.00
Propylene $/bbl	57.00
Sulfur $/metric ton	30.00

To the basic catalytic cracking refinery configuration, a SYDEC℠ process unit was added into the LP. Processing capacity downstream of the residue conversion process unit were added or expanded to accommodate the incremental light streams production. The configuration was set by the requirement to utilize existing capacity before adding new units. Utility systems and offsite requirements were increased for based on the additional demands of the new delayed coker complex. The impact of the delayed coker can be seen below in Figure 3:

Figure 3

BFD: Coker Case Refinery Configuration

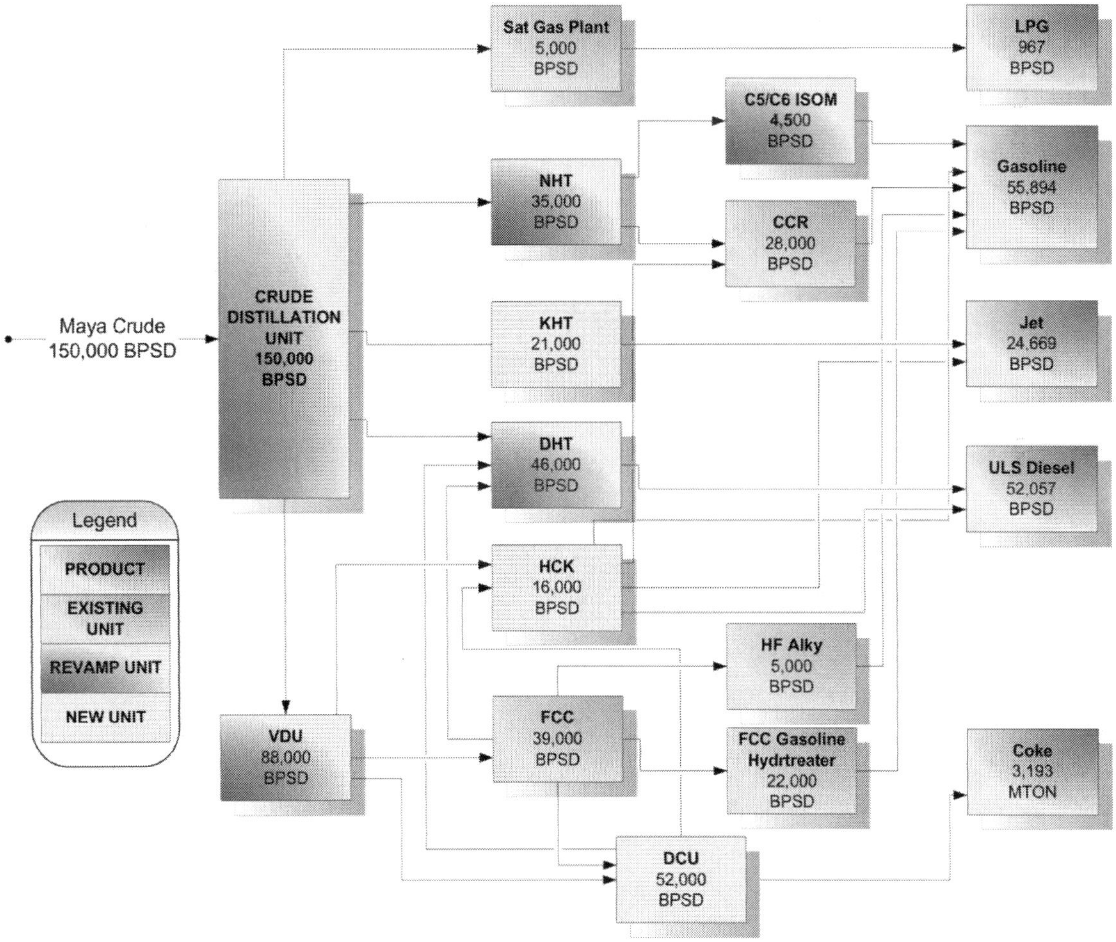

A comparison of the base and resulting unit sizes and refinery product yields are given in tables 5 and 6.

Table 5

Refinery LP Results for Evaluation Purposes – Process Units

Configuration Summary, BPSD except as noted			
Unit	Base Case: FCC only	With FCC + Coker Complex	
Crude Distillation Unit [CDU]	150,000	150,000	Revamp for high residue yield
Vacuum Flasher [VDU]	61,000	88,000	Revamp for high residue yield
Naphtha Hydrotreater [NHT]	35,000	35,000	Revamp for coker naphtha
C5/C6 Isomerization [ISOM]	3,000	4,500	Incremental capacity revamp
CCR Reformer [CCR]	28,000	28,000	
Kerosene Treater	20,000	-	Not Used
Kerosene HDT [KHT]		21,000	New unit
Diesel Hydroteater [DHT]	39,000	46,000	Incremental capacity revamp
Fluid Catalytic Cracker [FCC]	39,000	39,000	
FCC Gasoline Treater	22,000	22,000	
HF Alkylation Unit	5,000	5,000	
Saturate Gas Plant	5,000	5,000	
SYDECsm Delayed Coker [DCU]		52,000	New unit
Hydrocracker [HCK]		16,000	New unit
Sulfur Plant, MTPD	190	520	New incremental capacity
Amine Unit, GPM	590	1,610	New incremental capacity
Hydrogen Plant, MMSCFD		35	New unit

Table 6

Refinery LP Results for Evaluation Purposes – Refined Products

Feed / Product	Base Case: FCC only	With FCC + Coker Complex
Arab Light BPSD	150,000	-
Maya BPSD	-	150,000
LPG BPSD	2,452	967
Gasoline BPSD	60,448	55,894
Jet BPSD	4,984	24,669
Diesel BPSD	37,844	52,057
Fuel Oil BPSD	40,185	-
Sulfur MTPD	182	481
Coke MTPD	-	3,193
Propylene BPSD	3,182	4,968

The reduction of gasoline production from the Base Case to the Coker Case is attributed to the reduction in straight run naphtha from the Crude Unit (CDU). The Light Arabian straight run naphtha cut was approximately 22.4% LV while the same production from Maya is only 11.1% LV. A similar affect is seen in the LPG but it is also compounded by the use of LPG as feed to the Hydrogen plant. The non-extractive Kerosene caustic treater unit was replaced with a Kero Hydrotreater due to the increase in sulfur in the straight run kerosene. The LP optimized product yields based on product values. The result was lower LPG production with higher jet fuel and diesel productions. The resultant comparative economic benefits of residue conversion can be seen in Table 7:

Table 7

Economic Comparison of Adding Coking to Base Case

Economic Comparison (Basis: 15 Yr Project Life)			
	Base Case: FCC only	**With FCC + Coker Complex**	**Delta**
Feed Costs \$m/yr	2959	2516	-444
Product Revenues \$m/yr	3377	3538	160
Gross Margin \$/bbl	7.96	19.46	11.50
Operating Costs \$m/yr	-	-	75.70
Capital Costs \$m	-	-	780
Payout Period (no Tax) yrs	-	-	1.4
IRR % (no Tax)	-	-	49%
NPV @ 10% (no Tax), \$m	-	-	2391
Payout Period (35% Tax) yrs	-	-	2.2
IRR % (35% Tax)	-	-	35%
NPV @ 10% (35% Tax), \$m	-	-	1334

The economic comparisons above indicate the value that SYDEC^{SM} Delayed Coking adds to refinery margins and profitability. Foster Wheeler has also investigated the economic benefits of using Visbreaking and Solvent Deasphalting for Bottom of the Barrel conversion projects. While the IRR may look attractive when employing these processes they usually do not allow a refiner to exit the HSFO market and the NPV of those projects is significantly lower than for SYDEC^{sm} delayed coking.

Conclusion

Residue conversion and reduction of high sulfur fuel oil can be accomplished through many processing routes and the best solution is unique to each refinery. SYDEC^{SM} Delayed Coking offers a proven and reliable means to accomplish these goals relatively inexpensively while increasing refining margins and profitability in markets with declining High Sulfur Fuel Oil demand. In the simplified case study presented, SYDEC^{SM} Delayed Coking offered high economic returns with a strategic means for a complete exit from the HSFO market.

Effects of trace amount of noble metal (Pd/Rh) on the performance of Co-Clay catalysts for heavy oil upgrading

Mohammad M. Hossain
Chemical Reactor Engineering Center (CREC), The University of Western Ontario, London, Ont., Canada, N6A 5B9

Abstract

This communication reports the promotional effects of small amounts of Pd and Rh on the hydrodesulfurization (HDS) and hydrocracking (HC) activities of high porous saponite (HPS) supported Co catalysts. During catalyst preparation, HPS was first modified with the noble metal(s) using an incipient wetness impregnation technique followed by cobalt loading through an ion exchange method. Temperature-programmed reduction (TPR) studies showed that the reducibility of the catalysts was significantly improved by the noble metals (Pd, Rh). Higher metal dispersion and hydrogen spillover effects were considered to be the main reasons for the increased reducibility. Both the HC and HDS activities of noble metal-promoted Co/HPS catalysts were found to be higher than that of plain Co/HPS catalysts, when evaluated in a batch autoclave reactor using vacuum gas oil as feedstock. Furthermore, Rh displayed higher promotional effects as compared to Pd in both HC and HDS. This higher activity is consistent with the TPR results, which showed superior reducibility of the Rh promoted catalysts. A particular interesting finding of this study pertained to a synergic effects of bi-noble (Pd–Rh) metal promoted Co/HPS catalyst. It was observed that a trace amount of Rh and Pd exhibited a significant enhancement of the catalytic activity of Co/HPS. The activity of the bimetallic catalyst was comparable to that of the Rh promoted catalyst with higher Rh loading. It should be noted that this synergic effects was achieved with a lower overall noble metal content, which is crucial for the economic feasibility of the catalyst for industrial application. The coke deposition on the Pd–Rh promoted catalyst was also found be the lower than that of the plain and single noble metal promoted Co/HPS, which is a positive aspect for a longer catalyst lifetime. Therefore, this study revealed that trace amount of noble metal promoted Co/HPS catalyst holds great promise for HC and HDS of heavy oil.

Keywords: Heavy oil upgrading; Clay; Co; noble metals; bimetallic catalysts; TPR; HDS; HC

---

1. Introduction

The increasing demand for lighter engine fuel and recent escalation of crude oil prices have stimulated research focused on the processing of comparatively cheaper heavy crude oils and the heavy residues. The main problem associated with heavy oil fractions is their high organic sulfur, nitrogen and metal content, which must be removed due to their toxicity and environmental impact. In addition, environmental regulations are becoming increasingly stringent with regards to sulfur content in the fuels. Therefore, the need for more efficient hydortreating catalysts and process is growing. The key objective in developing a successful catalyst is to achieve maximum catalytic activity under milder reaction conditions. The duel functional hydrocracking catalysts are derived from a metal from Group VIB (Mo or W) or Group VII (Co or Ni) supported on an acidic support such as alumina. The hydrogenation-dehydrogenation functionality of these catalysts is due to the metal(s) while the acidic support provides the cracking activity. The optimum catalytic performance of the catalyst can be achieved by appropriate balancing of these two functionalities. Bimetallic or multimetallic catalysts have been found very useful because of their improved activity, selectivity or stability compared to monometallic catalysts [Savargaonkar et al, 1996]. Alumina supported Pd-Ru and Pt-Ir catalysts are possibly the best example of bi-metallic catalyst with superior performance, applied in reforming process. The introduction of the second metal provided a way of moderating the catalysts surface to maintain sustainable activity and stability for longer period of time [Sinfelt, 1983]. The success of bi-metallic naphtha reforming catalyst possibly encouraged researchers to apply the samilar concept on hydrotreating catalysts as an attempt to enhance catalyst activity, stability, and also to prevent coke deposition. Among the studied bimetallic catalysts, the noble metals promoted catalysts have been demonstrated excellent performance [Al-Saleh et al, 2003; Ali et al, 2002; Hossain et al, 2004; Sugioka et al, 1996, 1998; Fujimoto, 1999]. The presence of noble metal helps maintaining the balance between the cracking and hydrogenation functionalities. The cracking of heavy hydrocarbon molecules and the sulfur bearing molecules on the acidic sites produces unstable free radicals and olefins, which need to be hydrogenated to produce stable products. The inclusion of a noble metal (Pt, Rh, Pd etc) along with the main active metal (Co or Mo) on the acidic support provides extra hydrogenation-dehydrogenation activity of the catalysts [Ali et al, 2002]. Generally, the noble metals are excellent to dissociate hydrogen comparatively at lower temperature. Therefore, their presence provides the appropriate amount of reactive hydrogen to react with the cracking reaction intermediates. Consequently, the overall rate of the reaction becomes faster. The excess availability of reactive hydrogen species also prohibits the coke formation on the catalyst surface by reacting with the coke to form fluid product. As a result of that, the catalyst surface remains clean and active for longer time.

This study demonstrates the promotional effects of Pd, Rh and their combination (Pd-Rh) on the hydrocracking and HDS activities of a Co/HPS catalyst using vacuum gas oil (VGO) as feedstock. Keeping in mind about the rare availability and high price of the noble metals their concentration was kept at very small level in the catalyst. Before evaluation, the fresh catalysts were characterized by temperature-programmed reduction (TPR) to asses the noble metal effects on the reducibility of the catalyst. The spent catalysts were also characterized to find the amount of coke deposition during the hydrocracking reaction in a batch autoclave reactor.

2. Experimental methods

2.1. Catalyst preparation

In all catalysts the composition of the Co was maintained around 18 % (of the catalyst weight). For the single noble metal promoted catalysts the noble metal (Pd/Rh) loading was kept at 1 wt. % while in Pd-Rh promoted catalyst the Rh loading was reduced to 0.1 wt. % keeping the Pd loading still at 1 wt. %. During the catalyst preparation, the HPS was first impregnated with noble metal solution. The resultant material was then dried at 120 °C followed by calcinations at 600 °C for 4 hr. Co loading was achieved by an ion-exchange technique, which consisted of the following steps: $Co(NO_3)_2.6H2O$ solution was prepared in distilled water and aged at 80 °C for 1 hr. The noble metal(s) loaded HPS was added into the solution and the mixture was stirred at 80 °C for another 2 hr. The precipitated product thus obtained was filtered and washed with distilled water followed by ethanol. The washed product was dried at 120 °C and calcined at 600 °C for 4 hr and finally was pelletized for characterization and evaluation.

2.2. Characterization by TPR

The temperature programmed reduction (TPR) experiments were carried out in a Model TP-2000 TPR system, supplied by Ohkura Rikan Co. Ltd, Japan. Before analysis the sample was degassed by flowing dry air (30 ml/ min) at 400 °C. For catalyst reduction a mixture of 5 % hydrogen in argon gas was used at a flow rate of 20 ml/min. The temperature was programmed to increase linearly at 10 °C/min to 1030 °C and held at this temperature for 15 min. The degree of reduction during the TPR was determined by a TCD.

2.3. Catalyst evaluation in batch autoclave reactor

The batch reactor experiments were carried out using vacuum gas oil (VGO, a typical feedstock in commercial hydrocracking processes) as feedstock. The physical and chemical properties of the feedstock VGO are presented elsewhere [Al-Saleh et al, 2003]. For each batch, 3 g of pre-sulfided catalyst and 100 g of VGO were loaded into the reactor. The reaction was carried out at 400 °C and 13800 kPa for 90 min. The gas, liquid product and spent catalyst samples were collected when the reactor reached to the room temperature. Gas chromatography was used to analyze the gas sample, while the liquid product was analyzed for boiling point distribution using GC simulated distillation, hydrocarbon types by HPLC and elements by CHNS analysis.

3. Results and discussion

3.1. Temperature programmed reduction

In the fresh catalyst, the metals are present in their oxide forms. Given that the metals are more active in their metallic/sulfide state, it is very important to characterize the catalyst so as to investigate the reducibility of the catalyst. The TPR characterization can help measuring the reducibility of the metal oxides before putting into the reaction. Furthermore, TPR study also clarifies the promotional effects of noble metals on the reduction of the main active (cobalt) species. Concerning hydrogen spillover effects, TPR can also provide insight into this phenomenon through the measurement of hydrogen consumption and shifting of the reduction temperature [Al-Saleh et al, 2003]. Figure 1 illustrates the H_2 consumption profiles during the TPR of the different catalysts involved in this study. For un-promoted catalyst the first peak of H_2 consumption was evolved at 226 °C while the second peak started at 500 °C and completed at 850 °C with the maxima at 745 °C. The low temperature peak was assigned to the reduction of trivalent cobalt oxides (Co_2O_3/Co_3O_4) to divalent cobalt oxide. The second peak, appearing at 740 °C, (i.e., the temperature at which most of the reduction occurred) was attributed to the reduction of divalent Co oxide (CoO) to the metallic Co (Co^0) [Vob et al, 2002; Lango et al, 1997; Jacobs, et al, 2002; Noronha et al, 1996]. These observations are consistent with the following reaction scheme, which has been proposed based on the TPR measurement of pure Co_3O_4.

$$Co_3O_4 + H_2 \longrightarrow 3CoO + H_2O$$
$$CoO + H_2 \longrightarrow Co + H_2O$$

The TPR profiles of both Pd and Rh promoted Co-HPS were also similar to that of unpromoted Co/HPS, which suggests no alloy formation between cobalt and the noble metals [Yang et al, 1997]. However, a small peak was observed due to the presence of Pd/Rh, which was considered as the leftover of the second peak of Co/HPS, that was strongly bonded with the support and unaffected by the noble metal [Hossain et al, 2004]. It is interesting to note that the noble metal promoted catalysts exhibited a significantly low-temperature reduction. The shifting of the peak temperature, which occurred in the range of 140-215 °C, was clearly due to the promotional effects by the noble metals [Yang et al, 1997; Bianchi, 2001].

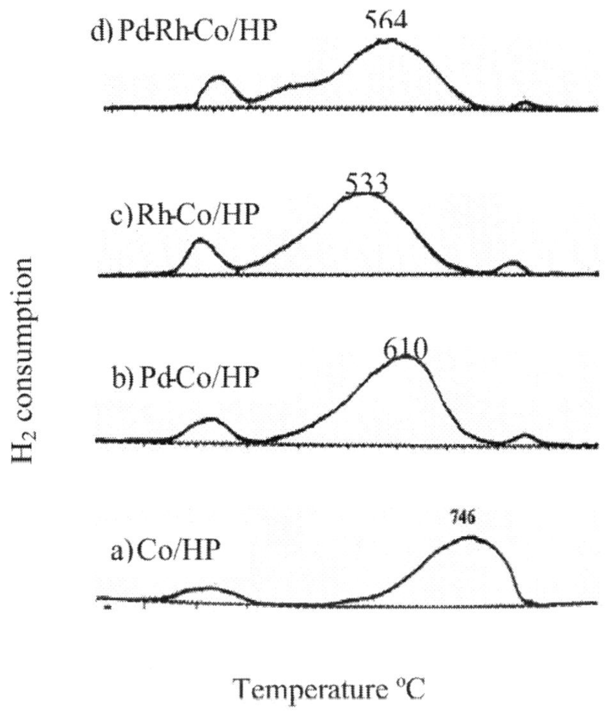

Figure 1: TPR profiles of (a) Co/HPS, (b) Pd-Co/HPS, (c) Rh-Co/HPS and (d) Pd-Rh-Co/HPS.

From a mechanistic point of view, this effect can be explained by so-called hydrogen spillover [Al-Saleh et al, 2003; Ali et al, 2002; Hossain et al, 2004] phenomenon, which can be described as follows: the noble metal can dissociate molecular hydrogen to highly reactive hydrogen species, i.e. H^+, H^-, H_3^+ or ion pair, and such reactive species can easily attack the cobalt oxide(s) and reduce them at comparatively lower temperatures [Kusakari et al, 2002].

Between the two noble metals used, Rh exhibited a more pronounced promotional effect compared to Pd in the reduction. In catalysts samples with equal loading, Rh shifted the peak temperature by 215 °C while Pd showed a shift of 140 °C, which suggested that Rh reduces cobalt oxide more readily than Pd. However, the price of Rh is 2.5 times the price of Pd. Bearing this in mind, the next objective was set so as to achieve the similar effects (as obtained with Rh at its higher amount) by employing a Pd-Rh bi-noble metal combination using a trace amount of Rh. For comparison purpose, the TPR profile of this catalyst is also included into Figure 1. It is clear that a trace amount of Rh along with Pd can provide a similar temperature shift as Rh did alone at a higher amount of its loading. The bi-noble metal effect can be reasonably explained as follows: the presence of Rh assisted in the reduction of Pd, which in turn assisted in the dissociation of more hydrogen molecules, thereby enhancing the reducibility of the cobalt oxide sites at a lower temperature. The increased reducibility of noble metal promoted Co/HPS catalysts could also be due to superior dispersion of cobalt species on the support surface in presence of noble metals [Kusakari et al, 2002; Hossain et al, 2004].

3.2. Catalytic performance in batch autoclave reactor

Batch reactor experiments were carried out at typical industrial hydroprocessing conditions (400 °C, 13800 kPa) and for 90 min in reaction. A standard $CoMo/Al_2O_3$ catalyst was also evaluated under identical reaction conditions and the activity data of this catalyst is also included in relevant figures of this report for comparison purpose. For noble metal promoted catalysts, the formation rate of gaseous hydrocarbon (C_1~C_5) was considerably lower than that of the unpromoted Co/HPS (Figure 2). Moreover, the Pd-Rh promoted Co/HPS showed the lowest rate of gas formation. This result was expected after analyzing the TPR data, which indicated the readily available reactive hydrogen due to hydrogen spillover introduced by the noble metal(s) that help to suppress the formation of lighter gases. The presence of Pd/Rh or Pd-Rh helped maintain a sufficient amount of reactive hydrogen needed for the hydrogenation reaction that determines the termination of the cracking reaction. Thus prevented the over cracking of the intermediates, which yields undesirable light gaseous products. In other literature reports it has been assumed that the addition of the noble metal introduces a better balance between the acidic function (for cracking) and hydrogenation function of the catalysts, which helps reducing the gaseous product formation by suppressing the over cracking of heavy hydrocarbon molecules [Elangovan et al, 2002; Blomsama et al, 1997]. This characteristic of a hydroprocessing catalyst is of great importance for its industrial application since deep cracking produces lighter gas, which is considered as waste. The liquid product analysis data showed that the HDS and HC activity of Co/HPS were significantly higher with noble metal modification, as illustrated in Figures 3 and 4. The order of activity was: Pd-Rh-Co/HPS > Rh-Co/HPS > Pd-Co/HPS > Co/HPS.

It is interesting to note that Pd-Rh promoted catalysts displayed the highest activity for both HC and HDS reactions. Such high activities were consistent with the reducibility data showed earlier in TPR studies of these catalysts. The higher HC and HDS activity of Pd/Rh and Pd-Rh promoted catalysts probably occurred due to the synergy effect introduced by the noble metals. This promotional effects can be explained as follows: Sulfur bearing heteroatoms, as well as the heavy hydrocarbon molecules, crack on the acidic HPS sites to produce unstable reaction intermediates.

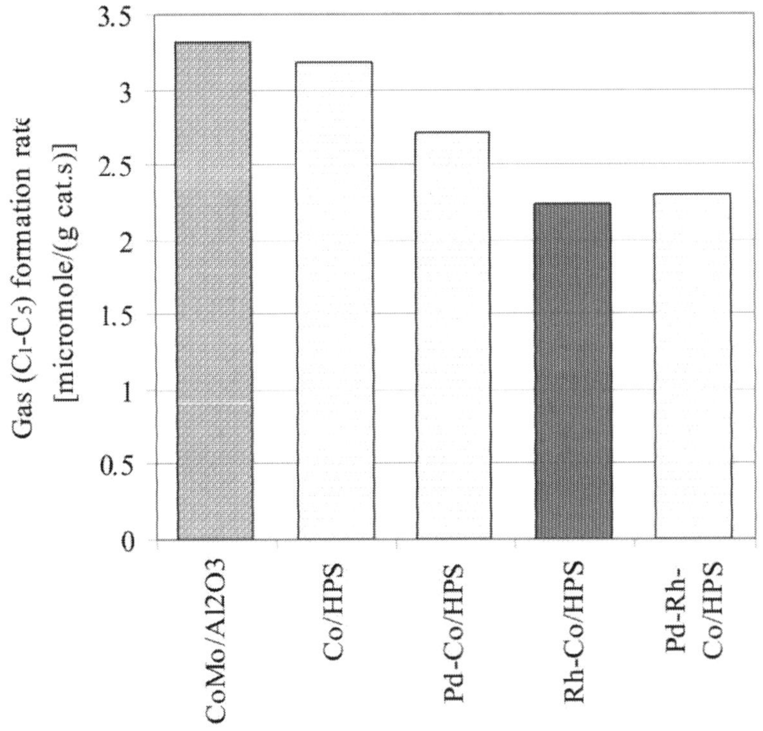

Figure 2: gas (C_1-C_5) formation rates in batch autoclave reactor with VGO as feedstock.

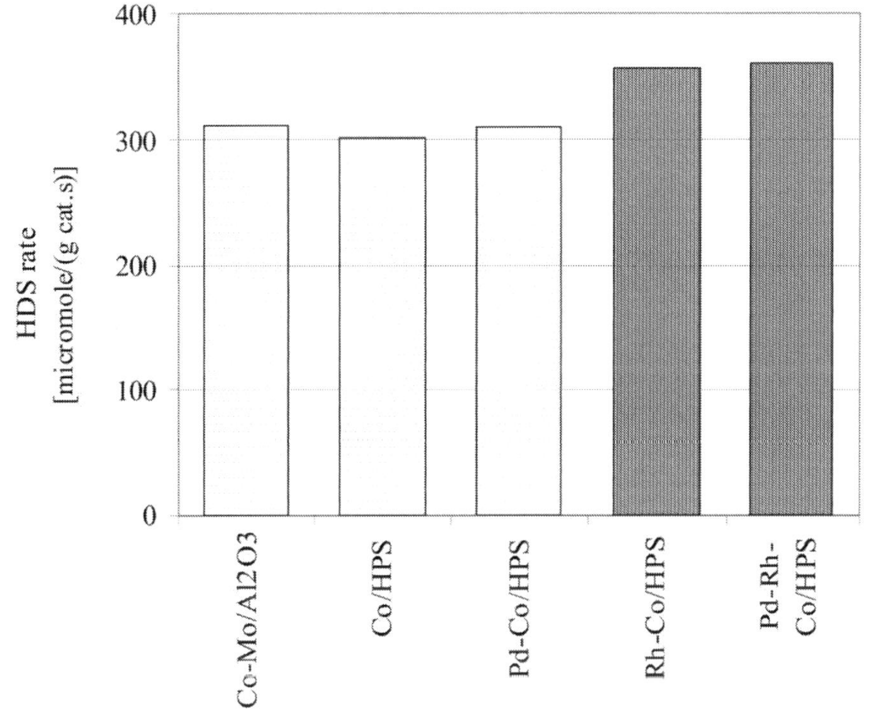

Figure 3: HDS rates of VGO using Co-Mo/Al_2O_3 and unpromoted, noble metal (Pd/Rh) and bi-noble metal (Pd-Rh) promoted Co/HPS catalysts in batch autoclave reactor.

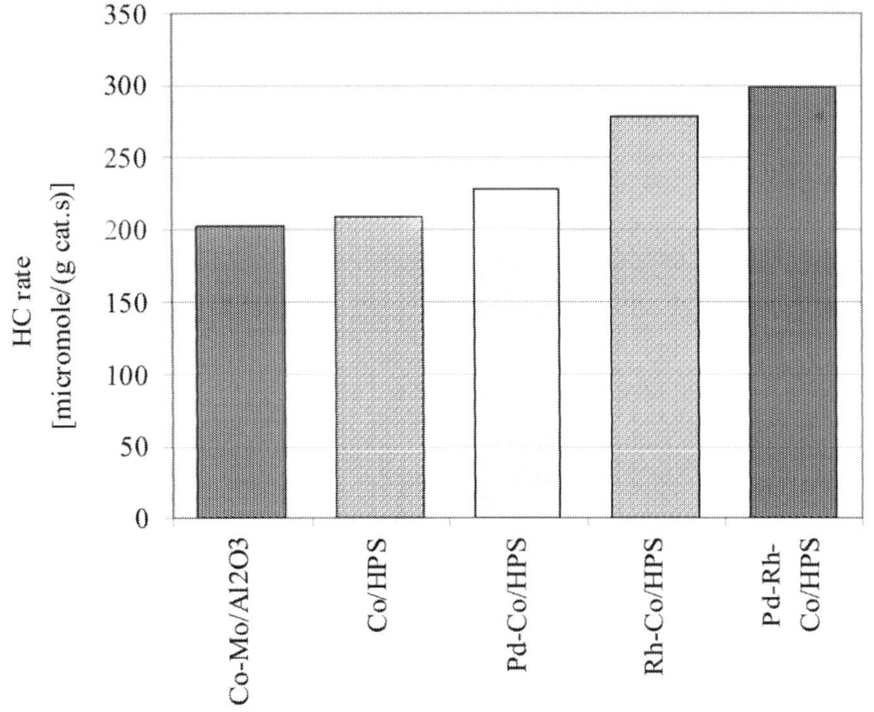

Figure 3: HC rates of VGO using Co-Mo/Al_2O_3 and unpromoted, noble metal (Pd/Rh) and bi-noble metal (Pd-Rh) promoted Co/HPS catalysts in batch autoclave reactor.

Meanwhile, the noble metal(s) supply reactive hydrogen via the dissociation of molecular hydrogen. The cracked reaction intermediates react with the hydrogen species to produce stable products. As a result the rate of the reaction and the activity become higher.

A mechanism consistent with the above explanation has been reported elsewhere [Al-Saleh et al; 2003]. A similar mechanism for hydrodesulfurization of Thiophene over USY zeolite supported Pt, Pt-Pd catalysts have also been demonstrated by other authors [Sugioka, et al., 1996, 1998]. The reasons of higher hydrogenation activity can be further confirmed from the H/C ratio and HPLC analysis data of liquid products, shown in Figures 5 and 6, respectively. The H/C ratios of the noble metal promoted catalysts were shown to be significantly high as compared to those of unpromoted Co/HPS catalysts. This result indicates the addition of hydrogen into the liquid products through subsequent hydrogenation of cracked intermediates, and also the removal of sulfur from the sulfur containing hydrocarbons [Ali et al, 2002]. The analysis for the types of hydrocarbons, also supports the higher hydrogenation activity of the noble metal modified Co/HPS. The presence of noble metal(s) resulted in higher saturates content in the liquid product, while yielding lower aromatics and polar content as compared to the plain Co/HPS. This result suggests that the addition of hydrogen to the liquid products during reaction due to the enhanced hydrogenation rate [Ali et al, 2002]. The higher HC and HDS activity of noble metal promoted catalysts can also be explained by the enhanced acidity introduced by the hydrogen spillover effects as observed in TPR [Fujimoto, 1999]. This phenomenon has been explained by considering the acidic nature of hydrogen species involved in hydrogen spillover, which suggests that the hydrogen species give rise to protons that generate brønsted acid sites on the Co sites, eventually leading to an increase in cracking as well as the hydrogenation activity. Other studies have also indicated such higher acidity and hydrogenation activity due to the presence of noble metal(s), although the term "hydrogen spillover" has not been used [Hossain et al, 2004; Blomsam et al, 1997; Wada et al 1996].

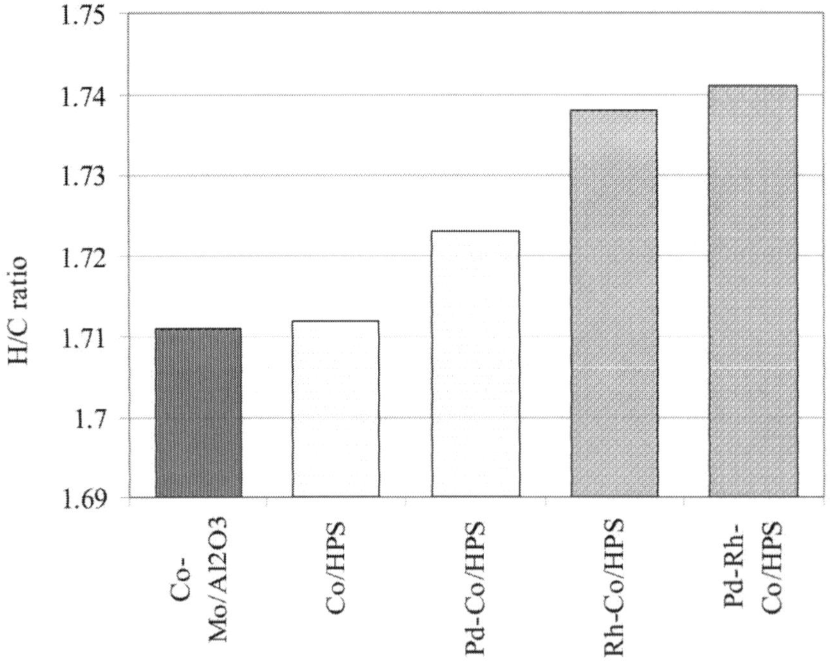

Figure 5: H/C ratio in liquid product from batch autoclave experiments.

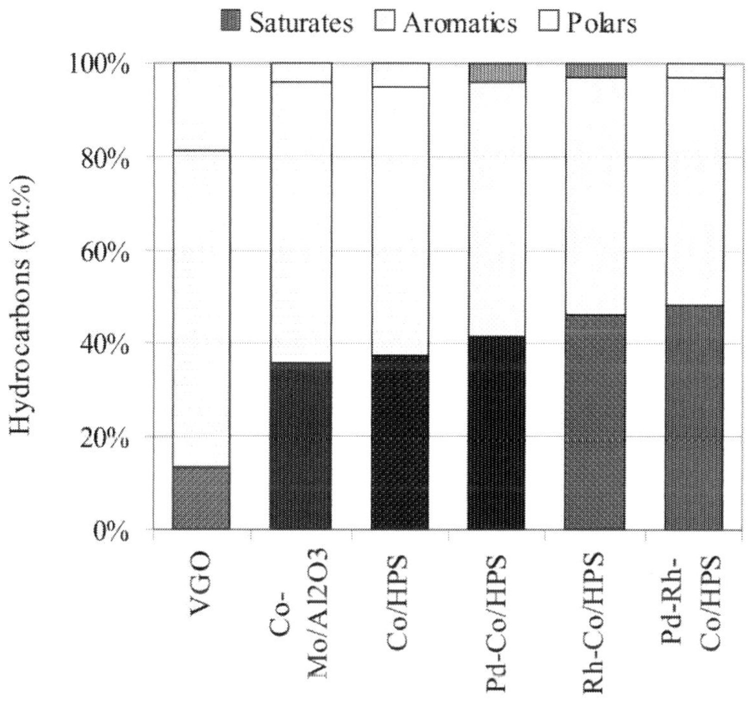

Figure 6: Hydrocarbon types in liquid product from batch autoclave experiments.

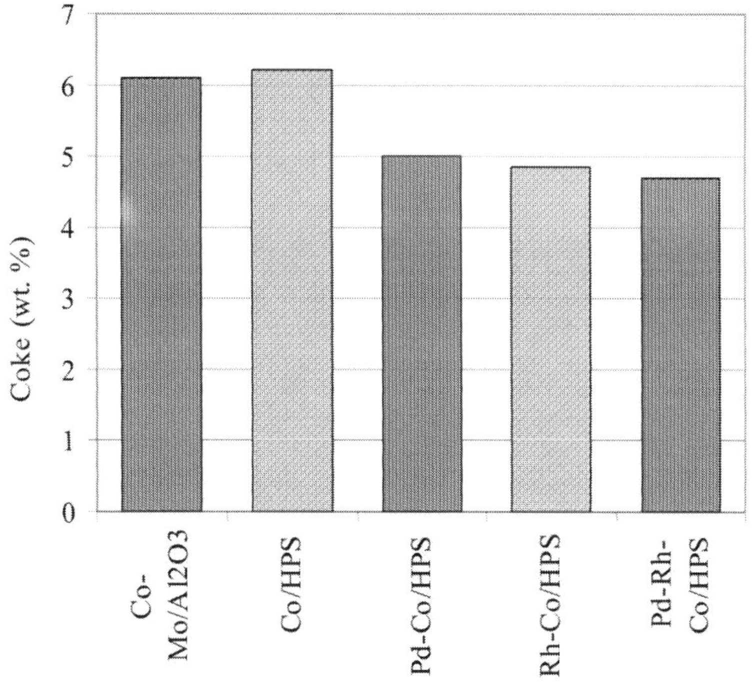

Figure 7: Coke deposition on spent catalysts in batch autoclave experiments.

The participation of spillover hydrogen was further demonstrated by the characterization of spent catalysts for the amount of coke deposition after the reaction. Figure 7 displays the amount of coke deposition on different catalysts after reaction time of 90 min. The result showed that the coke deposition was reduced significantly (18-25 wt. %) on Pd/Rh and Pd-Rh promoted Co/HPS catalysts; particularly Rh and Pd-Rh displayed more pronounced effects in suppressing coke deposition. This observation can be explained by considering the following: gas phase hydrogen molecules are adsorbed on the noble metal sites and then dissociate to produce reactive hydrogen species (H^+, H^-, H_3^+), which desorbs quickly. This reactive hydrogen species "spillover" to the Co sites (i.e., the main hydrogenation sites) as well as the acidic sites (HPS support) and react with the cracked intermediates to produce stable products. As a result, the deposition of the hydrocarbon intermediates on the active sites is significantly decreased, leading to a clean catalyst surface [Fujimoto, 1999; Al-Saleh et al, 2003]. Another explanation is provided by the so-called the remote control mechanism by hydrogen spillover on noble metal promoted catalysts [Hossain et al, 2004]. According to this mechanism the hydrogen involved in spillover influences the cracking of heavy hydrocarbon molecules by controlling the concentration of carbonium ions as well as the other unstable reaction intermediates, thereby hindering their deposition by converting them to stable products. Consequently, the catalysts remained active.

4. Conclusion

The reducibility of Co/HPS can be significantly enhanced by modifying the catalyst with noble metals (Pd, Rh) as well as their combination. Rh proved to be more effective than Pd in reducing the Co/HPS. A trace amount of Rh along with Pd showed a remarkable effect, which was much higher than that obtained with Pd alone. The enhanced reducibility of Co/HPS by noble metals also helped to increase the both HC and HDS activity of such catalysts, as shown by batch autoclave reactor experiments using vacuum gas oil as feedstock. Particularly, the Rh and Pd-Rh promoted catalysts were found to be very effective in both HC and HDS purpose. The analysis of the liquid products obtained from the batch reaction of VGO showed a considerable H/C ratio and higher saturates, which confirmed the higher hydrogenation activity of noble metal promoted Co/HPS catalysts. Furthermore, spent catalyst analysis also revealed a remarkable decrease in coke deposition on Co/HPS due to the addition of noble metals. Therefore, this study revealed that noble metal promoted Co/HPS holds great promise for hydrocracking and HDS of heavy oil.

Acknowledgements

The author wishes to acknowledge The Center for Refining & Petrochemicals, RI-KFUPM, Dhahran, Saudi Arabia for supporting this work.

References

Ali, M.A., Kimura,T., Suzuki, Y., Al-Saleh, M., Hamaid, H., Inui, T., *Appl. Catal. 227* (2002) 63.

Al-Saleh, M.A., Hossain, M.M., Shalabi,M., Kimura, T., Inui, T., *Appl. Catal. 253* (2003) 453.

Bianchi, C.L., *Catal. Lett. 76* (2001) 155.

Blomsma, E., Martens, J. A., Jacobs, P. A., *J. Catal.165* (1997) 241.

Elangovan, S.P., Cischof, C., Hartmann, M., *Catal. Lett. 80* (2002) 35.

Fujimoto, K., in: B. Delmon, G.F. Froment, P. Grange (Eds.), Hydrotreatment and Hydrocracking of Oil Fraction, Elsevier, Amsterdam, 1999.

Hossain, M.M., Al-Saleh, M.A., Shalabi,M., Kimura, T., Inui, T.,, *Appl. Catal. A 274 (2004) 43.*

Jacobs, G., Das, T.K., Zhang, Y., Li, J., Racoillet, G., Davis, B.H., *Appl. Catal. 233* (2002) 263.

Kusakari, T., Tomishige, K., Fujimoto, K., *Appl. Catal. 224* (2002) 219.
Lago, R., Bini, G., Peña, M.A., Fierro, J.L., *J. Catal.167* (1997) 198.
Noronha, F.B., Frydman, A., Aranda, D.A.G, Perez,C., Soares, R.R., Morawek, B., Castner, D., Campbell, C.T., Frety, R., Schmal, M, *Catal. Today 28* (1996) 147.
Savargaonkar, N., Khanra, B.C., Pruski. M., King, T.S., *J. Catal.* 162(1996) 277.
Sinfelt, J.H., Bimetallic Catalysts: Discoveries, Concepts, and Applications, John Wiley & Sons Ins., NY, USA, 1983.
Sugioka, M., Sado, F., Kurosaka, T., Wang, X., *Catal. Today 45* (1998) 327.
Sugioka, M., Sado, F., Matsumoto, Y., Maesaki, N., *Catal. Today 29* (1996) 225.
Voß, M., Borgmann, D. Wedler, G., *J. Catal. 212* (2002) 10.
Wada, T., Kaneda, K., Murata, S., Nomura, M., *Catal. Today 31* (1996) 113.
Yang, M.G., Nakamura, I., Fujimoto, K., Stud. Surf. Sci. Catal. 112 (1997) 221.

103d Salt Hydrolysis in Crude & Bitumen Refining

Paul E. Eaton, Technical Marketing, Champion Technologies Inc., 19827 Sunbridge Lane, Houston, TX 77094, Murray Gray, Chemical & Materials, University of Alberta, T6G 2G6, Edmonton, AB T6g 2G6, Canada, and Tuyet Le, Department of Chemical and Materials Engineering, University of Alberta T6G 2G6, Edmonton, T6g 2G6, Canada.
Abstract

Salt Hydrolysis in Crude and Bitumen Refining Paul Eaton, Champion Technologies, Fresno TX; Murray Gray University of Alberta; Edmonton, Alberta; Tuyet Le, University of Alberta, Edmonton, Alberta

Crude oil charged to the refinery, contains clay minerals, and may also contain chloride salts. The chlorides can undergo hydrolysis to form corrosive hydrochloric acid during the distillation process. Crude oil also contains significant concentrations of organic acid components known as naphthenic acids, which can cause corrosion in overheads and downstream equipment. In the presence of steam, the chloride salts, clays and organic acids may interact synergistically to promote formation of hydrochloric acid, causing significant corrosion in the downstream equipment. Despite the importance and the impact of these components to product quality and refinery operation, their behavior under upgrading conditions is not well understood.

The objective of this research is to investigate the rate and the extent of the hydrolysis reactions for salts contained in crude oil as a function of mineralclays and napthenic acid. Laboratory glassware is used to expose mixtures of bitumen, salts and organic acids to flowing steam at 100 oC to 400 oC under nitrogen purge. The vapors are condensed and the condensate is analyzed for chloride using ion chromatography. This apparatus promotes the formation of hydrochloric acid, and allows us to investigate the interactions of the salts with organic acids and clays and napthenic acid under upgrading conditions.

Mitigation of the hydrolysis reactions is shown to be possible with the use of chemical additives.

Light Cycle Oil Hydrotreating with Catalysts Containing W, Ni, Treated Zeolite, and Alumina

Lianhui Ding[a], Ying Zheng[a], Zisheng Zhang[b], Zbigniew Ring[c], Jinwen Chen[c]

[a] *University of New Brunswick, Fredericton, New Brunswick*

[b] *University of Ottawa, Ottawa, Ontario*

[c] *National Centre for Upgrading Technology, DEVON, Alberta*

Abstract

Zeolite Y was treated by the chemical approach using $(NH_4)_2SiF_6$ aqueous solution. Zeolite beta was prepared by hydrothermally treating NH_4-Beta. The series of WNi/Y, WNi/Beta, and WNi/Beta+Y catalysts were made using the treated zeolite Y, zeolite beta, and treated Y and Beta as the acidic components, respectively. The catalysts were characterized by BET, TEM, XPS, NH_4-TPD. The reaction performances on hydrodesulferization, hydrodenitrogen, and hydrodearomatics were evaluated with light cycle oil (LCO). The zeolite containing catalysts exhibited much higher HDN, HDS, and HDA activities than the WNi/Al_2O_3 catalyst. HDN activity among the three zeolite containing catalysts was identical. HDS activity of the WNi/Beta+Y and WNi/Beta catalysts was higher than that of the WNi/Y catalyst. The higher HDN, HDS, and HDA activities of the three zeolite-containing catalysts were mainly associated with the enhancement of hydrogenation activities and the increase in acidity.

Introduction

The more and more stringent environment legislation demands that the sulfur content in diesel fuel be reduced to lower than 15ppm[1]. Studies have revealed that HDS activity of refractory sulfur species left in the conventionally hydrotreated diesel fractions can be greatly improved by the hydroisomerization [2,3]. The hydroisomerization activity of catalyst can be imparted by the addition of acidic component zeolites into catalysts [4]. To improve the HDS activity of refractory sulfur species and avoid the rapid deactivation and overcracking of the diesel distillates, the surface acidity of zeolite should be properly modified. One of the effective ways to modify the acidity is to remove some tetrahedron Al atoms from the framework of zeolites. Among all the zeolites available, zeolite Y is widely applied and most extensively studied due to its appropriate pore structure, high acidity, and good thermal and hydrothermal stability [5,6]. As the same large-pore zeolite as zeolite Y, zeolite beta, a high SiO_2/Al_2O_3 ratio zeolite, exhibits higher hydroisomerization activity, a lower hydrogen transfer capacity, and lower catalyst decay by self-poisoning [7]. These features make zeolite beta a good catalyst component to convert refractory sulfur species and lower the deactivation rate of catalyst.

Previous studies mainly focused on the application and the comparison of the two zeolites in hydrocracking of heavy oil [7,8]. Little research has been conducted to study the

HDS, especially hydrotreating performance of real inferior diesel fractions using chemically treated zeolite Y and beta as an acidic component of catalyst. The main objective of this study is to investigate the hydrotreating performance of the W-Ni catalysts prepared by chemically treated zeolite Y and zeolite beta using LCO as feedstock.

Experimental

Catalyst preparation

NH_4-beta (Zeolyst CP814E) was calcinated at 540℃ for 10 hours, and then treated with 0.5M hydrochloric acid aqueous solution at 80℃ for 2 hours under stirring. After separation, the filtrated cake was treated in an autoclave reactor at 600℃, 0.2MPa, for 2 hours. Chemically modified zeolite Y was synthesized using the method described in the reference [9]. The supports were prepared by mixing the zeolites, large-pore alumina, and a binder, extruded, dried at 110℃, and then calcined at 550℃ for 4 hours. The NiW catalysts were prepared by co-impregnation of the above extrudates using the incipient wetness method with an aqueous solution of the appropriate amount of nickel nitrate hexahydrate and ammonium metatungstate, and then dried, calcinated. The catalysts prepared from the supports containing Al_2O_3 , 15wt% of modified zeolite Y, 15wt% of modified zeolite beta, and 15wt% of modified zeolite beta plus 15wt% of zeolite Y, were designated as WNi/Al_2O_3 , WNi/Y, WNi/Beta, and WNi/Beta+Y, respectively. The nominal contents of WO_3 and NiO in all catalysts were 24wt% and 6wt%, respectively.

Catalyst characterization

Nitrogen adsorption measurements were performed on a Quantachrome Autosorb-1. TEM was performed on a JEOL 2010 STEM, operated at 200keV. The spectra were collected using an EDAX Genesis 4000 system. XPS spectra of the samples were taken in a VG Microtech MultiLab ESCA 2000 system with a CLAM4 hemispherical analyzer. NH_4 Temperature programmed desorption (NH_4-TPD) was performed in a 5ml tubular reactor. Under 40ml/min flow of helium, the temperature was raised from 100℃ to 610℃ at 15℃/min. The desorbed ammonia was analysized by MS.

Catalyst activity evaluation

The catalyst activity evaluation was carried out in a 1L stirred autoclave (Autoclave Engineers Division of Snap-tite, Inc. Eze-Seal stirred reactor) with LCO. The catalyst was in-situ sulfided with dimethly disulfide (DMDS) at 593K for 2 hours and at 633K for another 2 hours. The reaction conditions were as follows: initial reaction pressure of 1300psi; reaction temperature of 375°C; weight ratio of LCO to catalyst 10:1. The nitrogen, sulfur, and aromatic were analysized according to ASTM D4629, ASTM D4294, and ASTM D6591, respectively. The boiling ranges of LCO and liquid products were measured with simulated distillation by gas chromatography.

Results and discussion

Textual and acidity properties

Textual properties of the four catalysts were listed in Table 1. The surface area of WNi/Beta was almost the same as WNi/Al_2O_3, and the decrease extent of the pore volume and average pore size was much lower than that of WNi/Y.

The acidities of the four catalysts are also listed in Table 1. Comparing the acidity of WNi/Al_2O_3, the addition of zeolite beta and Y enhanced the overall acidity. The increase in acidity for zeolite Y was higher than that for the zeolite beta. Compared with WNi/Beta and WNi/Y, though the content of zeolite doubled, the total acidity and the proportion of the strong acidity of WNi/Beta+Y only slightly increased, suggesting that almost half of acidic sites are occupied by hydrogenated metals or aluminium species.

Table 1. Textual and acidity properties of the catalysts

	W-Ni/Al_2O_3	W-Ni/Y+Beta	W-Ni/Y	W-Ni/Beta
Pore Volume, ml/g	0.56	0.38	0.37	0.42
Surface area, m^2/g	165	222	221	165
Average pore size, nm	13.6	6.8	6.7	10.2
Total acidity, mmol/g cat.	0.340	0.546	0.506	0.480
Average length of slabs*,nm	8.7	26.3	17.7	16.1
Average number of layers*	2.0	3.7	4.4	3.7

* Obtained from TEM.

XPS

Surface species of oxidic and sulfided catalysts

The binding energies of W, Ni, and S for oxidic and sulfided samples are summarized in Table 2. W 4*f*5/2 and W 4*f*7/2 electrons at a binding energy (BE) position of 37.8-37.9 and 35.7-35.8eV are assigned with W^{6+} species, most likely $W^{6+}O_3$ [10,11]. The two Ni 2p3/2 (BE 856.4 eV) and Ni 2P 1/2 (BE 873.8 eV) of the oxidic samples were assigned to the spin-splitting and the two broad peaks to the envelopes of the corresponding shakeup satellite(SS) lines [12]. After sulfidation, BE at 34.4 and 32.5 is typical for tungsten with a formal charge state of +4 as in WS_2 phase [13]. The Ni 2p3 at 856 eV is due to the presence of the non-sulfided Ni^{2+} species [14]. The Ni 2p3/2 at 853.1–853.7 eV, together with the presence of the S 2p3/2 at 162 eV, is associated with nickel sulfides [14].

The sulfiding degree of W and Ni

Preliminary evaluation of the Ni 2*p*3/2 emission line of the sulfided catalysts indicated that two contributions, an oxidic one and a sulfidic one, could be identified. Therefore, the sulfidation extent of Ni can be determined by the relative integrated area ratio of the peaks corresponding to the sulfided Ni and oxided Ni. The extent of sulfidation of W species was calculated based on the atomic concentration ratios of the corresponding sulfide species over the total oxide and sulfide species. The values are summarized in Table 3. The S/W+Ni ratios for all catalysts are higher than the theoretical ratio (1.42), suggesting that very small tungsten and/or nickel sulfide clusters contain more than stoichiometric amount of sulfur. The sulfided W and Ni species range from 84.4% to 92.0% and 77.1% to 93.5% respectively. The sulfidation degree of Ni in WNi/Al_2O_3 is much lower than that of other three zeolite containing catalysts, showing that more nickel species strongly interact with the support and can not be sulfided. Among the three zeolite containing catalysts, the sulfidation degree decreases in the order: WNi/Y>WNi/Beta+Y>WNi/Beta.

Surface atomic ratios

The surface atomic ratios of Ni/Si+Al and W/Si+Al derived from XPS for each catalyst are also given in Table 3. The W/Si+Al and Ni/Si+Al ratios of all oxidic catalysts are much lower than the bulk ratios (11.8-15.1 for W/Si+Al and 9.2-11.7 for Ni/Si+Al), suggesting that considerable amount of W and Ni has been migrated into the bulk after calcination. After sulfidation, the amount of the surface W and Ni further decreases. The reduction quantity of W and Ni for WNi/Al_2O_3 is much higher than that for the other three zeolite containing catalysts. The surface W and Ni decreases in almost the same proportion for the three zeolite containing catalysts.

Table 2. Core level binding energies of the components for oxidic and sulfided forms of W-Ni catalysts

Sample	WNi/Al_2O_3	WNi/Beta+Y	WNi/Y	WNi/Beta
Oxidic				
W 4f7	35.81	35.67	35.72	35.68
W 4f5	37.89	37.80	37.86	37.82
Ni 2p3	856.67	856.33	856.36	856.45
Ni 2p1	874.27	874.08	874.00	874.34
Sulfided				
W 4f7	32.51	32.66	32.48	32.27
W 4f5	34.70	34.83	34.66	34.45
Ni 2p3	853.79	854.17	853.90	853.47
Ni 2p1	871.04	871.42	871.15	870.66
Ni 2p3 oxy-S	856.25	855.89	855.89	855.68
S 2p	162.04	162.2	161.98	161.81

Table 3. XPS intensity ratios for oxidic and sulfided forms of W-Ni catalysts

Sample	WNi/Al_2O_3	WNi/Beta+Y	WNi/Y	WNi/Beta
Bulk W/Ni molar ratio	1.29	1.29	1.29	1.29
Bulk W/Si+Al, ×100	15.1	11.8	13.3	13.2
Bulk Ni/Si+Al, ×100	11.7	9.2	10.3	10.2
Oxidic ×100				

I_W/I_{Al+Si}	6.35	7.80	8.47	7.67
I_{Ni}/I_{Al+Si}	6.28	6.05	5.72	6.10
I_W/I_{Ni}	1.01	1.29	1.48	1.26
Sulfided×100				
I_W/I_{Al+Si}	4.15	5.78	6.52	5.50
I_{Ni}/I_{Al+Si}	3.55	4.72	4.39	4.40
I_W/I_{Ni}	1.17	1.22	1.49	1.25
S/W+Ni	1.58	1.49	1.53	1.47
Degree of sulfiding				
WS_2	91.3	87.8	92.0	84.4
NiS	77.1	93.5	90.4	87.3

TEM

The TEM images reveal the presence of typical structures of the layered WS_2 phase (Confirmed by EDX), which are homogeneously dispersed on the support. In the sulfided state, the WS_2 slabs were formed on the surface of all catalysts and exhibited thread-like layers. The statistical results of the average length and number of layers of the WS_2 slabs are given in Table 3. For WNi/Al_2O_3 catalyst, the average particle length is equal to 8.7 nm and the average layer number is equal to 2.0, which are close to the values of 7.3nm and 2.4 reported in the literature [15]. When 15wt% zeolites (beta or Y) were added in catalyst (WNi/Beta and WNi/Y), the average length of WS_2 slabs increases to 16.1-17.7 nm, and the average layer number increases to 3.7-4.4. When the zeolite content further increases to 30wt% (WNi/Beta+Y), the average slab length also further increases to 26.3 nm; whereas, the average layer number nearly unchanged (3.7).

Activity evaluation

The LCO hydrotreating results are summarized in Table 4 and Figure 1 and Figure 2. Compared with WNi/Al_2O_3, the zeolite containing catalysts exhibit much higher hydrodenitrogenation (HDN), hydrodesulferization (HDS), and hydrodearomatation (HDA) activities (Figure 1). HDS, HDN, and HDA activities increased 1.17-2.63wt%, 18.51-19.45wt%, and 3.3-6.9wt% respectively. HDN activity among the three zeolite containing catalysts is identical. HDS activity of WNi/Beta+Y and WNi/Beta is higher than that of WNi/Y.

As shown in Figure 2, after hydrogenation, majority of aromatics is converted to monoaromatics. The saturation of aromatics increases in the order: WNi/Beta +Y > WNi/Beta > WNi/Y > WNi/Al_2O_3. The higher HDN, HDS, and HDA activities of the three zeolite-containing catalysts are mainly associated with the enhancement of hydrogenation activities and the increase in acidity. From XPS results, it was found that no any correlation between the sulfidation degree of W and Ni and reaction activities can be established. Therefore, the difference in hydrogenation activity can not be resulted from the variation in sulfidation extent. The enhancement of hydrogenation activity is most likely related to the size of hydrogenation active sites, WS_2. It is noted that, though the sizes of the surface WS_2 slabs were similar for WNi/Beta and WNi/Y, WNi/Beta presented high HDS and HDA activities, which indicates that there must be other factors influencing the activities besides the hydrogenation performance of the catalyst. After the zeolites beta and/or Y were added into the catalysts, HDN and HDA activities increased much greater than HDS activity. From the mechanism of HDN and HDA, it is known that the incorporation of zeolites enhances hydrogenation performance of the catalysts more pronouncedly than the increase in the acidity.

Table 4. Evaluation in 1L autoclave with LCO

Catalyst	WNi/Al_2O_3	WNi/Beta+ Y	WNi/Y	WNi/Beta
Density, g/ml (15°C)	0.9264	0.9009	0.9136	0.9171
Nitrogen,ppm	97.88	1.42	1.84	6.14
Sulfur,ppm	507	165	355	232
>412.1 °C fraction, wt%	0.9	2.1	2.5	1.5
Liquid yield, w%	92.3	90.3	88.5	91.5
Conversion from distillation curve	0.85%	3.70%	0.80%	1.55%
Conversion, w%	8.55	12.40	12.30	10.05

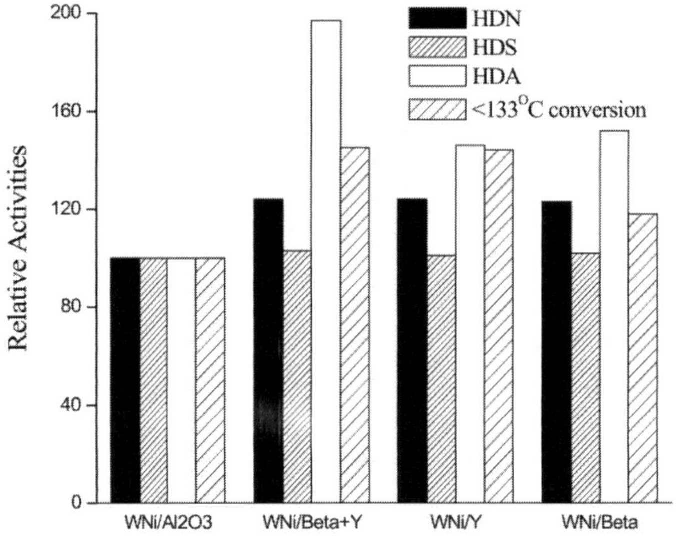

Figure 1. Relative HDS, HDN, and HDA activities, and <133 conversion

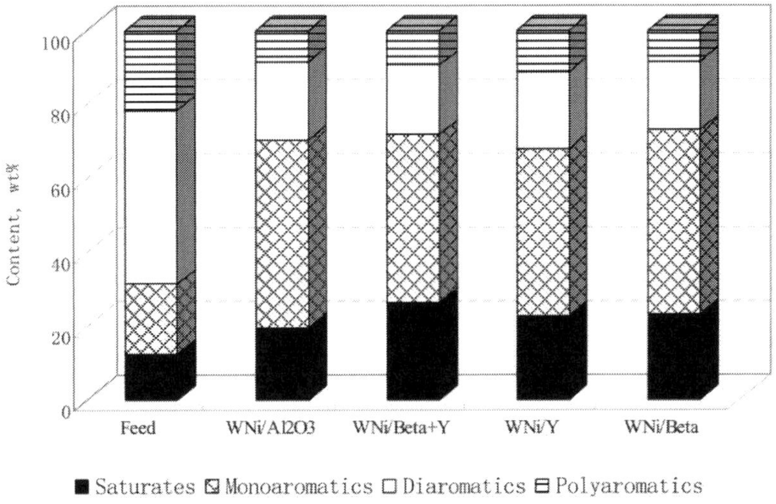

Figure 2. The saturates and aromatics distribution in feed and hydrotreated products on the four catalysts

Conclusions

The addition of modified zeolite beta and Y enhanced the overall acidity of the W-Ni catalysts. The acidity of the modified zeolite Y containing catalyst is higher than that of the zeolite beta containing catalyst. Very small tungsten and/or nickel sulfide clusters contain more than stoichiometric amount of sulfur. W and Ni in all catalysts present very high degree of sulfidation. Nickel species in WNi/Al_2O_3 more strongly interact with the support than they do in the zeolite containing catalysts. Among the zeolite containing catalysts, the sulfidation degree decreases in the order: WNi/Y>WNi/Beta+Y>WNi/Beta.

Considerable amount of W and Ni is migrated into the bulk after calcination. The W and Ni are further migrated into supports after sulfidation. WS_2 is present on the support in a typical layered structure for all catalysts. With the increase in the zeolite content, the WS_2 slabs become big and the number of the layers increases.

Compared with WNi/Al_2O_3, the zeolite containing catalysts exhibit much higher HDN, HDS, and HDA activities. HDN activity among the three zeolite containing catalysts is identical. HDS activity of WNi/Beta+Y and WNi/Beta is higher than that of WNi/Y. The higher HDN, HDS, and HDA activities of the three zeolite-containing catalysts are mainly associated with the enhancement of hydrogenation activities and the increase in acidity. After hydrogenation, majority of aromatics is converted to monoaromatics. The saturation of aromatics increase in the order: WNi/Beta +Y > WNi/Beta > WNi/Y > WNi/Al_2O_3. The higher acidity also leads to higher conversions. The conversions of the zeolite containing catalysts are much higher than that of WNi/Al_2O_3.

Acknowledgements

Financial support from Atlantic Canada Opportunities Agency is hereby acknowledged.

References

1. Song, C. (2003), Catalysis Today, **86**, pp.211-263.
2. Lecrenay, E., K. Sakanishi and I. Mochida (1997), Catal. Today, **39**, pp.13.
3. Isoda, T., S. Nagao, Y. Korai and I. Mochida (1996), Am. Chem. Soc., Prepr. Div. Petrol. Chem., **41**, pp.559.
4. Isoda, T., S. Nagao, X. Ma, Y. Korai and I. Mochida (2001), Energy Fuels, **1**0, pp.1078.
5. Bataille, F., J.L. Lemberton, G. Pérot, P. Leyrit, T. Cseri, N. Marchal and S. Kasztelan (2001), Applied Catalysis A: General, **220**, pp.191-205.
6. Isoda, T., S. Nagao, Y. Korai and I. Mochida (1998), Sekiyu Gakkaishi, **41**, pp.22.
7. Corma, A., V. Fornes, J.B. Monton and A.V. Orchilles (1987), J. of Catalysis, **107**, pp.288.
8. Marin, C., J. Escobar, E. Galvan, F. Murrieta, R. Zarate and H. Vaca (2004), Fuel Processing Technology, **86**, 391-405.
9. Breck, D.W., H. Blass and G.W. Skeels (1985), US Patent No. 4503023.
10. Ng, K. T. and D.M. Hercules (1976), J. Phys. Chem., **80**, pp.2094.
11. Salvati, L. Jr., L.E. Makovsky, J.M. Stencel, F.R. Brown and D.M. Hercules (1981), J. Phys. Chem., **85**, pp.3700.
12. Li, D., A. Nishishjima, D.E. Morris and G.D. Guthrie (1999), J. of Catalysis, **188**, pp.111-124.
13. Bendezu, S., R. Cid, J.L.G. Firro and A. L. Agudo (2000), Applied Catalysis A, **197**, pp.47-60.
14. Pawelec, B., J.L.G. Fierro, J.F. Cambra, P.L. Arias, J.A. Legarreta, G. Vorbeck, J.W. de Haan, V.H.I. de Beer and R.A. van Santen (1997), Zeolites, **218**, pp.250.
15. Vradmam, L., M.V. Landau and M. Herskowitz (2003), Fuel, **82**, pp.633-639.

103f Understanding the Mechanism of Iron Sulfide Induced Fouling in Upgrading

C.B. Panchal, Energy Systems Division, Argonne National Laboratory, 9700 S Cass Ave, Argonne, IL 60439
Formation of iron sulfide is recognized as a key step in formation of fouling in petroleum processing. Iron may be naturally present or generated by corrosion of pipes and process equipment by naphthenic acids. Limited data and analyses are available to quantify the effects of corrosion products on the propensity and threshold fouling conditions. Based on the experimental investigation and the literature data, an investigative analysis is performed to understand the underlying mechanisms of iron sulfide induced fouling/coking in upgrading processes.

103g Smart, Integrated Approach to Capturing Acid Crude Value

Mike Hodges, BP Refining Technology, Chertsey Road, Sunbury-on-Thames, TW16 7LN, England and Christopher Gould, Innovene Grangemouth Business Centre, Bo'ness Rd., Grangemouth, Stirlingshire, FK3 9XH, England.
Oil refineries are designed to run on a range of crude oils, so they can quickly respond to changing markets. However so called 'opportunity' crudes have a variety of properties which make them undesirable to the refiner. Acid crude is just one such 'opportunity' crude they may consider.

Acid crudes are found throughout the world, normally in shallow oil fields that have undergone microbial decomposition, accounting for about 10% of supply. However, if used as raw material, it can cause corrosion problems, leading to an elevated risk of line failure in certain areas of oil refineries. As a result, demand for acid crude is often lower than conventional crude oils, and the market discounts significantly, depending on the acidity level.

To handle the increased corrosion risk, refiners traditionally invest heavily in upgraded metallurgy of key refinery units, often using expensive stainless steels. This increased cost is not always economically justified given the variability of acid crude discounts. Therefore, if a way could be found to safely and economically increase the percentage of acid crudes processed, refiners could begin to use this less-expensive raw material to increase their margins.

In 1998, BP began a project aimed at understanding acid crude corrosion in the oil refinery at Grangemouth, Scotland. We have carried out significant work using on-line corrosion monitoring techniques, anti-corrosion chemicals, laboratory-based testing of main process streams, and detailed inspection during turnaround times. This smart, integrated approach to risk evaluation and refinery monitoring increased margins at the refinery while maintaining the integrity of the operating plant.

COMPARATIVE STUDY OF VAPOR-LIQUID EQUILIBRIUM DURING HYDROPROCESSING OF DIFFERENT PETROLEUM FEEDSTOCKS

Jinwen Chen, Wei Jiang, Hong Yang, Randall Hawkins, and Zbigniew Ring
National Centre for Upgrading Technology (NCUT), One Oil Patch Drive
Devon, Alberta, Canada, T9G 1A8

Prepared for Presentation at the 2006 AIChE National Spring Meeting, TA001: Advances in Hydroprocessing II, April 23-27, 2006, Orlando, FL, USA

March 2006

Abstract

VLE experiments were conducted in a continuous flow bench-scale unit at temperatures, pressures, and gas-to-oil ratios typically used for hydroprocessing. For comparison, three different feedstocks were used: light cycle oil (LCO) (highly aromatic), white oil (highly paraffinic) and a mixture of LCO and white oil. Under the same conditions white oil showed the highest volatility and LCO the lowest, although white oil had the highest IBP, FBP and averaged boiling point. The differences in volatility among these feedstocks were mainly attributed to their relative aromatics contents. A commercial flash program was calibrated by adjusting the interaction coefficients between hydrogen and hydrocarbons in the VLE model to match the flash calculation with experimental data. The interaction coefficients were further correlated with the boiling points and aromatics content of the feedstocks.

1. Introduction

Hydroprocessing of middle distillates, including hydrotreating and hydrocracking, involve oil and hydrogen flowing-through trickle-bed reactors under high temperatures and pressures (350-400°C and 50-100 atm.). Under such conditions, the reacting system (hydrogen and hydrocarbons) is close to thermodynamic vapour-liquid equilibrium (VLE) with both the vapour and liquid phases containing hydrogen and hydrocarbons (including sulphur and nitrogen compounds). VLE can significantly affect the flow rates of the two phases and the reactor hydrodynamics, also consequently affecting the reactor performance (Chen et al., 2005).

There have been only a few studies published considering VLE in three phase reactors (Akgerman et al., 1985; Smith and Satterfield, 1986; LaVopa and Satterfield, 1988; Jong, 1994; Tesser et al., 2003). There have been even fewer studies of the VLE effect on hydroprocessing (Kocis and Ho, 1986; Bellos and Papayannakos, 2003). Experimental data on VLE of hydrogen-petroleum fraction systems obtained under commercial hydroprocessing conditions are rarely published in the open literature because of the poor definition of the problem, stemming from the complex composition of those systems. Instead, systems consisting of hydrogen and a single hydrocarbon or a simulated hydrocarbon mixture are often studied (Simnick et al., 1978; Laugier et al., 1983; Ramanujam et al., 1985; Connolly and Kandalic, 1986; Wiegand et al., 1989). These studies have been briefly reviewed in our previous paper (Chen et al., 2005).

Continuing from our previously reported study on VLE with a light cycle oil (LCO), this paper presents our VLE experimental data obtained with a white oil, and a mixture of the white oil and LCO under the conditions similar to those used in the previous study. Comparisons between these VLE data sets are made. Also presented are the interaction coefficients between hydrogen and hydrocarbon boiling pseudo-components for the equation of state (EOS) used to perform flash calculations, which have been evaluated and correlated with pseudo-component boiling point and composition.

2. Experimental

The pertinent details of the experimental setup, procedures and validation experimental data were given in a previous paper (Chen et al., 2005). Here we give just a brief description.

A continuous-flow VLE cell was used to conduct the experiments. The oil feed was pumped from a feed tank and then mixed with hydrogen (99.99%) under the desired pressure. The resulting mixture went through a stainless steel coil pre-heater before entering the VLE cell where it was allowed to flash. Both the coil and the VLE cell were situated in a temperature-controlled furnace to ensure a constant flash temperature. The cell (I.D. of 2.54 cm and height of 30 cm) was arranged as a high-pressure, high-temperature phase separator in which a two-point thermocouple was positioned at the center to measure the temperatures of the liquid and vapour phase separately. The vapour phase from the top of the cell went into a low-temperature, low-pressure phase separator where the hydrocarbons in the vapour were condensed into the corresponding liquid while the uncondensed hydrogen went to a dry test meter (DTM). The liquid phase from the bottom of the cell entered another low-temperature, low-pressure phase separator where the hydrocarbons were condensed and the dissolved hydrogen was released and went into a high-precision gas meter (Ruska 2331-801, Technel Engineering Inc.) for accurate volume measurement. The condensed liquid samples were collected, weighed and analyzed separately for density and simulated distillation (SimDis). Overall mass balance calculation was performed to ensure that the experiments were properly conducted.

Table 1. Physical properties and chemical compositions of the materials used in VLE experiments

	LCO (original)	White oil	Mixture of LCO and white oil (50wt%-50wt%)
Density (15.6°C), g/mL (ASTM D4052)	0.9360	0.8152	0.8714
Paraffins (wt%)	6.8	31.1	20.8
Cyclo-paraffins (wt%)	8.9	67.7	40.2
Aromatics (wt%)	83.8	1.1	39.0
SimDis, °C (ASTM D2887)			
IBP	128.4	210.5	148.0
10wt%	225.0	237.5	233.0
30wt%	255.3	256.5	255.0
50wt%	288.5	272.5	277.0
70wt%	326.4	312.5	318.5
90wt%	373.9	395.0	381.0
FBP	430.8	490.0	483.5

Table 1 lists the key physical properties and chemical compositions of the three feeds used in the study – LCO, white oil and a mixture of white oil and LCO (50wt%-50wt%). The LCO has the highest content of aromatics, 84wt% (the balance being paraffins) while the white oil contains only 1wt% aromatics.

The temperatures and pressures used in this study ranged from 250°C to 400°C and 50 atm. to 100 atm, respectively. The liquid feed and gas rates were fixed at 10 g/h and 10 NL/h, respectively, resulting in a gas-to-oil ratio of 1000 NL/kg. All of these conditions have been used in our previous hydrotreating studies with the same LCO (Chen et al., 2003; Chen and Ring, 2004). Such conditions are routinely applied commercially with the exception of the gas-to-oil ratio, which is in the higher end of the commercial range.

3. Results and Discussion

3.1. Comparison of VLE Data for Different Feedstocks

One of the important objectives of this study was to investigate the relative volatility of oils of differing compositions under different operating conditions. Figures 1 and 2 show the wt% of oil in the vapour phase at different temperatures and pressures, respectively, for the three oils investigated in this study.

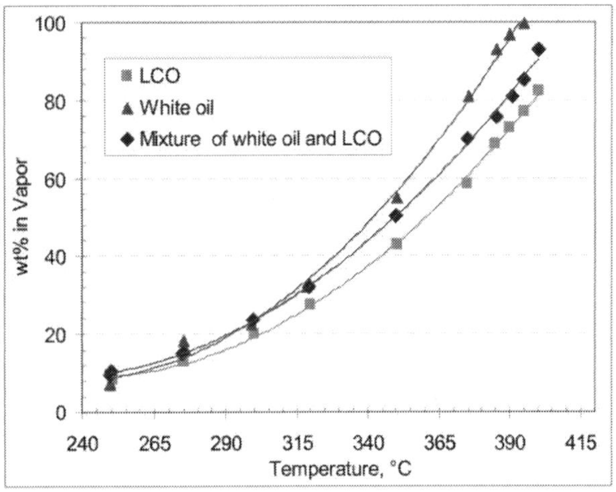

Figure 1. Wt% of oil in the vapour phase at different temperatures (P =70 atm)

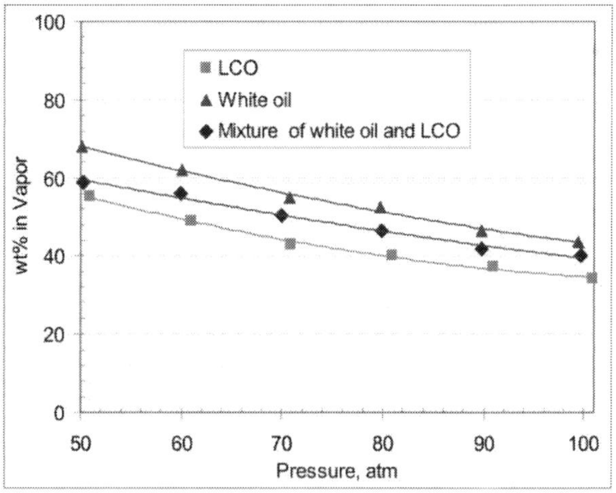

Figure 2. Wt% of oil in the vapour phase at different pressures (T = 350°C)

As shown in these figures, the wt% of the individual oils in the vapour phase increase significantly with temperature and decrease with pressure. At 250°C (P = 70 atm.), only about 10wt% of oil (regardless of composition) is vaporized. In this case, VLE does not have a significant effect on the flow dynamics in the reactor. However, at 400°C, over 80wt% of the LCO and 100% of white oil are found in the vapour phase, which would dramatically affect the flow dynamics in the reactor. In the pressure range of 50 to 100 atm. investigated in this study, the wt% of oil in the vapour phase changed from about 60wt% to about 40wt% at 350°C. It is interesting to note that, at the range of operating conditions used, and the same temperature and pressure, the wt% of oil in the vapour phase is always the highest for the white oil (except for the lowest temperature of 250°C). The wt% is always the lowest for the LCO and, as expected, the white-oil-LCO mixture is always in the middle. This observation indicates that the volatilities of the three feedstocks are in the order: white oil > mixture of white oil and LCO > LCO. From the physical and chemical composition data listed in Table 1, LCO has the highest density but much lower IBP and FBP, while the white oil has the lowest density but much higher IBP and FBP. The chemical composition (hydrocarbon type distribution) controls the relative volatility of the oil feedstocks. Therefore, it can be concluded that at a similar boiling point, paraffins vaporize more easily than aromatics.

3.2. *Evaluation of Interaction Coefficients*

As mentioned earlier, to accurately model a hydroprocessing trickle-bed reactor, VLE has to be accounted for, and the chemical composition of each phase critically affects VLE. Therefore, in principle, flash calculations are required at each position at which the reactor model differential equations are integrated to calculate the temperature and conversions of individual reactants. With petroleum systems, flash calculation is normally performed by solving the equation of state (EOS). The most frequently used EOS is the Peng-Robinson equation:

$$P = \frac{RT}{V-b} - \frac{a(T)}{V^2 + 2bV - b^2} \tag{1}$$

where a and b are equation parameters associated with critical properties:

$$a(T) = \frac{\Omega_a R^2 T_c^{\,2}}{P_c}\alpha(T) \qquad b = \frac{\Omega_b R T_c}{P_c} \tag{2}$$

For a mixture, such as petroleum fractions, mixing rules are applied to a and b:

$$b = \sum_{i=1}^{N} y_i b_i \qquad a = \sum_{i=1}^{N}\sum_{i=1}^{N} y_i y_j a_{ij} \tag{3}$$

$$a_{ij} = (a_{ii} a_{jj})^{1/2}(1 - d_{ij}) \tag{4}$$

In Equation (4), d_{ij} is called the interaction coefficient between components i and j. For a petroleum system, it is impossible to identify all the hydrocarbon components. Boiling pseudo-components are routinely used in VLE calculations instead.

The interaction coefficient d_{ij} between components i and j is critical to flash calculation. There are relatively reliable correlations available to estimate d_{ij} between hydrocarbons. However, there are no such correlations available for d_{ij} between hydrogen or sour gases (H_2S and NH_3), and hydrocarbons, which are required for flash calculation in hydroprocessing. The evaluation of the interaction coefficients between hydrogen and hydrocarbon pseudo-components was performed in this study by fitting the EOS – Equations (1) to (4) – to experimental data. These data included the wt% of oil in the vapour phase and the SimDis data of the top and bottom samples from the VLE cell.

To achieve this goal, a commercial flash program was purchased from Computer Modeling Group Ltd. and modified to perform the flash calculation. The flow diagram in Figure 3 shows the logic followed in the calculations. The feed oil was divided into 29 boiling pseudo-components according to its SimDis data. Hydrogen was treated as a separate component. To use the program for flash calculation, the density, molecular weight and boiling of each

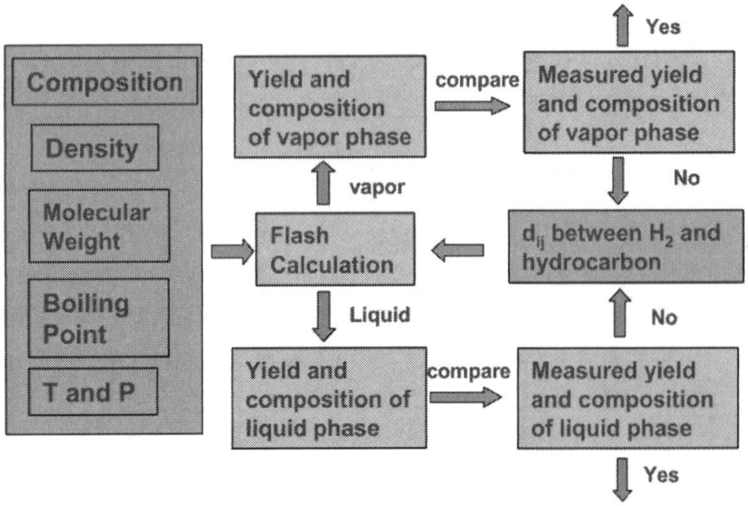

Figure 3. Computation diagram to evaluate interaction coefficients

component have to be known. All other properties, such as critical properties, are estimated by the CMG software based on these three properties. For hydrogen, all the data are available in the open literature. To get the molecular weight and density of the 29 hydrocarbon pseudo-components, distillations were performed to generate 8 fractions for each of the three feedstocks. Molecular weight, SimDis and density were then measured for each fraction, resulting in separate correlations between molecular weight and boiling point, and between density and boiling point for each feedstock. These correlations were subsequently used to estimate the molecular weight and density of the 29 pseudo-components. For each feedstock, a set of initial values of the interaction coefficients between hydrogen and the pseudo-components were assumed. Flash calculation was then performed under all the experimental VLE conditions to calculate the vapour and liquid yield (wt% of oil in vapour), and the SimDis of the vapour and liquid products of the flash calculation. These were compared with the experimental data. Minimum discrepancies between the calculated and measured data were achieved by adjusting the values of the interaction coefficients. The interaction coefficients were assumed to vary linearly with boiling point, and the coefficients A and B in the formula below were the outcome of the parameter estimation calculation.

$$d_{iH} = A + B \times BP_i \tag{5}$$

where d_{iH} represents the interaction coefficient between hydrogen and pseudo-component i, and BP_i represents the boiling point of pseudo-component.

3.3. Discussion

Typical matches between the experimental and calculated data are shown in Figures 4, 5 and 6. Figures 4 (a), (b) and (c) show the calculated and measured wt% of oil in the vapour phases at different temperatures for LCO, white oil and the mixture of LCO with white oil, respectively. Figures 5 (a), (b) and (c) show those at different pressures. Figures 6 (a), (b) and (c) show the calculated and measured SimDis data of the top and bottom liquid products for the three feeds. As shown in these figures, the calculated wt% of oil in vapour phase and

SimDiS data match the experimental data reasonably well except some deviations observed for white oil (Figure 4 (b)) at the temperatures where the oil is almost completely vapourized. This is where the experimental measurements would be expected to be much less accurate. It is worth noting that the flash calculations predict correct shapes of the SimDis curves (Figures 6 (a), (b) and (c)) which are quite different among the three feeds, indicating a high predictive capability of the flash program if the right interaction coefficients between hydrogen and hydrocarbons are used.

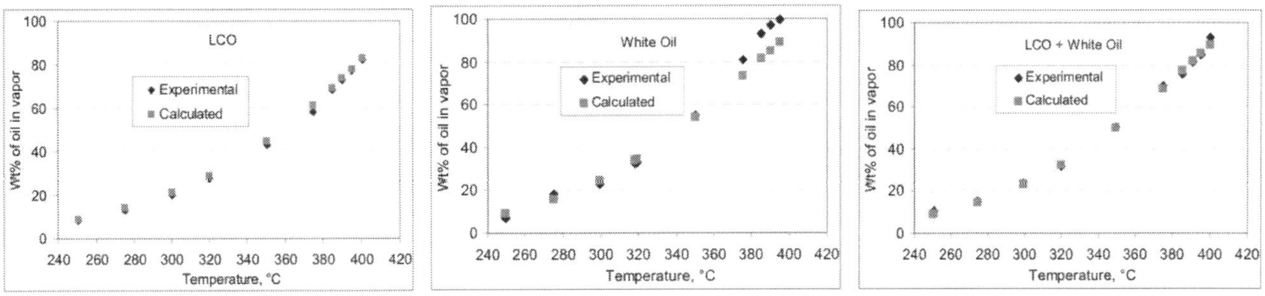

(a) LCO (b) White oil (c) Mixture of LCO and white oil

Figure 4. Calculated and measured wt% of oil in vapour phase at different temperatures

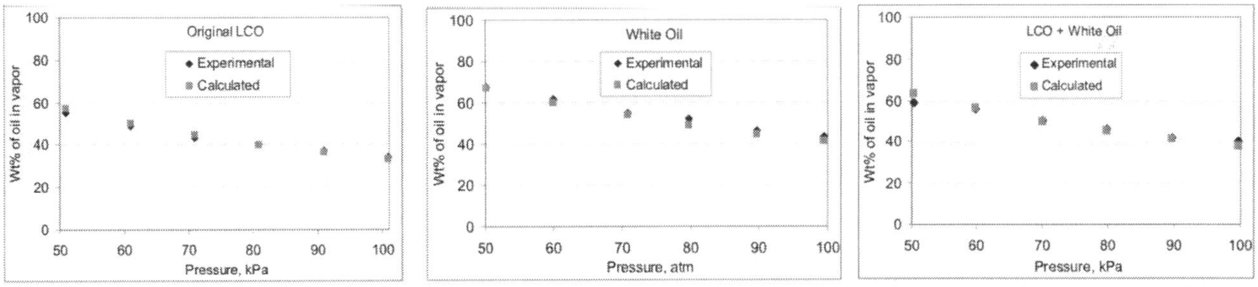

(a) LCO (b) White oil (c) Mixture of LCO and white oil

Figure 5. Calculated and measured wt% of oil in vapour phase at different pressures

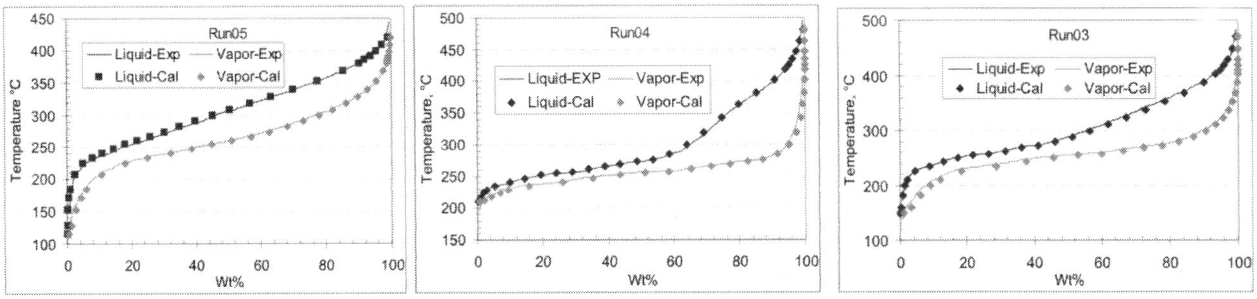

(a) LCO (b) White oil (c) Mixture of LCO and white oil

Figure 6. Calculated and measured SimDis

The fitted values of A and B for the three feedstocks are listed in Table 2. The interaction coefficients (d_{iH}) calculated using these A and B values at different boiling points and plotted in Figure 7, which shows that at the same boiling point, LCO has the highest d_{iH} and white oil has the lowest d_{iH}. As expected, the d_{iH} of the mixture of LCO and white oil is in-between. The values of A and B for the mixture of LCO and white oil are approximately equal to those averaged ones between LCO and white oil, as seen in Table 2. Figure 7 also shows that the interaction coefficients of the three feeds are practically identical for components boiling at about 120°C, but the differences among them increase for the higher-boiling components.

Table 2. Values of A and B in Equation (5)

	A	B
LCO	0.1865	0.00388
White oil	0.386	0.00226
Mixture of LCO and white oil	0.285	0.00302

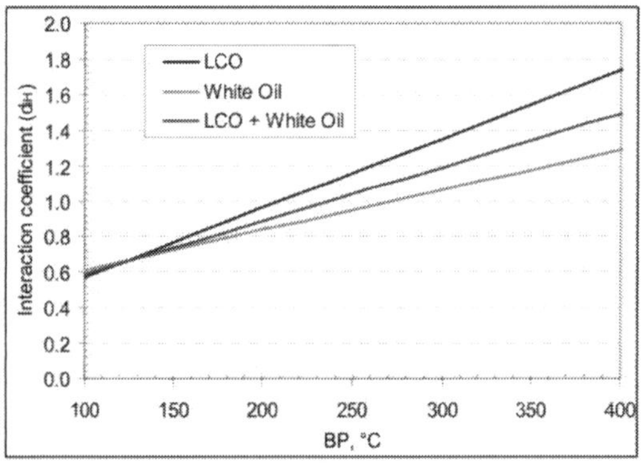

Figure 7. Interaction coefficients between hydrogen and hydrocarbons

The differences in the interaction coefficients for the three feeds can be attributed mainly to their chemical compositions. As discussed earlier, the high aromatics content in LCO make it less volatile compared to white oil, which contains mainly paraffins (Figures 1 and 2). The higher interaction coefficients for LCO (Figure 7) also suggest a stronger interaction between hydrogen and the oil. Therefore, in order to perform flash calculations with any type of middle distillate feedstocks, the parameters A and B (Table 2) were correlated with the aromatics content (wt%) in the feed. Two linear correlations were obtained:

$$A = 0.3852 - 0.00241 \times C_{Aromatics\%} \quad (7)$$

$$B = 0.002245 + 1.96 \times 10^{-5} \times C_{Aromatics\%} \quad (8)$$

Considering the modest sensitivity to the interaction coefficients, the simple preliminary correlations for A and B (Equations (7) and (8)) are expected to be sufficiently accurate to produce reasonable flash calculation results for any middle distillate feedstock.

It should be mentioned that in order to perform the flash calculation, the SimDis data, density and molecular weight distributions with boiling points have to be known prior to the flash calculation. SimDis analysis is easy to conduct and can be done in any petroleum laboratory. However, density and molecular weight distributions with boiling points normally need real distillation followed by the measurement of density and molecular weight for the individual fractions, which are expensive and time consuming. We have developed a methodology to measure the molecular weight distribution by boiling point using GC-FIMS (Gas Chromatography/Field Ionization Mass Spectrometry). The idea is to use the scanned mass of hydrocarbon ions and their abundance to calculate the molecular weight at different GC retention times, from which the boiling point of the ions can be calculated by using pre-established calibration data between retention time and boiling point. The details on this method development will be reported/presented in the future.

4. Conclusions

1) Comparative VLE experiments were conducted in a continuous flow cell with hydrogen and petroleum distillates, including LCO, white oil, and a mixture of LCO and white oil, under typical hydroprocessing conditions.
2) The relative amount of oil in the vapour phase significantly increased with the increase in temperature and decrease in pressure.
3) The volatilities of the three feedstocks have the order: white oil > mixture of white oil and LCO > LCO, which correlates with the relative aromatics contents in these feedstocks.
4) The interaction coefficients between hydrogen and hydrocarbon boiling pseudo-components were evaluated by fitting the CMG software calculations to the experimental data. These interaction coefficients linearly increased with the increase in boiling point.
5) LCO had the highest interaction coefficients, while white oil had the lowest, which is also attributed to their relative aromatics contents. The parameters A and B correlating the interaction coefficients with boiling point were further correlated to the aromatics contents in the feed. These correlations are expected to reasonably simulate flashing of any type of feedstock, provided that its SimDis, density, molecular weight distributions with boiling point and aromatics contents are known.

5. Acknowledgements

The authors are grateful to the following individuals for their help and support: Dennis Carson for conducting the VLE experiments and Norm Sacuta for editing this paper. Partial funding for NCUT has been provided by the Canadian Program for Energy Research and Development (PERD), the Alberta Research Council (ARC) and the Alberta Energy Research Institute (AERI).

6. References

1. Akgerman, A., Collins, G. M., Hook, B. D., "Effect of Feed Volatility on Conversion in Trickle-bed reactors", Ind. Engng. Chem. Fundam., 24, 398-401, 1985.
2. Bellos, G. D., and Papayannakos, N. G., "The Use of a Three Phase Microreactor to Investigate HDS Kinetics", Catal. Today, 79&80, 349-355, 2003.
3. Chen, J., Yang H., and Ring, Z., "Experimental Study on Vapour-Liquid Phase Equilibrium of LCO Hydrotreating Systems at High Temperatures and Pressures", Proceedings of 2005 AIChE National Spring Meeting, Paper 61b
4. Connolly, J. and Kandalic, G., "Gas Solubilities, Vapour-Liquid Equilibria, and Partial Molar Volumes in Some Hydrogen-Hydrocarbon Systems", J. Chem. Eng. Data, 31, 396-406, 1986.

5. de Jong, K. P., "Effect of Vapour-Liquid Equilibrium on Coke Deposition in Trickle-Bed Reactors during Heavy Oil Processing. 2. Modeling", Ind. Engng. Chem. Res. 33, 3141-3145, 1994.
6. Kocis, G. R. and Ho, T. C., "Effects of Liquid Evaporation on the Performance of Trickle-Bed Reactors", Chem. Engng. Res. Des., 64, 288-291, 1986.
7. Laugier, S., Richon, D., and Renon, H., "Vapour-Liquid Equilibria of Hydrogen-Hydrocarbon Mixtures Up to 40 MPa, 400°C: Measurements and Representation", Chemical Engineering Thermodynamics, paper presented at the Thermodynamics Sessions of the 2nd World Congress on Chemical Engineering, 1983, Montreal, Que., Canada.
8. LaVopa, V., and Satterfield, C. N., "Some Effects of Vapour-Liquid Equilibria on Performance of a Trickle-Bed Reactor", Chem. Engng. Sci., 43, 2175-2180, 1988.
9. Ramanujam, S., Leipziger, S., and Well, S., "Vapour-Liquid Equilibrium for a Hydrogen/ Simulated Coal-Derived Liquid System", Ind. Engng. Chem. Proc. Des. Dev., 24, 364-368, 1985.
10. Simnick, J., Sebastian, H., Li, H-M, and Chao, K-C, "Solubility of Hydrogen in Toluene at Elevated Temperatures and Pressures", J. Chem. Eng. Data, 23, 339-340, 1978.
11. Smith, C. M. and Satterfield, C. N., "Some Effects of Vapour-Liquid Flow Ratio on Performance of a Trickle-bed reactor", Chem. Engng. Sci. 41, 839-843 1986.
12. Tesser, R., Di Serio, M., and Santacesaria, E., "Influence of the Vapour-Liquid Equilibrium (VLE) on the Kinetics in Gas-Liquid and Gas-Liquid-Solid Systems", Catal. Today, 79&80, 323-331, 2003.
13. Wiegand, K., Strobel, B. and Hofmann, H., "Vapour-Liquid Equilibrium of Hydrogen/Coal Liquid and Methane/Coal Liquid Systems at Elevated Temperatures and Pressures", Chem. Engng. Technol., 12, 280-288, 1989.

Value Driven Catalyst Developments in FCC Pretreatment Service

<u>AIChE Spring Annual Meeting</u>
Orlando, Florida
April 2006

Vito Bavaro
Patrick Gripka

Criterion Catalysts & Technologies L.P.
Houston, Texas

Dr. Alexei Gabrielov
Dr. Changan Zhang

CRI/Criterion, Inc.
Westhollow Technology Center
Houston, Texas

Clean air regulations have been the driving force for the reduction of stationary and tailpipe emissions. The net result has been significant refinery infrastructure and technology changes in units that produce both gasoline and diesel fuel components. Refiners have relied on innovative technologies that more effectively remove sulfur and other resultant air pollutants from fuel components and have rapidly applied them within the refinery configuration. The technologies selected have included both hardware modifications and catalytic improvements to provide for essentially zero sulfur fuels. Refiners have applied these technologies both where the contaminants are removed before conversion occurs as well in a final step before being blended into the finished cleaner burning fuels of the future. Table 1 highlights these recent changes in on-road diesel and gasoline sulfur specifications.

Table 1

Fuel Sulfur Specifications

	Diesel		Gasoline	
	North America	EU	North America	EU
	Sulfur, ppm	Sulfur, ppm	Sulfur, ppm	Sulfur, ppm
1993	500	2000		500
2000		350		150
2005		50 / 10 *	30 **	50 / 10
2006	15 *			
2009		10		10

* Highway or On-Road Diesel. Different specifications and timing exist for the reduction in sulfur content of the other categories (Off-Road, Marine, Locomotive, Heating).

** Starting in 2004, maximum per gallon caps are reduced per schedule.

Most other nations are lowering diesel sulfur content to <50 or <15 ppm by 2010.

Most other nations are lowering gasoline sulfur content to <50 by 2011.

The drive to zero sulfur has provided the impetus among catalyst technology providers for significant expenditures in research and development that has resulted in dramatic gains in desulfurization and nitrogen removal activity. Among catalyst technology providers, most of the focus has been on specific catalyst development for sulfur removal via post treatment approaches for ultra low sulfur diesel or low sulfur gasoline production. Criterion Catalysts and Technologies L.P. and our partner companies have continued to demonstrate industry leadership in the innovation of catalysts and technologies that target these post treatment ultra low sulfur production operations. In a parallel path, Criterion has also recently completed an extensive R&D program and commercialized an extension to the ASCENT family of catalysts focused on contaminant removal prior to the FCC unit. This provides the ability to achieve the low sulfur product requirements with the added advantage of improved FCC conversion, while further improving the refiner's catalyst life cycle value compared to other available catalyst technologies.

65% of US refiners will utilize a combination of FCC Pretreatment (FCC PT) and various forms of post-treatment of the gasoline and diesel boiling range to produce the near zero sulfur fuels. Of the 65%, 25% will rely on FCC PT alone (no postreatment) for production of low sulfur gasoline, while relying on post treatment to meet the ULSD specification of 15 ppm. The demand for improved catalysis in the FCC PT service has driven Criterion's R&D program to continue to innovate. A recently completed Criterion FCC Pretreatment catalyst R&D program targeted the need for improved life cycle value for refineries that apply FCC feed pretreatment technology in their clean fuels production strategy. The specifically targeted areas were desulfurization, denitrification, aromatic saturation, stability, and regenerability. In addition, elements such as ease of handling, loading, storage, and startup were also considered.

Catalyst Technology Application into FCC PT

Catalyst technology commercially applied for FCC PT include catalysts that were developed for other services such as the newest Cobalt Molybdenum Ultra Low Sulfur Diesel and Nickel Molybdenum Hydrocracking Pretreat catalysts (Type II Active Sites with pore size optimized for these services), as well as catalysts that were specifically developed for higher boiling fractions (HVGO/CGO) including full range resid processing. The reason for the wide variety of catalyst technology for FCC PT service is due to the wide boiling range typically encountered and the specific requirements of each refinery configuration. It is not atypical to encounter feedstocks that contain as much as 10% naphtha, up to 50% diesel, with the remaining fraction comprised of vacuum gas oils and heavier fractions boiling above 1200°F. The feedstocks can be sourced from operations such as crude atmospheric and vacuum towers, cokers, ROSE/deasphalters, Fluid Catalytic Crackers, and Lube units. Environmental drivers, such as refinery FCC flue gas emission requirements and gasoline sulfur blending requirements, and/or economic drivers, such as obtaining maximum FCC conversion, gasoline, olefin, and gasoline octane, put more complex demands on FCC PT catalyst performance. The wide variety of feeds encountered and the requirement of multi functional catalyst performance present catalyst selection challenges for the refiner and as well challenges for catalyst technology developers.

Catalyst applied in this service must operate robustly within the wide boiling range of the feeds charged to these units. As opposed to the lighter and relatively narrow boiling range of ultra low sulfur diesel operation, catalysts that operate in FCC PT service must obtain a specifically optimized sulfur, nitrogen and aromatic saturation performance to drive the downstream FCC environmental and economic performance and be able to maintain their activity (stability) in an environment where contaminant metals such as nickel, vanadium, silicon, sodium, arsenic, and asphaltenes exist. Often, several catalysts in a tailored package are utilized to fulfill the demands placed on the FCC PT unit. A detailed understanding of how to apply several different functionalities in an optimized manner is what make this a complex challenge.

The Starting Point for New Catalyst Development for FCC PT Application

In 1997, Criterion introduced a new concept in FCC PT catalysis, which resulted in a step change in FCC PT catalyst performance. The pore size distribution was optimized from previous generation catalysts with a focus on making the catalyst more tolerant (stable) to the feed contaminants expected in this service and enhancing the aromatic saturation performance that is critical in maximizing the FCC conversion. After 60+ FCC PT unit applications, the 200 series catalysts (DN-200) became the industry benchmark with commercially demonstrated improvements in catalyst stability (up to 200%) in addition to increases in sulfur, nitrogen, and aromatic saturation activity. Figure 1 demonstrates the concept behind the stability improvement of DN-200.

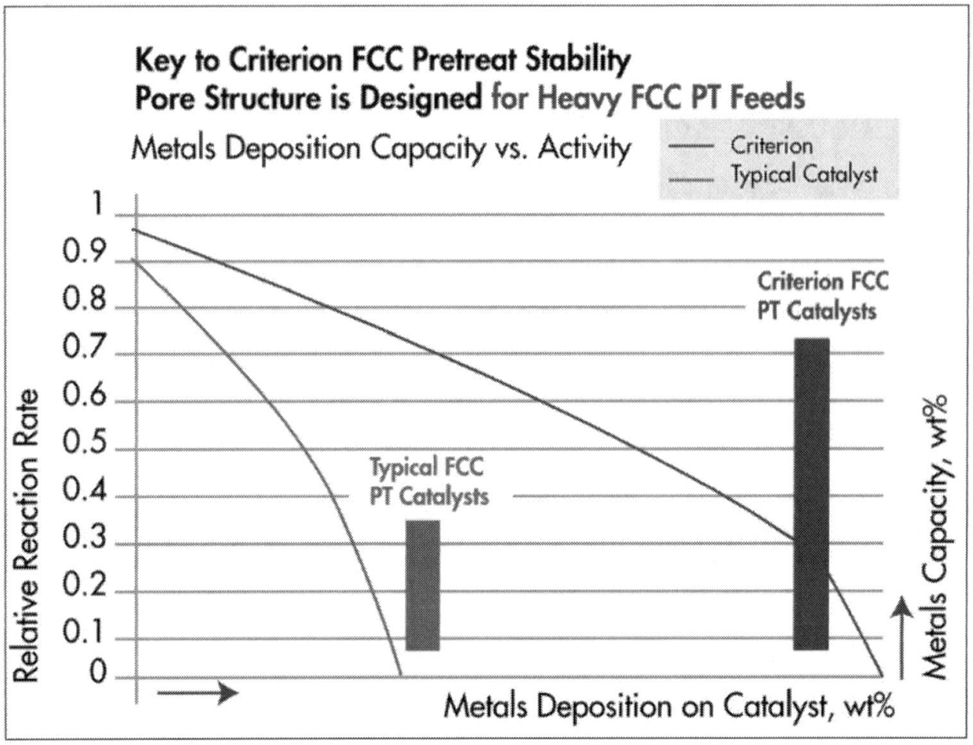

Figure 1: Improved Metals Tolerance of Criterion FCC PT Catalysts

DN-200 has since been succeeded by higher activity Criterion catalysts, including the CENTINEL family of catalysts, which employ catalyst manufacturing technologies that take the same optimized pore size distribution of the 200 series catalysts and create a catalyst that predominantly contains Type II active sites. These products have commercially demonstrated the value of the improved catalytic efficiency that Type II active sites provide with demonstrated FCC conversion improvements of up to 1.5 volume% and significant reductions in the FCC gasoline sulfur. Most importantly, the stability has been maintained as a result of the previously optimized pore size distribution.

Refiners have embraced and benefited from the Type II active site technology. However, new technologies usually place new demands, e.g. startup and handling issues, on the refiner in order to take full advantage of the benefit. Criterion's on-going R&D effort in FCC PT catalysis never rests and a new target was set to further extend FCC PT catalyst Life Cycle Value performance.

The New Targets

- Deliver improved FCC feed properties over the entire catalyst cycle
- Improve regenerability performance (90% of fresh activity by conventional regeneration)
- Provide flexibility in delivered state of catalyst (oxide or pre-sulfurized state)

Exceeding the performance of Type II active site catalysts, while maintaining conventional regeneration capability was the mountain that needed to be climbed. In addition, an added complexity was to provide flexibility in the delivered state, catalyst startup and handling. Even with these challenges in conjunction with the wide variety of feedstocks available and contaminants present, Criterion researchers achieved their targets in a 12-month concept to commercialization effort.

There are many places to begin in catalyst design and innovation, such as the metal function, surface chemistry function, or the pore accessibility (diffusion) function. These parameters have been the target of catalysis research for many years and new developments are not easily achieved in FCC PT service. FCC PT catalyst design demands a balance of all functions and overcoming the limitations as each bottleneck is observed.

Improving Diffusion Limitations

It was identified that the current FCC PT catalysts were bottlenecked by diffusion. Criterion researchers first had to overcome this obstacle. When hydroprocessing heavier boiling fractions, it is well known that diffusion limitations can interfere with active site accessibility. This affects overall catalyst performance regardless of the inherent high activity of the promoter metal. Researchers must strive for the proper balance of diffusion and inherent activity of the promoter metal in order to obtain optimized catalyst performance.

Reaction rates for contaminant removal are largely determined by the Thiele Modulus, which is a function of the ratio of the intrinsic rate constant (k_n) and the effective diffusion constant (D_{eff}):

$$\phi \sim \left(\frac{k_n}{D_{eff}} \right)^{1/2}$$

When diffusion limitations are determining the overall reaction rate, the effective diffusion constant can be affected by the following factors:

$$D_{eff} = \frac{D_{bulk}}{\tau} [\varepsilon_M (1 - \frac{r_{mol}}{r_M})^4 + \varepsilon_m (1 - \frac{r_{mol}}{r_m})^4]$$

Where:

τ is the pore tortuosity factor

D_{bulk} is the bulk diffusion constant

ε_M, ε_m are the catalyst's macroporosity and microporosity, respectfully

r_{mol} is the contaminant molecular radius

r_M, r_m are the catalyst's macropore radius and micropore radius, respectfully

FCC PT feedstocks contain a variety of contaminant species having different intrinsic reactivities and molecular dimensions. Optimizing the catalyst pore geometry and size will significantly improve the effective diffusion constant, which will in turn enhance the overall rates of hydrotreating reaction pathways previously limited by diffusion. The development of catalysts in applications where diffusion limitations are significant requires an iterative development process since the active metals, which provide the inherent catalyst activity, also influence diffusion when they are applied to the catalytic surface. Optimizing the diffusion characteristics does not just imply modifying the support. It is a combination of support and active metal interaction. In the final analysis it was discovered that modification of the pore shape design reduced the existing diffusion limitations. This will now allow the inherent promoter metal activity to be more fully utilized. An additional advantage is that the modified pore shape should be more tolerant to contaminant metals.

Improving Active Metal Efficiency

The next focus area is on improving metals utilization on the redesigned support. Criterion had previously developed the ASCENT oxide catalyst technology, which utilizes innovative support preparation methods and proprietary impregnation techniques. The ASCENT catalyst technology provides highly efficient active metals interacting with the alumina carrier and optimizes both hydrogenation and hydrogenolysis reaction pathways for heteroatom removal. ASCENT technology generates an active surface that is characterized by an increased number of adsorption sites. Figure 2 compares the capacity of conventional and ASCENT technology catalyst support surfaces, with a higher number of adsorption sites shown for the ASCENT support. These results indicate an increased surface heterogeneity for the ASCENT support, which generates a high metal efficiency and an optimal active (metal) site configuration on the alumina surface to catalyze the hydrotreating reactions.

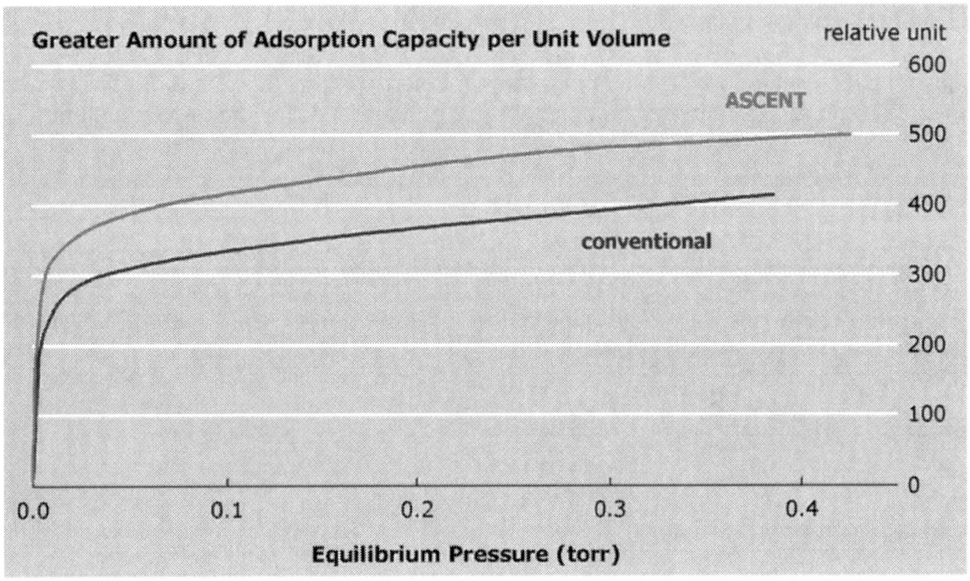

Figure 2: Adsorption Capacity for ASCENT and Conventional Catalyst Supports

Tailoring Catalytic Performance

The third area of focus is the promoter metal application. Again, incorporating the proprietary ASCENT manufacturing process allowed for further enhancement of the catalyst inherent performance characteristics through the application of proprietary impregnation techniques. A key result of the ASCENT technique is a significant increase in both the number and inherent activity of active sites as demonstrated in Figures 3 and 4. In order to provide for the wide range of feedstock and performance characteristics required in FCC PT service, both NiMo and CoMo promoter metals were explored.

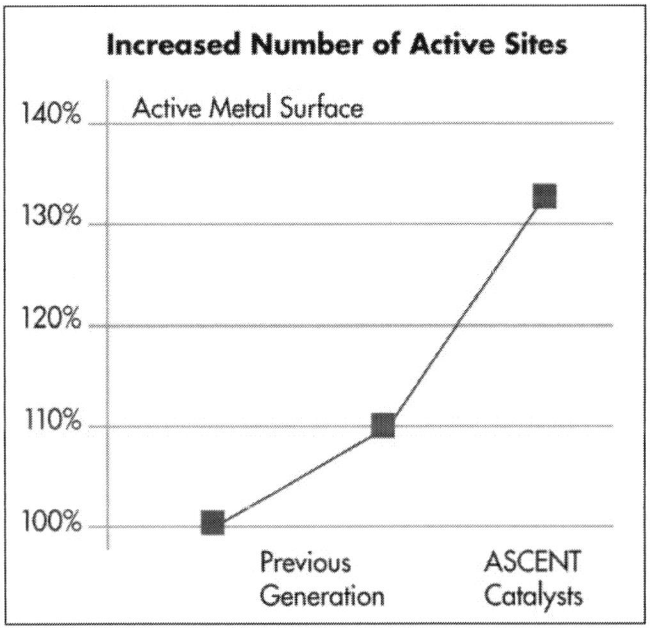

Figure 3: Increased number of Active Sites with ASCENT Technology

The net result of the combined interaction of modified pore geometry design, proper support surface functionality, and the proprietary impregnation techniques produce catalysts that more efficiently utilize the active metals that are available. Figure 4 demonstrates the more efficient metals utilization for the ASCENT products compared to conventional catalyst technologies.

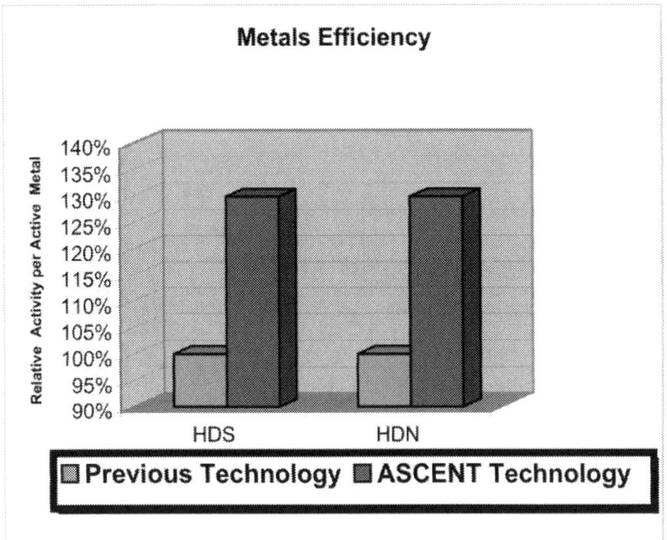

Figure 4: Relative HDS and HDN activities per active metal for conventional catalysts and ASCENT catalyst technology

Additional analysis has been performed to more fully understand the active metals configuration of NiMo and CoMo promoted catalysts produced by this new approach. In order to demonstrate differences compared to other catalyst technologies, TEM and NO(ad)/FTIR analyses were performed. The characteristics of the enhanced performance of an ASCENT NiMo catalyst (in it's sulfided form) produced by this novel approach are demonstrated in Figure 5. The TEM analysis allows quantification of average particle size (L) and the average degree of stacking (n) of the MoS_2 crystallites on the active catalyst surface. The latter method (FTIR) selectively detects the probe molecules of nitrogen oxide pre-adsorbed on the promoter atoms residing on the edges of MoS_2 crystallites.

Figure 5: High Resolution Transmission Electron Microscopy (HRTEM) photo of DN-3551 nickel molybdenum catalyst illustrates the high level of MoS_2 dispersion on the catalyst surface. Small crystallites of MoS_2 are prevalent, with some stacked slabs visible as thin parallel lines.

Table 2 below compares the relative average particle size and degrees of stacking for the conventional and ASCENT nickel molybdenum products. We found that the conventional catalyst (Type I sites) has the lowest degree of MoS_2 stacking compared to a higher degree of stacking for ASCENT DN-3551 NiMo. This can be an indirect indication that the ASCENT technology provides an optimal combination of metal-support interactions.

Table 2
TEM Results Comparison of Previous Generation and DN-3551

Previous Generation		DN-3551	
L (rel)	n (rel)	L (rel)	n (rel)
1	1	1	1.2

In order to assess the MoS_2 "edge occupation" by promoter metals, we have employed the NO chemisorption method in combination with the FTIR technique. The combination of these methods allowed detection of nitrosyl stretch vibrations of specific frequencies for the interaction of NO with the promoter on the edge sites (Figure 6).

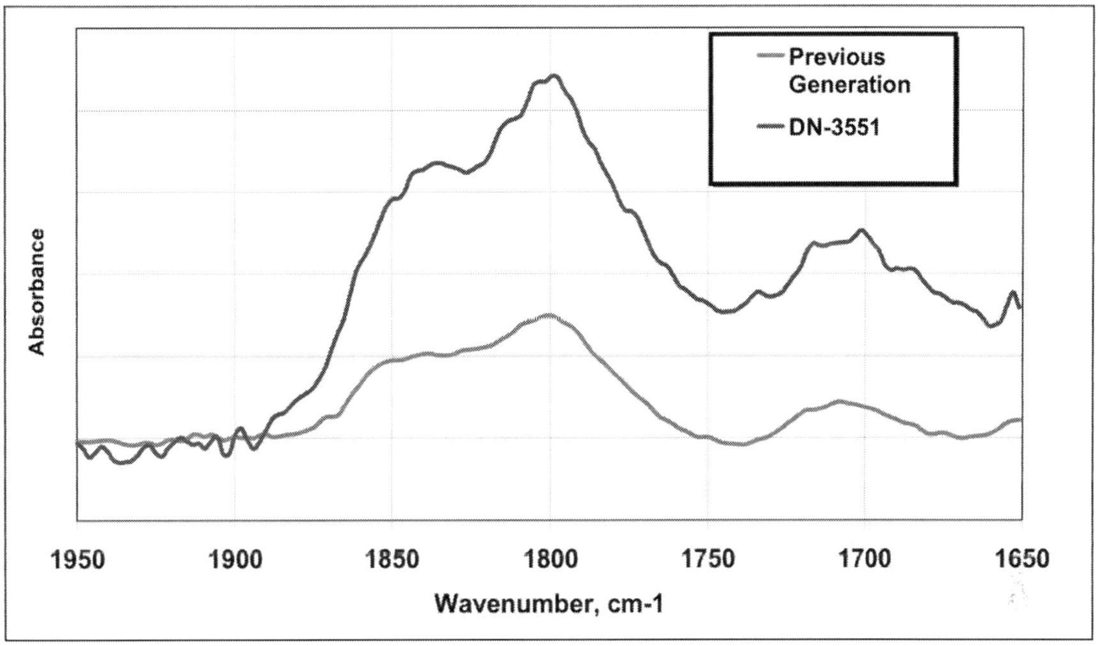

Figure 6: FTIR Spectra of NO chemisorbed on the promoter metal of the sulfided catalysts

As evidenced from the ratios of the integral intensities of the bands centered at ~1850 and ~1700 cm^{-1} in the spectra shown in Figure 6, the DN-3551 nickel molybdenum catalyst demonstrates the highest degree of MoS_2 edge occupation by the promoter. By providing the highest accessibility of active sites (particularly edges) and balanced interaction of the MoS_2 with the carrier (i.e. Type I and II sites), the result is that both the direct and indirect routes for heteroatom removal are catalyzed in the same catalyst composition.

Criterion's catalyst design methodology provides a new catalyst platform improving both the transport of reactants to active sites as well as enhancing the inherent activity of those sites. The combination of improved utilization of active metals, optimal mix of Type I and II active site morphologies and the highest degree of MoS_2 edge occupation results in a FCC PT catalyst technology with superior performance. This is demonstrated in Figure 7 by improved HDS, HDN, and hydrogen consumption for DN-3551 NiMo compared to previous technologies.

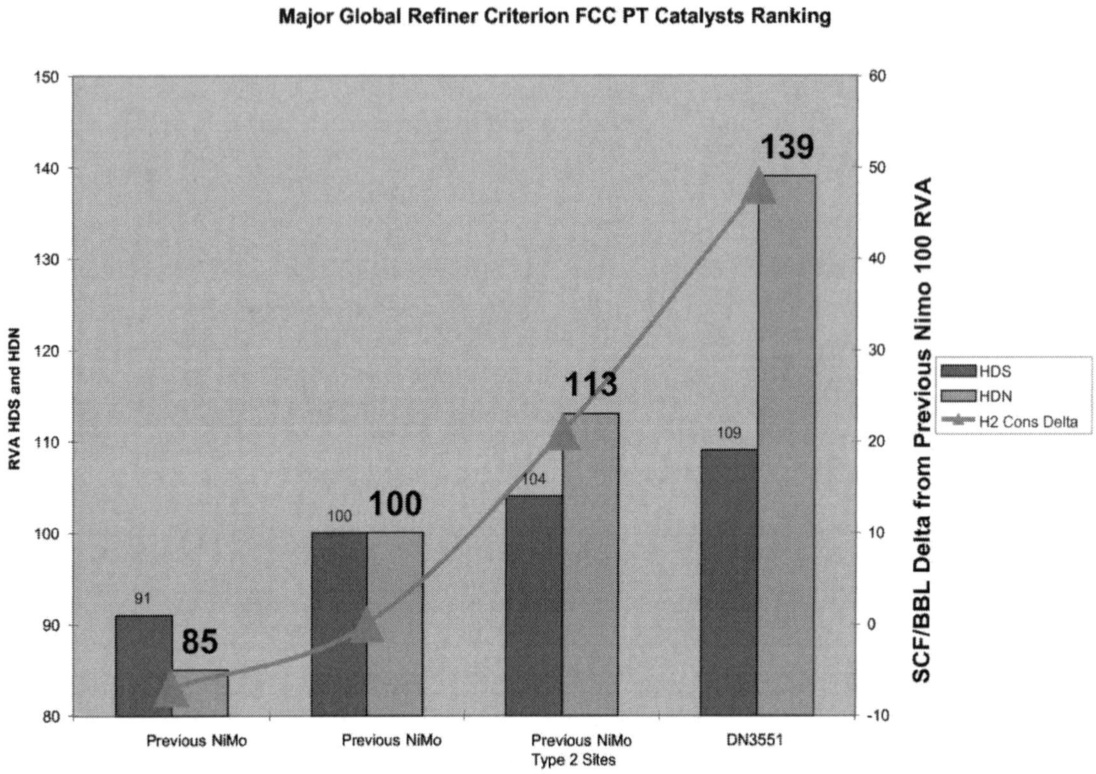

Figure 7: Increased activity as documented by a third-party major oil company laboratory during their evaluation of DN-3551.

Out of the Lab and Into the Field

Criterion R&D has produced yet another step out in NiMo and CoMo FCC PT catalyst performance. The activity improvements described so far are only interesting from a catalyst development standpoint. The real value will be demonstrated when these new high performance catalysts are properly applied to increase conversion in an FCC unit, reduce the sulfur content of the FCC products, and extend unit run lengths. The ultimate value of these new inventions will be highlighted when a complete catalyst life cycle value analysis is performed.

Catalyst life cycle value, when applied to FCC feed pretreatment, may be measured by a key set of objectives that include:

- Maximum FCC Conversion
- Maximize Product Quality
- Improved Cycle Life Economics
- Flexible Feed Stock Capability
- Multi-cycle Catalyst Usage Capabilities
- Minimized Handling and Turnaround costs

DN-3551 NiMo and DC-2551 CoMo target each of the items in the life cycle value equation.

Maximize FCC Conversion and Product Quality

The integration of the FCC PT unit with the FCC unit is where technology improvements demonstrate their significant value. A catalyst improvement, like the ones described, can be worth more than $10 MM/yr in net profit to a refinery when integrated effects are taken into account.

We previously identified that optimum FCC PT catalyst performance is a function of each refinery infrastructure, environmental drivers, and economic drivers. In this case, DN-3551 NiMo and DC-2551 CoMo catalyst technology improvements have targeted both the inherent activity and activity maintenance properties (stability). The enhanced sulfur, aromatic and nitrogen removal functions improve FCC conversion, reduce both FCC gasoline and LCO sulfur concentrations, and reduce SOx emissions in the flue gas. Our approach was to optimize both the NiMo and CoMo functionality separately rather than try to incorporate them together in one catalyst and potentially sub-optimize the performance. Previous work has demonstrated that stacking optimized NiMo and CoMo catalysts can result in a more optimized system compared to one catalyst that is promoted with both Ni and Co.

Catalyst selection is a complex evaluation that should ultimately be driven by a rigorous economic analysis of each catalyst's performance and corresponding FCC feed properties. The demonstration of the value of DN-3551 and DC-2551 is shown in the following examples of integrated FCC PT and FCC operation. The flexibility and potential that these new products can provide is the reason Criterion continues to innovate in the FCC PT area.

Figure 8 demonstrates the interaction between the FCC Pretreatment Unit and the FCC. Table 3 and 4 compares FCC feed properties and FCC response obtained by the new catalyst technology compared to previous generation technology. The response of the FCC is also shown to demonstrate the potential economic improvement and reduction in FCC gasoline sulfur that the new catalyst technology can provide. The analysis is based on calibrated FCC PT and FCC modeling in which a FCC delta analysis is used to demonstrate the change in FCC conversion and FCC properties as a result of the changing feed quality.

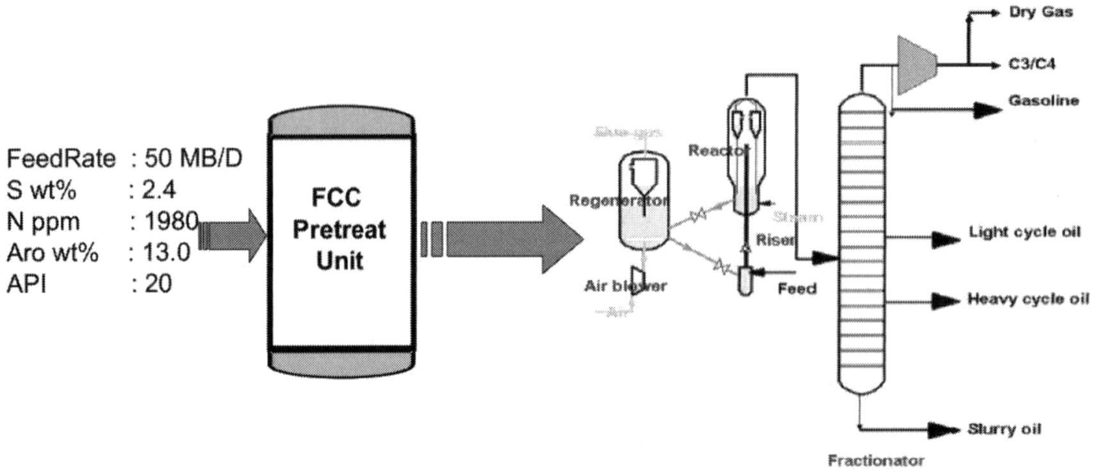

Figure 8: FCC PT and FCC Integration

TABLE 3 - Comparison of DN-3551 with other NiMo Catalysts

Catalyst	H2 Consumption SCF/BBL	FCC Pretreat Unit Product Properties S wt%	N ppm	di+ Aro wt%	FCC Conversion vol%	Economics $MM/yr	FCC Gasoline Yield vol%	LCO Yield vol%	Gasoline Sulfur ppm
Conventional NiMo	609	0.15	900	5.5	Base	Base	Base	Base	112
New Generation									
NiMo (Type II Sites)	625	0.14	800	5.2	+0.82	+4.5	+0.59	-0.40	100
DN-3551 NiMo	655	0.13	665	4.6	+1.56	+8.5	+1.15	-0.75	86

In Table 3 the change in FCC feed and the resulting FCC product properties combined with the economic benefits are demonstrated based on the use of DN-3551 NiMo in FCC PT. A potential $8.5 MM economic benefit is obtained in this example from the enhanced activity benefits of DN-3551 NiMo compared to previous and newer generation NiMo catalysts. As a result of significantly lower nitrogen and aromatic content of the FCC feed that DN-3551 NiMo produces, the FCC conversion and economics are further improved beyond that which was produced by even a Type II active site catalyst. Of significant note is the reduction of LCO with DN-3551 NiMo which has beneficial cycle life implications for ULSD production units. DN-3551 NiMo demonstrates additional FCC gasoline sulfur benefit as a result of previously documented interaction of nitrogen and aromatics content of the FCC feed and its impact on the FCC product sulfur distribution.

TABLE 4: - Comparison of DC-2551 with Previous NiMo and CoMo Catalysts

Catalyst	H2 Consumption SCF/BBL	FCC Pretreat Unit Product Properties S wt%	N ppm	di+ Aro wt%	FCC Conversion vol%	Economics $MM/yr	FCC Gasoline Yield vol%	LCO Yield vol%	Gasoline Sulfur ppm
Conventional NiMo	609	0.15	900	5.5	Base	Base	Base	Base	112
New Generation									
CoMo (Type II Sites)	614	0.10	900	5.5	0	0	0	0	75
DC-2551	620	0.09	900	5.3	+0.25	+1.4	+0.22	-0.12	67

In Table 4 the change in FCC feed and the resulting FCC product properties based on the use of DN-2551 CoMo are shown. Particularly noteworthy is the potential to reduce FCC gasoline sulfur by 45 ppm with DC-2551 CoMo without FCC conversion loss compared to the conventional NiMo FCC pretreat operation. As the industry moves further into clean fuels production, a CoMo catalyst that minimizes or eliminates the FCC conversion penalty will be of substantial value. The latest generation CoMo catalysts have achieved the nitrogen and aromatic saturation performance of previous generation NiMo catalysts due to significant advances in HDN and aromatic saturation performance. As we move into lower sulfur clean fuels requirement, DC-2551 provides the opportunity to extend FCC conversion beyond latest generation Type II CoMo catalysts and take advantage of the FCC sulfur distribution response discussed earlier. Many refiners will take advantage of the sulfur removal improvement demonstrated by avoiding post treatment capital requirements. Others will use it to minimize octane loss within FCC gasoline desulfurization processes or to extend run length of FCC PT units.

DN-3551 NiMo and DC-2551 NiMo produced by the ASCENT manufacturing process provide improved FCC conversion, improved economics and improved FCC product properties due to the increased HDS, HDN and aromatic saturation activity produced from this new catalyst technology development.

Improved Cycle Life Economics

Cycle life economics are driven by a catalyst's ability to tolerate high feed contaminant metal and con-carbon levels encountered in typical FCC pretreat service while achieving the required HDS, HDN, aromatics saturation at the desired cycle length. The stability of Criterion FCC pretreatment catalysts is built on our knowledge of optimized catalyst pore structure. This results in providing industry leading contaminant metals uptake capacity and feed con-carbon tolerance. Further optimization of the pore shape design applied to DN-3551 and DC-2551 continues this heritage providing the opportunity to process higher margin opportunity feeds while maintaining catalyst stability.

Innovations in catalyst active site chemistry and configuration have produced substantial activity gains in the last 5 years. Regeneration of top tier catalysts containing Type II active sites to 90% of fresh activity has generally required the use of special proprietary regeneration processes. DN-3551 NiMo and DC-2551 CoMo, which exceed the performance gains of the past 5 years, are made utilizing the ASCENT manufacturing process which produces oxide catalysts that can be regenerated to 90% of fresh activity using conventional regeneration processes as shown in Figure 9. Additionally, the increased physical strength of ASCENT manufactured catalysts have the added benefit of increased yield recovery during regeneration as per Figure 10. These features provide the opportunity to generate significant multi-cycle use economics.

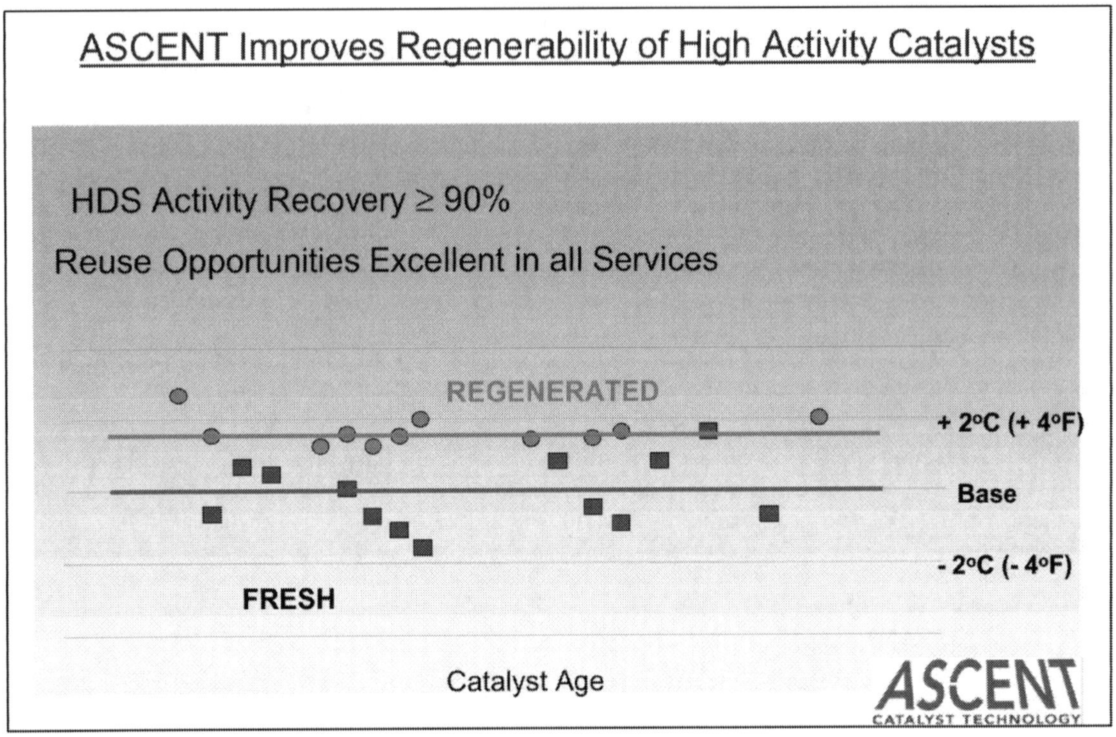

Figure 9: Regenerability of ASCENT Catalysts

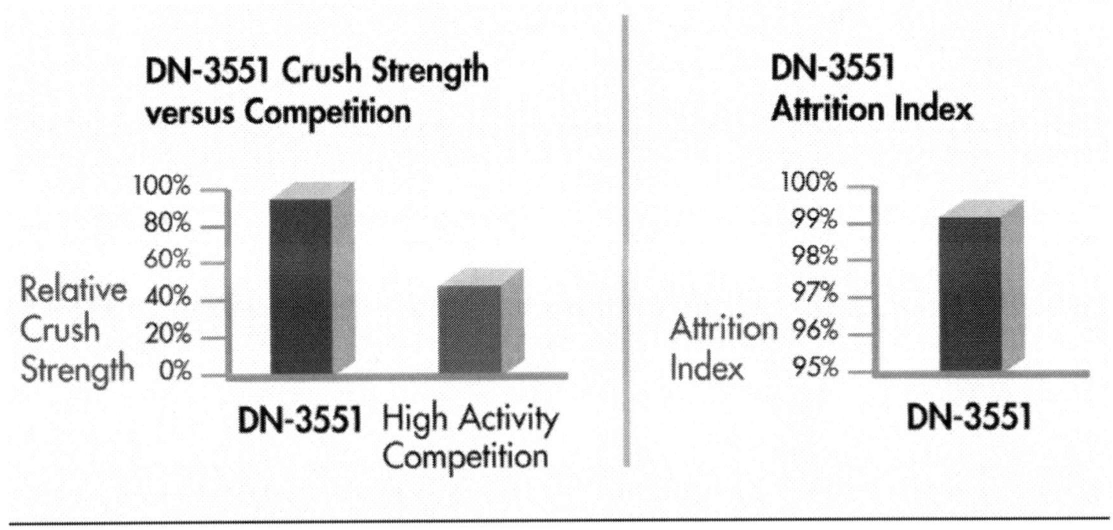

Figure 10: Strength of ASCENT Catalysts

An Example of Multi-Cycle Use Analysis

We have often observed that multi-cycle use capability is often overlooked. Often, this is because of the complexity of such an analysis. This analysis is different for each operating company as there are many possible scenarios and disposition for the re-use of regenerated catalysts. One must also consider the cost associated with performing the regeneration, the unique catalyst activity recovery capability, and the value/debit of the performance of the regenerated catalyst. However, it is not unusual to encounter opportunities for multi-cycle use, which approach 1 million dollars value justifying the effort to analyze a catalysts multi-cycle use value.

In order to demonstrate this concept, we will use a scenario that was encountered in one industrial application. A refinery had an alternate unit that could utilize regenerated catalyst from an FCC PT unit as long as the regenerated catalyst had at least 90% recovery of original desulfurization activity. In this scenario, regenerated catalysts that can achieve the activity recovery target are considered equal in performance value. The refinery was considering 2 catalysts, DN-3551 NiMo and an alternate high activity catalyst (Catalyst A). As previously mentioned, DN-3551 can achieve 90% activity recovery from conventional regeneration. Catalyst A can achieve 90% activity recovery only after undergoing a proprietary regeneration process which adds additional cost and results in lower regeneration yields. In addition, the unit utilizing the regenerated catalyst will need to purchase additional fresh catalyst to make up for regeneration yield losses.

Table 5

Multi-Cycle Use Cost Analysis

	Catalyst A	DN-3551
% Recovery from Regeneation	85%	95
Make-Up Catalyst Requirement, ft3	877	292
Cost of Regeneration, US $K	920	120
Make-Up Catalyst Cost, US $K	370	110
MultiCycle Use Cost, US $K	1,290	230
MultiCycle Use Cost Difference, US $K	1,060	

In the multicycle use cost analysis, DN-3551 is advantaged, as it does not require special regeneration processes to achieve the required activity recovery. In addition, DN-3551's excellent strength properties provide a higher yield recovery during regeneration, which reduces the amount of fresh catalyst that is required for makeup.

Minimize Handling and Turnaround Costs

DN-3551 NiMo and DC-2551 CoMo are provided in both oxide and presulfurized form. This may have the benefit of reduced turn-around cost or improved operating value associated with startup, transportation, storage, loading and general handling compared to other high activity catalysts. In one case, this extra cost was determined to be in excess of US $400,000.

An analysis of each component of a catalyst's life cycle value is essential to understand the true return on investment that a catalyst provides. Criterion Catalyst and Technologies' latest innovations in FCC Pretreat Service, DN-3551 NiMo and DC-2551 CoMo, not only cater to more severe operation but also to improve the life cycle value equation.

Commercial Application

Refiners have quickly accepted these new catalysts based on both in house pilot test results and cooperative development efforts. Both products are commercially operating and demonstrating the performance improvements identified in the pilot tests. As a demonstration of the flexible potential that these catalysts provide, two refiners have selected DN-3551 to produce ULSD while simultaneously producing very low sulfur VGO products.

Conclusions

Extending catalyst performance from the current high level requires a rigorous and specific approach to catalyst design for the targeted feedstocks. Criterion researchers have further optimized the key parameters required to enhance catalyst performance in the diffusion limited FCC PT processing arena.

Criterion Catalysts & Technologies L.P. continues to invest heavily in R & D to provide unique and innovative catalyst solutions to facilitate economic processing of opportunity crudes and reduce capital expenditures for environmental compliance. DN-3551 NiMo and DC-2551 CoMo are the latest Criterion R&D advancements in FCC pre-treatment catalysts offering significant opportunities for refiners.

Bibliography

1. Bavaro, Vito F, *et al,* “Improvement in FCC pretreatment catalysts”, PTQ Catalysis, Vol 11, No. 2, 2006.
2. Bavaro, Vito F, “Clean Fuels FCC Operation via FCC Pretreatment”, Fundamentals of FCC Pretreatment. Feb. 2005.
3. Bavaro, Vito F., “Produce Low Sulfur Gasoline via FCC Feed Pretreatment”, World Refining, Vol 10, No. 2, March 2000
4. Bavaro, Vito F, *et al,* "Maximize FCC Profitability with Pretreater Performance" Hydrocarbon Eng. Vol. 4 Number 9 October 1999.
5. Torrisi, Salvatore P., *et al*, AM-05-15, “Catalytic Advancements To Increase Reliability and Value of ULSD Assets”, NPRA Annual Meeting, March 2005.
6. Swain, Justin, *et al*, “Catalyst Innovations That Increase Reliability and Value of ULSD Assets”, ERTC 10th Annual Meeting, November 2005.
7. Pereira, C.J., Ind. Eng. Chem. Res., 1990, 29, 512-519
8. Tamm. P.W., Harnsberger, H.F. and Bridge, A.G., Ind. Eng. Chem. Process Des. Dev., 1981, 20, 262-273.
9. US EPA, "Tier 2 / Gasoline Sulfur Final Rule with Amendments", 8/29/2002 revision as posted on the NPRA website.
10. Kiuru, Liisa, "Worldwide Fuel Quality of Trends - Focus on Asia", International Fuel Quality Center, Better Air Quality in Asia and Pacific Rim Cities (BAQ 2002), December 2002.
11. Klein, Tammy, "Automotive Fuel Markets in Eastern/Central Europe and Former Soviet Union", Internation Fuel Quality Center, UN/EPA Partnership on Cleaner Fuels & Vehicles, November 2002.

Low emission diesel production through LCO upgrading

Roberto Galiasso Tailleur[1]

Artie Mc Ferrin Dep't of Chemical Engineering, TA&M University, College Station, Texas, galiasa@cantv.net

ABSTRACT

The production of high quality low emission diesel component has experimentally been studied in a hydrotreating micro-plant using a novel catalyst and compared to the commercial operation with a conventional HDT catalyst. Both catalysts were tested incrementing the operational temperature as function of cycle length to keep sulfur in product constant, while keeping the other conditions constant. The reaction products were characterized and tested in a diesel engine to measure NOx and PM emission. The spent catalyst was characterized using solid ^{13}CNMR spectroscopy and ammonia adsorption, and tested with probe molecules to verify its naphthenic ring opening capability. An important decline in ring opening activity was detected along the cycle length due to acid sites deactivation. The deactivation is attributed to LCO poly-aromatics adsorption on a particular type of acid sites.

Keyword: Hydrotreating, LCO hydrogenation, WNiPd/$TiO_2Al_2O_3$ catalyst, Coke, NOx and PM emissions

INTRODUCTION

The hydrotreating of LCO using high-pressure hydrotreating (HDT) commercial units to produce a low sulfur high cetane Diesel is a growing area of interest. Several authors (cf. Babich and Moulijn 2000[1]) including ourselves [2], have performed the analysis of different potential routes to revamp the available HDS units to produce a high quality diesel using a low value refinery streams as LCO and LKGO. The key point is the catalyst deactivation, which is associated to the type of catalysts and the hydrogen partial pressure used for the upgrading since some limitations in catalyst life had been commercially observed. The optimum hydrotreating cycle length depends on the composition of feedstock used and how it affects some active centers on the catalyst surface. In particular, when cracked material is used, the important issue is how to transform the high molecular weight poly-aromatics into n- and iso-paraffins with high volumetric yields and moderate hydrogen consumption to improve the cetane number and reduce the emission.

The standard commercial practice is to blend some LCO with other Light Gas Oils and perform a high-pressure hydrotreatment all together. The economy of doing that is strongly dependent on hydrogen availability, type of unit, and quality and proportion of LCO incorporated in the feed. With more than 50% of LCO in the hydrotreating feed, a very low space velocity is required (c.f. Lee and Mc Govern [3]) to obtain the 15ppm of sulfur, and 42 cetane number. Considerable attention has been devoted in recent years to the development of new catalysts and processes for LCO upgrading, with particular emphasis in a deep sulfur removal and aromatic saturation (c.f. Cooper and Donnis, 1996 [4]). However, the analysis of the Gulf Coast refinery indicated that some refineries still use a limited amount of LCO (less than 15% by weight) in the high-pressure hydrotreating units producing 500ppm sulfur diesel. At least in two of those refineries an important catalyst deactivation and bed plugging was observed. These HDT units were using conventional last generation catalyst that hardly achieves one-year operation. There is a high economical incentive to increase both the LCO and the cycle length.

---

[1] Correspondent author: 203 Jack Brown Bldg.,3122 TAMU, College Station TX 77843, Ph # (979) 218 9103

Recently, a new catalyst to treat LCO based on WNiPd/TiO2 Al2O3 was developed (Galiasso T. and Ravigli [5]). The catalyst seems to be able to reduce sulfur to a very low level and to improve the cetane number via hydrogenation and naphthenic ring opening. Deactivation of this type of catalyst occurred by adsorption of coke precursor on its metal and acid sites. The acid strength distribution in amorphous type structure is broader than in zeolytic type, according with the ammonia TPD studies [5]. In accordance with that, the adsorption energy of the heavy aromatic would follow the acid distribution function. In some of the strongest acid sites, the adsorbed material suffers dehydrogenation-cracking and dimerization reactions that generate the coke precursors. In other acid sites the same molecules can follow other reactions such as isomerization, cyclization and ring opening, helped by a nearby metallic site.

The LCO used in this study – 180-350oC boiling points range - contains high amounts of aromatics (70%) and sulfur (0.7wt %), and has an extremely low cetane number. In it the "aromatics" are formed by 15% of mono-ring-aromatics, 45% of di-ring-aromatics, and 10% by weight of tri-ring-aromatics. The LCO has been produced in the HDS-FCC complex in one refinery that uses highly naphthenic vacuum gas oil, and it can only be valorized as a diesel component if it is deeply hydrotreated and most of the "aromatics" converted (>80%) into paraffins.

In a previous communication (Galiasso T. 20005, [6]) it was shown that diesel engine emissions can be reduced if the cetane number of the LCO is increased above 42, the diesel product has extremely low content of tri- and di-aromatics and naphto-aromatics (less than 10%), and a high content of paraffins and so-paraffins . The poly-aromatics hydrogenation and ring opening reaction have recently been studied by many authors (cf. ref. [7, 8 and 9]) and by us [10]. In spite of their large scientific contribution little is known about the ring opening reactions. In a recent communication, Santana et al. 2006 [11] evaluate different strategies for cetane number improvement.

This paper is focused on studying the WNiPd/TiO_2-Al_2O_3 catalyst deactivation that occurred during the processing of LCO in a medium high pressure HDT unit. We will be concentrated in aromatic hydrogenation and ring opening reaction, as well as in evaluating the impact of deactivation in diesel engine emission. The emission of the pilot plant products are compared with the performance of a sample obtained from the refinery after one year operation with limited amount of LCO in the feed.

2. EXPERIMENTAL

2.1 Catalyst

Three identical charge of WNiPd/$TiO_2Al_2O_3$ catalyst were installed in one fixed bed micro-reactor. They were used to process LCO at standard HDT conditions during one, four and ten weeks on stream. Other sample of an unknown deactivated catalyst was obtained from a commercial plant that operated for one year using 15% of LCO in the feed. The spent and commercial catalyst samples were then characterized.

Physical method: Surface, pore volume, and average pore diameter were measured using standard nitrogen adsorption (Micromeritics 250) and mercury porosimetry methods. See results in Table 1.

Chemical characterization: Bulk metal analysis was performed using atomic adsorption spectroscopy (Varian Techtron analyzer) and coke measured by LECO carbon analyzer. Metals were reported in % by weight (bulk) of total metal in the support (W, Ni, / on Ti+Al support). See results in Table 1.

Solid ^{13}CNMR spectra: The analysis of the carbon deposited on solids (coke) was performed using a Varian spectrometer operating at a frequency of 100.58 MHz with cross-polarization. The technique allowed us to detect the type of carbon present on the surface (Figure 1) using the information published by Callejas et al., 2001[11].

.

Ammonia TPD: Ammonia thermal desorption analysis was used to characterize the total acidity strength at the start of run, and at the fourth and tenth week on stream catalysts. A Mc Bain microbalance was employed for that with 1 mg of sample. Catalysts were treated with argon in vacuum until a constant weight was reached; after that the temperature was increased to 300°C using a program of 10°C/m, kept constant for two hours and subsequently cooled down to 30°C at a rate of 10°C/m. At this temperature the ammonia was adsorbed and measured as mol of NH_3*10^2adsorbed/m^2 of catalyst. Then, the sample was heated using a ramp of 6°C/min and the remaining amount of NH_3adsorbed by m^2 was calculated at 200°C-300°C (Table 1). More than 50 different spent catalysts from commercial plants and laboratory were also tested using the same procedure for comparison purposes. The analysis was complemented by heat adsorption measurement (DTA) and by testing the sample with hexadecane (n-paraffin) in mild hydrotreating conditions. All of that allowed us to correlate isomerization and cracking reaction rate with ammonia adsorption at 200 and 300°C [13].

2.2 Feed and products characterization

Analysis of the feedstock and HDT products. The fresh feed (LCO) was fractionated in three cuts, each of which was analyzed using a Programmed-Temperature GC techniques coupled with a low intensity mass spectrometry method. In parallel, a preparative HPLC was also performed on each cut using hexane carrier and an ultraviolet detector. The fractions were analyzed using a LECO (elemental analyzer), a ^{1}HNMR spectroscopy (Varian - 400 MHz spectrometer - 10 mm Broadband Probe), alumina percolation method (using tetra-hydrofuran as solvent) and mainly a high-resolution mass spectrometry (HRMS-Philips 2341). With all of this information, a reconciliation procedure was used to fit the analysis of the full range LCO with the results cut by cut (Mendez and Carbognani, 2002 [14]). This special-in-house developed analysis was able to discriminated families of aromatic, naphthene, and paraffin (see results in Table 2). The cetane number was determined by using the D 613 ASTM method.

Microplant long run tests. The micro-plant is composed by a conventional isothermal down flow-fixed bed reactor [5]. It was operated at temperature ranging between 350 and 377°C, 0.2 h^{-1}, 20 (molar ratio) of H_2/HC, and 12.3 MPa. The WNiPd on $TiO_2Al_2O_3$ catalyst was charged in the reactor diluted with 50% of inert material, which was having 0.001 mm diameter particles, to improve the reactor fluid-dynamics (see the methodology in De Brujin,(1976) [15], among others). The catalyst was pre-activated in-situ with straight run light gas oil spiked with 1% of CS_2, at 300°C during 6 hours. Then the microreactor was started up with LCO and operated continuously either for one week, four weeks, or ten weeks on stream generating the catalyst and products called Spent 1, 2 and 3 respectively. Duplicate samples of hydrotreated LCO were taken daily and mass balance was performed to check the operation. Having achieved the period on stream, the catalyst was cooled down to 300°C, washed with toluene for half a day, and then by nitrogen at 200°C for another half a day. After that the deactivated catalyst was ready to be tested with synthetic feedstock. The procedure for the three samples of catalyst was identical.

Test with synthetic feed (probe molecules). The spent catalysts were tested using a blend of 30% of naphthalene, 10% of tetralin and 5% of cyclohexane in hexadecane. The operating conditions were 340°C 0.5 h^{-1} of LHSV 10 H_2/HC molar ratio and 12.5 MPa. In other runs and to recover the naphthalene conversion to the initial values (SOR), the operating temperature was risen up. Finally the catalysts were washed by continuous pumping of xylene at 200°C for half a day, dried in nitrogen for another half a day (120°C) and discharged in a dried-inert chamber. The three samples of spent catalyst were characterized together with the one coming from a commercial plant.

Diesel Engine Test. The specialized literature (cf. EPA report, 2003 [16]) indicates some dependency of exhaust gas emissions with cetane number in 2000 heavy duty diesel engines. More recently we had demonstrated the effect of diesel composition on 2002 EURO IV heavy duty diesel engine (Gutierrez el at., 2005 [17]). Here the NOx (nitrogen oxide) and PM (particulate) emissions were measured in a high speed Detroit Diesel Series 50 (2002), direct injection engine operating at 2000 rpm, 350 KW power, 1364 kPa BMEP, 1630 N-m torque and without EGR. To determine NOx and PM, a micro-tunnel device coupled with on line micro-filter and IR analyzer were used. The curve that measured the heat released in the piston chamber was analyzed at stationary conditions and during the operation in transient mode by using a EURO III type of driving cycle.

3. RESULTS

3.1 Catalysts characterization. Table 1 presents the catalysts properties. The catalyst contains W, Ni, and Pd as metal active phase. The support is based on TiO_2-Al_2O_3, a low acidity type of active support (see in reference [5] details about catalyst formulation and characterization). The analysis of the physical properties indicates that the solid has a high surface, medium pore volume structure with a large contribution of interconnected mesopores and an average pore diameter of 12 nm. The pre-activated (sulfided) catalyst (fresh sample) contains 5.6 % weight of sulfur and less than 0.1% of Carbon (Table 1). A previous study [18] showed that this catalyst had 30% higher hydrogenating and denitrogenating capability, and 15% lower desulfurization activities than a Ni-Mo/Al_2O_3 commercial HDT catalyst. The main contribution of the new catalyst is due to its higher (15%) ability to carry out naphthenic ring opening reactions and its acid center stability.

The spent catalyst properties after one, four and ten weeks in operation are shown in Table 1. It can be observed that at the start of run the catalyst contained 5.4% of C, and had lost 40% of the surface, 15% of pore volume and 48% of the micropore volume compared to the fresh catalyst. Then, at the fourth and tenth weeks of sustained operation, the physical properties of the spent catalyst continued to decline, showing additional losses in surface and pore volume, while the carbon and sulfur content were slightly increased. The differences in physical properties between spent samples are relatively small but noticeable. The analysis of spent catalyst obtained from commercial unit showed 6.3% by weight of carbon, 6.1% of sulfur. Its physical properties are similar to those of Spent 3 catalyst.

The pore size distribution analysis in all deactivated samples indicated that most of the microprobes structure was lost by coke deposition, which is responsible for the important reduction on activity. That reveals the (relatively low value of the information obtained during the first week on stream, when the initial activity and selectivity of the catalyst were still changing due to carbon deposition. In spite of different coke concentrations between the samples, a uniform coke deposition across the 1 mm particle diameter was observed by a SEM-XR micro-diffraction analysis, not shown here, that might indicate a "similar" deactivation mechanism is occurring in pilot plant and commercial operation.

Table 1 Fresh and spent catalyst properties

Catalyst	Fresh	Spent 1	Spent 2	Spent 3	Commercial
NiO (wt %)	4.2	--	--	--	--
WO3 (wt %)	12.5	--	--	--	--
PdO (wt %)	0.2	--	--	--	--
Surface (m^2/g)	245	145	132	112	--
Pore Volume (cm^3/g)	0.52	0.44	0.4	0.38	--
Micropore volume (cm3/g)	0.22	0.1	0.08	0.07	--
Average Pore diameter (nm)	12	14	15	16	--
Adsorbed NH3 (mmol/g) 200/300°C	0.4/02	0.1/0.08	0.08/0.05	0.08/0.05	0.03/0.01
Sulfur (wt %)	5.4	6.1	6.2	6.1	6.1
Carbon (wt %)	0.1	5.4.	5.8	6.2	6.3

One way to measure the remaining acidity is by adsorption and then desorption of ammonia, the former performed at 30°C and the latter at high temperatures (200 and 300°C). The fresh catalyst showed the highest retention of ammonia because it has the cleanest surface exposed to the adsorption (negligible coke content). The spent catalysts dramatically lost their acidity (more than 80% for all the samples). Their analysis indicated high acid sites blocking by coke, in spite of the help provided by the high metal loading and the presence of Pd cluster capable of hydrogenating coke precursors. It can be seen by the results of naphthalene and tetralin hydrogenation, showed later on, that the metallic function is less affected than the acid function. The microstructure of the catalyst, where most of the acid and metal sites are present, seems to be plugged by the coke formed, and the internal diffusion of reactant and products are reduced in large extent.

Figure 1 shows the solid ^{13}CNMR spectra for spent catalysts. The carbon deposit analysis allows us to understand the evolution of coke in the pores with time on stream and compare commercial sample with those from micro-plant. The lower two spectra correspond to the Spent 1, 2 and the upper two to Spent 3 and commercial samples respectively. The one-week sample presents a well-defined peak at 135 ppm that is assigned to characteristic aromatic polymeric structures [11 and 12], and other signals at 70 and 35 ppm are attributed to aliphatic carbon structures. The 10-week sample spectrum shows a signal at 130 ppm, which represents a shift of 5ppm with respect to that of the previous sample, it has lost the 80ppm signal, and the signal at 35ppm is augmented. All of these signals are in agreement with a more organized turbostratic structure due to more "aged" coke. Clearly, the internal carbonaceous deposit on the catalyst evolves as function of time on stream into a more condensed type of coke that

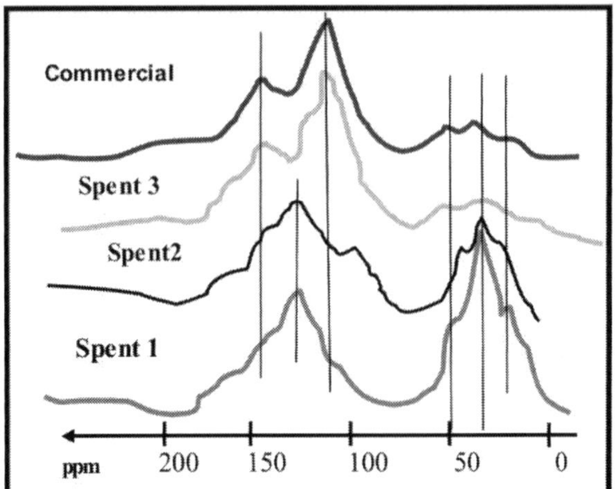

Figure 1 ^{13}CNMR analyis

isolated some of the surviving metallic sites. It seems that the adsorption of part of the heaviest poly-aromatics compounds present in the LCO seems to be responsible for the coke formation on the strongest acid sites. The FTIR analysis of the toluene used for washing the spent samples had confirmed the presence of adsorbed polyaromatics on catalyst.

The signals for the 10-week sample are quite similar to those observed in the spectrum of the commercial catalyst (lower spectrum in figure 4). In spite of differences in catalyst composition and structure, both catalysts seem to accumulate similar type and amount of coke.

3.2 Microplant results

LCO Product quality as a function of time on stream - Results from the micro plant operation with LCO. Table 2 shows the properties of the products after operating for one, four and ten weeks on stream. The temperature required to keep around 50ppm of sulfur in the products is indicated in the upper row. When the time on stream increases, the catalyst performance shows a clear deactivating pattern, and a higher temperature is needed to maintain constant the sulfur in the product.

. During the time on stream a change occurs in the hydrogenation and cracking reaction balance. The relative deactivation of metallic/acid sites functions on the catalytic surface is responsible for the change in the balance between aromatic ring hydrogenation and ring opening. This phenomenon has an impact in the yield and quality of

Table 2. Products quality as a function of time on stream. 12.5 MPa, 0.2 h^{-1} , 20 (mol) H_2/HC

Properties /weeks on stream	LCO	Spent 1	Spent 2	Spent 3	Commercial
Temperature °C	--	340	354	376	390
Sulfur (ppm wt)	12340	50	55	65	60
Nitrogen (ppm wt)	1230	30	55	75	100
Mono-aromatics (wt %)	15	12	15	17	15
Di-aromatics (wt %)	45	10	15	16	25
Tri-aromatics (wt %)	10	0	1	2	6
Mono-naphthene (wt %)	5	18	20	21	10
Di-napththenes (wt %)	10	12	14	16	12
Tri-naphthenes (wt %)	6	0	2	3	3
HDAr (% wt)	0	68	55	50	34
Cetane number	----	44	41	38	<38

the diesel produced during the cycle length. It can be observed that the hydrogenation capability is slightly lost, in spite of the increments on temperature performed to compensate for the sulfur removal. On the other hand, the products of the naphthentic ring opening reactions (alkyl-benzene + alkyl-cyclohexane) were more reduced than those of hydrogenating reactions along the cycle. As a result, it can be observed in the last rows of Table 2, an increment in aromatics and naphthenic-aromatic content that affect the main diesel property: the cetane number. The same result had been observed in commercial operation.

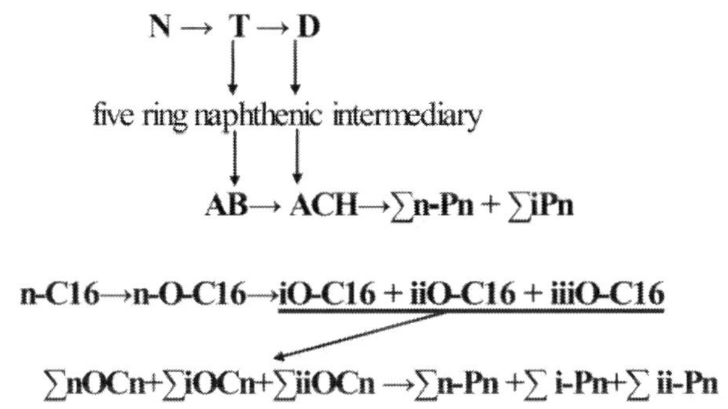

Figure 2 Hydrogenation of Aromatic and paraffins cracking

The aromatic ring conversion into paraffins occurs via a complex network of reactions that use both acid and metal sites through several steps of adsorption-reaction and desorption. It starts by aromatic ring adsorption on metal sites, where it is bounded by σ or π interactions. The hydrogenating reaction uses some hydrogen molecules adsorbed (σ bond) nearby in a metallic site (or by spillover in the border of the cluster). The adsorbed alkyl-poly-aromatics can also suffer some hydrodealkylation of the branches, as well as some hydrogenolysis of the side chains. These reactions reduce the "paraffinicity" of the products and their cetane number. In addition these reactions change the aromatic distribution along the boiling range by transferring the heavy aromatics into the lighter part of the diesel. The hydrogenated poly-alkyl-naphthenic-aromatics molecules desorb from the metal site and migrate into the acid sites where the naphthenic ring might be isomerized and then it is open to form another poly-alkyl-aromatic (Figure 2 shows, in a simplified scheme, how the Tetralin (T) and Decalin (D) go through the isomerization of their C_6-naphthenic-cycle into C_5-nathenic-cycle, which then are open into Alkylbenzene (AB) and Alkylcyclohexane (ACH). The last two types of molecules are cracked into n-paraffins (n-Pn) and iso-paraffins (iPn) (a detailed explanation about the potential mechanism can be found in ref [19], among other).

Table 3. Test of Spent catalyst by probe molecules. LCO, 340 °C, 0.8h^{-1}, 12.5 MPa, 10 (molar) H_2/HC

Chemical / Catalyst	Spent 1	Spent 2	Spent 3	Commercial
Naphthalene %	3	5	9	8
Tetralin % wt	10	12	15	15
Decalin % wt	11	12	14	15
Alkyl-benzene wt%	10	7	5	1
Alkyl-cyclohexane wt%	4	2	1	0
Benzene wt%	0.1	0.1	0.1	1
Cyclohexane wt%	2.5	2.7	2.9	4
n-C_{16} conversion wt%	25	14	10	12

There are many discussions about the role of the acid and the metal sites in the conversion of an isolated naphthenic ring or naptheno-aromatics compounds into alkyl-aromatics and alkyl-naphthenic molecules. No attempt is made here to clarify the point, but our previous studies about mildhydrocracking [9] using probe molecules (1-methyl-naphthalene, 1-methyl-tetralin and n-C_{16}) to characterize the sites indicated the benefits of having nano-clusters of -WPd- and Pd to protect nearby acid sites. It is suspected that some of the acid sites are used for isomerization and ring opening reactions while others are employed to crack the paraffins. Here, to analyze the effect of deactivation, the spent catalysts were tested with naphthalene, tetralin, cyclohexane and n-C16 molecules to check what kind of acid sites survive after the deactivation.

Test with synthetic feed. Table 3 presents the results obtained in hydrotreating a synthetic feed with the three deactivated samples at constant operating conditions. It can be seen that the start of run catalyst is able to produce a quite fast conversion of the naphthalene (N) into tetralin (T) and then into decalin (D). In parallel these two products were converted into alkyl-benzene and alkyl-cyclohexane compounds. The latest reactions are responsible for the shift in the hydrogenating equilibrium (N=T=D, Figure 2)) toward decalin, and then into paraffins, in agreement with the Rautanen (2002 results [19]. Both factors help the cetane number improvement. At the above mentioned operating conditions, the start of run catalyst generated 25% of hexadecane conversion. The mass balance of the paraffins formed indicated that most of the cyclohexane conversion occurred by ring opening reaction into C_6 compounds.

Table 3 shows that catalyst deactivation is responsible for the lost of ring opening capability. The third and fourth columns depict a progressive increment of naphthalene and tetralin content in detriment of the other products. The conversion of cyclohexane is slightly reduced and the cracking of hexadecane with Spent 2 an 3 catalysts fall down to 14% and 10% respectively at 340°C. In other test, the testing temperature was increased to 348°C and 355°C respectively for Spent 2 and 3 catalyst samples to maintain constant the naphthalene conversion measured with Spent 1 sample. At these conditions the amount of product generated by ring opening reaction did not follow the naphthalene-hydrogenating increment. It can be observed that alky-benzene + alkyl-cyclohexane compounds were incremented to a lower value (15%) than observed with Spent 1 catalyst. At the same time hexadecane conversion is only augmented to 18 and 15% at the above mentioned temperatures. The low increments in cracking conversion confirm the reduction in the number acid centers responsible for the cracking reactions, while the others used for the ring opening reactions seems to survive in much higher proportion.

The analysis of the type of paraffins produced by cracking provided information about the intensity of the isomerization and the cracking reactions that take place during the hydrotreatment of the synthetic material tested. Figure 2 shows a simplified scheme of the paths of reactions that starts by the conversion of n-paraffins (n-Pn) into an oleffinic intermediary (o-P), followed by an in series isomerization into nono- di- and tri-branched oleffinic intermediaries (io-P = iio-P = iio-P), of which the latter can be cracked and hydrogenated toward low molecular weight n-paraffins and iso-paraffins. It can be seen that light paraffins (C_1-C_4) can be formed by cracking the alkyl-benzene and alkyl-cyclohexane as well as by cracking the n-C_{16}, but the C_7^+ paraffins are preferentially produced from C_{16} cracking. For that the ratio between iso/n-C_7 and iso/n-C_8 paraffins is used here to analyze the ability of the acid sites to isomerize paraffins. Cyclohexane (CH) is converted into paraffins, but it is also produced in minor proportion during the cracking and dealkylation reaction of ACH. Cyclohexane conversion shows in Table 4 is calculated taking in account the first reaction.

Table 4 shows the C_7 and C_8 iso/n-paraffin ratios. There is a smaller reduction in the iso/n-paraffin ratio and in cyclohexane conversion as a function of time on stream than in n-C_{16} conversions. The results confirm that the deactivation of the isomerizating acid site was different than that for the cracking acid site. It is suspected that the same isomerizating acid site is responsible also for the C_6 into C_5 naphthenic ring opening, and that they might be "protected" from coking by the nearby metal site hydrogen adsorption. The fact that there is not noticeable formation of benzene, indicates a deep reduction of alkyl-branches cracking activity in spent catalysts. This also supports the hypothesis that negligible amount of cyclohexane is formed from alkyl-cyclohexane or decaline, and that its disappearance from the feed is mainly due to ring opening reaction.

Table 4 Iso/n-C7 and C8 ratio of paraffins in the synthetic hydrotreated product *

Ratio	Spent 1	Spent 2	Spent 3	Commercial
iC_7/ nC_7 paraffin	1.6	1.52	1.49	1.5
iC_8/n-C_8 paraffin	1.8	1.74	1.62	1.4
i/ii C_7 paraffin	1.4	1.45	1.51	1.8
i/ii C_8 paraffin	1.3	1.34	1.42	2
Cyclohexane conversion %**	50	46	42	20
n-C_{16} conversion %	25	14	10	12

*at the operating conditions shown in Table 3 ** mol disappeared/ initial mol

The iso/n-paraffin ratio in the product obtained with the commercially deactivated catalyst is lower than that shown for the pilot plant deactivated catalysts. Notice that commercial catalyst, at the operating conditions mentioned in Table 3, has a higher cracking activity than Spent 3 catalyst and it mainly formed mono-branched paraffins and benzene. Table 3, 4 and 5 shows that the remaining acid sites in commercially deactivated catalyst are performing better for cracking and dealkylation than for isomerization and ring opening reactions that explain the low cetane number and high aromatics content of LCO hydrotreated product.

3.3 Diesel engine test

To demonstrate the effect of catalyst deactivation in diesel emission, the three hydrotreated LCO samples obtained with Spent 1, 2 and 3 catalyst were tested in stationary conditions using a Detroit diesel engine. In parallel the performance was confirmed averaging the emission obtained by performing a EURO III driving cycle simulation. The stationary tests were performed in the conditions mentioned above to compare the emission on particulate (PM) and nitrogen oxides (NOx). Since the sample had different cetane number and composition along the cycle length, the test shows the changes in engine emission associated to both factors. In this respect the lowest emission is produced by the SOR sample, which is the most hydrogenated product (lower total aromatics and mono-ring-aromatics). The higher the time on stream, the higher the NOx and PM diesel emissions (Table 5). The effect is attributed to combustion delay and the shape of the energy release. When the fuel is injected into the chamber after the engine is crancked on, the liquid is atomized and the ignition begins. In this phase, called uncontrolled burning or flame propagation period, most of the PM precursors are formed by thermal cracking in which the aromatics play an important role. Then, the second phase starts with full combustion resulting in peaks of pressure and temperature. From that point up, most of the NOx is formed associated to the maximum temperature attained. The pressure-volume curve for SOR product is flatter than those observed for the four- and ten-week samples. That behavior is equivalent to an earlier (and better) vaporization and a lower maximum temperature for the SOR product than for the other. Table 5 shows NOx and PM increments as a function of time on stream. It was speculated that the SOR product has also a lower light aromatics in the most volatile part of diesel and because of that it achieves a better engine emission behavior. The diesel produced at the end of cycle with commercial catalyst is used in table 4 as reference. It has a higher NOx and lower PM emission than those diesel sample obtained in pilot plant after 2 month on stream. Both diesel has a similar total aromatics content and cetane number, but different type of aromatic and aromatic distribution along the cut. The effect of cetane and aromatics on emission is noticeable at this low level of cetane number- high naphthenic type diesel and the observation could be valid for vehicles 2002+.

Table 5 Diesel engine emissions (Engine EPA 02)/ 2000 rpm/ 350 Kw / 1364 kPa BME No EGR in use

Emissions	Sample I (Spent 2)	Sample II (Spent 2)	Sample III (Spent 3)	Commercial Sample*
NOx g/bhp h	6	9	17	12
PM g/bhp h	0.1	0.4	1.1	0.7
Cetane Number	45	42	40	41

* Commercial sample at the end of the operation (50 ppm S, 30% aromatics)

CONCLUSION

LCO contains high amounts of polyaromatics and sulfur, and this fuels valorization depends on converting their main component into paraffinic type material boiling in the diesel range. The commercial catalyst is able to hydrogenate aromatics but it has shown a very limited capability to transform them into paraffins at the end of cycle. The commercial experiences with limited amount of LCO in the feed indicated that the catalyst suffers a strong deactivation as a function of time on stream. The present hydrotreating study was performed using pure LCO and a particular WNiPd on $TiO_2Al_2O_3$ catalyst in an attempt to understand catalyst deactivation. It was observed that:

- Acid function deactivated much faster than metallic functions due to coke deposition.
- Coke deposited on catalyst after ten weeks on stream had a similar structure than that observed in conventional last-generation HDT catalyst operating in a commercial plant with 15% of LCO in the feed.
- The catalyst ability to open naphthenic ring molecules (ca. tetralin and decalin) was partially lost due to deactivation of some acid sites, while the cracking of normal paraffins was less affected.
- The deactivation produces a loss in cetane number, and an augment of aromatic content, which generated higher NOx and PM emissions in diesel engine operation.
- It seems that the mechanism of ring opening active center deactivation will be a critical point in future LCO valorization.

ACKNOWLEDGEMENT

The author wants to thank the experiments performed by R Albertti Tissano, and the support provided by the Universidad Simon Bolivar (Venezuela), Instituto de Tecnología Química and Departamento de Ingeniería Mecánica, Universidad Politécnica de Valencia (Spain).

REFERENCES

1. Galiasso T. R; "Diesel options to meet future US and European specification", preprint of World Chemical Engineering Congress, Melbourne. Paper 13 Sep (2001)
2. Babich I.V., Moulijn J., "Science and technology of novel process for deep desulfurization of refinery stream". A review Fuel, 82, (2003), 607-631.
3. Lee C.K., Mc Govern S., "Comparison of clean diesel production technologies"; PTQ, 35-39 (2001).
4. Cooper B.H., and Donnis B.B.L., "Aromatics saturation of distillates. An overview, *Appl. Cat. A: General* 137, (1996), 203-223.
5. Galiasso T. R., and Ravigli J., "Catalyst for ultra low sulfur and aromatic", *Appl. Catal. A: Gen.* 282, 1-2, (2005), 227-235.
6. Galiasso T. R, "Diesel upgrading into a low emission fuel", preprint CHISA 2005, accepted for publication in *Fuel Processing l* (2006).
7. Toppinen S., "Liquid phase hydrogenation of some aromatic compounds". *Doctoral thesis*, Department of Chemical Engineering, Abo Akademi University, (1996)
8. Chowdhury R., Pedernera E., and Reimert R., Trickle-Bed Reactor Model for Desulfurization and Dearomatization of Diesel, *AIChE J.* 48, 1, (2002), 126-135.
9. Arroyo J., and Galiasso T. R., "Kinetics model for n-C_{16} hydrocracking in presence of di-aromatics and DBT" sent to *Ind. Eng. Chem. Res. & Dev.(*2006)
10. Santana R., Do T., Santikunaporn P., Alvarez M., Taylor W., Sughrue J.D., Resasco E.D. "Evaluation of different reaction strategies for the improvement of cetane number in diesel fuels", *Fuel*, 85, (2006), 643-656.
11. Callejas MA., Martinez MT., Blasco T., Sastre W., "Coke ^{13}CNMR analysis" *Applied Catalyst*, 218, (2001) 181-188.
12. Martinez M.T., Jimenez J.M., Callejas M.A., Gomez F.J., Rial C., Carbo E.; Hydrotreatment and Hydrocracking of oil fraction , Elsevier (Ed G.F. Froment, B. Delmon and P. Grange), (1997) p. 311.
13. Albertti Tissano R., and Galiasso T.R., "HDTspent catalyst characterization", unpublished report (2004).
14. Mendez A., Carbognani E.; "GC-MS and HTMS method to characterize LCO", internal report Int-2356-02, unpublished results, (2002).
15. De Brujin A., "Trickle bed pilot plant testing", *The 6th. Int. Congress on Catalysis, London*, paper B34, (1976), 121-32.
16. EPA report, "The effect of cetane number increase due to additive on NOx emissions from Heavy-Duty Highway Engines" EPA420-R-03-002 February (2003).
17. Gutierrez A, Paliza J., Galiasso T. R, "Effect of aromatics on diesel emissions on a EURO III engine" PST report 1246XZ (2002). Unpublished results (2002).
18. Galiasso T. R, and Prada S. R., Effect of catalyst pretreatment in Milhydrocrackino operation" preprint of AICHE Spring Meeting New Orleans (2003), 201-210.
19. Rautanen P.A., Lylykangas M.S., Aittamaa J.R., Krause A.O.I., "Liquid-Phase Hydrogenation of Naphthalene and Tetralin: Kinetic Modelling", *Ind. Eng. Chem. Res.*, **41**, (2002), 5966-5975.

Effect of Promoters on Structural and Chemical Properties of Hydrotreating Catalysts

R.G. Egeberg, K.G. Knudsen, A. Carlsson, M. Brorson, H. Topsøe
Haldor Topsøe A/S, Nymøllevej 55, DK-2800 Lyngby, Denmark

P.G. Moses, J.K. Nørskov
Center for Atomic-scale Materials Physics (CAMP), Nano DTU, Department of Physics, Building 307, Technical University of Denmark, DK-2800 Lyngby, Denmark

J.V. Lauritsen, F. Besenbacher
Department of Physics and Astronomy, Interdisciplinary Nanoscience Center (iNANO), University of Aarhus, DK-8000 Aarhus C, Denmark

Introduction

In the years to come, the demand for transportation fuels will increase. Simultaneously, a legislative drive to remove sulphur will mean a higher demand for cost-effective desulphurization processes and catalysts [1]. In order to develop improved catalyst technology it is mandatory to gain better understanding of the structure of the active phase and reaction mechanisms. For instance, several years of practical experience have shown that both Ni and Co dramatically promote the activity of MoS_2, and that NiMo and CoMo catalysts respond differently to changes in e.g. pressure, feed composition and space velocity [2-3]. Although different theories have existed, there have been no direct observations of changes induced by the promoters in the atomic scale structure. In this paper we present recent observations of MoS_2 nanostructures as well as the Ni- and Co-promoted phases.

Structure of Catalyst Nanoparticles

The atomic structure of the active phase in hydrotreating catalysts was only recently directly observed with the introduction of Scanning Tunneling Microscopy (STM) studies [4-5]. The preferred shape of the MoS_2 nanoclusters deposited on an Au substrate was observed to be triangular. This means that one of the two low-index edges, denominated the Mo edge and the S edge, respectively, is more stable than the other. By combination with Density Functional Theory (DFT) calculations it was deduced that only the Mo edge was exposed. The STM pictures further revealed a pronounced bright brim located on top of the triangular MoS_2 nanoparticles adjacent to the edge itself. The brim reflects regions of high electron density, and by analysis of the electronic band structure it was found that so-called one-dimensional edge states existed. These brim sites were found to play an essential role in the hydrogenative (HYD) route of the hydrodesulphurization reaction. Upon adsorption of hydrogen and formation of SH groups on the edge, the adjacent brim sites were able to adsorb and react with thiophene [5]. The brim sites were found not only to facilitate the hydrogenation of thiophene but also to cleave the first C-S bond producing a *cis*-but-2-ene-thiolate intermediate. The final step of cleaving the second C-S bond and removing the sulphur most likely occurred at the edge by reaction with a sulphur vacancy. The brim sites thus play a very important role in the HYD pathway of sulphur molecules. The sulphur vacancies seem to play a role in both the direct desulphurization (DDS) and the HYD pathway; a feature supported by the observation that

H_2S inhibition of the HYD route is caused by poisoning of the final hydrogenolysis step and not by poisoning of the hydrogenation step [6].

Promotion with Co and Ni is known to increase the activity significantly [7-8]. Whilst NiMo catalysts in model feed studies typically display higher activity for the HYD route, CoMo catalysts are often seen to outperform NiMo catalysts at high space velocities and low hydrogen pressures. This is partly due to the inhibition by aromatic and especially basic nitrogen compounds present in real feeds, which severely poison the HYD route [9]. A DFT study [10] showed that pyridine was adsorbed on the metallic-like brim of MoS_2 thus blocking these sites. The sterically hindered molecules such as 4,6-dimethyl-dibenzothiophene were thus unable to interact via π-bonding with the brim.

The enhanced activity when adding promoters can be explained by the changes in electronic structure caused by the incorporation of Ni or Co at the edge of the MoS_2 structure facilitating the formation of sulphur vacancies at the edge [11]. Furthermore, the formation of highly active type II structures has been related to the support interaction of the catalyst nanoparticles. The formation of Mo-O-Al linkages occurs preferentially at the S edge, and these linkages significantly increase the energy required to form sulphur vacancies [12]. Since the promoter atoms are observed to substitute Mo at the S edge, the promoters will weaken the support interaction thus creating type II structures. Still, the detailed atomic structure of the Ni-Mo-S and Co-Mo-S has not yet been determined, and may also influence the reactivity. Therefore, a STM study of the structures on an Au substrate was undertaken.

Addition of Co results in structures similar to those of the MoS_2 phase. However, the nanoclusters have a truncated triangular (or almost hexagonal) shape as opposed to the predominant triangular morphology of the unpromoted MoS_2 nanoclusters. This implies that both the Mo edge and the S edge are exposed, the morphological change being induced by the affinity for Co to be located only at the S edges. A brim state was also observed on the S edge – significantly brighter than that of the Mo edge. Simulated STM pictures from DFT calculations confirmed that Co fully substituted Mo on the S edge and that this edge had a 50% S coverage.

For the Ni-promoted model system, the changes in particle morphology were even more dramatic. In contrast to the Co-promoted case, two types of coexisting Ni-Mo-S clusters were observed that are distinguishable in terms of size and shape. The first type is characterized by a truncated triangular shape similar to that of Co-Mo-S, whereas the second type shows a more complex form resembling a dodecagonal shape. The distribution of the two structures depends on annealing temperature; the truncated triangular shapes are larger at higher annealing temperatures, whilst the size of the dodecagonal structures is more or less independent of annealing temperature. This phenomenon does not seem to be caused by kinetic limitations during the particle growth. Rather, the equilibrium particle shape seems to be intimately linked to the particle size.

The larger truncated triangular Ni-Mo-S particles are very similar to their Co-promoted counterparts. The Ni atoms fully substituted the Mo atoms on the S edge, and a very bright brim was observed along this edge. The dodecagonal structures were quite different and cannot be interpreted by a simple edge truncated triangular model. A new model that includes a higher index $(11\bar{2}0)$ edge in addition to the low index Mo edge and S edge was found to match the experimental data closely. Such high index edges have usually been considered energetically unfavourable, but here it can be directly observed that such edges can be exposed under equilibrium conditions. Typically, these edges only spanned a single unit cell, but in some cases they were two or three unit cells wide. This high index edge termination must thus have a free edge energy comparable to that of the more closed packed Mo and S edge. The dodecagonal structures also displayed a bright brim behind the Ni substituted high index edges and the S edges.

HAADF-STEM Observations

The morphology of Ni-promoted and unpromoted MoS_2 and WS_2 nanoclusters supported on graphite was also investigated with High Angular Annular Dark Field Scanning Transmission Electron Microscopy (HAADF-STEM) [13-14]. The graphite support only interacted weakly with the sulphide structures and allowed the formation of type II sites. It was observed by analyzing the image intensity levels and from edge-on imaging that most of the nanoclusters present were single layer structures (S-Mo-S or S-W-S). Type II sites are thus not necessarily linked to multi-layer slabs.

The most common MoS_2 and WS_2 nanoparticle regular shape is triangular but with a slight truncation. Whilst MoS_2 particles are generally smaller than WS_2 particles prepared under the same conditions, both displayed a similar degree of truncation.. In both systems, the presence of Ni promoter atoms changed the particle shape. This indicates that the Ni-Mo-S and Ni-W-S structures were formed. The truncations of the triangles became less sharp and the particle shape more dodecahedral-like. This is also in very good agreement with the STM observations and supports the suggestion to the effect that high-index planes are present in the Ni-Mo-S phase.

BRIM™ Catalyst Technology

The fundamental insights provided by the STM and HAADF-STEM observations combined with the DFT calculations and our long experience in carrier and catalyst manufacturing have enabled Haldor Topsøe A/S to develop improved catalysts for the production of ultra-clean transportation fuels. By using the BRIM™ technology, we are able to increase the frequency and strength of both type II activity sites and brim sites.

The catalysts produced with the Topsøe BRIM™ technology have been recognized by refiners as top-tier catalysts for ULSD and FCC P/T service. More than 70 high severity hydrotreater units have been loaded with BRIM™ catalysts over the past three years. The proven performance in commercial units regarding both activity and stability has helped refiners meet specifications, while processing more difficult feeds at higher rates and longer operating cycles.

TK-575 BRIM™ is a new top-tier NiMo catalysts designed specifically for use in ULSD service. Production of ULSD requires removal of the refractory sulphur compounds that proceed via the hydrogenating desulphurization pathway. TK-575 has a high hydrogenation activity and is 5-7°C more active than its predecessor depending on feedstock type, operating conditions and unit severity. Already, the first commercial reference for TK-575 BRIM™ has been obtained at a major U.S. refinery.

For pretreatment of hydrocracker feed, the new TK-605 BRIM™ NiMo catalyst offers superior HDN activity. The effective removal of refractive organic nitrogen compounds means that the performance of the hydrocracker is significantly improved. TK-605 BRIM™ exhibits about 5°C higher HDN activity as well as a higher HDS activity than the previous generation catalyst and has the same high stability.

Conclusions

By understanding the desulphurization mechanisms and the atomic structure of the active catalytic phases, the development of new catalysts and processes for the refining industry can be advanced. By assessing both electronic and structural factors influencing the desulphurization reactions, the link between catalyst formulation and macroscopic reactivity can be rationalized. The well-known changes in activity and selectivity of MoS_2 catalysts upon promotion with Ni or Co can now be directly linked to morphological changes on the atomic level. Such insights have aided the development of new high activity catalysts.

References

[1] C. Song, Catal. Today 86 (2003) 211.
[2] K.G. Knudsen, B.C. Cooper, H. Topsøe, Appl. Catal. A 189 (1999) 205.
[3] F. van Looij, P. van der Laan, W.H.J. Stork, D.J. DiCamillo, J. Swain, Appl. Catal. A 170 (1998) 1.
[4] S. Helveg, J.V. Lauritsen, E. Lægsgaard, I. Stensgaard, J.K. Nørskov, B.S. Clausen, H. Topsøe, F. Besenbacher, Phys. Rev. Lett. 84 (2000) 951.
[5] J.V. Lauritsen, M. Nyberg, J.K. Nørskov, B.S. Clausen, H. Topsøe, E. Lægsgaard, F. Besenbacher, J. Catal. 224 (2004) 94.
[6] M. Egorova, R. Prins, J. Catal. 225 (2004) 417.
[7] H. Topsøe, B.S. Calusen, F.E. Massoth, in:J.R Andseron, M. Boudart (Eds.), Hydrotreating Catalysis – Science and Technology, vol. 11, Springer Verlag, Berlin, 1996
[8] R. Prins, V.H.J. de Beer, G. Somorjai, Catal. Rev.-Sci. Eng 31 (1989) 1.
[9] P. Zeuthen, K.G. Knudsen, D. D. Whitehurst, Catal. Today 65 (2001) 307.
[10] A. Logadottir, P.G. Moses, B. Hinnemann, N.-Y. Topsøe, K.G. Knudsen, H. Topsøe, J.K.Nørskov, Catal. Today 111 (2006) 44.
[11] H. Topsøe, B.S. Clausen, N.-Y. Topsøe, J.K. Nørskov, C.V. Ovesen, C.J.H. Jacobsen, Bull. Soc. Chim. Belg. 104 (1995) 283.
[12] B. Hinnemann, J.K. Nøskov, H. Topsøe, J. Phys. Chem. B 109 (2005) 2245.
[13] A. Carlsson, M. Brorson, H. Topsøe, J. Catal. 227 (2004) 530.
[14] A. Carlsson, M. Brorson, H. Topsøe, J. Microsc., *in press.*

130e Hydrocracking for Clean Fuels Production

Suheil F. Abdo[1], Vasant P. Thakar[2], and Bart Dziabala[2]. (1) Catalysts, Adsorbents and Specialties, UOP LLC, 25 E. Algonquin Rd, Des Plaines, IL 60017, (2) PT&E, UOP LLC, 25 E. Algonquin Rd, Des Plaines, IL 60017

Hydrocracking for Clean Fuels Production

Suheil F. Abdo Vasant Thakkar Bart Dziabala

UOP, LLC Des Plaines, Illinois, USA

Abstract

Hydrotreating and hydrocracking are two of the best suited upgrading technologies to help meet the increasingly challenging clean and reformulated fuels challenge. The wide variety of hydrocracking catalyst and process schemes currently available offers a fair degree of leverage which can be utilized to meet the challenge. The most effective approach to meet the clean fuels challenge will greatly benefit from enhanced knowledge of the chemistry taking place in under varied process conditions. However, due to the very complex nature of typical feed stocks and the simpler but still complex product composition, detailed assessment and control of the required molecular transformations has eluded technologists in the field.

New developments will require increasingly specific tailoring of new catalysts to optimize the desired reaction pathways which vary with process design and the consequent environment the catalyst must operate in. Development efforts to meet clean fuels production objectives must start by establishing a detailed knowledge of the required molecular transformations and then coupling this knowledge with catalyst design principles and process conditions. This presentation will illustrate the benefits achieved through application of new tools for detailed catalyst and hydrocarbon characterization can help in enhancing the reaction pathways in the hydrocracking unit to achieve these goals. Examples of catalyst and process schemes offered for the upgeading of cycle oils into a mix of clean, high-value gasoline and diesel products under hydrocracking conditions will be presented to illustrate these principles.

Thermochemistry of Coking in Hydroprocessing Units: Modeling Competitive Naphthalene Saturation and Condensation Reactions

Paul R. Robinson and Lawrence S. Kraus
Albemarle Catalysts Company LP
Houston, Texas

Abstract

The condensation of polyaromatic hydrocarbons (PAH) often is cited as a mechanism for coke-induced catalyst deactivation in fixed-bed hydroprocessing units. Saturation of PAH inhibits condensation reactions. One purpose of this paper is to reiterate the fact that PAH chemistry in hydroprocessing units involves more than just a single entity called "naphthenes" and four entities called 1-, 2-, and 3+ "aromatics." Another is to show that understanding this chemistry can have practical value.

The simplest system involving PAH saturation comprises naphthalene, tetralin, decalin, and hydrogen. We modeled this system thermochemically, adding as a proxy for naphthalene condensation a reaction between naphthalene and o-xylene to form chrysene. The calculations show the change in equilibrium concentrations as conditions change from 100 to 3000 psig and 400 to 1500 °F. As expected, decalin is favored by high pressure and low temperature, and naphthalene is favored by high temperature and low pressure. But thermochemistry also predicts small but significant amounts of chrysene, even at moderate pressures and temperatures.

Examples are presented showing how an understanding of this kind of information can be used to explain, and in some cases improve, the performance of hydroprocessing units.

PART I. BACKGROUND AND LITERATURE REVIEW

Coke Induced Deactivation in Fixed-Bed Hydroprocessing Units

In many oil refining processes, coke lay-down is responsible, to a lesser or greater extent, for catalyst deactivation. For hydrotreating and hydrocracking units, coke formation is slow due to the presence of excess hydrogen. For example, after a 2 or 3-year cycle in which straight-run gas oil with an API gravity of 23 is processed at an LHSV of 0.8, a typical amount of coke on the spent catalyst is (at most) 10 wt%. This means that less than 0.0005 wt% of the carbon in the fresh feed ends up as coke. Practically speaking, coke formation is too slow to measure.

The above-described calculation is crude in part because the rate of coke formation increases as a catalyst cycle proceeds. Consider an operation in which catalyst activity is a function of the temperature required to meet a certain process objective (such as product sulfur content, product nitrogen content, or conversion). For most of the cycle, the rate of increase in required temperature is slow, but it accelerates at high temperatures near end-of-run (Figure 1). This can be attributed to accelerated coke formation and/or catalyst pore mouth blockage.

The spent catalyst at the top of a bed tends to contain more coke than bottom-of-the-bed material. Also, according to some industry pundits, different kinds of coke can be formed. According to the "hard coke, soft coke" postulate, "soft coke" comprises unconsolidated hydrocarbons and therefore can be removed by increasing hydrogen partial pressure (ppH_2) or by flushing with light oil at relatively low temperature. Over time, "soft coke" turns into "hard coke," which cannot be removed until the catalyst is unloaded and then regenerated. However, this hypothesis is hard to test directly because coke-on-catalyst can't be sampled while a unit is running, even if that unit is a pilot plant.

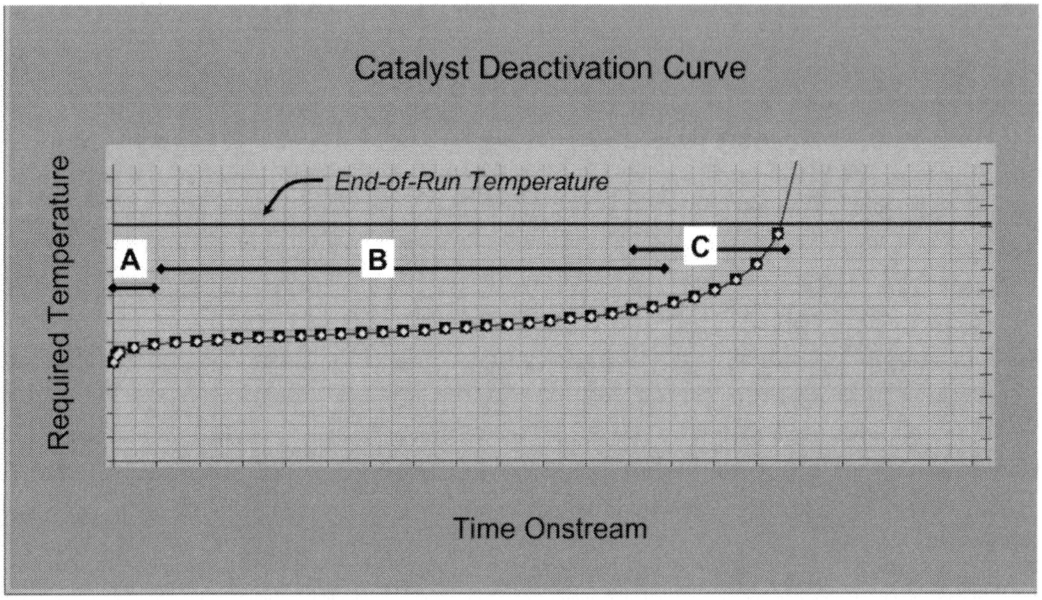

Figure 1. Catalyst deactivation curve showing three main regions. **(A)** Conditioning period in which the required temperature rises rapidly but tapers off. **(B)** Long middle-of-run period in which the required temperature rises slowly and almost linearly. **(C)** End-of-run period in which the required temperature rises exponentially.

Despite its insignificance from a mass-balance standpoint, coke formation is significant due to its considerable effect on catalyst activity. Understanding coke formation and the processing conditions that favor it are valuable for managing hydroprocessing catalyst deactivation. Consequently, it continues to be a major focus of catalyst research.

Coke Formation Mechanisms

The literature proposes at least two classes of coke-formation reactions – sequential growth and polymerization; the latter is also called condensation.

The sequential growth mechanism in Figure 2 is based on a paper by Van Speybroeck, et al.,[1] which discusses the formation of soot in flames. It involves alkylation of condensed aromatics by small free radicals followed by cyclization and dehydrogenation to form larger condensed-ring aromatics.

The so-called "naphthalene zigzag" pathway for the formation of large PAH from small PAH, as described by Sullivan, et al.,[2] is illustrated in Figure 3. The black arrows indicate peri-ring closure in which a 2-carbon moiety adds to an accessible "bay." The white arrows indicate ortho-ring closure in which a 4-carbon chain is added. The authors believe that in PAH growth, ortho- and peri-addition may alternate.

Other investigators[3] offer an explanation for the stability of coronene and ovalene, which can build up in the recycle oil of full-conversion hydrocrackers with the potential to foul heat exchangers and accelerate catalyst deactivation. They believe that 3-carbon bays (Figure 4) are less reactive than 4-carbon bays, which in turn are less reactive than 5-carbon bays. Coronene and ovalene accumulate because they contain only 3-carbon bays, and hence are less reactive than PAH with similar hydrogen/carbon ratios.

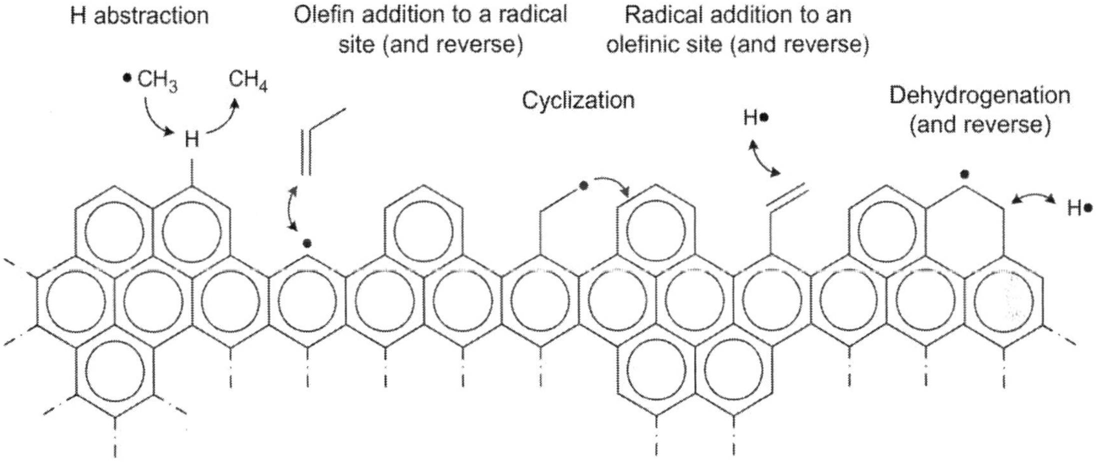

Figure 2. Coke growth from alkylation of condensed aromatics by small free radicals

Hydrogenation versus Condensation

For fixed-bed hydroprocessing units, kinetic data imply an alternative to sequential-growth pathways.[4] Measured rates of catalyst deactivation are consistent with the following mechanism:

(1) $PAH + xH_2 \leftrightarrow x/2\text{-hydroPAH}$
(2) $PAH + PAH' \rightarrow PAH'' + 2H_2$
(3) $PAH'' \rightarrow coke + xH_2$

This sequence shows a competition between PAH saturation (Reaction 1) to form a naphthene or naphthenoaromatic, and PAH condensation (Reaction 2) to form a larger molecule (PAH") and hydrogen. Reaction 1 is reversible, but Reactions 2 and 3 are not. As mentioned above, in fixed-bed hydroprocessing units accumulation of coke is very slow. Therefore, either Reaction 2 or 3 – or both – is/are significantly slower than Reaction 1.

The model described in Reference 4 is based on Reactions 1-3, where coke precursors include PAH, poly-ring naphthenes, and naphthenoaromatics. This is because at high end-of-cycle temperatures, naphthenes and naphthenoaromatics in the feed can undergo significant dehydrogenation (the reverse of Reaction 1) to form PAH.

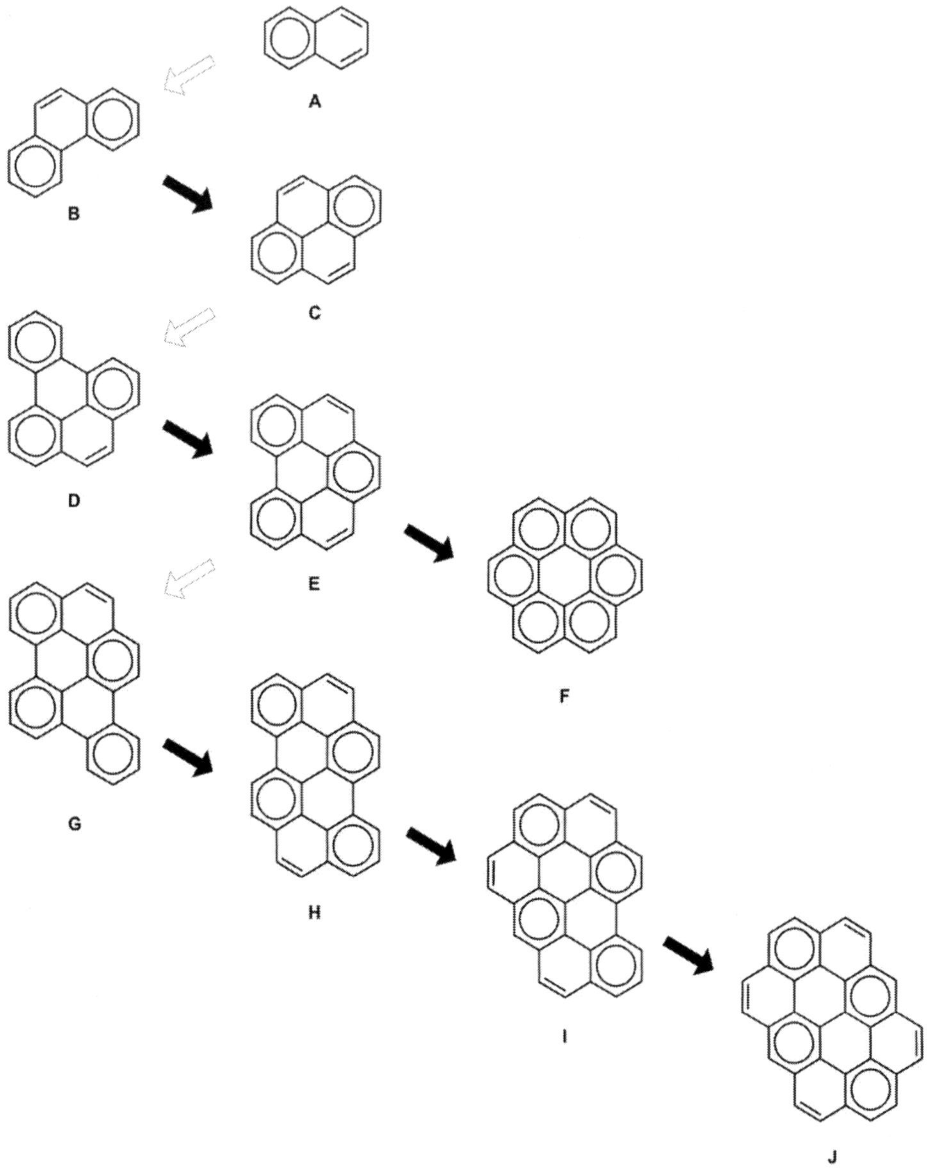

Figure 3. "Naphthalene Zigzag" pathway for forming heavy PAH from light PAH. A = naphthalene, B = phenanthrene, C = pyrene, D = benzo(e)pyrene, E = benzo(ghi)perylene, F = coronene, G = dibenzo(e,ghi) perylene, H = benzo(pqr)naphtho(8,1,2bcd)perylene, I = naphtha(8,2,1abc)coronene, J = ovalene.

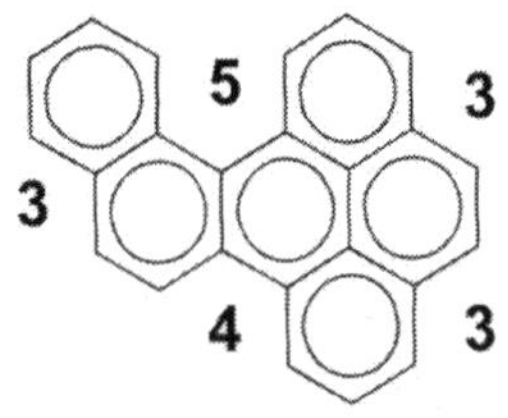

Figure 4. Illustration showing 5, 4, and 3-carbon bays in a PAH molecule

Saturation of Aromatics: Thermodynamics vs. Kinetics

The literature of hydroprocessing frequently cites the competition between hydrogenation and condensation. "Aromatics crossover" (Figure 5) refers to the temperature where the saturation of aromatics starts to become difficult. This phenomenon is attributed to a competition between kinetics and thermodynamics, with the latter dominating at higher temperatures. Often, the inflection point of the "crossover curve" is referred to as the temperature at which the system reaches equilibrium. That is incorrect. Strictly speaking the inflection point is the temperature at which dehydrogenation becomes significant. With respect to aromatics, the system reaches equilibrium when there is no net saturation, i.e., the concentration of aromatics in the product is equal to the concentration of aromatics in the feed.

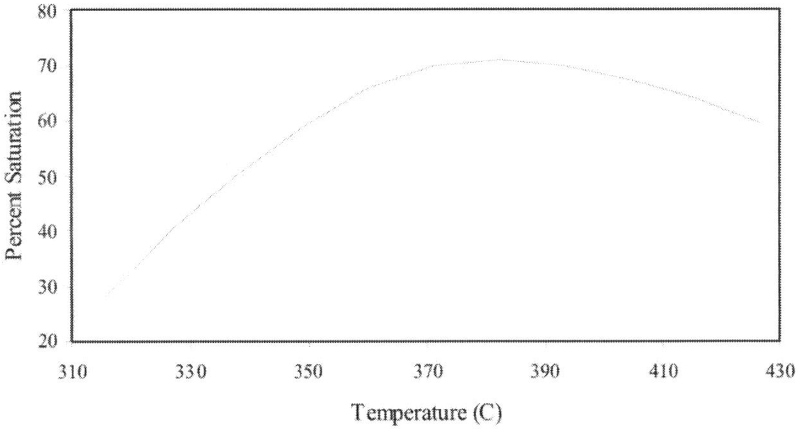

Figure 5. Crossover curve for saturating aromatics in a VGO hydrotreater at 1500 psig

Noble metals such as Pd and Pt are excellent catalysts for hydrogenation of pure aromatic compounds, but they are sensitive to H_2S and sulfur. Therefore, it's expensive to use them in commercial hydrotreaters. In the world of alumina-supported hydrotreating catalysts, sulfided cobalt-molybdenum (CoMo) is highly sulfur tolerant and catalyzes saturation at relatively low temperature and pressure. Sulfided nickel-molybdenum (NiMo) catalysts also are sulfur tolerant. At moderate pressures, they are better for saturation than CoMo catalysts. NEBULA™ catalysts are new materials, which employ base-metal supports. For saturation of aromatics, they are among the most active non-noble-metal catalysts available today. For saturation of aromatics in clean feeds, i.e., feeds with nil sulfur and nitrogen, the following trend is widely accepted:

Pt, Pd > NEBULA™ > NiW on alumina > NiMo on alumina > CoMo on alumina

If saturation of coke precursors inhibits condensation reactions, then catalysts with higher saturation activity should be less susceptible to coke-induced deactivation. It's hard to check this independently, because the chemistry of hydrotreating is rather complex, involving olefin saturation, aromatics saturation, hydrodesulfurization (HDS), and hydrodenitrogenation (HDN).

In conventional fixed-bed units, oil and hydrogen are mixed, heated, and sent down through the reactor(s) under conditions that approach trickle-bed flow through the catalyst(s). Such units are adiabatic, and (except for the afore-mentioned dehydrogenation of naphthenes and naphthenoaromatics) all major reactions consume hydrogen and release heat. In units with multiple catalyst beds and/or multiple reactors, temperatures are controlled by the injection of relatively cold, hydrogen-rich recycle gas. From the inlet to the first catalyst bed to the outlet of the last bed, temperatures rise and the partial pressure of hydrogen (ppH2) falls. The amount of variation from inlet to outlet depends on several factors, including feed rate, feed quality, ppH_2, H_2 circulation rate and H_2 availability. Low ppH_2 and low H_2 availability decrease the capability of the unit and the lifetime of the catalysts, especially those with relatively high activity for aromatics saturation and (consequently) high consumption or hydrogen.

"What Is an *Aromatics*?"

To this point, we have been talking about "aromatics" as if they were a single entity. The truth is, there are thousands. According to Briker, et al.,[5] straight-run gas oils (diesel fuels) contain 15 wt% to 30 wt% aromatics, either with one ring or 2-5 condensed rings. Robbins and Hsu[6] report that, in heavy distillates, most of the ring compounds are attached to alky groups, which can contain 1 to 17 carbon atoms. The average is 3 carbons. Light cycle oil (LCO) from an FCC unit can contain 60 – 90% aromatics and 10 – 20 wt% olefins. In LCO, more than 90% of the alkyl groups attached to aromatic rings are methyl groups. Under a given set of process conditions, saturation rates will be different for each compound and the temperature response of those reaction rates will differ significantly.

Different analytical methods give different results for bulk concentrations of aromatics. Table 1 compares analytical results for several different diesel fuels[5] as measured by different methods. Due to its relatively low cost, HPLC is the method most commonly used by petroleum refiners. The largest differences are for polyaromatics, where results can vary by a factor of six. Even with diaromatics, variations can be significant. This is important in hydroprocessing, because polyaromatics are much more likely to form coke than compounds with fewer rings.

Naphthenoaromatics are not reported in the table, although they probably are lumped in with "aromatics." In hydrotreated material, concentrations of naphthenoaromatics can be significant. With HPLC in particular, naphthenoaromatics with a given number of rings can elute with fully unsaturated aromatics having an equal number of rings, especially when the compounds have long alkyl side chains. The eluted band may be broader when partly unsaturated compounds are present, but in our experience, technicians (and computers) in refineries laboratories seldom consider this when reporting results.

Even if we know the "true" composition of a feedstock, the picture is not simple. Consider the reactions shown in Figure 6. At constant pressure and temperature over the same non-acidic catalyst, the relative extents of the reaction vary substantially as a function of residence time.[7] Figure 7 shows that the 1st reaction, the saturation of naphthalene, is always the least favored. For phenanthrene saturation, the 2nd reaction is preferred if the LHSV is higher than 1.8, but the 3rd reaction is favored if the LHSV is 1.0 or less. At less than 0.5 LHSV, the 4th reaction is more preferred than the 2nd.

Sample	*LC/GC-MS*	*SFC*	*SPE/GC-MS*	*GC-FIMS*	*HPLC*
Ref2					
Aromatics	29.5	29.8	30.6	29.7	-
Monoaromatics	23.8	23.4	24.1	23.8	-
Diaromatics	5.4	5.7	5.1	5.5	-
Polyaromatics	0.3	0.7	1.4	0.4	-
SWRII					
Aromatics	33.9	32.6	33.3	33.3	-
Monoaromatics	23.5	21.3	22.0	23.4	-
Diaromatics	9.8	10.4	8.7	11.9	-
Polyaromatics	0.6	1.0	2.6	0.7	-
S20B					
Aromatics	22.3	23.5	-	26.4	22.8
Monoaromatics	19.2	20.3	-	23.5	20.9
Diaromatics	2.8	2.7	-	2.7	1.3
Polyaromatics	0.3	0.5	-	0.1	0.6
C20B					
Aromatics	19.4	20.2	-	20.1	19.8
Monoaromatics	15.7	16.8	-	16.3	17.4
Diaromatics	3.6	3.2	-	3.6	2.2
Polyaromatics	0.1	0.3	-	0.1	0.3

Table 1. Breakdown of "aromatics" in selected diesel fuels

In short: Just as "aromatics" are more than one (or four) entities, relative saturation rates depend on several factors, including process conditions, feed composition, and catalysts.

PART II. THERMOCHEMICAL CALCULATIONS

In fixed-bed hydroprocessing units, under typical hydroprocessing conditions, coke formation from PAH precursors is too slow to measure. Therefore, when trying to gain a fundamental understanding, it is helpful to work with models.

Methodology

In this paper, we report the results of a *thermochemical* model designed to illustrate the competition between the saturation and condensation of naphthalene – the simplest of all polyaromatic hydrocarbons. We emphasize *thermochemical* because we did not address reaction rates. Rather, we calculated the equilibrium concentrations of 6 components – naphthalene, tetralin, decalin, chrysene, ortho-xylene and hydrogen.

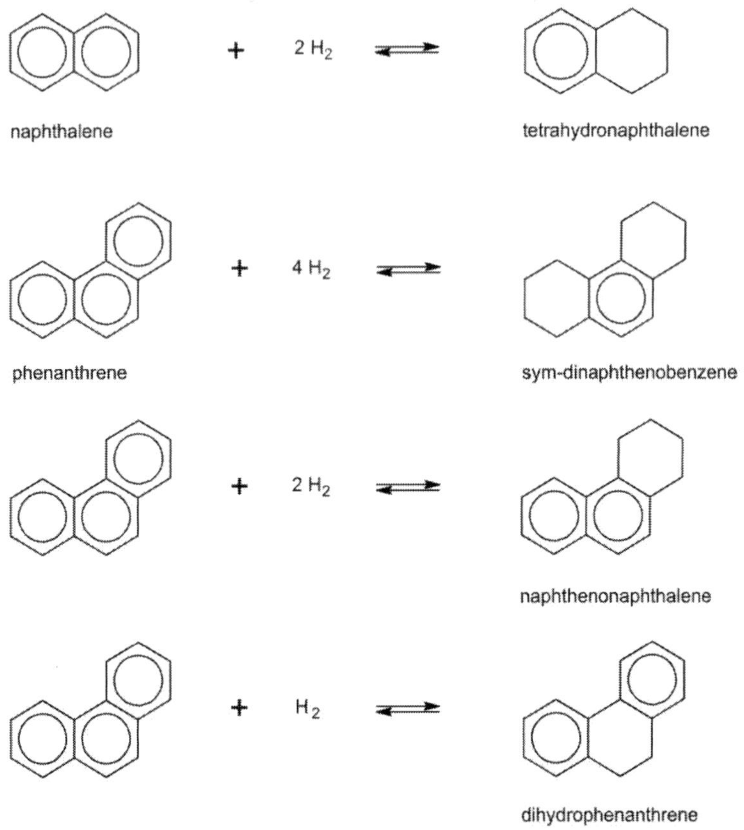

Figure 6. Saturation reactions for naphthalene and phenanthrene

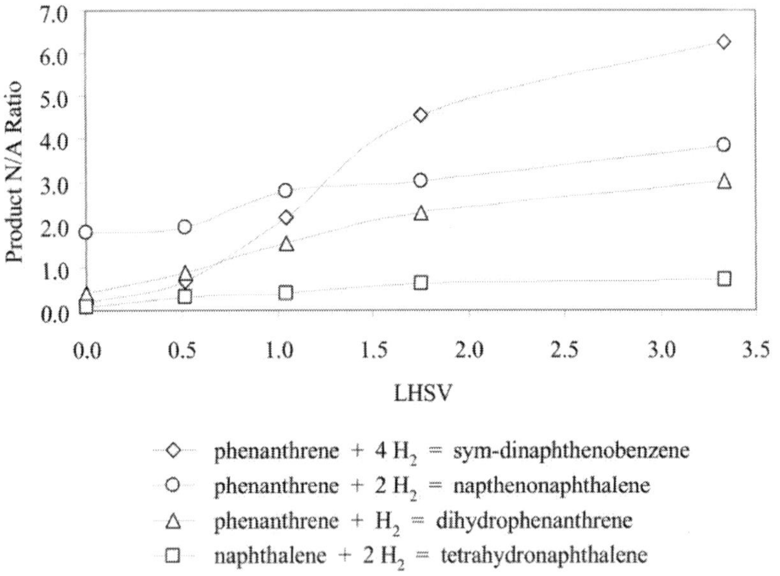

Figure 7. Product naphthene/aromatic (N/A) ratios for the saturation of naphthene and phenanthrene at 800°F, 2000 psig (426°C, 13,891 kPa) over a non-acidic catalyst. N/A ratios at zero LHSV (infinite reaction time) come from equilibrium calculations, not measurements.

For our model, we used the reaction scheme shown in Figure 8. For the last reaction, we wanted to use the dimerization of naphthalene to a C_{20} compound, but the standard physical property database associated with our modeling tool – Aspen Plus 12.1 – didn't contain data for any C_{20} PAH with 4 or 5 condensed rings. As a substitute, we used o-xylene and allowed it to undergo only one reaction – condensation with naphthalene.

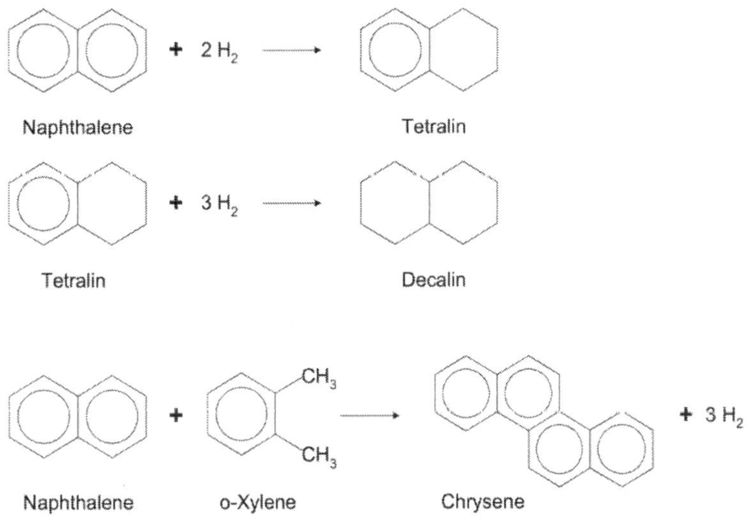

Figure 8. Components and reactions used in the thermochemical model

An Aspen Plus flowsheet was used, as shown in Figure 9. Hydrogen and hydrocarbon streams were mixed into a feed stream (MIX1-EX) and sent to Gibbs reactor in which only the three above-shown reactions were permitted. The reactor overhead and bottom streams were combined into a total product stream (SUM-PROD). The composition of this stream is reported below for different sets of temperatures and pressures.

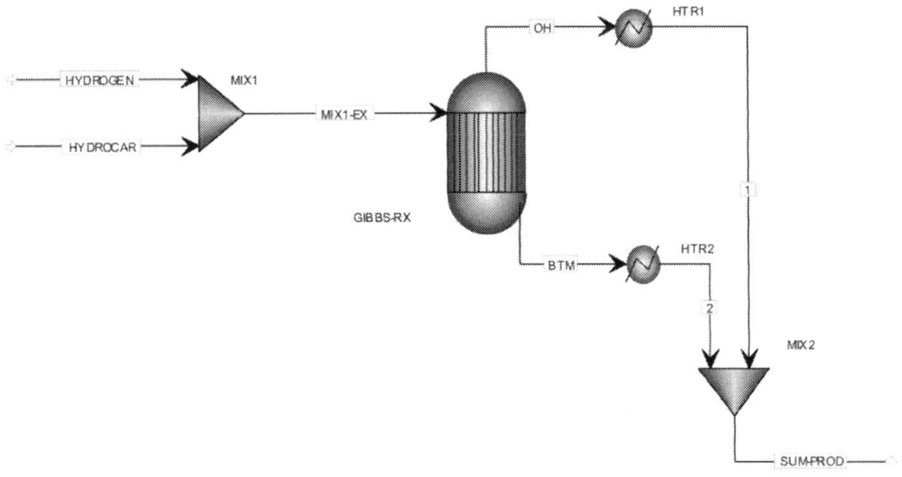

Figure 9. Aspen Plus Flowsheet

Table 2 shows results at 3000 psi and 1000°F. As with all other cases, the feed comprised 50 lb-moles/hr each of o-xylene and naphthalene, and 1000 lb-moles/hr of H_2.

3000 psi	HC	H2	MIX-RX	OH	BTM	SUM-PROD
H2	0	1000	1000	1000.79549	0	1000.79549
O-XYLENE	50	0	50	49.3163291	0	49.3163291
CHRYSENE	0	0	0	6.84E-01	0	6.84E-01
DECALIN	0	0	0	1.30E-05	0	1.30E-05
TETRALIN	0	0	0	6.28E-01	0	6.28E-01
NAPHTHL	50	0	50	48.6885899	0	48.6885899
Total Flow lbmol/hr	100	1000	1100	1100.11182	0	1100.11182
Total Flow lb/hr	11717.046	2015.88	13732.926	13732.9264	0	13732.9264
Total Flow cuft/hr	692.089837	13334.1093	14289.04	8133.61657	0	19748.2856
Temperature F	1000	1000	955.89444	1500		1505.57208
Pressure psi	1214.69595	1200	1200	3000	3000	1200
Vapor Frac	1	1	1	1		1

Table 2. Results of Calculations at 3000 psi and 1000°F

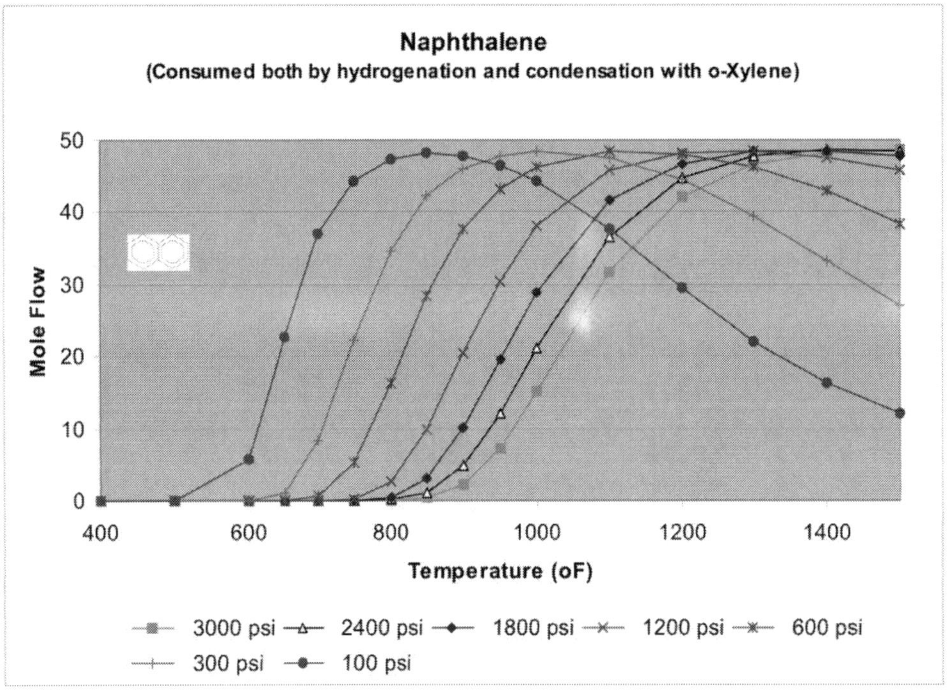

Figure 10. Equilibrium Mole Flow of Naphthalene as Function of Temperature and Pressure

Figures 10-15 show equilibrium mole flows for individual components. Figures 16-19 show all components except hydrogen at a single pressure and various temperatures. These are more informative because they show the relative concentrations of the organic components. Note how tetralin changes on each graph as a function of temperature. Under conditions typically used for hydroprocessing 400-2000 psi and 600-800 °F, tetralin always is present in significant amounts.

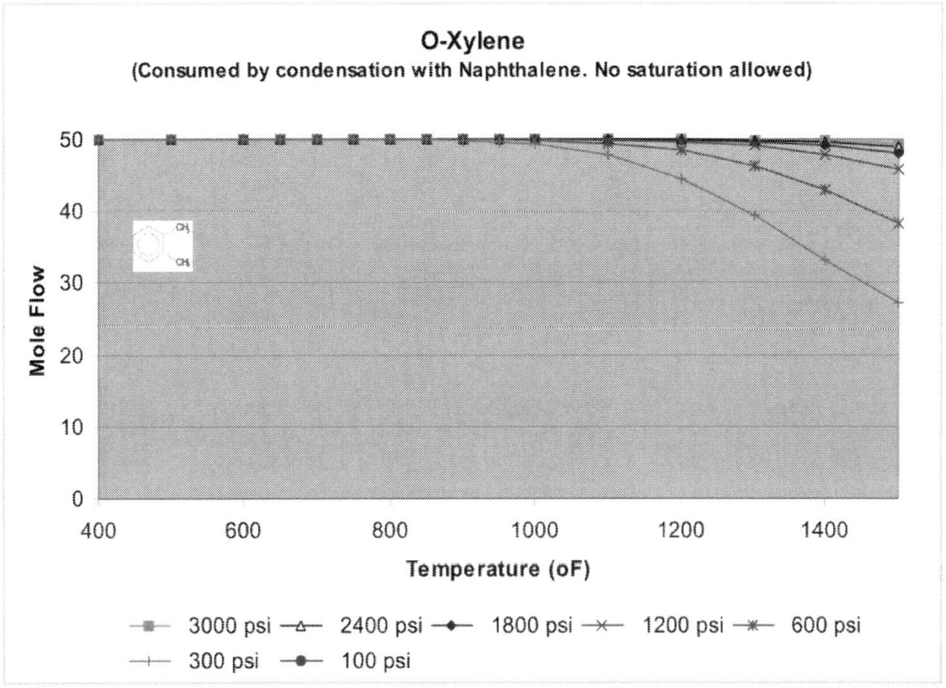

Figure 11. Equilibrium Mole Flow of O-Xylene as Function of Temperature and Pressure

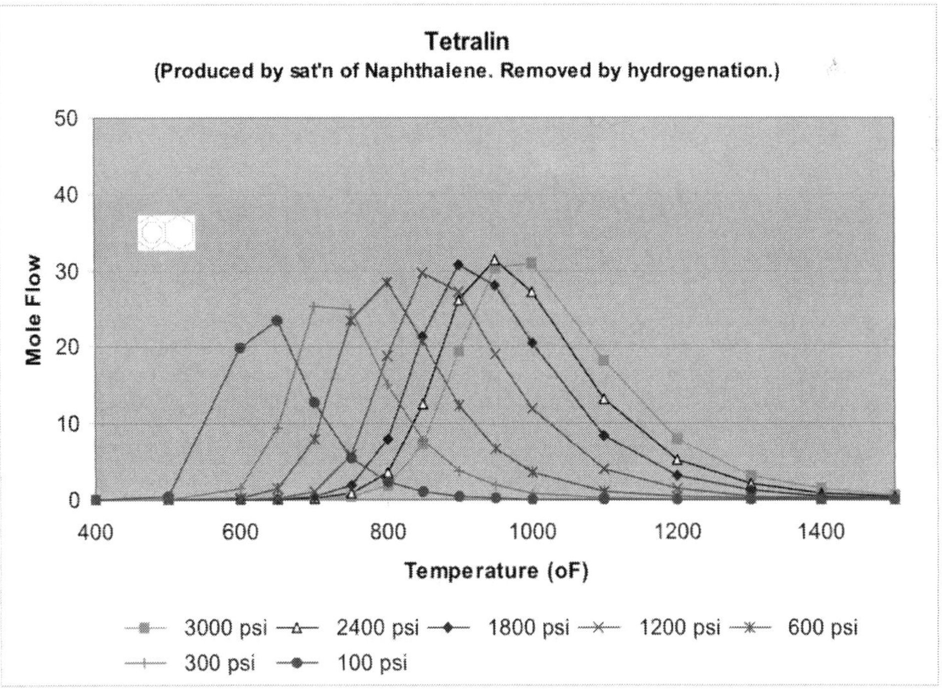

Figure 12. Equilibrium Mole Flow of Tetralin as Function of Temperature and Pressure

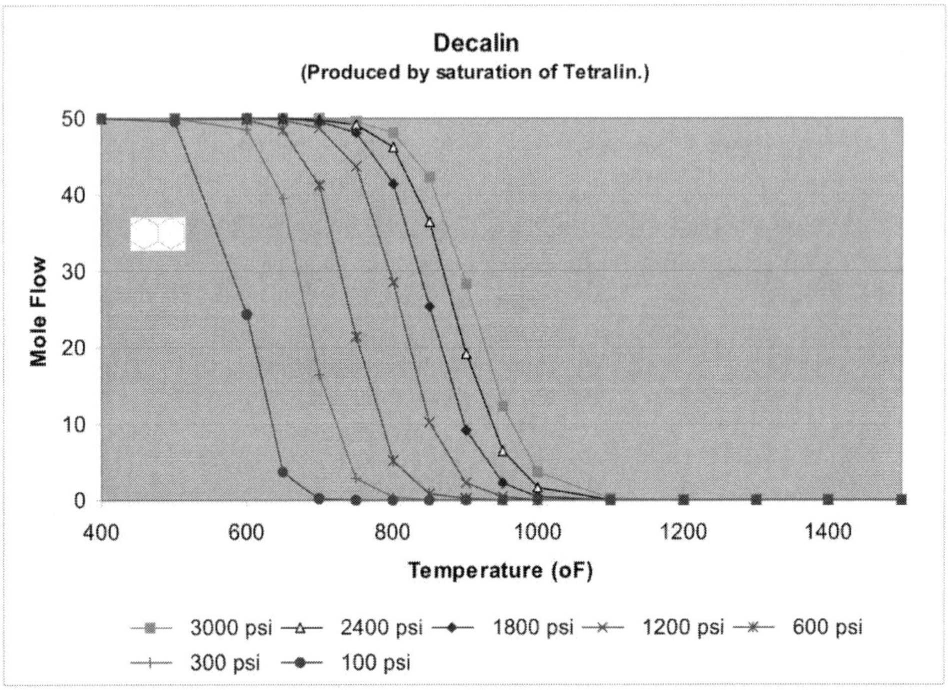

Figure 13. Equilibrium Mole Flow of Decalin as Function of Temperature and Pressure

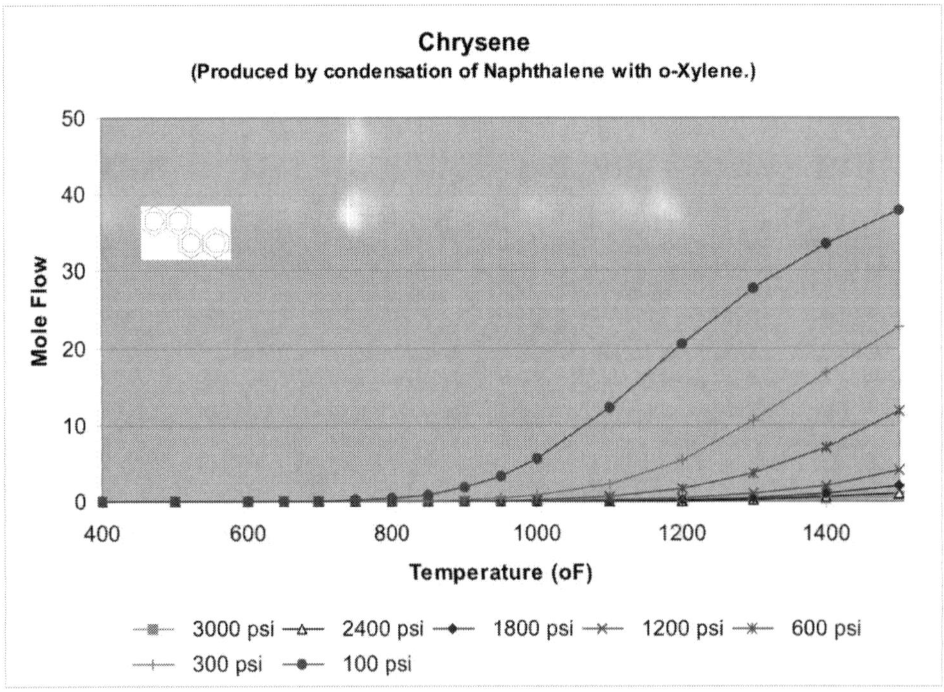

Figure 14. Equilibrium Mole Flow of Chrysene as Function of Temperature and Pressure

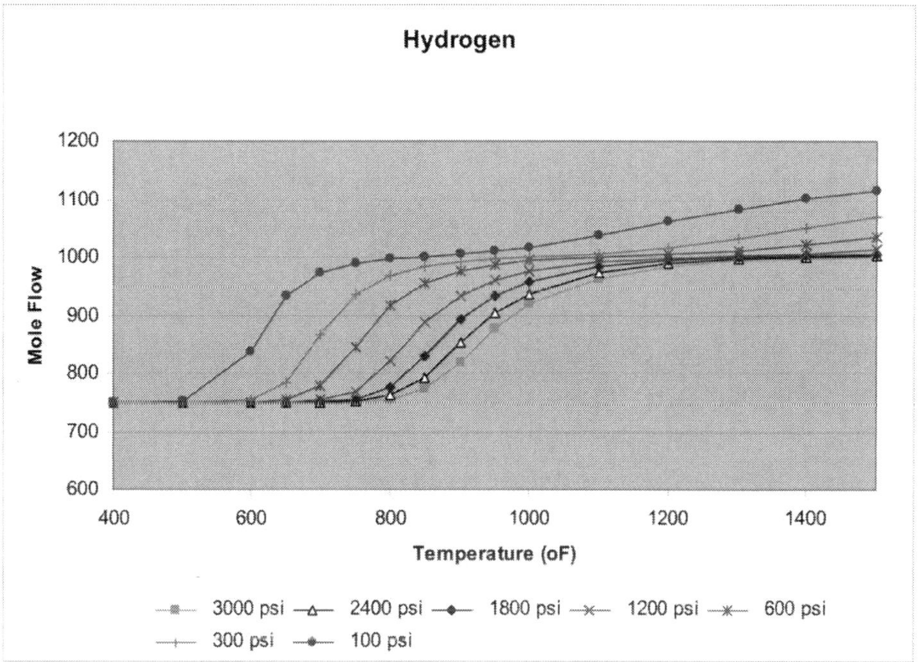

Figure 15. Equilibrium Mole Flow of Hydrogen as Function of Temperature and Pressure

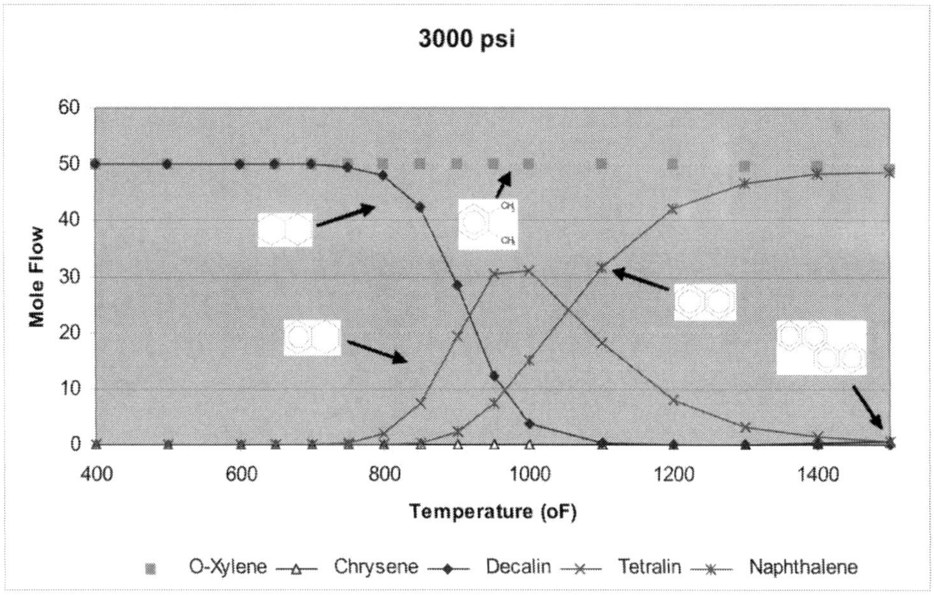

Figure 16. Equilibrium Mole Flows at 3000 psi

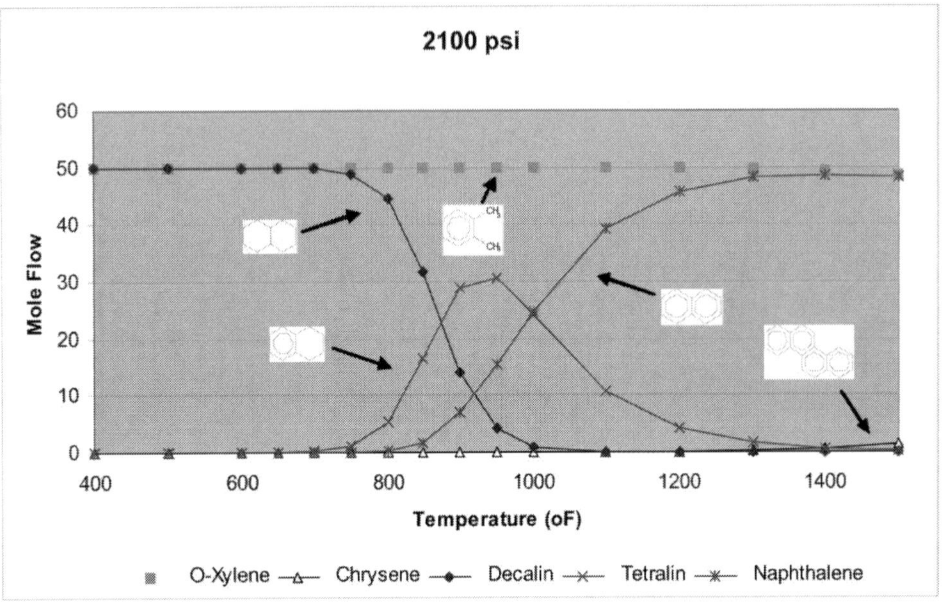

Figure 17. Equilibrium Mole Flows at 2100 psi

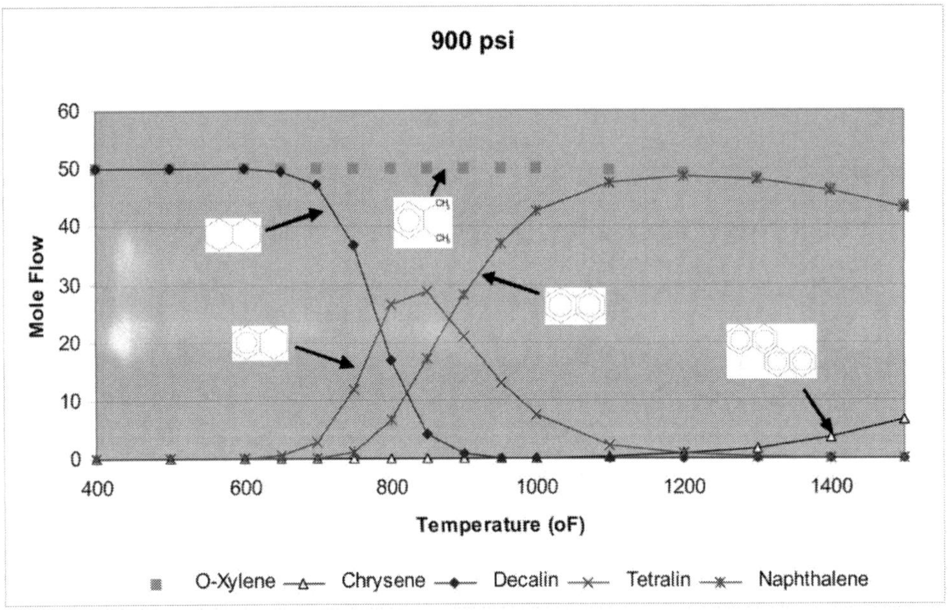

Figure 18. Equilibrium Mole Flows at 900 psi

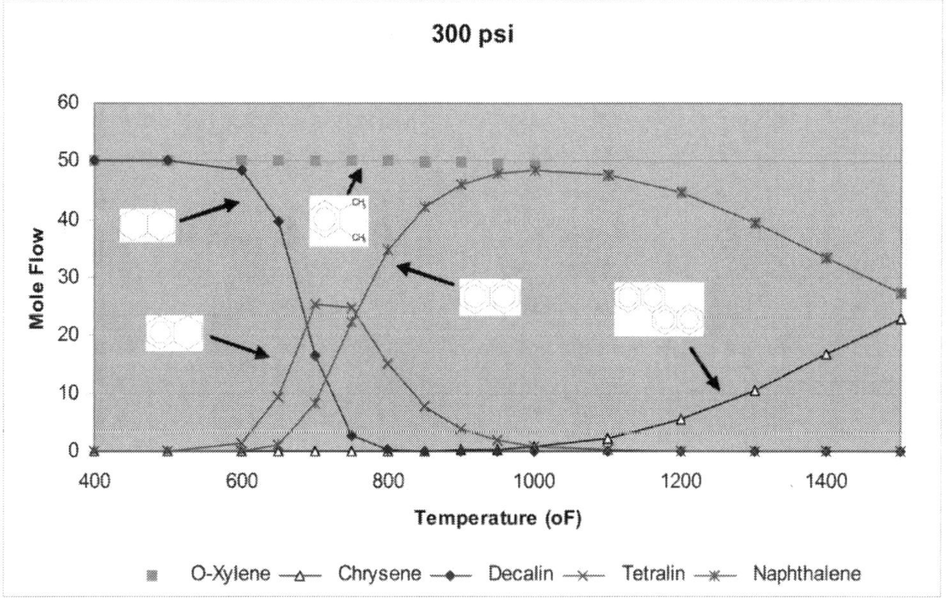

Figure 19. Equilibrium Mole Flows at 300 psi

As implied by Figures 16-19, poly-ring naphthenes start to dehydrogenate at 700°F, generating 4-ring and 5-ring PAH susceptible to coke generation and subsequent catalyst deactivation.

According to Van Speybroeck, et al.,[1]

> "...the nature of coke formation is autocatalytic: the larger the macroradicals, the faster the subsequent reactions that lead to further growth of the polyaromatic surface."

To illustrate this thermochemically, we calculated free energies for several PAH condensation reactions as a function of temperature.[8] As shown in Figure 20, even for relatively small 20-carbon PAH, free energies of reaction are significantly negative, even at very low temperature. Our calculations are for gas-phase reactions under standard conditions, in which the concentrations of all reactants are 1 molar. With higher hydrogen partial pressure, the reaction potential is smaller, but not significantly so. Even at initial concentrations $<10^{-6}$ molar and pressures >10,000 bar (147,000 psi), these reactions tend to go to completion.

PART III. IMPLICATIONS FOR COMMERCIAL OPERATION

There are several ways in which this chemistry can be used to explain or even improve commercial operations. The best way to illustrate this is with real-world examples.

Example 1. A Canadian refinery was processing a synthetic crude oil, which had been severely hydrotreated at the production site and delivered through a pipeline. Unsurprisingly, the crude had a high gravity and contained low amounts of sulfur and nitrogen. Pilot plant studies predicted that VGO derived from the crude was easier to hydrocrack than conventional VGO. The refiner built a hydrocracker based on the pilot plant study, and indeed, at low start-of-run temperatures, the unit performed as expected, easily meeting all process guarantees. But during the cycle, as the required temperature rose past 700°F, deactivation of the 1st-stage treating catalyst accelerated dramatically, far more than expected. One explanation was that blocks of synthetic crude became contaminated as they flowed for 30 days through a common-carrier pipeline. Indeed, analysis of the crude at the refinery gate confirmed this, but analysis of the VGO produced by the crude distillation unit (CDU) did not show contamination, probably because the contaminants were rejected into the CDU residue. Another explanation came from a GC/MS analysis of the VGO. It revealed high concentrations of poly-ring naphthenes, with significant amounts of 4-ring and 5-ring moieties. It is probably that, at T >> 700°F, these compounds underwent dehydrogenation to form condensable PAH.

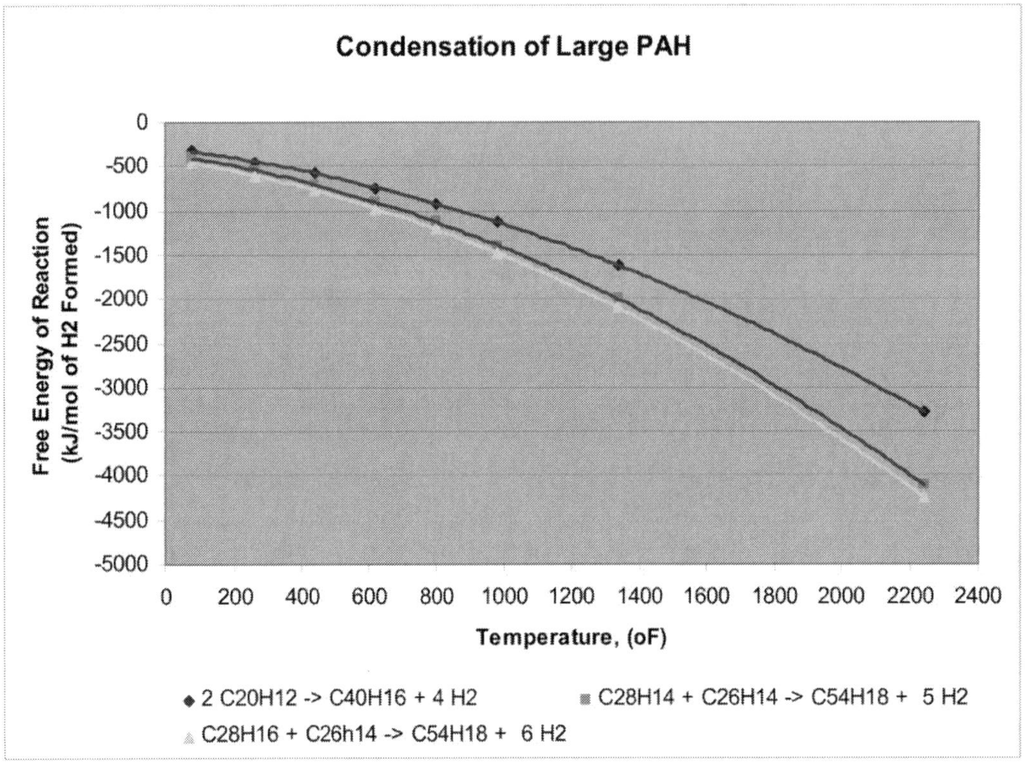

Figure 20. Free Energy of Reaction for the Condensation of Large PAH

Example 2. A high-pressure hydrocracker became heater limited due to feed rates 30% higher than design. To run at these rates, the refiner maximized heat recovery from the 4-bed pretreat reactor by operating with an ascending temperature profile, with no quench between the 2nd, 3rd, and 4th catalyst beds. These beds were loaded with a high-activity hydrotreating catalyst to remove organic nitrogen. Toward end-of-run, the Bed 4 inlet temperature was 765 °F and outlet was 759 °F. In other words, instead rising across this bed, the temperature dropped. One explanation was that the catalyst was dead (or nearly so) due to high-temperature operation since start-of-run. Another was that the feed to that bed, after being saturated thoroughly by the first 3 beds, was dehydrogenating at such high temperatures. To test these assumptions, the refiner lowered the Bed 4 inlet temperature to 750°F. In response, the temperature drop turned into a temperature rise, producing an outlet temperature of 756°F and making a product with an organic nitrogen content equal to that obtained with the higher inlet temperature. With a revised (but still ascending) temperature profile, the refiner is able to get performance out of the 4th bed while maintaining high heat recovery.

Example 3. Figures 21 and 22 show data from a commercial FCCU feed hydrotreater. Table 3 shows some bulk properties of the feed and the process objectives for two catalyst cycles. For Cycle A, a conventional CoMo/KF-757 catalyst system was used in (approximately) a 70/30 volume ratio. KF-757 is a high-activity STARS catalyst, where STARS stands for Super Type II Active Reaction Sites.[9]

Feed Bulk Properties	
Feed Type	LVGO
T95, °F	970
Gravity, °API	22.0
Sulfur, wt%	1.7
ppH_2, psia	650
Process Objectives	
Cycle A (Conventional CoMo + KF-757, 70%/30%)	
Product sulfur, wppm (WABT < 700 °F)	2000
Product sulfur, wppm (WABT > 700 °F)	500
Cycle B (KF-757, 100%)	
Product sulfur, wppm (WABT < 700 °F)	500

Table 3. Hydrotreater Feed Properties and Process Objectives for Example 3

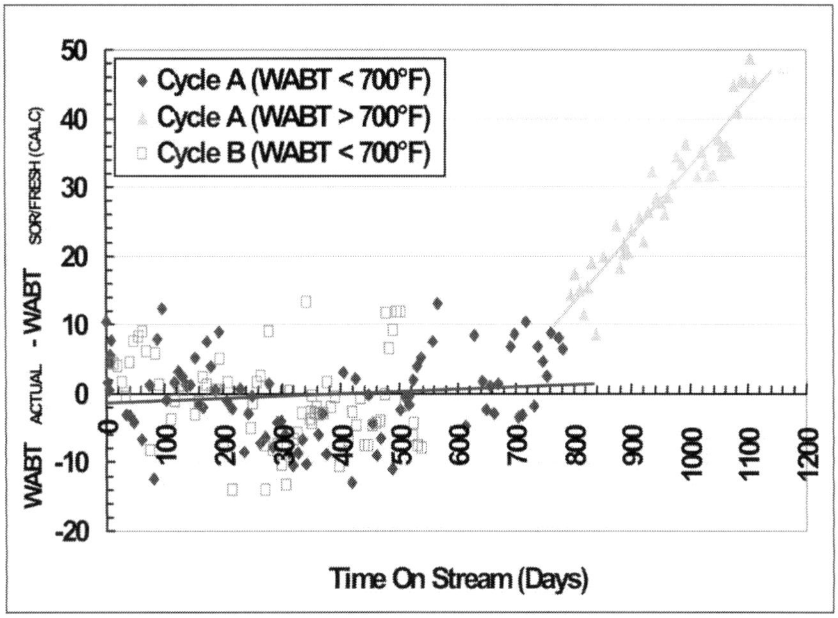

Figure 21. Required WABT versus Time on Stream

During the first two years of Cycle A, the product target was <2000 wppm S. This target could be achieved at temperatures below 700°F. After two years of operating with this target, the product objective was changed to <500 wppm S. To achieve this target, the required WABTs were raised to >700°F. The normalized WABT vs. TOS (time on stream) is shown in Figure 21. Below 700°F, deactivation was very low: <0.3°F/month. The deactivation rate at WABTs >700°F was 10 fold higher than that observed when the target was <2000 wppm S. This can be explained by evaluating the impact of WABT on reactor axial ΔT, as shown in Figure 21. At WABTs below 700°F, the axial ΔT is in the 50-60°F range. At WABTs above 700°F, the axial ΔT steady decreased with TOS (and increasing WABT) from the 45-50°F range to just above 35°F. The decrease in axial ΔT is caused by reduced aromatics saturation at temperatures above the crossover temperature (per Figure 5). The increase in aromatics results in higher levels of coke precursors and an increased rate of deactivation.

For Cycle B, a complete load of KF-757 was used. The significant increase in HDS activity with this catalyst system allowed product with <500 wppm S to be produced at WABTs <700°F. This resulted in an operating environment favorable for aromatics saturation, as shown in Figure 22 by axial ΔT values in the 50-55°F range and a deactivation rate similar to that of the Cycle A catalyst system at product target of <2000 wppm S.

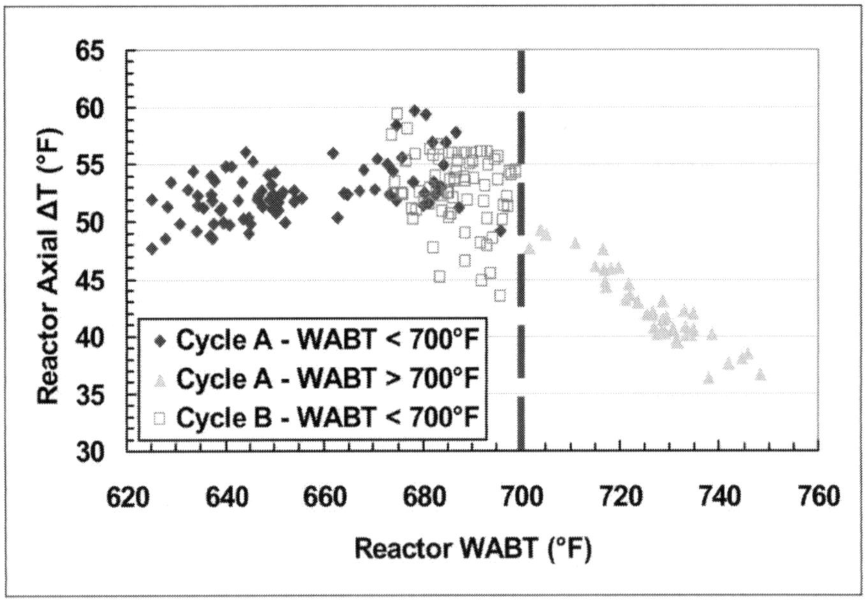

Figure 22. Cycles A and B: Temperature Rise versus Reactor WABT

CONCLUSIONS

1. When considering the impact of different feeds on the coke-induced deactivation of hydroprocessing catalysts, it is overly simplistic to talk about aromatics as if they are three single entities with 1-, 2-, and 3+ rings. Naphthenes and naphthenoaromatics also must be considered, due to their tendency to dehydrogenate above about 700°F.
2. In the thermochemical model presented herein, condensation is favored by high temperatures and low pressures. However, some condensation occurs under conditions typical for fixed-bed hydroprocessing.

3. Looking at the naphthalene-tetralin-decalin group, we see that, for any given pressure, decalin predominates at low temperatures and naphthalene predominates at high temperatures. Tetralin appears at moderate temperatures. In real systems, at temperatures below about 700°F, the opposite is observed. That is, aromaticity increases with temperature. This is consistent with the thermodynamics-versus-kinetics explanation for the "aromatics cross-over" effect.
4. Larger PAH have a much greater tendency to condense than smaller ones. It is safe to say that our naphthalene-based model represents the most optimistic scenario for PAH saturation. Free energies of reaction for the condensation of PAH containing 20+ carbon atoms are very negative, even at low temperatures and high pressures. Therefore, to minimize coking due to PAH in commercial units, it's best simply to keep such molecules out of the feed.
5. Understanding this chemistry can help to explain and improve the operation of commercial hydroprocessing units.

REFERENCES

1. V. Van Speybroeck, M.-F. Reyniers, G.B. Marin, and M. Waroquier, "The Kinetics of Cyclization Reactions on Polyaromatics from First Principles," *Chem. Phys. Chem. 2002*, 3, 863-870
2. R.F. Sullivan, M.M. Boduszynski, and J.C. Fetzer, *Energy and Fuels*, 3:603 (1989)
3. A. Keller, R. Kovacs and K.-H. Homann, *Phys. Chem. Chem. Phys., 2000*, 2, 1667-1675
4. P.R Robinson, "Catalyst Life Management with a Predictive Deactivation Model," 2004 NPRA PADS Conference, PD-04-171, September 19-21, 2004
5. Y. Briker, Z. Ring and H. Yang, "Intergrated Methodology for Characterization of Petroleum Samples and Its Apllication for Refinery Product Quality Modeling," Chapter 4 in *Practical Advances in Petroleum Processing*, C.S. Hsu and P.R. Robinson, eds., Springer: New York, 2005.
6. W.K. Robbins and C.S. Hsu, "Composition," in the *Kirk-Othmer Encyclopedia of Chemical Technology (4th ed.)* John Wiley & Sons: New York, 1996
7. W.L. Wu and H.W. Haynes, "Hydrocracking Condensed Ring Aromatics over Nonacidic Catalysts," in *Hydrocracking and Hydrotreating*, J.W. Ward and S.A. Qader, eds., American Chemical Society Symposium Series 20, 1975, 65-81
8. Calculations based on data taken from: J.Y.R. Sumathi and W.H. Green, Jr., "Accurate and Efficient Method for Predicting Thermochemistry of Polycyclic Aromatic Hydrocarbons - Bond-Centered Group Additivity," and references contained therein. (http://web.mit.edu/sdl/www/Accurate%20and%20Efficient%20Method.pdf)
9. D.A. Pappal, F.L. Plantenga, W.J. Tracy, R.A. Bradway, G. Chitnis, and W.E. Lewis, "Stellar Improvements in Hydroprocessing Catalyst Activity," National Petroleum Refiners Association Annual Meeting, San Antonio, Texas, March 2003 (AM-03-59)

ENERGY CONSERVATION AND INNOVATION OF BASIC CHEMICAL PROCESSES: DRIVERS AND BARRIERS

Tao Ren, Department of Chemistry, Utrecht University, Utrecht, The Netherlands and Dutch Energy Research Center, Petten, The Netherlands (ECN)

Abstract

Energy conservation of chemical processes is a key option for the chemical industry to reduce greenhouse gas emissions. This paper is a study of drivers and barriers to energy-efficiency related process innovations in the petrochemical industry. For improving existing processes, main drivers are process energy saving, tight supply of gas feedstocks and personal commitment. Main barriers are shortage of staff and time, competition from other prioritized projects and existing configurations. For developing new processes, main drivers are low-cost feedstocks, competition and knowledge spillovers. Main barriers are unfavorable business environment, insufficient decision-making tools and concerns on job security. These drivers and barriers can be understood in terms of knowledge on reward/risk ratios.

Introduction

In 2004, approximately 16 EJ process energy use (in primary terms) and one billion tons of CO_2 equivalent greenhouse gas emissions were estimated to have been generated by basic chemical processes. About 40-80% of operating expenses of the basic chemical industry are process energy cost, which was roughly 60 billion dollars globally in 2004. Therefore, it is of significance to study innovation of basic chemical processes from the perspective of energy efficiency.

Much previous work on the innovation in the basic chemical industry was done from either the economic perspective [1, 2] or the environmental perspective [3, 4]. However, there is little work to bridge these two perspectives and to help understand what encourages or hinders the innovation of such processes. Therefore, our research question is: What drivers and barriers to innovations are there in the case of petrochemical processes as far as energy conservation is concerned? The aim of this paper is to provide readers with some insight in innovation of chemical processes. Petrochemical processes are used as the case study because of the large amount of energy use involved. Over 30 interviews were conducted during 2003-2005 in the US and Europe. This paper presents the findings on the barriers and drivers to two types of activities respectively, improving existing processes and developing new processes. Implications for energy conservation will be also discussed.

Definitions

In this paper, barriers are defined as those factors that negatively affect a firm's intentions for innovation and therefore hinder activities. Drivers are defined as those factors that positively affect a firm's intentions for innovation and therefore encourage activities. Barriers and drivers are inextricably interlinked. A driver, e.g. an economic opportunity, could lead to a barrier, such as fear for making a wrong choice. Innovation activities consist of two types of activities, improving existing processes and developing new processes as defined in the widely used Oslo Manual issued by OECD [5]. See examples in Table 1.

Improving Existing Processes

Drivers

First, the most important driver is cost saving, which is the result of reducing process energy use per ton of products in existing processes. Process energy use can be reduced through installing steam and power cogeneration (improving overall energy efficiency). Such improvement could save up to 5-25 million dollars annually out of an annual process energy costs of several hundreds of millions of dollars in a large firm, or up to 15% of total annual energy costs in a single plant [6-8]. Process energy use can also be reduced through installing improved furnace coils (with higher or more flexible yields so less energy use per ton of product), more efficient distillation columns (with better heat integration), more effective catalysts (with lower activation energy), etc. Such improvement could save up to 10-20% of the total process energy use in comparison with conventional processes. Process energy savings directly increase profits, or earnings before income tax, a key indicator for performance. In fact, process energy cost is one of a few reducible production costs excluding costs of feedstock and maintenance. Process energy costs (or the costs of combusted fuels and electricity) is roughly 20% of the production cost of a naphtha-based steam cracking process in 2000 [9]. During 1999-2003, process energy costs per ton ethylene more than doubled in the US and rose over 50% in Europe and Japan [10].

Second, another strong driver for improving existing processes is the tight supply of gas feedstocks (i.e. ethane and propane derived from natural gas and LPG). This is mainly due to increasing use of natural gas for electricity production and natural gas liquids (containing LPG) for transportation fuel production. This is especially relevant for the US producers who depend on ethane and propane as 70% of feedstocks. Over the past 30 years, ethane prices in the US were around $1-3/GJ. As a result, the average US steam crackers were 25% less energy-efficient than those in Japan and 16% less than those in Europe [10]. However, in 2005, the prices of gas feedstocks in the US increased by 200-300% in 2000-2005. This has caused an increase of the five-year-average spending on energy conservation in US petrochemical plants (including both innovation and non-innovation) in 1999-2003 by 4.5 times from that in 1994-1998 [10].

Third, commitment of individuals is also a driver for initiating R&D projects and adopting new or improved technologies. Some process engineers and managers are committed to energy saving due to incentives of bonuses that are tied to energy savings [7]. Others are committed to energy savings due to fear for losing local employment to competition from overseas [11]. Commitment from the top leadership or a leading coordinator for energy efficiency issues can also be an important driving force for adopting new or more energy-efficient technologies in the refinery and petrochemical plants [12, 13]. This confirms an earlier finding that innovations are triggered by people inside firms [1].

Barriers

First, shortage of staff and time is a general barrier to implementation of any type of innovative technologies. This was agreed by all of our interviewees. Employment in the US basic chemical industry was reduced by 4% in 1994-2004 [14] (No comparable data for Europe and Japan is available). This occurs despite the fact that labor cost is less than 1% of petrochemical production costs [9, 15]. Many engineers in petrochemical plants are daily occupied with putting-out-fire type of troubleshooting and other operational priorities [16]. Little time is left for understanding the current energy use or getting information on innovative energy-efficient technologies [15]. This is especially true when market demand is strong and plant utilization is high. Even in some of the world's largest firms, there are no more than ten people globally specializing on energy-efficiency improvement related coordination and management [7]. Labor-shortage problems are also occurring to contractor and engineering firms. It is even difficult to find experienced engineers to meet labor demand [12, 13] or to replace the retired personnel [17].

Second, other projects that could lead to higher sales and internal rate of returns often take investment funds away from less-profitable projects on improving existing processes and energy efficiency. In 2004, ExxonMobil posted return on capital investment of 23% [18], which is much higher than the return from average energy efficiency improvement projects such as cogeneration (with returns up to 15%) [19]. The situation in other firms is similar. Further, R&D on basic chemical processes, even if it leads to better energy efficiency and higher profits, does not necessarily increase the sales of basic chemicals, which are among the major criteria for investment [20]. Another interviewee remarked that due to the expected diminishing returns in the long term, spending on improving existing processes such as is unlikely to be extremely high or to increase forever [6]. The effect can be seen in a 2005 Solomon report that capital spending on energy conservation in US petrochemical plants (including both innovation and non-innovation) planned for the five-year-average level in 2004-2008 was reduced by 66% from that level in 1999-2003 [10].

Third, reliability and flexibility of existing configurations and operation control are considered more important than the potential benefits from improving energy efficiency. One example is adopting new high-performance furnace coils, which may cause problems in the other parts of the steam cracking processes (possibly incompatible with new coils) or product purity [11]. Another example is integration of refinery and petrochemical processes, which inevitably add complexity and reduce flexibility. Either linked directly or indirectly, an upset in one unit can lead to unexpected disruptions at the other unit (or even a shutdown of the whole plant) [21]. One additional example is that combined distillation and evaporation columns could slow down the throughput, then the benefits of energy efficiency can be reduced or offset [15]. In fear of such "hidden costs", plant operators are often reluctant to adopt any kind of process changes [21]. Economic loss due to disruptions and shutdowns are usually in the range of hundreds of thousands of dollars to millions of dollars per month for a plant [15]. Severe loss of human lives and liabilities of environmental damages that could be caused by leak and explosions can be devastating to any firm.

Explanations

Technologies, drivers and barriers to improving existing processes are summarized in Table 1. The links between drivers and barriers can be explained in terms of *rewards* and *risk*. *Rewards* refer to potential benefits as perceived, such as energy cost savings. *Risks* refer to the potential costs as perceived. Such costs also include opportunity costs, or failures to obtain rewards. The degree of a reward/risk ratio is perceived on the basis of knowledge on such rewards and risks.

For improving existing processes, technical and economic data are often already present or can be calculated. Therefore, improving existing processes are perceived as having moderate rewards and low risks. In other words, the reward/risk ratio is considered as being relatively well known. If the risks from the adoption of an innovative, energy-saving furnace, such as disruptions in the separation processes, are assessed and then are found to be small and controllable, such project would have a high chance to go ahead. Also, sometimes such risks are similar to every firm, or being "shared" or "equally distributed" [15]. This occurs, for instance, when most of large US firms recently are adopting cogeneration or new furnaces during revamps and in fact have similar risks. When the reward/risk ratios are known or similar among competitors, drivers such as energy savings can outweigh barriers and improving existing processes will go ahead. As a general remark by interviewees, given the importance of energy costs, energy efficiency improvement is carried out as a routine practice and is always a highly desirable and possible outcome of improving existing processes.

Developing New Processes

Drivers

First, the strongest driver for both producers and engineering firms is the possible economic gains from converting low-cost feedstock to high-value chemicals. One example of such feedstocks is so-called "stranded gas", which is found in remote locations in the Middle East or Eastern Europe where available natural gas far exceeds local demand but is too costly to be transported [22]. The cost of stranded gas is 50-80% cheaper than conventional feedstocks in $/GJ terms [23]. Such low cost of stranded gas is the main driver for UOP (a major engineering firm) to develop natural gas-to-olefins via methanol [24]. Another example of such feedstocks is heavy oils and $C_{4\text{-}9}$ byproducts from refinery and steam cracking that newly developed catalytic olefin processes produce high-value petrochemicals, which is about 50% cheaper than naphtha in $/ton feedstock terms [25]. An additional example is that production cost of coal to ethylene via methanol is about half that from weighted average US feedstocks in 1997-2005 [26]. These new processes were promoted on the ground of stabilizing profit performance in case of high relative prices of conventional feedstocks to alternatives. As said earlier, more than half of the total petrochemical production costs is feedstock cost.

Second, fierce competition among process licensors and engineering firms for new projects is also a strong driver to developing new processes. The survival of these firms, especially independent engineering firms such as ABB Lummus and UOP, primarily depends on contracts for new projects. It is widely expected that more new projects will be announced in the coming years. The global petrochemical capacity is expected to grow about 20% in 2001-2007 (based on known start-ups and planned projects; annually about 2%; all in physical terms), but global market demand for petrochemicals is expected to grow 25-40% (annually 3-5%) during the period. In particular, the annual demand growth of propylene is expected to be 5-6% while that of ethylene is 4% [27]. Since most new projects in the Middle East are ethane-based and will only produce ethylene, a possible propylene shortage or tight supply has been expected. Targeting at potential propylene projects, several firms have recently offered their own catalytic technologies with high yield propylene for commercialization. These processes, such as low-value feedstock based catalytic cracking, metathesis and methanol-to-propylene, are even referred to as propylene-on-purpose processes [28]. These firms intend to use new processes to compete with conventional processes (and among new processes themselves) on several frontiers: energy use, costs of capital investment and equipment, flexibility of feedstocks and products, effectiveness of catalysts, yields, length of construction time and maintenance costs [28, 29]. Our data show that catalytic cracking of heavy feedstock are also (besides its cheap feedstock advantage) promoted on the ground of better energy efficiency of 10-15% in comparison with the conventional steam cracking [25].

Third, knowledge spillover from catalytic processes to the non-catalytic steam cracking process helped engineering firms and producers to advance petrochemical processes. This effect was observed by both interviewees and our earlier techno-economic feasibility studies [25]. Many interviewees remarked that currently well-understood refinery processes, Fischer-Tropsch (natural gas, coal and biomass-to-syngas, fuels and basic chemicals), partial oxidation and related catalysis knowledge were either transformed or stretched into new processes that convert heavy or alternative feedstocks to basic chemicals. Those firms that are active in such field of new processes, such as those listed in Table 1, already possess extensive knowledge and experiences in refinery and catalytic processes in the past decades [24, 28, 30, 31]. We characterize such phenomena as "knowledge spillover".

Barriers

First, increasingly globalized supply-demand market has led many petrochemical producers to consider the future of business environment as being extremely volatile and unfavorable and, therefore, they feel very reluctant to invest in developing new processes. A strong concern among firms is the uncertain outlook of the world market with respect to demand and prices of feedstocks and products [32]. Market demand has been weak or growing slow in the US, Europe and Japan and the threat of overcapacity is constantly present. Today's business cycles become shorter and more volatile than what is normally known as the seven-year cycle, observed by one interviewee [30]. Global prices of naphtha and light olefins have fluctuated more than 100% (recently above 200%) since 1993. Profit margins fell by 2% per year in 1986-1996 reportedly because some new players from the Middle East and Asia have strong feedstock cost advantages. Also, there is constant fear for overcapacity. An actual increase of demand could be smaller than expected and could be easily met by de-bottlenecking or even by expanded refineries. For example, new processes that produce a single product, e.g. propylene and ethylene, are more vulnerable to market volatility than the conventional naphtha-based steam cracking with higher product flexibility [28]. One interviewee estimated that 95% of innovation failure is due to overcapacity and subsequent losses [33]. As said earlier, new processes cost billions of dollars of investment in supply logistics of new feedstocks and subsequent basic chemical processes. Given these high risks, even the advantages of low-cost feedstocks and expected market growth cannot easily convince producers to use new processes. Recently, firms in major industrialized countries have expanded relatively little while rather preoccupied with restructuring, consolidation and regulatory compliance. Only one or two new plants or major expansion projects in the US or Europe were announced in recent years. In fact, many big chemical firms have divested all or part of the petrochemical business from their portfolio–in favor of expanding in faster-growing, more value-creating businesses such as life science, transportation fuels and electricity related.

Second, conventional tools for making decisions and modeling processes are seen by producers and engineering firms as insufficient to deal with high uncertainties that are inherent in long-term development of new processes. It is well known that lack of sufficient tools can discourage firms from engaging in innovation. Conventional decision-making tools typically consist of extrapolative methods such as history-based forecasting, net present value analysis and utility theory [34]. These tools are suitable for analyzing relatively regular business cycles and then forecasting market demand for the coming five years. They typically give quantitative recommendations aiming for small rewards, such as improving existing processes. However, these tools are unable to address the complexities or the likelihood of market entry by new processes using alternative feedstocks in the long term [31]. Similarly, conventional process-modeling engineering tools often have problems to simulate the complexities of radically new processes e.g. cogeneration of petrochemicals, heat, electricity and fuels from coal [15]. In fact, it is even difficult to simulate a conventional process such as a distillation column to an advanced level [35]. Small firms, which often do not possess reliable and understandable software tools to (at least to some extent) compensate for the lack of operational experiences and economic certainty, will find it difficult to access the overall cost/benefit of any innovation project [35]. Things become even more complex when sensitive issues such as regulations and sustainability are involved, e.g. market deregulations, CO_2 emission costs and energy tax, according to a number of interviewees.

Third, concern for job security could be a significant barrier too. This is especially valid for producers due to the large R&D and plant investment costs involved. Engineers and managers at big firms have worked on natural gas-to-chemicals and fuels since the 1970s—all via the Fischer-Tropsch (FT) process. Major firms involved have spent over two billions dollars in R&D in the field since early 1980s. ExxonMobil claims to have spent over 600 million dollars and acquire over 3,500 patents worldwide on FT and gas-to-liquid technologies. Others firms, e.g. Shell, Sasol, ConocoPhilips, BP, Statoil, Rentech and Syntroleum, together have over 1,000 patents on FT processes. Managers and engineers from some big firms ignored a non-Fischer-Tropsch process ECLAIRS due to the fear of losing their jobs. Or, in other words, it is the familiar "not-invented-here syndrome", according to some interviewees [36]. According to another interviewee, however, is that the syndrome is often used as a marketing technique and those who cite such syndrome are concerned of their own jobs as well [31]. A similar observation is that executives find it safer to develop the same technologies as others are working on--if everyone is wrong, then no one is better than us [37]. The danger of losing jobs is indeed significant since the cost of innovation is high.

Explanations

Technologies, drivers and barriers to developing new processes are summarized in Table 1. Similar to improving existing processes, we explain the links between drivers and barriers in terms of rewards and risks. However, rewards and risks involved in this case differ from those in improving existing processes in three aspects. First, the magnitude and content of rewards and risks involved in developing a new process is much greater in comparison with those involved in improving the existing processes. Rewards and risks involved in developing a new process, or the reward/risk ratios, are relatively unknown. Second, rewards and risks involved in developing a new process in fact concern innovation strategies and scenario planning of a firm. Essentially, developing new processes is of strategic importance. Third, in contrast to improving existing processes, energy efficiency is not always considered as a top relevant issue for developing new processes. New processes compete over relative feedstock costs, not over energy efficiency. For a new synthetic pathway using a new feedstock (e.g. methane or coal), the specific pathways and relative feedstock economics are the first issues to be considered.

Conclusion

Improving existing processes and developing new processes are subject to different sets of drivers and barriers respectively. They are related to each other, but for different activities they differ in terms of knowledge on the reward/risk ratio. Improving existing processes is rather a routine matter while developing new processes is of strategic importance. In both cases, energy efficiency is a desirable and possible outcome, especially in the case of improving existing processes. However, because its importance is determined by a wider technological and economic context, it is in itself not necessarily a top objective for innovation of new chemical processes. For developing new processes such as natural gas and coal-to-chemicals, specific synthetic pathways and relative feedstock economics are considered first.

Based on these findings, we propose three points for policymakers to consider. First, the relative importance of energy efficiency improvement for the innovation of basic chemical processes indicates that firms have many more considerations other than just energy cost savings. Without taking these important considerations into account, policymakers will find it difficult to stimulate energy efficiency improvement. Second, the strong interest in low-cost, alternative feedstocks and knowledge spillover-resulted new processes should be seen as an opportunity for battling climate change. Many opportunities exist for clean use of recycled waste, renewable biomass feedstocks and alternative fossil resource (including coal). Governments should sponsor and coordinate R&D partnerships among current users and suppliers of alternative feedstocks/processes to improve the reward/risk ratios of new processes. Third, external costs (i.e. costs of CO_2 emissions and sequestration) were not found to be a driver for innovation. Policymakers and firms alike should explore the business opportunities and consequences of adding external costs as part of scenario planning in major chemical firms.

References

1. Albach, H., et al., Innovation in the European Chemical Industry. (1996), Wissenschaftszentrum Berlin: Final report and discussion paper (FS IV 92-26) prepared by Research Area Market Processes and Corporate Development for the European Commission DG XXIII-D-4, "Innovation and Technology Transfer".
2. Fleischer, M., S. Kelm, and D. Palm, Regulations and Innovation in the Chemical Industry. The Impact of EU Regulation on Innovation of European Industry. (2000), Institute for Prospective Technological Studies at the European Commission DG Joint Research Center: Seville.
3. Dijkema, G., Process system innovation by design: towards a sustainable petrochemical industry (PhD thesis), M. Weijnen and J. Grievink, Editors. (2004), Delft University of Technology.
4. Patel, M., et al., Medium and long-term opportunities and risks of the biotechnological production of bulk chemicals from renewable resources (The BREW Project). (2005), Department of Science, Technology and Society at the Utrecht University: Utrecht, The Netherlands.
5. OECD, Oslo Manual, Proposed guidelines for collecting and interpreting technological innovation. (1997), Paris: OECD/Eurostat.
6. Anonymous-1, (2005), Personal communication, R&D department of a major petrochemical company located in the Westhollow Technology Center, Houston, Texas, April 18, 2005.
7. Theising, T., (2005), Personal communication, BASF, Freeport, Texas, April 19, 2005.
8. Burchmore, S., N. Ayscough, and A. Gendebien, Techno-Economic Study on the Reduction Methods, Based on BAT of Industrial Emissions (Air, Water, Wastes) from the Basic Petrochemical Industry. (1993), European Commission.
9. ChemSystems, Process Evaluation/Research Planning: PERP 2002/2003 Program. (2002), ChemSystem/Nexant Inc.: Houston, Texas.
10. Cagnolatti, C. Global Ethylene Plant Performance Comparison. in 2005 AIChE Spring National Meeting and 17th Annual Ethylene Producers Conference. (2005). Atlanta: American Institute of Chemical Engineers (AICHE).
11. Anonymous-2, (2005), Personal communication, Process control engineers at a major petrochemical company in the US Gulf Coast, Baytown, Texas, April 26, 2005.
12. Ferland, K., (2005), Personal communication, Texas Industries of the Future at the University of Texas, Austin, Texas, May 2, 2005.
13. Zarker, K., (2005), Personal communication, Pollution Prevention and Industry Assistance at Texas Commission for Environmental Quality, Austin, Texas, May 3, 2005.
14. CEN, Employment: Another Decline. Chemical & Engineering News. Also available at www.cen-online.org, (2005)(July 11).
15. Burchmore, S., N. Ayscough, and A. Gendebien, Technico-Economic Study on the Reduction Methods, Based on BAT of Industrial Emissions (Air, Water, Wastes) from the Basic Petrochemical Industry. (1993), European Commission.
16. Tallman, M., (2005), Personal communication, KBR, Houston, Texas, May 5, 2005.
17. Thornley, S., (2005), Personal communication, ABB Lummus, Houston, Texas, April 26, 2005.

18. ExxonMobil, Annual Report 2004. See also at www.exxonmobil.com.corporate/files/corporate/ar_2004_chemical.pdf, E. Chemical., Editor. (2005).
19. Deason, D., (2005), Personal communication, ExxonMobil, Baytown, Texas, April 25, 2005.
20. Strohrman, M., (2005), Personal communication, BASF, Ludwigshaven, Germany, April 16, 2005.
21. Herzog, C., (2005), Personal communication, Mustang Engineers and Constructors, Atlanta, Georgia, April 12, 2005.
22. Burke, B. Emerging Technologies to Develop Stranded Gas. in 2001 Annual Meeting of NPRA (National Petroleum & Refinery Association). (2001). New Orleans: NPRA.
23. Ren, T., M. Patel, and K. Blok. Steam Cracking and Natural Gas-to-Olefins: A Comparison of Energy Use and Economics. in 2005 Spring National Meeting of American Chemical Engineering Association (AIChE). (2005). Atlanta: New York: American Institute of Chemical Engineers.
24. Vora, B., (2005), Personal communication, UOP LLC, Des Plaines, Illinois, August 11, 2004 and September 27, 2005.
25. Ren, T., M. Patel, and K. Blok, Olefins from conventional and heavy feedstocks: Energy use in steam cracking and alternative processes. Energy, (March 2006). 31(4): p. 425-451.
26. D'Auria, J. The Role of Innovation in USGC Competitiveness. in UH-GEMI Conference on The Future of the Gulf Coast Petrochemical Industry. (2005). Global Energy Management Institute (UH-GEMI) at the University of Houston in Houston.
27. SRI. The Chemical Industry and Technology Opportunities in China. in The 9th Annual China Chemical Industry Conference. (2003). Shanghai, China: Menlo Park, California: SRI International Consulting Co.
28. Laghate, A., (2005), Personal communication, Technip, Zoetermeer, The Netherlands, April 13, 2005.
29. Blom, H., (2005), Personal communication on November 1, 2005, Shell Global Solutions, Amsterdam,
30. Richman, P., (2005), Personal communication, UOP LLC, Houston, Texas, April 27, 2005.
31. Calis, H., (2005), Personal communication, Shell Gas & Power, Amsterdam, The Netherlands, June 14, 2005 and July 14, 2005.
32. Arbogast, S. A Conference Report on the Future of the Gulf Coast Petrochemical Industry. 2005 Conference on the Future of the Gulf Coast Petrochemical Industry. (2005). Houston.
33. Madden, J., (2005), Personal communication, Vice President of Chemical Marketing Resources, Houston, Texas, April 18, 2005.
34. Schoemaker, P., Profiting from uncertainty. (2002), New York: The Free Press. 9.
35. Anonymous-3, (2005), Personal communication, Research and Technology Center of a major petrochemical company, Amsterdam, Netherlands, March 22, 2005.
36. Hall, K., (2005), Personal communication, Professor, Dep. of Chemical Engineering at A&M University, College Station, Texas, May 4, 2005.
37. Clark, W. Internal Growth and Risk Taking in the Chemical Industry. in R&D Meets M&A, 2003 Conference on Innovation and Creativity in Chemical R&D. (2004). Philadelphia: Chemical Heritage Press.

Table 1 Barriers and Drivers to Innovation Activities

	Improving Existing Processes		Developing New Processes	
Commercial Status	Widely commercialized; very often implemented	Mostly commercially available; some implementations	Mostly commercially available; rarely implemented	Some commercially available; a few implementations completed or planned
Feedstocks	Conventional feedstocks from crude oil and ethane/propane/butane-rich natural gas or petroleum gases		Conventional feedstocks and low-value feedstocks from crude oil	Alternative feedstocks (e.g. methanol from natural gas or coal)
Potential for process energy saving	5-15% saving on conventional processes	10-20% saving on conventional processes	Up to 20% savings on conventional processes	Methane-based processes use twice as much energy and coal/biomass-based ones use four times as much as oil-based ones
Actors and technologies that are going through various stages of an innovation procedure	Lockwood Green for BASF/Atofina: cogeneration at a steam cracker site GE/ExxonMobil: residue gasification for cogeneration; Air Liquide for Dow and Shell: combined-cycle cogeneration ASPEN Tech: real-time plant control and energy cost management system UOP: integration of petrochemical and refinery complex	Hamworthy: burner design FM: coated furnace coils Linde: short coil tubes ABB Lummus: online decoking; gas turbine integration Stone & Webster: advanced product recovery gas turbine integration KBR: short residence time Technip: distillation schemes Shell/ECN: integrated distillation columns for separation	ExxonMobil, Shell, ABB Lummus, KBR, UOP, DSM Linde and S&W/Sinopec: low-value hydrocarbons (e.g. $C_{4\text{-}9}$ refinery or steam cracking byproducts) to petrochemicals in FCC-like reactors at lower temperatures than steam cracking; KRICT, LG, AIST, VNIIOS and ASAHI: catalytic cracking of conventional, C_4/C_5 streams and heavy feedstocks Dow and ExxonMobil: catalytic oxidative dehydrogenation of ethane and propane BP/MTR: membrane separation	ExxonMobil, BP, ConocoPhilips and Shell: via syngas to liquid fuels and petrochemical feedstocks UOP: via methanol to olefins Lurgi: via methanol to propylene and fuels DSM: via ethane; partial oxidation; ethylene and electricity cogeneration Synfuel/A&M University: via acetylene; partial oxidation; ethylene or fuels Bechtel: coal to fuels, electricity and petrochemical feedstocks Eltron: ethylene and electricity in a fuel cell
Drivers	Process energy cost saving Tight supply of gas feedstocks Personal commitment		Low-cost feedstocks Competition among process licensors and engineering firms Knowledge spillovers	
Barriers	Shortage of staff and time Competition from other prioritized projects Existing configuration		Unfavorable business environment Insufficient tools for decision making and process modeling Concerns on job security	

133b Energy Savings for Refinery and Petrochemical Industries

Robin Smith, Centre for Process Integration, The University of Manchester, Sackville Street, PO Box 88, M60 1QD, Manchester, United Kingdom

The state-of-art in process integration will be presented for energy conservation in refining and petrochemical industries. The presentation will not only review well-established ideas, concepts, techniques, and methodologies of process integration in the area of energy savings, but also provide recent developments, extensions and applications in the design and retrofit of heat-integrated refinery and petrochemical systems. The presentation will cover how process integration methods can be applied to achieve high energy efficiency and low emissions by providing the optimal set of operating conditions and/or configurations when increased throughput, changes in feedstocks, and changes in product specifications are to be accommodated.

133c Energy Integration - a Fresh Look for a Changed Energy World

Bill Townsend, Alan Ryder, and Cyril Collins. Petrochemicals & Advanced Chemical Engineering, M W Kellogg Ltd., Kellogg Tower, Greenford Road, Greenford, UB6 0JA, United Kingdom

The fundamental world shift in energy prices has a structural basis, and leading industry analysts are forecasting a long term oil price above $40/bbl. At the same time, environmental pressure to reduce CO2 emissions is growing steadily. Consequently the incentive to invest in technology to reduce energy consumption in process and energy plant has also fundamentally shifted, and a fresh look at capital schemes for energy conservation is warranted. However, this is against a background of the 'low hanging fruit' energy schemes having already been 'plucked' over the last 10 years.

Consequently there are two strategic directions in which it is logical to look for new energy projects. The first is to look to new ideas for heat integration schemes between adjacent processes and sites. An example is the use of refinery waste heat to evaporate LNG, substituting fuel firing in the LNG terminal's vaporisers. M W Kellogg Ltd. has developed a study and design methodology for such schemes and demonstrated their economics. The approach and results are presented in outline, and implementation issues discussed against the background of the operational difficulties perceived independently by the two centres.

The second strategic direction is to re-visit those 'old' ideas which in the past have been discounted on the grounds that they were uneconomic, but which in the new energy world could pay out. These can be given added incentive by adding new design ideas. Examples include the use of heat pumping in above-ambient distillation schemes, and some dividing wall column applications. A worked example is presented in which heat pumping and use of a dividing wall column for C4 isomer separation are re-visited as energy-saving alternatives.

133d Combined Energy and Water Analysis for the Oil Sands Industry

Alberto Alva-Argaez, Luciana E. Savulescu, and Abdelaziz Hammach. CETC-Varennes, Natural Resources Canada, 1615 Lionel-Boulet Blvd, Varennes, QC J3X1S6, Canada
In oil sands processing water and energy are closely interrelated. Water is the main carrier of energy either as hot water or steam. The various site configurations that are in place or in the planning stages for the production of synthetic crude oil in the Canadian oil sands face several technological challenges, one of which is reduction of water and energy demands. Integration between steam and power systems has been shown to bring about economic and efficiency benefits for various industrial sites. Further expansion of the system boundary to include the site water system should also reveal interesting opportunities for improvement. The links between steam generation, cooling water demands and process water should be explored in more detail as these have usually been evaluated independent from each other. CETC-Varennes is working in extending the boundaries of the utility system in order to produce a comprehensive understanding of the trade-offs involved in the design and operation of an efficient system. In a number of scenarios it is possible to reduce energy consumption and water simultaneously with significant economic gains from greenhouse gas emissions reductions and fuel savings.

Paper No. TA017/05

Enhanced Heat Transfer Technology Application in Crude Units for Saving Energy

Kirtan K. Trivedi and Tom Rudy
ExxonMobil Research & Engineering Company
3225 Gallows Rd.,
Fairfax, VA 22037
Phone: (703) 846-2179
Fax: (703) 846-7906 (office)
Fax: (262)-313-4675 (fax received on email account)
Internet Email: kirtan.k.trivedi@exxonmobil.com

Presented at AIChE 2006 Spring National Meeting April 23-27, 2006
Orlando, FL

Session: Energy Conservation

9th Topical Conference on Refining Processing

Unpublished

Extended Abstract

A number of applications of enhanced heat transfer (EHT) have been identified that have enabled energy savings with attractive discounted cash flow (DCF) returns. Enhanced heat transfer technology is maturing with a significant increase in applications identified, or such applications are in the process of being installed, or are already installed. This paper discusses a systematic way to identify exchangers in a Crude Unit where application of appropriate enhanced heat transfer technology will result in an increase in the fired heater coil-inlet temperature and thus result in reduced energy consumption.

The heat transfer duty of an existing exchanger can be improved by increasing the UA value of the heat exchanger i.e. increase the overall heat transfer coefficient or increase the area of the heat exchanger or both.

Multiple technologies are available for heat transfer enhancement. These include enhancements to tube side, shell side, or both simultaneously in a shell and tube type heat exchanger. Inside the tube can be bulk mixing devices such as twisted tapes and coiled wires, or near-wall devices such as springs or fins. Outside of the tube can be fins or special baffle designs. Other configurations may twist the tubes or apply enhancement to both sides. Additionally other types of exchangers such as plate or spiral heat exchangers are also available.

For a given network topology, it is possible to generate a relation between the % increase in UA of an exchanger and its impact on coil inlet temperature (CIT) to the furnace. UA sensitivity tables can be generated using the UMIST SPRINT program. Comparing the UA sensitivities for all the exchangers will identify the candidate heat exchangers that can significantly increase the CIT by increasing the UA value of that exchanger.

For these heat exchangers, application of appropriate EHT technology can be screened by incorporating the appropriate heat transfer and pressure drop correlations in rigorous heat exchanger simulators.

Enhanced Heat Transfer Technology Application in Crude Units for Saving Energy

AIChE Spring National Meeting, Orlando, FL
April 23-27,2006

Kirtan Trivedi & Tom Rudy
ExxonMobil Research and Engineering Company,
Fairfax, VA

ExxonMobil

Background

- Crude Units are one of the largest consumers of energy in a refinery
- Low cost techniques required to enable additional heat recovery
- Heat transfer enhancement techniques enable heat recovery at low cost
  - EHT can be installed as a new bundle into an existing shell
- Low cost retrofit of Crude Unit Heat Exchanger Networks require
  - Minimal topology changes
  - Minimal downtime

ExxonMobil

Enhanced Heat Transfer

- Current belief
  - Additional pressure drop
  - Reduce surface area
- Recent developments
  - Enhancement is obtained with little or no pressure drop increase
  - Sometimes it is possible to achieve surface area increase and pressure drop reduction
- Other benefits
  - Capital cost reduction (grass roots projects)
  - Low capital requirement (revamp projects)
  - Improve exchanger performance in terms of operability, mitigation of fouling and improved flow distribution within exchangers

ExxonMobil

Enhanced Heat Transfer in Crude Units

- Attractive for revamp
- Most appropriate application of EHT is on the side of high thermal resistance
- Typically occurs on the tube side where the crude is flowing
  - Low film heat transfer coefficient and fouling
- Ideal application for tube inserts
  - may increase pressure drop but average pressure drop over run length is likely to decrease
  - Easy implementation during exchanger cleaning
  - No facilities modification

ExxonMobil

Available EHT Technologies for Crude Unit Applications

- Twisted Tapes
- Static Helically Coiled Wires e.g. TOTAL "FIXOTEL"
- Wire Matrix Inserts e.g. HiTran
- Flexible Metal Springs e.g. SPIRELF
- Low-finned (Integral fin tubes)
- Twisted tubes
- Helical Baffles
- Rod Baffles

ExxonMobil

Enhancement Techniques

- Exchanger heat transfer duty can be increased by
  - Increasing heat transfer coefficient
  - Increasing area
  - Both
  - Decreasing fouling
- For Crude Units
  - Heat transfer normally controlled by tube side film coefficient and fouling
  - Additional area that can be added to an existing shell is limited
  - Enhancement can most likely be obtained by increasing heat transfer coefficient and decreasing fouling

ExxonMobil

Sensitivity Analysis

- Hohmann
- Kotjabasakis, E. and Linnhoff, B.
- Kays and London
- Pretty, Gibbard & Polley
- Based on effectiveness-NTU approach

$$Q = CP_c(t_2 - t_1)$$

$$Q = CP_h(T_1 - T_2)$$

$$Q = F_{t_{min}} UA(\Delta T_{LMTD})$$

$$(1 - RB)T_2 + (B - 1)Rt_1 + (R - 1)T_1 = 0$$

$$(1 - RB)t_2 + (B - 1)T_1 + (R - 1)Bt_1 = 0$$

$$R = \frac{CP_c}{CP_h}$$

$$B = \exp[(UA / CP_c)(R - 1)]$$

ExxonMobil

Example – Sensitivity Analysis

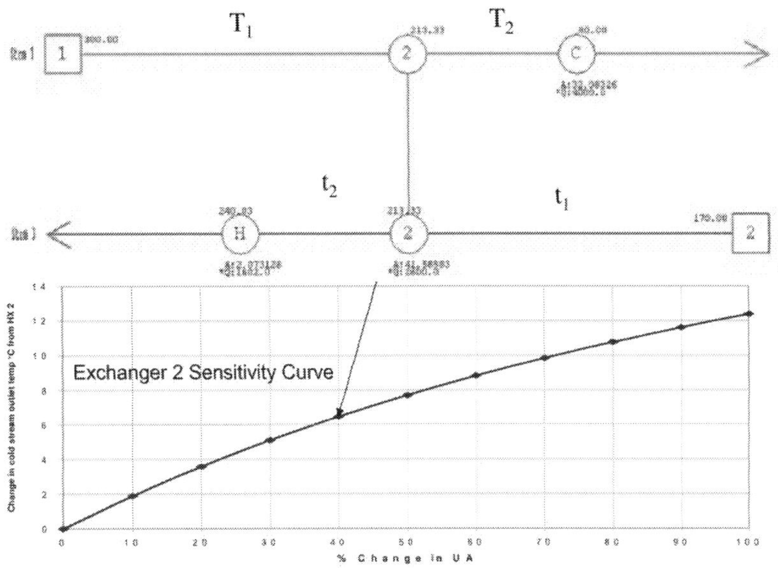

ExxonMobil

Example – Sensitivity Analysis

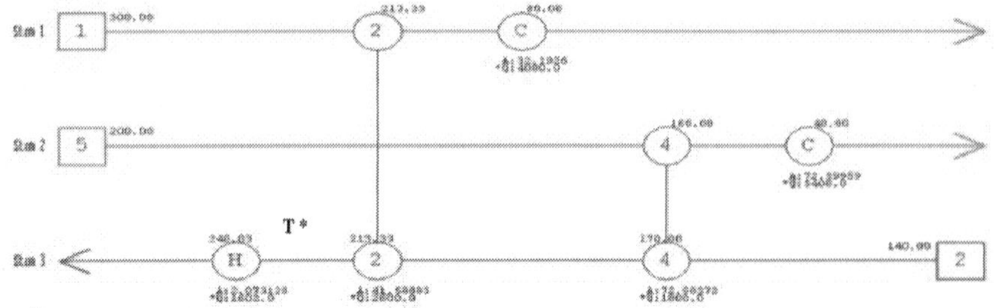

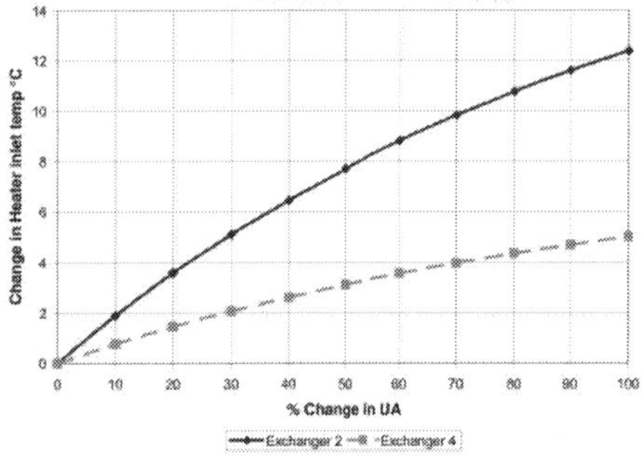

- Assume 50% change in UA of Exchanger 2 and 4
- Change in heater inlet temperature
  = 8 + 3 = 11°C

ExxonMobil

Comments on this approach

- Limitations
  - F_{tmin} = 1 and all exchangers are assumed to be pure counter current
  - Heat transfer equations for heat exchangers are non-linear
  - Impact of change in UA is additive
- Objective is to do a 80-20 type analysis
  - Quickly identify energy recovery targets
  - Likely candidates for application of enhanced heat transfer
  - This approach may result in 10 % errors
  - Good enough for doing scoping / screening analysis
  - Confirm attractive options later with rigorous simulation

ExxonMobil

Case Study

- Simple Crude Unit
- Identify candidate exchangers for application of EHT
- Max UA increase 40%
- Total Duty of PA circuits is fixed
- Bottom PA circuit does not have any coolers. E5 is the only exchanger
- All other hot streams have coolers
- Data taken from Zhu, Zanfir, and Klemes, "Heat Transfer Enhancement for Exchanger Network Retrofit", Heat Transfer Engineering, 21, 7-18, (2000)

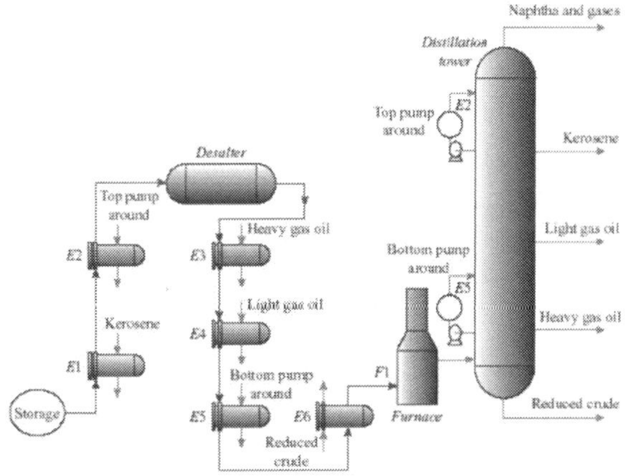

ExxonMobil

Grid Diagram for the Crude Unit

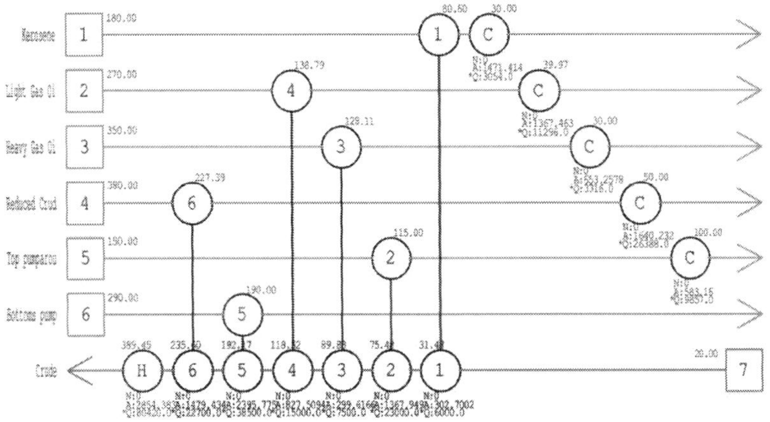

ExxonMobil

UA Sensitivity Plots

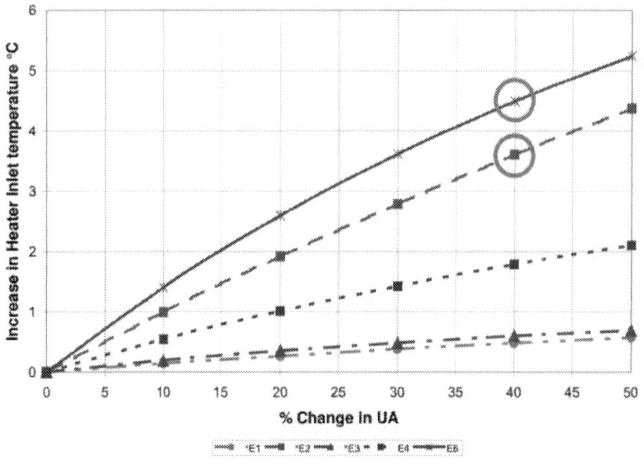

- Increasing the UA of E2 and E6 by 40% will increase the heater inlet temperature by 3.6°C and 4.5°C respectively
- Increasing the UA of E1, E3 and E4 is not attractive
- Increasing UA of E2 will increase the crude temperature to E5 and thus will decrease the heat recovery in E5 and hence the BPA outlet temperature will increase

ExxonMobil

UA Sensitivity for E2 and E5

- Increasing UA of E2 by about 40% will result in increasing the outlet temperature from E5 by about 5°C. Hence we need to increase the UA value of E5 by about 10% to restore the outlet temperature to the original value
- Increasing the UA value of E2 and E6 by 40% and E5 by 10% the heater inlet temperature can be increased by about 8°C or about 4200 kW or 5% of the current duty
- Simulation within SPRINT shows the heater inlet temperature is increased by 10.5°C and the energy savings is 4920 kW or 6%

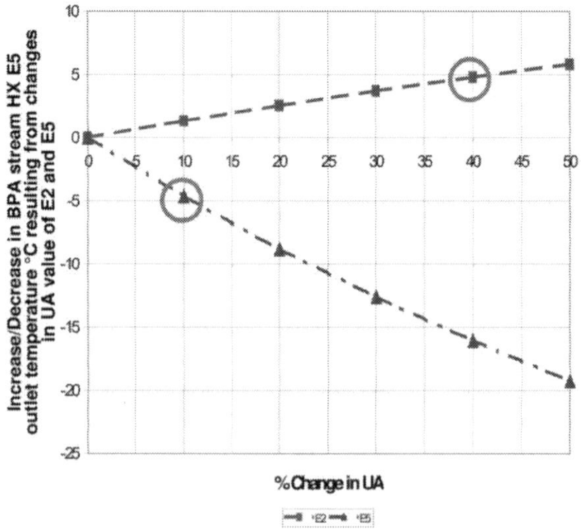

ExxonMobil

Comparison of Network Performance – Before and After revamp

		Hot Stream		Cold stream			
Exchanger No	Description	Tin °C	Tout °C	Tin °C	Tout °C	Duty	UA kW/°C
E1	Kero/Crude	180.0	80.6	20.0	31.5	6,000	61.18
		180.0	*80.6*	*20.0*	*31.5*	*6,000*	*61.18*
E2	TPA/Crude	150.0	115.0	31.5	75.5	23,000	291.38
		150.0	*106.8*	*31.5*	*85.8*	*28,394*	*407.90*
E3	HGO/Crude	350.0	128.1	75.5	89.8	7,500	57.75
		350.0	*136.2*	*85.8*	*99.6*	*7,227*	*57.94*
E4	LGO/Crude	270.0	138.8	89.8	118.5	15,000	165.26
		270.0	*145.7*	*99.6*	*126.8*	*14,214*	*166.01*
E5	BPA/Crude	290.0	190.0	118.5	192.2	38,500	458.53
		290.0	*190.0*	*126.8*	*200.5*	*38,500*	*509.29*
E6	Reside/Crude	380.0	227.4	192.2	235.6	22,700	293.36
		380.0	*219.5*	*200.5*	*246.1*	*23,874*	*405.50*
Heater						80,420	
						75,500	

- Duty of E3 and E4 decreases. LGO and HGO streams go to storage tank and the new temperature needs to checked if it is acceptable. If not modifications will be required to these exchangers
- Use EHT scoping tools to select the right technology

ExxonMobil

Conclusions

- Simple procedure to identify exchangers that can be candidates for application of Enhanced Heat Transfer Technology
- 80-20 rule followed
  - Potential errors acceptable for the scope of the procedure
- Encourage exploring the application of this procedure to revamp of complicated Crude Unit Networks
- Potential application for analyzing impacts of fouling and optimizing cleaning cycles for heat exchangers in a Crude Unit pre-heat train

ExxonMobil

133f Extraction of Benzene from Wastewater Using Refinery Liquids by L-L Extraction Instead of Distillation

Ashok V. Naimpally and Marcia G. Zimmermann. Chemical Engineering, California State University, Long Beach, 1250 Bellflower Blvd, Long Beach, CA 90703
Benzene in waste water is a major problem at refineries and cannot be released to the county treatment plants.

This paper will show how to separate benzene from water using L-L extraction , using the extractant solvent alkylate, a a common liquid hydrocarbon produced in refineries.

Data has been collected and plotted on a ternary-diagram. The solvent alkylate is treated as a pseudo-one component system for this purpose.

A typical computational exercise is provided, showing the amount of benzene that can be extracted and the savings obtained over an energy-intensive process like distillation.

DEBOTTLENECKING OF HEAT EXCHANGER NETWORKS USING OPTIMUM PRESSURE DROPS

Mohammad Hassan Panjeshahi, Nassim Tahouni,
Department of Chemical Engineering, University of Tehran, Iran

Abstract

Process Integration technology is now widely applied in grass-roots design, energy saving retrofit and debottlenecking of Heat Exchanger Networks. This technology has been used in a variety of industries and approved to be reliable and applicable in engineering design. Debottlenecking may apply to a specific part or entire unit, whether it is due to increased throughput or process modifications.

One of the advanced methods for debottlenecking, which is currently used, is based upon fixed allowable pressure drops, through which a retrofit can be achieved without a need for pump and/or compressor replacement. However, this method does not consider optimum pressure drops for the process streams, and hence capital investment for new exchangers, operating cost of the network and final topology would not be optimum;

This research is trying to resolve these problems, by development of a new procedure for pressure drop optimisation in debottlenecking. This procedure enables the designer to study pump and/or compressor replacement whilst at the same time optimizing additional area and operating cost of the network. A new economic function has been introduced, which is the summation of heat exchanger area investment, pump capital cost and power consumption. This generalized investment function should be minimized to obtain optimum stream pressure drops.

The new procedure has been effectively applied to a Crude Oil Pre-heat Train, which was subject to some 20% increase in throughput, and the corresponding results proved to be accurate enough. Namely, the results of targeting stage showed a very good agreement with network synthesis and final detailed design.

Introduction

The world population is increasing and energy consumption paradigm is changing as well. On the other hand, establishing new plants is very expensive. Hence, increasing the capacity of existing plants and removing the subsequent bottlenecks is considered as the best way to maximize profitability of the plants.

Generally, the best debottlenecks are those, which avoid the purchase of additional capacity for the more expensive items of equipment (furnaces, compressors, refrigeration units, etc.) through the manipulation of the heat recovery system.

The design of heat recovery networks is generally concluded in a number of stages. First, there is a targeting stage where the economics of heat recovery are evaluated in order to set the recovery level. This involves the trading-off energy and network capital costs. The second stage involves network synthesis. Here the topography of the heat recovery system required to realize the targeted energy recovery is determined. Finally, the detailed design stage is entered. Here the detailed design of the exchangers forming the recovery system is undertaken.

Delay in considering pressure drops in targeting and design of heat recovery networks can lead to serious mistakes in both grass-root and retrofit designs. These mistakes can be summarized as follows:

1. Inconsistency between network synthesis and detailed design, with respect to surface area of the exchangers;
2. Incorrect capital-energy trade-off and network optimization;
3. Misleading targeting results during retrofit projects.

It has, therefore, made clear that stream pressure drops should be considered during network targeting and synthesis, but existing methods do not consider optimum pressure drops for the process streams, and hence suffer from the following drawbacks.

1. Capital investment for new exchangers would not be optimum;
2. Operating cost (i.e., Hot utility, cold utility and Electricity) of the network tends to be sub-optimum;
3. Wrong targeting results cause wrong design initialization and hence wrong network topology.

Therefore, in addition to considering pressure drop instead of film heat transfer coefficient, we have to make some decisions regarding pressure drop optimization. So, pumps and/or compressors are taken into account as well as heat exchangers.

This research deals with the problem of optimal debottlenecking of heat exchanger networks considering minimum total cost.

New Method to Target Using Optimum Pressures

Debottlenecking retrofit differs from energy saving retrofit in a number of important ways. The first difference is, of course, in the objective of the exercise. In debottlenecking, increased throughput is often not only the principal objective but an overriding one. The second difference is in the economic criterion applied. This is generally just an investment ceiling. Project payback is not generally considered.

Definition of New Targeting Function

It is clear that during pressure drop optimization, the stream pressure drops should be addressed as key variables and optimized according to unit specifications. The more stream pressure drops, the more turbulence the higher the film heat transfer coefficients. When heat transfer coefficients increase, the required heat exchanger area installed over that stream reduces, and therefore the purchasing capital cost of new exchangers reduces. But, on the other hand, new pumps and/or compressors investment and running cost increase due to higher pressure drops. Hence, there would be a trade-off between the capital cost associated with the new heat exchanger surface areas and capital cost plus additional power cost associated with pumps and compressors.

In this research, an economic objective function is introduced as following:

Total cost = $f(\Delta P_1, \Delta P_2, \ldots, \Delta P_n) = CC(\Delta A) + CC(\Delta P) + NPV(\Delta P)$

= Additional Area + New Pumps/Comp. Cost + Pumps/Comps. Running Cost

Where:

$CC(\Delta A)$: Capital investment for new exchangers.
$CC(P)$: Capital cost for purchasing new pumps & compressors.
$NPV(\Delta P)$: Net Present Value, Operating cost for the pumps and compressors.
Note that this function is independent of ΔT_{min} due to fixed energy consumption. Now, the objective function should be minimized, in order to optimize stream pressure drops. There are two possible cases:

a) When we use bigger pumps instead of existing pumps.
b) When we replace existing pumps with smaller ones.

For optimization in case a, we face with a 2-way trade-off between network additional area investment on one hand and capital and operating cost of the pumps on the other hand.
i.e.: $[CC(P) + NPV(\Delta P)] \Leftrightarrow CC(\Delta A)$
But, in case b, we have a trade-off as below:
$[CC(P) + CC(\Delta A)] \Leftrightarrow NPV(\Delta P)$

It is necessary to say that the minimum value of objective function is found by direct searching methods.

Determination of Retrofit Curve

The stages of newly established method in targeting for debottlenecking are as follows:

1. simulation of the existing process operating at the desired increased throughput. It will often be found that the required process temperatures are not met. It is then assumed that additional utility is used to restore the temperatures.
2. Specifying the area efficiency in existing network after increasing throughput using area-energy plot
3. Calculating the pseudo network area
4. Determination of optimum targeting point
5. Evaluation of required additional area

Area-energy plot is shown in Figure 1. According to this figure, if targeting point has exactly the same position as point B,

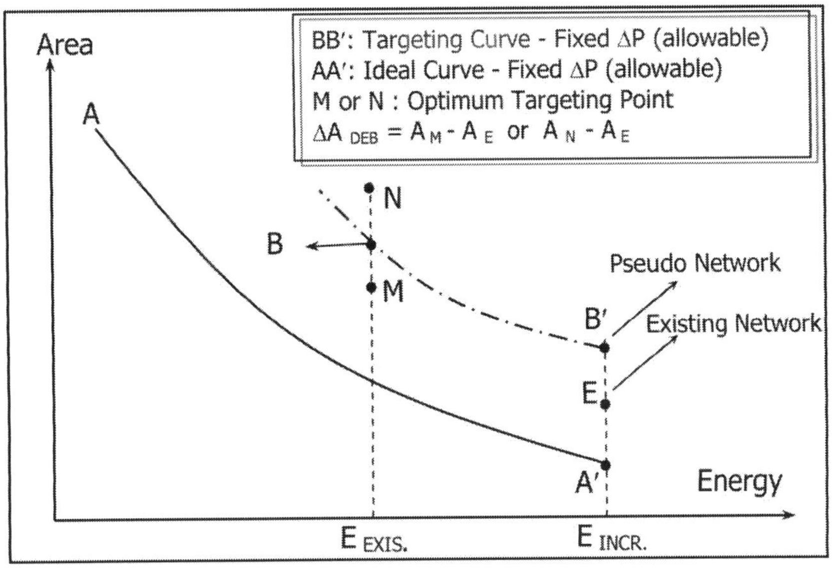

Figure 1. Evaluation of Required Additional Area Based on Optimum Stream Pressure Drops

- We do not need to invest for new pumps.
- We do not need to pay for additional running cost (i.e.: electricity).
- Required additional area is the same as fixed pressure drop situation.

If targeting point is located below point B (for example in point M):

- We need to buy bigger pumps.
- We have to pay for additional electricity cost.
- The additional heat exchanger area would be less than before (i.e.: fixed-Δp situation, $\Delta A_{DEB} = A_M - A_E$).

If targeting point is located above point B (for example in point N):

- We need to replace our pumps with smaller ones.
- Consequently, we have to pay less running cost.
- However, the required additional area would be more than fixed-Δp situation ($\Delta A_{DEB} = A_N - A_E$).

Case Study

The case study is the Crude Pre-heat Train. As can be seen in Figure 2, the crude oil feed stream is pre-heated in two sections by interchange with the hot fractions resulting from the distillation columns. The first section runs from storage to a desalter unit and the second from the desalter to the crude tower. Process heating is provided by a furnace, which pre-heats the crude into the crude tower.

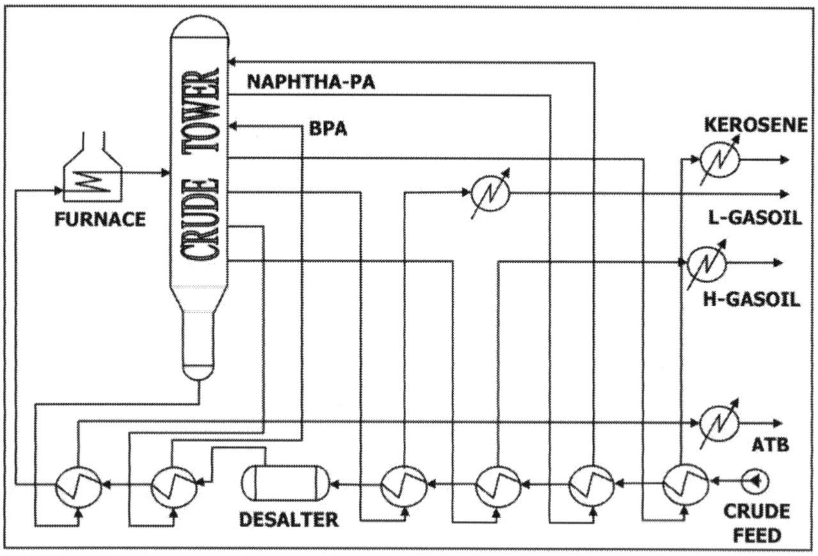

Figure 2. Flowsheet for the Crude Pre-heat Train

The heat exchanger network shown in flow sheet form is reproduced in grid form in Figure 3, with stream temperatures, match heat loads and specific heat mass flowrates marked. The hot streams are grouped together in the top part of the grid and numbered from H1 to H6. The only cold stream is drawn below and numbered C1.

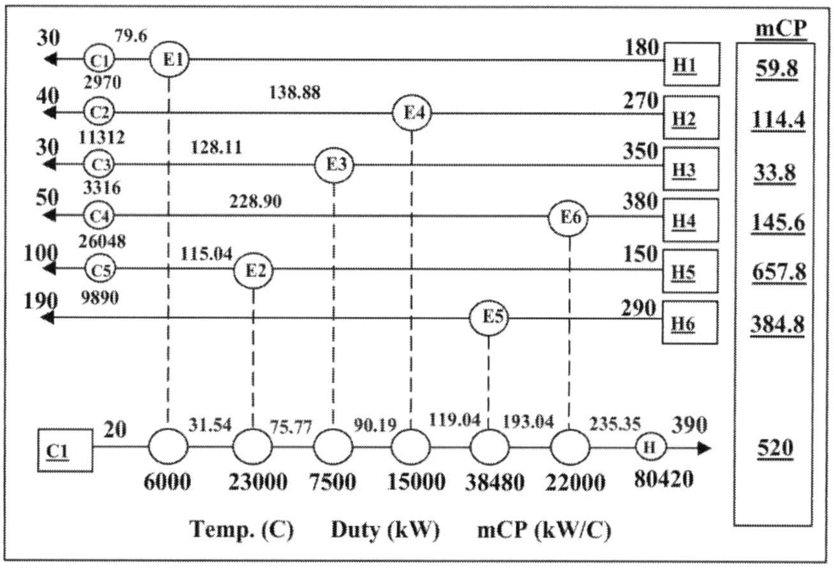

Figure 3. Existing Network for the Crude Pre-heat Train

Now, we would like to investigate this case study subject to some 20% increase in throughput.

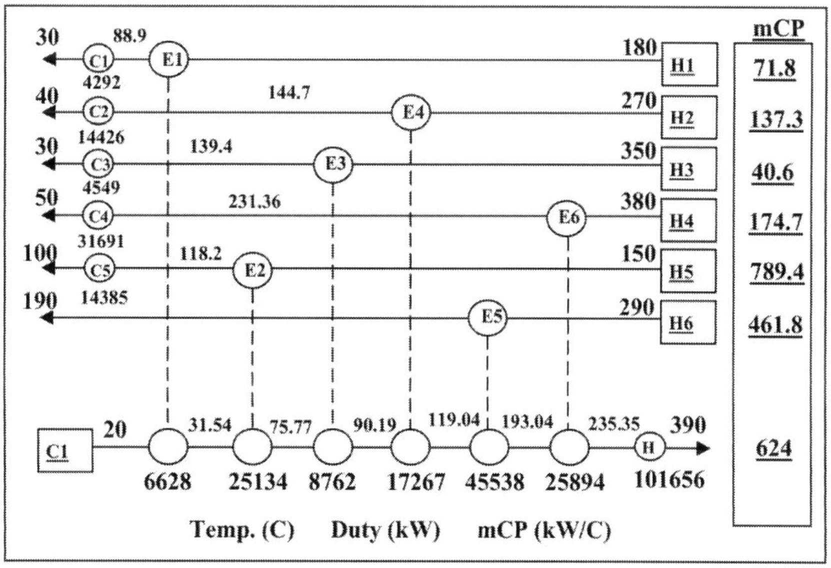

Figure 4. Existing Network for the Crude Pre-heat Train after Increasing Throughput

In order to estimate the stream pressure drops available for heat transfer, we have to analyze the performance of the existing exchangers after increasing throughput. The analysis used an exchanger-rating program based on Bell-Delaware method. The new data for exchangers and streams have been shown in Table 1 and 2. Having obtained exchanger output temperatures, loads and heat transfer coefficients the existing network after increasing throughput has been drawn in Figure 4.

As it can be seen, the load of heater is higher than the maximum capacity and hence, creates a bottleneck. Now, it is time to evaluate the required additional area for debottlenecking of existing network, and its associated capital cost.

Table 1. Performance of Exchangers after Increasing Throughput

	E1	E2	E3	E4	E5	E6
h_{io} (W/m^{2}°C)	2252.96	1039.01	2252.96	1646.03	852.29	1646.03
h_o (W/m^{2}°C)	1931.88	2756.49	1318.00	1685.94	1788.89	1603.84
ΔP_t (kPa)	45.589	16.051	45.589	43.420	14.964	73.816
ΔP_s (kPa)	67.591	94.372	44.624	86.868	108.177	123.052
$R_{f,io}$ (m^{2}°C/W)	0.00144	0.00137	0.00140	0.00142	0.00157	0.00142
$R_{f.o}$ (m^{2}°C/W)	0.00144	0.00137	0.00140	0.00142	0.00157	0.00142

Table 2. Stream Data after Increasing Throughput

St. No.	Stream Name	Supply Temp. (°C)	Target Temp. (°C)	Flow (kg/s)	HTC (W/m^{2}°C)	ΔP (Kpa)
1	Kerosene	180.0	30.0	28.0	510.822	67.591
2	LGO	270.0	40.0	53.0	496.736	86.868
3	HGO	350.0	30.0	16.0	463.236	44.624
4	ATB	380.0	50.0	67.0	489.356	123.052
5	Naphtha	150.0	100.0	304.0	577.107	94.372
6	BPA	290.0	190.0	178.0	469.703	108.177
7	Crude	20.0	390.0	240.0	372.900	239.429

Targeting

In this stage, the targeting is carried out based on optimum pressure drops. The cost data and results of targeting are in Table 3 and 4 respectively. Also, Table 5 shows the stream heat transfer coefficients, the optimum pressure drops and the contact areas corresponding to targeted ΔT_{min}.

Design

In design stage we aim to minimize the number of changes we make to the network, within the pressure drop constraints, in order to achieve the targeted goal. This procedure consists of three main steps, analyzing the existing exchangers, improving the inefficient exchangers and revolution of network.

Given the targeted ΔT_{min} and hence the Pinch temperatures, exchangers E3 and E6 and heater H are identified as transferring heat across the Pinch and hence must be corrected if energy is to be saved. Table 6 shows that the total amount of heat being transferred across the Pinch is equal to the targeted value of the energy reduction. Also,

Table 7 shows the thermal and pressure efficincies in existing exchangers. According to this table, we can make a right decision to achieve improvement. Table 8 shows the results of Remaining Problem Analysis for corrected exchangers.

The complete design grid, which involves all the corrected units as well as new matches, is illustrated in Figure 5.

Table 3. Cost Data General Data for Crude Oil Pre-Heat Train

Hot Utility (£/kw.yr) : 70 Cold Utility (£/kw.yr) : 7 Power Cost (£/kw.yr) : 60	Plant Life Time (yr) : 20 Interest Rate (%) : 15 Max Area Per Shell : 580 Number of Tube Passes : 2 X-Factor for Multiple Pass Shells: 0.8
Exchanger Capital Cost : $8600+670\ (\text{Area})^{0.83}$ Pump Capital Cost : $8600+7310\ (q\sqrt{H})^{0.2}$ Pump Efficiency (%) : 70 Compressor Capital Cost : $8600+83.66(W)^{0.62}$ Compressor Efficiency (%) : 70 Annualization Factor : $i(i+1)^N / \{(i+1)^N - 1\}$	

Table 4. The Targeting Results for Specification of Required Additional Area

Debottlenecking Point Target			
Min.Temp.Diff.	58.00 °C	Minimum 1-2 Area	10983.89 m^2
Cold Utility	48751.79 kw	Target Area	12313.25 m^2
Hot Utility	80419.79 kw	Existing Area	6960.00 m^2
Psedud.Hot Ut.	101656.00 kw	Additional Area	5353.25 m^2
Hot Ut.Enhanc.	21236.21 kw	Constant Alpha	.892
Power Consum.	26554. £/kw	Area Investment	1374618. £
Energy Saving	1635188. £/kw	Flow Investment	0. £
	Generalized Investment 1226065. £		

Stream	Rout	Velocity	H.T.C.	Pres.Drop	Cont.Area	ID/Phase
H 1	S	.691	513.36	67295.8	256.38	P(L)
H 2	S	.764	514.34	244815.5	1307.01	P(L)
H 3	S	.592	489.58	207721.7	625.86	P(L)
H 4	S	.726	505.16	409157.4	2804.84	P(L)
H 5	S	.866	562.79	49675.9	1291.39	P(L)
H 6	S	.683	473.99	153234.0	3347.84	P(L)
H 7	S	.019	1000.00	.0	.28	C(L)
C 1	T	2.158	393.78	624152.9	9633.46	P(L)
C 2	T	.086	1000.00	.0	.14	C(L)

Table 5. Heat Transfer Coefficients and Optimized Pressure Drops at ΔT_{min}=58°C

α_{tar}=0.892					
St. No.	**Stream Name**	**Exchanger Side**	**Targeted-h ($w/m^2 {}^\circ C$)**	**Opt. Pressure Drop (kPa)**	**Contact Area(m^2)**
1	Kerosene	Shell	513.36	67295.8	256.38
2	LGO	Shell	514.34	244815.5	1307.01
3	HGO	Shell	489.58	207721.7	625.86
4	ATB	Shell	505.16	409157.4	2804.84
5	Naphta	Shell	562.79	49675.9	1291.39
6	BPA	Shell	473.99	153234.0	3347.84
7	Crude	Tube	393.78	624152.9	9633.46

Table 6. The Total Amount of Heat Being Transferred Across the Pinch

Unit	**Calculation**	**Heat (MW)**
E3	$(350-290)*41.6*10^{-3}$	2.496
E6	$(380-290)*174.2*10^{-3}$	15.678
Furnace (H)	$(232-227.09)*624*10^{-3}$	3.064
Total		21.238

Table 7. Remaining Problem Analysis

$\Delta T_{min,tar}$=58°C $Q_{H,tar}$=80418 kW		Constant Coefficient $A_{1\text{-}1,min,tar}$=9145.88 m^2			Fixed Pressure Drop $A_{1\text{-}2,min,tar}$=9634.16 m^2		
Exchanger Name	**Rem. ΔT_{min}**	**Exis. $A_{1\text{-}1}$**	**Rem. $A_{1\text{-}1,min}$**	**α_{max} (h)**	**Exis. $A_{1\text{-}2}$**	**Rem. $A_{1\text{-}2,min}$**	**α_{max} (ΔP)**
E1	58	285.87	8925.21	0.993	280	9438.45	0.991
E2	58	1302.66	7851.07	0.999	1480	---	---
E3	54	276.36	9415.89	0.944	280	10052.23	0.932
E4	51.85	767.69	8664.53	0.970	800	9549.23	0.931
E5	14.72	2323.73	8603.93	0.837	2760	9251.59	0.802
E6	29.04	1318.38	9008.19	0.886	1360	42326.19	0.721

Table 8. Remaining Problem Analysis for Correction of Existing Exchangers

$\Delta T_{min,tar}$=58°C $Q_{H,tar}$=80418 kW		**Constant Coefficient** $A_{1\text{-}1,min,tar}$=9145.88 m^2			**Fixed Pressure Drop** $A_{1\text{-}2,min,tar}$=9634.2 m^2		
Exchanger Name	**Rem. ΔT_{min}**	**Exis. $A_{1\text{-}1}$**	**Rem. $A_{1\text{-}1,min}$**	α_{max} **(h)**	**Exis. $A_{1\text{-}2}$**	**Rem. $A_{1\text{-}2,min}$**	α_{max} **(ΔP)**
E6	56.113	1287.19	7882.27	0.997	1360	8183.43	1.000
E6+E1	56.113	1573.06	7660.92	0.990	1640	7983.96	1.000
E6+E1+E5	56.113	3828.21	5408.45	0.990	4400	5476.54	0.975
E6+E1+E5+E3	56.113	3947.97	5289.38	0.990	4680	5348.04	0.961
E6+E1+E5+E3 +E2	56.113	5235.35	4009.05	0.989	6160	4039.84	0.945
E6+E1+E5+E3 +E2+E4	56.113	5866.75	3411.73	0.986	6960	3279.47	0.941

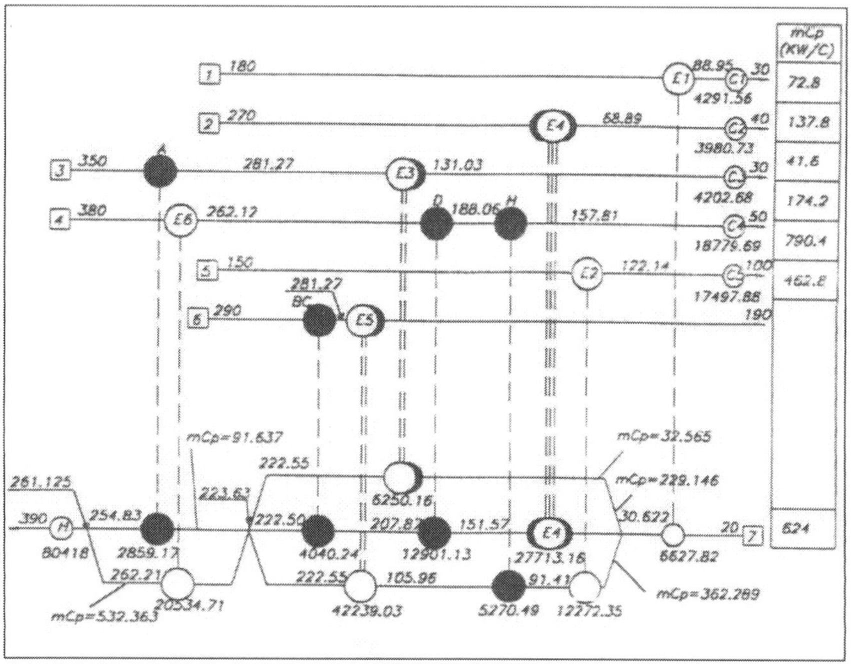

Figure 5. Complete design grid showing the new matches

Comparison

The resulting final network saves 1.238 MW of energy, which is exactly the target value. It requires the addition of 3818.75 m^2 of new area, which is 0.6% less than the targeted value of 3840.63 m^2. The results of targeting and design are compared in Table 9.

Table 9. Comparison Between Targeting and Design

	Targeting	Design	Difference
Energy saving (MW)	21.238	21.238	0.0%
Area requirement (m^2)	3840.63	3818.75	-0.6%
Overall efficiency	0.892	0.894	+0.2%
Power Consumption (£/yr)	26554	26554	0.0%
Area investment (£)	969653	965392	-0.4%
Flow investment (£)	90459	90459	0.0%
Generalized investment (£)	1226282	1222021	-0.3%

References

1. Smith, R., "Chemical Process Design", McGraw-Hill , New York, 1995.
Linnhoff, B. et. al., "User Guide on Process Integration for the Efficient Use of Energy, The Institution of Chemical Engineers, England, 1982.
2. Polley, G.T., and Panjeshahi, M.H., "Process Integration Retrofit Subject to Pressure Drop Constraints", Process Technology Proceedings, 9, Elsevier, 1990.
3. Polley, G.T., Panjeshahi, M.H., and Jegede, F.O., "Pressure Drop Considerations in the Retrofit of Heat Exchanger Networks', Trans. I.Chem.E., 68A, pp. 211-220, 1990.
4. Polley, G.T., Panjeshahi, M.H., and Nunez, M., "Rapid Design Algorithms for Shell-and-Tube an Compact Heat Exchangers", Trans. I.Chem.E., 69A, PP. 435-444 , 1991.
5. Polley, G.T., and Panjeshahi, M.H., "Interfacing Heat Exchanger Network Synthesis and Detailed Heat Exchanger Design", Trans. I.Chem.E., 69A, PP. 445-457, 1991.
6. Panjeshahi, M.H., "Pressure Drop Consideration in Process Integration", Ph.D. Thesis, UMIST, 1992.
7. Panjeshahi, M.H., and Ghafoorian M., "Debottlenecking of HEN's with Fixed Pressure Drop Constraints-I) Targeting", Journal of Faculty of Engineering, Tehran University, No. 57, 1996.
8. Panjeshahi, M.H., and Ghafoorian M., "Debottlenecking of HEN's with Fixed Pressure Drop Constraints-II) Design", Journal of Faculty of Engineering, Tehran University, No. 58, 1997.
9. Fallahi, H.R., "Pressure Drop Optimisation in Process Integration", Ph.D. Thesis, Tehran University, 1999.
10. Panjeshahi, M.H., and Fallahi, H.R., "Pressure Drop Optimization in Grass-roots Design of Heat Exchanger Networks", Proceeding of the Forth National Congress on Chemical Engineering, Sharif University of Technology, Tehran, 1999.
11. Panjeshahi, M.H., and Fallahi, H.R., "Pressure Drop optimization in Retrofit of Heat Exchanger Networks", Journal of Faculty of Engineering, Tehran University, Vol. 33, No. 4, 2000.
12. Panjeshahi, M.H., and Tahouni, N., "Debottlenecking of Heat Exchanger Networks with Optimum Pressure Drops", Proceeding of the Sixth National Congress on Chemical Engineering, Isfahan University of Technology, Isfahan, May 2001.
13. PILOT, Software for Process Integration, 2002.

133h Systematic in-Process Modification Approach for Enhanced Waste Energy Recovery in Gas Plants

Mahmoud Bahy Noureldin[1], Ahmed S. Aseeri[1], Ali H Al-Qahtani[2], and Saleh Al Hashimi[3]. (1) Consulting Services Department, Saudi Aramco, Dhahran, Saudi Arabia, (2) Consulting Services Dept., Saudi Aramco, Dhahran, Saudi Arabia, (3) Chemical Engineering Department, Petroleum Institute, Abu Dhabi, United Arab Emirates
Oil and gas industry is currently one of the most growing industrial sectors in the world. World demand for hydrocarbons is forecasted to grow rapidly during this decade and the one after. The growing demand for energy consumption have initiated another wave of new research focusing on new cost-effective technologies and systems for increased productivity and product quality while improving energy efficiency, preserve sustainability, reduce greenhouse gas emissions and other environmental consequences as well. Many methods and tools for energy conservation have been developed during the last two and half decades. These methods used mainly mathematical programming technique and pinch technology to synthesize and operate processes optimally from efficient use of energy point of view. Several challenges still face the industrial community in adopting these methods efficiently and stretching the envelope of knowledge beyond it towards reaching global optimal energy-consuming processes. This paper demonstrates the benefits obtained in form of more energy savings in gas industry through the utilization of systematic technique for in-process modification approach in a typical gas plant. It shows a new methodology for systematic targeting of energy consumption under all possible combination of in-process modifications and the automatic selection of the optimal process operating conditions that achieve desired targets. The application of this approach "In-process modification for energy efficiency optimization" is not only beneficial in maximizing profitability but also in reducing the energy-based air pollution in gas processing plants. Key Words: In-process modification, Heat Integration; gas processing plant; TEM_icons™ software

145a Compositional Analysis of Opportunity Materials: Characterization of Heavy Crude Oil and Bitumen by Fourier Transform Ion Cyclotron Resonance Mass Spectrometry

Ryan P. Rodgers[1], Donald F. Smith[2], Geoffrey C. Klein[2], Tanner M. Schaub[3], Alan G. Marshall[2], and Parviz Rahimi[4]. (1) Department of Chemistry and Biochemistry, National High Magnetic Field Lab at Florida State University, 1800 E. Paul Dirac Dr., Tallahassee, FL 32310, (2) Department of Chemistry and Biochemisty, National High Magnetic Field Lab at Florida State University, 1800 E. Paul Dirac Dr., Tallahassee, FL 32310, (3) National High Field FT-ICR MS Facility, National High Magnetic Field Lab at Florida State University, 1800 E. Paul Dirac Dr., Tallahassee, FL 32310, (4) National Centre for Upgrading Technology, 1 Oil patch drive, suite A202, Devon, AB T9G 1A8, Canada

Introduction

Previous efforts in the detailed characterization of crude oils and bitumen, their associated production deposits and isolated subfractions have been hindered by the oil's inherent complexity. The use of conventional analytical techniques requires extensive fractionation procedures and even then, the complexity of the isolated fractions typically exceeds the capabilities of most commercially available analytical equipment. Furthermore, the lack of suitable standard compounds often precludes identification based on retention time comparisons. For example, the detailed characterization of naphthenic acids requires a pre-isolation step followed by chemical derivitization in order to make the acids gas chromatography (GC) amenable. However, the complexity often leads to coeluting peaks that hinders identification. The addition of a mass spectrometer may aid in the analysis; however, as with GC, coeluting peaks complicate the mass spectrum and often prevent individual component identification. The recent rapid development of FT-ICR mass spectrometry allows for the detailed compositional analysis of complex petroleum samples with little or no sample fractionation.1 Sample preparation is minimal, and simply involves dilution of the sample in a toluene:methanol prior to introduction. The attainable mass-resolving power allows baseline resolution of isobaric species (i.e. C3 vs. SH4, both with a nominal mass of 36 but an exact mass difference of 3.4 mDa) differing by as little as 1 mDa, and the high mass accuracy allows for individual component identification at the level of elemental composition assignment. The ability to assign elemental compositions by class, type, and carbon number to all peaks in the mass spectrum by FT-ICR affords three different levels of compositional information. First, the heteroatom content or class of all peaks is determined. For example, carboxylic acids are members of the class, O2. Second, for determination of the aromaticity range of a specific class, the type or number of rings plus double bonds (double bond equivalents, or DBE) is determined. Third, for every class and type, the carbon number distribution provides a measure of the degree of alkylation for every class and type of species identified. Based on its ability to provide class, type and carbon number distributions for all identified species, FT-ICR mass spectrometry is currently unmatched in its ability to provide detailed compositional information on petroleum derived materials. Here we present the latest results from the ESI FT-ICR MS analysis of heavy crude oils and their associated production deposits as well as Athabasca bitumen samples subjected to common upgrading and fractionation processes.

Experimental

Sample preparation for ESI FT-ICR MS. Crude oil and bitumen samples were each prepared by dissolving ~ 30 mg of material in 30 mL of toluene. A one mL aliquot was then diluted with 0.5 mL of methanol. Ten microliters of either acetic acid (positive ion ESI) or ammonium hydroxide (negative ion ESI) was added to 1.5 mL of the final solution to facilitate protonation or deprotonation in the electrospray ionization process.

Mass spectrometry. Mass analysis was carried out with a homebuilt FT-ICR mass spectrometer equipped with a 22 cm horizontal room-temperature bore 9.4 T magnet (Oxford Instruments America, Inc., Concord MA). Data were collected and processed by a modular ICR data acquisition system (MIDAS). Two hundred co-added time domain data sets were Hanning apodized, followed by a single zero-fill before fast Fourier transformation and magnitude calculation.

Data processing. Mass values for singly charged ions between 225-1000 Da with FT-ICR mass spectral magnitude greater than 5-sigma of baseline noise were converted from IUPAC mass to Kendrick mass as previously described.1 The Kendrick masses were then sorted based on Kendrick mass defect and nominal-z value and imported into an Excel spreadsheet. A molecular formula calculator program limited to molecular formulas consisting of up to 100 12C, 2 13C, 200 1H, 5 14N, 10 16O, 5 32S, and 1 34S assigned elemental compositions. Because members of a homologous series differ only by integer multiples of CH2, assignment of a single member of such a series usually sufficed to identify all higher-mass members of that series. Due to the wealth of information provided by FT-ICR MS analysis of petroleum derived materials, several data reduction methods are useful in quickly determining class, type, and carbon number differences between samples of interest. Kendrick, van Krevelen, and DBE vs. Carbon number plots are commonly used to summarize compositional information provided by FT-ICR MS analysis of petroleum derived materials. Information on all three may be found elsewhere.1-3

Results and Discussion

Detailed characterization of the acidic species in Athabasca bitumen identified the suite of acidic species present. The negative ion ESI FT-ICR MS analysis of the bitumen identified 13 abundant classes of interest. The DBE and carbon number variations in the naphthenic acid class were of special interest. However, a similar level of composition detail is available for all other classes identified in the analysis. Abundant acidic classes include N, NS, NOx, Ox and SOx. The most abundant species are the DBE=3 naphthenic acids that range from 25-40 carbon numbers. A successively decreasing abundance is shown for both increasing carbon numbers and DBE values. The presence of highly abundant SOx species was surprising, and was further investigated through distillation cuts.

The compositional information provided by the FT-ICR MS analysis of the bitumen sparked interest in the behavior of these species in common refining operations. The fate of these species in distillation was investigated through a simulated vacuum distillation performed at NCUT in Devon, Alberta, Canada. Fractions (8) were collected from ~300 – 500+ °C and subsequently analyzed by FT-ICR MS. The TAN of the collected fractions increased as temperature increased for all but the 500+ °C distillation cut. However, the FT-ICR MS results showed a gradual decrease in the naphthenic acid (O2 class) profile. Interestingly, as distillation temperature increased the levels of multifunctional acidic (NO2, O3-4 and SxOx) species increased. Results from the 500+ °C cut showed a reduction in SO2 and O4 species compared to the 475-500 °C cut, suggesting that these species may be responsible for the decrease in TAN number for the highest boiling fraction. As expected, the carbon number and DBE values progressively increased as the distillation temperature increased. As distillation temperature increases the O2 species shifted from DBE 2-8 (C# 20-35) to DBE 2-12 (C# 25-45) and finally 2-14 (C# 30-50) in the highest temperature cut. All three samples share similar "hot spots" in the images, indicating that there is one preferred core structure. The SO2 species follow a similar trend with observed shifts from DBE 2-8 (C# 15-35) to DBE 2-12 (C# 20-40) and finally 2-16 (C# 25-50) in the highest temperature cut. Interestingly, the SO2 class has a pair of "hot spots" in the DBE vs. Carbon Number images, suggesting that there is a change in core structure as distillation temperature increases. Similar details will be presented for all other classes identified in the analysis. Thermal conversion of acidic species in Athabasca bitumen was carried out in a 1L autoclave at 300-400 °C and a residence time of 60 minutes. The reactions were carried out in the presence of slow flow of nitrogen to remove water formed in the condensation reactions of naphthenic acids. Besides small gaseous products and light end products collected in the condenser, major products (heavy ends) were collected from the reactor and submitted for different analyses. Under the reaction conditions employed the amount of gaseous products was negligible. At low temperatures, including 300 °C and 325 °C, the light ends were low in abundance, and only at higher temperatures of 375 and 400 °C were the light ends collected in the condenser at 3 and 6 wt% respectively. The analysis of these products showed a large amount of low DBE, low carbon number naphthenic acids. The mass spectra showed that there is little change in the species identified as the temperature increased from 300 °C to 375 °C. However, there was a significant decrease in oxygen content at 400 °C. The FT-ICR MS analysis of the thermally treated samples showed very little difference in the naphthenic acid (O2 class) distributions for the four samples, although the TAN of the heaviest fraction was substantially lower than for the other 3 lower temperatures. However, the FT-ICR MS analysis did show substantial differences in the O4 class for the different reaction temperatures. As the temperature increases, the number and relative abundance of the O4 species rapidly declines. The depletion of these species could account for the TAN reduction. Other studies presented include the compositional analysis of a heavy crude oil and its associated deposits and compositional variations in pressure drop and solvent drop asphaltenes. Examples and the implications of self aggregation will be presented for both bitumen and heavy crude oil.

1. Marshall, Alan G.; Rodgers, Ryan P., Accounts of Chemical Research (2004), 37(1), 53-59. 2. Wu, Zhigang; Rodgers, Ryan P.; Marshall, Alan G., Analytical Chemistry

(2004), 76(9), 2511-2516. 3. Kim, Sunghwan; Kramer, Robert W.; Hatcher, Patrick G., Analytical Chemistry (2003), 75(20), 5336-5344.

THE STATE-OF-THE-ART OF NAPHTHENIC ACID AND SULFIDIC CORROSION EVALUATION AND PREDICTION

Russell D. Kane, Elizabeth Trillo and Sridhar Srinivasan
Honeywell Process Solutions[(1)]
14503 Bammel N. Houston Rd., Suite 300
Houston, Texas USA 77014

[(1)] Recently acquired InterCorr International, Inc.

Abstract

Conventional rules for assessment of naphthenic acid and sulfidic corrosion and operating risk have evolved over the past 40 years, primarily based on experience at individual plants processing certain crude streams. This is the basis for the commonly used rules-of-thumb for conditions that generally produce elevated corrosion rates in fractionation systems. Currently, API RP 581 Base Resource Document provides guidelines that are based on process conditions of TAN, sulfur, and temperature, and alloy content. However, it has been found that these guidelines are, in many cases, overly conservative, leading to excessive alloy requirements. This paper reviews these factors for both sulfidic and naphthenic acid corrosion, and indicates a path forward for a more rational basis of assessing crude oil corrosivity. This revised procedure includes analysis, testing and a corrosion model that takes into account critical parameters for evaluation of operational risk, unit upgrading requirements and inspection/maintenance planning.

Keywords: Naphthenic acid, refining, total acid number, hydrogen sulfide, inhibition, corrosion prediction

Introduction

Corrosion in crude oil distillation units and associated feed and transfer lines continues to be a major concern in refinery operations. In a few locations in the refining circuit, high vaporization and multiphase flow results in very high wall shear stresses (>1000 Pa). In most cases, these areas have now been alloyed to high levels with Cr, Mo and Ni (i.e. UNS S31600 or UNS C31700 or higher) to prevent corrosion. The remainder of the circuit is comprised of primarily liquid filled lines with generally lower wall shear stresses (<100 Pa). These circuits are commonly made from lower alloy materials (carbon, 5Cr-0.5Mo, 9Cr-1Mo and 12Cr steels). However, only limited data is available for making these alloying decisions. Potential for unit deterioration, leaks and fires still persists. Industry experience and the recent API RP 581 Base Resource Document[1] provide guidelines for assessing corrosivity and operational risk, but these have been found to have substantial conservatism. Studies are being initiated under joint industry sponsorship to improve the guidelines found in the API RP 581.

Background

To remedy the current situation described above, more data is required to base engineering decisions for crude selection and blending strategies, and for unit inspections, repairs and/or upgrades. There is a need for expanded definition of boundaries between corrosive and non-corrosive service conditions over a broader range of refinery process conditions. This paper reviews the subject of corrosion in refinery distillation units including the effects of sulfur species and naphthenic acids, and offers a path forward taking into account critical parameters for evaluation of operational risk, unit upgrading requirements and inspection/maintenance planning.

Refinery Experience

For over 40 years, most of the guidelines for processing corrosive crude oil feedstocks were based plant experiences and related heuristics. This situation has resulted in only limited understanding of the interactions between corrosive species commonly found in crude oil feedstocks and those produced in the process stream. Additionally, the current understanding of the interaction between chemical and mechanical effects in high temperature oil systems is not widely utilized. However, based on service experience, two types of corrosion have been identified: (a) Sulfidic corrosion based on the presence of sulfur and a variety of organic and inorganic sulfur species in crude oil, (b) Naphthenic acid corrosion derived from certain types of organic acids present in the oil. Other effects of organic and inorganic chlorides are substantially different in nature and are consequently not addressed herein.

Sulfidic Corrosion. The most apparent cause of sulfidic corrosion in refinery distillation units is due to the presence of hydrogen sulfide naturally in the oil or that produced during heating. This latter source is from the thermal decomposition of organsulfur compounds (See Figure 1).[2] It is not entirely clear if these compounds act as a corrosive in the RSH (protonated) form, or if they are converted by hot metal surface and thermal degradation to H_2S as the active species. In most cases, the production of H_2S is thought to predominate, a fact that can be valuable in proper assessment of corrosivity.

Use of alloys containing chromium (usually 5 to 18 percent) can reduce sulfidic attack to acceptable levels. The extent to which Cr need to be present as an alloying element is a function of the level of available sulfur species. Guidelines are given in the modified McConomy Curves (See Figure 2).[3] These values are generally representative of conditions found in the gas phase rather than in flowing liquid or multiphase environments. Due to the influence of viscosity and density on increasing levels of wall shear stress, higher corrosion rate can be experience in some cases where flow conditions predominate.

Even in the absence of "significant" levels of naphthenic acids, corrosion has been observed in sulfidic environments that is morphologically similar to naphthenic acid corrosion (localized attack). These results in velocity accelerated sulfidic corrosion, producing higher than expected corrosion rates and requires materials of construction with both a high chromium and molybdenum content in excess of that indicated in the McConomy curves.

Naphthenic Acid Corrosion. Naphthenic acids are also present in many crude oils in varying amounts from around the world and provide another major class of corrosion-inducing contaminants. Those areas especially noted for producing crude oils containing elevated concentrations of naphthenic acid have historically included some areas in California, Venezuela, India, China, and Russia. More recently, these areas have also included some regions not historically noted for having high levels of naphthenic acid which include new fields in the North Sea, West Africa, Mexico, Canada (tar sands) and offshore Brazil.

Until recently, little quantitative information was known regarding, variations in molecular weight, boiling point and ring structure and their relationship to corrosivity in petroleum refining systems. As with most organic molecules, these variables can influence both their fractionation characteristics and chemical reactivity. This is particularly true of naphthenic acids as shown in Figure 3.[4] These effects will be discussed in more depth later in this paper.

Commonly, the naphthenic acid content in oils is assessed in terms of total acid number (TAN) per ASTM D664. TAN is given in terms of milligrams of KOH needed to neutralize a 100 gram sample of oil. It is also possible to obtain a slightly different value using an alternative method of TAN assessment using UOP method 565 where some of the non-naphthenic species are removed prior to analysis. Depending on the crude source, this can result in values that are from 0.1 to 0.5 units lower than that obtained by the ASTM method leading to somewhat better correlation with corrosivity.

Values of TAN in the range of 0.1 to 3.5 are common in crude oils, but particularly severe situations have been found for particular hydrocarbon fractions that can have TAN values in excess of 6. The complexity of the situation can be realized when considering that organic acid components in crude oils can vary considerably with their source. Furthermore, these acidic components can include both naphthenic acids (those typically referred to as having two or more cyclopentane rings) and lower molecular weight straight chain organic acids further complicating the picture.

Much of the practice of corrosion control for naphthenic acid containing crude oils has come from experiential data gained from plant operations. This has helped in the development of qualitative guidelines and heuristics ("rules of thumb" based on experiential information) primarily developed for long standing plant operation using specific feedstocks. The important questions being asked today relate to the question of how well these general rules apply to refining operations that now vary their feedstock mix, sometimes significantly, on a daily, weekly or monthly basis based on their price on spot market.

Naphthenic acid corrosion is differentiated from sulfidic corrosion by the corrosion mechanism and the morphology of the attack. Sulfidic corrosion is usually described are being a general mass loss or wastage of the exposure surface with the formation of a sulfide corrosion scale. By comparison, naphthenic acid corrosion is typically characterized as having more a localized character particularly at areas of high velocity and, in some cases, where condensation of concentrated acid vapors can occur in crude distillation units. The attack is oftentimes described as lacking surface corrosion products exposing fresh metal. Consequently, damage is in the form of unexpectedly high corrosion rates and localized attack on alloys that would normally be expected to resist sulfidic corrosion (particularly steels with more than 9% Cr). In some cases, even very highly alloyed materials have been found to exhibit sensitivity to corrosion. For these cases a minimum alloy of 18Cr-2.5Mo (AISI 317) has been used.[5]

As a rule, whole crudes with a TAN value greater than 0.5 and refined hydrocarbon fractions with a TAN higher than 1.5 have been generally considered to be potentially corrosive. The difference in these two numbers comes from the concentration of naphthenic acid that occurs in certain hydrocarbon fractions during refining (See Figure 3). One problem is that there are many naphthenic acids with varying structures, molecular weights and boiling points. Therefore, it is a better practice to assess the potential corrosivity based on the TAN values in the hydrocarbon fractions. Furthermore, it is important to also review the sulfur content of these fractions are well. Both types of information are available in assay information based on a standard fractionations.

There have been several recent developments in the quest for prediction and control of naphthenic acid corrosion. These developments include identification that:

- The presence of sulfur-containing compounds in hydrocarbon fractions can inhibit naphthenic acid corrosion. This understanding has come from both experiential information and, more recently, experimental data on high TAN fractions that are also high in sulfur species, and from blending of high TAN crude with low TAN crude with significant sulfur content (See Figure 4).[6] The most important factor in this effect is that the form of sulfur and it ability to provide reactive sulfur (H_2S) to inhibit the naphthenic acid corrosion under the process conditions.

- Some recently discovered crude oil sources produce hydrocarbon fractions with high TAN in combination with low sulfur (<0.5 wt%) can be particularly corrosive since. This effect results since the fractions do not have enough reactive sulfur to counteract the highly corrosive naphthenic acids. This situation can also occur when the sulfur species present, as a result of their structure, are not reactive to the point where they can participate in corrosion reaction to provide their inhibiting effect.

- Some sources of crude oil and refined fractions have been found that have high TAN values, but appear not to be as highly corrosive as one might expect based solely on their TAN value. These crudes appear to have a very beneficial combination of conditions: (a) the presence of sufficient sulfur in the fractions, and (b) the presence of certain naphthenic acids with particular structures that are less corrosive.[6]

There is also evidence to indicate that both sulfidic and naphthenic acid corrosion can be accelerated by velocity of the flowing process environment or by local turbulence (See Figure 5).[7] The wall shear stress produced by the flowing media contributes an added mechanical means to remove the normally protective sulfide films. The wall shear stress is generally proportional to velocity in single phase systems but also takes into account the physical properties and flow regime of the flowing media.

One of the properties included in the wall shear stress is the density of flowing medium or media (in the case of multiphase flow) which is, in turn, affected by the degree of vaporization and temperature. The reason for the strong emphasis on quantifying wall shear stress is because this parameter reflects the mechanical forces acting on the surface of operating equipment. This requires an emphasis on flow modeling to obtain values of wall shear stress.

API RP 581 RBI Base Resource Document

The American Petroleum Institute (API) Risk Based Inspection (RBI) Project was initiated in May 1993 by an industry sponsored group to develop practical methods for Risk Based Inspection. API RP 581 – "Risk-Based Inspection Base Resource Document". This document outlines a method for using risk as a basis for prioritizing and managing the efforts of an inspection program. The petroleum industry lacks a specific experience database of failure frequency categorized by equipment type and specific process environment. As a result, the RP 581 Base Resource Document provides assessment methodologies and reference data for evaluating in-service degradation occurring in particular refinery systems and environments. It provides "Technical Modules" to describe the assessment methodologies for many types of corrosive degradation including sulfidic and naphthenic acid corrosion found in Appendix G under the general category of thinning.

The general flow chart for addressing sulfidic and naphthenic acid corrosion is shown in Figure 6.[1] It provides for four primary input parameters: (a) Material (carbon steel through AISI 317), (b) Maximum service temperature, (c) Sulfur content (wt%), (d) Naphthenic acid content (TAN), and (d) stream velocity. To apply this methodology, the input data needs to be specific to the stream(s) under consideration. Therefore, whole crude values are of little value in this analysis. The first three parameters are used to estimate the corrosion rate based on presentation of tabular data for commonly used alloys: Carbon, 5Cr, 7Cr, 9Cr steels, AISI 410 (12Cr), AISI 316 (<2.5% Mo) and AISI 317 (>2.5% Mo).

Table 1 provides an example of the data format for 9Cr-1Mo steel.[1] It shows corrosion rates for various conditions as defined by the sulfur content, TAN value and temperature. The data sets for <0.2, 0.21-0.6 and >3 wt% sulfur from this table are shown graphically in Figure 7a-c. These figures show the current understanding of the relationships discussed previously in this paper, which include: (a) severe conditions for naphthenic acid corrosion at very low sulfur levels, (b) lower corrosion rates with nominal sulfur levels (0.2-0.6 wt%), (c) increasing corrosion rates at higher sulfur levels, (d) increasing corrosion rate with increasing TAN levels.

As mentioned above, the APR RP 581 methodology for assessment of sulfidic and naphthenic acid corrosion also includes provisions for velocity. As shown in Figure 5, once the baseline corrosion rate is determined for the appropriate stream conditions and material, the flow chart then describes application of a velocity factor 5X for flow rates of 100 ft/s (32 m/s) or greater. Figure 8 shows the impact of this factor for 9Cr-1Mo steel at 0.2-0.6 wt% sulfur. These are much higher corrosion rates that would apply for certain vacuum column heat tube and transfer line applications where velocities can be in the range of 100 to 300 ft/s (32 to 97m/s).

While the corrosion model in API RP 581 is simplistic, it is based on a significant data base of sulfidic corrosion based on the modified McConomy curves. While some data exists on the influence of naphthenic acid on corrosion rate, the base corrosion data for naphthenic acid corrosion is less well established. But, the simplicity and structure of the methodology provide an effective and easily applied method for corrosion prediction and a state-of-the art basis for RBI in crude oil distillation units and piping. This said, many oil companies have gone further by generating corrosion data for specific crude oils, blends and their fractions used in their refining operations. However, the extent of this data is currently limited and has not been generally applied across the industry. Further work to develop an expanded crude corrosivity database and corrosivity model incorporating flow modeling is starting in 2006.

Specific Crude Oil Properties Influence Corrosivity

One of the areas of growing interest is the understanding of conditions that either promotes or helps minimize naphthenic acid corrosion. As shown in the previous section, most major factors are taken into account in the API RP 581 method for assessment of sulfidic and naphthenic acid corrosion. This method provides guidance for RBI procedures and materials selection. As experience with this methodology progresses, it has been noted that the methodology while beneficial, has a tendency for being overly conservative particularly with respect to certain naphthenic acid conditions.

Naphthenic Acid Type

One of the key findings from a major research study was that the acidic components commonly found in crude oils which contribute to the TAN value can vary substantially and have a major impact of the resulting corrosivity. As shown in Figure 9, the acidic components present in crude oils consists of both straight chain molecules (fatty acids that have no ring structures characteristic of naphthenic acids) and true naphthenic acids (with one or multiple ring structures).[8] The fatty acids while contributing to the total TAN value are thermally unstable and are decomposed quickly during process heating in the refining operation. Consequently, they contribute little to the resultant corrosivity. Secondly, those acidic components that do have naphthenic acid character have chemical structures involving a R value of 1, 2, 3, 4 or possibly more in the formula $R(CH_2)_nCOOH$ where R is the number of cyclopentane rings and the value of N is >12.

In order to further interpret this type of information, Table 2 shows data obtained using the Fast Atom Bombardment –Mass Spectroscopy (FAB-MS) technique for three “naphthenic acids” obtained from crude oils that where were characterized by their 50% off boiling point as low, medium and high boiling point acids.[9] These data are also compared to those obtained for a reagent grade naphthenic acid of know corrosivity. The data presented in Table 2 are generally similar to those shown in Figure 7 for commercial crude oils having varying amounts of straight chain and ring structures.

In the present case, two pieces of information were correlated. Data from this study showed that the higher boiling point acids have greater percentages of higher ring structures, whereas the low boiling point naphthenic acid had a higher percentage of the single ring components similar to that found in the reagent grade naphthenic acid. Furthermore, the results from the corrosion testing program showed that in the range of typical process temperatures where naphthenic acid corrosion is commonly observed, the most corrosive acids were the reagent grade acid and the low boiling point acid. The reagent grade naphthenic acid and low boiling point samples had a similarly high percentage of single ring structure naphthenic acid. In summary, it was determined that the overall aggressivity of the acids in a hydrocarbon fraction (at constant TAN value) was proportional to the percentage of their single ring structure acidic components. Therefore, it appears that FAB-MS data indicating the nature of the acidic structures can be used as a method to qualify the corrosivity of acid containing fractions.

The abovementioned data also help to explain, to some degree, the reason for the potential conservatism in the API RP 581 method for naphthenic acid corrosion in certain applications. The cause for this effect is the inability for the TAN value alone to take into account the distribution of naphthenic and non-naphthenic acid components in the process stream.

H_2S versus Naphthenic Acid Domination

In a recent study, a particular source of crude oil was evaluated for addition to a feedstock mix. It was characterized as having high TAN and sulfur, but it would be blended with a sweet crude mix with low sulfur and low TAN. Laboratory tests were conducted to assess corrosivity in various fractions of this new feedstock for comparison to the corrosion rates obtained with the API RP 581 methodology and in an effort to develop less conservative guidelines.

To illustrate the importance and use of laboratory test data, consideration was given to the HVGO stream in the temperature range 500 F and 675 F. The stream conditions consisted of 3.4 wt% sulfur and 4.0 TAN. Based on this condition, the API RP 581 methodology indicated corrosion rates on 9Cr-1Mo steel of 20 mpy (0.5 mm/y) at 550 F (C) and 30 mpy (0.7 mm/y) at 675 F (C). These corrosion rate values are in excess of the design cut-off value of 5 mpy for piping and would require upgrading to a higher alloy material.

Corrosion autoclave tests were conducted in the HVGO stream for assessment of corrosion rates at these two temperatures. The methodology for the corrosion test was to run the HVGO sample at minimal pressure so the autoclave included the use of a pop valve to release the pressure from off-gassing of the HVGO sample during the test.

The data from these tests were further evaluated to obtain representative corrosion rates for use in design of plant piping. This analysis consisted of first consideration of the baseline corrosion rates for 9Cr-1Mo steel in the tests along with naphthenic acid corrosion index (NACI) values produced using carbon steel coupons under the same conditions. The NACI is determined from the corrosion rate of carbon steel (in mpy) divided by the scale weight (in mg/cm^2). High values (>30) of NACI indicate naphthenic acid dominated corrosion conditions whereas low values (<10) of NACI indicate sulfur dominated corrosion conditions. In this case, the NACI values were observed to be low indicating domination by sulfidic corrosion. Therefore, sulfide scaling would be expected which tends to reduce corrosion rates with time.

A post-test analysis procedure can be applied to laboratory derived corrosion rates for use in refinery materials selection decisions. The protocol for this procedure is provided below:

- **Sulfur Dominated Conditions.** A reduction factor for cases of low NACI value where sulfide scale would be expected to lower long term corrosion rates. This is based on short and long term corrosion rates conducted under sulfur dominated conditions. See Figure 10.[9]

- **Naphthenic Acid Dominated Conditions.** Where high NACI values are obtained from laboratory tests, no reduction factor is applied since naphthenic acid corrosion because this form of corrosion does not usually involve filming and does not involve a reduction in corrosion rate with time of exposure as exhibited by sulfidic corrosion.

- **High Wall Shear Stress Conditions.** Where high velocity, multiphase flow conditions result in high wall shear stress a correction factor is incorporated to test results conducted at low flow conditions that effectively increases the expected corrosion rate. This factor varies depending on the NACI index and allows for velocity accelerated sulfidic corrosion at low NACI, and velocity accelerated naphthenic acid corrosion at high NACI values, as well as intermediate conditions. See Figure 4.

- **Safety Factor.** After the abovementioned adjustments are made to the laboratory corrosion rates, a safety factor can be added depending on the confidence in the test data and the plant operating conditions. This factor is usually 2X for most cases but can be adjusted higher, if necessary for specific cases.

Due to the low NACI values in the present case, a reducing factor was applied the 9Cr-1Mo corrosion rates based on Figure 10 to account for the sulfide scaling effect expected in the HVGO stream. In addition a safety factor of 2X was applied to the adjusted corrosion rate data. No additional factor was used for high wall shear stress conditions, since the application involved liquid filled pipe flow with wall shear stress levels generally less than 100 Pa. The adjusted corrosion rates for 9Cr-1Mo steel were 2 mpy at 550 F (C) and 5.3 mpy at 675 F (C). As can be seen, these values are substantially lower than those predicted by the API RP 581 methodology for an HVGO stream with 3.4 wt% sulfur and 4.0 TAN. The results were actually closer to those predicted by the API RP 581 procedure assuming no contribution from the naphthenic acid at all. For this case, the corrosion rates predicted by API RP 581 were 5 mpy at 550 F and 10 mpy at 675 F.

The reason for the lack of naphthenic acid effect in the corrosion rate appears to be related to the type of acidic structures in the base oil and concentrated in the HVGO for the present case. The results of the FAB-MS analyses on the base oil and the HVGO are shown in Table 3. The unexpected lack of naphthenic acid character in this case is caused by the predominance of acidic components that are not highly corrosive over the temperature range used in the crude oil refining unit. Furthermore, the presence of enough sulfur in this stream to provide a protective corrosive scale would also be expected to impart further corrosion protection over time. Therefore, this study shows that the API RP 581 methodology provided a sound but very conservative basis for assessment of corrosivity. However, when this method is combined with a laboratory test data simulating process conditions, it can provide a more effective tool for both implementing RBI in crude oil refining units as well as for material selection in design and revamping these units as the feedstock changes over time.

Path Forward

Prediction and evaluation of crude oil corrosivity for refinery operations continues to be an involving field. In the past ten years, major strides have been made to separate and understand the roles the corrosive components commonly found in crude oil. As discussed herein, laboratory testing (with high temperature autoclave and flow loop systems) augmented by analytical methods such as FAB-MS have provided new insights and directions for improved assessment and predictive capabilities. A new joint industry sponsored research effort has been initiated by Honeywell InterCorr involving major oil companies to further extend technology in this area.

The technical emphasis of this program is the integration of an experimental corrosion test program with supporting analytical data on naphthenic acid and sulfur species. This information will be used to define improved, quicker test methods and better define the safe operating envelopes and conditions that pose a risk of accelerated corrosion to commonly used alloys. The results will be integrated with a new software tool to aid in the prediction and assessment of refinery hot oil corrosivity. The software tool will utilize the new test data to provide an improved corrosion prediction and risk assessment than currently available with the API RP 581 RBI Resource Document using similar methodologies as highlighted in this paper.

An accelerated test method for corrosivity assessment would be valuable to be able to quickly evaluate limited quantities of actual hydrocarbon fractions and benchmark them against other fractions of known corrosivity. This procedure would also help to characterize if the corrosion was dominated by sulfidic or naphthenic acid attack which would aid in better predictive capabilities for long term corrosion rates as discussed herein. Additionally, an corrosivity database will be developed based on levels and types of sulfidic and naphthenic corrosive constituents and flow conditions quantified in terms of wall shear stress for proper linkage to actual plant operating conditions. Finally, the data from this latter portion of the study will be assembled in a software tool to accommodate flow modeling and rapid interrogation of the corrosivity database and relationships developed from this data.

Conclusions

Based on the information provided herein the following conclusions were made:

(1) Sulfur and naphthenic acids are corrodants in refinery crude oil distillation units that can act separately or in combination with each other to define the severity of corrosion in particular process streams.

(2) It is important to evaluate corrosion in these units based on conditions specific to the process stream of interest and not use whole crude properties.

(3) The presence of sulfur-containing compounds in hydrocarbon fractions can inhibit naphthenic acid corrosion. This understanding has come from both experiential information and, more recently, experimental data on high TAN fractions that are also high in sulfur species

(4) High TAN in combination with low sulfur (<0.5 wt%) can be particularly corrosive. This effect predominates when the fraction does not have enough reactive sulfur to counteract the highly corrosive naphthenic acids.

(5) Some sources of crude oil and refined fractions have been found that have high TAN values, but appear not to be as highly corrosive as one might expect based solely on their TAN value. These crudes appear to have a very beneficial combination of conditions: (a) the presence of sufficient sulfur in the fractions, and (b) the presence of certain naphthenic acids with particular structures that are less corrosive.

(6) FAB-MS techniques are useful in characterizing acidic constituents in refinery crude oil feedstocks.

(7) The API RP 581 methodology provided a sound but very conservative basis for assessment of corrosivity.

(8) The API RP 581 methodology can be augmented with the use of laboratory testing and a protocol for evaluating the test data for use in materials selection based on certain conditions indicated below:

- **Sulfur Dominated Conditions.** A reduction factor for cases of low NACI value where sulfide scale would be expected to lower long term corrosion rates.

- **Naphthenic Acid Dominated Conditions.** Where high NACI values are obtained from laboratory tests, no reduction factor is applied.

- **High Wall Shear Stress Conditions.** Where high velocity, multiphase flow conditions result in high wall shear stress a correction factor is incorporated to test results conducted at low flow conditions that effectively increases the expected corrosion rate.

- **Safety Factor.** After the abovementioned adjustments are made to the laboratory corrosion rates, a safety factor is added depending on the confidence in the test data and the plant operating conditions.

References

1. RP-581, Risk Based Inspection, Base Resource Document, American Petroleum Institute, Washington, D.C. (2003)
2. L.F. Hatch, S. Matar, Hydrocarbon Processing, June (1977), p. 189
3. J. Gutzeit, et.al., Metals Handbook, Vol. 13 - Corrosion, ASM International, Materials Park, OH, p. 1270, (1987)
4. R.L. Piehl, "Naphthenic Acid Corrosion in Crude Oil Distillation Units", Corrosion/87, Paper No. 196, NACE International, Houston, TX (1987)
5. R.D. Kane, "Understanding Critical Factor Influencing Refinery Crude Corrosiveness", Materials Performance, NACE International, Houston, TX, July (1999), p. 48-54
6. B. Messer, B. Tarleton, M. Beaton, T. Phillips, "New Theory For Naphthenic Acid Corrosivity of Athabasca Oil Sands Crudes", Corrosion/2004, Paper No. 04634, NACE International, Houston, TX (2004)
7. R.D. Kane, M.S. Cayard, "A Comprehensive Study on Naphthenic Acid Corrosion", Corrosion/2002, Paper No. 02555, NACE International, Houston, TX, (2002)
8. F.H. Behar, P. Albrecht, Organic Geochemistry, Vol. 6 (1981), p. 597
9. R.D. Kane, "Predicting Crude Oil Corrosivity: Influence of Velocity, Interactions of Oil Composition, Temperature and Alloying – Final Report", Honeywell Process Solutions (acquired InterCorr International), Houston, TX, Feb. (2000)

TABLE 1 – TABULAR DATA FROM API RP 581 FOR 9CR-1MO STEEL

Table G-21—Estimated Corrosion Rates for 9% Cr Steel (mpy)

Sulfur (wt.%)	TAN (mg/g)	Temperature (°F) ≤ 450	451–500	501–550	551–600	601–650	651–700	701–750	> 750
≤ 0.2	≤ 0.7	1	1	1	2	3	4	5	6
	0.71-1.5	1	2	2	4	4	5	6	8
	1.6-2.1	2	4	5	8	10	15	15	20
	2.1-4.0	3	6	10	12	15	20	20	25
	> 4.0	5	8	12	15	20	25	30	30
0.21–0.6	≤ 0.7	1	1	2	3	4	6	7	8
	0.71-1.5	1	1	2	4	5	7	8	10
	1.6-2.1	2	2	3	5	8	8	10	10
	2.1-4.0	3	3	5	8	10	10	12	15
	> 4.0	4	5	8	10	10	12	15	15
0.61–1.0	≤ 0.7	1	1	2	3	5	8	9	10
	0.71-1.5	1	2	3	5	8	10	10	10
	1.6-2.1	2	3	5	8	10	10	10	15
	2.1-4.0	3	5	8	10	10	15	15	15
	> 4.0	5	8	10	10	15	15	20	20
1.1–2.0	≤ 0.7	1	1	2	4	6	10	10	15
	0.71-1.5	1	2	3	5	7	10	15	15
	1.6-2.1	2	4	4	6	8	12	15	20
	2.1-4.0	3	6	5	8	10	15	20	20
	> 4.0	5	8	10	12	15	20	20	25
2.1–3.0	≤ 0.7	1	1	3	5	7	10	15	15
	0.71-1.5	1	2	4	6	8	10	15	15
	1.6-2.1	2	4	5	8	10	15	15	20
	2.1-4.0	3	6	10	12	15	20	20	25
	> 4.0	5	8	12	15	20	25	30	30
> 3.0	≤ 0.7	1	1	2	5	8	10	15	15
	.71-1.5	2	3	5	8	10	15	15	20
	1.6-2.1	3	5	10	12	15	20	20	25
	2.1-4.0	5	8	12	15	20	25	30	30
	> 4.0	7	9	15	20	25	30	35	40

TABLE 2

Naphthenic Acids Structures from Crude Oil & Reagent Naphthenic Acid Using FAB-MS

Oil Sample	Straight Chain	1 Ring	2-Ring	3-ring	4-ring	Other
Reagent Nap Acid	12	45	31	6	3	3
High B.P. Nap Acid	9	15	24	19	14	19
	(8)	(18)	(28)	(20)	(12)	(14)
Medium B.P. Nap Acid	7	27	37	17	6	6
Low B.P. Nap Acid	20	42	25	6	3	4

*Double values indicate replicate FAB-MS rings on same sample.

TABLE 3

Structures of Naphthenic Acids from Base Oil and HVGO vs. Reagent Naphthenic Acid using FAB-MS

Oil Sample	Straight Chain	1 Ring	2-Ring	3-ring	4-ring	Other
Reagent Nap Acid (corrosive)	12	45	31	6	3	3
Refinery Base Oil Sample	9	15	23	20	14	19
Refinery HVGO Sample	8	22	30	20	10	10

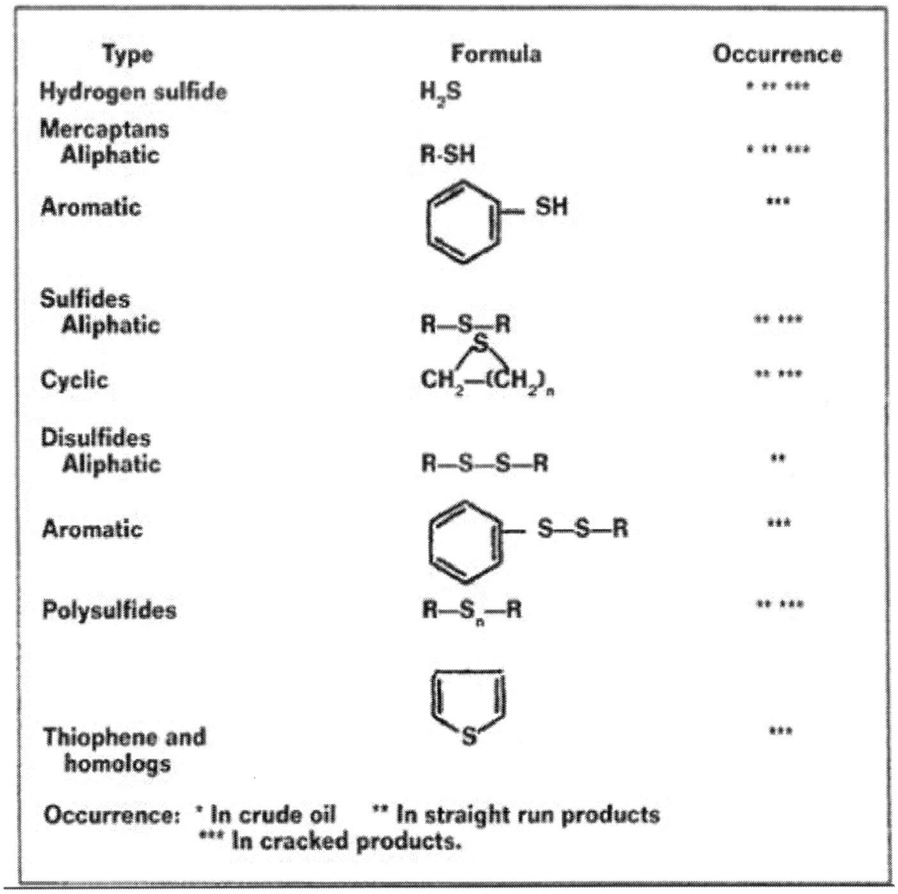

Figure 1 - Major sulfur compounds and chemical structure found in crude oil and it refined fractions.

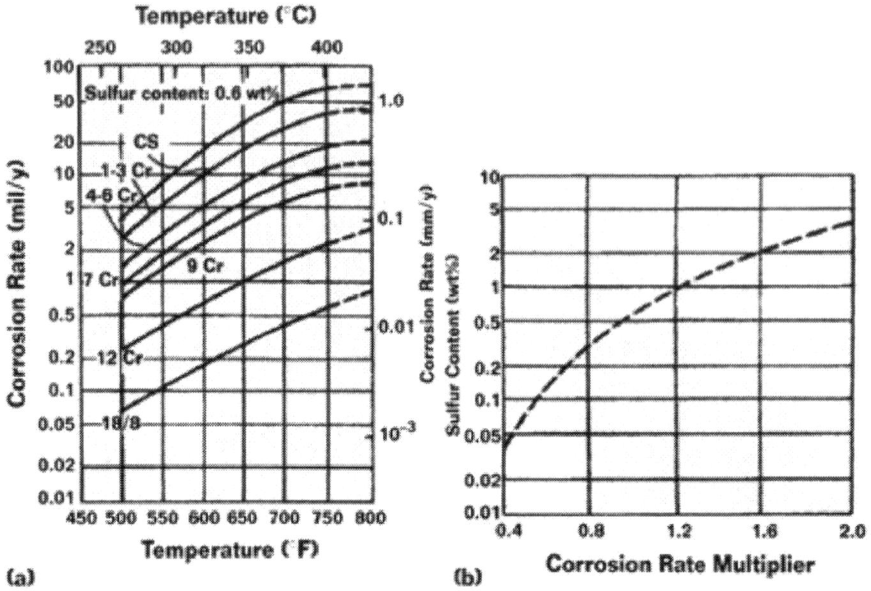

Figure 2 – Modified McConnomy curves for selection of alloys versus sulfidic corrosion

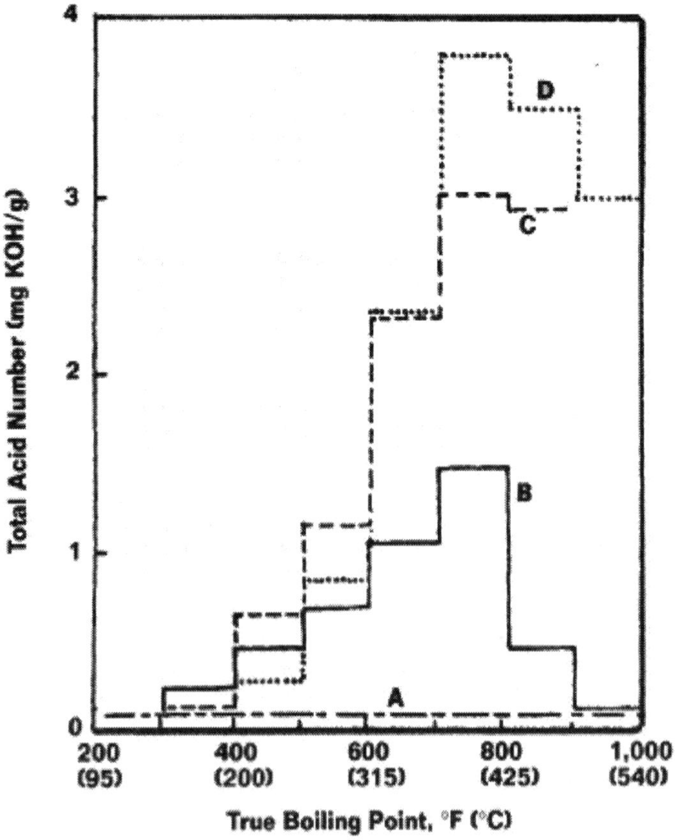

Figure 3 – Variation in TAN with temperature for four base oils (A-D).

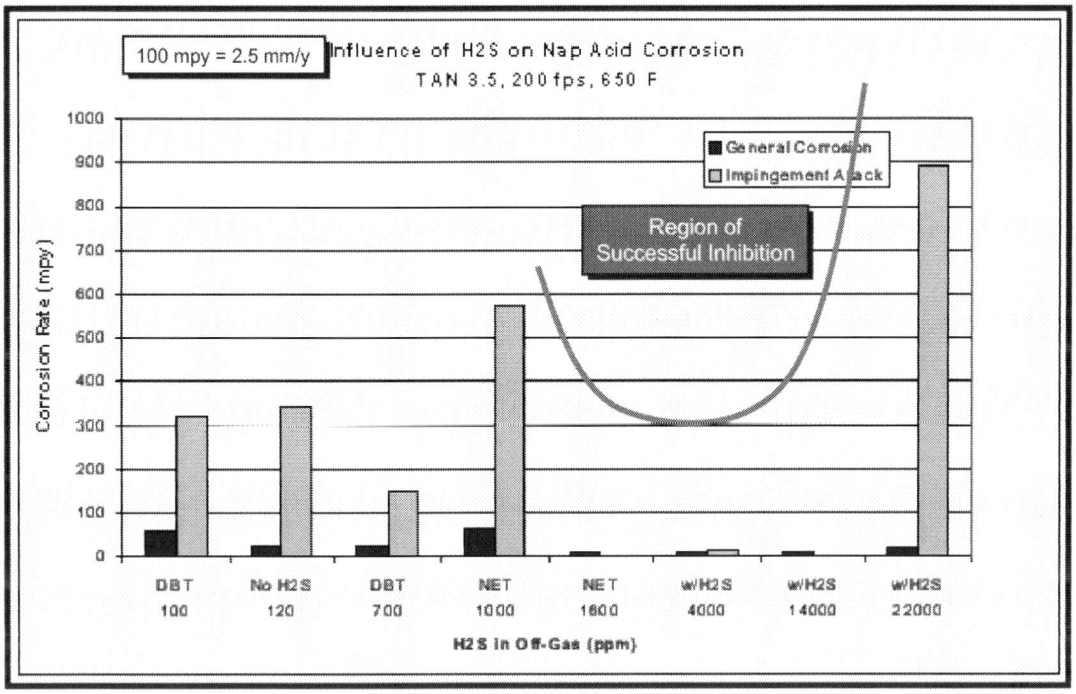

Figure 4 – Corrosion of 9Cr-0.5 Mo Steel at TAN 3.5 and high velocity as a function of H_2S in the environment.

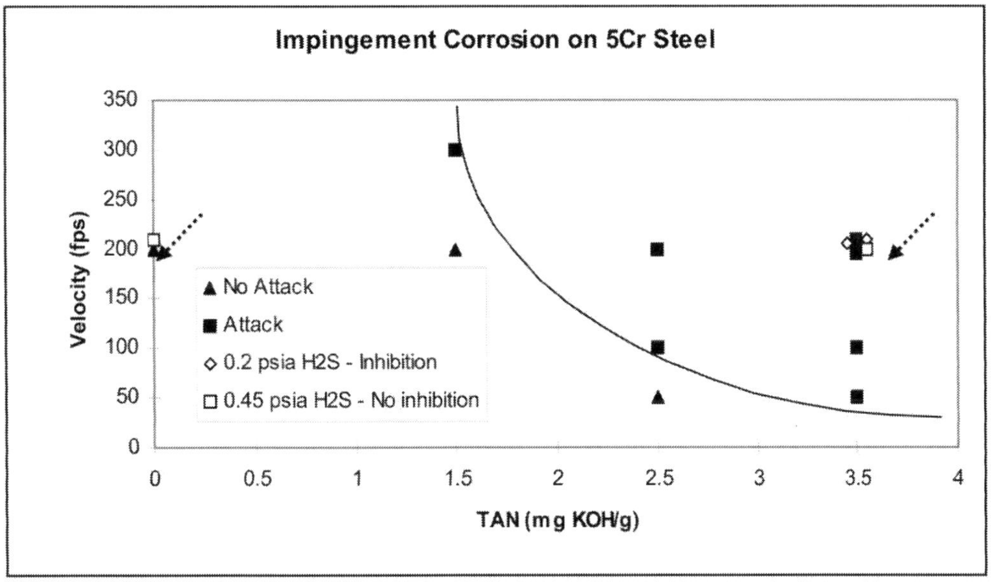

Figure 5 – Example of conditions were velocity accelerated corrosion can occur due to naphthenic acid corrosion and sulfidic corrosion, and inhibition of this attack can take place on 5Cr-0.5Mo Steel (See arrows).

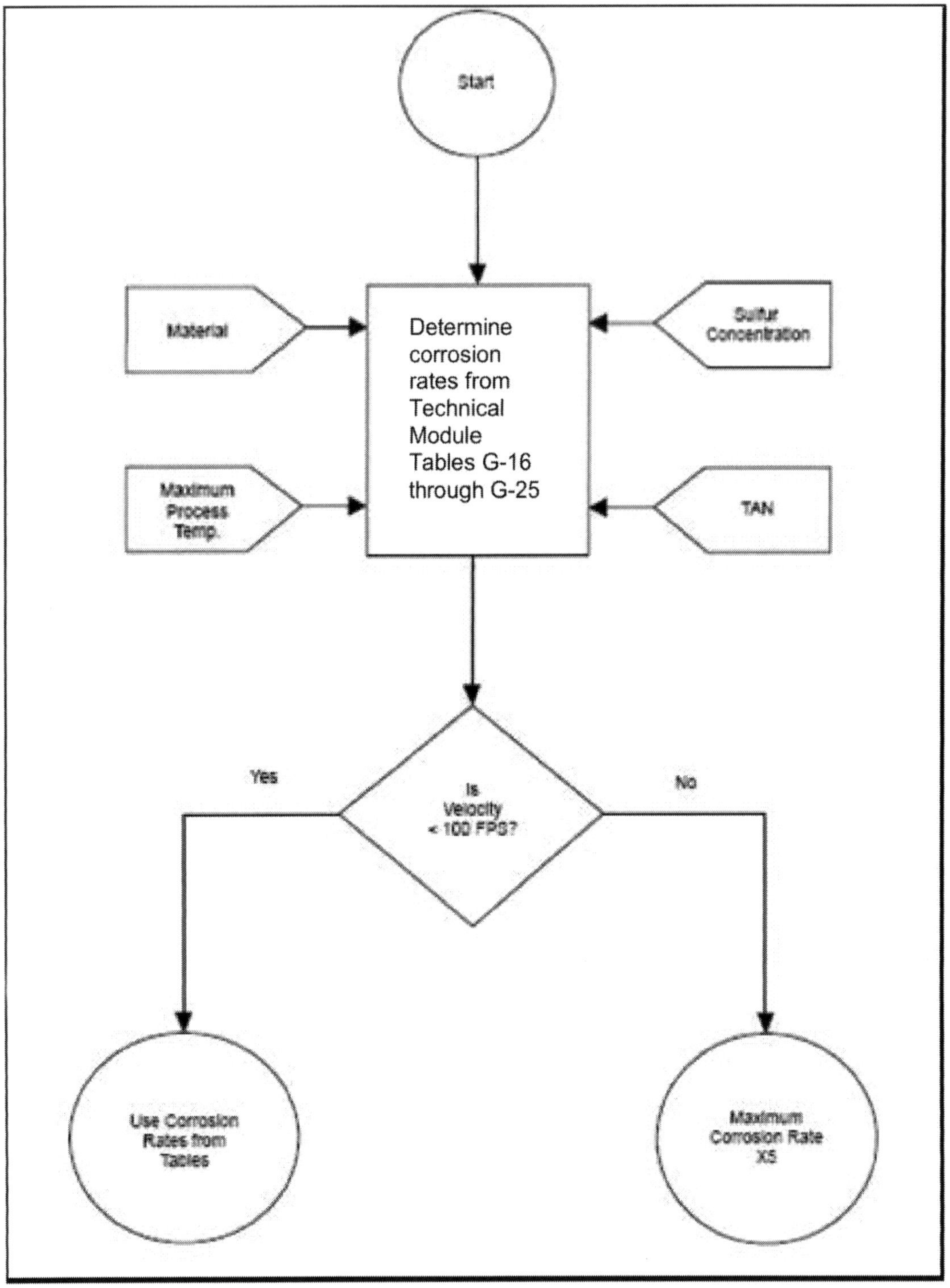

Figure 6 – API RP 581 assessment of high temperature sulfidic and naphthenic acid corrosion.

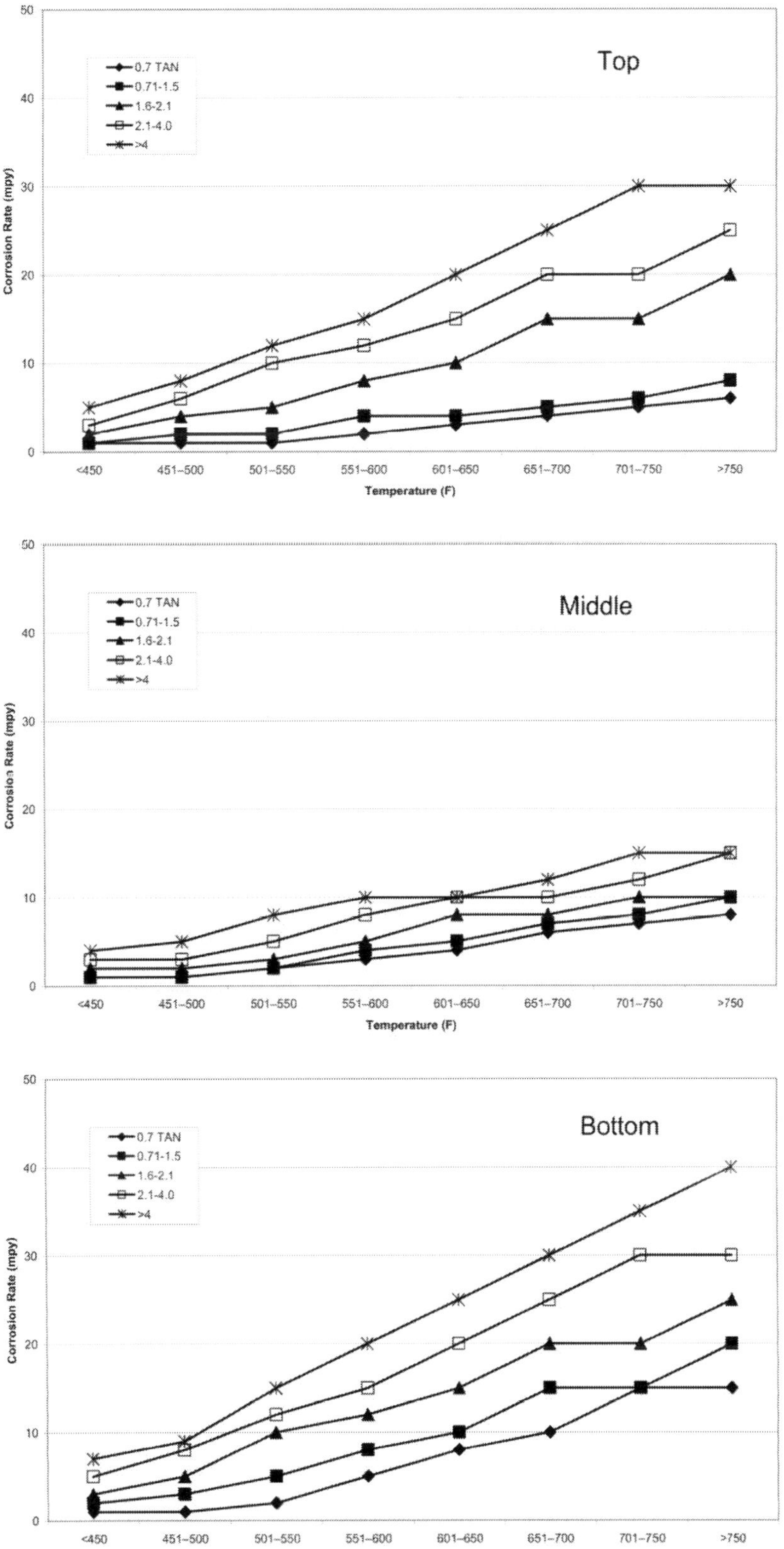

Figure 7 – Corrosivity assessment chart for 9Cr-1Mo Steel – Top - <0.2% S, Middle – 0.21 to 0.6% S; Bottom - >3% S; per API RP 581 (Conversion: 40 mpy = 1 mm/y; 600 F (C).

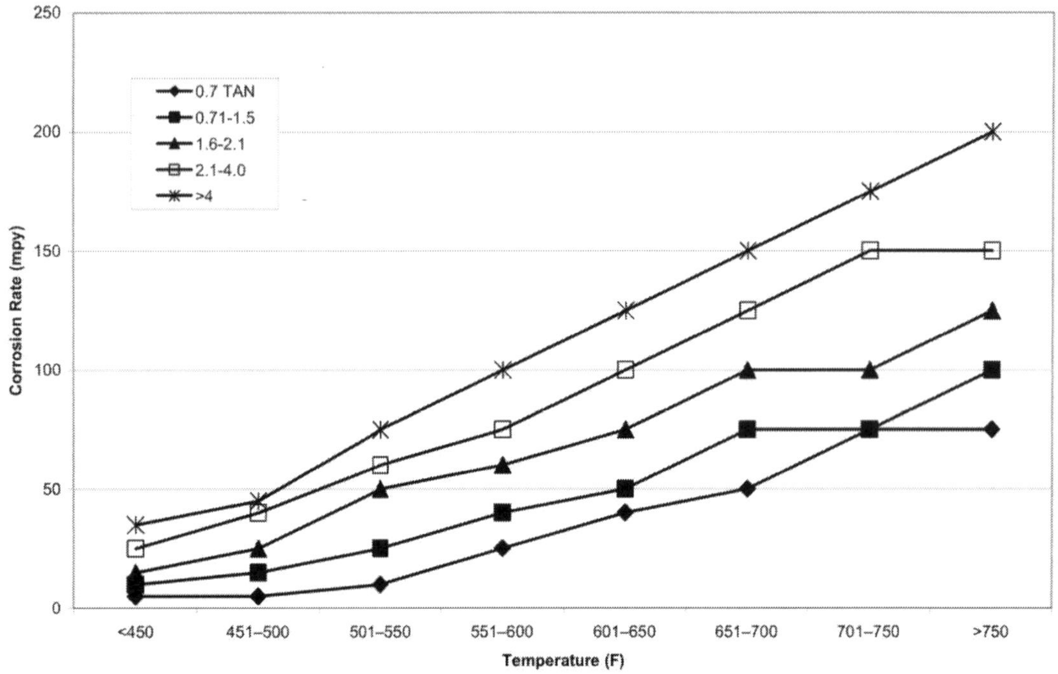

Figure 8 - Corrosivity assessment chart for 9Cr-1Mo Steel at >3% S and >100 ft/s per API RP 581
(Conversion: 40 mpy = 1 mm/y; 100 ft/s = 32 m/s; 600 F (C).

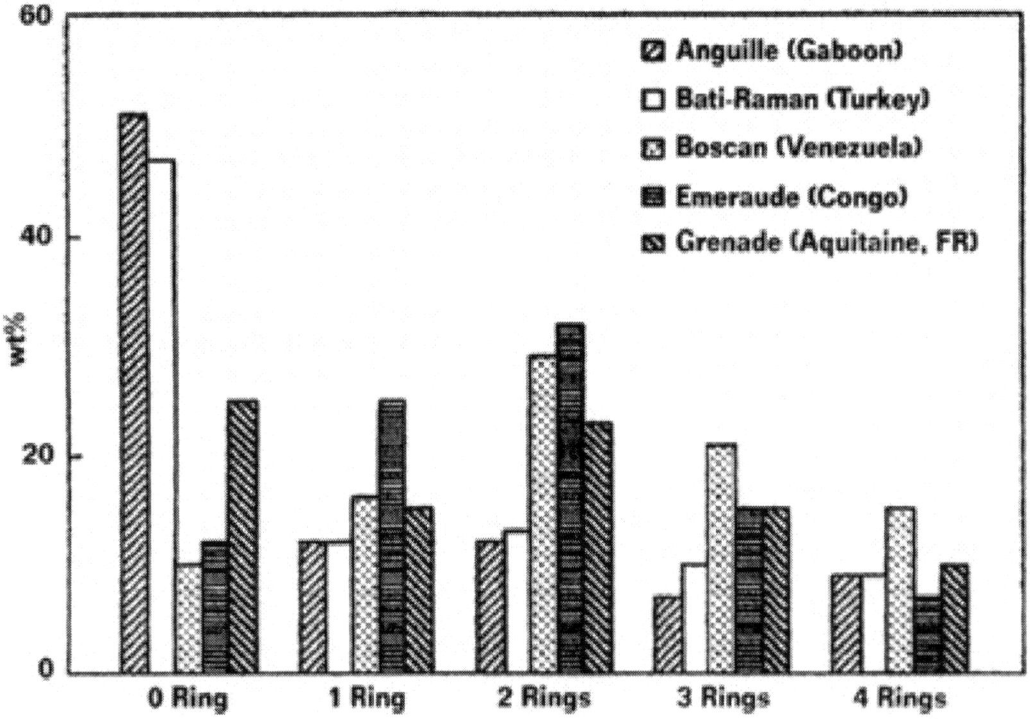

Figure 9 – Variations in organic ring structures in selected crude oils.

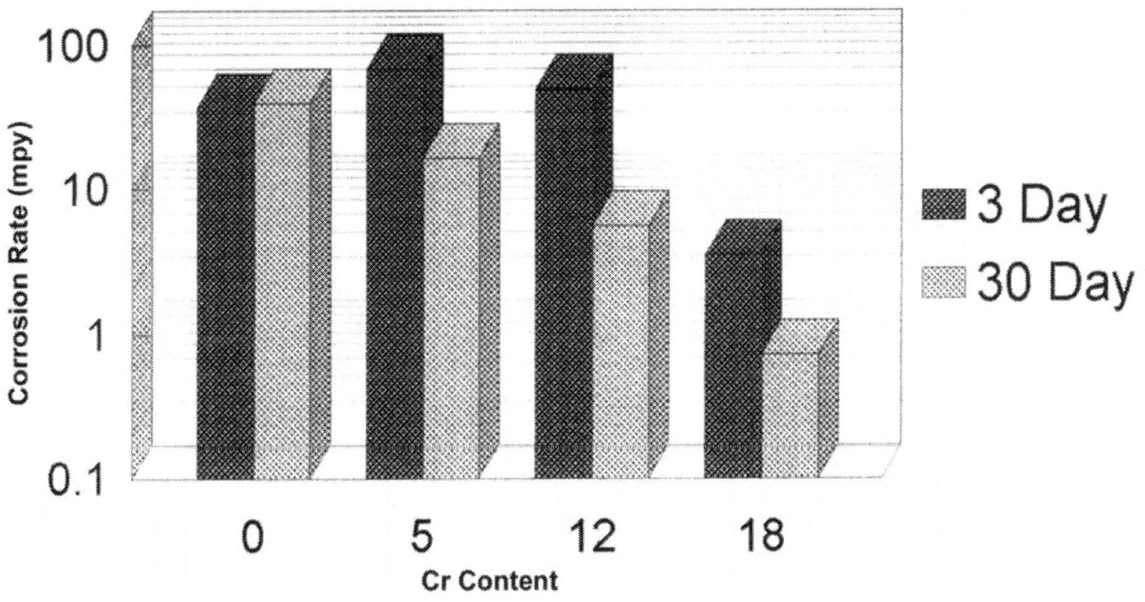

Figure 10 – Corrosion rates versus time for exposure on commonly used alloys in sulfur dominated hot oil environment (9Cr Steel values interpolated from those of 5Cr and 12Cr steels).

145c Crude Oil Corrosivity Measurement Using Radioactive Tracer Technology

Douglas C. Eberle, Craig M. Wall, and Martin B. Treuhaft. Engine, Emissions and Vehicle Research, Southwest Research Institute, PO Drawer 28510, 6220 Culebra Road, San Antonio, TX 78228
New sources of crude oils with unknown corrosion potential present significant opportunities and uncertainties for refiners. Traditionally, total acid number (TAN) measurements are often used to gauge corrosivity, but with poor results. Loss-in-weight corrosion test methods suffer from poor sensitivity, long exposure times and low reproducibility. Gravimetric coupons can sometimes gain weight due to adhesive product buildup. Radioactive tracer technology has been used for more than fifty years by the automotive industry to measure real-time wear in operating engines. This technology has been adapted to develop two techniques to measure high temperature crude corrosivity in a timescale short enough to avoid thermal degradation of reactive acids. The first technique allows for high sensitivity real-time measurement of alloy corrosion when exposed to high-temperature crude oil in a flow loop under high shear stress conditions. The second utilizes a pressurized stirred reactor. This latter technique has been used to measure both dissolved corrosion product buildup in the oil as well as adhesive corrosion product buildup on the test coupon. In both cases, corrosion product buildup sensitivity is in the 100 ppb range, with measurements made as quickly as one hour at temperature.

Effect of Naphthenic Acid Structure on Carbon Steel Corrosion

Priya Rangarajan, Fred Holmes and Bruce Randolph, ConocoPhillips Company

ABSTRACT

This work investigates the effect of structure on the naphthenic acid corrosivity towards carbon steel. Three acid mixtures with a comparable carbon number range were selected for this study: two containing predominantly acyclic structures (Acids A and B) and one containing a predominantly cyclic type structures (Acid C). Structural information for these acid mixtures was obtained initially by liquid chromatography/mass spectroscopy. In the temperature range of 425-500 °F, acids A and B showed comparable corrosion rates (32 mpy at 425 and 55 mpy at 500 °F). However, acid C showed somewhat higher corrosion rates (42 mpy at 425 °F and 87 mpy at 500 °F). It is presently unclear why the cyclic structures are more corrosive. No evidence for excessive thermal breakdown or volatilization was seen in any of the three acid mixtures under the conditions studied. Presently, these structural differences have much smaller influences on the corrosion rate than the overall TAN and the temperature.

INTRODUCTION AND BACKGROUND

Acid crudes contain naphthenic acids, a generic term for a broad range of organic acids present in the crude oils, which can cause corrosion in downstream processing equipment. Naphthenic acids are believed to have the chemical formula $R(CH_2)_nCOOH$ where R is a cyclopentane or cyclohexane ring [1]. Areas like California, Venezuela, India, China, Russia, West Africa, North Sea and Brazil are known to supply crudes containing naphthenic acids [2].

The acidity of the crudes is measured in terms of TAN or the total acid number, which is referred to in terms of milligrams of KOH needed to neutralize one gram of oil. TAN values can range from essentially 0 to as high as 10. Testing the whole crude and side cuts determines the region where different acids would concentrate during distillation. Crudes containing a TAN >0.5 or refined fractions with TAN >1.5 are found to be more corrosive in general. In addition to corrosion, fouling problems can also occur due to the corrosion products (iron naphthenate salts). Desalting problems can occur due to the emulsification tendencies of acid crudes; as the pH of the desalter wash-water increases, metal naphthenate salts can form very stable emulsions.

Naphthenic acid corrosion is localized particularly in areas of high velocity or where condensation of concentrated acid vapors occurs. It is typically more prevalent in carbon steel equipment. Also, corrosion is most severe in crude distillation units in the 220-400 °C (425-825 °F) range. In high temperature, high velocity applications like furnace and transfer lines, naphthenic acids can accelerate sulfidic corrosion. Thus, high TAN crudes have tremendous implications from a downstream processing perspective.

An important concern that has not been sufficiently addressed from a corrosion standpoint is the nature of the acids present in the crude oils. Although generally grouped together as "naphthenic acids", the acids present in crude oils can vary widely in terms of the molecular weight and their side chain groups, both of which can potentially have an impact on their corrosivity. Although the nature of the acid type has long been speculated to contribute to different corrosivities, the concept has never been proven definitively. Literature reports have shown data that indicate higher boiling acids tend to be less corrosive than lower boiling acids [2, 3].

Previous researchers have speculated on the possibility of acid structure influencing corrosivity but never definitively established the result [4]. Our work explores this possibility and is designed towards understanding whether the nature of the hydrocarbon chain attached to the acid group can influence acid corrosivity. Three different types of acid mixtures were investigated: two containing mainly acyclic type acids and the other containing predominantly cyclic acids.

EXPERIMENTAL

Materials: Naphthenic acid mixtures were purchased from three different sources and are named acids A, B, and C respectively.

Setup: The glassware apparatus used is depicted in Figure 1. The process employed a glass stirring reactor of 500 ml fitted with two metal test coupons (carbon steel) screwed on to a metal rod, a diffuser tube with fritted glass to disperse inert nitrogen gas (50 ml/min) through the reaction medium, and a condenser for condensing any lights volatilizing from the reactants. The vessel was filled with mineral oil containing known amounts of the desired acid compounds described below in Table 1. The coupons were placed such that there was sufficient space to allow the reaction mixture to flow freely between them. The oil was heated to the desired temperature by a 270 watt external heat mantle and stirred at a speed of 640 rpm. The oil was also analyzed for soluble iron content to get a measure of the corroded iron that was soluble in the oil phase.

The reaction temperature was maintained at 425 °F, and about 400 ml of mineral oil acidified to a TAN of 5 was purged constantly with 50 ml/min of N_2. The TAN was composed of a selected naphthenic acid mixture and the reaction times varied from 4 h to 24 h. The corrosion rate of the coupons immersed in the oil mixture was evaluated by weight loss technique and inductively coupled plasma (ICP) measurement of soluble iron. This sequence was also repeated for different reaction temperatures (425 °F, 450 °F, 475 °F and 500 °F) for a 24 hour run for all three naphthenic acid mixtures. Standard runs for the experimental protocol were performed by running acid A in mineral oil (TAN=5) at 425 °F in a nitrogen environment for 24 h every two weeks. The corrosion rates obtained were assessed for reproducibility. Average corrosion rates and standard deviation values were computed.

Analytical: The three acid mixtures were characterized in order to determine their carbon number and type distribution. Flow injection mass spectral analysis was performed using a Waters Micromass QuattroLC triple-quadrupole mass spectrometer interfaced to an HP1100 HPLC. Thermal gravimetric analysis (TGA) was done on all three acid samples in order to determine their boiling point distributions. A TA Instruments 2950 TGA was used for this purpose.

Figure 1: Experimental Set up in Lab

The corrosion rate was monitored using the following relation:

Corrosion rate (mpy)=(K x ΔW)/(A x T x D)

Where K=3.45 x 10^6, a unit conversion factor
A=exposed surface are (cm^2), area
ΔW=weight loss of iron (mg)
T=exposed time (h)
D=density of material (g/cm^3)

RESULTS AND DISCUSSION

Analytical studies on naphthenic acid characterization: Mass Spectroscopy

The molecular weight range of the acids was investigated by atmospheric chemical ionization (APCI) mass spectrometry (MS) and the spectra are shown in Figure 2 and selected ions of variable mass/charge (m/e) ratio are given in Table 1.

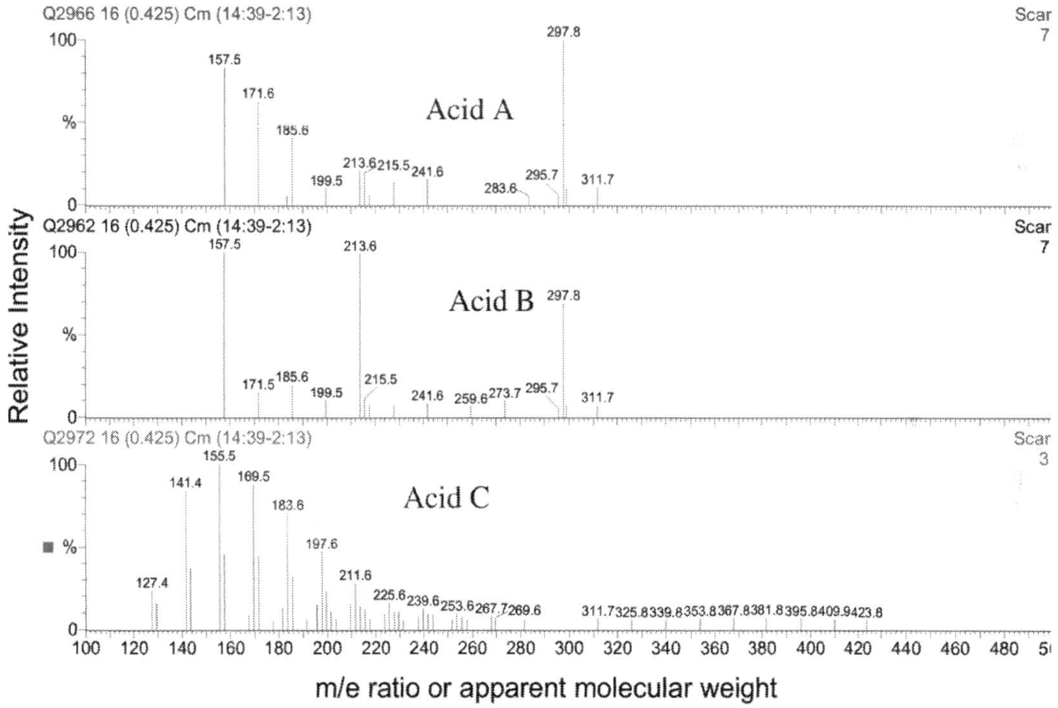

Figure 2: APCI data for the three different acid mixtures studied

Under the conditions of the analysis, the pattern of mass spectral peaks may be used as a fingerprint indicative of the carbon number and degree of unsaturation of acids in the sample. Table 1 shows the molecular weights of the anions obtained by the extraction of a proton from the acidic species. As is evident from the data, the molecular weights are distinct for a given carbon number and serves to distinguish the species in the APCI-MS spectrum shown in Figure 2. Note that this approach relies on the assumption that all ions observed are from acids, requiring isolation of the acids from other organics that produce negative ions.

Table 1: Molecular weight of anion formed by proton abstraction of acidic species

# C	Diacid or Acyclic	One DBE	Two DBE	Three DBE	Four DBE	Five DBE	Other (Higher DBE)
atoms	$C_nH_{2n}O_2$	$C_nH_{2n-2}O_2$	$C_nH_{2n-4}O_2$	$C_nH_{2n-6}O_2$	$C_nH_{2n-8}O_2$	$C_nH_{2n-10}O_2$	$C_nH_{2n-12}O_2$
7							
8	143	141	139	137	135		
9	**157**	**155**	153	151	149		
10	**171**	**169**	167	165	163	161	
11	**185**	**183**	181	179	177	175	
12	**199**	**197**	195	193	191	189	
13	**213**	**211**	209	207	205	203	201
14	227	**225**	223	221	219	217	215
15	**241**	**239**	237	235	233	231	229
16	255	**253**	251	249	247	245	243
17	269	**267**	265	263	261	259	257
18	283	281	279	277	275	**273**	271
19	**297**	**295**	293	291	289	287	285
20	**311**	309	307	305	303	301	299
21	**325**	323	321	319	317	315	313

(Figures in **bold** represent most of the m/e ratios depicted in Figure 2)

In the above table and subsequent graphs, the following definitions apply:

Acyclic—carboxylic acid with alkane chain containing no rings or double bonds.
One DBE (Double Bond Equivalent)—carboxylic acid with alkane chain containing one ring.
Two DBE—carboxylic acid with alkane chain containing two rings.
Three DBE—carboxylic acid with alkane chain containing three rings.
Four DBE—carboxylic acid with phenyl group on the alkane chain.
Five DBE—carboxylic acid with phenyl group and one additional ring on the alkane chain.
Higher DBE: Diacid/acyclic acid molecules containing two carboxylic acid groups on an alkane chain with no rings.

It can be seen that the range of molecular weights of the acids in all the three acid mixtures are comparable, although there are variations in the distribution and types of acid constituents. Acids A and B contain similar distributions although each displays one or more constituents which vary in composition from the other. Furthermore, it is instructive to compare the molecular weights corresponding to the spectral peaks shown in Figure 2 with the species corresponding to these molecular weights shown in Table 1. Both acids showed little unsaturation and are predominantly acyclic as compared to acid C which contained higher amounts of components with one DBE, or a higher degree of unsaturation. This is most likely due to the presence of cyclic components in the hydrocarbon chain, rather than olefinic structures.

Analytical studies on naphthenic acid characterization: TGA Studies

Thermal gravimetric analysis (TGA) was done on the three acid samples in order to gauge their relative volatility. Figure 3 shows the comparison of three acid samples. The volatilization temperatures for two acids, A and B, were primarily in the temperature range of 250-350 °C. Although both weight loss profiles appeared to be similar, some differences were noted. Acid A showed an early weight loss region prior to the main weight loss region. Also, acid A appeared to have a slightly narrower weight loss range than acid B. It should be cautioned that the TGA is not able to distinguish volatilization due to boiling from decarboxylation and subsequent volatilization of decomposition products.

On the other hand, acid C showed a temperature increase of about 20-30 °C in the onset of weight loss as compared to that displayed by acids A or B. The increase can be attributed to the nature of the hydrocarbon chain attached to the acidic species, the higher boiling point of the acid species, or better stability towards thermal breakdown of the acid constituents in this acid mixture. Higher temperatures (>300 °C) did not show any variation in volatilization behavior of the three acid mixtures.

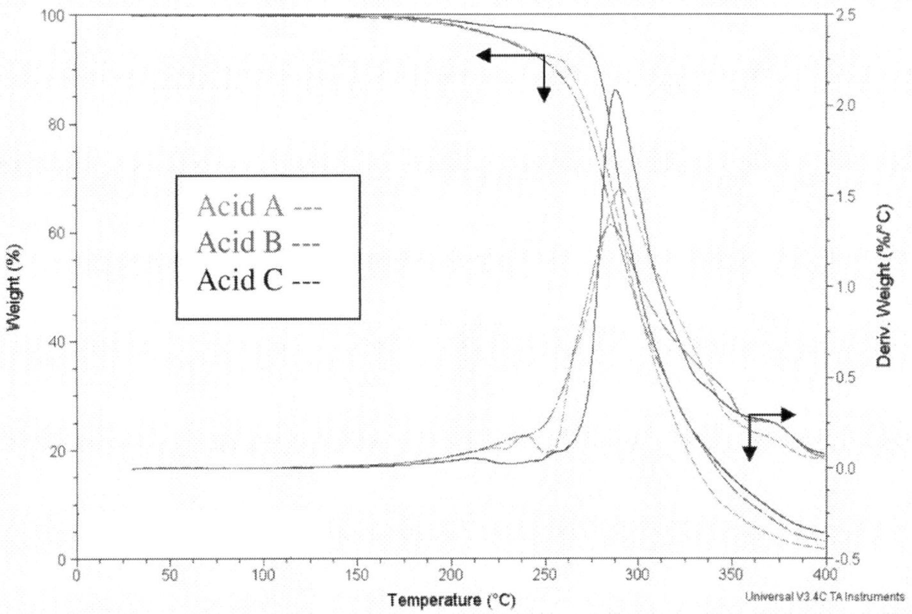

Figure 3: Weight loss curves as a function of temperature and the derivatives for the three acids

Based on the results from LC/MS and the TGA, it was concluded that acids A and B are similar in the type of the acids (acyclic), although there are slight variations in the composition of the acid constituents. On the other hand, acid C displayed markedly different characteristics from the other two acid types, mainly in the form of varying hydrocarbon chain structure (cyclic) and higher temperature for onset of volatilization.

Corrosion Tests: Full Range Naphthenic Acid Mixtures

Initial standard tests at 425 °F for acid A in mineral oil (TAN of 5) gave a baseline corrosion of 29.5±5.8 mpy for carbon steel. Five sets of data were used for performing the error analysis. Each set of data consisted of the average corrosion rate of two coupons used in each experiment. Less than 10% variation in corrosion rates was observed for two coupons used in the same experiment. The overall error in reproducibility (19.6%) is well within the range of typical errors associated with corrosion experiments (20-25%) [2]. The standard test data was in good agreement with Piehl's data of 29 mpy for a TAN of 5 [5]. A control was also run with no coupons, and only the coupon holder and stirrer, to determine whether any iron dissolution occurs due to corrosion effects from the coupon holder. Less than 5 ppm Fe was obtained by ICP (Inductively Coupled Plasma) atomic absorption spectroscopy, indicating that the corrosion of the metal coupon holder was negligible under the typical run conditions.

In order to study the effect of the nature of acid side chain on corrosion, corrosivities of the three naphthenic acid mixtures were investigated. The effect of exposure temperature on corrosion rate (for an initial TAN of 5 with the acid mixtures) is shown in Figure 4.

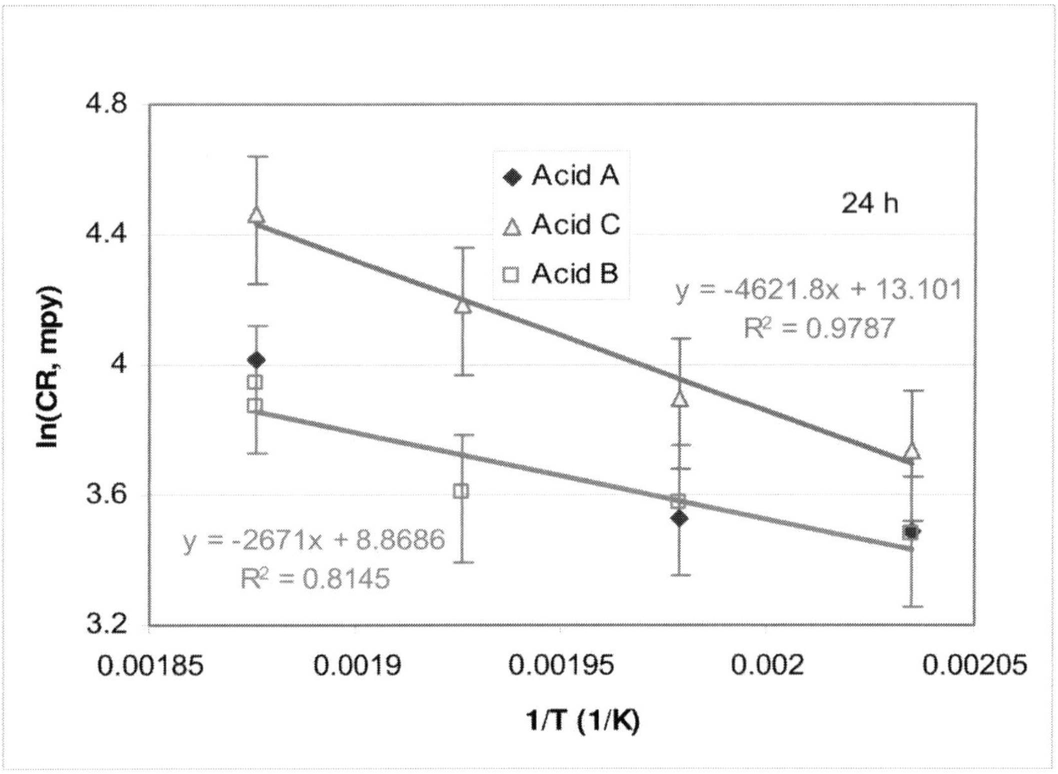

Figure 4: The effect of temperature on Fe weight loss for different acid types

The corrosion rates are seen to be comparable for acids A and B within experimental error. The temperature profile (ln(corrosion rate) vs. 1/T did not give a good linear fit (R^2=0.814). Since a mixture of naphthenic acids instead of a single model compound was studied, the effective corrosivity of the mixture at any reaction temperature would depend on the molecular weight ranges, concentration and boiling points of the acids. This could provide at least a partial explanation of why the linear fit did not show a higher correlation coefficient. Also, the experimental reproducibility has an error of $\pm$20%, which could also contribute to the lower value of correlation coefficient. The current set of experiments shows that in the temperature range studied (425-500 $^\circ$F), both types of acid mixtures were most corrosive at 500 $^\circ$F (i.e., 1/T=0.000188 K). It can be seen from Figure 3 that this temperature of 500 $^\circ$F (or 260 $^\circ$C) is near the region of initiation of rapid weight loss representative of volatilization or vaporization of a large number of acid components present in the two acids. This could potentially explain why both acid types displayed higher corrosion rates at 500 $^\circ$F than those predicted by the linear profile (Figure 4), if experimental variations can be discounted. This observation supports the theory expounded by other publications that the acids are most corrosive near their boiling points [3]. Furthermore, the linear correlation (Figure 4) predicted a corrosion rate of 76 mpy at 600 $^\circ$F for acids A and B. Acid B yielded an experimental corrosion rate of 70 mpy, which varies less than 10% from the predicted value.

On the other hand, acid C displayed significantly higher corrosion rates which remained exponentially related to the temperature. A good correlation coefficient was obtained for the linear plot of ln(corrosion rate) vs. (1/temperature) (R^2=0.978). It was also seen that the slope of the line was greater than that seen for acid B, suggesting that the increase in corrosion rates with temperature is higher for acid C.

Figure 5 shows the corrosion rates of the three acid mixtures at various exposure times and at 425 °F. Acids A and B show very similar corrosion rates as a function of reaction time except for the short time (4 h) run which showed a significant difference. The discrepancy between the corrosion rates for acids A and B for the 4 h run is not clear, and may simply be indicative of a bad data point.

A more discernible difference in corrosion rates was seen when comparing acids A and/or B with acid C. Ideally, the corrosion rates for all reaction times for an acid mixture should be constant. However, acid C showed a gradual decline in corrosion rates with time, as is sometimes observed in experimental runs. Compare this with the more closely aligned values obtained for acid B at various experimental times. Thus, we see definitive differences in the corrosion rate trends for acid A and B versus acid C.

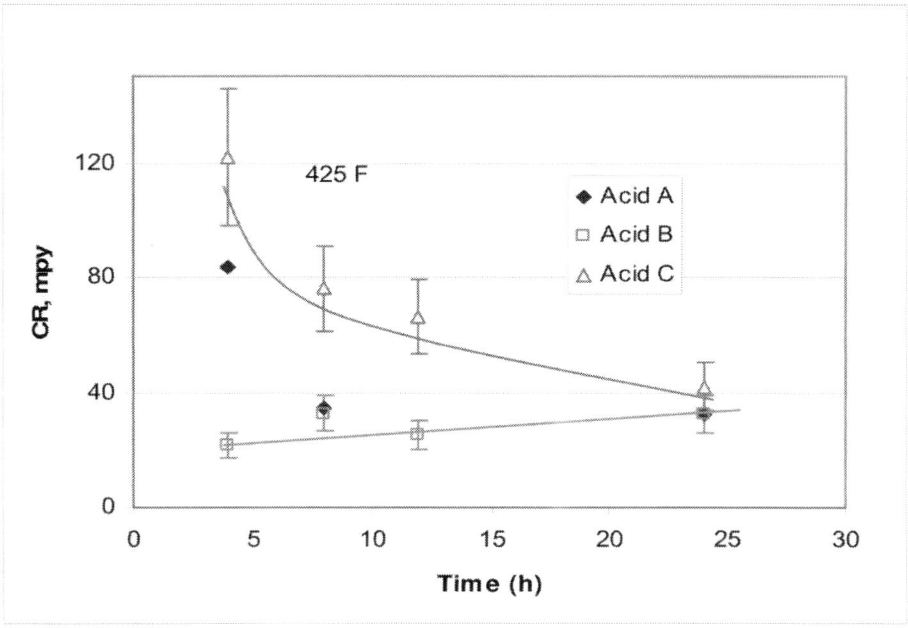

Figure 5: The effect of reaction times on Fe weight loss for different acid types

From these studies, it is not apparent whether this difference in corrosion rates can be attributed solely to structural differences between the acid species. Aromatic acids are known to be more acidic than acyclic acids in an aqueous environment and at room temperature due to better stabilization of the negative charge upon loss of a proton. However, whether such an effect with alicyclic acids would translate to higher acidity at a much higher temperature in a non aqueous environment is debatable and still unresolved. Other factors, such as varied boiling points and volatilization behavior, distribution of acidic constituents (molecular weight), solubility of the acid species in the base oil, and thermal stability of the acidic mixture could also play a role.

CONCLUSIONS

Glassware experiments performed at 425 °F showed significant differences in corrosion rates of carbon steel for a predominantly acyclic naphthenic acid mixture (Acids A and B) versus a predominantly cyclic type naphthenic acid mixture (Acid C) having similar carbon number range. This pattern was repeated at temperatures of 450, 475 and 500 °F. It is unclear why cyclic structures would be more corrosive. Other parameters that could be important are the carbon-number distribution, concentration and the boiling points of the acid components.

Based on this study, it appears that the structure of an acid could have a discernible effect on the corrosivity of the acid at the conditions studied.

ACKNOWLEDGEMENTS

We are thankful to ConocoPhillips Company for granting us the permission to present this paper. We also acknowledge Tom Baugh, Principal Scientist, for TAN analysis, Laura Allison, Senior Scientist, for naphthenic acid analysis by LC/MS, Dave Krottinger, Principal Scientist, for TGA analysis, and Jeff Larmer, Analytical Technician, for ICP analysis.

REFERENCES

1. E. Slavcheva, B. Shone and A. Turnbull, "Review of naphthenic acid corrosion in oil refining", British Corrosion Journal, 34(2), 1999, pp. 125.
2. R. Kane, "A Comprehensive study on naphthenic acid corrosion", Corrosion 2002, Paper no. 025555.
3. B. Messer, B. Tarleton, M. Beaton, and T. Phillips, "New theory for naphthenic acid corrosivity of Athabasca oilsands crudes", Corrosion 2004, Paper no. 04634.
4. K.R. Lewis, M.L. Duane, and R. Schelling, "Processing corrosive crude oils", Corrosion 1999, Paper no. 377.
5. R.L. Piehl, "Naphthenic acid corrosion in crude distillation units", Materials Performance, January 1988, pg. 37.

145e Biological Upgrading of Petroleum – Recent Developments and Research Needs

Abhijeet P. Borole, Oak Ridge National Laboratory, 1-Bethel Valley Rd, Oak Ridge, TN 37831

Biological upgrading of petroleum is a potential alternative for improving quality of crudes which are high in sulfur and nitrogen and potentially for heavy crudes as well. This presentation will give an overview of the current status of various technologies including biodesulfurization, biodenitrogenation, and viscosity reduction. Progress in biodesulfurization will be discussed in detail. Research in biocatalyst development has resulted in identification of several new microbial catalysts with properties different from that of Rhodococcus sp. IGTS8, especially in terms of substrate specificity and thermostability. While most work has been in identifying new strains, a few strategies to improve initial biocatalyst activity and cell density have also been reported. This can result in up to an order of magnitude improvement of initial conversion rates, however, little progress has been made in sustaining biocatalyst activity. What can be done to improve biocatalyst stability? This and other biocatalyst related questions will be addressed. Developments in biocatalyst immobilization, bioreactor design, oil-water-biocatalyst separation can assist in pushing the rates and process efficiency further. Biocatalyst as well as bioprocess issues will be discussed with assessment of potential for commercialization and evaluation of research needs.

145f Market Developments for Opportunity Crude Processing

Stephen L. Jones, Purvin & Gertz, Inc., 600 Travis Street, Suite 2150, Houston, TX 77002

Crude oil supply patterns are changing rapidly as production from traditional supply basins decline and new sources of supply come on-stream in the Atlantic Basin. The presentation would discuss and quantify the issues and impacts that these new crude oil streams will have on refiners' crude oil supplies, qualities and related operating challenges. The specific areas that would be covered include.... A continuing build-up of new and developing supply sources from the deep water Gulf of Mexico, production from the Caspian region, expansion of exports from the Former Soviet Union (FSU), new streams from West Africa and China, expansion of synthetic crude oil and bitumen supplies from Canada and developments in Latin America.... Prospects for production and quality shifts from traditional U.S. crude oil suppliers such as Mexico, Venezuela and Middle East producers.... Developing market pressures that change crude oil trade patterns, pricing dynamics and regional refinery expansions.... The historical and anticipated changes in crude oil qualities and the potential impact of these changes in consideration of refinery operating strategies.... Highlight opportunities and issues that changing supply patterns and qualities would imply for the major refining markets, including N.W. Europe, U.S. and Asia refiners. The presentation would be of broad interest to refiners (as well as upstream market participants) and would best fit the agenda of the Upgrading and Processing of Opportunity Crudes Session (TA 017) at the 2006 Spring National Meeting.

NEW TECHNIQUES FOR MEETING NEW PRODUCT SPEC AND DEMAND UNDER CATASTROPHIC FAILURE THROUGH MULTI OBJECTIVE MULTI REFINERY OPTIMIZATION

Ahmed Sirag Khogeer, Research & Development Center, Saudi Aramco, PO Box 5836, Dhahran, 31311, Saudi Arabia(ahmed.khogeer@aramco.com) and M. Nazmul Karim, Chair, Department of Chemical Engineering, Texas Tech University, Lubbock, Texas

Abstract

With the political instability and increased risk of war and terrorist attacks in which refineries and fuel transportation grids are possible targets, pressures are exerted on refiners to explore new techniques to meet product demand under the case of partial or total failure of a refinery or multi refineries system. Since we are assuming a sudden failure or a need to promptly meet a new product spec, thus major upgrading of refineries is not an option but the option is to optimize the operations and the process streams within a refinery and intra refineries in order to produce salable products meeting the required product spec. A provision of importing high quality steams intended for sell at international markets was allowed with constraints. Simulation runs were carried out using ASPEN PIMS utilizing nonlinear techniques to solve the optimization problem. The results addressed the need to debottleneck some refineries or transportation media to meet the demand for essential products under partial or total failure scenarios. They also addressed the benefits of importing high spec products to help recover some of the losses and what is needed in order to accomplish this. In addition, the results showed nonlinear relations with the global objective for some refineries. In addition, crude processing capacity has more effect on products supply than the treating and upgrading capacity.

Introduction

During past years, many catastrophic failure happened world wide leading to sever interruption to products supplies. For example, morocco, Kuwait, and Texas City. With the increased risk of war and terrorist attacks, refineries and fuel transportations grids are targeted since they supply fuels to essential nodes such as power plants and war efforts and thus they need to secure supply in case of a catastrophic failure event. In these cases, the aim will be to meet the supplies to essential nodes first, such as power plants and then produce sellable products with the remaining operational units. It is not possible under these cases to recover from the failure without some investment in the process units that should take time to rebuild. Another case if is the refining system is required to meet a new product spec that the system of refineries was not originally designed for.

The way to fulfill the above mentioned objectives is through multi refinery multiple objectives optimization, where catastrophic failure effects are quantified and incorporated into the objective function. This way, catastrophic failure effects could be accounted for in the economic objective and alternatives from other refinery or the process units within the refinery can be suggested. In literature, not much attention has been paid to this field until recently. Most work in literature is concentrated on the effects of process accidents on refinery emissions and safety. For example, for some refiners, the scope might be instability and phase equilibria research rather than economic consideration [2, 3, 4, 5].

Model formulation

Below are the major components of the model that was formulated to address the optimization problem. The effect of a catastrophic failure resulting from a process accident or a deliberate action causing partial or total failure should be divided into two groups [7, 8]:

a) Catastrophic failure having on or off (0/1) operations. 0/1 indicates that a particular equipment or process unit is either working or not. For example, in the case of pipeline rapture, the pipeline cannot usually be used and should be shutdown. For this group, constraints were treated as binary (0/1) constraints in the model.

b) Relative increased risk of operations due to process or non process reasons. This includes war risks and increased risk of terrorist acts such as hijacking of tankers or trucks. For example, in case of war, tankers are penalized $1.0 per barrel for shipment from Arabian Gulf but not from the West Coast of the Arabian Peninsula.

Supply-Demand Chain

A Supply–Demand chain was constructed which included the major refinery feed, processing, storage and consumptions nodes. This supply demand chain will be the heart of any reallocation and inter refinery transportation in the future in order to maximize the sellable products. Fig 1 shows a schematic example of a simplified supply demand chain which gives an example of feed pools, refineries as blocks, internal units in refineries, major consumption nodes and interconnections among refineries and within the chain. To have a good model of a refining system Supply –Demand chain, the following should be identified:

a. Different feeds such as crude, fuels, condensate, reduced crude, MTBE, etc.
b. Different refineries with each refinery having different process units.
c. Process limitations and capacities.
d. Different transportation media such as tankers, pipelines and trucks.
e. Major consumer nodes such as desalination plants, power plants, etc.
f. Markets and products requirements such as gasoline for summer & winter modes.
g. Supply/demand growth per year.

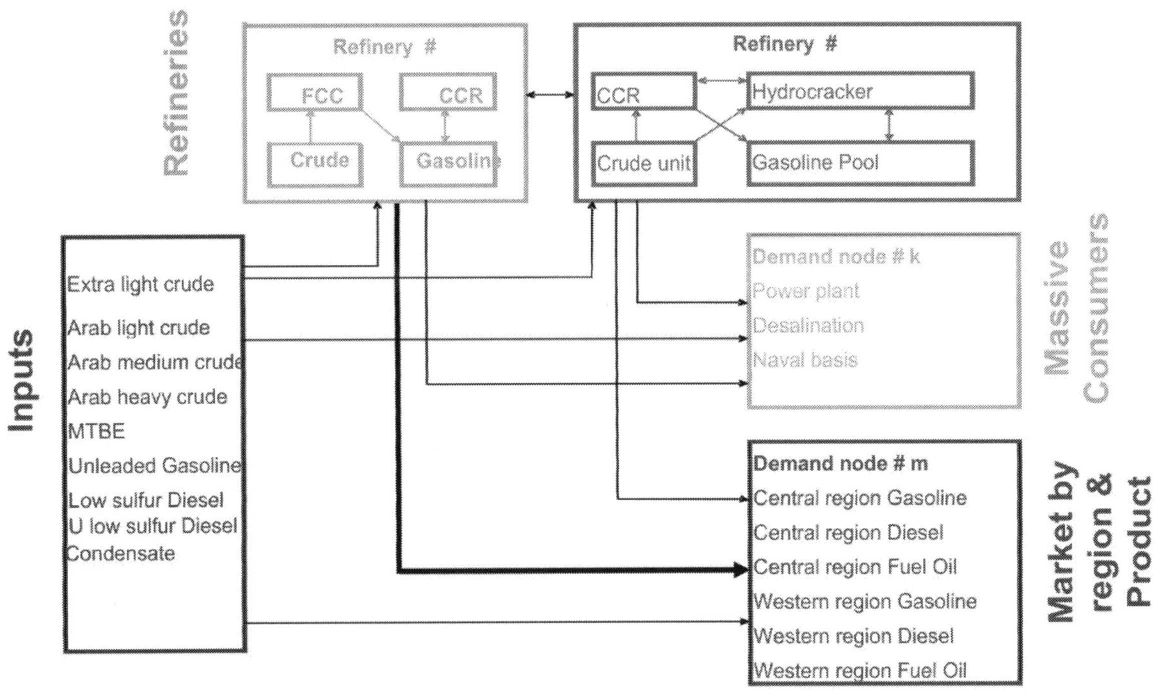

Figure 1: Refining-Supply / Demand Chain

Objective Function

The objective function can be expressed mathematically as follows [1]:

Minimize $\Phi = -(P_1 - P_2 - P_3 - P_4)$ ---(1)

Where:

- Φ is the profit-based objective function,
- P_1 is the selling price of the products which is the major income of a refinery
- P_2 is the price of the raw materials,
- P_3 is the operating costs,
- P_4 is the product-property related parameter and is usually set to zero.

P_1 is given by the following equation:

$$P_1 = \sum (Prod_j \,.\, Price_j \,.\, wt_j) \quad (2)$$

Where

- j = 1, ..., 7 is the product for the refinery such as LPG, Gasoline, etc.
- $Prod_j$ is the production rate of $product_j$, (BBL/Day)
- $Price_j$ is the price of a unit quantity of product j, \$/BBL
- wt_j is the weighting on the jth term of the objective function

P_2 is given by the following equation:

$$P_2 = \sum (Feed_k.\, Price_k \,.\, wt_f) \quad (3)$$

Where

- F is the feed or imported stream flow rate (BBL/Day). This can be crude oil, intermediate product (such as vacuum gas oil VGO for FCC feed) or finished product (such as ultra low sulfur diesel) imported from other source.
- Price k is the cost of the feed, \$/BBL
- wt_k is the weighting on the term related to the cost of raw materials.

P_3 is given by the following equation:

$$P_3 = \sum (utility_n \,.\, UC) \quad (4)$$

Where

- $Utility_n$, (m^3/Day) is the utility consumed in process unit n;
- UC is the cost of the utility (\$ /m^3).

Constraints

Refinery constraints are usually capacity limits, process variable constraints and product specifications. Other constraints such as environmental constraints can be added as limit or penalty constraints. Table 1 gives an example of such constraints for a typical mid-size, medium-complexity refinery.

Table 1: Typical Refinery Constraints

	Unit	Constraint Name	Lower limit	Upper limit
Capacity constraints	CDU	Throughput, KBPD	40	100
		Naphtha, %	18	26
	FCC	Throughput, KBPD	15	25
Process constraints	CDU	Heater outlet temp, C	NA	370
	FCC	Max reactor temperature, C	NA	530
Product constraints	Gasoline	RVP, Psia	7	11
	Kerosene	Flash Point, C	38	--

Optimization Algorithm

Simulation runs were done using ASPEN PIMS utilizing nonlinear techniques to solve the optimization problem. The nonlinear algorithm is Successive Linear Programming SLP which has a wide acceptance and success in solving mildly nonlinear term in large refining system linear programming models [6]. SLP uses a Taylor series expansion to approximate the nonlinear terms in the constraints and the objective function around a feasible point.

Software selection and features

Simulation runs were carried out using both MATLAB and ASPEN PIMS. Since the model was based on actual refineries data, there was no need for detailed simulations of refinery units. It was preferred to continue the runs using PIMS since it is the primary operations optimization tool for many refineries. PIMS uses several tools to perform the optimization, including a third party optimization engine. In addition, PIMS uses the same basic chemical engineering equations needed for mass and material balance.

Description of the Refining System

The model consisted of seven refineries; in which five are domestic refineries with variable capability of exporting excess products to other refineries or international markets. These refineries where built with the vision of supplying local markets so there are limitations on the type and quality of products they can produce. The other two are export refineries and are state of the art refineries with capabilities of meeting stringent products specifications.

Table 2: Summary of Refinery Processes

	Crude Capacity, KBPD	Condensate	Reformer	CCR	FCC	HC	DHT
Export refinery # 1	200		N	Y	Y	N	Y
Export refinery # 2	200	N	N	Y	N	Y	Y
Domestic Refinery # 1	300	200	Y	Y	N	Y	Y
Domestic Refinery # 3	80	N	Y	N	N	N	N
Domestic Refinery # 4	450	N	N	N	N	N	Y
Domestic Refinery # 5	210	N	Y	N	N	N	Y

Cases

Different scenarios were done to simulate the consequences of catastrophic events on the products supply and the cases for a new product spec upgrade. Different refineries had different scenarios depending on the location, proximity to oil supply channels (such as seaport or pipelines), complexity factor of the refinery (which is a factor of presence of treating and conversion units) and availability of alternatives. All of the cases presented are based on real refining situation with some figures and constraints modified for confidentiality purposes. Following are the major cases in the model:

1. Total & Partial Capacity Loss as a Result of a Catastrophic Failure

These cases were simulated on Refinery # 1, which is a large refinery with capability of processing different crudes and condensate. This refinery is considered as the work horse of the supply for distillate and jet fuels. Note that the products from each refinery are derived from the following equation:

$$\text{Product}_i = \Sigma\, \text{Product}_{ij}\, Tr_k\,, \qquad Tr_k \in [0,1] \text{ binary}$$
$$= \text{Product}_{i1}\ (0 \text{ or } 1) + \text{Product}_{i2}\ (0 \text{ or } 1) + \ldots.\text{Product}_{in}\ (0 \text{ or } 1)$$

Where:

- Product is any petroleum product such as low sulfur diesel, JET-A1, etc.
- i is product number.
- J stands for the refinery number.
- Tr_K stands for the unit or train abbreviation and K is a binary variable to show that the selected unit or train in a unit is either on or off.

Three scenarios were done for this refinery:

1. Partial loss in the condensate splitting capacity, which can be due to one or more of the following considered reasons:
   a. Logistics of supplying the refinery with condensate such as an attack on the pipeline or constraint on the flow due to pumping limitation as a result of an attack, mechanical failure or cooling limitations.
   b. Processing capacity loss in condensate splitter column, products coolers, condenser or downstream product treatment of the condensate section of the refinery.
2. Total loss in the condensate capacity along with decreasing crude capacity to the minimum running capacity. This can be a result of a partial outage in the crude unit such as loss of one or more crude trains.
3. Hydrocracker unit flexibility loss. In this scenario, no constraint was imposed on the refinery capacity but a constraint was put on the Hydrocracker unit decreasing its capacity of running in diesel mode. This is important in producing high quality distillates which is needed not only for production of diesel but also for blending with straight run high sulfur diesel from the crude unit to make it salable. A loss in the hydrocracked diesel of 5000 BPD may cause more than double this amount of loss in total diesel available for sale. In addition, very low sulfur diesels are usually produced from hydrocrackers and this constraint can affect the supply of low sulfur diesel to the whole country as discussed in the results.

Figure 2 shows the effects of condensate capacity loss on margins & products yield for different products in scenario # 1. This is due to the availability of alternatives and the composition of the condensate. The effect was most clear on the production of unleaded gasoline (regular). The second product was low sulfur diesel. In both cases the effect was mainly observed after the capacity loss exceeded 40 %.

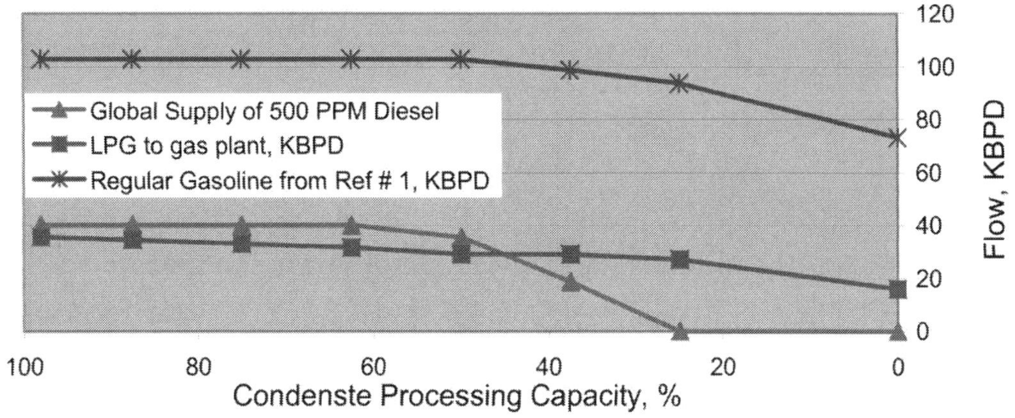

Figure 2: Effect of Condensate Capacity Loss on Objectives & Products Flow

For scenario # 2, the effect was clearer on the objective function for both local and global objectives. Figure 3 shows the effect of simultaneous crude and condensate capacities loss on selected light products yields.

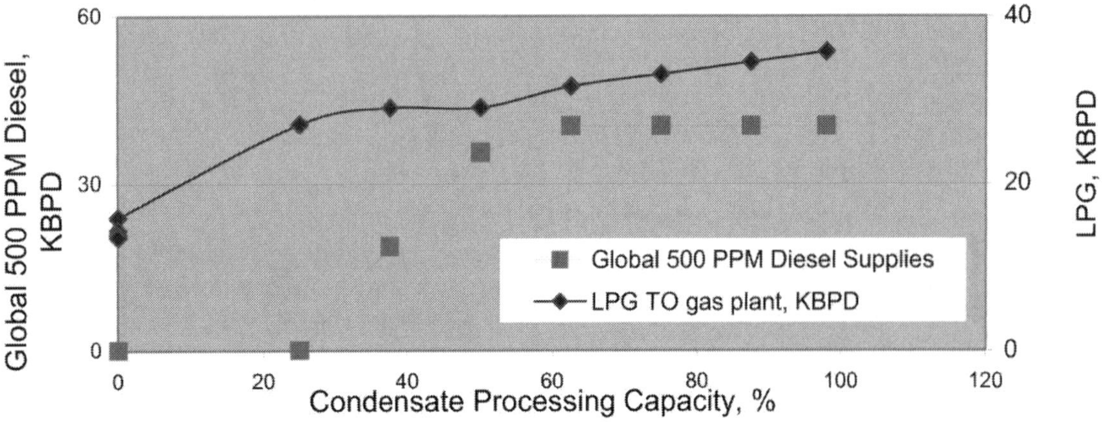

Figure 3: Effects of Condensate Processing Capacity Loss on Selected light Products

Scenario # 3 showed interesting results as shown in Figure 4 that both global and local objectives were not significantly affected by the loss of 50 % of Hydrocracker flexibility even thought this had a significant effect on the overall refinery low sulfur diesel (LSD) production. This is mainly due to the fact that the total Hydrocracker output is not affected but it is the quality and composition of products which are affected. That is, the Hydrocracker charge in all case is about 100 % of the base case, but due to limitations on the diesel mode capacity, the unit can not run towards maximum distillate but rather can go to the maximum constrained limit and have a point between maximum naphtha and maximum distillate modes. The main issue is that the amount of LSD available to the country is critically affected even thought it does not appear to effect neither the local nor the global objective functions.

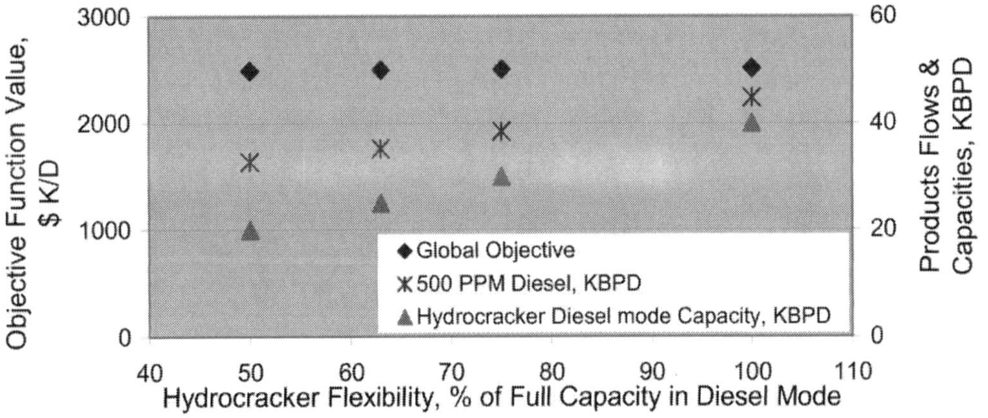

Figure 4: Effect of Hydrocracker Flexibility Loss on Objectives and Capacities

2. Effect of Product Specifications upgrade on a Specific Refinery-Market System

The purpose of this case is to examine the effect product specifications upgrades on a specific refinery-market system. This case is based on an actual situation for an urban market system consisting of north and south bulk plants. The south bulk plant provides products to other cities, as well as to a naval base, a desalination plant and a power plant. Due to design and logistic limitations, some products are only provided to this bulk plant by a nearby refinery, even though some other products are shipped from other refineries to the bulk plant through that refinery. Considering the consequences of the location of this market, it was assumed that the current medium sulfur diesel having a spec of 4000 PPM sulfur would be upgraded to same low sulfur diesel level of 500 PPM. Even though the new level is high compared to that required in some countries, it is a great achievement, considering that in many developing countries the diesel sulfur spec is 10,000 PPM. For this case, Refinery # 3 was selected. This refinery has the limitation of processing only light crudes. It also has cooling problems, which are a serious issue in countries of the Middle East especially during summertime. However, no other refinery in the model produces this type of diesel, even though there are other diesel types which are even lower in sulfur.

The refinery demonstrated a need to import 500 PPM sulfur diesel to be blended with the refinery-produced streams to make them salable. However, due to logistic reasons, this amount was limited to 10 KBPD. Consequently, with the lowered sulfur content ceiling, the total amount of refinery-made diesel that could be blended became less and less until it reached a point at which continuing at the same refinery output level meant having an unblendable stock of diesel. Some of this diesel was dumped to fuel oil, even though this constituted degradation for diesel and could limit the flash point spec of the fuel oil at some point. This process is clearly shown in Figure 5. At this point, the refinery's throughput has to be decreased to avoid having imbalanced product slates, as shown in Figure 6.

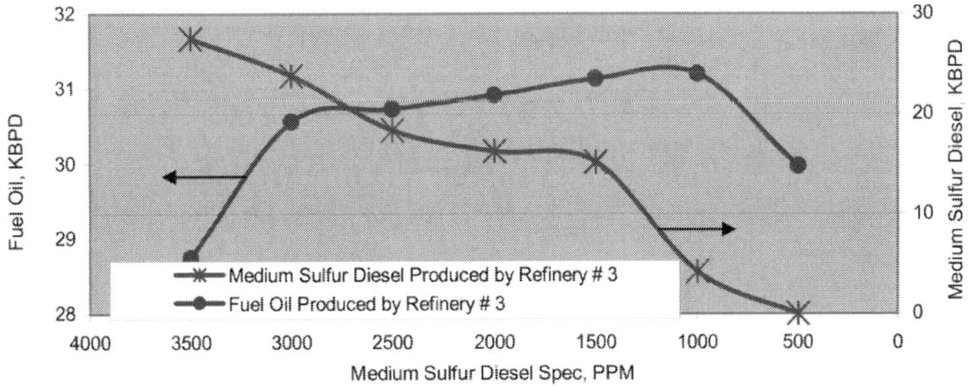

Figure 5: Un-salable Diesel Being Degraded to Fuel Oil

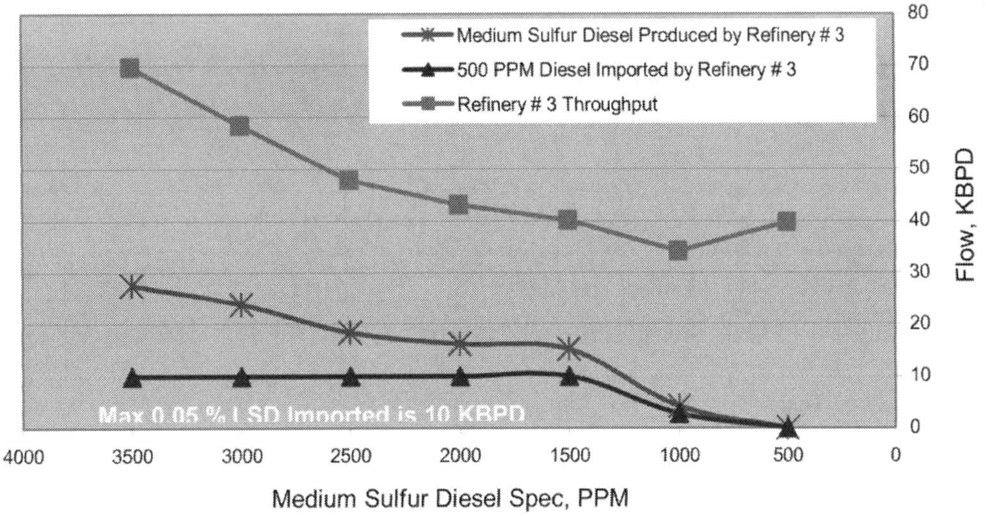

Figure 6: Unblendable Diesel Limits Refinery's Throughput

3. Global product specs changes

In this case, it was assumed that the national diesel sulfur spec for low sulfur diesel would change from 0.05% wt, which is 500 PPM to 0.005 % wt or 50 PPM. This is termed severe cutting, which requires deep desulphurization and affects all refineries that produce low sulfur diesel. To meet this low sulfur diesel level internally, refineries must have one or more of the following tools: distillate Hydrotreater, kerosene Hydrotreater, or Hydrocracker unit. In addition, if distillate from a fluid catalytic cracking unit FCC is to be used in diesel blending, either the FCC unit feed or distillate products should be hydrotreated. Building a Hydrotreater unit is a significant investment that can cost anywhere from two hundred million to more than half a billion dollars, depending on the unit's capacity, location and processing flexibility. As diesel-powered public transportation is the main means of transportation in most developing countries, diesel prices can be seriously affected by such a move toward a lower diesel specification, requiring a huge investment. On the other hand, many large cities and capitals in developing countries are suffering from pollution generated by diesel-fueled vehicles such as buses and trucks. This makes it even more essential to implement a sulfur phase-out program, starting from the areas most affected. A similar case in the US is the Los Angeles basin area, where the pollution is severe but unique, thus necessitating to implement more stringent environmental limits than those at the national level.

The results demonstrated a reverse exponential decrease in the global objective with the reduction in sulfur level, as shown in Figure 7. This is usually due to the higher cost of intermediate blending stocks required to produce the lower sulfur diesel, in addition to the higher operating cost of process units such as Hydrotreaters. The effects of environmental objectives represented as upgrades in product specs in this case are clearly nonlinear and closer to an exponential effect.

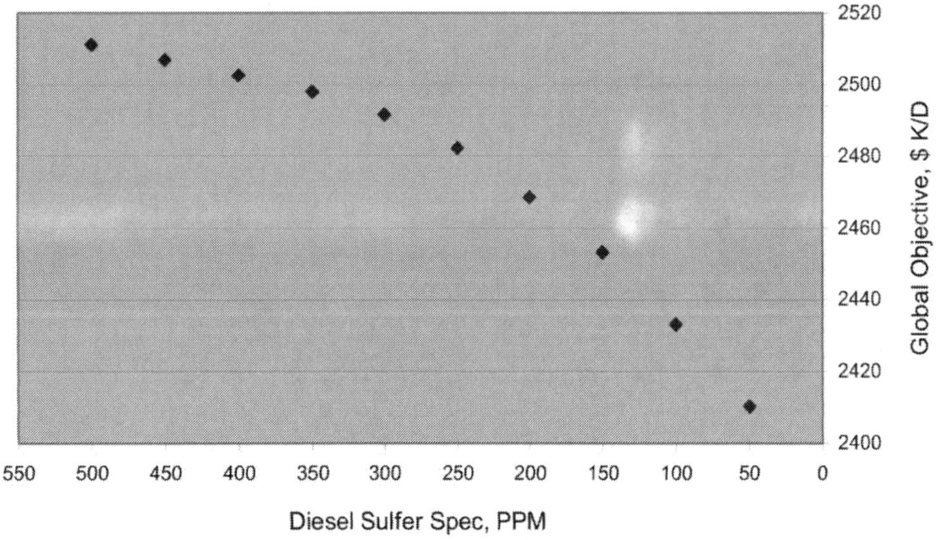

Figure 7: Nonlinear Effect on Global Objective

It was noted that each refinery experienced different effects depending on its process limitations. For example, for Refinery # 4, the extra hydrotreating capacity requirements increased linearly up to the 150 PPM level. After that, the extra hydrotreating requirements increased considerably but in a linear manner with a high sloping line, as can be seen in Figure 8. This suggests an optimum desulphurization level for that particular refinery of 150 PPM sulfur. This may be taken into consideration when deciding to upgrade the product spec. This refinery can be upgraded only to produce 150 PPM sulfur diesel. If the national spec is higher than this in sulfur, the product of this refinery can be sent as a blending component to other refineries lacking hydrotreating capacity. If the national sulfur spec requires a lower sulfur level, then the refinery may operate at 150 PPM sulfur and study the benefit of importing some lower sulfur diesel as a blending component.

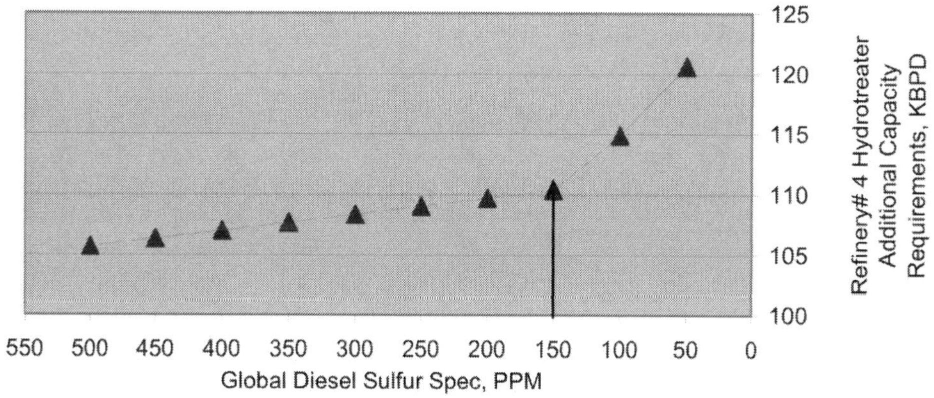

Figure 8: Optimum Desulphurization Level at 150 PPM

Conclusion

1. Multi objective local optimum may be different than the local optimum for the economic objective case. The deviation from the economic optimum depends on the degree of flexibility of the refinery in both selecting different feed / product scheme and in having different operating modes. These flexibilities act as a relief on the imposed multi-objective constraints by providing more alternatives to the model which increases the feasible region and thus may lead to a better optimum solution. Transportation costs of feeds and other intermediate products is also a strong factor.
2. For most refineries especially the large ones, the change in the capacity or the local objective as a result of imposed constraints usually follows a discrete behavior in contrast to the global objective which normally follows a smooth curve. This discrete behavior is due to the fact that refinery operations are hard to change and there are minimum turndown ratios that cannot be exceeded. In addition, the operating cost of refinery unit is partially fixed. That is, if a unit's capacity is reduced by 20 %, the operating expenses such as cooling costs and manpower requirements will not drop by the same percentage. In actual life, changing the operating conditions requires time to stabilize the unit in order to be able to produce on spec products.
3. For catastrophic failure effects, crude processing capacity is the main factor affecting the objectives. The loss of feed processing capacity such as crude or condensate has more effect on the objectives than the loss of upgrading and treatment capacities such as gasoline reforming. In addition, the results show the benefit of having more than one crude unit, which gives higher flexibility in running the refinery in less than the 60 % minimum turn down ratio in case of a catastrophic failure event. That is, for two refineries of the same size, if a refinery has one crude unit, it can usually run at 60 % of the maximum capacity. However, if the same refinery has 'm' crude units, it can run as low as 0.6/m of the total capacity.
4. The results show the importance of examining the effects of loss of internal processing flexibility on the product supplies which may not be clear in either local or global objective functions but it will appear at a point where the supplies cannot be

met from all local refineries and thus imports are needed. Proper import tiers penalties should be imposed on imports in order to clearly address the shortage of some products under the event of partial or total loss.

5. Different refineries had different levels beyond which the effect of the environmental objective increased sharply. For example, one refinery demonstrated an optimum level at 150 PPM sulfur diesel spec, while another showed a huge drop in the objective and capacity after 300 PPM. This should be taken into consideration when deciding to upgrade refineries.
6. The results demonstrated that the increase in Hydrotreating requirements is not linearly related to the decrease in product sulfur level. In some cases, the relationship was exponential, while in others it was close to exponential but having two different linear regions. This should be taken into consideration when revising product specs.

Nomenclature

ABBREVIATION	Description
LP	Linear Programming (also known in industry as linear planning)
SLP	Sequential Linear Programming
NLP	Nonlinear Programming
PIMS	Process Industry Modeling System
HYSYS	Steady State or Dynamic Process simulator
RVP	Raid Vapor Pressure
RON	Research Octane Number
MON	Motor Octane Number
ON	Octane Number
BPD	Barrels per day
KBPD	Thousand barrels per day
FCC	Fluidized Catalytic Cracking unit
CCR	Continuous Catalyst Regeneration (Type of Catalytic reforming)

References

1. Basak, K, Abhilash, K, Ganguly, S and Saraf, D. On-Line optimization of a crude distillation unit with constraints on product properties. Industrial and engineering chemistry research, v 41, 2002.
2. Bloss, K. Dynamic process optimization through adjoint formulations and constraint aggregation; PhD dissertation. Lehigh University, 2001.
3. Chaskunchuensaken, S. A combined optimization algorithm for stability, phase, and chemical equilibria (Phase separation). PhD Dissertation, Polytechnic University, 2000, ISBN 9-599-75785-X.
4. Hartmann, J. Distinguish between scheduling and planning models; Hydrocarbon Processing, Vol. 77 Issue 7, 1998.
5. Li, X. Refinery-wide optimization. Ph.D. dissertation. Texas Tech University. 2000.
6. Pike, R. Optimization for Engineering Systems. Louisiana State University 2001. http://www.mpri.lsu.edu/tcontentsindex.html.
7. Khogeer, A. Nonlinear analysis of catastrophic failure effects on a multi objective intra refinery optimization. AIChE spring national meeting. Mar 2005.

8. Khogeer, A. Multi Objective Multi Refinery Optimization with Environmental and Catastrophic Failure Effects Objectives. PhD Dissertation, Colorado State University. 2005.

Optimization of the benzine and diesel fuels blending

Lisitsyn N.V., Kuzichkin N.V.

The mathematical model of petroleum products blending consists individual typical operations for each product. The entire set of operations simulates the blending process as a whole.

The individual typical operation is characterized by several input streams, x_i and one output stream, y. The general equation for the blending is:

$$x_i = f_i(\theta, y), i = 1,...,n \tag{1}$$

The blending system can be obtained by combining individual operations. Different models, having various types of f-functions, can be considered. Inserting blending ratios of the streams of $u_i = f(\theta)$ types into the operation equation (1), we can rewrite this expression as follows:

$$x_i = u_i \cdot y \tag{2}$$

The set of specific values of the variables $u_1^*,...,u_n^*$ is regarded as an optimal composition mixture.

The admissible mixtures $U=\{u_1,...,u_n\}$ is determined by the equality $\sum_i u_i = 1$ and the constraints on the final product-quality index:

$$\varphi_k\left(u_1,...,u_n, d_{k_1},...,d_{k_n}\right) \geq b_{k_S} \quad , \tag{3}$$

where b_{k_S} is the extreme value of the k-th principal parameter with the quality standard S; d_{ki} is the k-th qualitative characteristic of the i-th component involved in the blending.

Furthermore, the constraints $u_i \cdot y \leq R_i$ must be observed, where R_i is the limits of the i-th component.

The principal quality indices are as follows:

- for diesel fuels:

density, sulfur content, diesel fuel distillation, cetane number, kinematic viscosity, flash point, pour, cloud and filterability points;

- for gasoline fuels:

density, gasoline distillation, octane units (engine and research methods), vapor pressure.

The blending of components is normally regarded as a problem of linear programming. Major impediment to using linear programming to solve the optimization problem is the non-additivity of the most of product-quality indexes. There is a non-linear dependence between product-quality indices and component-quality indices, which has been shown empirically. Normally the index method is used to solve the blending process simulation problem. This method assumes the existence of such functional transformation which makes property dependences linear. Then, determining the properties of the mixture components requires inverse transformation, i.e. the calculating the value of the inverse function. The index method is used in the well-known systems for simulating of industrial processes, like Aspen PIMS, BLEND-2000, etc.

However, the required function cannot be guaranteed to solve the problem. Generally, this function can be derived statistically with the help of the experimental data, but this functional dependence will change with feedstock quality.

Therefore, constructing the functional dependence $\varphi_k(u_1,...,u_n,d_{k1},...,d_{kn})$, one should take into consideration that some variables are additive, and thus , linear, e.g. distillation, density, sulfur content etc. The choice of the directly measured variables should show the chemical character of the variable studied. For instance, aniline point, density, distillation, and viscosity may be chosen for cetane number determination.

The density and aniline point can be used to determine the ratio between the aromatic hydrocarbons and paraffin hydrocarbons The boiling temperature and viscosity indicate the molecular mass. For example, if the viscosity is comparatively high, there is a considerable quantity of heavy molecules in the fuel, such as aromatic hydrocarbons, naphthenes, and the heaviest paraffins.

Let $r_{i_1},...,r_{i_m}$ be the designated additive quality indices of the i-th component, and the functional dependence is given by:

$$d_{k_i} = f(r_{i_1},...,r_{i_m}) \tag{4}$$

If the values $r_{i_1},...,r_{i_m}$, meeting the quality standard requirements are substituted, into equation (4), the condition $d_{ki} \geq b_{k_s}$ is complied with and the constraint on d_{ki} may be excluded from the model as being redundant. Otherwise the quality standard requirements for $r_{i_1},...,r_{i_m}$ may be corrected:

$$d_{k_i} = f(r_{i_1} + \Delta r_{i_1},...,r_{i_m} + \Delta r_{i_m}) = b_{k_s} \tag{5}$$

By directly measuring the quality indices r_{i_1} r_{i_m}, the dependencies for cetane number, flash point, pour, filterability and cloud points can be derived and corrections can be made for octane unit deviations from non-additivity.

The structure of the refinery determines the choice of the optimal blending problem solving procedure.

Nowadays the following blending structures are utilized:

- in stream blending of components ;
- blending by means of blending stations;

- combinatious systems (using the base components).

Combinatious systems are of the greatest interest. On the one hand, as compared to employing blending stations only, they decrease power usage by reducing the pumping of components between tanks. On the other hand, as compaired to in-stream blending quality standarts are much easier to meat.

If the blending is done using blending stations and there is a relatively small number of the components (≤ 5), non-linear programming methods can be used in calculations. For more complicated systems, where several types of products are obtained and the base components are used at blending, it is expedient to decomposition methods.

It is worth noting saying that in general the decomposition algorithm depends on the system structure for the blending. First, the quality indices for the base components are defined based on the blending operation, and the problem of the base component redistribution between petroleum products is solved (upper optimization level). Then the problem of cost minimization is solved for each petroleum product (lower optimization level). In the latter case, the components of the tank farm are independent variables.

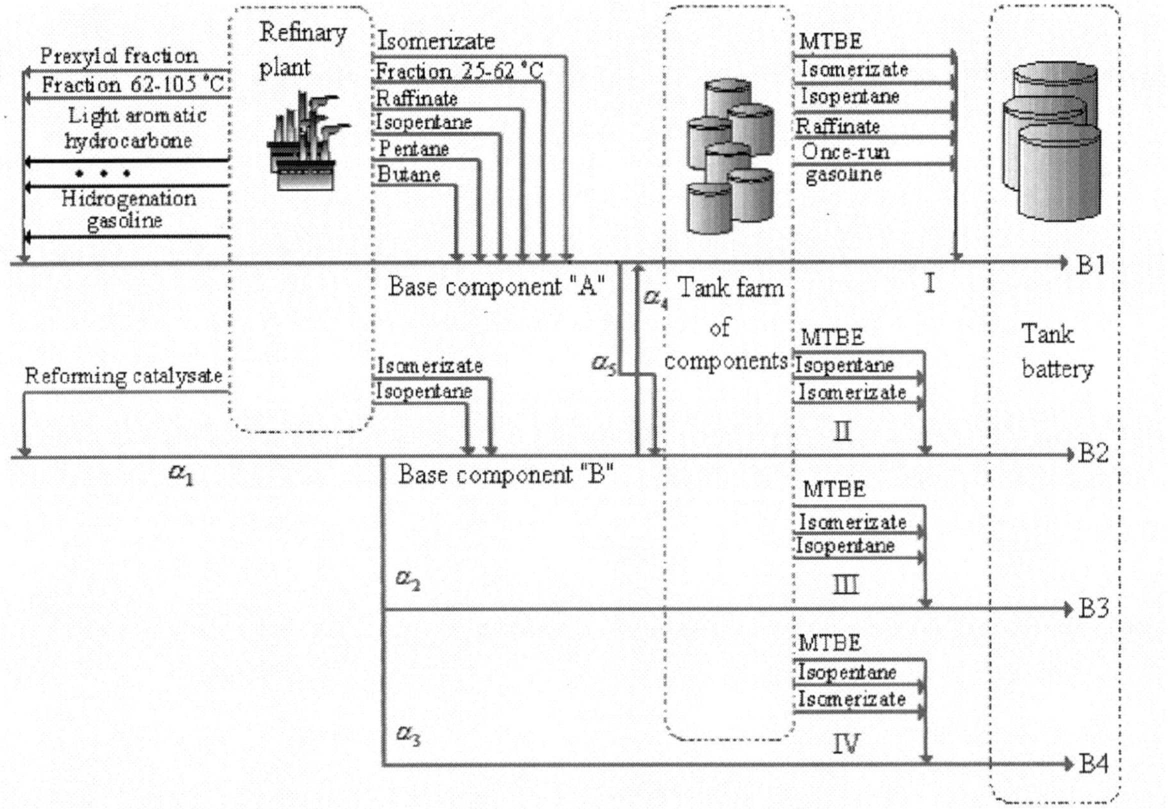

Fig.1 Gasoline blending diagram

The above is an example of a blending diagram for the simultaneous production of four types of gasoline (Fig. 1).

The coefficients of separation of the streams of reforming catalysate and basic components ($\alpha_1,...,\alpha_5$) were chosen as the optimized variables at the upper decomposition level.

Profit was used as the optimization criterion F:

$$\max_{\alpha_1...\alpha_5} F = \sum c_{0_j} \cdot y_j - \sum a_j \qquad (6)$$

Where c_{0_j} -price of the j-th product, a_j- price of the used components in the j-th product.

Upon setting the separation coefficients $\alpha_1,...,\alpha_5$, the flow rates of the base components and quality indices were calculated for the blending operation model. After that, flow rates and quality indices of the first through fourth streams were determined. Next the minimum cost of each gasoline type was calculated by the optimization procedures of the lower lever, and the components from the tanks farm were the independent variables u_{ij} (Fig. 1).

$$\min_{u_{ij}} a_j = \sum c_j \cdot u_{ij} \cdot y_j , \qquad (7)$$

In doing this, each optimization procedure was an independent blending operation.

The approach discussed above with use of the developed blending operation model has allowed the problem to be reduced from one of high dimensionality to a series of sub-problems of lower dimensionality.

The above calculation method can be utilized for both blending stations and in-stream blending.

175c Energy Management

Peter Stanley, Asset Optimization Division, Emerson Process Management, 12603 Southwest Fwy., Ste. 100, Houston, TX 77477
Utility complexes at refinery sites are a critical part of the operation – they supply the necessary steam and electricity that make the salable products. However, within the dynamic nature of the refining environment continually meeting site power and steam demands, balancing available fuels and staying within emission limits is a challenging task.

Energy Management Systems (based on optimization technology) continually monitor the dynamics of the operation and determine the set of "optimum" operating setpoints that will allow the utility operation to continually run at its economic optimum. Additionally, these systems are helping refiners cope with their increasing operating costs driven by rising energy prices (Natural Gas and Oil).

This paper will review a number of recent projects that highlight how Energy Management Systems are being deployed by refining companies to enable them to achieve substantial benefits through improved automation. Benefits in the order of 2 to 5% reduction in annual fuel bill expenditures per year are not uncommon for these types of applications.

The paper will also address a number of key elements that refiners should consider in the development of a successful Energy Management Project.

175d Development and Implementation of Process Analytical Toolbox

Fangwei Xu[1], **Edgar Tamayo**[2], and Biao Huang[1]. (1) Chemical Engineering, University of Alberta, Department of Chemical and Materials Engineering, Edmonton, AB T6G 2G6, Canada, (2) Process Automation Services, Syncrude Canada Ltd., P.O. Bag 4009, M.D. 8009, Fort McMurray, AB T9H 3L1, Canada

Advanced Process Control (APC) applications, such as Model Predictive Control (MPC) technology, have been widely accepted and applied in process industries. Monitoring of APC performance especially economic performance has been a great interest both in academia and in industry. This motivated us to develop a state-of-the-art software that can be used to monitor the performance of APC applications and facilitate the task of APC maintenance. The software package that we have developed during the last few years is called Process Analytical Toolbox (PAT).

The package of PAT was written in MATLAB, which is widely used in academia by researchers. MATLAB itself is a very powerful language in computation with large amount of speciality toolboxes in different research areas. It also provides a compiler that can compile most MATLAB codes directly into executable stand-alone applications. Moreover, the MATLAB OPC Toolbox makes it possible to retrieve real-time process data directly from DCS system via standard OPC interface, which is available in almost all different kind of DCS systems in process industries. We took full advantage of these features in our development of PAT package. On the other hand, it is possible for researchers to run their applications on line with direct real-time process data. This makes it a very convenient shortcut for the technology transfer from academia to process industries.

PAT is composed of application components and interface components. Two interfaces, PI Driver and Data Collector, are provided to retrieve process data, either historical process data off-line from the PI server or real-time process data on-line from the DCS system. They are designed to provide process data to the application components from difference data sources. The main application components include Cluster Analysis, Process Model Identification, Laguerre Network On-line System Identification, Univariate Time-Varying Controller Performance Assessment, Multivariate Controller Performance Assessment, Robust Minimum Energy Control with Output Covariance Constraints, MPC Economic Performance Analysis and Tuning Guidelines, and many others. The output results of these application components will tell process engineers useful information on the investigated process or controller directly based on the process data, most of which are routine operating data. The outcome will be process models, controller performance of time-varying processes, multivariate controller performance, optimal designed controller, MPC economic performance, MPC tuning guidelines, and so on. The most updated version of PAT includes some new algorithms developed very recently such as model-free subspace approach to multivariate control performance assessment. The synthesis of variety of results makes this toolbox a powerful package for process and control engineers. For example, the controller performance indices obtained from multivariate controller performance monitoring component can be imported into MPC economic performance analysis and tuning component. It can then give the possible

MPC economic benefit potential as well as some corresponding MPC tuning guideline. These two application components together with Data Collector will make a plant oriented solution for APC performance monitoring.

In this presentation, we will discuss the software architecture, implementation platform, mathematical principles behind this package, and its application potentials. Case studies will be provided to demonstrate some of its important functionalities.

177a Meeting the Challenge of Delayed Coker Naphtha Hydroprocessing

Terry A. Reid, Albemarle Catalysts Co. LP, 2625 Bay Area Blvd., Suite 250, Houston, TX 77058

Coker naphtha hydroprocessing is challenging because it combines highly exothermic reactions, quantitative poison removal, and essentially complete removal of sulfur and nitrogen. Feed diolefins can polymerize and create pressure drop problems. Accurately predicting silicon breakthrough is necessary to prevent silicon damage to the downstream reforming catalyst. Si Breakthrough is dependent on catalyst properties and processing conditions and is frequently the limiting factor in catalyst cycle length. This paper examines current and future coker naphtha processing trends and discusses the impact of improved catalyst technology and processing feeds with higher levels of metal contaminants (arsenic and silicon), nitrogen, and sulfur on coker naphtha hydrotreating.

177b Developing Knowledge from Spent Hydroprocessing Catalyst Chemical Analysis

Terry A. Reid, Albemarle Catalysts Co. LP, 2625 Bay Area Blvd., Suite 250, Houston, TX 77058

The refining industry uses hydroprocessing to remove metallic poisons from oil during upgrading. Every hydroprocessing catalyst has a specific capacity for individual metal poisons and a collective capacity for multiple metallic poisons. Catalyst metal poison capacity knowledge is developed from spent commercial hydroprocessing catalyst samples. This paper discusses the development of catalyst silicon and arsenic capacity knowledge from delayed coker naphtha hydroprocessing spent catalyst samples and the application of this knowledge in catalyst system design. Methodologies for determining catalyst metallic poison capacities and reactor metal deposition profiles from the chemical analysis of spent catalyst samples are discussed.

177c Recombination: A Complicating Issue in FCC Naphtha Desulfurization

Laura E. (Jones) Leonard and Peter Kokayeff. Refining Process Development, UOP LLC, P.O. Box 163, Riverside, IL 60546

The aim of cracked naphtha selective hydrodesulfurization (SHDS) technologies, such as the SelectFining™ process licensed by UOP, is to reduce the sulfur content of the naphtha while simultaneously maintaining its octane value. Both of these objectives are complicated by recombination, the term applied to the reaction of H_2S and olefins to form mercaptans. A solid understanding of recombination reaction kinetics is critical in the design and operation of a selective hydrodesulfurization unit as the lowest achievable sulfur level is often dictated by the extent of recombination. Furthermore, recombination contributes to the conversion of the valuable high octane olefins. Recombination is a very rapid reaction which frequently, but not always, approaches equilibrium. The extent of reaction is governed either by the equilibrium relationship or kinetics depending on the process conditions, feedstock composition, and reaction rates. If the rate of desulfurization is slow relative to mercaptan formation, the extent of recombination will be driven by the mercaptan-olefin equilibrium. Predicting the extent of recombination is a non-trivial challenge as it is impacted not only by the total olefin and H_2S content, but also the carbon number, double bond position, and branching of the olefins. These issues will be discussed in the context of the results of a series of numerical simulations and pilot plant experiments.

A Novel Oxidative Desulfurization (OxyDS) Process for Diesel and VGO

Tzong-Bin Lin, Project Manager
Hsun-Yi Huang, Research Engineer
Jyh-Haur Hwang, Research Engineer
Hung-Chung Shen, Director
Refining & Manufacturing Research Institute (RMRI)
Chinese Petroleum Corporation (CPC)
217, Min-sheng South Road
Chia-yi, Taiwan 60036

Karl T. Chuang, Professor
Department of Chemical Engineering
University of Alberta
Alberta, Canada

and

Fu-Ming Lee, Ph.D., P. E.
Technical Consultant
Katy, Texas

Abstract

The OxyDSTM process being developed by CPC can oxidatively desulfurize petroleum oils under non-aqueous and single phase condition, using such an active oxidant that oxidation of sulfur and nitrogen occur at low temperature and short residence time without a catalyst. As a result, essentially all the valuable hydrocarbon components in the oils are preserved. No troublesome waste acid regeneration associated with most other oxidative desulfurization processes is required either. Furthermore, after oxidation step the oxidant is converted into a "built-in" extractive solvent for removing oxidized sulfur and nitrogen compounds from the oxidized petroleum oils. The extraction step is so effective, that no complicated adsorption step (associated with all other oxidative desulfurization processes) is required for final polishing. For example, sulfur in diesel was reduced from 300 to 10 ppm through this simple but effective OxyDS process.

More significantly, this is the only known oxidative desulfurization process that can process heavy petroleum oils without difficulties, because the oxidation reactions occur in a single phase, non-aqueous environment without catalyst. The OxyDS treatment can significantly increase the quality of the FCC feed to not only substantially improve the conversion and product distribution, but also reduce the product sulfur contents. Consequently, no post desulfurization treatment is required for FCC gasoline and light olefins.

CPC is currently installing a continuous pilot plant at its Research & Manufacturing Technology Center in Chia-Yi, Taiwan, for commercializing of the OxyDS process. Start-up of the pilot plant is expected forward the end of this year.

Environmental Regulations on Road Diesel Sulfur

Stringent environmental regulations on sulfur content in petroleum fuels have been continuously enacted worldwide. In the United States, gasoline sulfur level will be reduced by 90% from current 300ppm down to 30ppm and CalRFG3 called for a level of 15ppm by 2006, while European regulations set a requirement of 50ppm by 2005(1). On the road diesel fuel, most of industrialized countries have adopted for a level of 50ppm or even as low as "zero sulfur" in the future. Other countries have progressively followed up the program to reduce engine emission to improve air quality. The EPA in Taiwan more closely follows European regulations and sets a 50ppm sulfur diesel by 2007. CPC has started to provide 50ppm diesel, since September 2004 to the local market, and the other company, Formosa Petrochemical Corporation (FPC) followed immediately. Therefore, the EPA in Taiwan revised the regulation demanding 50 ppm sulfur in diesel by 2005. This is a dramatic reduction from 500ppm level only six years ago. Both oil companies in Taiwan are also actively revamping or planning for new hydrotreating units to produce 10ppm sulfur diesel. The Refining & Manufacturing Research Institute (RMRI) of CPC began oxidative desulfurization research program few years ago with a great success and the first of many U. S. patent applications has been made in late 2004.

Low Sulfur Diesel Production Technologies

There are reports on alternatives other than the conventional hydrotreating technology to produce ULSD. For example, biodesulfurized diesel (2) or Gas-To-Liquid diesel (3) is inherently sulfur free and can be an excellent blending stock when producing in enough large quantity. There are also reports on enhancing performance of existing conventional hydrotreater operation by addition of a pretreater unit. Such a pretreater system has been designed as reactors in series (4,5) or multiple adsorbers (6). Biodesulfurization study has been the main objective for many publications (7-9). The process has limitations in relatively lower reaction rate, bacteria stability, oxygen supply, and the commercialization in the near future is in doubt. S-Zorb by ConocoPhillips developed a process similar to FCC with adsorption and desorption cycles at 500psig and reported a diesel product less than 10ppm sulfur (10).

Removal of sulfur compounds by oxidation reaction (11-15) has drawn attention from refining industry recently. Hydroperoxide was the major oxidant with various catalysts to enhance reaction rate. In this paper, we would like to report the use of a non-aqueous organic peroxide oxidant that has several advantages over the traditional hydrogen peroxide or other organic peroxide oxidant.

Sulfur Compounds and Oxidation Reaction

Determination of detailed sulfur compound distribution in diesel fraction is possible with the development in atomic emission detection (AED) technology (16). Relative HDS activity for typical sulfur compounds has been reported (17) and is shown in Table1. Due to stereo hindrance, alkyl-dibenzothiophene compounds are difficult to be removed by hydrotreating. Reaction mechanism includes ring saturation and therefore catalysts with high saturation activity and operation at high pressure are required to achieve ultra low sulfur level.

Reduction of LHSV, installation of amine scrubber to maintain higher hydrogen purity in the recycle stream and increase of H_2/Oil ratio or use of dense catalyst loading may be required to assure the production of ULSD in a traditional HDS unit (18). A polishing reactor may also be required to deal with the color problem that occurs due to higher temperature operation (19).

Literature had shown that alkyl-dibenzothiophene could be preferentially reacted over dibenzothiophene with an oxidant. A generalized oxidation route is the formation of sulfoxide and then to corresponding sulfone. Alkyl group and the ring structure are essentially intact as can be seen from Figure 1 (15). And this places oxidation process in an attractive position for ULSD production. Unipure Corporation made a presentation on oxidative desulfurization for ULSD production at AIChE meeting in 2002 (20). Conventional stirred tank reactor was chosen for the oxidation reaction. Straight run diesel was mixed with hydrogen peroxide, in proportional to the sulfur content, water and recycled organic acid catalyst. At the completion of reaction, spent acid extracted nearly 50% of sulfur compounds presented as sulfones. The diesel, being reduced in sulfur contents, was neutralized, washed, dried and then passed through an alumina bed to adsorb the remaining sulfones. Water-like diesel containing less than 5 ppm sulfur could be produced. The alumina bed was regenerated by light hydrocarbon washing and hot methanol extraction. Water balance loop had to be carefully organized to accommodate the water formed as oxidation by-product, water presented in the commercial peroxide solution, water stripped from the spent acid regeneration and water used to wash off residual acid. No further advance of the technology was publicly reported since 2002.

Instead of using relatively expensive hydrogen peroxide, Eni/UOP developed the process using organic peroxides and solid catalyst. It was also claimed that the change in polarity of sulfur compounds would facilitate the separation of sulfur from diesel stream (21). It claimed that a demonstration plant is foreseen in 2006.

CPC Process and Technology Development

Hydrogen peroxide, as in the aqueous phase, has a drawback of poor mixing with the oil phase. Phase transfer agent (or catalyst) has to be developed in order to break mass transfer barrier. Organic peroxide, readily mixed well with petroleum fraction would be an ideal candidate for selective oxidative reaction.

The CPC process uses peracetic acid (PAA) as the oxidative agent to take the advantages of its high oxidation activity without a catalyst. PAA could also be generated at high concentration and therefore could be, without further concentration, directly used in oxidation reaction. High concentration PAA is also beneficial as the oxidation activity would be higher.

Figure 2 is the schematic diagram of the developed diesel oxidation process. It consists of three major steps: (1) generation of PAA, preferentially from acetaldehyde with a homogenous catalyst system; (2) oxidation of sulfur and nitrogen compounds by the non-aqueous PAA without catalyst; and (3) removal of sulfones and nitrogen oxides by extraction, adsorption or combination of the two. As the major by-product of the current oxidation reaction is acetic acid (AA), sulfones and nitrogen compounds will preferentially present in the AA phase over original petroleum phase. In another words, the byp-roduct AA can also serve as the extraction solvent. Acetic acid, with a high market value, is then distillated to remove sulfones and purified. Low purity acetic acid stream could be recycle to the sulfone extractor as the extractive solvent.

One of the distinguish differences between the CPC process and other announced processes is the elimination of phases transfer agent during oxidation reaction. As the non-aqueous PAA is extremely active and completely miscible with the petroleum oils, design of oxidation reactor is very simple. AED analysis results, shown as Figure 3, indicted that DMDBT completely reacted with PAA in a few minutes. We selected a tubular reactor, possible with multiple injection points, to keep PAA at high concentration throughout the reactor and to provide a better temperature control in the reactor.

Generation of the peroxide at relative low cost also offers the competition advantage of the CPC process. Trace amount of homogeneous catalyst in low ppm level is enough to produce an oxidant containing 20 wt% PAA from acetaldehyde (AcH) at mild condition.

Diesel Product Quality

As an example for oxidation reaction, a diesel sample with 198 ppm total sulfur was treated with PAA. GC with sulfur detector was employed for the detailed analysis of sulfur compounds. Sulfur peaks of the diesel, before and after oxidation, were shown in Figure 4. The chromatograms clearly show the shift of peaks after oxidation indicating the formation of heavier sulfone compounds. Also shown in Figure 4 was the chromatogram of samples after water wash and adsorption by active alumina. As sulfones were insoluble to water, sample sulfur content simply by water wash was practically unchanged. Adsorption by alumina would nearly remove all sulfones as shown by the last chromatogram in Figure 4. As mentioned in the previous section that removal of sulfones could also be done by AA extraction. To facilitate the design of the extraction column in the future, a correlation was also developed to predict the sulfones' distribution between oil and AA phases. Prediction of the amount of residue sulfur in diesel phase matched quite well with the actual analytical results as shown in Figure 5. This confirmed that production of ultra low sulfur diesel could be realized by oxidation reaction and followed by extraction.

Color of all samples was presented as Figure 6. The color of the sample changed from light yellow to deep orange after oxidation reaction. This should be reasonable as sulfones were compounds with darker color. The transparent appearance of the oxidatively desulfurized oil might be more comfortable to customers. Other important properties of diesel samples were essentially unchanged and were listed as Table 2. Most noticeable of all were the slight improvement of cetane index and Rams Bottom Residues.

Treating Vacuum Gas Oil (VGO) with OxyDS Process

To take the advantage of high activity (without phase transfer agent or solid catalyst) of the PAA oxidant and non-aqueous nature of OxyDS process, heavy fraction of the petroleum oils such as vacuum gas oil (VGO) was also treated by the process. We demonstrated that the OxyDS treatment can significantly increase the quality of the FCC feed to not only substantially improve the conversion and product distribution, but also reduce the product sulfur contents. Consequently, no post desulfurization treatment is required for FCC gasoline and light olefins. However, detailed information on VGO treatments can not be disclosed at this time.

Pilot Plant Development

A one-barrel per day oxidative desulfurization and denitrogenation pilot plant for upgrading VGO and heavy oil as well as producing ULSD is being constructed at the Refining & Manufacturing Research Institute of CPC.

Summary

As an energy and petroleum refining company, CPC seeks sustainable growth. Continuous commitment to reduce fuel product sulfur contents to meet or even beyond the government regulations would need the involvement of extensive research activity. In the past few years the Research Institute of CPC had various research achievements successfully implemented in refineries. The development of this novel process for the production of low sulfur fuels is actually one of the many achievements of our oxidation research project. As detailed described in previous sections, there are several unique advantages of this novel process, including:

(1) Novel and efficient oxidant production method minimizes the overall investment cost.
(2) High by-product value benefits greatly the overall process economics.
(3) High reactivity oxidant eliminates the requirement of catalyst or a phase transfer agent.
(4) By-product could serve as the extraction solvent.
(5) Non-aqueous system minimizes unexpected phase separation or solid precipitation after oxidation reaction.
(6) Great oil affinity nature makes it viable to extend the system for VGO treatment.

Reference

(1) G. Fredrick; "Sulfur Reduction: What are Options?", Hydrocarbon Processing, Feb 2002, p.45.
(2) D. Alder, "Introduction to Biodiesel", 2002 presentation materials, searched from website.
(3) B. McCormick and T. Alleman, "Properties and Performance of Gas-to-Liquids Fisher-Tropsch Diesel Fuels" 2003 presentation materials, searched from website.
(4) Ron D. Key, Michael D. Ackerson, M. Steven Byars, John Laurent and Jim Hallock, "Isotherming – A New Technology for Ultra Low Sulfur Fuels", NPRA AM 03-11
(5) Michael D. Ackerson, M. Steven Byars and J.B. "Bob" Roddy, "Revamping Diesel Hydrotreater for Ultra-Low Sulfur using Isotherming Technology", NPRA AM-04-40.
(6) Seung-Woo Lee, Jae Wook Ryu and Whasik Min, "SK Hydrodesulfurization (HDS) Pretreatment Technology for Ultra low Sulfur Diesel (ULSD) Production" Catalysis Surveys from Asia 7 (4), 271, 2003, searched from website
(7) Je Hwan Chang, Yong Keun Chang, Kyung-Suk Cho and Ho Nam Chang, "Desulurization of model and diesel oils by resting cells of Gordona sp", Biotechnology Letters, 22, 193, 2000
(8) Abhijeet P. Borole, Choo Y. Hamilton and Karen Miller, "Improving Enzyme Activity and Broadening

Selectivity for Biological Desulfurization and Upgrading of Petroleum Feedstocks", 2003, DOE Report DE-FC07-00ID1400.
(9) Julia R. Acero R., "Development of DBT-Monooxygenase Mutants by Directed Evolution for Organic Sulfur Removal", 2002, searched from website
(10) World Fuels Conference 2001 Information from website
(11) J.M. Campos-Martin, M.C. Capel-Sanchez and J.L.G. Fierro, "A novel methodology towards deep desulfurization of fuels by chemical oxidation", searched from website
(12) Richard Hickey et. Al. "OST Technical Progress Report FY2001 Results, Carnegie Mellon University, searched from website
(13) Robert E. Levy, Alkis S. Rappas, Stephen J. Decanio and Vincent P. Nero, "UniPure's ASR-2 Desulfurization Process Provides Cost-Effective Solution for Ultra-Low-Sulfur Refined Products", World Refining, searched from Website
(14) C.D. Gosling, V.A. Gembicki, R.M. Gatan, Agostino Cavanna and Daniele Molinari, "The Role of Oxidative Desulfurization in an Effective ULSD Strategy", AM-04-48, NPRA 2003 Annual Meeting
(15) Frank J. Liotta and Yuan Z. Han, "Production of Ultra-Low Sulfur Fuels by Selective Hydroperoxide Oxidation", AM-03-23, NPRA 2003 Annual Meeting
(16) Bruce D. Quinby and Vince Giarrocco, HP Application Note 228-394
(17) Michael E. Leister, "Diesel Desulfurization Technologies", 2001 presentation materials, searched from website.
(18) Yataka Mukai and Yoshihiro Mizutani, "Survey of Petroleum Refining Technology for High-quality Diesel Fuel, Japan Energy Research Center Report, 2001, searched from Website
(19) P. Marion, "Axens Clean Fuels Technologies", 2002 presentation materials, searched from website.
(20) Robert E. Levy, Alkis S. Rappas, Fu-Ming Lee and Vincent P. Nero, "Unipure's Oxidative Desulfurization Process Creates New Market Opportunity for Supply of Ultra-Low Sulfur Fuels", AIChE Spring Meeting, New Orleans, March 13, 2002.
(21) Franco Baldivaghi et. al., "New approaches for ultra low sulphur diesel" Tpoint 3/2004, searched from Website.

Table 1. Typical sulfur compounds and relative HDS reactivity

Sulfur Species	Boiling Point, F	Relative Reaction
Thiophene	185	100
Benzothiophene	430	30
Dibenzothiophene	590	30
Methyl dibenzothiophene	600~620	5
Dimethyl dibenzothiophene	630~650	1
Trimethyl dibenzothiophene	660~680	1

Table 2. Diesel properties of example samples

Diesel Properties	BEFORE	AFTER	Method
Density @ 15.5° C	0.826 g/ml	0.824 g/ml	ASTM D5002
Flash Point	95° C	95° C	ASTM D93
Pour Point	-15° C	-15° C	ASTM D97
Viscosity @ 40° C	2.847 cSt	2.812 cSt	ASTM D445
Water & Solids	0.00 vol%	0.00 vol%	ASTM D1796
Cetane Index	54.9	55.5	ASTM D976
Corrosivity 3hr (at 50° C	1a	1a	ASTM D130
Rams Bottom Residue	0.09 wt%	0.04 wt%	ASTM D524
Ash	0.002 wt%	0.001 wt%	ASTM D482
Residue	1.6 vol%	1.6 vol%	
Sulfur	198 ppm*	< 5ppm**	*ASTM D2622
Boiling Range (° C)			*ASTM D5453
IBP	218.4	217.3	
10 vol%	235.1	235.2	
20 vol%	243.2	243.1	
50 vol%	264.8	264.4	
90 vol%	319.7	317.8	
End Point	358.4	359.1	

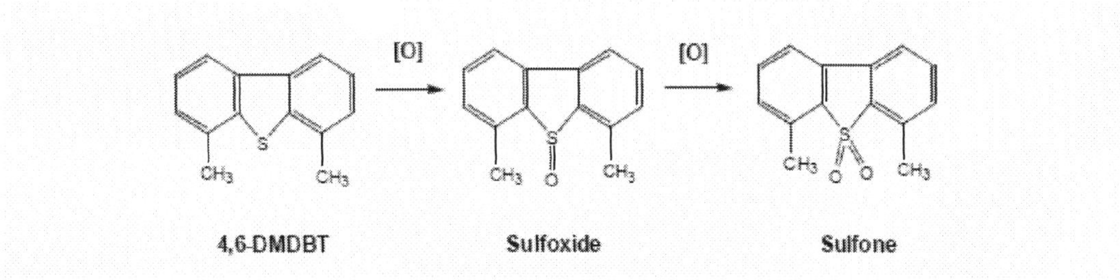

Figure 1. Typical Representative Diagram of Oxidative Desulfurization

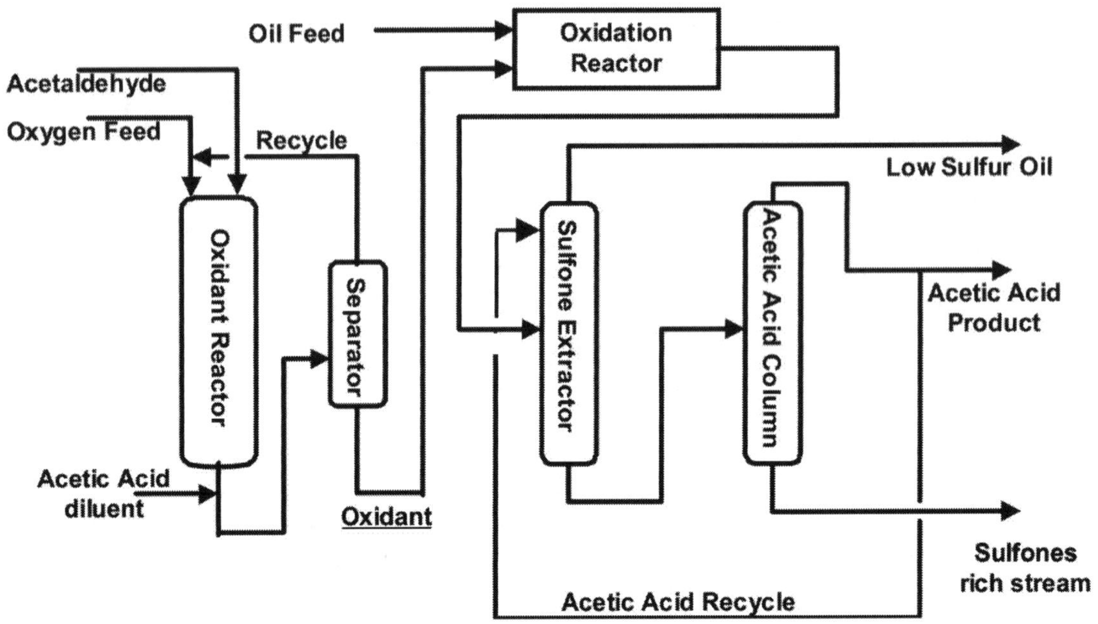

Figure 2. Schematic Diagram of CPC Diesel Oxidation Process

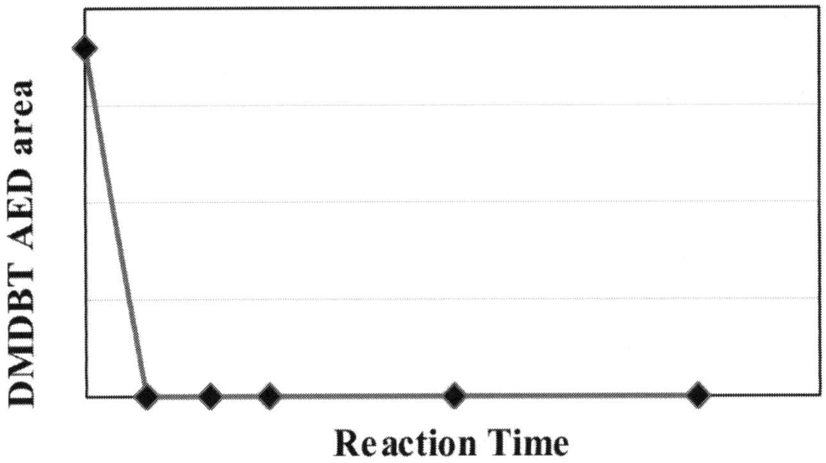

Figure 3. Fast reaction observed for DMDBT with PAA

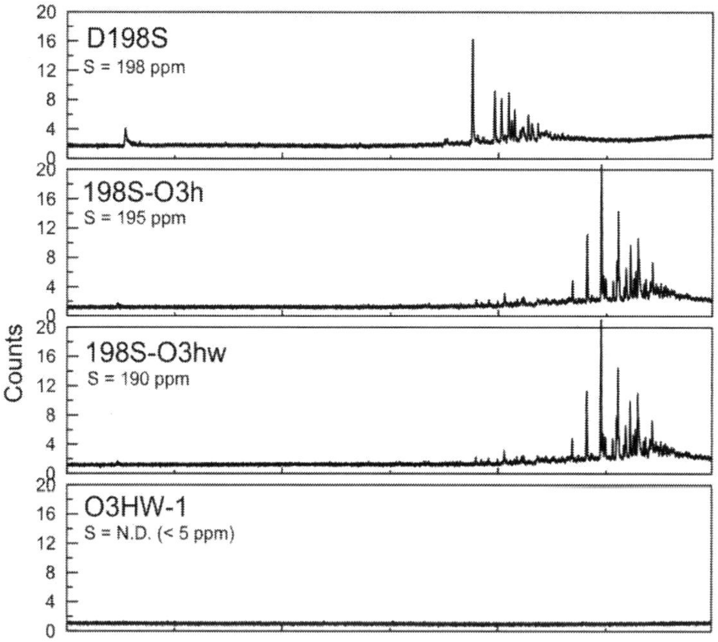

Figure 4. GC AED results of sulfur peaks for example samples

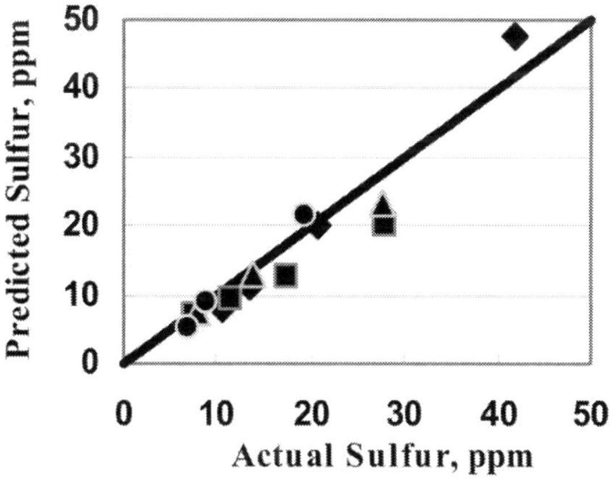

Figure 5. Correlation diagram on AA extraction of sulfones

Figure 6. Color appearance of diesel samples

Mathematical Modeling of a Trickle Bed Bio-desulfurizer of Hydrotreated Diesel for the Production of ULSD (Ultra-low Sulfur Diesel)

M. Mukhopadhyay, R. Chowdhury and P. Bhattacharya*
Chemical Engineering Department, Jadavpur University
Kolkata-700032
India
*Author to whom correspondence may be addressed
E-mail: ftbe_bon@yahoo.com Telefax: 91-33-24146378

Abstract

Biodesulfurization of hydrodesulfurized diesel has been carried out in a 600mm high and 66mm diameter trickle bed reactor using *Rhodococcus Sp.* in immobilized form. The initial concentration of organo-sulfur compounds I diesel has been varied in the range of 200 to 540 ppm. LHSV has been varied in the range of 1.838 to 3.676h^{-1}. Recycle ratio of 0.25 to 2 has been used at LHSV of 1.838h-1. A deterministic mathematical model has been developed using all kinetic and hydrodynamic parameters including μ_{max}, maximum specific growth rate of Monod type, K_s, saturation constant, Y, yield coefficient , LHSV, biofilm thickness, bed porosity and recycle ratio. Experimental trends have been observed to be well explained by the model predictions.

Key words: Bio-desulfurization, diesel, uninhibited Monod model, intrinsic kinetic parameters, immobilized, trickle bed reactor.

Introduction:

Distillates such as diesel contain significant quantities of organically bound sulfur compounds, namely, benzothiophenes (BT), dibenzothiophene (DBT) and napthothiophenes. These are very difficult to be removed by the conventional hydrodesulfurizaton process practiced in the refineries and without a suitable removing step, their use as fuel results in the formation and emission of polluting sulfur dioxide and other oxides during combustion, which in turn leads to service environmental hazards [1]. Due to stricter environmental legislation, the desulfurization of diesel is becoming increasingly important. The diverse sulfur containing hydrocarbons present in diesel react differentially in the conventional refinery practice of hydrodesulfurization (HDS) [9,11] and alkylated DBTs are among the most resistant [4,12] compounds due to steric hindrance. Thus conventional deep desulfurization should be followed by a complementary desulfurization process to achieve ultra low sulfur diesel (ULSD) (10-15ppm or less) [4]. In this direction, biodesulfurization (BDS) serves as a potential alternative for reduction of sulfur level from 100 to ultra low level of 15ppm [2,3,5]. From the literature review it is clear that although a number of pioneering works have been carried out on BDS, many of them are restricted to batch mode of operation for the desulfuriztion of model organosulfur compounds, like BT, DBT etc. [2,4,6,7,8,10,12-15]. Only a few works have been reported on the biodesulfurization of diesel [3,5]. Moreover, most of the studies have been undergone using reactors of CSTR type [19,20]. However, the usage of usual stirred tank reactors are faced with thorough downstream processing of diesel for the separation of biomass from diesel. Keeping this disadvantage in mind, a trickle bed reactor using immobilized bacterial strain has been used under the present study as an extension of the previous work of the group [16,17]. In addition to the process parameters, namely, initial substrate concentration, volumetric flow rate, used I the previous studies [16,17], recycle ratio and liquid hourly space velocity (LHSV) have been incorporated as new input parameters and the progressive change of axial distribution of biofilm thickness and bed voidage have been considered as some of the important response parameters, playing a vital role during the performance of the reactor. A deterministic mathematical model incorporating the effect of recycle ratio and LHSV has been developed for the present study.

Materials:

Beef extract GR (E. Merck, Mumbai), peptone GR (E. Merck, Mumbai), Nacl (Ranbaxy, Mumbai), methanol GR (E. Merck, Mumbai), acetone GR (E. Merck, Mumbai), benzothiophene (Lancester, England), N_2 (Prakash traders, Kolkata), dithiozone GR (E. Merck, Mumbai), NaOH GR (E. Merck, Mumbai), acetic acid GR (Process chemical industries, Kolkata), mercuric oxide GR (E. Merck, Mumbai), HCl GR (E. Merck, Mumbai), isopropyl alcohol GR (Process chemical industries, Kolkata) and nickel aluminum alloy GR (E.Merck, Mumbai) were used during the present investigation.

Microorganism: The pure bacterial strain of *Rhodococcus sp.* (NCIM.2891) was purchased from NCIM (National Collection of Industrial Microorganisms), India. Cells were cultivated and enriched using sulfur free medium supplemented with diesel oil in 50ml Erlenmeyer flasks.

Diesel used: Hydrodesulfurized diesel samples were purchased from IOC (Indian Oil Corporation, Kolkata) having the characteristics given in the Table1.

Table 1.Characteristics of diesel used

Compound	I.B.P. ^{o}C	F.B.P. ^{o}C	Sp. Gravity (Basis:density of water=1000 kg/m^3)	Density kg/m^3	Viscosity Kg/m^3/s	Sulfur ppm	Aromatic %(w/w)
Diesel	140	370	0.8216	832.4	0.002081	200 to 540	27.16

Analytical Method:

Dry weight Method for the determination of Bacterial Mass: The biomass concentration in the reaction broth was determined by dry weight method. In this method the broth was centrifuged at the rate of 10,000 rpm for 15 minutes at -15 ^{0}C. The bacterial mass was then transferred to a pre-weighed aluminum cup and dried at 50°C overnight. The exact weight of the bacterial mass was determined by subtracting the weight of dry cup from that of the cup containing dry bacterial mass.

Sulfur analysis: UOP 357-80 (Trace Sulfur in Petroleum Distillates by Nickel Reduction) method and X-ray fluorescence (XRF) (ASTM-D4294-03) method have been followed to determine the concentration of sulfur in the diesel samples.

Experimental:

Description of the Trickle bed bio-reactor and its operation:

The trickle bed reactor has a diameter of 66mm and height of 600mm.The reactor was initially packed with 12mm diameter pith balls. Bacterial medium containing *Rhodococcus sp.* having the biomass concentration of 20g/dm^3 was circulated through the packed bed until the bio-film thickness on the sphere became 0.1mm. The initial bed porosity was measured to be 0.6. The reactor was operated under atmospheric pressure. Diesel having different organo-sulfur concentration namely, 200, 330, 430 and 540 ppm was fed into the trickle bed reactor at a rate varying from 0.25 to 0.5 dm^3/hr in downward direction. The LHSV of diesel was varied in the range of 1.838-3.676 h^{-1}. The reactor was continuously sparged with air at 480 dm^3/hr in upward direction. The recycle ratio was varied in the range of 0.25 to 2 to ensure high conversion of sulfur. Effluent stream coming from the bio-reactor was analyzed for biomass using dry weight method and for organo-sulfur compounds using Nickel reduction method (UOP357-80) and X-ray fluorescence (XRF) method respectively.

Theoretical analysis:

The trickle bed bio-reactor has been schematically represented in Figure 1.

The mathematical model of the system has been developed on the basis of the following assumptions,

1) Influent stream of the bio-reactor is sterile.
2) Organo-sulfur compounds are the only growth limiting substrates.
3) Microbial reaction occurs only at the outer surface of the bio-film.
4) There is no external mass transfer resistance present in the system.
5) Microbial growth follows the Monod kinetics [(19)].
6) The immobilization matrices, namely pith balls are perfect spheres (radius R). Some of the spheres, n, are always in contact with one spherical particle and it leads to loss in biofilm surface area (A_L) and volume (V_L) per unit sphere in contact.
7) Additional biomass produced due to biochemical reaction at the packing surface get adhered to the biofilm on packing itself and is not transported with the liquid stream.
8) No biochemical reaction occurs in the recycle stream.

The system equation based on differential material balance for the organosulfur compounds of diesel is as follows:

$$\frac{dC_B}{dZ} = -\frac{1}{F_A(1+R)} \cdot \frac{C_B\, C_{A0}.L_f}{(k_s + C_B\ ).Y} \cdot \mu_{max} \frac{\left[4\pi(\frac{d}{2}+L_f)^2 - nA_L\right]}{\frac{4}{3}\pi(\frac{d}{2})^3} . (1-\varepsilon\ ) A.L_f \qquad (1)$$

Where $A_L = 2\pi(d/2+L_f).L_f$ (2)

Following assumption 7, the variation of biofilm thickness due to bacterial growth on the surface of the packing material is,

$$\frac{dL_f}{dt} C_A = \eta \frac{\mu_{max}.C_B.C_A}{(k_s + C_B)}.Lf - b_d.C_A.L_f \qquad (3)$$

The variation of bed porosity within the reactor may be correlated to biofilm thickness using C. Alsono, *et.al.'s* (1997)[18] concept of and is represented as follows,

$$\varepsilon_f = 1-(1-\varepsilon_0)\left[\left(1+\frac{L_f}{d/2}\right)^3 - \frac{n}{4}\left(\frac{L_f}{d/2}\right)^2\left(\frac{2L_f}{d/2}+3\right)\right] \quad (4)$$

The boundary conditions are as follows,
For $t \leq 0$, Z=0,
$C_A = C_{A0}$, $C_B = C_{B0}$, $L_f = L_{f0}$
For t>0, Z = 0,
$C_A = C_{A0}$, $C_B = \frac{C_{B0} + R.C_B}{(1+R)}$

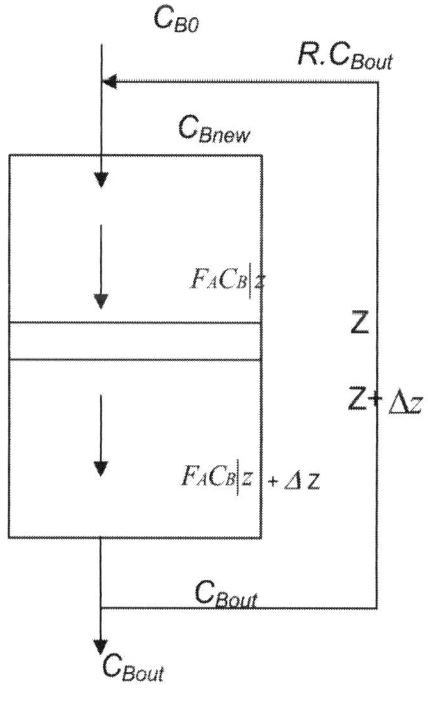

Figure 1: Schematic representation of the trickle bed bioreactor and the packing materials. F_A: Volumetric flow rate of the liquid stream; C_B: Substrate concentration; Z: axial position within the reactor.

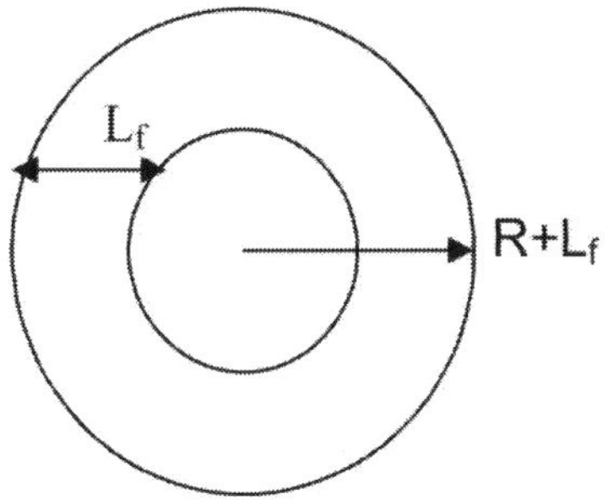

Figure 1a: Characteristic sphere. R: Radius of each packing sphere; L_f: Bio-film thickness over each packing material.

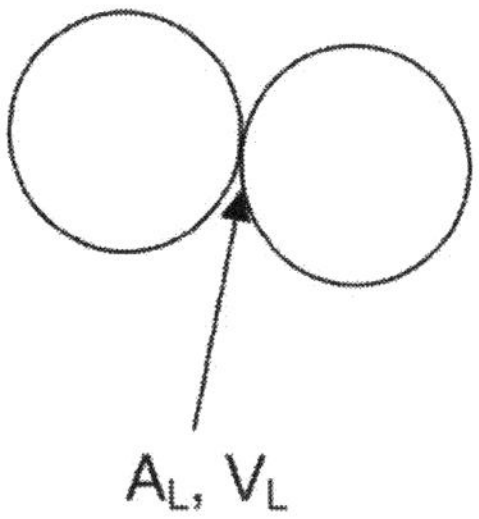

Figure 1b: Contacting pattern of the spheres.
A_L: Bio-film area loss per unit packing sphere in contact; V_L: Bio-film volume loss per unit packing sphere in contact.

The equations have been solved by 4th order Runge-Kutta technique using a suitable C-program. The values of kinetic parameters, μ_{max} and K_s and $Y_{A/B}$ have been taken from the previous publication [17].

Results and discussion:

Figure 2 represents the plot of simulated time history of biofilm thickness over the packing material at maximum LHSV of $3.676h^{-1}$ against time with initial sulfur concentration as a parameter. With due course of reaction time, the biofilm thickness has been found to increase with the increase in the initial sulfur concentration. This is because of the fact that with the progress of reaction time the quantity of biomass, adhering to the packing material, increases at the expense of substrate, i.e., organosulfur compounds present in feed diesel. In figure 3 time history of simulated and experimental values of bed porosity have been plotted with initial substrate concentration as the parameter. The figure reveals that the bed porosity decreases with the increase in initial substrate concentration. This may be accounted by the fact that the higher the substrate concentration the more is the substrate degradation and hence the more is the biomass formation. This results in the greater rate of increase in biofilm thickness which is reflected by the decrease in bed porosity. The agreement between simulated and experimental data is satisfactory. In figure 4 the simulated values of substrate

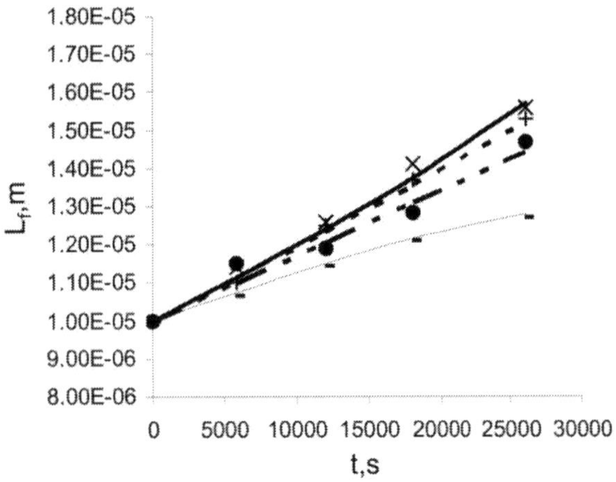

Figure 2: Simulated (Lines: —— 540ppm; — . . — 330 ppm; .... 430ppm; —— 200ppm) Experimental(Points: × 540ppm; • 330ppm; + 430ppm; -200ppm) Simulated and experimental profiles of biofilm thickness against time with initial substrate concentration as parameter.

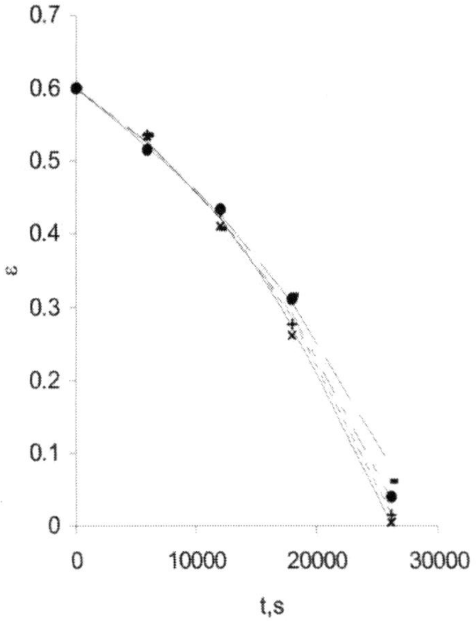

Figure3: Simulated (Lines: —— 540ppm; ----- 330ppm; .... 430ppm; — · · — 200ppm) Experimental (Points: × 540ppm; •330ppm; + 430ppm; -200ppm) Simulated and experimental profiles of bed porosity against time with initial substrate concentration as a parameter.

concentration in the liquid phase at the minimum LHSV of $1.838h^{-1}$ have been plotted against axial length of the reactor with initial sulfur concentration as a parameter. The experimental data have been superimposed on the same plot. As expected, the substrate

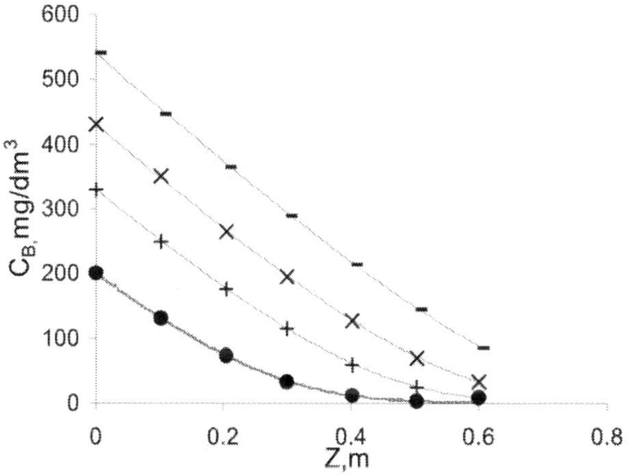

Figure 4: Simulated (Lines: .... 540ppm; .... 330ppm; ——— 430ppm; — · · — 200ppm)Experimental (Points: × 430ppm; •200ppm; + 330ppm; -540ppm) Simulated and experimental profiles of substrate concentration against axial length of the reactor.

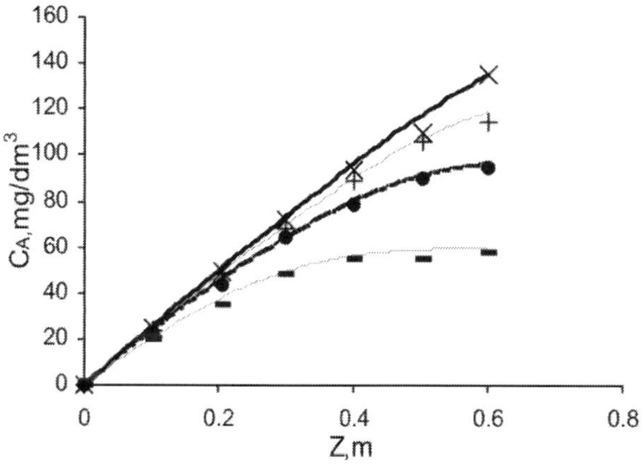

Figure 5: Simulated (Lines: ——— 540ppm; — . . 330ppm; .... 430ppm; ——— 200 ppm) Experimental (Points: × 540ppm; •330 ppm; + 430ppm; -200ppm) Simulated and experimental profiles of biomass concentration against axial length of bioreactor.

concentration decreases with the axial length of the trickle bed reactor. Moreover, the trend shows that the lower the initial substrate concentration, the better is the degradation efficiency. The simulated data are in good agreement with the experimental ones. Figure 5 represents the simulated and experimental profile of biomass concentration against the axial length of the reactor at LHSV of 1.838h^{-1}. The figure reveals of sulfur in diesel lie above those corresponding to lower ones. This is quite expected and is that the axial profiles of biomass concentration corresponding to higher initial concentrations due to the higher rate of formation of biomass under the condition of greater availability of diesel sulfur. Figure 6 is the plot of simulated and experimental concentration profile of substrate against axial length of the reactor at different LHSV of diesel for initial sulfur concentration of 200 ppm. LHSV has been varied in the range of 1.838h^{-1} and 3.676 h^{-1}. Close analysis of the figure reveals that with the increase in the LHSV the value of sulfur concentration in the reactor exit stream increases and the ultimate sulfur conversion decreases. The sulfur removal efficiency of 95.46% has been predicted by the mathematical model at the lowest inlet LHSV of diesel of 1.838h^{-1}, where as the corresponding experimental value is 95%. Similarly, the simulated

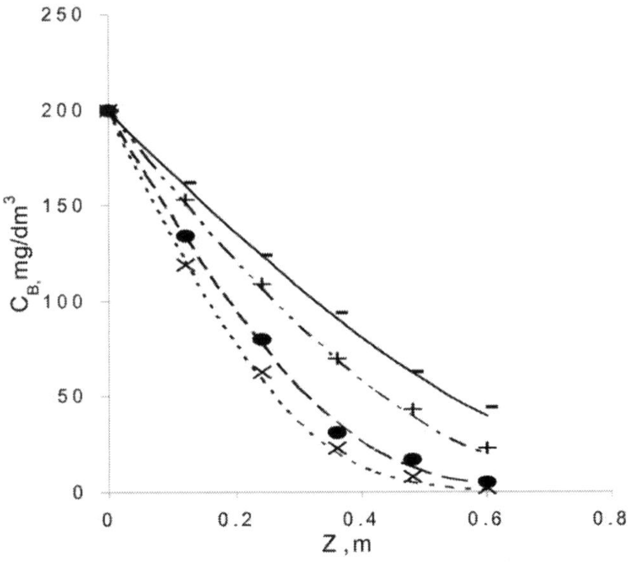

Figure 6: Simulated (Lines: .... 1.838h^{-1}; - - - - - - 2.158h^{-1}; ______ 2.94h^{-1}; .... 3.676h^{-1}) Experimental (Points: × 1.838h^{-1}; • 2.158h^{-1}; + 2.94h^{-1}; -3.676h^{-1}) Simulated and experimental profiles of substrate concentration against axial length with volumetric flow rate as a parameter with initial substrate concentration 200ppm.

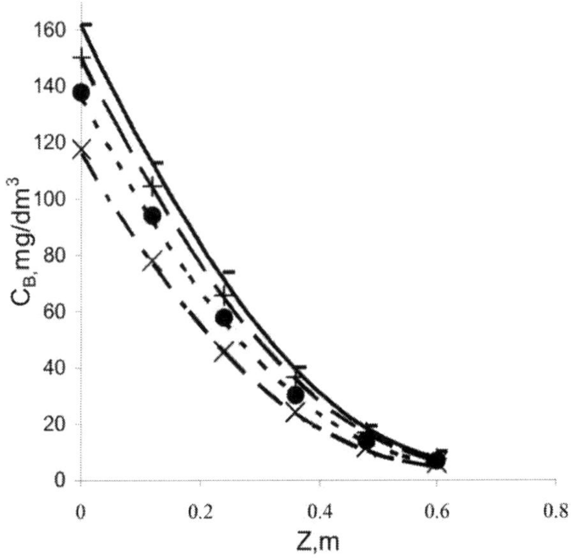

Figure 7: Simulated (Lines: ______ 0s; — — 180s; •••••• 5940s; — • •12060s; _ . . 18000s; ______ 26280s) Experimental (Points:+ 0s; ♦180s; -5940s; ▲12060s; ×18000s; •26280s) Simulated and experimental profiles of substrate concentration against axial length with time as a parameter at recycle ratio 0.25 and initial substrate concentration200 with ppm and at LHSV of 1.838h^{-1}.

and experimental values of sulfur removal efficiency at LHSV of 3.767h^{-1} are 80% and 79.8 % respectively. This is, however, expected as the increase in LHSV implies the decrease in reactor residence time causing drop in the ultimate conversion of the reactant. The simulated trend can explain the experimental one also in this case. In figure 6 the simulated concentration profile of substrate against axial length of the reactor at different recycle ratio ranging from 0.25 to 2 for initial sulfur concentration of 200 ppm diesel has been represented. The experimental values have been superimposed on the same plot, and they have been found to be well explained by the simulated results. The figure reveals that the higher the recycle ratio, the more is the reduction in sulfur concentration. At lowest recycle ratio of 0.25 with LHSV of 1.838h^{-1}, the simulated sulfur removal efficiency is 96.16%, while the corresponding experimental value is 95%. Again at recycle ratio of 2 with the same LHSV, the simulated result shows 97.87 % sulfur removal efficiency against its experimental counterpart of 96.4%. This can be easily realized from the fact that, as the recycle ratio increases, the scope for the substrate to come in contact with the immobilized microorganisms becomes more resulting in greater chance of substrate degradation reaction.

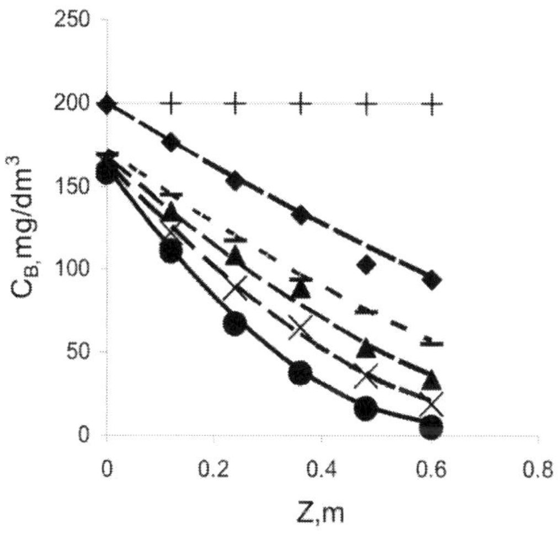

Figure 8: Simulated (Lines: — · — R= 0.25; - - - - R= 0.5; ——— R= 0.75; — — - R= 01.25) Experimental (Points: × R= 0.25; • R= 0.5; + R= 0.75; - R= 0.1.25) Simulated and experimental profiles of substrate concentration against axial length with recycle ratio as a parameter with initial substrate concentration 200ppm and at LHSV of 3.676h^{-1} at 26820s.

Conclusion:

Rhodococcus sp. (NCIM 2891) has been used to reduce the sulfur level in diesel for the present study. The initial sulfur concentration has been varied in the range of 200-540ppm. A trickle bed reactor was studied with the variation of LHSV and recycle ratio as parameters. Biofilm thickness over the packing material and biomass density have been measured time to time to get their profile during the course of substrate degradation reaction. A mathematical model capable of describing the biodesulfurization of diesel in a trickle bed reactor has been proposed. The model predictions are able to explain the experimental observation satisfactorily. However, there is scope for refinement of the model by the incorporation of more realistic characteristics such as presence of external mass transfer resistance etc. prevailing in the actual system.

Acknowledgements

The authors acknowledge the financial support rendered by C.S.I.R by providing Senior Research Fellowship to the first author.

Nomenclature

A	Cross-sectional area of the reactor (m);
A_L	Bio-film area loss per unit packing sphere in contact (m);
C_{A0}	Initial biomass concentration (mg/dm^3);
C_A	Biomass concentration at a particular time (mg/dm^3);
C_{Af}	Final biomass concentration at the reactor outlet (mg/dm^3);
C_{B0}	Initial substrate concentration (mg/dm^3);
C_B	Substrate concentration at a particular time (mg/dm^3);
C_{Bnew}	New inlet substrate concentration at a particular time during recycle (mg/dm^3);
C_{Bout}	Final substrate concentration at the reactor outlet (mg/dm^3);
F_A	Volumetric flow rate of the liquid stream (dm^3/hr);

Ks Saturation constant;
L Length of the reactor (m);
L_f Bio-film thickness over each packing material (m);
N Number of Pressure drop;
D Diameter of each packing sphere (pith ball) (m);
R Recycle ratio;
V Volume of the TBR (m^3);
V_L Bio-film volume loss per unit packing sphere in contact (m^3);
$Y_A/_B$ Yield coefficient = Mass of biomass produced /mass of substrate consumed;
Z Axial position within the reactor;

Greek letters
μ_{max} Maximum specific growth rate of biomass;
ε_0 Initial bed porosity;
ε_f Porosity in bed with bio-film;

Abbreviations
I.B.P. Initial Boiling Point;
F.B.P. Final Boiling Point;
LHSV Liquid hourly space velocity= volumetric feed flow rate / catalyst bed volume;
UOP Universal oil products;
ASTM American standard for testing materials;

References

(1) Singh, H.; Malhotra, M; Geroge, N.; Ghosh, S.K. Biocatalytic sulphur removal from fossil fuel.; *Petrotech* **2001**.
(2) Marcelis, C.L.M.; Leeuwen, M.V.; PoldermaN; H.G.; Janssen, A.J.H.; Lettinga, G., Model description of dibenzothiophene mass transfer in oil/water dispersion with respect to biodesulfurization. *Biochem. Engg.Jr.* **2003**, 16, 253.
(3) Grossman, M.J.; Lee, M.K.; Prince, R.C.; Minak-Bernero, V.; George, G.N.; Pickering, I.J., Microbial desulfurization of a crude oil middle distillate fraction: analysis of the extent of sulfur removal and the effect of removal on remaining sulfur, *Appl. Env. Microbiol.* **2001**, 67, 1949.
(4) Prince, R.C.; Grossman M.J.; Substrate preferences in biodesulfurization of diesel range fuels by Rhodococcus sp. strain ECRD-1, *Appl. Env. Microbiol.* **2003**, 69, 5833.
(5) McFarland, B.L.; Boron, D.J.; Deever, W.; Meyer, J.A.; Johnson, A.R; Atlas, R.M., Biocatalytic sulfur removal from fuels: applicability for producing low sulfur gasoline, R.M. *Crit Rev Microbiol.* **2001**, 24, 99.
(6) Bharami, A.; Shojaosadatia, S.; Mohebali, G., Biodegradation of dibenzothiophene by thermophilic bacteria, *Biotechnol.* **2001**, 11, 899.
(7) Okada, H.; Nomur, N.; Nakhara, T.; Maruhashi, K., Cultivation of desulfurizing bacterium, *Mycobacterium* strain G3, *Biotechnol.* **2001**, 24, 2047.
(8) Matsui, T.; Hirasawa, K.; Koizumi, K.; Maruhashi, K.; Kurane, R., Optimization of the copy number of dibenzothiophene desulfurization genes to increase the desulfurization activity of recombinant Rhodococcus sp., *Biotechnol.* **2001**, 11, 1715.
(9) Gregory, W.M.; Mc.Caffrey, D.S., Deep desulfurization to meet ultra low sulfur levels in distillate products, *Hydrocrbon Tech.* **1994**, 23, 15

(10) Kirimurak, K.; Furya, T.; Sato, R.; Ishii, Y.; Kino, K.; Usmani, S., Biodesulfurization of dibenzothiophene and benzothiophene through selective cleavage of carbon-sulfur bonds by Rhodococcus sp. strain WU-K2R, *Appl. Environ. Microbiol.* **2002**, 68, 3867.
(11) Chowhury, R.; Pedernera, E.; Reimert, R., Trickle-bed reactor model for desulfurization and dearomatization of diesel, *AICHE Jr.* **2002**, 48, 126.
(12) Maghsoudi, S.; Vossught, M.; Kheirolomo, M.; Tanaka, A.E.; Katoh, S., Biodesulfurization of hydrocarbons and diesel fuels by Rhodococcus strain P32C1, *Biochem Eng. Jr.* **2001**, 8, 151.
(13) Matsui, T.; Hirasawa, K.; Konishi, J.; Tanaka, Y.; Maruhashi, K.; Kurane, R., Microbial desulfurization of alkylated dibenzothiophene and alkylated benzothiophene by recombinant Rhodococcus sp. strain T09, *Appl. Microbiol Biotechnol.* **2001**, 56, 196.
(14) Chang, J.H.; Kim, Y.J.; Lee, B.H.; Cho, K.S.; Rya, H.W.; Chang, Y.K.; Chang, H.N., Production of a desulfurization biocatalyst by two-stage fermentation and its application for the treatment of model and diesel oil, *Biotechnol.*; **2001**, 17, 878.
(15) Furuya, T.; Ishii, Y.; Noda, K.; Kino, K.; Kirimura, K., Thermophilic biodesulfurization of hydrodesulfurized light gas oil by Mycobacterium phlei WU-F1, *Microbiol Lett.* **2003**, 221,137.
(16) Mukhopadhyay, M.; Chowdhury, R.; Bhattacharya, P., Mathematical modeling of a trickle bed reactor for the ultra low desulfurization of hydrodesulfurized diesel, in IICHE conference (CHEMCON, Mumbai) 28-30 December **2004**.
(17) Mukhopadhyay, M.; Chowdhury, R.; Bhattacharya, P., Two phase mathematical model for a bio-trickling reactor for the production of ULSD (ultra low sulfur diesel) from deeply hydrodesulfurized diesel, *Catal. Today*, **2005**, 06, 80.
(18) Alosono, C.; Suidan, M.T.; Sorial, G.A.; Smith, F.L.; Biswas, P.; Smith, P.J.; Brenner, R.C., Gas treatment in trickle-bed biofilters: Biomass, how much in enough?, *Biotechnol. And Bioengg.* **1997**, 54, 583.
(19)Guchhait, S.; Biswas, D.; Bhattacharya, P.; Chowdhury, R., Bi-phasic Bio-conversion of sulfur present in model organo-sulfur compounds and hydrotreated diesel, *Catal.Today*, **2005**, 106, 233.
(20) Guchhait, S.; Biswas, D.; Bhattacharya, P.; Chowdhury, R., Bio-desulfurization of model organo-sulfur compounds and hydrotreated diesel—experiments and modeling, *CEJ*, **2005**, 112, 145.

OPPORTUNITY CRUDES
- ADDRESSING THE CHALLENGES OF MECHANICAL RELIABILITY

Jerold I. Danis
The Equity Engineering Group, Inc.
Dallas, TX

Reliable unit operations are key to the success of petroleum refiners. The introduction of opportunity crudes, typically those containing high levels of naphthenic acid and sulfur, present challenges to maintaining high unit reliability. These components in opportunity crudes can contribute to accelerated corrosion, both localized and general, often in areas that have been free of corrosion when processing less aggressive crudes. This paper will address steps that can be taken by refiners, such as the use of equipment monitoring envelopes, to minimize the potential for unexpected and unplanned incidents associated with processing of opportunity crudes.

Introduction

Diminishing quantities of historical crude sources coupled with economic incentives have encouraged refiners during the past 5- 10 years to consider processing crudes that at one time may have been unattractive. These "opportunity crudes" present a variety of challenges which could include issues such as differences in yield distribution, more demanding requirements for efficient desalting and increased potential for fouling. Along with such process related changes these crudes also can expose the refiner to corrosion related issues that previously were not experienced.

Corrosion issues associated with opportunity crudes may occur in the overhead of the atmospheric and vacuum towers in crude units along with their associated high temperature circuits. This paper will address only hot circuit issues, i.e. high temperature sulfidic (sulfidation) corrosion and naphthenic acid corrosion. A brief review of these mechanisms will be discussed followed by an approach that can be used by refiners to address these corrosion related issues.

Sulfidic and Naphthenic Acid Corrosion – Two Forms of High Temperature Corrosion

Sulfidic Corrosion

As reflected in API Recommended Practice 571, Damage Mechanisms Affecting Fixed Equipment in the Refining Industry[1], sulfidic corrosion of iron based alloys (e.g. carbon steel, low alloy steels and stainless steels) is generally considered to be a concern for metal temperatures in excess of about 500F. Industry typically evaluates the potential for sulfidic corrosion using guidelines referred to as the Modified McConomy Curves – see NACE TCR 34103[2]. These curves reflect the relationship between corrosion rate and metal temperature as a function of hydrocarbon total sulfur and the alloy content of the base material. Some refiners use proprietary technology reflecting only the contribution of the reactive sulfur (most corrosive) species to corrosion. For example, thiophenic sulfurs, generally considered to be non-corrosive[3], may be ignored by some refiners.

Typically the corrosion rate increases as the temperature increases for a given sulfur level in a hydrocarbon stream. Corrosion can be mitigated or reduced by increasing the chromium content of the material. Most refiners "alloy up" from carbon steel to 5% Cr, then to 9% Cr and finally to stainless steels (both the 300 series austenitic grades and the 400 series martensitic and ferritic grades) to address progressively more severe corrosion concerns.

Sulfidic corrosion has been observed in both piping and pressure vessels. Affected processing units include not only the primary crude unit, but also portions of downstream units, such as the feed circuits of the FCCU, Coker and hydroprocessing units. Corrosion is considered to be primarily uniform although some variations in corrosion rate do exist across piping systems. Corrosion rates can be accelerated by turbulence if the semi-protective iron sulfide scale is removed. The general nature of this type of corrosion makes it suitable for monitoring via ultrasonic thickness measurements. Note: while corrosion due to high temperature sulfur will occur in hydroprocessing units in the presence of hydrogen, the corrosion mechanism is different and is not addressed in this paper.

Naphthenic Acid Corrosion

The term naphthenic acid refers to a family of carboxylic acids of varying molecular weights. As reflected in API Recommended Practice 571, naphthenic acid corrosion has been noted to occur in processing streams operating at temperatures above 425F although reports of corrosion at temperatures as low as 350F have been reported. The corrosion potential typically increases with increasing temperature up to about 750F-800F. Corrosion has been noted to be accelerated in regions of high velocity or turbulence, but has also been noted in areas where condensation is likely to occur, such as beneath trays in vacuum columns. Pure vapor systems are considered to be free of naphthenic acid corrosion issues.

While widely recognized to occur in crude units, the potential for naphthenic acid corrosion also exists in downstream processing units such as the feed systems for the FCCU, Coker and hydroprocessing units as well as lubes units. It is important to recognize that the distribution of the acids vary with the crudes being processed. However, in crude units the greatest concentration of the acids is typically seen in vacuum tower side streams. Quantification of the acids is usually expressed in terms of total acid number, TAN. This is measured typically by ASTM Test Methods D974[4] and D664[5].

Much of industry's knowledge of naphthenic acid corrosion has been based on experience processing crudes originating from Venezuela or California. Many years of processing these crudes resulted in the adoption of various rules of thumb, e.g. corrosion is not an issue with TAN < 0.5 or corrosion can be severe at TANs > 2.0. These rules have resulted in many refiners blending their crudes to low TAN levels in order to have manageable corrosion rates. However, experience over the past ten years with crudes originating from other locations has shown that these old "rules" are no longer valid. For example, corrosion has been experienced at very low TAN levels (< 0.5)[6] and severe corrosion has been observed at TAN levels much less than 2.0[7].

In general, refiners have found that while "low levels" of naphthenic acid can be managed at times with low chrome alloys, i.e. 9%Cr, the most effective alloys for mitigating or reducing the corrosion potential are 316 / 316L or 317 / 317L stainless steels. However, even these alloys have experienced naphthenic acid corrosion at high acid levels.

While some refiners have developed proprietary technology to enable improved predictive capability, industry in general is relying on studies and experiences that have been either published in the open literature or discussed in open industry forums. Such information includes the following:

a. A relationship exists between sulfidic corrosion and naphthenic acid corrosion[8]. As reflected in API RP 571, sulfidic corrosion promotes the formation of iron sulfide which can have an inhibitive effect on naphthenic acid corrosion. This helps explain the corrosion experienced with very low sulfur crudes (<0.25 wt %) that exhibit naphthenic acid corrosion at low TAN (<0.5) values.
b. The introduction of hydrogen to a stream in a hydroprocessing unit will eliminate the potential for naphthenic acid corrosion downstream of the mix point.
c. Some studies have reported the greatest corrosion potential to occur just below the boiling point of the acid. The molecular weights of the naphthenic acids in a crude or fraction can influence stream corrosivity. In general, the lighter the molecular weight of the acid the more aggressive the corrosion[9]. This supports the observation that while two different crudes can produce an identical TAN, the corrosion rates experienced can be different, i.e. the TAN may be the same, but the molecular weight of the acids may be different leading to differences in corrosion rate.
d. The use of external hydrogen probes may potentially be capable of detecting areas of high hydrogen flux associated with localized metal loss due to naphthenic acid corrosion[10]. There has been considerable discussion in industry meetings concerning this technology which continues to evolve.

What is Required to Start an Evaluation?

The first step in evaluating an opportunity crude is to set a basis for the evaluation. Typically the refinery's Supply organization will propose to refinery management that opportunity crudes be introduced into the site. Usually a change in the operating target will be proposed, e.g. a change from a vacuum side stream limit of 0.5 TAN to a vacuum side stream limit of 1.2 TAN. The new limit may be based on an opportunity identified by Supply to enter into a long term contract for purchase of a specific crude. However, it may also be based on a general target that would enable the site to process a variety of crudes to that specified limit. While the first approach permits detailed evaluations of the crude's assay and its distillation behavior / distribution both approaches can lead to a meaningful assessment.

Once the desired target is established, the next step in evaluating an opportunity crude is to have good documentation of what has already been processed at the site and what has been experienced when running these crudes. This is often referred to as a "baseline" determination. Baselines need to be defined for the applicable process variables as well as the applicable inspection issues. Controls on a variety of process related variables are necessary to ensuring unit reliability. Such controls are often referred to as the Materials Operating Envelope (MOE)[11,12].

Establishing a Process Baseline

A review of key segments of a unit's Materials Operating Envelope should be conducted as part of the baseline evaluation. If data is not readily available from an existing MOE it will typically be necessary to generate new data to define the baseline. Typical items which should be documented include:

1. Historical Crude Oils Processed: A tabulation of the historical crude slate that the refinery has successfully processed is recommended. A snapshot of crudes typically representing a 3-5 year window should be tabulated. While some refiners may process in excess of 50 crudes per year, the tabulation can be "fine tuned" to reflect only those crudes that represented a significant portion of the annual crude slate.

2. TAN: The TAN should be documented for the whole crude as well as all of the relevant atmospheric and vacuum tower side streams and bottoms streams. If such data is not readily available this often requires that data be gathered over a representative period (often for a minimum of 3 months) when processing crudes that ideally resemble the baseline crude slate. Data gathering may require the installation of sample coolers in order that accurate hot samples can be obtained. Sampling at key locations downstream from the crude unit is also recommended.

3. Sulfur: Similar to TAN, documentation of the sulfur distribution for the whole crude and relevant side streams is also helpful. It is important to note, however, that knowledge of total sulfur does not necessarily mean that one will have a good understanding of the potential

for sulfidic corrosion. As reported in industry meetings and in the open literature some sulfur species (reactive sulfurs) are more likely to lead to corrosion than others [13,14]. As mentioned above, while some refiners may have analytical procedures to evaluate these reactive sulfur species others may have to rely on less precise methods to assess the corrosion potential associated with sulfur.

4. Other Variables: Variables such as operating temperature and velocities in piping systems should also be reviewed as part of the baseline evaluation.

Establishing an Inspection Baseline

There are a variety of inspection related issues that also need to be defined before embarking on the processing of opportunity crudes. Issues to be addressed include:

1. Reviewing Piping Inspection Isometrics: Up to date and accurate piping isometrics are important when trying to assess the current condition of a unit. These isometrics, containing the Thickness Monitoring Locations (TMLs), are useful in determining which specific regions have been stewarded in the past. Due to the general nature of sulfidic corrosion, typically only a small sampling of the piping needs to be performed. However, the localized nature of naphthenic acid corrosion often means that additional TMLs need to be established. Once these regions are identified, additional monitoring to establish the baseline for the new TMLs is required. Note: due to the highly localized nature of naphthenic acid corrosion the use of radiographic testing (RT) or wide area scanning with ultrasonic testing (UT) is recommended for these suspect locations. Typical areas for which baseline readings should be obtained include turbulent areas such as:

a. Downstream and often upstream of control valves.
b. Reducers / swages
c. Piping downstream of orifice plates
d. Piping immediately adjacent to and downstream of thermowells and corrosion probes due to eddy currents which can lead to highly localized damage
e. Piping adjacent to the downstream side of butt welds used to fabricate mitered connections
f. High velocity areas such as those associated with pump discharge, bypass and warm-up lines. Note; there have been references in the literature indicating that wall shear stress (a relationship linking factors such as velocity to hydrocarbon properties such as density) plays a role in corrosion by contributing to the removal of the protective iron sulfide layer on the ID of piping systems[8]. It has been suggested that this factor plays a more significant role than velocity with respect to the potential for corrosion due to naphthenic acids.
g. The collection header for the piping exiting the individual passes of the vacuum furnace prior to entering the transfer line to the tower.

2. Evaluating Piping and Fixed Equipment Thickness Data: A review of existing thickness data and identification of the remaining corrosion allowance allows the remaining life of the component to be determined based on the historical crude slate. Most inspection data

management software programs will automatically generate this data and subsequently identify components that are close to retirement or require additional inspection based on historical corrosion rates. This is a useful step since it flags any "weak links" in the unit prior to introduction of the opportunity crude. Note: it is not unusual for ultrasonic thickness data compiled in data management programs to be misleading. Assessments of the data and recommendations based upon it need to be conservative due to the variability in the quality of the data.

3. Reviewing Corrosion Probes: Corrosion probes are the most common tools used to provide rapid feedback on the corrosivity of a given stream. A review of the location and data generated by existing corrosion probes in hot circuits is encouraged. However, often few if any probes are present in circuits subjected to high temperature sulfidic corrosion (typically > 500F) or naphthenic acid corrosion (those operating at temperatures > 425F). Consequently, it is suggested that probes be installed in all major circuits subjected to naphthenic acid or elevated temperature sulfidic corrosion. A data gathering period of 3 months minimum is suggested to ensure proper functioning of the probes and to gather sufficient data to accurately define the baseline corrosion rate for the piping circuits in question. Typically, the following items need to be addressed during probe selection:

a. A probe designed specifically for high temperature applications.
b. A sensing element metallurgy that matches the piping into which the probe is placed.
c. A tubular rather than wire loop sensing element. This typically is more robust than a wire loop and can potentially provide data for a longer period of time.
d. A means to continuously monitor the probe's output signal. This can be accomplished by either hard wiring the probe's signal into the central process control computer or by the installation of a remote data logger in the field. Output from the data logger can be periodically downloaded for review. The use of manually monitored probes is discouraged since corrosion events may occur during the interval between manual readings.
e. A documented and site endorsed procedure for change out of probes on the run. Corrosion probes have a finite life and as such, probes will require replacement. Process technicians should either be trained in the safe removal and insertion of probes or provisions should be made with an outside vendor to come into the plant on a contracted basis to change probes as required.

4. Identifying the Need and Location for Corrosion Coupons: The use of corrosion coupons can provide useful long term information regarding corrosion. Coupons can be installed on the end of corrosion probes in piping or equipment if the probes have been fitted with the appropriate holder. Or, if equipment will be opened and entered prior to the introduction of the opportunity crudes, corrosion coupon racks can be installed in suspect areas of key vessels. These will require another downtime to remove the coupons and as such are only useful for verifying long term corrosion rates. Some refiners have installed corrosion racks in the downcomer regions of atmospheric and vacuum towers to assist in the determination of the appropriate material for subsequent metallurgical upgrades.

Once the Baseline is Established, What is Next?

Once the existing operating envelope is confirmed (via the baseline measurements) a new operating envelope can be established. Often times the desired target is set by Operations due to their interest in replacing existing "historical" barrels with equivalent barrels of an opportunity crude. The number of "opportunity" barrels that can be introduced will be influenced by the existing metallurgy in the suspect circuits, the remaining available corrosion allowance and in the case of naphthenic acid corrosion, whether or not provisions were put in place to inject a corrosion inhibitor[15,16].

A strategy needs to be developed for the introduction of the opportunity crudes. One approach is to incrementally introduce the new crude in stages, e.g. in a step-wise manner over a number of weeks. During these stages, monitoring of the key process variables, e.g. TAN, sulfur and corrosion probe rate, should be conducted. The duration of the various steps will be influenced by the amount of the new crude that is available and any limitations or constraints associated with off-site tankage. It is important that the steps are of sufficient duration such that changes in corrosion rate can be noted. Note: adequate controls need to be in-place at the refinery's tank farm to ensure that the proper amount of opportunity crude is being blended into the site's base slate. Uncontrolled additions of opportunity crude can potentially lead to undesirable corrosion rates.

Mitigation Methods

Along with the development of a targeted operating envelope it is necessary to evaluate the options for mitigation of high temperature corrosion. Three standard approaches are normally followed: A) alloy upgrading, B) use of corrosion inhibitors for naphthenic acid, or C) a combination of both.

Many refiners today are using the provisions of API RP 580[17], and the associated Publication 581[18] (dealing with Risk Based Inspection) to assess their equipment and piping for risk. Such an evaluation addresses not only the probability (or likelihood) of incurring an incident, but also the consequence of the incident. A risk assessment of areas on the unit(s) susceptible to naphthenic acid corrosion or elevated temperature sulfidic corrosion can define the type and extent of mitigation that is needed. The options for mitigation are discussed below.

A. Alloy Upgrading

A unit risk assessment may identify the need for some selective alloy upgrading in order to minimize the possibility of an unplanned unit outage. As mentioned above, sulfidic corrosion is normally addressed by progressing from carbon steel to 5% Cr to 9% Cr and then to the stainless steels. For naphthenic acid concerns the first step in upgrading from carbon steel or low alloy material is often 316 / 316L stainless steel. This molybdenum containing alloy (2-3% Mo per ASTM specifications) has historically been the alloy of choice for addressing all but the most severe cases of naphthenic acid corrosion. However, due to changes in steel making practices over the past 15+ years the level of molybdenum found in this alloy is nearly always at the lowest end of the specification, i.e. essentially at 2.0%. Consequently, performance of the alloy has been mixed, especially at moderate TAN levels, forcing many purchasers to request this material with a guaranteed minimum molybdenum level of 2.5%. This normally leads to special mill orders which can be costly and have long delivery times. To avoid delivery constraints, 317 / 317L stainless steels (containing 3-4% Mo) are being more frequently specified by refiners.

B. Use of Naphthenic Acid Corrosion Inhibitors

The use of inhibitors to minimize the potential for naphthenic acid corrosion has been used in industry over approximately the past 10 years. Their use has been widely discussed in industry meetings and reported / documented in the open literature. Some refiners expecting to process high TAN opportunity crudes on an intermittent basis have relied on inhibitors to manage corrosion. Their use can provide a reliable means of managing corrosion without incurring the expenses associated with extensive metallurgical upgrades.

If inhibitors are being considered, a detailed review of the piping systems and tower tray configurations is required to identify the optimum locations for the installation of the inhibitor injection points. For those piping systems fabricated from low alloy materials (typically 5% Cr or 9% Cr), the installation of the necessary access fittings can only be done during a unit downtime to enable the subsequent use of post weld heat treatment (PWHT) as required by the B31.3 Piping Code[19]. The need for a downtime in many cases has been found to be a short term obstacle to the application of naphthenic acid inhibitors. Close interaction and involvement of the chemical vendor providing the inhibitor is critical to the success of its application. The vendor should be a key member of any decision making group involved with use of such an inhibitor. They should be able to provide guidance on the need for prepassivation of equipment (typically at high inhibitor dosages) as well as the required dosage for maintenance of a protective layer of the inhibitor.

C. Use of Alloy Upgrading plus Inhibitor

Many times refiners are reluctant to commit to major metallurgical upgrades due to the uncertainty of future crude supplies. However, the uncertainty with inhibitor performance in high velocity areas also has been expressed as a concern. This generally leads to some partial upgrading along with injection of a corrosion inhibitor. Metallurgical upgrading is normally reserved for those circuits where inhibitors may not be totally effective, e.g. high velocity vacuum unit furnaces and transfer lines and / or locations where periodic inspection of the piping system is difficult due to access limitations.

Corrosion in Downstream Units

A high level of attention is normally applied to the crude unit when evaluating opportunity crudes. However, issues such as sulfidic corrosion and naphthenic acid corrosion can be experienced in downstream units as well. Consequently, baseline evaluations for these units are also required. Typical areas to be addressed include hot feed streams in units such as the FCCU, gas oil hydroprocessing units, lubes units and cokers. Note: as mentioned earlier, naphthenic acid corrosion in hydroprocessing units is not expected to occur downstream of the hydrogen injection point. Since some of these units may receive feed streams from units other than the crude unit under consideration, there may be some beneficial effects associated with stream dilution. Corrosion probes may be required in the susceptible areas of these units if dilution is not sufficient to address potential corrosion concerns.

Other Issues for Consideration

In addition to the issues discussed above, it is necessary for the refiner to ensure that they have an adequate Positive Material Identification (PMI) program in place. PMI needs to be performed for existing installed materials (often called retroactive PMI). Additionally, a program needs to be in place to ensure that all new materials being installed to address the opportunity crudes are in fact what were specified. For example, for both a retroactive and new PMI program it is necessary to ensure that there is no 304 SS in a piping system that has been identified as 316 SS. API RP 578[20] provides guidance on PMI programs.

Another important, but often overlooked issue involves the increased sulfidic corrosion potential for low silicon (<0.15 wt %) containing carbon steels. For services in excess of 500F, there have numerous reports of incidents in industry where carbon steel components corroded at unpredictably high corrosion rates when exposed to a high temperature sulfur environment. The industry has recognized that piping materials such as ASTM A106 is produced with sufficient silicon. However, ASTM A53 piping materials may be produced with low silicon and hence are susceptible to this corrosion mechanism. The refiner needs to assess their vulnerability to this degradation mechanism since processing of a high sulfur opportunity crude may result in some unforeseen corrosion issues.

Conclusion

The processing of opportunity crudes presents challenges to the refiner due to the potential for increased corrosion. However, these challenges can be adequately addressed first by a review of the historical baseline information on the units in question and secondly by the implementation of appropriate monitoring and mitigation strategies.

References

1. API Recommended Practice 571, "Damage Mechanisms Affecting Fixed Equipment in the Refining Industry", First Edition, December 2003.
2. NACE TCR 34103, "Overview of Sulfidic Corrosion in Petroleum Refining", 2004.
3. ASM Metals Handbook Volume 13 Ninth Edition, "Corrosion", pg 1271.
4. ASTM D974, "Standard Test Method for Acid and Base Number by Color – Indicator Titration", Annual Book of ASTM Standards.
5. ASTM D664, "Standard Test Method for Acid Number of Petroleum Products by Potentiometric Titration", Annual Book of ASTM Standards.
6. Nugent, M.J. and J. D. Dobis, "Experience with Naphthenic Acid Corrosion in Low TAN Crudes", NACE Corrosion98, Paper No. 577.
7. NACE RefinCor – Minutes from STG 34 Refining Industry Corrosion.
8. Kane, R.D. and M. S. Cayard, "Understanding Critical Factors that Influence Refinery Crude Corrosiveness", Materials Performance, July 1999.
9. Lewis, K.R. et al, "Processing Corrosive Crude Oils", NACE Corrosion99, Paper No. 377.
10. Etheridge, A.M. et al, "Examples of Use and Interpretation of Field Data Using a Portable Hydrogen Permeation Monitor", NACE Corrosion2004, Paper No. 04477.
11. Danis, J.I. and A. C. Gysbers, "Creating Materials Envelope Statements for Petroleum Refining Operating Units", NACE Corrosion98, Paper 589.
12. Buchheim, G.M., J.I. Danis and D. M. DeHart, "Construction Materials Operating Envelopes: Safe and Reliable Operations within Boundaries", presented at NPRA Conference, 2005.
13. Slavcheva, E. et al, "Review of Naphthenic Acid Corrosion in Oil Refining", British Corrosion Journal, Vol. 34 No 2, 1999.
14. Kapusta, S. D. et al, "Safe Processing of Acid Crudes", NACE Corrosion2004, Paper No. 04637.
15. Babaian-Kibala, E., "Phosphate Ester Inhibitors Solve Naphthenic Acid Corrosion Problems", Oil and Gas Journal, Feb 1994.
16. Zetlmeisl, M.J., "A Laboratory and Field Investigation of Naphthenic Acid Corrosion Inhibition", NACE Corrosion95, Paper No. 334.
17. API RP 580, "Risk Based Inspection", 2002.
18. API Publication 581, "Base Resource Document-Risk Based Inspection", 2000.
19. ANSI/ASME B31.3 Piping Code.
20. API RP578, "Material Verification Program for New and Existing Alloy Piping Systems", 1999.

Crude Oil Management: Reduce Operating Problems While Processing Opportunity Crudes

By Larry Kremer
Baker Petrolite Corporation

It is important to evaluate opportunity crude oils not only for cost savings but also for potential impacts on desalting, corrosion, environmental discharges, fouling, catalyst poisoning, and product degradation. Once problem areas have been identified, mitigation strategies should be developed and implemented to make the processing of the crude oil as profitable and trouble free as possible. There can be considerable batch to batch variation in some quality parameters so continued monitoring is required to minimize the potential for upsets. Examples will be given to illustrate monitoring and mitigation strategies.

Introduction

There are several quality parameters in crude oil that can be used to predict potential processing problems in a given crude oil. Some of these parameters are specified when the crude oil is purchased, but others must be determined separately. Most contracts specify gravity, sulfur, and BS&W. Producers blend crude oils to meet gravity and sulfur specifications so that refineries will have a consistent product from shipment to shipment. Many of the other factors can vary significantly from batch to batch, which can cause intermittent problems or unanticipated upsets. There are also constituents that are rarely in crude oil shipments, but when present in the crude oil can have major negative impacts on the refinery. The source of many of these quality variations has been described elsewhere.[1] Examples of these different types of quality issues will be discussed in terms of recognizing potential problems and developing mitigation strategies to deal with them.

Table I. Selected crude quality parameters related to potential processing problems

Commonly Monitored Parameters	Other Significant Parameters
Gravity or Density	Organic Chlorides
Sulfur	Methanol
BS&W	Naturally Occurring Radioactive Material (NORM)
Salt (usually as water soluble chlorides)	Calcium Naphthenates
Total Acid Number (TAN)	Crude Incompatibility
Filterable Solids	Asphaltenes and Resins (SARA)
Metals (Ni, V, Fe, Hg, Se, *etc.*)	Ease of Desalting
Crude Distillation Properties	Volatile Amines

Table I lists some of the crude oil quality parameters that might indicate problems when processing a specific crude oil. The commonly monitored parameters in the first column are typically measured on a daily basis for every batch of crude oil received by the refinery. When an opportunity crude oil is evaluated with quality parameters outside the range of crude oils typically processed by the refinery, it is useful to conduct a more in depth evaluation of potential problems.

Some crude quality parameters such as organic chlorides, methanol, and NORM occur infrequently, so are not monitored by many refineries on a routine basis. Unfortunately the problems that can be caused by the presence of these compounds in crude oil can be severe. In many cases if the refinery is aware of the presence of these contaminants, they can develop strategies to handle them.

Baker Petrolite Crude Oil Management

The identification of potential problems that can arise from opportunity crude oils and development of mitigation strategies is a very broad topic. Baker Petrolite has programs to address issues such as fouling, corrosion, catalyst poisoning, and finished product issues that might arise when processing opportunity crude oils. This paper will concentrate on Baker Petrolite's Crude Oil Management approach to handling crude oil issues from receipt in the tank farm through potential issues around the desalter and the impact on the waste water handling facilities. How crude oil is handled in the tank farm impacts the desalter. The performance of the desalter impacts the waste treatment plant. Handling slop oil by the waste treatment plant affects desalter operation.

Case History: ASIT^{SM} Test Indicates Asphaltene Instability

When considering an opportunity crude it is critical to make sure that it is compatible with the existing crude oils in which it will be blended. Aspahltene destabilization from crude oil blending can result in fouling from the crude tank through the rest of the refinery as well as stabilize emulsions that might manifest as desalter upsets. In the most severe cases asphaltene precipitation can lead to rapid sludge build up in crude oil storage tanks. Several methods have been used to monitor crude incompatibility including SARA (saturates, aromatics, resins, and asphaltenes) testing, asphaltene to aromatic ratio, and various precipitation tests. Baker Petrolite has developed the Asphaltene Stability Index Test (ASIT^{SM} test) to predict asphaltene instability. The ASIT test results cannot only be used to predict problem crude blends, but also to set up crude blending schemes and to select proper treatment programs to control problems associated with incompatible blends.

An example taken from a paper presented at the 2002 AIChE conference[2] involves an opportunity crude oil (designated Crude 2 in Table II), which produced desalter upsets when blended with other crude oils currently processed by the refinery. There was a build up of a rag layer that led to the shorting out of the bottom grids. Samples of the various crude oils were evaluated in the ASIT test equipment to determine stability with respect to asphaltene precipitation. Crude 2 was the least stable of those tested, and would be considered an unstable oil. The blend that caused the desalter upsets was the one indicated as # 4 in Table II. This blend had an asphaltene stability index in the unstable range; therefore asphaltene destabilization was believed to be a source of the problems. It was also learned in the ASIT testing that an asphaltene stabilizer could be added to the crude blend and shift the titration curve towards the stable crude oil region.

The refinery took several actions so that they could continue to run the economically favorable crude. Asphaltene stabilizer was injected into the crude oil #2 as it was transferred into storage tanks. The desalter was optimized for better performance by increasing the wash water rate from 4.0% to 5.3%. A wetting agent was also injected into the desalter wash water to further minimize oil under carry. The refinery was able to continue to run blends containing crude oil #2 without operational problems.

Table II: Stability Characteristics of Crude Oils and Blends

#	Oil / Blend	ASI	% Asphaltenes	% Resins	Asph./Res.
1	Crude 1	1.75	3.4	7	0.49
2	Crude 2	0.96	5.4	10.8	0.50
3	Crude 3	1.5	6.53	-	-
4	55% (1) 20% (2) 25% (3)	1.24	-	-	-

Case History: Impact of Opportunity Crude Oil Processing on Desalter Operations

Often the first indications of problems with processing opportunity crude oils are desalter upsets. When new crude oils are blended into the normal crude slate, there is the possibility of increased conductance, poor dehydration and desalting, and increased levels of hydrocarbon in the effluent water. Baker Petrolite has developed the EDDA (Electrical Desalting and Dehydration Apparatus) for simulating desalter performance in the laboratory. The testing procedure uses the EDDA device, which has high voltage electrodes which accurately simulates the desalting process. New crude oils can be screened to see if they pose any desalting issues compared to the normal crude slate. If potential problems are suspected, changes in operating parameters (such as wash water rate and mixing energy) and chemical treatment programs can be evaluated and recommended.

A US refinery was considering the purchase of a newly developed Canadian crude oil. Blends containing this new crude oil were screened with the EDDA equipment using the existing emulsion breaking chemical product (BPR 27140 emulsion breaker) and other potential treatment schemes. As shown in Figure 1, the incumbent product worked as well or better than other treatment programs evaluated. The last cluster of water drop data shows that with the BPR 27140 emulsion breaker treatment, virtually all of the water was released from the emulsion within the first 10 minutes. Other treatments did not show such rapid resolution of the emulsion. It was found however, that it would be necessary to treat these blends at higher dosage levels than were required for the normal crude slate.

Figure 1. EDDA Demulsification Test Results with Proposed Opportunity Crude Blend

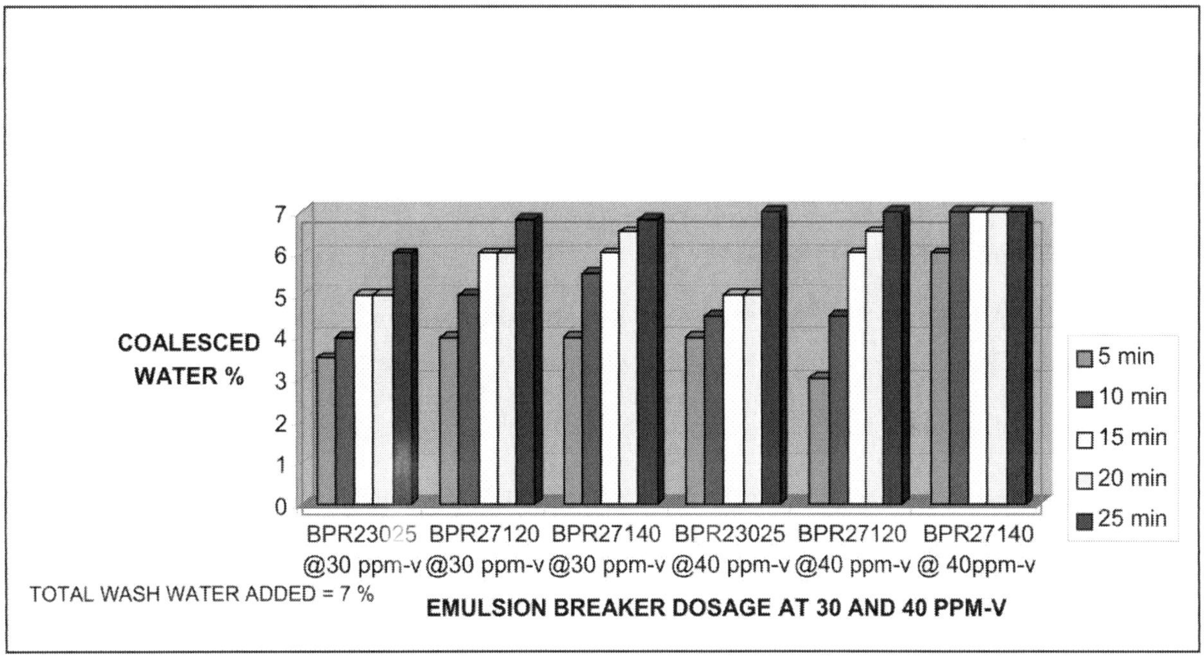

Case History: Doba Crude Oil

In recent years several crude oils containing calcium naphthenate have come onto the market. Calcium naphthenates cause severe fouling issues in oil field production equipment and can also lead to problems in the refinery. Crude oils with high levels of calcium naphthenate can stabilize emulsions in both the oil field and the refinery. Calcium naphthenate is not water soluble and is non-volatile, so it tends to stay with bottom products in the refinery. Depending on the refinery configuration, the calcium can end up in the coke, the residual fuel, asphalt, or on the FCC catalyst. Calcium deactivates FCC catalyst and will shorten the useful life of the catalyst. If the refinery is producing anode or metallurgical grade coke, then the calcium can impact the coke specifications and price. One refiner has observed that high levels of calcium in residual streams promote plant boiler and process heater tube slagging when these resids are blended into heavy fuel oils.

In 2004 a new crude oil, Doba, came on the market with favorable pricing that encouraged many refineries to consider running this crude oil. While the crude oil had many useful characteristics, it had the highest level of calcium naphthenate (200-400 ppm Ca) of any crude oil yet to be marketed on a large scale. It would be necessary for refineries with limitations on calcium in their bottoms streams to reduce the amount of calcium in the crude oil to an acceptable level. Baker Petrolite had developed EXCALIBUR metals removal technology to solubilize calcium and other metals into water to remove the metals from the crude oil in the desalting process. It was decided that this technology would be employed by one refinery to enable them to process the Doba at about 15% of their crude charge.

Several tests were performed during the planning phase to determine the impact of running this crude oil on the refinery and to develop a mitigation strategy. The BS&W of the Doba was believed to be in excess of 1.0%. Often high BS&W in an incoming crude oil indicates the potential to form stable emulsions, leading to poor dehydration and rag build up in the desalter. For this reason it was decided to treat the Doba with BPR 23025 dehydration aid while transferring the crude oil into tankage. ASIT^{SM} asphaltene stability tests were also run to make sure there were no crude incompatibilities. All of the crudes and crude blends that the refinery anticipated running were compatible. Laboratory simulations were conducted with the EDDA test equipment and procedures. A developmental emulsion breaking chemical product was found to give better dehydration on the expected crude slate. However, the existing desalting aid was judged to be sufficient to meet the needs of the refinery.

Preliminary testing by Baker Petrolite indicated that the impact of desalter effluent water on the waste water treatment plant should be minimal. While the refinery was worried that components of the EXCALIBUR calcium removal additive would add to the COD and BOD loading in the effluent water, the actual result of the treatment is to drive water soluble organics in the desalter effluent into the oil phase, so there is no net increase in COD and BOD. The environmental group was also concerned about the large amounts of calcium being removed having an impact on scale formation down stream. The EXCALIBUR metals removal program includes scale control agents which prevent the formation of scale in the effluent water. In addition we provided the waste water plant with computer based algorithms to calculate the Langelier Index (LSI) and Ryznar Index (RSI) so scale forming tendency could be monitored during the initial evaluation.

The EXCALIBUR additive package also includes corrosion inhibitor to protect the metal surfaces in the desalter wash water circuit. As part of the monitoring program, the refinery installed pH meters and ER probes in the wash water line to monitor the injection of the EXCALIBUR additive package. Prior to arrival of the Doba shipment, the EXCALIBUR treatment was activated to make sure all of the equipment performed properly and to verify that there would not be a negative impact on the waste water plant.

Once Doba arrived in the refinery, the mitigation strategy was implemented. The refinery was able to handle the crude with excellent desalting results. Calcium removal with the EXCALIBUR metals removal program met and exceeded expectations. Indeed it was shown that a calcium removal efficiency of 95+% could be achieved, although it was not required to meet refinery specifications. The desalter effluent water proved to be no problem for the waste treatment plant. Attention to detail on many of the potential problems enabled the refinery to process this crude oil with little disruption to operations. The implementation of the EXCALIBUR metals removal program has enabled the refinery to continue to meet desired calcium levels in their bottoms streams while processing this opportunity crude oil.

Conclusions

Proper screening of opportunity crude oils can help anticipate operational problems in the tank farm, desalter and waste water treatment plant. If needed, mitigation strategies can be developed and implemented to result in smooth operation when the opportunity crude oil arrives.

EXCALIBUR and ASIT are trademarks of Baker Hughes Incorporated

References

1. Lawrence N. Kremer, "Crude Oil and quality variations" Petroleum Technology Quarterly, Autumn 2004, pp 87-97.

2. Joseph L. Stark, Joe Nguyen and Lawrence N. Kremer, "Crude Stability as Related to Desalter Upsets," 5th International Conference on Refinery Processing, 2002 AIChE Spring National Meeting, New Orleans, LA, March 11-14.

METALLURGY AND EQUIPMENT UPGRADES TO PROCESS "OPPORTUNITY CRUDES"

Timothy John Ruggles - ConocoPhillips
Sweeny, TX

Biography

I graduated in 1986 with a Bachelor of Science Degree in Mechanical Engineering from Brigham Young University in Provo, Utah. I started my professional career working for BETZ Industrial in West Texas and Eastern New Mexico in the water treatment business. In 1988, I started working for Phillips Petroleum at the Sweeny Refinery in Sweeny, TX. I spent four years in the Design Department, two years as a Shutdown Planner and five years as a Reliability Engineer in both rotating and fixed equipment. For the past seven years, I've been working as the Senior Reliability Engineer for fixed equipment in the Crude, Cat Cracker, Reformer, Sulfur, HF and ARDS units.

Overview

To process "Opportunity Crudes" through the sour crude unit at the ConocoPhillips' Sweeny Refinery, many equipment modifications have been made during the past 11 years. The unit has been in the process of debottlenecking over its entire lifetime but during the last 11 years, a stronger emphasis has been made to process more severe crude slates at higher rates. During the period of 1995-2000, the Top Pump-Around exchangers were experiencing severe corrosion due to chloride salt deposition. These were upgraded to 2507 duplex stainless steel bundles. New Desalter internals were installed during the 2000 T/A and operations began single stage desalting. In August 2000, a new Coker was brought online. With the Coker came the "opportunity crudes" and the problem associated with them – ***increased corrosion***. The Crude Tower was revamped in 2001 to be able to process even more crude. This revamp included tray, packing and metallurgy modifications. With the increased charge rate and the "opportunity crudes" came equipment failures due to corrosion. During the 2004 T/A, the O.H. line and Top Pump-Around lines were replaced with Inconel 625 overlaid pipe because of severe corrosion. In 2005, a VAT (Vulnerability Assessment Team) was formed to study the increased corrosion problems, to identify equipment vulnerabilities throughout the entire crude processing system and develop procedures to reduce the corrosion and eliminate the vulnerabilities. This Team continues to study the corrosion problems and identify vulnerabilities that need to be corrected.

Top Pump-Around System

Top Pump-Around Exchangers

During the mid-1990's, multiple tube failures began occurring in the top pump-around exchangers on an unacceptable frequency. Unit operations had to cut crude rates to be able to blind, pull, clean, inspect and repair/plug the tubes that had failed. Sometimes a demurrage fee would have to be paid to hold ships waiting to deliver their cargo of crude.

Failed tube samples were analyzed to determine the cause of the frequent failures. The failure analysis indicated severe corrosion from increased chloride attack. Under deposit corrosion was also a contributor due to fouling caused by amine-chloride salt and iron sulfide deposition.

The life expectancy of carbon steel tubes was only 9 months. The majority of the tube failures were on the lower rows of the bundle. Heavy hydrocarbon sludge was always found packed in between the tubes and baffles and was thought to be causing the failures. Several bundles were retubed with 2507 duplex stainless steel in the lower half of the bundle. Unfortunately, the failures continued in the upper half of the carbon steel tubes.

When the failures continued, the upper half of the bundles were retubed with 2507 or new bundles were purchased with the upgraded metallurgy. All bundles but one have carbon steel baffles, tie-rods and spacers. Hindsight indicates that baffles, tie-rods and spacers should have been made out of 2205 duplex stainless steel because the carbon steel is continuing to degrade while the tubes are doing fine.

These bundles are operating near the temperature and chloride concentration limits of the 2507 Duplex Stainless Steel. The "Pitting Resistance Equivalent" (PRE = 41) and "Critical Pitting Temperature" (CPT) characteristics of 2507 Duplex Stainless Steels have a proven track record in this service; but, the more severe service bundles have had some pitting on the tubes.

These tubes have also had some cracking. The cracking has been traced back to the fabrication process. As the tubes are pulled through the baffles, the carbon steel baffles gall (scratches) the Duplex Stainless tubes and embeds iron into the surface as well as adds stress to the surface of the tubes. As the bundles are left outside (waiting to be installed), the embedded iron begins to rust in a linear fashion. This disturbs the natural passivation process of the Duplex Stainless Steels. These stressed and scratched areas are also vulnerable to cracking. A valuable lesson learned is to make the baffles, spacers and tie-rods out of the same (or equal) material as the tubes or to "pickle" the bundle after fabrication to remove the iron embedment.

After seeing the cracking and pitting in the 2507 tubes, Sandvik was contacted to investigate the problem. In the investigation process, a new "proprietary" metallurgy, 2707, was suggested as a possible solution. One test bundle was fabricated with the 2707 tubes with 2205 baffles, spacers and tie-rods and is showing great performance.

During the long process of upgrading the metallurgy on 20 bundles, one shell was left without a bundle. This shell was left open to the process. It was opened during the turnaround and had accumulated a significant amount of sludge which was made up of amine-chloride salts, hydrocarbon and iron sulfide.

The carbon steel exchanger shells are also experiencing significant pitting on the bottom inside surface of the inlet side of the exchanger. Whenever a bundle is pulled, the pits have to be repaired by welding. Ideas are still being considered to prevent further shell attack. The shells are regularly monitored by UT inspection and corrosion probes have also been installed.

Top Pump-Around Pumps

Two pumps are used to circulate the Top Pump-Around through the system. One operates while the other is idle. Both of these pumps have experienced severed degradation during their service. The damage has been seen as: severe corrosion to the pump case which even holed-through at one location in the top of the case, the "cut-water" was damaged from its original location to about 90 degrees away and several drain lines/nipples have failed. To reduce the corrosion in the critical areas of the pump, an Inconel 625 overlay was put on all "registered fits" and Belzona was applied to all wetted surfaces. After about 4 months, the pump was opened for seal repairs. At this time the Belzona coating was inspected and found to be in good condition except for a few areas that had popped off. This was probably due to poor surface preparation before applying the Belzona. One new duplex stainless steel pump case has been purchased and will be installed at the earliest opportunity.

Both pumps run with a solid Monel 400 impeller. The impellers are performing very well at the present time. The "boss fits" on the head are also overlaid with Inconel 625.

Top Pump-Around Piping and Valves

In order to select an appropriate metallurgy, a corrosion coupon spool was fabricated and installed in the discharge piping of the pump. This spool consisted of 4 thredolets with screwed-in plugs. The plugs were a special design that extended into the process flow. The plugs were made from different metallurgies with some being carbon steel that had specialty coatings applied to them. These plugs have been pulled and inspected during pump outages and others put in to take their place. Carbon steel, Inconel 625, Hastelloy C-276, carbon steel coated with various Belzona products, boronized and nickel plated carbon steel plugs have been used as test plugs. The Inconel, Hastelloy and Belzona have been performing the best. Continued plug testing will be done to determine the best corrosion resistant product.

The suction and discharge valves in the Top Pump-Around pumping circuit have been replaced with Alky-valves (Monel trim) because of previous valve failures (valves leaked when blocked in for maintenance). New Hastelloy C-276 butter-fly valves are being installed on the suction side of the pumps to help isolate the pumps when maintenance is required on the pumps.

The by-pass block valve, the DMV and downstream piping have experienced failures. The by-pass valve had a through wall hole and was replaced with an Alky-valve (Monel trim). The DMV's internals were severely damaged and the valve was replaced with a new Hastelloy C-276 overlaid valve. The piping downstream of the DMV was also severely damaged from the product flashing across the valve and was replaced with Hastelloy C-276 overlaid pipe.

The Top Pump-Around draw line coming directly off of the tower experienced severe corrosion from the tower flange to the first 90 degree elbow off of the tower. The corrosion mechanism has been identified as a heavy liquid amine salt. It is a very viscous material and very corrosive. It appears that the weld seam on the I.D. of the piping caused some of the liquid amine salt to "pool" on either side of the weld and corroded through the pipe. It is also evident from photographs that this liquid amine salt accumulates in the bottom of the pipe and leaves a trail of corrosion (on the bottom of the pipe) behind as it travels horizontally through the pipe. Once it reaches a vertical section of pipe, it disperses into the rest of the product stream and becomes less aggressive. Equipment inspection found the corrosion on the bottom of the pipe by UT inspection. Once the severity was known, new Inconel 625 clad pipe was ordered fabricated and installed. While this pipe was being fabricated, the existing pipe was wrapped with a carbon composite material called Diamond Wrap.

A two inch drain valve on the bottom of a horizontal run of the Top Pump-Around line is being looked at to be used as an amine chloride draw. Piping will be installed to draw the amine-chloride off the bottom of the pipe to a collection pot. This pot will be piped up to the oily water sewer. The intent is to drain the amine chloride from the system to reduce the concentration of amine chloride salts and reduce corrosion, similar to an intermittent blowdown system in a boiler.

Crude Tower

Top Pump-Around Draw on Crude Tower

It appears that the "dew point" of the amine chloride salt is near or at the same temperature as the draw tray. The liquid amine salt is accumulating inside the tower at the same elevation as the draw tray. This is evident from the "through wall" hole that occurred on the west side of the tower at the same elevation. Not only was there a hole but there was a thin area at the same elevation as the draw tray about 75% around the circumference of the tower. The amine chloride salt appears to be concentrating at this location and becomes very aggressive. The repair consisted of a Hastelloy C-276 grooved band that went 90% around the tower. The grooved band was supplied with injection, vent and drain ports. A Belzona product was injected into the grooved area until it flowed out the vents and drains. These ports were later plugged with Hastelloy C-276 plugs.

Crude Tower O.H. Line

There have been several failures in the Overhead line. These include: carbon steel fittings welded to the Monel 400 piping, carbon steel slip-on flanges, areas of piping where flow direction changed (elbows & tees) or velocities were high.

The Overhead line is Monel 400 down about 40' from the top of the tower. The upper section of this pipe had carbon steel fittings branching from it and had severe corrosion. During the last turnaround, the fittings were replaced with Monel 400 fittings. After the 40' of Monel piping, the remaining carbon steel pipe was replaced with Inconel 625 and 316 s.s. clad carbon steel pipe to the Overhead condensers.

The O.H. condensers have 2205 duplex stainless tubes but the header boxes are carbon steel. Each bank of O.H. condensers has it own water wash system to reduce the possibility of forming salts. Infrared thermography is used to monitor the tubes to see if plugging is occurring from salt formation. Corrosion coupons have also been installed around the condensers to monitor their operating condition.

Top Head of Crude Tower

The top head of the Crude Tower is clad with Monel 400 (.079" thick). There are several specific areas with accelerated corrosion where the process has actually penetrated the cladding to the carbon steel base metal. During previous turnarounds, these areas have been weld overlaid with Inconel or Monel. An Inconel 625 strip liner was installed over the most severely damaged areas of the top head during the last turnaround with Belzona being applied to other areas not strip-lined.

History Time-Line
1978 – Unit was commissioned

1995-2000 – Top Pump-Around exchangers experiencing severe corrosion. Put in 2507 Duplex Stainless Steel bundles.

2000 – Started up Coker Unit to start processing Venezuelan crudes. Currently processing Maya and Merey.

2000 - Started using Baker-Petrolite as chemical vendor.

2001 – Revamped the Crude Tower: added AL6XN packing and trays, added new AL6XN and Inconel 625 clad nozzles.

2004 – Modified the water wash injection in the O.H. line and at each of the O.H. condensers.

2004 – Replaced two O.H. condensers.

2004 – Replaced cracked AL6XN distribution piping with 2205 Duplex Stainless Steel.

2004 – Increased charge from 160,000 bpd to 230,000 bpd.
May 2005 – Organized VAT (Vulnerability Assessment Team) - decided that the Hamaca crude with MEA (used as a H2S scavenger) was forming Amine-chlorides. These Amine-chlorides were the primary corrosion mechanism in our system. The VAT team is continuing their investigation of the corrosion issues in the Crude Unit.

Set a new monthly crude throughput record of 255,711 B/D in July 2005 and then broke it again with 261,877 B/D in August!

October 2005 – Started using G.E. BETZ as chemical vendor with their LoSalt technology. Injecting Citric acid/emulsion breaker into the Desalter water, LoSalt into the Crude Tower O.H. Have had some injection tubing failures due to the HCl in the emulsion breaker and/or environmental chlorides. Swapped to different emulsion breaker without the HCl.

Impact of H_2S Scavengers and Other Tramp Amines on Refinery Operations and Equipment Integrity

Lawrence White
GE Water and Process Technology
9669 Grogans Mill Road
The Woodlands, Texas USA

M. Craig Winslow
GE Water and Process Technology
9669 Grogans Mill Road
The Woodlands, Texas USA

James Rue
GE Water and Process Technology
9669 Grogans Mill Road
The Woodlands, Texas USA

Abstract

Amine and nitrogen based compounds are commonly used in the petroleum industry to treat a wide variety of acidic contaminants found in crude oils, gas, petroleum intermediates and finished products. In oil and gas production, the principal function of these compounds is to reduce the vapor phase concentration of H_2S so that the product can be safely shipped and stored. In the refining industry these compounds are principally used to control the pH of the aqueous environment in the crude overhead condensing equipment, control the pH of condensed steam and to sweeten or remove H_2S and CO_2 gases from various refinery gas streams. The presence of these compounds, their reaction products and their chloride salts can have potentially serious impact on some refinery operations, equipment integrity and refinery profitability.

The principal refinery operations that have been impacted by these compounds have been the desalter, crude fractionation and condensing equipment and the wastewater treatment plant. It has also been well established that the source of these compounds can be traced to many different points. They can be present in the crude oil, refinery slops, stripped sour water, as desalter wash water, overhead accumulator water (as desalter make up or as wet reflux to the tower) or in some cases in the stripping steam used in the towers.

The impact of some of these amines, like those being added to control pH in the overhead condensing system, can be controlled by properly selecting an amine[1,3,4] that is suitable for the process conditions, equipment and environment or by applying a properly designed and implemented overhead corrosion control program[2]. The other amines like triazine that is used to scavenge H_2S from crude oils or MEA, DEA, MDEA and DGA used to sweeten refinery gas streams, are much more difficult to control. These amines are more difficult to control because they emanate sporadically from uncontrolled sources and have a much wider impact area. Controlling their impact is significantly more challenging.

This paper presents laboratory data and information on the impact of "tramp" amines on refinery operations.

Key Words: amine, triazine, amine chloride salt,

Introduction

The majority of sour crude oils imported into the United Sates or shipped within the US are treated to limit the amount of hydrogen sulfide in the vapor space above the crude oil. Many other crude oils and petroleum intermediates, such as Heavy Vacuum Gas Oil may also be treated similarly in order to control H_2S in the vapor space. This need to control H_2S in the vapor space is required by federal, state and local regulations; and the local authority often controls the target vapor space specification and test method. Chemicals used to reduce the concentration or volatility of H_2S in crude oil includes a variety of amines and nitrogen based compounds.

While many chemicals can be used to control H_2S volatility, amines and nitrogen based compounds represent the vast majority of the commercially available choices. Amines react with H_2S to form a water-soluble salt that can dissociate and liberate H_2S as a result of changes in temperature, pressure or pH. For this reason, triazines are usually preferred because they react irreversibly with H_2S to form an organic sulfur compound and because they react more specifically with the H_2S than amines.

Triazine is used as a generic term for a small class of compounds that contain a six-member ring with 3 nitrogen and 3 carbon atoms. The atoms bonded to the nitrogen but not included in the ring structure define the differences between various triazine compounds.

The most commonly used triazine in the petroleum industry is made by reacting para formaldehyde with (mono) ethanolamine (MEA). The proper chemical name for this compound is 1,3,5-Tris (2-hydroxyethyl)hexahydro-s-triazine, and it is usually referred to as 1,3,5 triazine or simply triazine. The synthesis of the compound is depicted in Figure 1.

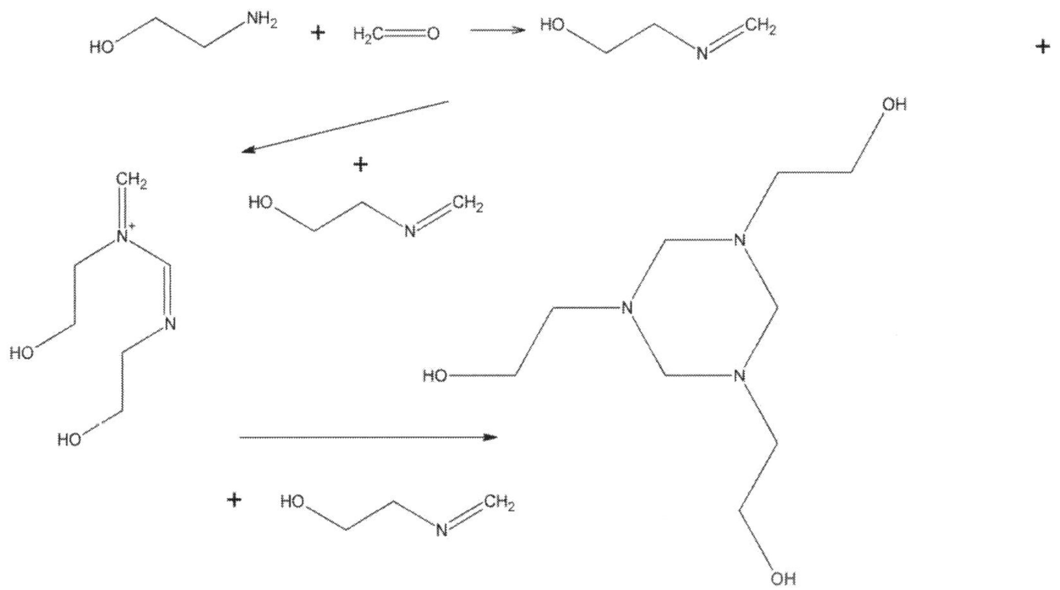

Figure 1. One mole of Triazine from 3 moles each of MEA and formaldehyde

References in literature on the reactions of this compound with H_2S indicate that up to 3 moles of H_2S could react with 1 mole of triazine and that 3 moles of MEA will be produced as a by-product. A recent paper by Jan M. Bakke, Janne Buhaug and Jaroslav Riha titled "Hydrolysis of 1,3,5-Tris (2-hydroxyethyl)hexahydro-s-triazine and its reaction with H_2S" (American Chemical Society, 11/21/2001) provides additional insight into the actual reaction mechanism, kinetics and efficiency. One significant conclusions of their study reveals that only with 2 moles of H_2S, not the 3 moles theoretically possible H_2S will react . This conclusion agrees with field treatment experiences and confirms that additional mixing and reaction time do not improve performance above the 2/1-reaction stoichiometry. A summary of the reaction mechanism between 1,3,5-Tris (2-hydroxyethyl)hexahydro-s-triazine and H_2S is shown in Figure 2.

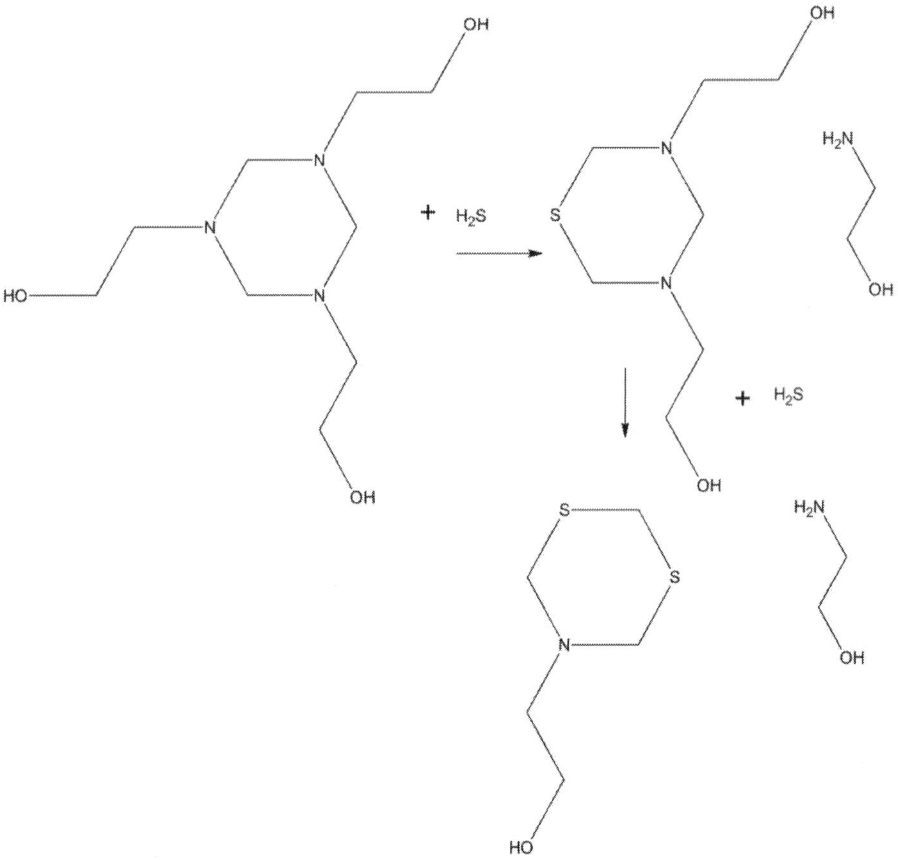

Figure 2 React MEA Triazine with hydrogen sulfide yields MEA by-product

This MEA byproduct from the triazine reaction with H_2S is only one of many sources of amines that can enter the refinery desalter crude fractionation processes. Other sources include the recycle of amines used to control pH in the in crude distillation overhead systems. They can enter the desalter as desalter wash water and can also return to the atmospheric tower in the naphtha reflux. Additionally, MEA, DEA, MDEA and DGA used in the amine sweetening process have been reported in the desalter and crude processing units. They generally enter the system as a contaminant of slop oil or stripped sour water. Amines used for controlling pH in the steam condensate system also enter the crude units as stripping steam. Regardless of the source, these amines are generally referred to as tramp amines.

Impact of Tramp Amines on Desalting

In the desalting process, water soluble contaminants in the crude are removed by washing the crude with water and then separating the water from the crude. In this manner, the contaminants are extracted to the water and dehydrated crude is sent to the fractionation process. Since amines partition in both the water and hydrocarbon phases, understanding the partitioning coefficient (the ratio of tramp amine in the water to the tramp amine in the hydrocarbon) is important to defining the impact.

A North American refinery processing a high sulfur crude oil experienced desalter dehydration problem and a corrosion failure in the atmospheric column naphtha pump around. The crude oil was treated at the production site with a triazine H_2S scavenger at the high end level of treatment. The crude tendered was processed in the normal manner, through the desalters. Wash water to the desalter was a nominal 4% of crude charge and temperature was in the range of 250-280° F. BS&W content of the desalted crude had increased from a nominal concentration of 0.5 wt.% up to 1.2 wt.%. There was no oil in the effluent brine, but the dehydration efficiency of the desalter had been seriously impaired. Chloride concentration in the overhead accumulator was analyzed to be in the range of 70-120 ppm.

A sample of this heavy sour crude oil was specially treated in the laboratory to remove the H_2S reaction by-products. Analysis of the crude oil indicated that 21 ppm of MEA was present in the crude oil. Analysis of the same crude oil sample by acid extraction and GC/MS did not detect MEA, but ion chromatography revealed 95 ppm MEA.

Further laboratory testing of the crude oil sample determined that at the typical operating conditions of this refinery desalter, 52% of the un-reacted triazine and roughly 80% of the MEA by-product partitions to the desalter wash water. Further testing of other crude oil samples has indicated that the partitioning values vary from crude to crude and as a function of the concentration of un-reacted triazine and MEA. Regardless of the desalter efficiency some triazine and MEA will be present in the desalted crude.

Further investigation of the desalter operations indicated that the presence of triazine and the MEA byproduct, as shown in Table 1, have a strong negative impact on desalter performance. In laboratory simulation, adding 600 ppm of triazine to the crude oil reduced oil and water separation rates by about 50%. Even though the triazine is not necessarily surface active, the basic triazines form an acid/base slat complex with species such as the naphthenic acids, and these can be a very effective hydrophile. They exhibit a high degree of surface activity that can be a powerful stabilizing agent for oil and water emulsions. Table 2 lists the estimated concentration of triazine and MEA in the desalted crude and desalter tail water for a 100 ppm triazine treatment.

Table 1 Study of affecting of 600 ppmv MEA-triazine in water separation, chloride and amines concentration. SamplesRaw crude (log # D-03-232 2). Wash water (log # W-03-231). Procedure: Add 90 ml crude in the tube first, treat the crude with 50 ppm Emulsion Breaker A @ ambient temperature, shake for 5 minutes at high by a shaker. Add 10 ml wash water, cap, heat to 137C, and mix @ 10,000 rpm for 2 sec (= 10 psi), grids on. Record the water drop. After 64 minutes collect top 70 ml oil for total chloride by TOX and the separated water for amines & chloride analysis.

	Water Drop reading in ml at time indicated							Oil	Water	Water
Test	1 min	2 min	4 min	8 min	16 min	32 min	64 min	Cl ppm	MEA ppm	Chloride ppm
Blank	3.5	5.5	8	9	10	10	10			
	3.5	6	8	8.5	10	10	10	55.2	158	63
Triazine	0.8	2.5	5	7	8	9.5	10			
	0.6	2.5	4.5	6.5	7	9	10	43.3	2450	56

Table 2 Estimate of MEA and triazine concentrations in water and desalted crude oil
Basis, 100 ppm triazine treatment, 50% reacted 1:1 with H_2S
100,000 BPD Crude Rate (18 API), 5% Desalter Wash water

	Before Desalter	Partitioning	Post Desalter	
	(ppm in oil)	% to water	(ppm in oil)	(ppm in water)
Un-reacted Triazine	50	52%	24	492
MEA	50	80%	10	757

Not only do a percentage of the triazine and MEA remain in the crude oil after the desalter, but their presence can also directly impact the rate of emulsion resolution. Reducing the rate of resolution results in increased water in the desalted crude, and the water carries more chlorides and tramp amines in the fractionation section.

A review the operating conditions showed that there was a significant impact on the desalter operations. Higher injection rates of demulsifier were required to control the interface. It was also noted that the periods of excess water carryover correlated with a significant increase in corrosion activity (which will be discussed in more detail in the Corrosion Impact section of this paper).

Impact of Triazine and Tramp Amines on Waste Water Treatment

Even though only a portion of the triazine and MEA partition to the water in the desalter is removed in the effluent brine, there have been occurrences of impact to wastewater plants associated with their use. In one example, the waste treatment plant of a North American refiner processing a high sulfur crude experienced a significant impairment to performance. In this case the refinery received a tender of high sulfur crude treated with approximately 500 ppm of H_2S scavenger. This crude was blended at varying percentages of the total crude and was charged to the unit.

The most significant impact observed in the wastewater plant was the increase in ammonia content in the plant outfall as depicted in Figure 3. In evaluating this impact, refinery personnel looked into the concentration of ammonia leaving the desalter and the DAF. This analysis shown in Figure 4 indicates that the level of ammonia leaving the DAF had not increased and was, in fact, exhibited a decreasing trend. The analyses of the desalter effluent brine pH and ammonia concentration in Figure 5 and Figure 6 indicate a slight increase with increasing percentages of high sulfur crude.

Using the data, it was determined that the ammonia in the feed to the wastewater facility had not changed significantly and was not the source of the impact. Further analysis (taking into account the lag time differences resulting from changes in wastewater inventory and flows to the unit), clearly established the only plausible explanation for the high level of ammonia in the outfall was processing the high sulfur crude containing the H_2S scavenger.

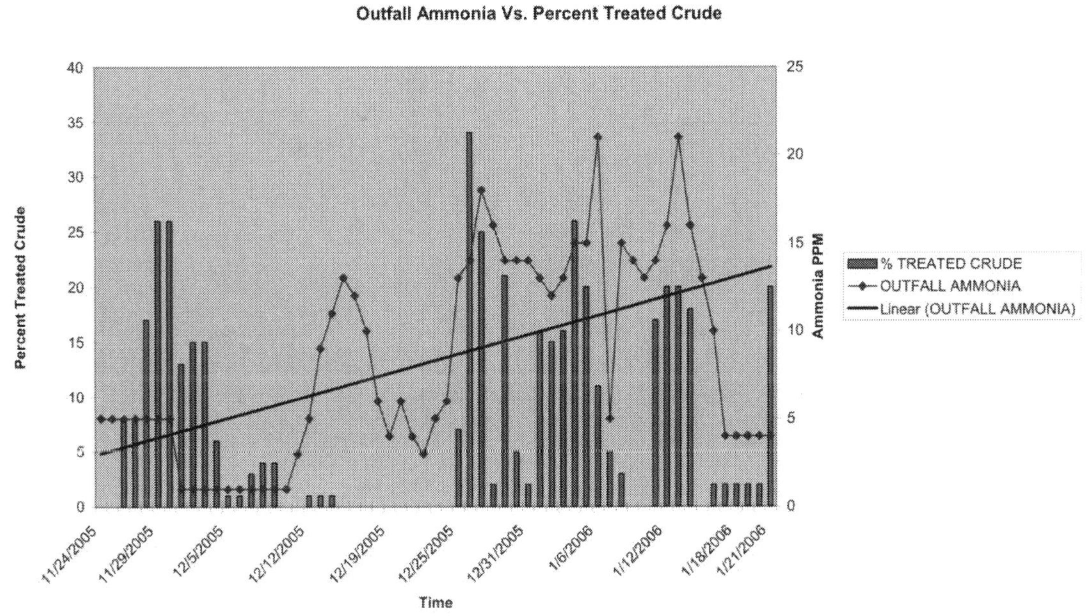

Figure 3. Refinery wastewater outfall ammonia concentration vs.. % High Sulfur Crude

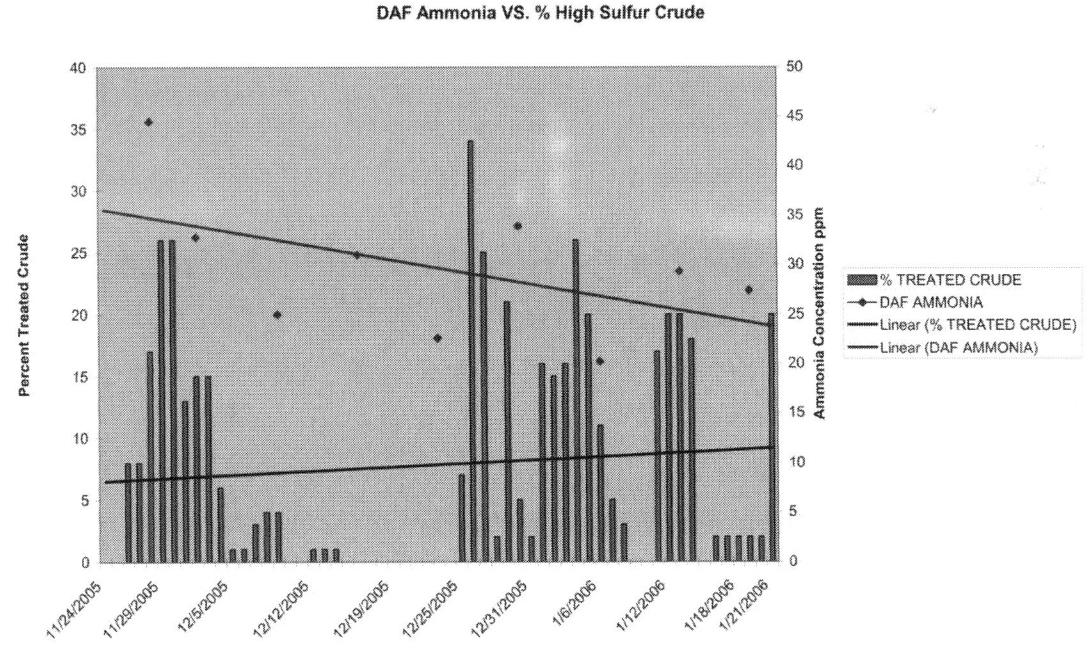

Figure 4. Refinery DAF Ammonia Concentration vs. % High Sulfur Crude

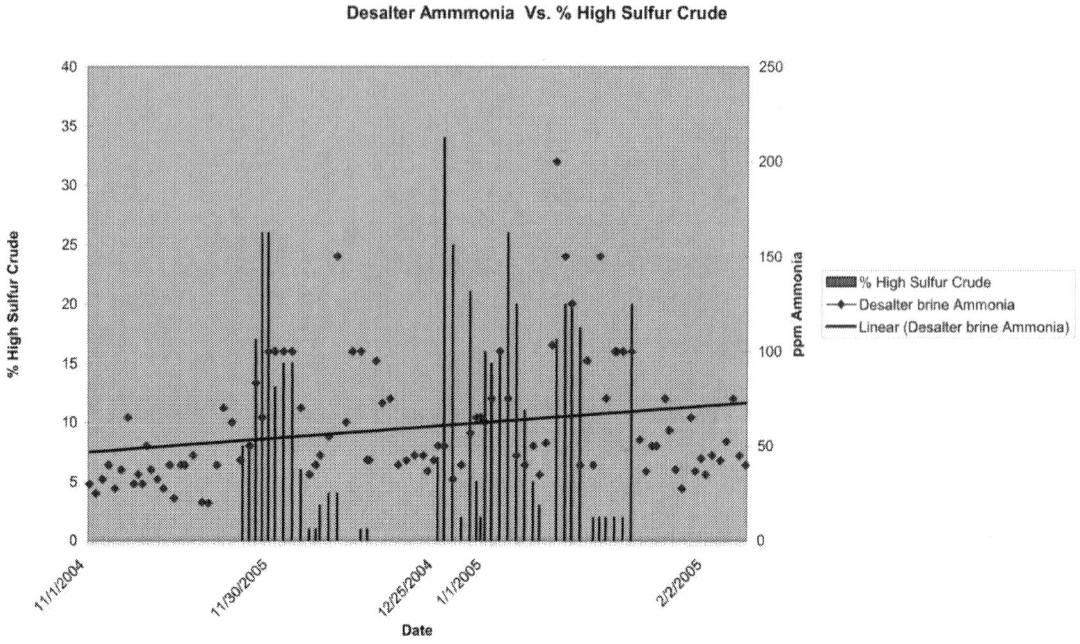

Figure 5. Refinery Desalter effluent brine ammonia concentration vs. % High Sulfur Crude

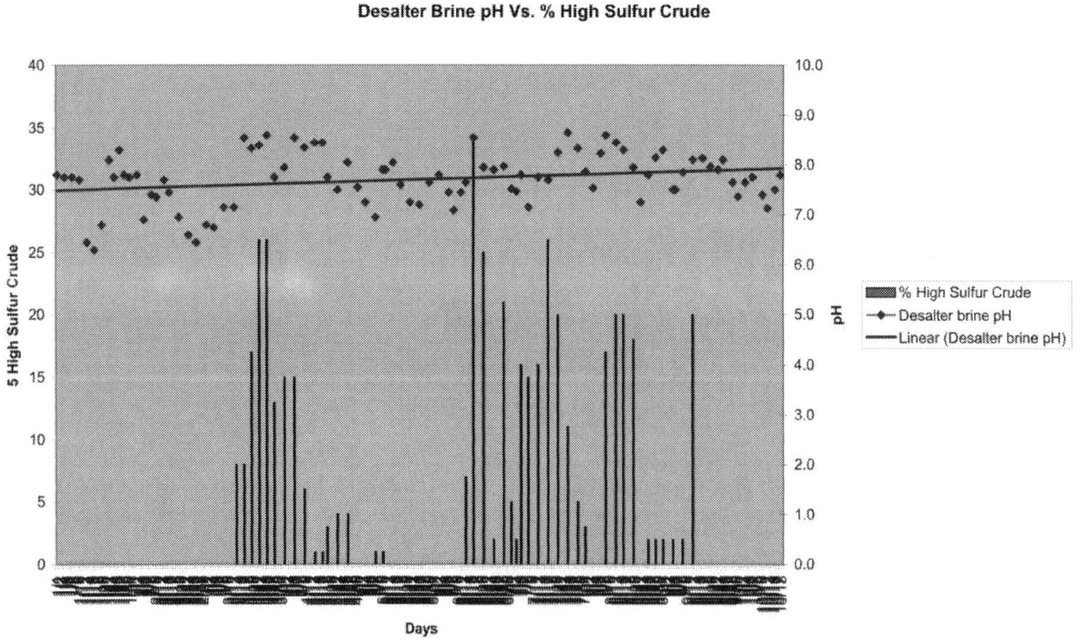

Figure 6. Refinery Desalter effluent brine pH vs. % High Sulfur Crude

In analyzing this impact, refinery personnel determined that the level of the amine byproduct had been high enough to inhibit the nitrification process of the biological population in the aeration basin. In order to bring the system back to equilibrium, refinery personnel re-seeded the system with fresh nitrifying bacteria and reduced the flow rate to the unit.

Impact of Triazine and Tramp Amines on Crude Distillation

Research done in the early 1990s by Edmondson and Lehrer[5] and confirmed by refining corrosion control operations[1,2,3,4] during the last 15 years has clearly established that amines react with chlorides to form an amine chloride salt, and that in some cases the resultant amine chloride salt can precipitate from the vapor phase at a temperature above the temperature of water dew point. From these works, it has also been well established that these salts are mostly hydrocarbon insoluble, hygroscopic and corrosive.

The deposition of these amine chloride salts in locations of the crude fractionation and overhead condensing system at temperatures above the water dew point is another critical impact of tramp amines. This condensation temperature of the amine chloride salt is called the salt point temperature. It has been well established that with some amines like MEA, the temperature of MEA chloride salt point can often be significantly higher than the water dew point. When salt point above water dew point occurs, there is no liquid water to wash the salt away, resulting in deposition of the salt will result and severe corrosion to the underlying equipment.

Corrosion Impact Compounded by multiple "tramp" amines

Any "tramp" amine entering the crude fractionation tower reacts with hydrochloric acid at the salt point temperature, to produce strong acid-weak base salt. This salt formation temperature is a strong function of the concentration of HCl and MEA, but it is also influenced by the presence of other components like hydrocarbons, acids and bases.

As previously mentioned, a North American Refinery experienced a failure to a naphtha section of piping in the naphtha draw line on the atmospheric crude column. A detailed analysis of the deposited material in and around the corrosion failure indicated that the corrosion had been caused by multiple "tramp" amine chloride salts.

Because of the rapid and extreme onset of the corrosion, refinery personnel were able to review the available data and piece together a comprehensive understanding of the event. The refinery had been processing a high sulfur crude that contained approximately 500 ppm of triazine H_2S scavenger. Review of other operating data showed that the refinery had also been experiencing MDEA losses from their alkanol amine units. The source of these loses was traced back to knock out pots that were being dumped to the refinery slop system and from their to crude charge to the affected crude unit.

Table 3 shows the presence of the key components of interest in the deposits removed from the area of the failure. Chlorine, the principal corrosion precursor, is listed as are corrosion by-products. Since the piping in the area of the corrosion was carbon steel the presence of nickel and chromium in the analysis suggest that corrosion may also have been occurring in the tower and that these corrosion by-products were migrating from the tower to the area of the failure. Table 4 lists the "tramp" amines found in the various deposits removed from the area of the failure.

Table 3

Constituent	Area of Corrosion	Adjacent Top	Adjacent Bottom	Pipe Top	Flange
Iron	66.7	87.8	87.0	81.6	69
Nickel	10.9	2.3	ND	ND	8.1
Chromium	8.6	2.1	0.8	<0.5	6.9
Sulfur	9.5	2.1	3.5	11.3`	13.1
Chlorine	1.3	4.6	8.1	5.7	0.9

Table 4 Concentration (ppm) of various "tramp" amines found in the deposits in and around the area of corrosion.

Sample	Free Ammonia	MEA	DMAPA	MDEA
Top of Pipe	68	216	ND	ND
Gasket	612	12790	130	1755
Corroded Area	219	10819	ND	ND
Adjacent	117	2170	ND	ND

Because of previous corrosion in this area of piping, the segment had been monitored closely for corrosion. Weekly UT surveys were being conducted. The monitoring showed that the period of onset of corrosion to piping retirement occurred in 30 days. The corrosion rate in mpy during this period of time was calculated to be approximately 1,000 mpy.

Further analysis of the corrosion event, using standard laboratory corrosion testing technique conducted in conjunction with a major refiner resulted in the data presented in Table 5. This data corroborates that corrosion rates could exceed 900 mpy.

Table 5: Laboratory Corrosion test Salt: Monoethanolamine Hydrochloride
2:1 Salt/HydrocarbonMetallurgy: Low Carbon Steel
Temperature: 300 FTest Duration: 16 - 17 hours

Coupon Type	Water (%)	Initial Wt.	Final Wt.	Mg Wt. Loss	MPY
Flat	0	9.4407	9.0639	376.8	1012.6
Flat	3	9.447	9.0847	362.3	973.7
Flat	3% HCl	9.4313	8.7196	711.7	1912.7
Flat[1]	0	21.1412	20.9159	225.3	327.8
Flat[1]	3	21.0889	20.8474	241.5	286.7
Pipe[1]	0	22.0715	21.4741	597.4	1066.6
Pipe[1]	3	20.4276	19.6023	825.3	1651.4

[1] Customer prepared coupons

Further investigation at the refinery, including sampling of the naphtha stream, revealed an aqueous sample collected by decanting small amounts of water from the hydrocarbon of a naphtha reboiler. Over 2.5% MEA and over 4.5 % chloride were reported as noted in Tables 6 and 7. High levels of free and fixed ammonia confirm a mixture of ammonium and amine salts were present.

Table 6 Naphtha draw line and reboiler sample analysis for nitrogen compounds

Sample	MEA ppm	DMAPA ppm	MDEA ppm
Top	216		
Gasket	12790	130	1755
Corroded	10810		
Adjacent	2170		
Naphtha Reboiler	2730		
Naphtha Reboiler	27200		

Table 7. Water analysis from sample removed from Naphtha Reboiler

Ammonia,		Chloride, as Cl, ppm	45000
Free as N, ppm	1690	Iron, (total) as Fe, ppm	8760
Fixed Organic as N, ppm	9510	Manganese, as Mn, ppm	130
Free & Fixed as N ppm	11200	Molybdenum, as Mo, ppm	16.5
pH	4.4		
Sulfate, as SO4, ppm	1480		

Corrective actions put in place following this event included identifying all possible sources of in plant tramp amines that could enter the desalter wash water system and limiting the migration of these materials via wash water and slop oils into the crude processing unit.

Another Gulf Coast Refinery was also processing crude slates containing high sulfur opportunity crude oils, and it was experiencing corrosion failures in both the top pump around to raw crude heat exchangers and the kerosene to raw crude heat exchangers. The heat exchangers were being taken off line, the leaking tubes were being plugged and the heat exchangers were being put back on line without empirical data on the root cause of failure. It was believed that he root cause of the failures was attributable to ammonium chloride corrosion, amine chloride corrosion or a combination of both. Salt point modeling of the system confirmed that salt formation from any number of amines could be occurring, but there was no conclusive evidence implicating any specific amine.

To better understand the root cause, GE engineers designed and installed a special sampling device to sample the water in the top pump around loop. The aqueous sample collected in the sampler was submitted for analysis to determine the presence and concentration of amines. Gas chromatography with nitrogen specific detection was used to identify and quantify the amine constituents. The identification concentration and source of these amines are detailed in Table 8.

Table 8

	Top Pumparound (ppm)	Amine Source
Ammonia	100	Sour Water Stripper
MEA	72	H_2S Scavenger
MMEA	161	MDEA degradation product
DMAPA	485	Boiler neutralizing amine
Cyclohexylamine	298	Boiler neutralizing amine
DBA	185	H_2S Scavenger, Oil field raw material
MDEA	2310	Amine Sweetening unit bleed to slop

The results from this analysis indicate that the sample contained a number of "tramp" amines including Monoethanolamine (MEA), Monomethylethanolamine (MMEA), Dimethylaminopropanolamine (DMAPA), Cyclohexylamine, Dibutylamine and MonoDiethylEthanolAmine.

The amine data from Table 8 was then used in the LoSALT™ modeling program to define temperatures where the various amine chloride salts could exist in the atmospheric crude tower. The modeling (Table 9) indicated that the "tramp" amine chloride salts could have existed at temperatures as far down the column as the diesel section.

Table 9. Amine chloride salt formation

Amine	300° F Top Tray	450° F Kerosene Draw	525° F Diesel Draw
MEA Cl		Salt Condensation	No Salt Formation
MMEA Cl		Salt Condensation	No Salt Formation
DMAPA Cl			Salt Condensation
Cyclohexylamine Cl	Salt Condensation	No Salt Formation	No Salt Formation
DBA Cl	Absorbed	Salt Condensation	No Salt Formation
MDEA Cl			Salt Condensation
Ammonium Chloride		Salt Condensation	No Salt Formation

Using Table 8 and Table 9, it was determined that the most likely corrosion precursors were the MDEA, which came from the acid gas sweetening units, the DMAPA from the boiler steam and the ammonia from the sour water stripper. Reducing the corrosion that was being observed in the heat exchangers would require a combination of actions including optimization of the sour water stripper operations, control of MDEA losses and changing the boiler water steam condensate neutralizing amine. All of these issues were individually addressed, and no further corrosion issues have occurred.

Summary

Amines play an essential role in many hydrocarbon processes. As a consequence of these many roles, amines can enter the refinery crude unit operations. In the crude unit, the presence of these amines has been shown to contribute to poor desalter performance, fouling and corrosion.

Controlling the negative impacts of these amines is a complex challenge but is vital to the reliable operation of the crude unit. Controlling the impact begins with a thorough assessment of the amines and their potential impacts. Once the assessment is completed, procedures can be implemented to control the predicted impacts. These procedures may include minimizing or eliminating the ingress of amines into the crude unit, enhancing their removal in the desalter, or may require process changes.

References:

1. "Advances in Neutralizing Amine Technology," Edmonson and Lehrer, NACE 1994.
2. "Control of Salt Initiated Corrosion," Rue and Edmondson, NACE 2001
3. "Crude Unit Corrosion and Corrosion Control," Bagdesarian, Feather, Hull, Stephenson and Strong,, NACE 1996
4. Corrosion and fouling in crude units using low base strength neutralizing amines," Clarida, Johnston, McConnell and Strong
5. Lehrer and Edmondson, US Patent 5,211,840, 1993.

188e A Comprehensive Approach to Assessing Opportunity Crudes

Doug Milton, James Feather, Kenneth Bagnoli, and James McLaughlin. Materials Engineering, ExxonMobil Research & Engineering, 3225 Gallows Road, Fairfax, VA 22015

Recently produced crudes can be purchased at attractive discounts, largely because the crudes possess an increased corrosion potential, present increased processing concerns, contain problematic contaminants and can provide reduced product value. An evaluation of processing these crudes must include all potential downside aspects to effectively determine if the margin discount is adequate to justify the purchase. As a result, a systematic approach has been developed to evaluate opportunity crudes in order to permit processing them. The major focal points of this assessment are the following: sulfidation and naphthenic acid corrosion, processing effects, contaminants, product value, and downstream effects.

OPTIMUM CRUDE BLENDING AND OPERATION

Kiyomi Enomoto Meason, Valero Energy Corp., Sunray, TX

Biography

She is originally from Tokyo, Japan and moved to United State in 1984. Since 1984 she has lived in San Diego CA, Saudi Arabia, Fayetteville AR, and various locations in Texas. She attended Texas A & M at Kingsville earning Chemical Engineering degree and graduating in 1999. She taught mathematics at a local high school for year and was then offered work at a Valero Refinery as an Engineering Representative for Catalyst Platinum Recovery in 2000. She was responsible for witnessing the weighing, screening, and sampling of catalyst for platinum recovery. In 2001-2002, she worked as a Process Engineer and later moved to Economics and Planning. She provided instructions to the crude tank terminal. She was responsible for daily crude pipeline composition (crude blending) and monitoring crude qualities. She tracked different crude impacts on the refinery such as sulfur production, Kerosene smoke point and cetane numbers. She also maintained and updated day by day operation data to predict gas oil, diesel, and naphtha production to meet refinery targets.

Required knowledge before crude blending

With increasing crude prices over the last decade, a refinery has to run the optimum crude to increase its profitability. However, just because a model provides a prediction of an optimum crude, feeding that crude may upset the refinery operation and reduce profitability. To be able to meet Refinery demands without upsetting operations, exceed environmental limits, and violate safety regulations requires:

- Crude Quality (Assay data)
- Crude Evaluation
- Terminal operation and blending
- Pipeline operation and blending
- Daily refinery unit operation and product demand.

Following the steps below when planning daily crude blending can make the process simpler and refinery operations smoother.

1. Update current crude inventory and quality data
2. Plan crude receipts into crude terminal tanks
3. Blend crude into pipeline
4. Estimate the crude feed quality and production
5. Adjust unit rates and endpoints to meet the refinery demand
6. Writing an Order for Terminal Crude Receipt

7. Prepare a weekly Refinery Operating Plan

These steps may need to be repeated several times to find the most optimum crude blend.

An actual case study was used to demonstrate how a crude blend was prepared at a refinery on the Gulf Coast. This refinery has four crude tanks at the crude terminal and there is a crude pipeline between the terminal to the refinery. The refinery has one crude/vacuum unit, Rose unit, FCC unit, Hydrocracker unit (HCU), CCR and Semi-Regen Reformers units, Benzene, Toluene, and Xylene unit (BTX), and an SRU. The most optimum crude feed will change based on economics such as gasoline or distillate mode. However, typically the optimum feed is to maximize the sour crude up to the SRU limit and produce enough gas oil and naphtha to fill conversion units and BTX units in summer. During winter, typically the optimum crude feed is to maximize distillate to fill the Diesel Hydrotreater (DHT) and also BTX. Also balancing the intermediate and product tanks to ensure not to exceed the minimum and maximum tank levels. The goal is to blend crude to meet the refinery demand without any refinery upsets. If a particular crude charge is not desirable for the refinery and will upset the refinery operation, it is crude blender's job to notify management.

Crude Quality

The crude quality is typically determined by assay. The assay contains information about the crude such as gravity, weight percent sulfur, viscosity, pour point, cloud point, and distillations. Table 1 is a summary of a micro-cut assay. The blender may have to use lab data if the assay is not available. For example, the Purchased Naphtha in the assay table had no assay and all of the data was generated in the refinery lab.

There are assay's that run micro cuts which is a distillation based on an interval of 20 degrees instead of 40 degrees. However, additional lab analysis needs to be completed so a comparison between the assay and the actual crude quality can be made. For each crude receipt, a composite sample should be taken at the crude terminal from each crude compartment on the transport ship and sent to the lab for least an API, percent Sulfur, and BS&W for actual crude quality data. Further lab analysis such as distillation might be requested if the difference between the assay and actual are great. Verifying the actual crude quality data help to estimate the final crude feed quality and production rates.

For this paper, Light and Heavy Crudes are defined by their gravity (see Table 1). If a crude has a gravity of more than 30 or greater such as Santa Barbara or Oso Condesate they are considered light crude. Crude with an API less than 30 like Troll and Basra are considered heavy crudes. Light crude typically contains a high percentage of LPG, Naphtha, and a less percentage of Distillate. On the other hand, Heavy crude contains a high percentage Distillate and a low percentage Naphtha and LPG. Depending on what crude is currently in the crude tanks, the current demand and how the refinery is operating will determine what crude needs to be requested from a crude purchaser. For example, if the current crude tank quality is light and there is a high diesel demand and low diesel inventory, a heavy crude may be the optimum choice. However, if the Diesel Hydrotreating Unit is shutdown due to maintenance for several days and the diesel inventory is high, then a light crude may be the optimum choice. Another example is that if the VTB inventory is low and the gasoline demand is high, the optimum choice may be a heavy crude which contains more VTB than a Light crude. This will maintain the PDA or ROSE unit charge and increase the gas oil inventory. Choosing a Light or Heavy Crude is dependent on different economic requirements.

Table 1. Crude assay data supplied by refinery

Crude Name	Troll	Purch Naphtha	Santa Barbara	Oso Condensate	Zuata Sweet	Petro Zuata CGO	Basra
API	27.2	36.8	37.9	51	32.2	14.9	29.5
wt% Sulfur	0.26	0.01	0.53	0.03	0.5	3.2	3.05
RVP, lbs	0.01	na	1.54	2.282	0.01	0	5.124
Vis, cSt	8.23	na	2.21	1.25	5.78	na	4.65
Pour, F	-27.4	na	-57	65	-16.6	na	5
Acid Number	1.21	na	0.06	0.16	0.05	na	0.13
LPG	0.6%	0.0%	1.1%	3.0%	2.1%	0.0%	2.6%
C5-150 LSR	1.2%	5.0%	5.4%	3.5%	3.2%	0.0%	5.2%
150-380 Naphtha	14.6%	95.0%	27.9%	25.5%	13.6%	0.9%	15.2%
380-500 Kerosene	12.5%	0.0%	13.7%	13.8%	15.9%	3.4%	10.4%
500-650 Distillate	21.5%	0.0%	18.3%	15.9%	26.0%	17.2%	13.5%
650-750 AGO	10.1%	0.0%	9.7%	8.6%	15.4%	24.4%	8.5%
750-1050 VGO	27.7%	0.0%	18.8%	20.8%	24.5%	51.6%	22.1%
1050+ VAC Resid	11.8%	0.0%	5.1%	9.0%	0.0%	2.5%	22.3%

Sweet and Sour crudes are classified by the percent sulfur in a crude. A crude such as Santa Barbara, Oso, Purchased light naphtha, and Troll contain less than 1 percent sulfur and are considered sweet crudes and Basra which contains around 3 percent sulfur would be considered a sour crude. Typically sour crude is a more economic crude than sweet crude. The amount of sour crude a refinery may process is determined by environmental permits such as total sulfur production limits set by the EPA or appropriate state agency and the capacity of the Sulfur recovery unit in a refinery. The metallurgy of the operating units also effect the amount of sour crude that can be run. Blending sour crude up to a sulfur production limit will require monitoring theoretical and actual percent sulfur in the crude and daily sulfur production. If conversion unit rates in the HCU, FCCU, and PDA or ROSE are at maximum and the percent sulfur is too high the result can fines penalties and reduced economics. In today's environment, prison to be is also a possibility. The way to control sulfur production while reducing the percent sulfur in the crude charge is to reduce conversion unit rates. However, reducing conversion unit rates will reduce profitability by not maximizing production.

Crude Evaluations

Evaluating a new crude before allowing it to be processed in the refinery has to be coordinated through several departments.

Planning and Economics

The crude quality will be evaluated by Planning and Economics based on the price, crude availability, and production demand. Some refineries use linear programs such as PIMS to determine the optimum crude at the refinery. Depending on where a refinery is located, such as inland or coastal, opportunity crudes may be limited. There are various foreign crudes available if a refinery is located close to coast with less transportation costs. On the other hand, if a refinery is located in mid-continent transportation costs will be higher than those of coastal refineries, and that is assuming there is a pipeline available.

Refinery Operations

Operations will also need to review the proposed crude based on operating limits. For instance, if the crude is too light, the crude overhead production will exceed the maximum feed rate and will not be able to process all the LSR unless there is a cut in crude charge. If the feed is too heavy with a high percent VTB production that exceeds the PDA or ROSE maximum feed rate, then the VTB inventory will increase. If a crude contains a high acid number such as Troll, a chemical injection system with a neutralizer will be needed to protect the crude unit if the metallurgy is not of proper quality to run high acid crudes. The cost of operation will increase due to adding neutralizer, and that will need to be considered as part of the economics.

Environmental and Safety

The Environmental Department will also need to evaluate each crude based on environmental and safety regulations. For example, if the sulfur production exceeds the limits permitted, the environmental department would have to report to the EPA and other appropriate government authorities. Also safety issues have to be evaluated based on handling the crude such as crudes containing high H2S. Operators at the terminal and the refinery may require special personal protection equipment and additives to minimize H2S exposure.

Terminal Operation and Blending

Knowledge of terminal operations including crude tanks, crude ships and barges, and pipeline is necessary for terminal crude blending. Information such as a ship size and speed, minimum and maximum pump capacity of the ship and barges, and the terminal schedule of ships or barges to estimate the arrival time at the crude terminal and to estimate the receiving time are critical. For example, a large ship can hold approximately one million barrels crude and if all the crude compartments are full, it will take more time to get to the terminal than a small barge. Knowledge of high and low tides is also a part of determing when a ship may be able to offload. If the refinery is low on crude, the refinery may need to reduce crude or shutdown if offloading is not well planned.

Update current crude inventory and quality data

There are some instances when the quality of a particular crude is different than the assay value and subsequently a good estimate of tank and pipeline quality is lacking and will adversely effect production. Testing for and verifying the difference between the assay and lab data can be used to gain concessions on future crude purchases. Monitoring crude into the crude tanks will also provide a better estimate of downstream products.

The volume of each tank and its quality can be estimated by using a spreadsheet and actual data. Table 2 presents seven different crudes in Tank 1 starting with 75.6% Zuata sweet and 21% Rabi Light on February 10th. On February 11th, the tank received Bonny Light which raised the percent Bonny light up to 92% from 0% and the percent Rabi Light and Zuata percentage went down. Also the gravity increased from 31.4 to 33.4 after the Bonny light receipt which indicates Bonny Light is lighter than the original tank quality. However, it is necessary to check the final tank quality by taking a sample from the tank and running a lab analysis for least API and sulfur. Typically terminal operators will have a daily tank volume report with tank quality which can be used to compare the estimate and modify the quality in the spreadsheet.

Table 2. Tank Composition supplied by refinery

Date of Receipt	Rabi Light	Purch Naph	BASRA Heavy	BONNY LT	FORCADOS	Zuata Sweet	QUA IBOE	TK COMP/API	TK COMP/% sul
2/10/2004 0:00	21.0%	1.3%	0.0%	0.0%	0.0%	75.6%	1.4%	31.4	0.1683
2/10/2004 8:00	21.0%	1.3%	0.0%	0.0%	0.0%	75.6%	1.4%	31.4	0.1683
2/10/2004 16:00	21.0%	1.3%	0.0%	0.0%	0.0%	75.6%	1.4%	31.4	0.1683
2/11/2004 0:00	3.4%	0.2%	0.0%	83.9%	0.0%	12.1%	0.2%	33.2	0.1697
2/11/2004 8:00	1.7%	0.1%	0.0%	91.8%	0.0%	6.2%	0.1%	33.4	0.1699
2/11/2004 16:00	1.7%	0.1%	0.0%	91.8%	0.0%	6.2%	0.1%	33.4	0.1699

Plan crude receipt into crude terminal tanks

The crude receipt process needs to take into account the following items:

1. Minimize Crude Pipeline shutdown while receiving crude
2. Receiving crude into a dead tank
3. Segregate heavy, light, sour, and sweet crudes
4. Minimize demurrage costs
5. Mixing Time

This process is done by trial and error (see table 3). For case 1, first receive crude into the tank 1 (T1) while blending the pipeline crude from Tank 2 (T2), Tank 3 (T3), and Tank 4 (T4). After the crude receipt is completed into T1, change the pipeline blend using T3 and T4, and receive the crude into T2. This gives time to mix T1. After the crude receipt into T2 has been completed, change the pipeline blend using T1, T3, and T4. This gives time to mix T2. Case

1 will not require the pipeline to shutdown while receiving crude into the dead tanks which are T1 and T2. However, it requires blending crude by transferring crude from one tank to the other before the receipt. Case 2 transfers crude from T1 to T2 before the receipt for two reasons such as emptying T1 before the receipt so that T1 will fill with one type of crude and blend T1 to T2 to make a desirable pipeline crude while receiving crude into T1, T3, and T4. Although there are multiple cases that can be blended; the goal however, is to make sure not to upset the refinery by causing the refinery have to reduce crude charge or exceed sulfur production.

There are many cases that require receiving into undesirable tanks. For Example, if a sour crude tank is full, it may be necessary to receive sour crude into sweet tanks to minimize demurrage costs.

The most desirable way to receive crude is to segregate each crude type by tank, such as heavy, light, sour, and sweet. However, sometimes that luxury may not be available.

Table 3. Trial and Error Case Study for Crude Receiving Order

Case 1		T1	T2	T3	T4
	1	Receive	P/L	P/L	P/L
	2	Mix	Receive	P/L	P/L
	3	P/L	Mix	P/L	P/L
	4	P/L	P/L	P/L	Receive
	5	P/L	P/L	P/L	Mix
	6	P/L	P/L	P/L	P/L
Case 2		T1	T2	T3	T4
	1	transfer to M2	receive crude from M1	P/L	P/L
	2	Receive	Mix	P/L	P/L
	3	Mix	P/L	Receive	P/L
	4	P/L	P/L	Mix	Receive
	5	P/L	P/L	P/L	Mix
	6	P/L	P/L	P/L	P/L

Samples can be taken by operators for a final lab analysis of tank quality after each crude receipt. To get the final tank quality estimate correct the blend order will need to include the mixing time. If the tank is not mixed well, the final tank quality from top, middle, and bottom samples will show different qualities than the estimated crude quality and when it is pumped to the refinery can result in swings on the units.

Pipeline Crude Blending

After the process of terminal crude blending, the pipeline blend will be done by the percent crude draw from each tank with the specified pipeline charge. Here it is important to know the minimum and maximum pipeline capacity. If the refinery increases the crude charge, the pipeline rate has to be increased to feed the crude unit. The pipeline crude blend is also done by a simple spreadsheet (See Table 4). Table 4 shows that the percent draw is based on the meter reading.

Table 4. Pipeline Crude Blending

	Crude Receipt	T1	T2	T3	T4	API	Sulfur	% Naphtha	% Gasoil Distillate
2/7/2004 0:00	18.9	25.0%	35.9%	25.6%	13.5%	33.0	0.568	23.7%	63.3%
2/7/2004 8:00	0.0	25.0%	35.9%	25.6%	13.5%	33.0	0.568	23.7%	63.3%
2/7/2004 16:00	37.2	0.0%	74.0%	26.0%	0.0%	34.4	0.550	21.7%	66.3%
2/8/2004 0:00	0.0	0.0%	74.0%	26.0%	0.0%	34.4	0.550	21.7%	66.3%
2/8/2004 8:00	0.0	0.0%	74.0%	26.0%	0.0%	34.4	0.550	21.7%	66.3%
2/8/2004 16:00	111.1	0.0%	73.2%	26.8%	0.0%	34.4	0.563	21.7%	66.3%
2/9/2004 0:00	65.3	0.0%	74.1%	25.9%	0.0%	34.4	0.549	21.7%	66.3%
2/9/2004 8:00	0.0	0.0%	74.1%	25.9%	0.0%	34.4	0.549	21.7%	66.3%
2/9/2004 16:00	0.0	0.0%	74.1%	25.9%	0.0%	34.4	0.549	21.7%	66.3%

The quality of a pipeline blend is based on refinery economics and should include the intermediate tank inventories and production demands. If the refinery needs more naphtha to maintain the maximum Reformer rate and increase gasoline or BTX production, the crude blend will be light with a high naphtha percent. If there is no light crude available, a request to purchase naphtha to a crude purchaser may be an option if the cost is economical. After the optimum blend is determined, the pipeline rate has to be determined based on how fast the refinery needs the specific blend. For example, if the SRU is upset, the sulfur in the crude may have to be reduced to minimum as soon as possible to dilute the existing crude in the refinery crude tank with a tender of sweet crude.

The final check for the pipeline blend is to check the pipeline sample at the blend header at the terminal, refinery pipeline, and the crude charge feed quality. The samples are sent to the lab for at least an API, BS&W, and percent sulfur and compared to the results with the calculated or estimated numbers.

Refinery Crude Tank Blending

The refinery production can be estimated by the pipeline composition. The production includes LPG, LSR, Naphtha, Distillate, AGO, VGO and VTB from Crude/Vacuum unit, and can be estimated by using the equation below.

$$\Sigma[(\text{Total Crude Charge}) \times (\% \text{ Crude Comp}) \times (\% \text{ Distillation})]$$

The total crude charge is based on the daily actual and the predicted charge rate. The percent crude composition is the pipeline crude compostion which is estimated by multiplying total crude charge and the percent crude. The percent Distillate in crude is estimated using an assay table with cut point data. The cut point on LSR, Naphtha, and Diesel has to take into account the estimated production for final product estimation (See Table 5). For instance, if the naphtha cut point decreases, the naphtha production decreases and Kerosene production increases. If the diesel cut point decreases, the diesel production decreases and there is a corresponding increases in AGO, LVGO, and HVGO.

Table 5. Crude/Vacuum Unit Production

	RM Chg	Slop	Isr CP	Nap CP	Dsl CP	LPG	LSR	Naphtha	Kero	Diesel	AGO	LVGO	HVGO	VTB (1050+)
Date/Unit	MBBL	MBBL	F	F	F	MBBL	MBBL	MBBL	MBBL	MBBL	MBBL	MBBL	MBBL	MBBL
09/08/04	154.1	0.8	145	366	650	4.9	10.3	43.9	21.8	21.8	12.0	5.2	21.8	13.3
09/09/04	154.4	0.8	145	366	630	4.9	10.2	43.7	21.9	19.2	15.0	5.3	21.9	13.2
09/10/04	154.2	0.8	145	340	600	4.9	10.1	39.0	26.2	15.2	19.6	5.3	22.0	12.9
09/11/04	154.2	0.8	145	340	600	4.9	10.1	39.0	26.2	15.2	19.6	5.3	22.0	12.9

The total sulfur in the crude feed can also estimated by the equation below. The equation is based on mass since the weight percent sulfur is available in the assay.

$$\Sigma[(\text{Total Crude Charge}) \times (\%\ \text{Crude Comp}) \times (\text{wt\% Sulfur})]$$

The daily total sulfur production in this refinery was calculated by evaluating total sulfur to the refinery which included total crude sulfur and the intermediate's sulfur if intermediate tanks are drafting to maintain the conversion unit feed. The product sulfur leaving the refinery and the intermediates to storage must be accounted for in the refinery sulfur balance. If the HCU reduces rate, the excess gas oil will goes to storage reducing the sulfur production in the refinery. Below is a block flow diagram on how daily sulfur production was evaluated.

Block Flow Diagram 1. Refinery Sulfur Balance

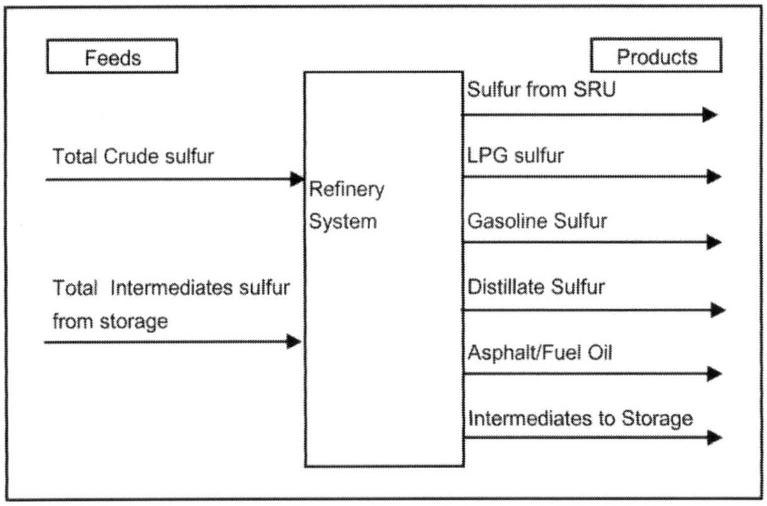

The equation for total sulfur production is therefore:

$$\Sigma[\{(\text{Crude (mass)} \times (\text{wt\% sulfur in distillation cut})\} + \Sigma[(\text{intermediate from storage (mass)} \times (\text{wt\% sulfur})] - \Sigma[(\text{product (mass)} - (\text{wt\% sulfur})] - \Sigma[(\text{Intermediate to storage (mass)} \times (\text{wt\% Sulfur})]$$

Updating the daily lab analysis on feeds and products including weight percent sulfur and specific gravity are required to predict the sulfur production, usually reported in long tons per day.

Additionally unit data needs to be updated such as charge rate, percent naphtha, percent distillate, percent conversion, and the volume of LCO/HCO to HCU. For the FCCU, the charge rate, volume of ATB to the FCCU, percent catgas, and percent LCO yield, and percent slurry are good monitoring points. Similar data is required for the ROSE unit and Reformers to balance the intermediate tank inventories. The crude production was added to the previous day's tank actual inventory to estimate the tank loss/gain for the future. The intermediate tank inventories such as naphtha, distillate, gas oil, and VTB are also updated before adding in the production. Those tanks will be the feed tank to the conversion units and the hydrotreating units that need to be balanced based on production rates and the unit charges. A simple spreadsheet will provide loss/gain of each tank. An example of a naphtha inventory balance is in the Table 6.

Table 6. Naphtha Inventory Balance

	Production	Production		Actual/Estimate	Loss/Gain
	Crude Naphtha	HCU Naphtha	Reformer Rate	Inventory	Inventory
9/11/2004	32000	9000	44000	30000	
9/12/2004	32000	9000	44000	27000	-3000
9/13/2004	35000	8000	44000	26000	-1000
9/14/2004	37000	8000	46000	25000	-1000

The total naphtha production should include the crude naphtha and HCU naphtha. However, if the naphtha hydrotreater (NHT) does not have the capacity to process all of the sour naphtha, the total naphtha production will be the sweet naphtha from NHT unit plus the HCU naphtha. The total distillate production can be separated into Kerosene and diesel. The Kerosene production should include crude Kerosene and HCU Kerosene. Also diesel production should include crude diesel, HCU diesel, and LCO make. The same principle will apply for distillate and VTB production. If the Merox unit does not have the capacity for the total sour Kerosene, the total Kerosene production will be limited by the Merox rate plus HCU kerosene, and if the DHT does not have the capacity for all the sour diesel, the total diesel production will be the DHT rate plus HCU production.

If the HCU and FCCU charge rates are higher than the total gas oil production, the gas oil inventory will be drafting. If the intermediate tank is not in balance with the production rate, the refinery has to cut the unit charges or crude feed to balance the intermediate tank which is an undesirable operation. The crude composition will affect the entire refinery and requires repeating the process from crude receipts through the tank balances until the most optimum crude composition is achieved.

The crude blender's job is to keep operations as smooth as possible by meeting the most optimum and consistent crude blend.

Writing an Order for Terminal Crude Receipt

After the different cases have been analyzed and a plan is finalized, blend orders need to be issued on how to receive crude at the terminal. The order should include:

1. Which crude to receive into which tank and approximate volume
2. Percent draw from each tank for pipeline blending while receiving crude
3. How long to mix after the receipt
4. When and which tank to sample of (bottom, middle, and top)
5. When the pipeline must be shutdown and how long to be shutdown

It is very important to write an order very clearly to the terminal operators. However, it also requires contacting operations as needed to clarify any questions about the order to minimize any misunderstanding.

Writing the Weekly Refinery Operating Plan

The weekly refinery operating plan is also integral to the refinery and is based on the crude blend. This report contains a daily estimation of the crude feed quality, composition, production, products and intermediate inventory targets, unit charges, and end points. Publishing the plan to operations including the gasoline blender, Product Handling Manager, and Lab Manager, and Environmental and Safety Managers to review and provide input to finalize the weekly plan is a good practice. This process helps prepare operations when the crude feed changes and guides them on how to react to those changes.

Conclusions

There are some computer programs such as Orion or Simto that will do many of the same calculations and provide a more detailed analysis for crude blending. However, the process of daily updates and monitoring the unit processes is still necessary whether using an off the shelf program or when using spreadsheets. The most optimum crude blend will minimize refinery upsets with stable operation that increases reliability and profit. Probably the most importance skills for a crude blender are 1) good communications 2) knowing why and how to target the specific crude feed quality 3) good decision making within a limited time frame and 4) maintain the crude blending tool as accurate as possible. The author spent time visiting crude ships to see how crude samples are taken, sitting in a control rooms to see what are the unit operating capacities, minimum and maximum rate, or operating control variables for each unit, and even simply watching how an operator opens and closes a valve. After all, whatever is blended, has to come out somewhere. Depending on what is blended, the refinery will reduce or increase profit.

215a Apriori Alarm Management during Project Feed Stage Using the Results from the Hazop

E. Michael Stafford and **Harry West**. Shawnee Engineers, London, United Kingdom
1 ABSTRACT A recent Front End Engineering Design (FEED) on a major plant expansion included a significant effort in apriori alarm management which represents innovative engineering. This consisted in implementing alarm classifications directly after the HAZOP, using the results of HAZOP prioritization and grouping to suggest Alarm priorities

User Approval of Safety Instrumented System Devices

Angela E. Summers, Ph.D., P.E
Susan Wiley
SIS-TECH Solutions, LP
Houston, Texas 77034
Phone: 281-922-8324

Prepared for Presentation at
American Institute of Chemical Engineers
2006 Spring National Meeting
8th Process Plant Safety Symposium
Orlando, Florida
April 24-26, 2006

February 2006

UNPUBLISHED

User Approval of Safety Instrumented System Devices

Angela E. Summers, Ph.D., P.E
Susan Wiley
SIS-TECH Solutions, LP
Houston, Texas 77034
Phone: 281-922-8324

Abstract

User approval is a simple process that supports design, engineering, configuration management, and management of change. User approval assesses manufacturer information and installed performance data to justify that a device is suitable for use in a safety instrumented system (SIS) application. The process supports configuration management and management of change activities by providing a means to determine whether a new or revised device is acceptable. This paper explains the concept of user approval as documented in ANSI/ISA 84.00.01-2004, ANSI/ISA TR84.00.04, and the Center for Chemical Process Safety book, Guidelines for Safe and Reliable Instrumented Protective Systems.

1.0 Introduction

This paper is distributed "as is." All warranties, either expressed or implied, are disclaimed as to quality, performance, or merchantability, whether expressed, implied or statutory, hidden defects, or fitness for any particular purpose. The author makes no representations about the suitability of the information in this paper for any purpose. Reader bears the entire risk relating to the use of this paper.

A paragraph similar to the above is contained in many end-user license agreements accepted during the installation of user software and with the hardware manufacturer's terms and conditions. The underlying concept is that the "user" of the product bears the risk associated with determining whether a product is fit for a particular purpose.

The manufacturer limits its liability associated with effects beyond its control, such as the operating environment. Other limitations are generally addressed in the installation and maintenance manual (or in the safety manual for safety instrumented system devices). Product manuals provide warnings concerning the installation, commissioning, maintenance, and testing requirements. These agreements, terms, and warnings establish the boundary for the manufacturer's claims. Deviation from the manufacturer's recommended practices generally invalidates all warranties, either expressed or implied.

No manufacturer's claim, third-party analysis, or certification report reduces the user's responsibility for determining the product is fit for purpose. This user responsibility was acknowledged in ISA 84.01-1996, which used the term "user approved" as "hardware, software, procedures, etc., that the user has evaluated and determined to be acceptable for the application."

2.0 The User Approval Process

A user approval process should be established to examine evidence of the suitability of devices for the application and operating environment. A device can be user approved and considered suitable for use when it can be demonstrated that the device meets the core attributes associated with its protection layer. The device boundary includes the hardware and software elements necessary for the device to perform its design intent.

The user approval process examines the device using analysis and testing and demonstrated performance in the operating environment. The outputs of the process should be an approved manufacturer's list and an installation, commissioning, and maintenance plan (e.g., the safety manual for the SIS) that outlines how the approved devices should be implemented.

User approval requires a close relationship between the manufacturer and the owner/operator as a product moves from early development through obsolescence. Technology evolution results in some devices moving rapidly toward obsolescence. For example, the lifespan of some programmable devices is less than 10 years due to the rapid development of computer technology, the increasing demand for interconnectivity, and a desire for uniform configuration tools. Such rapid evolution results in the manufacturer selling products that have some of the desired hardware and software features with a migration path for future upgrades. In some cases, significant changes in the product may be required to correct problems or to provide requested functionality.

The owner/operator should have a process to control the replacement of components in any way other than "like" for "like." When the manufacturer offers a new version of an approved component or recommends a replacement, these new components should go through the user approval process. For new versions of approved products, this process can be simplified when the manufacturer provides documentation explaining changes to the products and highlights any configuration changes.

Configuration management can be challenging, but is essential to achieve the core attributes. Success requires that SIS product specialists serve as an information resource for the owner/operator. These SIS specialists should be highly familiar with SIS requirements, receive and review advisory notices and product information from the manufacturer, and make decisions regarding the addition or removal of devices from the approved manufacturer's list. It is recommended that the SIS product specialist be independent of the project team seeking approval of the device.

Field performance and lessons learned in applying a device should be shared with the manufacturer to assist in future product development and in the dissemination of information regarding critical failures. Monitoring field performance and sharing failure reports with project and plant personnel is important for performance improvement.

3.0 Operating Environment

The operating environment can be identified by drawing an imaginary bubble around the device as installed in the process. The bubble establishes the boundary of the device. The operating environment may include a variety of items that affect device operation:

- External environmental conditions,
- Process operational conditions,
- Communications and interconnectivity,
- Human interfaces,
- Access security,
- Support systems, e.g. instrument air and electricity.

For Programmable Electronic (PE) logic solvers, the operating environment also covers the embedded software, the hardware architecture, application software, I/O configuration, communication to other systems, operator and engineering interfaces, and access security.

4.0 Analysis and Testing

Devices should be subjected to analysis and testing to evaluate the device's design, manufacture, and validation procedures, as well as the manufacturer's quality and change management systems. The rigor of the analysis and testing is related to the complexity of the device and the maturity of its technology. When completed, there should be a better understanding of how the device functions and fails. This analysis should consider the device boundary and should result in the following:

- Description of the analysis boundary.
- Identification of potential dangerous failures.
- Identification of means to detect dangerous failures by diagnostics or testing.
- Description of assumed diagnostic coverage factor and any user requirements necessary to achieve the diagnostic coverage, e.g., configuration requirements and the need for external diagnostics.
- Statement of the probability of safe and dangerous failures with any assumptions regarding maintenance activities and testing intervals.

Analysis and testing should also include evaluation of the device's conformance to applicable codes, standards, and practices. Meeting the area classification (e.g., hazardous environments) and following applicable electrical codes is important for all instrumented system applications. Devices that do not meet the area classification can become ignition sources should a loss of containment occur. For SIS, the requirements of IEC 61511 Clause 11.5 should be met. Guidance on complying with this clause is also provided in ANSI/ISA TR84.00.04 Annex L.

Manufacturers, who wish to advertise their products as safety products, analyze and test their devices for compliance with the intent of IEC 61508. An IEC 61508 evaluation may yield a "certification" if the product is evaluated by a recognized certification body. A certified device can be labeled with an "SIL (Safety Integrity Level) Claim Limit." In the United States, certification to IEC 61508 is issued by Nationally Recognized Testing Laboratories (NRTL), such as those listed on the US-OSHA website (www.osha.gov). The OSHA website has a complete listing of NRTLs. Other countries establish certification requirements by policy or regulation acceptable certification bodies.

When a device is certified according to IEC 61508, the manufacturer supplies a safety manual providing the criteria for its implementation. Successful implementation of SIS devices often requires a specific configuration, the addition of external diagnostics, the provision for inspection and maintenance, mandated proof test intervals, and specific installation details. Deviation from the safety manual may invalidate the SIL Claim Limit.

It is important to note that IEC 61508 is a generic functional safety standard that applies to safety-related systems used in many different industry sectors, including process, rail, machinery, and medical. Since this standard applies to many sectors, there are devices that meet the requirements of IEC 61508, but are not sufficiently robust for the operating environment in many process industry facilities. A device developed for use on a manufacturing floor may not be appropriate for use in a refinery SIS.

Manufacturers sometimes provide field performance reports based on an installed base at other owner/operator sites. More often, manufacturers supply analysis and testing results and documented predictive calculations of their product's probability to fail on demand and spurious trip rate based on the product's "shelf-state" design and manufacture. For any new technology, the analysis and testing is generally based on very limited field operating experience. Thus, the manufacturer evaluation represents the highest performance that can be expected from the device when it is implemented in accordance with its safety manual.

Field device and non-PE logic solver performance is generally dictated by the operating environment and the owner/operator's inspection and maintenance practices. For example, a sampling of data for pressure transmitters from various manufacturers show 600 to 800 year mean time to failure dangerous (MTTFD), however, owner/operator prior use data show a range of 75 to 200 year MTTFD.

Many important failure modes are excluded from the manufacturer's boundary. This includes the process connections, manner of installation, power supplies, and communication interfaces. Failures due to the operating environment can be significantly greater than failures due to device manufacture. Failures in the interfaces between the device, the process, and other protective systems should be considered when determining its suitability for addition to the approved manufacturer's list.

Some owner/operators develop installation details for each type of technology. These owner/operators then analyze the installation itself to ensure that the core attributes of a protective system are achieved, e.g., independence, functionality, integrity, reliability, auditability, access security, and management of change. This provides a consistent operator and maintenance interface and allows the analysis and testing for device approvals to be limited to the device and its ability to work within the existing approved installations.

For PE logic solvers, performance is generally dictated by how well the safety manual is followed, how recommended upgrades are executed, and whether the manufacturer's recommended operating environment is maintained. When implemented according to manufacturer's recommendations, PE logic solvers tend to achieve the reported level of safety. This does not necessarily mean that the PE logic solver has the robustness to provide the trouble free service that may be desired. In general, PE logic solvers tend to fail-safe more frequently than reported by analysis. This is largely due to violation of the operating environment requirements or human error.

5.0 Prior Use History

In general, manufacturer failure rate data is 3 to 10 times better for programmable devices than the actual performance observed in the process industry. For mechanical devices, the ratio is 30 to 100 times better. Programmable electronic systems are not immune, with some certifications assuming diagnostics coverage factors that are simply not achievable using current practices. Because of the discrepancy between manufacturer and owner/operator environments, devices should be selected based on demonstrated history in a similar operating environment.

The intent of prior use is to gain an understanding of the device's failure modes so that the design and operating basis can take them into account. Previous operating experience provides valuable information for the selection of field devices, because it identifies how the operating environment degrades the theoretical performance claimed by the manufacturer. For some devices, especially field devices, evidence of successful operating experience in similar process applications is very important. In IEC 61508, this is called "proven-in-use;" while in IEC 61511, it is called "prior use."

There are specific requirements and limitations regarding the prior use evaluation in IEC 61511 Clauses 11.4 and 11.53 through 11.5.6. The requirements vary depending on the device type (sensor, logic solver, final element) and whether the device uses programmable elements. If the device is programmable, the requirements also vary dependent on whether fixed programmable language or limited variability language is used to configure the device. These requirements are further discussed in ISA TR84.00.04 Annex L.

For SIS, IEC 61511 requires devices be selected based on their expected performance in the operating environment. Maintenance records are a valuable information source. The plant maintenance tracking system can be used to flag devices that have repeatedly failed so that more detailed analysis can be performed. In general, it takes approximately three years of operating time to gain sufficient understanding of the failure modes of a field device. This operational time can be obtained in the actual or similar operating environment in process control, other non-SIS applications, or at other owner/operator sites.

Operational time can be gathered during alpha and beta testing where owner/operators work in close association with the manufacturer. Alpha testing is conducted to demonstrate the basic functionality of the device and general compatibility with the service. A successful alpha test leads to beta testing, which is usually conducted using multiple installations to gather information on the device failure modes in a variety of operating environments.

However, alpha and beta testing results are not sufficient for approving devices for SIS service.

Manufactures often provide references to other owner/operators who use their product. These references can often provide valuable insight into product application and use. In addition, the manufacturer user's group can provide valuable networking opportunities to gather information concerning product performance. The owner/operator may choose to rely on the manufacturer and other owner/operator's field experience or may choose to supplement this evidence with more direct experience. This may include bench testing and field trials in process control or low hazard services. Generally, the more unfamiliar or complex the technology is, the more time should be spent understanding how it works and how it fails.

Proven performance takes on special significance for SIS, because most operate as dormant systems that only take action when a process demand occurs. In contrast, control system devices are expected to operate frequently, if not continuously, so failures in control system are rapidly detected. If these failures cannot be managed by the control system or operator, an independent SIS is often implemented to mitigate any unacceptable process risk.

One of the strongest arguments for separation of the SIS and the control system is the common practice of using the latest technology in control applications. Many owner/operators who strive to use the newest control technology often find that the products are not as well developed as originally thought. New control system implementation can turn into an unplanned product development project. While necessary for production or quality reasons, this can be a hazardous practice for SIS. Proven operating history cannot be underestimated.

6.0 Summary

User approval is an important aspect of complying with ANSI/ISA 84.00.01-2004. User approval relies of predictive evidence, such as analysis and testing reports, balanced with field operating history. While some manufacturers are now providing devices that are "certified for use" in safety instrumented systems (SIS), these devices are relatively limited in their range of application and technology. Many are not field proven and some are not demonstrating the robustness necessary to survive a chemical plant environment. Within an IEC 61508 analysis, an unreal operating environment is often assumed: one that does not include the process impact, ambient conditions, stress, manufacturing defects, software errors, installation issues, electrical disturbances, instrument air or other support system quality problems, etc.

Consequently, owner/operators in the process sector should rely heavily upon operating experience for selection of devices for SIS, especially for those devices installed in the process environment. User approval seeks to get the proper balance of analysis/testing information and field experience. The amount of information or experience required and the degree of rigor associated with their evaluation may vary, depending on the nature of the device and the required safety integrity level. The key point to the user approval process is to establish sufficient evidence to justify that the device can and does achieve the required performance in the intended operating environment.

7.0 References

[1] Instrumentation, Systems, and Automation Society (ISA), ANSI/ISA S84.01-1996, "Application of Safety Instrumented Systems (SIS) for the Process Industry," Research Triangle Park, NC (1996)

[2] Occupational, Safety and Health Administration (OSHA), "Process Safety Management of Highly Hazardous Chemicals; Explosives and Blasting Agents, 29 CFR Part 1910." Federal Register 57,36, Washington, DC (1992).

[3] International Electrotechnical Commission (IEC), IEC 61508, Functional Safety of Electrical/Electronic/Programmable Electronic Safety Related Systems, Parts 1-7, Geneva, Switzerland (1999-2001).

[4] International Electrotechnical Commission (IEC), IEC 61511, Functional Safety: Safety Instrumented Systems for the Process Sector, Geneva, Switzerland (2003).

Advances in Abnormal Situation Prevention in Refineries and Petrochemical Plants

Ahmad Hamad and Ravi Kant
Emerson Process Management
Eden Prairie, MN 55344

Abstract

Abnormal situation prevention is of high importance in the operation of refineries and petrochemical plants. Abnormal situation may result from equipment malfunction, process upsets, operator errors, etc. There have been several approaches to deal with those situations. The proposed abnormal situation prevention approach in this work targets the timely and early detection of abnormal situations to prevent undesired upsets, shutdowns and catastrophic consequences. The approach uses advanced statistical control techniques, smart field devices, and multivariate analysis to analyze trends of process variable and detect faults or deviations. The approach has three phases: learning, model development, and monitoring. All these phases can be implemented online or by using off line data. The capabilities of the proposed abnormal situation prevention approach will enable refineries and petrochemical plants to run safer and cleaner while achieving higher economic benefits.

Introduction

Abnormal situations can be defined as process disturbances, upsets, and events that can cause the process to deviate from its normal operating state and may lead to damaging consequences. Abnormal situations can vary in severity and consequences from off-spec product to catastrophic failure of equipment. Cost of abnormal situation consequences increases as the severity increases. Marsh Risk Consulting (2001) identified 379 abnormal events in the Hydrocarbon industries with individual losses exceeding $10 billion in the period from 1970 to 1999. Average annual losses were approximately $700 million with twenty-five of these events having individual losses of $150 million or more. Unfortunately, these economical losses are sometimes also accompanied by loss of lives and environment incidents. Till now, the main approach for preventing damaging consequences of abnormal situations is preventive maintenance. However, this approach has major disadvantages. First, this approach is costly since maintenance is done on regular basis regardless of the status of the process or equipment. This approach may include unnecessary plant or unit shut down.

Second, this approach does not detect abnormal situations. This is a major disadvantage because abnormal situation can occur between scheduled maintenance periods. Another disadvantage includes the direct and complete dependence on plant operators. To overcome these disadvantages, several efforts have been done by various industries and researchers to come up with diagnostic techniques to detect and prevent abnormal situations and reduce the cost of maintenance. Roemer and Kacprzynski (2000) presented a real-time, integrated health monitoring of gas turbine engines that can detect, classify, and predict developing engine faults. Kubiak et al. (2004) introduced a hybrid fault pattern approach for diagnosing gas turbine component degradation. Reising et al. (2005) discuss how abnormal situation management (ASM) is used by plant operators to recognize the presence of failures before alarms are issued. Eryurek et al. (2005) discusses how smart field devices are used for abnormal situation prevention. This paper presents a novel on-line methodology for abnormal situations prevention in the process industries. The proposed methodology builds-on several tools to come up with a novel abnormal situation prevention structure that is robust and operator-friendly. The methodology is embedded in Emerson's Plantweb technology and relies on using smart devices, high frequency transmitters, statistical-based analysis methods, on-line data gathering and analysis, and on-line process automation using operating software such as DeltaV. Examples of applying this approach to refineries and petrochemical plants will be presented.

Process Control and Abnormal Situation Prevention

Before talking about the proposed abnormal situation prevention methodology, it is important to differentiate between process control and abnormal situation prevention. Abnormal situation prevention technology is intended to provide early detection and timely corrective action. Consider Figure (1) where abnormal event severity is presented as a function of time. The severity of the abnormal situation increases as time progresses. It is traditional ASM approach that when the value of the process variable reaches unacceptable region an alarm is issued to alert the operator. The operator takes an action in an attempt to return the process into the normal operation region. The operator is doing process control in this case based on prior knowledge or rules. In many instances this approach is ineffective because the detection is too late and may lead to catastrophic situations or unplanned plant/equipment shutdown. In addition, this approach fails in most cases to identify the source of faults. The operator gets trapped in fixing the symptoms of the fault by manipulating process variables. The main goal of the operator is to bring the process back to operate within the acceptable region. The operator may fix the fault symptoms but fails to identify and eliminate the fault that caused these symptoms.

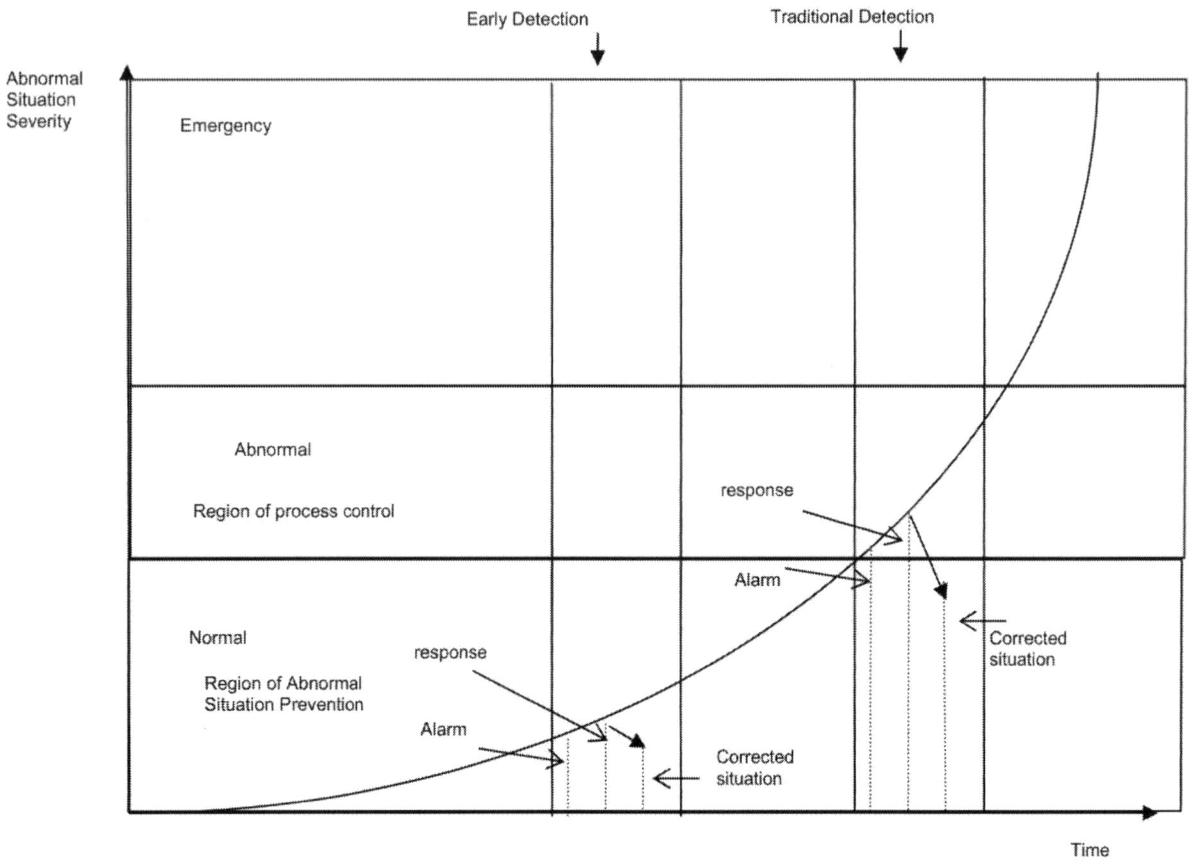

Figure 1: Typical Response Diagram for Abnormal Situations

Process control in this situation may take the process variable back to a point in the normal region that is not far from the abnormal situation. The proposed abnormal situation prevention approach, on the other hand, identifies the abnormal situation at an earlier time (Figure 1) and identifies the source of the fault. The abnormal situation is identified as it starts to develop. The whole diagnostic, detection, and corrective actions take place in the normal operating region. How is this done? This is done by comparing the actual real time operating conditions to the "ideal base." Ideal base here refers to the normal process operating conditions without any deviation or abnormal situation. The proposed approach uses advanced pattern recognition technologies, signal analysis, statistical process control, as well as expert system based solution with the smart devices for improved process operation, and early abnormal situation detection and prevention.

Technology Advances In Abnormal Situation Prevention

The Fluid Catalytic Cracker (FCC) unit is used to illustrate advances in abnormal situation prevention technology development. Figure 2 presents the cost of abnormal situation consequences resulting from slide valve failure in the catalyst transfer lines of the FCC unit. Valve failure is due to irregularity in catalyst circulation (abnormal situation) in the transfer pipes between reactor and regenerator. The abnormal situation consequences in this case progress can be as follows:

1. Stage 1: represents pressure irregularity due to catalyst erratic circulation in transfer pipes
2. Stage 2: represents loss of production due to regenerator fouling
3. Stage 3: Valve malfunction because of catalyst build up in transfer pipes
4. Stage 4: FCC unit shut down due to pressure loss across transfer pipes
5. Stage 5: Plant shut down

As the stage order increases, the cost severity associated with the abnormal situation increases and may result in major damaging consequences. Circle (A) in Figure (2) represents two corrective approaches, preventive maintenance and knowledge-based diagnostic and fault detection. The preventive maintenance takes a corrective action with respect to physical entity such as cleaning the valve or the transfer pipe. The knowledge-based diagnostic uses historian data to develop a multivariate model to detect the behavior of the process. This approach can get very complex. It uses low frequency filtered data and relies on historian data. It does not have the capability to gather data on-line to update the monitoring models. This approach fails to detect abnormal situations if the abnormal situation is very fast or with a high frequency unless a damaging consequence is evolved. Furthermore, this approach fails to predict health of machinery, devices, and transmitters used in the process. It can, though, give a good indication of process variable deviation from normal operation condition. The response time domain for circle A is slow and can be too late.

Circle B represents the use of smart field devices and field instruments. Eryurek et al. (2005) illustrates that improved process measurements and real time analysis/detection could have prevented or at least substantially reduced the damage of approximately 25-40% of reported abnormal situations. With advances in computing and measurement device technologies and advances in communication such as the digital Fieldbus technologies, today's smart field devices and transmitters are capable of providing much more information regarding the processes and its conditions. Smart field devices and instruments manufactured by Emerson Process Management have built-in ASP™ blocks (Figure 3) which captures high frequency process data (22 Hz) and have data analysis capabilities. Abnormal situation prevention algorithm now can have access to data that was lost with a low frequency analysis (Figure 4).

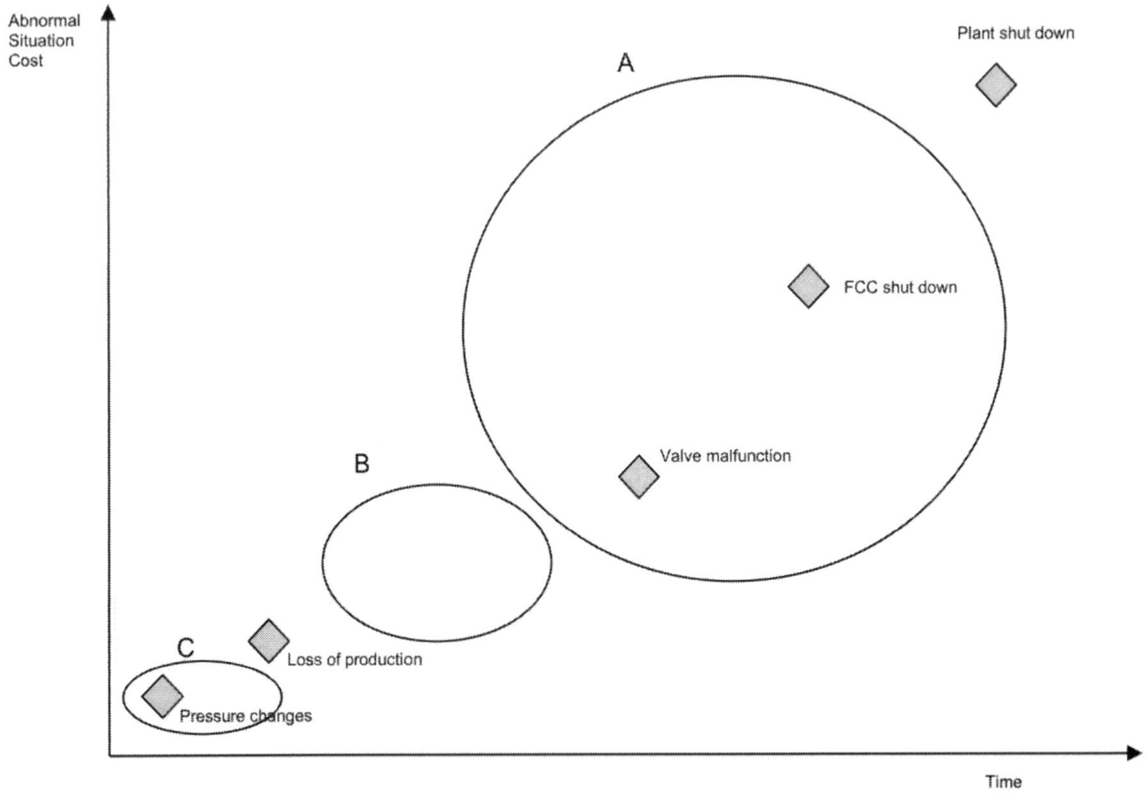

Figure 2: Abnormal Situation Cost For Slide Valve in FCC Unit

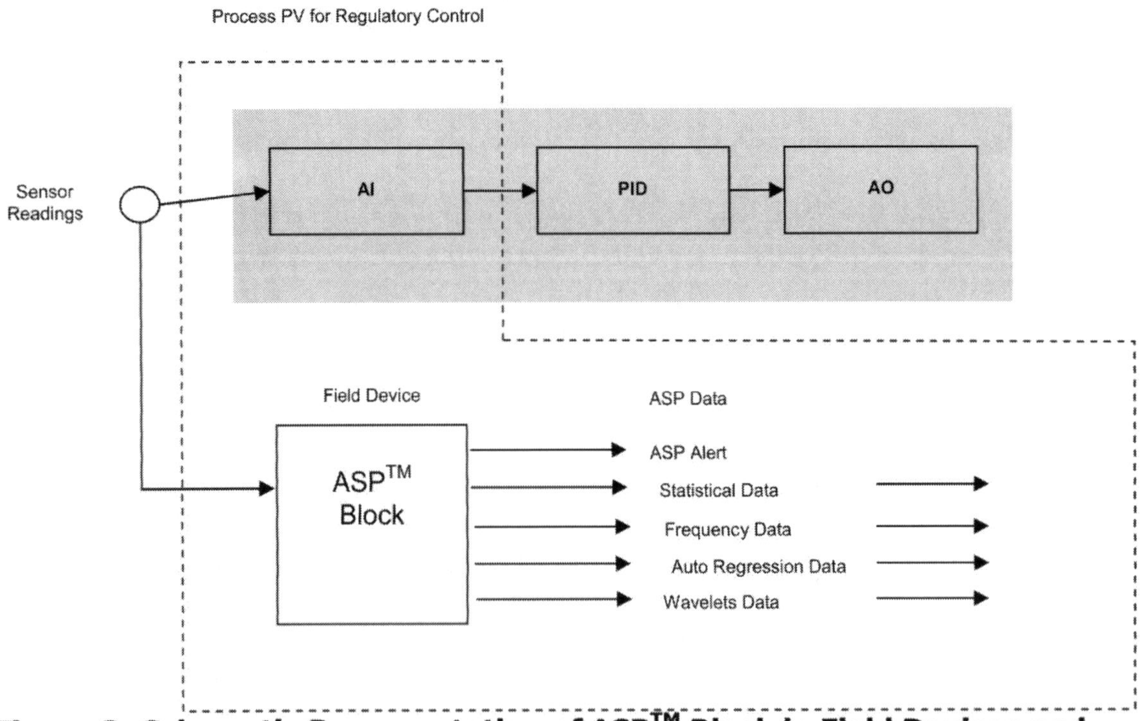

Figure 3: Schematic Representation of ASP™ Block in Field Devices and Instruments

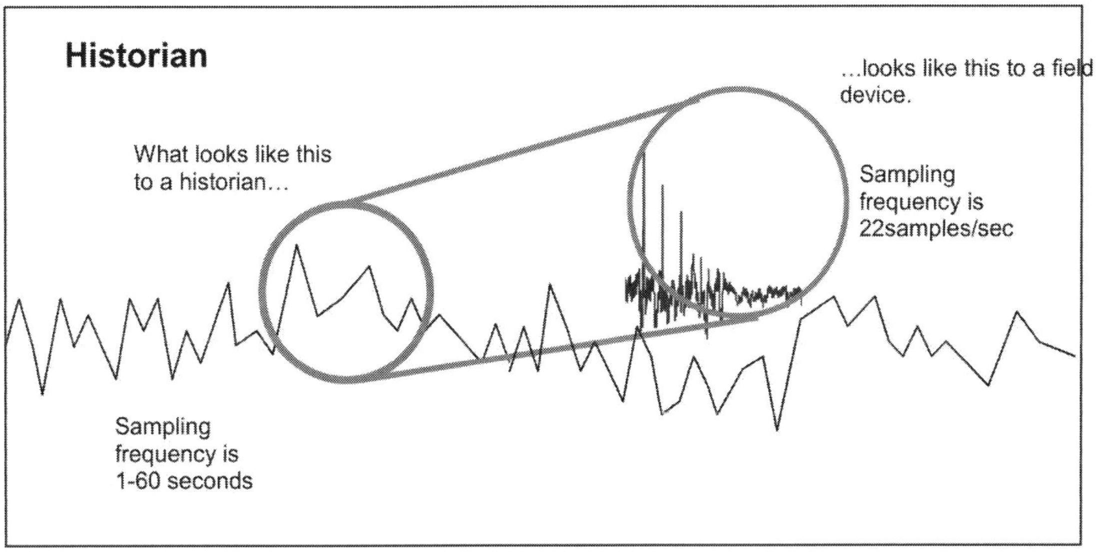

Figure 4: Data (Pressure, Temperature, etc.) Frequency Sampling in Smart Field Transmitters

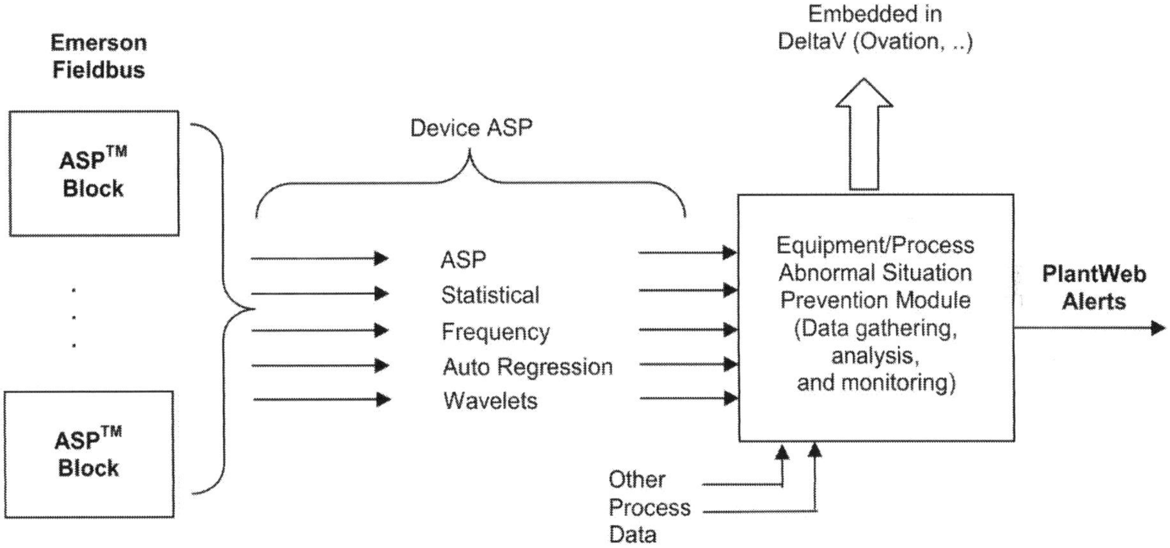

Figure 5: Equipment and Process Abnormal Situation Prevention Approach Using Plantweb Technology

This data could provide information about process performance and abnormal situations. Smart devices and instruments can provide the operator with an assessment of their own health by providing high frequency statistical data such as the variance and mean. This methodology still cannot go to detect abnormalities due to process equipment fouling or process-related abnormalities.

Circle (C) in Figure 2 represents the proposed equipment and process abnormal situation prevention approach in this work (Figure 5). This approach utilizes the benefits of knowledge-based, smart field devices and instruments, advanced process control techniques, and other diagnostic techniques to detect and prevent abnormal situation in early stages while the process is running in the normal operation region (see Figure 1). The proposed abnormal situation prevention approach extracts high frequency real time plant data using Plantweb components and applies on-line diagnostic techniques to analyze this data and identify any abnormal situation or behavior in the plant. Plantweb is a digital plant architecture that delivers proven results, and the DeltaV system makes it easy. Designed with this digital architecture in mind, the DeltaV system fully integrates "smart plant" capabilities including HART®, FOUNDATION™ fieldbus, high-speed discrete buses, embedded advanced control, and advanced unit and batch management. Such tools can be used in the proposed approach to make online analysis and data extraction easier and faster. The following section illustrates the phases of the proposed ASP approach.

Phases of Abnormal Situation Prevention Approach

The proposed abnormal situation prevention approach has three phases to detect faults and abnormal situations in the process industry. The phases include learning, model development, and monitoring. In the learning phase, the module collects data (input and output) as needed and applies various statistical techniques to relate input (load variable) and output (monitored variable) data. The models relating outputs and inputs can be as simple as variance with a threshold, or linear, or polynomial, or any other form of interest. Multivariate analysis techniques can also be employed to relate inputs and outputs. Selected model type is user defined. During the learning phase, data are collected for both input and output variables. Two time domains are important here and both are user specified. The first domain is time interval between data collected. It could range from fractions of seconds to minutes or hours. The shorter the interval the more data is collected. The second domain is the length of the learning phase. It could be hours, days, weeks, or months. It is recommended to have a learning period long enough to capture all anticipated variances in process operation. It is important to realize here, that the models developed during a learning phase are good enough for the range of data used in the regression process. If the process shifts to a new region of operation, then new models are needed. This means that a new learning phase is needed. The models are developed during the learning phase and then used in the monitor phase to detect and prevent abnormal situations. Operator may monitor point or pattern deviation of the process variable. Smart field devices and instruments are necessary to have robust data gathering and analysis of process variables.

Implementation of Abnormal Situation Prevention Approach

One major advantage of this approach is that it is easy to implement. The developed abnormal situation prevention algorithms or modules can be implemented in any of the following platforms:

1. Field Device Interface Module (e.g. Rosemount 3420)
2. Control System (e.g. DeltaV, Ovation)
3. Standalone software application

Various abnormal situation prevention modules can be developed which may be process equipment based or process unit based. Operators or plant engineers will have the flexibility to tailor ASP modules to fit the needs of their process unit application.

Case Studies

Various case studies are presented below to illustrate the effectiveness of the proposed abnormal situation prevention methodology.

1. Coke Detection in Refineries

The Delayed Coker is used in refineries to process the heavy residue (Coke) from the vacuum distillation unit. Although there are many components to the Coker unit, a component with great value to Abnormal Situation Prevention is the Coker Heater. This fired heater takes the feedstock from the vacuum bottoms, and heats it prior to sending it to the Coke Drums. The feed enters the unit at the left, and is divided into multiple passes (number of passes is user defined). A load-balancing control regulates the flow though each of the passes. One problem that can occur in the Coker Heater is the build-up of coke on the inside of one or more of the passes. When this happens, the performance of the furnace begins to degrade, and the other passes need to take more loads, in order to compensate. If coking continues to build up even more, eventually the affected passes, or even the entire unit, will need to be shut down. Thus, it is desirable to know when coke buildup begins to happen, in order that a corrective action can be taken.

Module Summary

The Coker abnormal situation prevention algorithm monitors each pass in the furnace of the Delayed Coker to check for high coking. Generally, a decrease in either the flow or heat transfer rate for a given pass indicates that there is high coking in that pass. During the learning phase, data is collected for each pass.

At the end of the learning phase, a regression model is developed which can predict the value of each monitored variable as a function of the load variable. The value predicted by the model is considered the normal operating condition for that monitored variable. When the actual value differs from the predicted value by more than a certain threshold, an abnormal situation (high coking) is detected. This algorithm is implemented in DeltaV (Figure 6) and is used in industrial trials.

2. Catalyst Circulation In FCC Units

Catalyst circulation in the FCC unit is extremely important. Without "proper" catalyst circulation the FCC unit deteriorates significantly. Detecting circulation problems requires good understanding of pressure balance around the reactor-regenerator system. It is important to detect erratic circulation; otherwise FCC unit may experience unscheduled and costly shutdowns.

Module Summary

A lower or higher differential pressure across the slide valve than "normal" slide valve opening is a good indication of abnormal situation development in the catalyst transfer pipe. Catalyst circulation can also be limited by mechanical problems with the slide/plug valve. The abnormal situation prevention module consists of two steps:

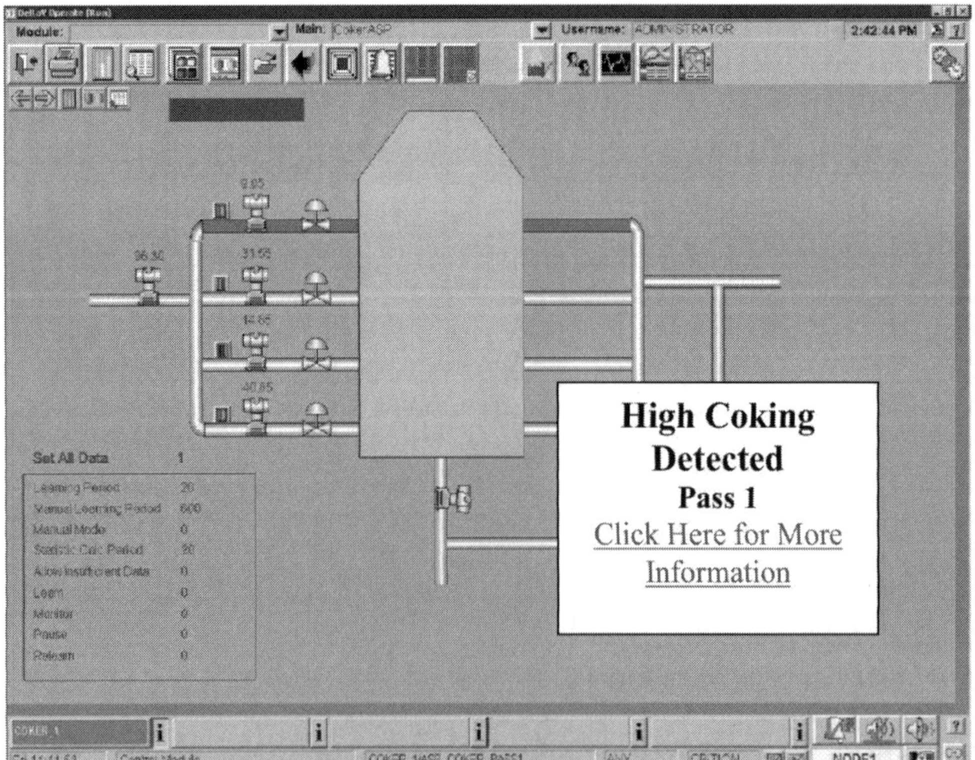

Figure 6: High Coking Detection as Implemented in DeltaV

a. Using smart field valves such as DVC6000f. This valve will keep operators informed of the health of the valve to avoid any unplanned shut down.
b. Pressure balance (drop) across the valve will be monitored using smart pressure transmitters to detect abnormality. With this information, operators will be able to take corrective action early in the problem development before reaching the abnormal situation failure (Figure 1).

3. Maltrays Detection in Crude Columns

One of the major problems that face crude columns and distillation columns in general is maltrays. Maltray could result from flooding, foaming, lost trays, etc. The major technology that is used currently for column diagnostics is Gamma ray. This technology is not handy and is expensive to use. As an example of maltrays, let's discuss column flooding. Flooding is excessive accumulation of liquid inside the column. This accumulation is generally caused by various mechanisms such as downcomer backup flooding, spray entrainment flooding, etc. Flooding can be recognized by some symptoms including excessive column differential pressure, loss of bottoms, loss of separation, etc.

Module Summary

The objective of this algorithm is to detect conditions in column operation that lead to flooding or to predict flooding conditions as early as possible to avoid entire column or tray flooding. Different algorithms can be developed to identify flooding conditions depending on existing instrumentations and potential future instrumentations.

4. Gas Turbines Abnormalities

One of the major problems in gas turbines (GT) is fuel supply and efficient usage of fuel. Uncontrolled and/or unburned fuel in the combustor could lead to various abnormal situations in gas turbines and could cause sudden shut downs. Abnormal situations include loss of fuel efficiency, damaged blades (due to high temperatures, localized explosions), and damaged nozzles (due to high temperatures). The objectives of the proposed abnormal situation prevention method in this work are:

1. Control and monitor fuel supply to prevent abnormal situations in GT
2. Detect abnormalities in fuel system to prevent abnormal situations in GT
3. Detect abnormalities in expander to prevent abnormal situations in GT and prevent unscheduled shutdowns

Module Summary

In the learning phase, the method collects data and applies various modeling techniques to relate input and output data. The models developed in the learning phase are used in the monitor phase to identify abnormal situations. Each model is assigned a region of confidence for operation. The region of confidence is user specified.

Conclusion

Preventing abnormal situations in the manufacturing processes (refineries, petrochemical plants, etc.) can lead to substantial operational, economic, safety and environmental benefits. The proposed abnormal situation prevention approach diagnoses and analyzes process variables to detect abnormal situations as they start to evolve. This advantage gives plant operators/engineers the opportunity to eliminate the source of the abnormal situation before it causes process variables to deviate from the normal operating region. The proposed abnormal situation prevention approach has three phases: learning, model development, and monitoring. The developed modules can be implemented in field devices, DeltaV, or stand alone software.

References

1. Eryurek, E. P. Sharpe, and D. White (2005), "Abnormal Situation Prevention through Smart Field Devices," presented at the NPRA 2005 Annual Meeting (AM-05-49), San Fransisco, CA
2. Kubiak, J. S., A. G. Garcia, G. B. Urquiza, and G. R. Gonzalez (April 2004), "Hybrid fault patterns for diagnosing gas turbine component degradation," Hydrocarbon Processing, p. 71 – 74
3. Marsh Risk Consulting (Feb. 2001), "Large Property Damage Losses in the Hydrocarbon Industries, A Third Year Review
4. Reising, D., T. DeMaere, J. Errington, K. Harris, and P. Bullemer (October 2005), "Establishing Operator Performance Improvements and Economic Benefit for an ASM Operator Interface, presented NPRA Plant Automation & Decision Support Conference, Grapevine, TX
5. Roemer, M. J., and G. J. Kacprzynski (May 2000), "Advanced Diagnostics and Prognostics for Gas Turbine Engine Risk Assessment," presented at IGTI/ASME Turbo expo, Munich, Germany

"Strategic Proof Testing"

Dr. William M. Goble
Exida
Sellersville, PA 18960
wgoble@exida.com
+1-215-453-1720

Introduction

Safety Instrumented Systems (SIS) are designed to provide automatic protection functions. A typical design includes a sensor to detect the potentially dangerous condition, a safety PLC and a remote actuated valve capable of taking action to prevent an accident. In a well designed chemical plant the potentially dangerous conditions occur very infrequently. Hence these equipment sets may be idle for long periods of time and operate only once every few years. In such a situation failures of the equipment that prevent a safety response may go undetected for years without a periodic "proof test."

Proof testing is an important part of the maintenance activities needed to assure safety. Proof testing is done on a periodic schedule using procedures specifically developed during the design. If the designer is following a safety lifecycle (SLC) per IEC 61511, an international functional safety standard, the periodic proof test interval is calculated based on a safety integrity level (SIL) design target. A SIS design process is shown in Figure 1.

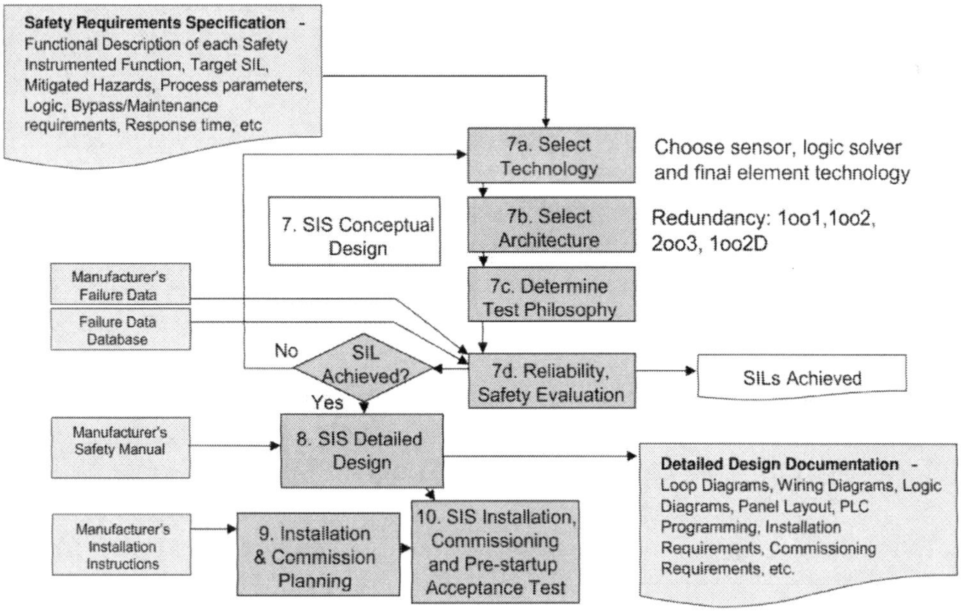

Figure 1: SIS Design Phase Activities (copyright exida 2001, from Safety I course notes)

During the design verification analysis (step 7d) the probabilistic calculation includes a number of variables including the proof test interval and the proof test effectiveness. The traditional approach to proof testing is to perform some action which simulates the dangerous condition and observe to the extent possible that the safety function performs its duty. A typical test (simplified) for a single channel design might be:

Procedure:
1. Procure all management of change permission forms
2. Block valve from closing.
3. Move input signal above trip point.
4. Verify that valve attempted to close.
5. Move input signal back to normal below trip point.
6. Remove valve block.
7. Document test results and restoration to normal.

An assumption is often made, especially when using simplified equations for the analysis, that the proof test is 100% effective. But how effective is that test?

FMEDA Analysis Techniques
In the late 1990's techniques were developed to determine the diagnostic coverage capability of smart instruments with automatic diagnostics (Ref. 1). These techniques were created from the need to more precisely determine the quality of proposed diagnostic tests in PLC product (Ref. 2). The objective of the analysis is to measure how many of the potential failures are detected by a diagnostic. This is measured by a metric called the "coverage factor." A number of 100% means all failures are detected and that is the goal.

The FMEDA technique involves studying every failure mode of every part and recording if a particular test can detect that part failure mode. The method is very systematic and complete. The same methods can be used to determine the effectiveness of any proposed manual proof test. Given a list of all components and failure modes one can reasonably determine which items are detectable. Using a weighted average based on failure rates the effectiveness is calculated.

The techniques have evolved and are now being used on instruments (pressure transmitters, temperature transmitters, etc.), PLC products and even mechanical products (solenoids, actuators and valves). An example from a ball valve is shown in Table 1.

Table 1: Partial FMEDA of ball valve partial stroke test (PST).

Valve component	Failure mode and effect	Lambda (FITS)	PST Credit	Lambda (detected
Ball	Damage Pin slot resulting in failure to close	31.50	0	0
Ball	Surface Damage resulting in failure to close	126.00	0.8	100.8
Ball	Surface Damage resulting in Internal leak, major	189.00	0	0
Drive dowel pins	Break/damage resulting in failure to close	25.00	0	0
Stem	Break/deformation resulting in failure to close	18.75	0	0
Stem	Damage/Deflection resulting in failure to close	112.50	0.8	90
Trunnion	Damage/Deflection resulting in failure to close	210.00	0.8	168
Seat Seals	Damage resulting in Internal leak, major	20.00	0	0
Stem Bearings	Deformation resulting in failure to close	80.00	0.8	64
Thrust washers	Deformation resulting in failure to close	20.00	0.8	16
Adapter Bolting	Break resulting in failure to close	1.00	0	0
Adapter Plate Pins	Break/deformation resulting in failure to close	2.00	0	0
Actuator Adapter Plate	Break/deformation resulting in failure to close	1.50	0	0
Seat Ring	Damage resulting in Internal leak, major	180.00	0	0
Seat Ring	Damage resulting in failure to close	60.00	0.8	48
	TOTAL DANGEROUS FAILURE RATE FOR VALVE	**1077.25**		
	DANGEROUS DETECTED FAILURES			**486.8**
	TEST EFFICIENCY			**45.19%**

Note that in this case the result shows an effectiveness of 45%, far from the 100% assumption often made.

Instrument manufacturers are proposing a range of proof test techniques as part of their IEC 61508 safety certification. For example, a page from the safety manual of a pressure transmitter is shown in Figure 2. This shows a relatively simple test that can be remotely with a handheld communicator. It has a proof test effectiveness of 65%. Another more extensive test is specified with an effectiveness estimated to be greater than 989%. One may wonder why the first test is even mentioned. Remember that the first test is far simpler and less costly to run. It is likely that a combination of the two tests, each run on a different schedule can provide good coverage with optimal cost.

SAFETY INSTRUMENTED SYSTEMS

Additional Safety Instrumented Systems information is available in the Rosemount 3051S Series Pressure Transmitter reference manual (document number 00809-0100-4801). The manual is also available electronically on www.rosemount.com or by contacting an Emerson Process Management representative.

Installation

The 3051S SIS can be used in safety instrumented functions up to SIL2 with a single instrument and SIL3 with hardware fault tolerance of 1. No special installation is required. Follow standard installation practices as outlined in this document.

Set-up

The security switch must be in the **ON** position during normal operation.

Configuration

No special configuration is required. Verify configuration as outlined in this document. Use any HART-compliant master to communicate with and configure the 3051S SIS. DD revision 3051S SIS Dev. 1 Rev 1 is required.

Operation and Maintenance

Proof Test and Inspection:

The following proof tests are recommended. Proof test results must be documented with corrective actions taken in the remote possibility that safety functionality is not correct.

To perform a HART Loop Test follow the "Loop Test" instructions available in the 3015S reference manual (document number 00809-0100-4801).

Five-Year[(1)] *Proof-Test*

This proof test, when combined with a HART Loop Test, will detect more than 65% of DU failures not detected by the automatic diagnostics programmed into the 3051S SIS.

1. Enter the milliampere value representing a high alarm state
2. Check the reference meter to verify the mA output corresponds to the entered value.
3. Enter the milliampere value representing a low alarm state
4. Check the reference meter to verify that the mA output corresponds to the entered value.
5. Execute the Master Reset command to initiate start-up diagnostics.

Ten-Year[(1)] *Proof-Test*

This proof test, when combined with a HART Loop Test, will detect over 98% of DU failures not detected by the automatic diagnostics programmed into the 3051S SIS.

1. Perform a minimum two point calibration check using the 4-20mA range points as the calibration points.
2. Check the reference mA meter to verify the mA output corresponds to the pressure input value.
3. If necessary, use one of the "Trim" procedures available in the 3051S reference manual to calibrate.
4. Execute the Master Reset command to initiate start-up diagnostics.

(1) May be a longer proof test interval as justified by PFDavg calculation.

Figure 2: Safety Manual page showing recommended proof tests along with effectiveness (used with permission of Rosemount)

Using FMEDA results to pinpoint effective proof tests

The FMEDA results can identify the component failure modes that cause the failure of the safety function. Given that these are now clearly identified why not design a proof test focused on discovering that failure?

This is effectively a reversal of thinking on the objective of a proof test. Conventionally a test is devised to show that the safety function works properly. It is assumed that any equipment failure that prevents the safety function will be discovered during the test. This is generally true but what about transient failures? What about marginal failures? What about the failure of an automatic diagnostic? Rarely are these revealed during a conventional proof test.

The reverse method of thinking says that a test should be focused on finding a functional failure. The assumption then is that if there are no functional failures then the safety function works. There are some advantages to this approach. One can be more precise in the testing. The tests can be designed to detect failures of any automatic diagnostics. And tests can be designed to reveal any weak or marginal parametric failures. One significant advantage that seems to be appearing is that the testing can be done with less time and effort if designed correctly.

Conclusions

If proof testing is designed NOT to see if a function is working but instead to see any component of a function has failed then it appears possible to create tests that are more effective and cost less to perform. In addition a side benefit is that more precise estimates of proof test effectiveness can be done allowing more conservative SIL verification calculations.

The FMEDA technique can be used for this purpose and several instrument manufacturers are beginning to provide more effective proof test recommendations.

References

1. Cheddie, H.L. and Goble, W.M., *Safety Instrumented System Verification – Practical Probabilistic Analysis*, Chapter 7, ISA, 2005.
2. Goble, W. M., Bukowski, J. V. and Brombacher. A. C., "How diagnostic coverage improves safety in programmable electronic systems," *ISA Transactions*, Vol. 36, No. 4, The Netherlands: Amsterdam, Elsevier Science B. V., 1998.

FIRED HEATER SAFEGUARDING SURVEY

Edward M. Marszal, Kenexis, Columbus, OH

Abstract

Refiners use a wide array of process heaters. The operation of these heaters poses a number of hazards that must be controlled, and are typically controlled using safety instrumented functions. The number and type of safety instrumented functions that are employed vary to a great degree from refiner to refiner and from site to site. This paper presents a summary of some of the results of a benchmarking study that was performed to determine the most prevalent practices in industry and establish the degree of adoption for some of the more common safety instrumented functions.

1.0 Introduction

Most refining organizations are very concerned with meeting the requirements of safety related standards and best practices in addition to their overall commitment to continuous improvement of safety and reliability. One common area for standardization among refining companies is the engineered safeguards that are implemented on fired heaters. There is a great benefit in standardizing the design of these systems because there are so prevalent. Like many refinery standard practices, standards developed to address fired heater safeguarding are often extensions of a related industry consensus standard – in this case American Petroleum Institute Recommended Practice 556 (API RP-556). API standards are extremely well adopted in the refining industry, and respected by both practitioners and regulators. Although the API RP-556 document is an excellent foundation, like most industry consensus standards it does not provide a completely definitive basis for design. API RP-556 contains many clauses where multiple options are available for meeting requirements and other clauses where the recommendations are not as strict as others (i.e., should instead of shall). Refiners typically address this limitation of consensus standards by issuing corporate standards that supplement other industry practices instead of developing a completely separate practice.

The author's organization performed a benchmarking study of the standard practices of refining organizations with respect to fired devices. The study was developed so that it would not only address common and well understood failure modes, but also address potential new approaches required as a result of new conditions and equipment, such as low NOx burners, and recent incidents and accidents related to fired heaters. Over the past several years, a number of accidents have occurred in the refining industry that were in whole or in part related to the following conditions.

- Improper use of bypass around heater fuel shut down valves
- Individual burner valves being left in the open position after occurrence of a shutdown resulting in inadvertent gas admission during subsequent light off
- Introduction of excess air into a fuel rich firebox subsequent to an operating anomaly

1.1 Benchmarking Survey Questions

The benchmarking was performed in accordance with a survey form that contained the information presented in *Table 1*.

Table 1 – Survey Questions

Which Automatic Trips are Implemented?
Fuel Gas High / Low Pressure Shutdown
Pilot Gas High / Low Pressure Shutdown
Loss of Flame on Main Burner - If yes, what method? (Flame rod, scanner, other)
Loss of Flame on Pilot Burner - If yes, what method? (Flame rod, scanner, other)
Low Process Flow (Pass Flow)
Low Excess Air / High Combustibles / Improper Fuel/Air Ratio - method for action?/detection?
Combustion Air Low Flow / Fan Not Running
Alarms
High Draft Pressure
High Bridgewall Temperature / Stack Temperature
High Process Outlet Temperature
Low Oxygen / High CO or High Combustibles
Permissives
Purge flow permissive – Automtic purge? – Material Used for Purging?
Fuel Gas Shutoff valve position permissive
Individual burner valves proved closed prior to light off (how?)
Pilot flame proven
Shutoff Valves
Double block and bleed on main / pilots
If a single valve, control valve or special purpose valve, what type?
Are bypasses used around shutoff/control valves for light-off or minimum firing
Pilots
Continuous pilots - Pilots sourced from a separate supply (i.e., natural gas)
Post Purge
Automatic Steam addition for purge upon initiation of a heater trip

The survey also developed information with regards to both new equipment projects and existing equipment installations, and the content of standards versus application of those standards in the field. Information regarding the content of corporate standards was requested in a yes or no question. Summary results for standards questions were then presented in terms of percentage of yes responses to the question. Information regarding the degree to which a practice has been implemented (both in new construction and existing equipment) was ranked from 1 to 3 where 1 corresponds to 'almost never' implemented, 2 corresponds to 'sometimes' implemented, and 3 corresponds to 'almost always' implemented. Summary results for implementation questions were then presented as the arithmetic mean (average) of all responses (one response per operating company).

While the form was oriented toward natural draft heaters, several of the questions were related to forced draft furnaces, specifically the questions related to combustion air flow shutdowns. The response from the survey indicated that there are some significant differences in how forced draft heaters are safeguarded as opposed to natural draft heaters. In some cases the author made revisions to survey forms returned by respondents, based on their comments, to ensure the focus of the study was natural draft heaters.

It is important to note that the commentary to the surveys is as important as the statistical data that was obtained. The rationale for why some companies chose certain practices over others gives a deeper understanding of how to obtain overall functional safety. Some statistical figures may be deceiving at first glance and should not used outside of the context of a full survey including all participant commentary. For instance, one respondent organization does not have fired heater standards at the corporate level, leaving all individual sites to set their own practices. As a result, a reported figure of less than 100% of companies require a certain safeguard as a standard practice, may actually mean that 100% of the companies that <u>have</u> standard require the safeguards – the remainder do not have standards, and thus would not require the safeguard as a standard practice. Furthermore, some hazards can be addressed in multiple ways. For instance, loss of combustion air flow can be detected using an air flow measurement, fan motor status, or fan shaft speed. If the survey reports that only fraction of respondents have a low combustion air flow shutdown that does not mean that the balance of respondents do not care about loss of combustion air, it means that other ways of detecting loss of combustion air are utilized at their sites.

1.2 Survey Participants

In order to understand the survey and put it in proper context, it is important to understand the demographics of the surveyed companies. Several operating companies responded to the survey (in some cases, multiple responses per company were received – in this case the responses were averaged for that company). All of these companies operate at least one refinery in the United States and are in the business as either pure-play refiners, or integrated oil and gas production, refining, and petrochemical companies. The survey team selected companies that have large market capitalizations, operate multiple refineries and have a large crude oil processing capacity. Significantly smaller companies with very low oil processing capabilities were not included in the study due to the potential that their practices may not reflect the best practices in industry.

It should be noted that the results presented in this survey are based on a somewhat limited data set. Although the survey includes companies that represent the preponderance of crude oil processing in the United States, typically only one or two respondents per operating company completed survey forms where the organization might have more than a dozen refineries. Furthermore, many refining organizations have recently been built up through a series of mergers and acquisitions, leading to a wide variety of safeguarding practices within a single operating company. While most refiners are working diligently to standardize their operations, that task is not complete yet. While the author has made great efforts to provide a comprehensive analysis of operating company practices, data from every site in every organization could not be obtained. For operating companies where the author knew that practices varied to a great degree from site to site, we made an extra effort to survey multiple sites to get an average response.

3.0 Survey Results Summary

A summary of some of the results of the survey is shown below. The information presented below is simply a statistical compilation of the responses to the survey forms and notes containing general commentary on the results and the comments that the respondents provided in their survey forms. While the original survey analyzed a number of design scenarios, the summary results presented here only represent the application of these design features on new equipment.

3.1 Shutdowns

Fuel Gas Low Pressure Shutdown

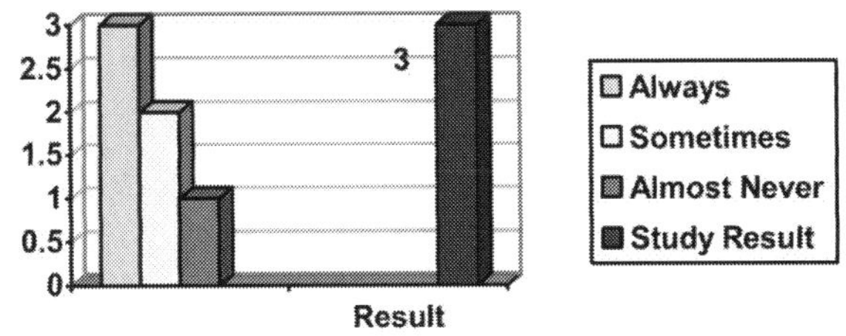

The low fuel gas pressure is the most common shutdown employed on fired heaters. All companies 'almost always' implement this shutdown on their new and existing equipment. Of the surveyed companies that have corporate standards, only one does not require that low pressure trip be implemented, but even that standard requires that the shutdown must be 'considered' based on a risk assessment and further requires that the low pressure trip shall be implemented unless continuously operated separately fuel sourced pilots are utilized on the heater.

Fuel Gas High Pressure Shutdown

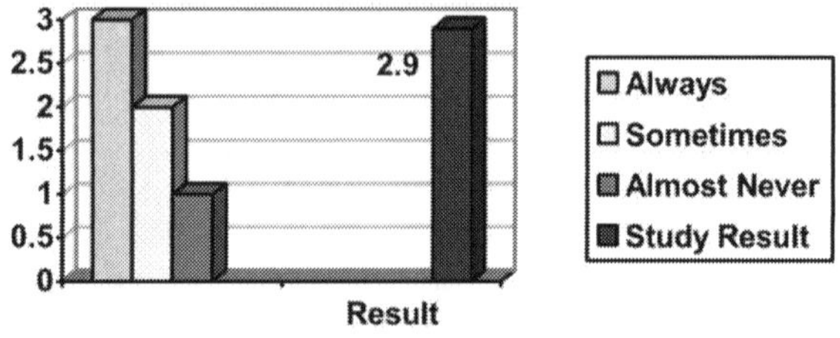

Of the operating companies that have fired heater safeguarding standards, all require implementation of a high pressure trip in their most current form. The high level of adoption of the high pressure trip is based on the knowledge that hazards resulting from high pressure fuel gas situations can not (in all cases) be safeguarded by other means such as flame scanners and continuously operating pilots. Even though standard practice dictates that this shutdown be implemented, existing equipment and even in some cases new equipment may not employ this shutdown. One respondent noted that even though their standard requires the high pressure shutdown, some projects may choose not to implement the trip. If this decision is made it is based on knowledge (and testing) of the burners to determine their stable operating range and verification that the fuel supplied to the burner can not reasonably exceed that pressure based on the design of the fuel supply system.

Pilot Gas High/Low Pressure Shutdown

High Pressure

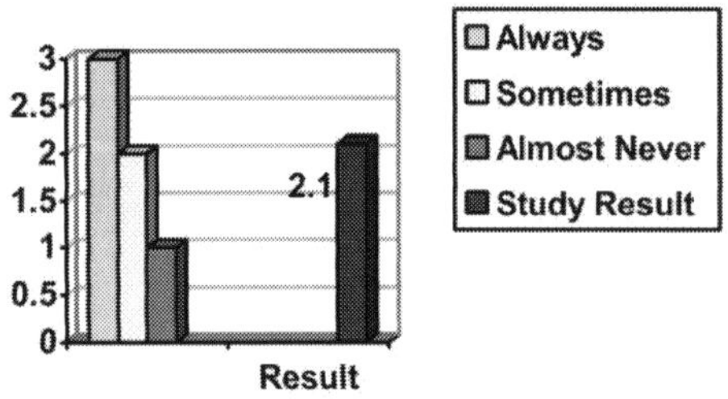

Low Pressure

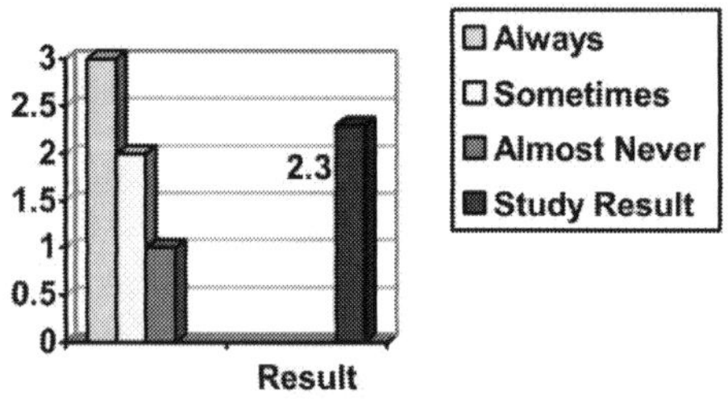

The figures reported for adoption of high pilot gas pressure shutdowns are difficult to interpret without considering the context in which they were generated and the respondent notes regarding how pilots are utilized at the various operating company sites. While the figures show that the preponderance of operating companies do not require shutdown due to high pilot gas pressure, that should not be taken to imply that high pilot gas pressure has been deemed to be an insignificant risk. Not all companies employ continuously operating pilots, and as such their standards would not require a high pressure pilot gas shutdown, not because the risk is insignificant, but because the pilots are not typically in operation. Of the operating companies that elect not to employ a high pilot gas pressure shutdown, their rationale is that the risk is tolerably low. A high pressure event in the pilot during normal operation would typically be mitigated by action of the main flame, leaving only startup and pilot-only operational modes as situations where the plant is at risk. In addition, many deem the likelihood of generating a pressure high enough to extinguish the pilot flame as very unlikely.

Many of the same considerations in analyzing the response to the question of high pilot gas pressure shutdowns apply to low pilot gas pressure shutdowns. Low pilot gas pressure shutdowns are more common for users of continuous pilots than high pressure shutdowns due to the increased likelihood that a low pressure pilot gas condition can occur and generate a hazard.

Loss of Flame on Main/Pilot Burner Shutdown

Main

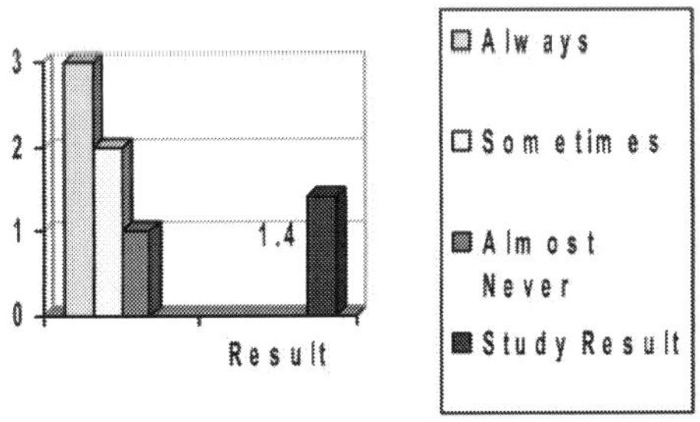

Pilot

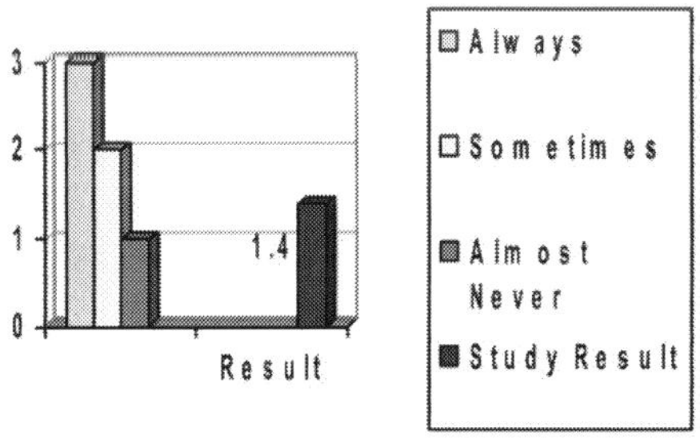

Shutdown on loss of flame at the main burner (as detected by flame scanners, flame rods, pattern recognition cameras, or other device) is not a very common shutdown, and many refiners feel that the hazards that would be prevented utilizing this function are adequately safeguarded using other means. It was noted by several participants that a loss of flame on the main burners' shutdown is not required to be measured or cause a shutdown if continuously operated pilots that are sourced with a supply of fuel that is independent of the main burners.

It should be noted that the figures presented above represent natural draft process heaters. Use of flame scanners is much more prevalent on induced and forced draft process heaters, and almost always done on boilers (true purpose-built boilers, as opposed to process heaters where steam generation is incident to the process purpose of the heater).

Shutdowns based on flame detection on pilots are also not extremely common. Although not common, one operating company employs pilot burner loss of flame shutdowns as a standard practice and almost always implements this shutdown in practice. The rationale of this operating company is that if the pilot burner is in operation, then a lot of the hazards that might occur as a result of main burner fuel system anomalies will be effectively mitigated by the pilot. As a result, they place a high degree of importance on ensuring that the pilot flame is always available to perform its safety purpose.

Low Process Flow (Pass Flow) Shutdown

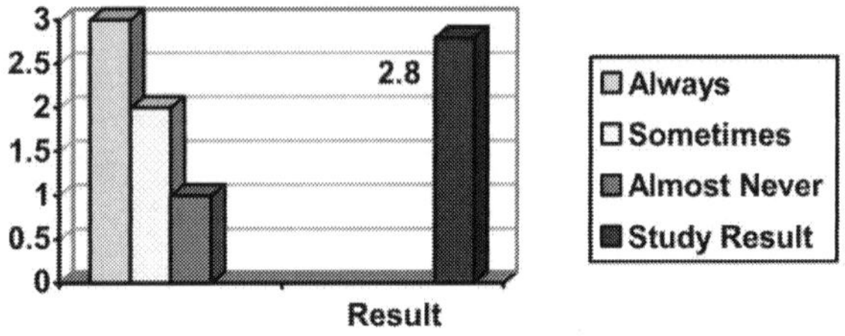

Heater shutdown based on low flow of process material through heater passes is very common. For those respondents whose standard practice does not require this trip, analysis is required to ensure that loss of flow through the heater tubes does not pose a significant hazard and are still recommended in coking services. Even for those whose standards require the shutdown, in some cases exceptions are allowed if an analysis of the type of heater and the process fluid is performed to determine the detrimental effects of loss of flow is performed that shows the risk posed by loss of flow is tolerable.

Low Excess Air / High Combustibles / Improper Fuel/Air Ratio Shutdown

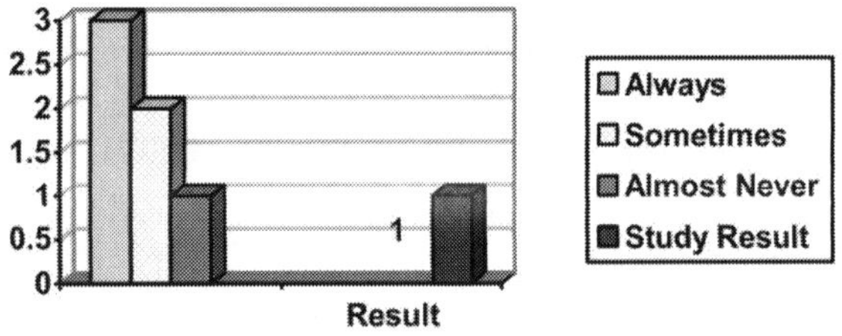

None of the operating company respondents employ any automatic actions to mitigate the hazards posed misoperation of a fired heater resulting in an oxygen deficient fuel rich atmosphere in the heater firebox. The basis of safety for preventing realization of fuel rich firebox hazards is through good control of furnace draft, in some cases employing closed loop control based on oxygen analyzer measurement, and thorough operating training to ensure that if an oxygen deficient firebox condition were to occur that the proper control action of cutting fuel gas, as opposed to adding air, is used.

Combustion Air Low Flow / Fan Not Running Shutdown

Low Air Flow

Fan Not Running

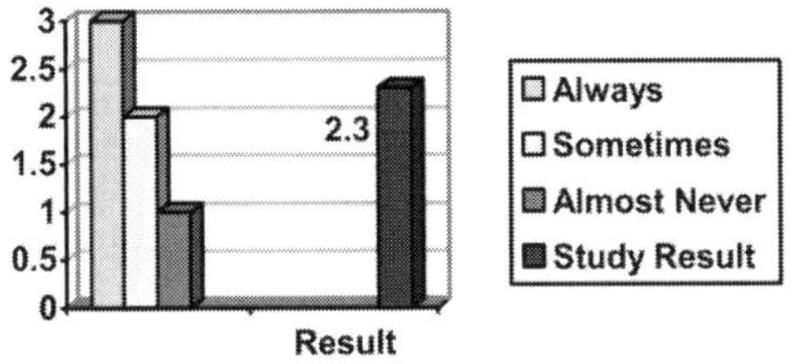

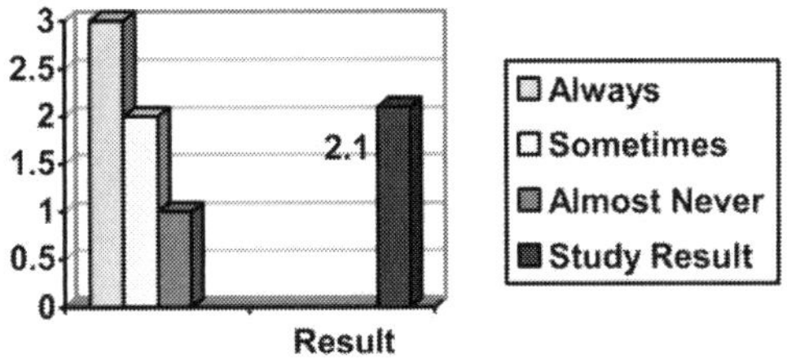

It was noted that in almost all cases, if a heater is equipped with a forced or induced draft air supply system, some sort of shutdown was employed to detect a loss of air condition. While the reported numbers for standard practice and implementation of either low combustion air flow trip or air fan not running trip were near 50%, these figures are not indicative of their importance to process safety, but to the preference of what type of measurement is used for causing the shutdown. The two primary means of measurement were direct measurement of combustion air flow or an auxiliary contact on the combustion air fan motor starter. For new installations there is a strong preference to move toward direct measurement of combustion air flow due to the more comprehensive and direct indication of a hazardous condition. In some cases, both fan not running and low air flow were used as measurements that can result in a heater shutdown. In the case of one of the respondent operating companies, their preference for detecting insufficient combustion air was the measurement of the speed to the fan motors.

3.2 Alarms

Alarms are very consistently applied across all operating company respondents, and very frequently applied. The only alarm where application was more represented by 'sometimes' than 'almost always' is the High CO or High Combustibles alarm. It was noted that application of this alarm is becoming more prevalent as regulatory requirements increase.

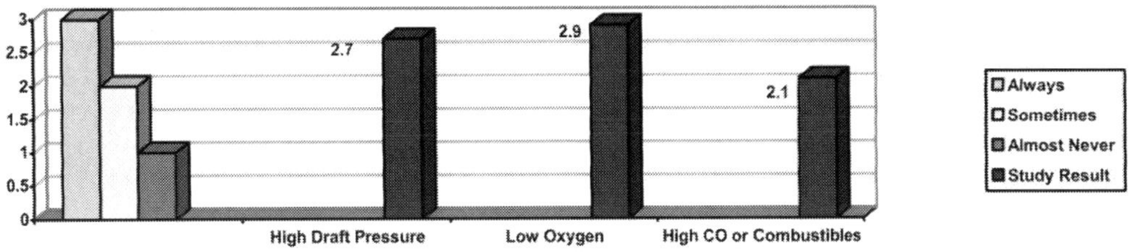

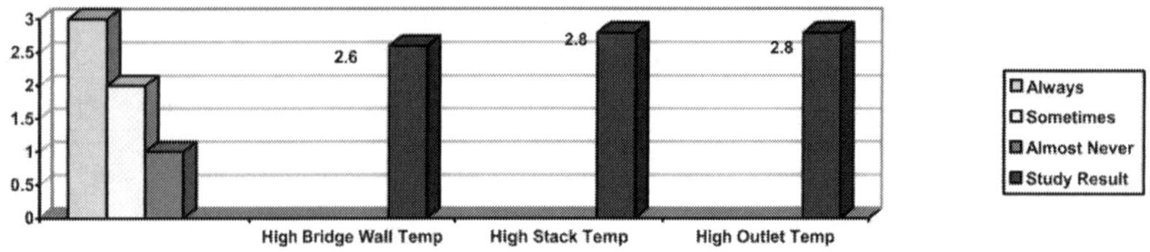

3.3 Permissives

When reviewing the permissives section it is important to note that the intention of the survey was to primarily review the operation of natural draft heaters, as opposed to forced or induced draft heaters. It was noted that the degree to which permissive functions are implemented is much greater for heaters with forced or induced draft fans and also for boilers.

Purge Flow Permissive

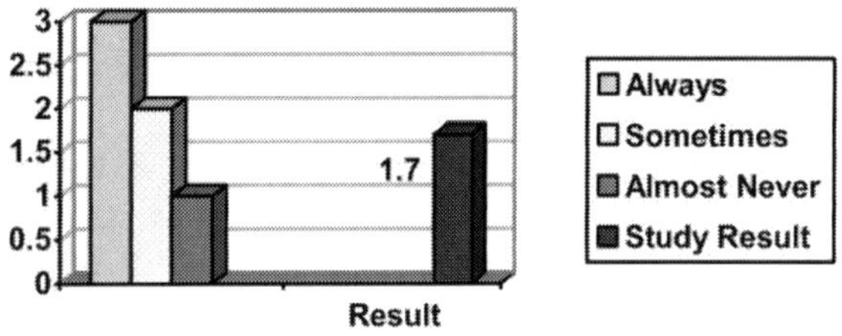

Operating company respondents that employ purge flow permissives were in the minority, but application was significant enough to warrant consideration. It is important to draw a distinction between this function and a subsequent question regarding automatic firebox purge. If automatic firebox purge is employed, then a purge flow timing permissive is virtually always used, but automatic purging is not required for a purge flow timer permissive. Some operating companies employ this permissive by measuring the heater draft or purge steam flow and ensuring that the measured variable is of a sufficient quality for the preset time interval.

Fuel Gas Shutoff Valve Position Permissive

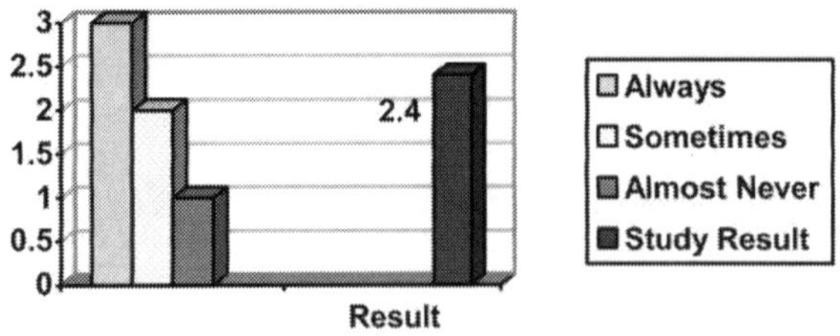

The fuel gas shutoff valve permissive is quite common in practice while not as often directly required in corporate standards. Where this shutoff is not employed in an automatic fashion it is always performed through operator procedure. When the permissive is employed in an automatic fashion it is typically performed using limit switches. No use of valve positioner signals for this purpose was noted.

Individual Burner Valves Closed (Seal Test) Permissive

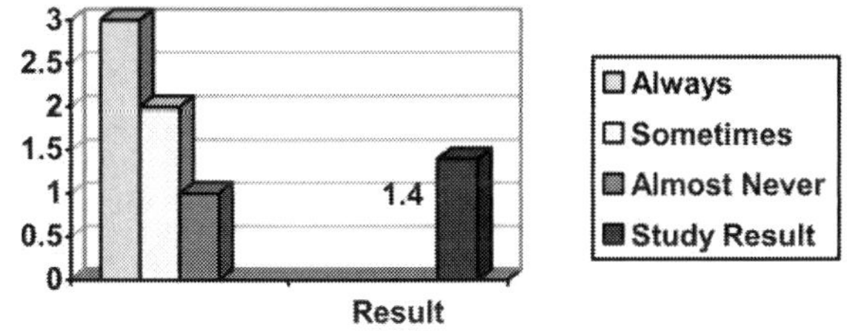

Closure of individual burner valves is always ensured prior to light off of the heater, but it is rarely performed automatically, unless the heater is of a certain type that has a very large number of burners, such as a steam reformer or an ethylene cracker. Only one respondent has historically been doing this type of permissive on a regular basis and only two respondents employ this type of permissive in the standards for new equipment (on all heaters). Of these companies, neither requires a 'seal test' type permissive on all heaters – and typically does not use the seal test methodology unless there is an extremely large number of burners (e.g., more than 100).

Automatic Firebox Purge

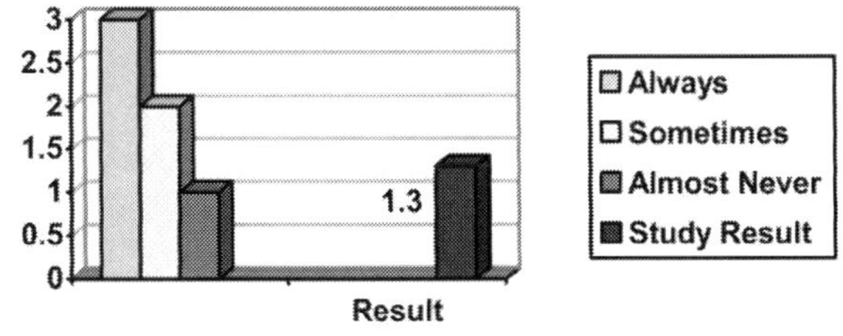

The results presented above show that implementation of automatic purges on natural draft heaters is very rare; in fact, only one site of one respondent organization is known to regularly utilize this technique.. Only one respondent incorporates this as standard practice, and most of their existing heaters do not employ the permissive. Where forced draft heaters are used, the automation of purging is more frequent, but still would only be characterized as "sometimes" for existing equipment. More new installations of forced draft heaters implement automatic purging of the firebox utilizing the forced/induced draft fans, but not enough to categorize the implementation as "almost always".

The medium used for purging is extremely consistent with the type of heater being used. For natural draft heaters, steam was virtually always used. For forced/induced draft heaters, combustion air was almost always used. No significant usage of any other purging medium (i.e., nitrogen) was identified.

Flame Proven Permissive

Pilot

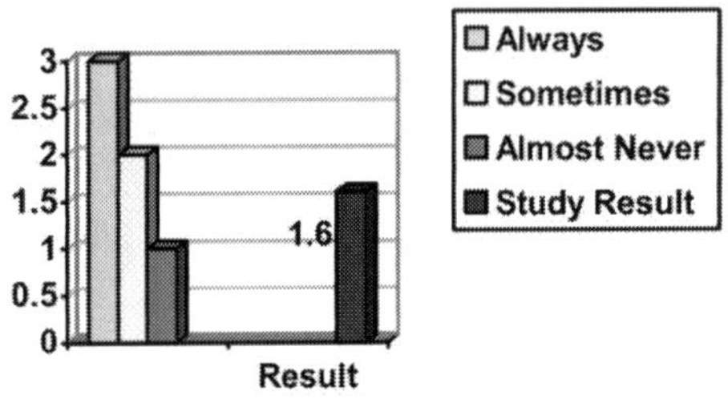

Main

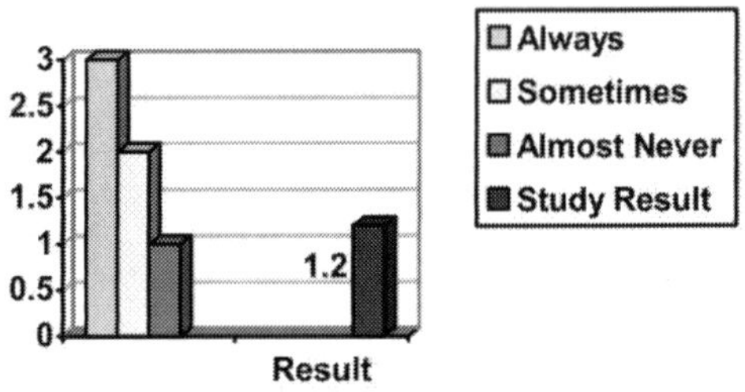

The survey indicated that flame proven permissives are not common. Flame proven permissives are typically performed automatically when there is a loss of pilot flame trip, but not performed if there is no associated trip.

3.4 Shutoff Valves

Double Block and Bleed Shutoff Valves

Main

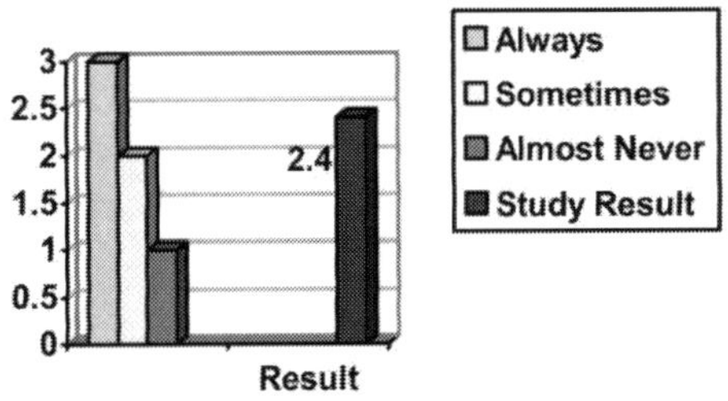

Pilot

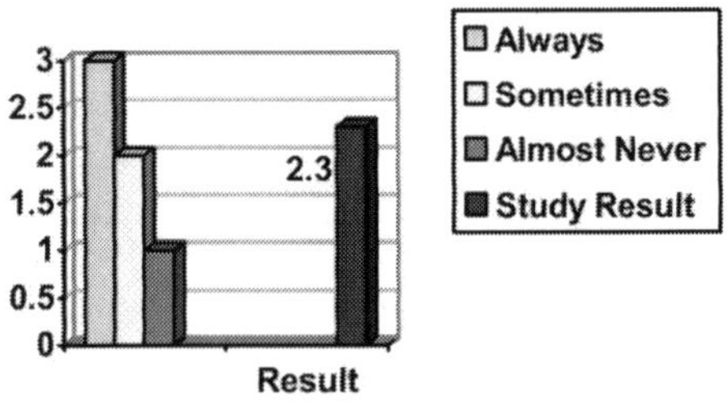

Use of double-block-and-bleed arrangements on heater fuel gas supplies is not uncommon now and is becoming even more common. Most of the respondents are requiring double block and bleed as standard practice, and many require in their standard that double block and bleed be considered during the SIL analysis process. In actual practice, use of double-block-and-bleeds is also prevalent.

Use of a double-block-and-bleed arrangement provides two benefits over a single valve. First, a positive isolation is ensured, preventing hazards resulting from fuel gas leaking across a single valve in to the firebox when the valve is closed. Second, the use of two devices to perform the shutdown (in a 1oo2 vote) decreases the probability of failure on demand and increased the safety integrity level (SIL) that can be achieved by the functions. These two goals can and are being achieved by measures other than use of double-block-and-bleeds by a number of respondent companies.

For positive isolation, some refiners isolate the fuel when not in use as a standard practice by blinding and rolling a spool piece in the gas supply. The main fuel source is typically not reconnected to the main burners until after the pilot burners have been lit, virtually eliminating the risk of leaking main shutdown valves and open individual burner valves for the case of a cold start. One refiner noted a move to quick disconnects to facilitate the practice of isolation.

To decrease the probability of failure on demand of gas shutoff some refiners are electing to use increased diagnostics and more frequent testing in lieu of increasing the number of valves. Where required to meet elevated SIL targets, partial stroke testing is sometimes utilized. Partial stroke testing is performed commonly either using manual jamming devices or with purpose built automatic testing devices – no strong preference between the testing types emerged in the survey.

In some cases, the use of double block and bleed arrangements of pilot gas systems was noted to be a matter of consistency between shutoff practices between the main burners and the pilot burners, and not for risk reduction in and of itself.

Bypasses Valves around Shutoff Valves (for Light-Off or Minimum Firing)

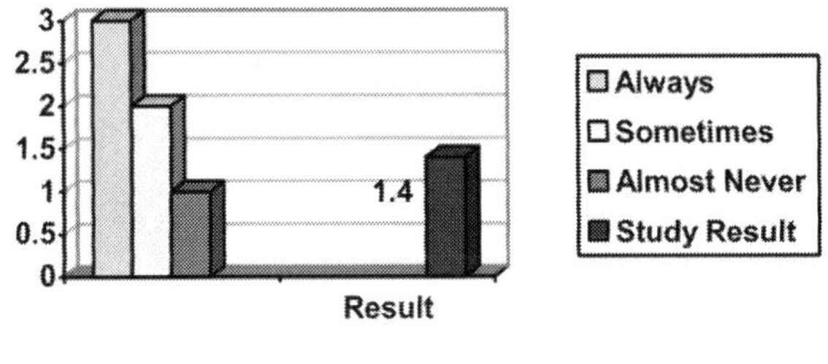

Bypasses around shutoff valves are not very frequently used and are falling out of favor in situations where they were traditional used. One respondent noted that the only instance where bypasses are allowed are for 'seal testing' and in this case the bypass was a manual 'dead man' valve and the bypass itself was only 1/4" tubing. The preferred design is moving toward either ensuring that SIL targets can be met with equipment that only needs to be tested at a turnaround interval (i.e., double block and bleed) or use of partial stroke testing.

3.5 Pilots

Continuously Operated Pilots

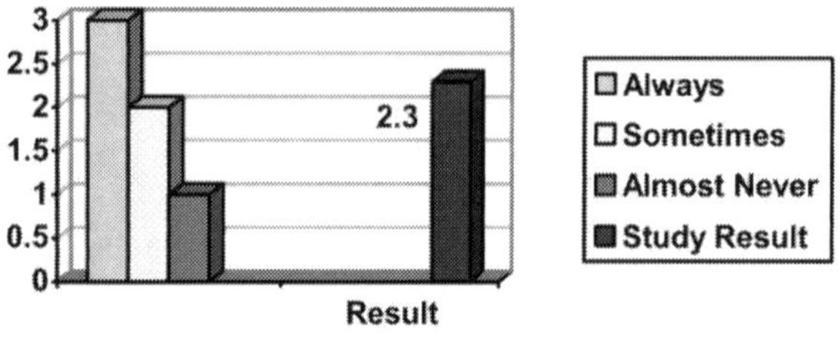

While the use of continuously operated pilots is not required by the preponderance of operating companies, actual use of continuously operating pilots is very common. Many of the standards that do not strictly require that this practice be implemented do require that it be considered. These standards require that if continuous pilots are not used that other means of enhancing safeguarding be employed, such as loss of flame detection.

Some operating companies tend to have very strong opinions about the use of pilots that permeate through their organizations. One operating company relies on continuous pilots as a primary means of safeguarding, including heater shutdown if loss of pilot flame is detected. Two operating companies that responded to the survey do not employ continuous pilots as a standard practice. One of these companies uses pilot burners to light the main burners, but then takes them out of service. Another company does not install pilot burners at all; instead, they light the main burners directly. This company stated that they find pilot burners to be very unreliable and hard to maintain due to the increased propensity toward clogging due to the smaller orifice size and lower flow rates. Their preferred means of ensuring safe operation is through strict control of burner pressure and well established testing and knowledge of the safe operating limits of the burner system.

Pilots are Source from a Separate Supply (i.e., Natural Gas)

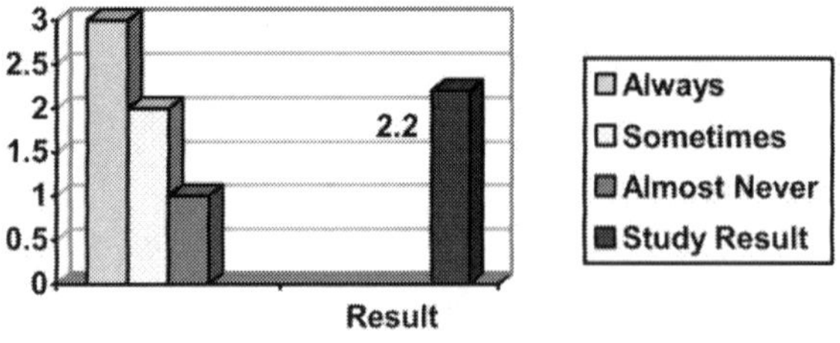

Use of a separate source of pilot gas is somewhat common where practicable, but not to the degree where it can be considered to be 'almost always'

employed. It should be noted that the figures shown here are somewhat skewed because those respondents who do not use pilots at all are also indicated as not using a separate supply. This means that use of a separate supply of pilot gas for those respondents who use pilots is actually somewhat higher than what is shown.

3.6 Post Purge

Automatic Steam Addition upon Natural Draft Heater Shutdown

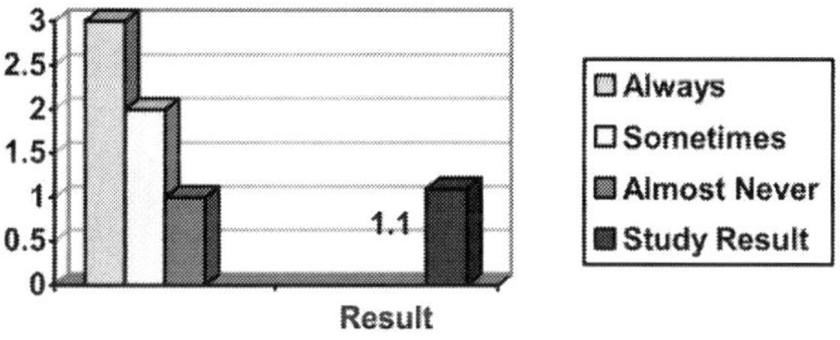

The surveys indicate that this action is virtually never done automatically. Only one respondent noted that this is done automatically in some situations, and another respondent noted that this practice is employed regularly at one of their sites, but never at the others. In fact, some respondents noted that this practice is intentionally avoided due to the potential that spurious operation of the function or use of the function in non-optimal conditions may actually cause damage to the heater.

4.0 Conclusions

Benchmarking is an effective tool in assisting refiners to determine how to implement portions of standard practices where the requirements are ambiguous, open to interpretation, or multiple options are available. In many cases the practices that are implemented are remarkably consistent and form the basis for "recognized and generally accepted good engineering practice".

216a A Self-Tuning Regulator for Fluid Catalytic Cracking Units

Felix Severino Farias Junior and Rubens Maciel Filho. Laboratory of Optimization, Design and Advanced Control (LOPCA), UNICAMP, Faculty of Chemical Engineering. State University of Campinas (UNICAMP), P.O.Box 6066, CEP 13081-970, Campinas, SP, Brazil

The fluid catalytic cracking converter is considered one of the most important units in petroleum refinaries since it converts heavy residues of oil, without commercial value, into valuable products. Because of its large thoughtput and nonlinear behaviour, improvements or better control strategies and operation of the process can generate expressive improvements in profits. Because of its complex multivariable, nonlinear and nonstationary behaviour, the FCC process control is a challenging problem in chemical engineering. Several linear and nonlinear advanced control approachs were reported in literature for this case: DMC, GMC, Neural Networks, among others. But one common problem with these designs is the limited range of process operating conditions in which tunning update is not necessary. So, they require periodic online tuning to accompaing process operation changes in real plant, which is not desirable or sometimes possible in real time plant operation. Adaptative controllers are know to be more robust to process operating conditions changes because of their intrinsical self-regulating nature. In this work, an adaptive control of adjustable auto regulating type (Self-tuning Regulator) for a FCC unit is proposed. This controller uses a recursive estimator to achieve the process parameters and these are then used for calculation of the desired control action through the controller design module. The STR controller is composed by two parts: an external block, formed by a conventional controller directly connected to the real process, and an internal one, that takes into account the changes in the process and provides the new set of controller parameters to the conventional controller. This internal block is formed by two parts: an estimator of parameters/states and another one that calculates the best parameters for the internal controller, the controller designer. The developed adaptive controller performance is compared with the DMC controller one. The robustness of each approach is compared thought theirs responses behaviour under varying FCC unit operating conditions.

ROBUST DYNAMIC PRINCIPAL COMPONENT ANALYSIS FOR PROCESS PERFORMANCE MONITORING

Daniel A. Castro-Rodriguez, Elaine B. Martin and Gary A. Montague

School of Chemical Engineering and Advanced Materials
University of Newcastle, Newcastle upon Tyne, NE1 7RU, England

Abstract

Multivariate statistics are being increasingly used for batch and continuous process monitoring and fault detection. Of the variety of methods available, Principal Component Analysis (PCA) has been one of the most widely employed. Different approaches have been formulated to take account of the dynamics of the process being monitored and the presence of outliers and/or missing data in the information gathered. However, outlier detection approaches have failed to address the situation when outliers exist in dynamic data. The present work proposes a robust dynamic PCA approach to address this problem. The technique is tested with data from a simulation of a polyvinyl acetate CSTR operating under normal and faulty conditions. Results are compared with those obtained using PCA, Dynamic PCA and Robust PCA. It is found that the proposed technique is more sensitive for fault detection than the other approaches considered when using data from the simulated process.

Introduction

Applications of multivariate statistical process control methodologies to linear, non-linear, steady-state and dynamic, batch and continuous systems have been widely reported in the literature. As a consequence of the on-line, real time implementation of these technologies, it is possible to detect changes in process behavior that may otherwise not have been identified if univariate statistical process control methodologies had been applied. However, for dynamic systems that possess a large number of highly correlated variables, the univariate methods of CUSUM and EWMA and their multivariate counterparts, MCUSUM and MEWMA are not ideally suited. This limitation has resulted in the development of methodologies formulated around the statistical projection techniques of principal component analysis (PCA) and projection to latent structures (PLS). Consequently, the most commonly implemented monitoring metrics are those of Hotelling's T^2 and the squared prediction error (SPE).

More recently, to address the dynamic nature of processes, Ku et al. [5] proposed dynamic principal component analysis (DPCA). A data matrix is formulated, similar to that used in time series modeling and PCA is then applied to the augmented matrix. However, a fundamental issue is the sensitivity of the projection based methods to the presence of outliers. A consequence of this additional complication is that a number of robust principal component analysis approaches have been proposed including a methodology proposed by Hubert et al. [3] based on projection pursuit and the minimum covariance determinant.

In this paper, robust and dynamic principal component analysis are combined thereby drawing together the dynamic nature of the process and recognising the potential presence of outliers. The approach is tested using a simulation of a polyvinyl acetate CSTR where known faults are introduced. In the case of the polyvinyl acetate CSTR, the faults pertain to a random perturbation on the controller manipulated variable simulating the impact of a faulty transducer. In addition, step changes of 10% in the flow rates of the feed, initiator and temperatures of feed and coolant are considered. This fault results in an undesirable change in the molecular weight of the product. Following these investigations, more severe outliers in the data are simulated and the performance of the fault detection strategies reassessed. The variables measured are the flow rates of feed and initiator, the temperatures of the reactor, the feed and cooling water and the output of the controller. All these variables are measurable with instrumentation present in any plant. The two metrics, Hotelling's T^2 and the SPE obtained from the application of PCA, robust PCA, dynamic PCA and the proposed robust dynamic DPCA are used to determine the presence of faults. The premise of the paper is that robust dynamic DPCA is more sensitive for the detection of abnormal operation than previous approaches that have been adopted.

Principal Component Analysis

One of the most widely used classical multivariate statistical techniques is Principal Component Analysis. Proposed originally by Pearson in 1901 [7] its objective is, given a set of correlated variables, to find a new set of "variables" with correlation minimized and variance maximized. This is performed by solving a dual objective optimization problem using the Lagrange multipliers procedure. The Lagrange multipliers that solve the problem of finding the scores with minimized correlation and at the same time maximized variance are the eigenvalues of the correlation matrix [8]. The new set of variables obtained is termed the scores. The scores are linear combinations of the original variables, with the weighting matrix given by the eigenvectors of the correlation matrix (termed the "loadings"). This leads to:

$$\mathbf{T} = \mathbf{XP} \qquad \text{Equation 1}$$

where **T** is the matrix of scores, **X** is usually the standardized original data matrix and **P** is the loadings matrix. In most cases the data is standardized, by centering with respect to its mean and dividing it by its standard deviation. This is performed to avoid variables with large absolute values masking the effect of variables with smaller absolute values. As many scores as original variables can be found. However, the scores associated with the smaller eigenvalues of the correlation matrix usually provide little information about the data and can be discarded. Many criteria can be used to select the number of scores to be calculated. Among the criteria are PRESS, Cross Validation Index, Percent variance test and Scree test. [1,2]. Cross validation was the chosen method in the present work as it was found to provide the most reliable results.

Estimates of the data can be obtained from the scores using:

$$\hat{\mathbf{X}} = \mathbf{TP}^T \qquad \text{Equation 2}$$

with $\hat{\mathbf{X}}$ being the estimates. Using these estimates is possible to obtain E, the residuals [9], according to:

$$\mathbf{E} = \mathbf{X} - \hat{\mathbf{X}} \qquad \text{Equation 3}$$

Scores and residuals are used to monitor the process as will be explained later. Scores are used to explain the variation within the PCA model whereas the residuals can indicate a change in the correlation structure of the process variables [12].

Fault detection using principal component analysis

T^2 index

One of the indices used to detect the presence of a fault is the T^2 index. With the scores obtained using Equation 1, it is possible to calculate and index termed T^2 [4] according to:

$$T^2 = \sum_{i=1}^{a} \frac{t_i^2}{s_{t_i}^2} \qquad \text{Equation 4}$$

where *a* refers to the number of retained scores. The threshold for fault detection [2] is given by:

$$T_\alpha^2 = \frac{(n-1)^2 (a/(n-a-1)) F_\alpha(a, n-a-1)}{n(1+(a/(n-a-1))) F_\alpha(a, n-a-1)} \qquad \text{Equation 5}$$

where *n* the number of observations and α an appropriate level of significance for performing the test. The variation within the PCA model is measured by the T^2 index. A violation of the T^2 limit can indicate the presence of a fault, but sometimes can simply mean that a change in the operation region has occurred [12].

SPE Index

Another index used to detect the presence of faults in a process is the Q index, also termed SPE. The residuals given by Equation 3 can be re-written as:

$$\mathbf{r} = (\mathbf{I} - \mathbf{PP}^T)\mathbf{x} \qquad \text{Equation 6}$$

The residuals obtained are used to calculate the Q index with:

$$Q = \mathbf{r}^T \mathbf{r} \qquad \text{Equation 7}$$

The fault detection threshold for the Q index [2] is given by:

$$Q_\alpha = \theta_1 \left[\frac{h_0 c_\alpha \sqrt{2\theta_2}}{\theta_1} + 1 + \frac{\theta_2 h_0 (h_0 - 1)}{\theta_1^2} \right]^{1/h_0} \qquad \text{Equation 8}$$

where $\theta_i = \sum_{j=a+1}^{n} \sigma_j^{2i}$, $h_0 = 1 - \frac{2\theta_1\theta_3}{3\theta_2^2}$ and c_α is the normal deviate corresponding to the (1-α) percentile for a level of significance of α.

The Q index is a measure of the variation not explained by the PCA model. This means, when the Q index is larger than the limits, it can be assumed that the covariance structure is no longer the same as that of the model and the relationships between the variables has changed.

Dynamic Principal Component Analysis

Ku et al. [5] proposed the use of PCA to analyze a data matrix similar to the matrix used to build an ARMAX model. The data matrix is then given by:

$$\mathbf{X} = [\mathbf{X}(k) \quad \mathbf{X}(k-1) \quad \mathbf{X}(k-l)] = \begin{bmatrix} \mathbf{x}^T(1) & \mathbf{x}^T(0) & \dots & \mathbf{x}^T(1-l) \\ \mathbf{x}^T(2) & \mathbf{x}^T(1) & \dots & \mathbf{x}^T(2-l) \\ \vdots & \vdots & \ddots & \vdots \\ \mathbf{x}^T(m) & \mathbf{x}^T(m-1) & \cdots & \mathbf{x}^T(m-l) \end{bmatrix} \qquad \text{Equation 9}$$

In this matrix delayed values of the variables are included and consequently the dynamics of the process are taken into account and the presence of autocorrelation is addressed. This strategy is proposed as applying PCA to the data matrix without adding the delayed variables will only provide a static approximation therefore failing to reveal true relationships between the variables. As a consequence, changes might not be detected or false alarms may arise. The lag l is determined by an iterative procedure to obtain the number of linearly independent scores. Once the matrix to be analyzed is obtained, scores and residuals are obtained. T^2 and SPE indices are used to determine the presence of a fault, in the same manner as when PCA is used to analyze the static data matrix. This technique has been found to be effective for modeling and process monitoring [6]. However, it is not resistant to outliers, the presence of which can lead to false alarms or the failure to detect abnormal process conditions.

Robust Principal Component Analysis

Classic PCA is very sensitive to the presence of outliers and/or missing data due to its dependence on the covariance/correlation matrix. Several approaches have been proposed to avoid these problems. Some are based on the use of robust estimates of the covariance or correlation matrix. A disadvantage of these methods is that they can be applied only when the number of observations is more than the number of variables [11]. Others make use of projection pursuit to obtain the principal components [10, 11]. Among the projection pursuit based techniques are the methods of Croix and Ruiz-Gazen and RAPCA, a modification proposed by Hubert et al [10, 11]. A disadvantage of these methods is that they can be slow [11]. In the present work, an algorithm termed ROBPCA proposed by Hubert and Engelen [3] was chosen which combines projection pursuit ideas with the use of the Minimum Covariance Determinant (MCD) estimator. As a consequence, this algorithm has interesting advantages; it is fast, allows the use of either low or high dimensional data and also yields accurate estimates when applied to data sets either containing outliers or not.

The robust principal component analysis approach is based on measuring the level to which an observation is categorized as an outlier. The degree by which an outlier is present is quantified according to:

$$outl(\mathbf{x}_i) = \frac{|\mathbf{x}_i'\mathbf{v} - m_{MCD}(\mathbf{x}_i'\mathbf{v})|}{s_{MCD}} \qquad \text{Equation 10}$$

where $\mathbf{x}_i'\mathbf{v}$ is the projection of an observation onto the univariate space defined through the minimum covariance determinant, m_{MCD} is the univariate minimum covariance determinant estimators of location and s_{MCD} is the corresponding scale estimator. The observations with the smallest value of $outl(\mathbf{x}_i)$ are retained and the resulting covariance matrix is used to calculate the loadings. The scores and estimates of the retained observations are then attained from the loadings calculated previously.

Robust Dynamic Principal Component Analysis

To take into account the process dynamics and at the same time deal with the presence of outliers and/or missing data, a robust dynamic principal component analysis technique is proposed. This technique is based on dynamic PCA proposed by Ku et al [5] and the ROBPCA Robust PCA approach proposed by Hubert et al [3]. In the present work the delay l was chosen not by the iterative procedure followed by Ku et al but by using an information criterion proposed by Akaike [16]. The number of scores to be retained is chosen using cross validation.

Application to a Poly vinyl acetate CSTR simulation

In the present work a simulation of a continuous solution polymerisation is used to generate the process data. DeCicco wrote the Matlab and Simulink routines used to simulate the process from a detailed mathematical model developed by Teymour and Ray [13,14] of a laboratory scale polymerisation reactor. The polymerisation occurs in a continuously stirred tank reactor. The simulation considered a solution polymerisation to produce polyvinyl acetate using vinyl acetate monomer of 99% purity, t-butanol as solvent and AIBN [2,2'-azobis(2-methyl-propionitrile)] as initiator. Before entering the reactor the monomer feed is cooled and the t-butanol is heated so their temperature is at all times above 25° C, which is the t-butanol freezing point. Then, the temperature-conditioned streams are pumped into the reactor where the reaction occurs in a nitrogen atmosphere. The temperature of the reactor is regulated by means of a blower that circulates air with the purpose of removing any excess of heat. The air temperature is controlled by a heater that is turned off or on depending on the need to remove or provide heat to the reactor. In the simulation the control scheme used is a single loop PI control to manipulate temperature of the input to the cooling jacket instead of the on off scheme used by Teymour and [13,14]. The controller parameters where calculated using the Ziegler – Nichol's open loop tuning procedure [15].

The faults considered consist of a random noise disturbance in the measured position of the manipulated variable. This simulates the effect of a faulty transducer producing a noisier signal than normal. In addition, perturbations can be introduced in the process, to simulate step changes of in the flow rates of the feed, initiator and temperatures of feed and coolant. In this paper disturbances are limited to be 10% changes in the feed temperature.

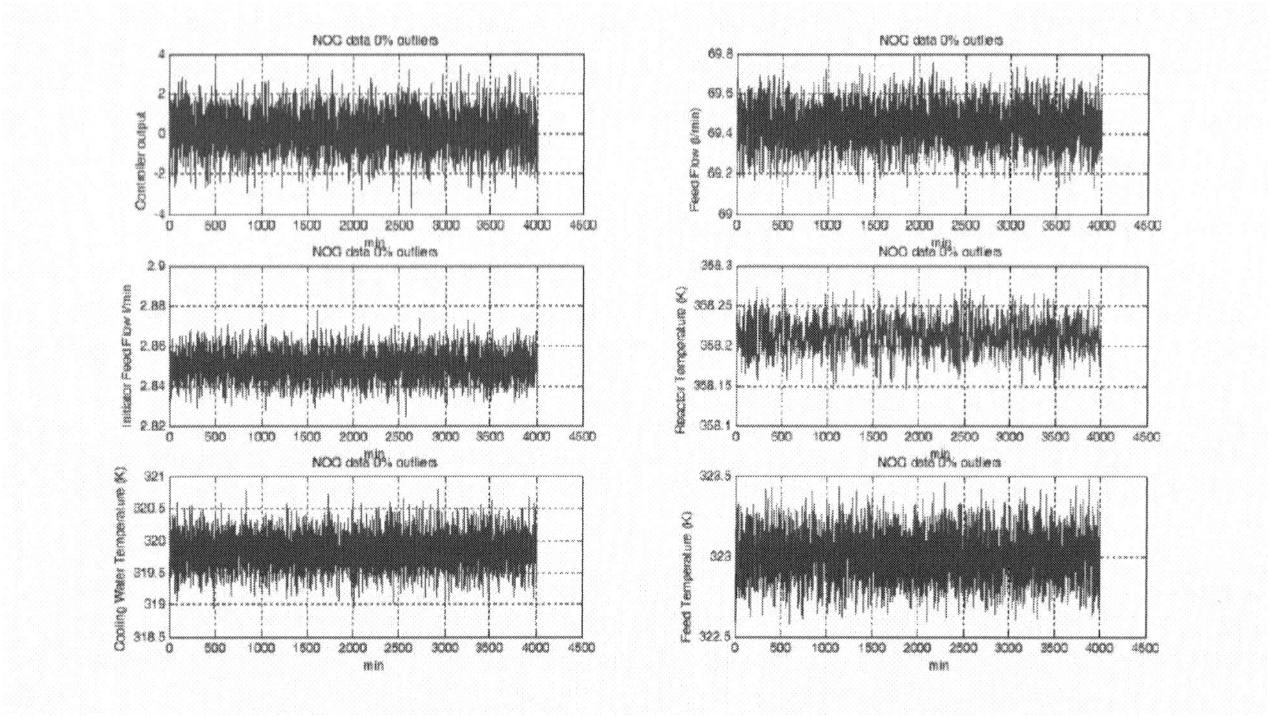

Figure 1. Process variables under normal operating conditions

Figure 1 shows the process variables behavior in normal operating conditions. The x axis scale is in minutes .Temperatures are in K, flow rates are in l/min and the controller output (PID) is in percentage. From these graphs it can be observed that the noise level is relatively low.

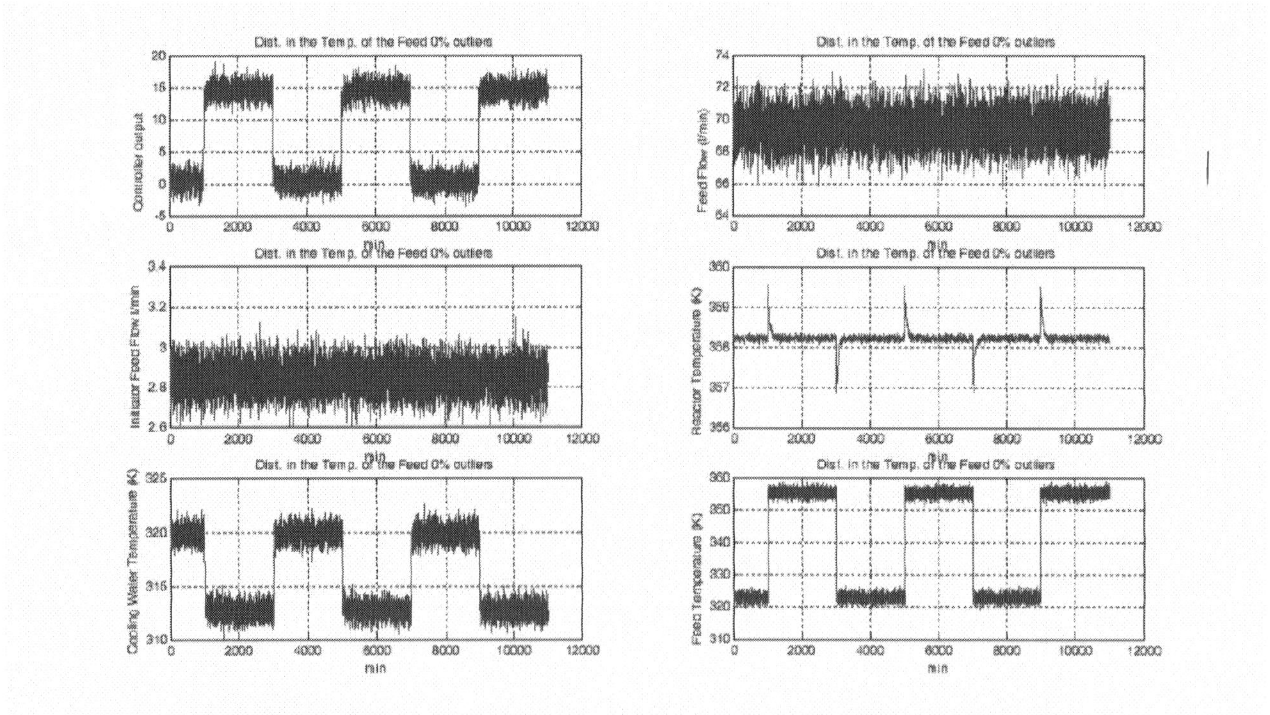

Figure 2. Process responses with 10% changes in feed temperature

Figure 2 shows the effect on the process of 10% step disturbances in the feed temperature. It can be observed that the reactor temperature is brought back to normal very quickly by the feedback controller.

To investigate the performance of the monitoring methods when the process is subject to fault conditions, the process is simulated for 4000 minutes under normal operating conditions. At 4000 minutes the transducer fault occurs and remains for the remainder of the simulation. From 5000 minutes disturbances in the feed temperature occur. It can be observed with all the four monitoring methods tested that the disturbances register the temperature change in the form of spikes in both, T^2 and SPE, indices. The SPE index registers the disturbance because the correlation structure of the variables changes briefly until the controller action has caused the temperature of the reactor to return to the desired value, as can be seen in the Figures 3, 4, 5 and 6.

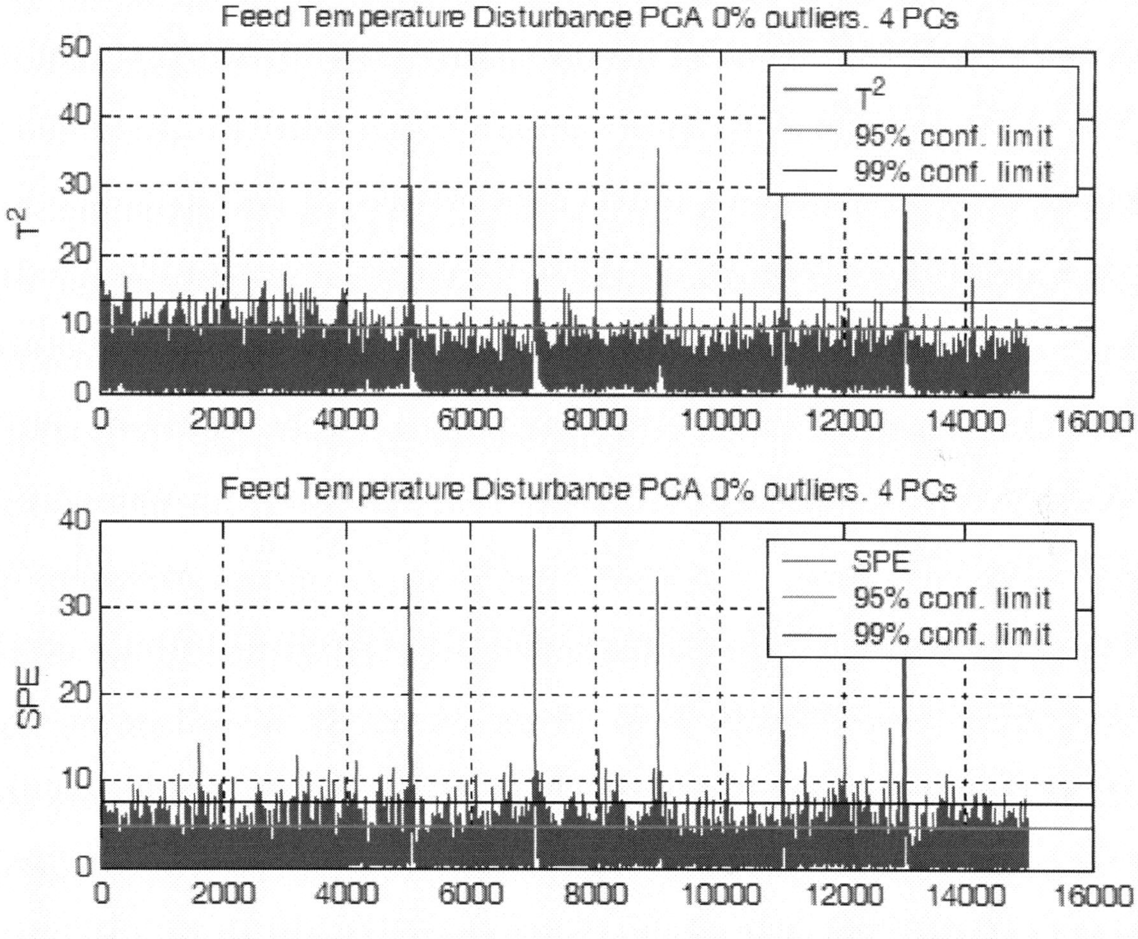

Figure 3. SPE and T^2 for PCA applied to fault / disturbance data

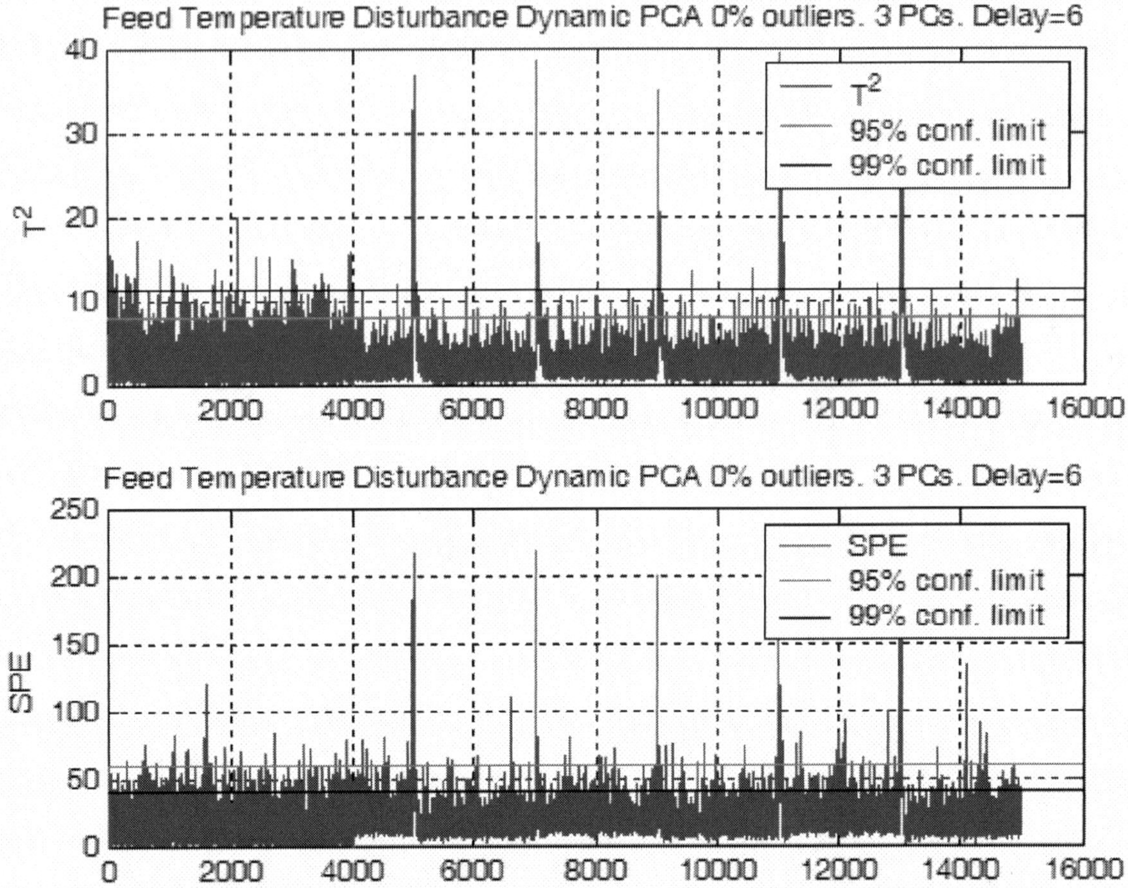

Figure 4. SPE and T^2 for Dynamic PCA applied to fault / disturbance data

It is observed in Figures 3 and 4 that, PCA and Dynamic PCA fail to capture the presence of the faults as the SPE and T^2 plots indicate that the signals predominantly stay within the 95% confidence limits. The methods fail to distinguish the presence of feed temperature changes and transducer fault to any extent other than the spike at the onset of the disturbance. This is because the level of the faults that occur are such that they fail to register above the normal operating regime window of performance.

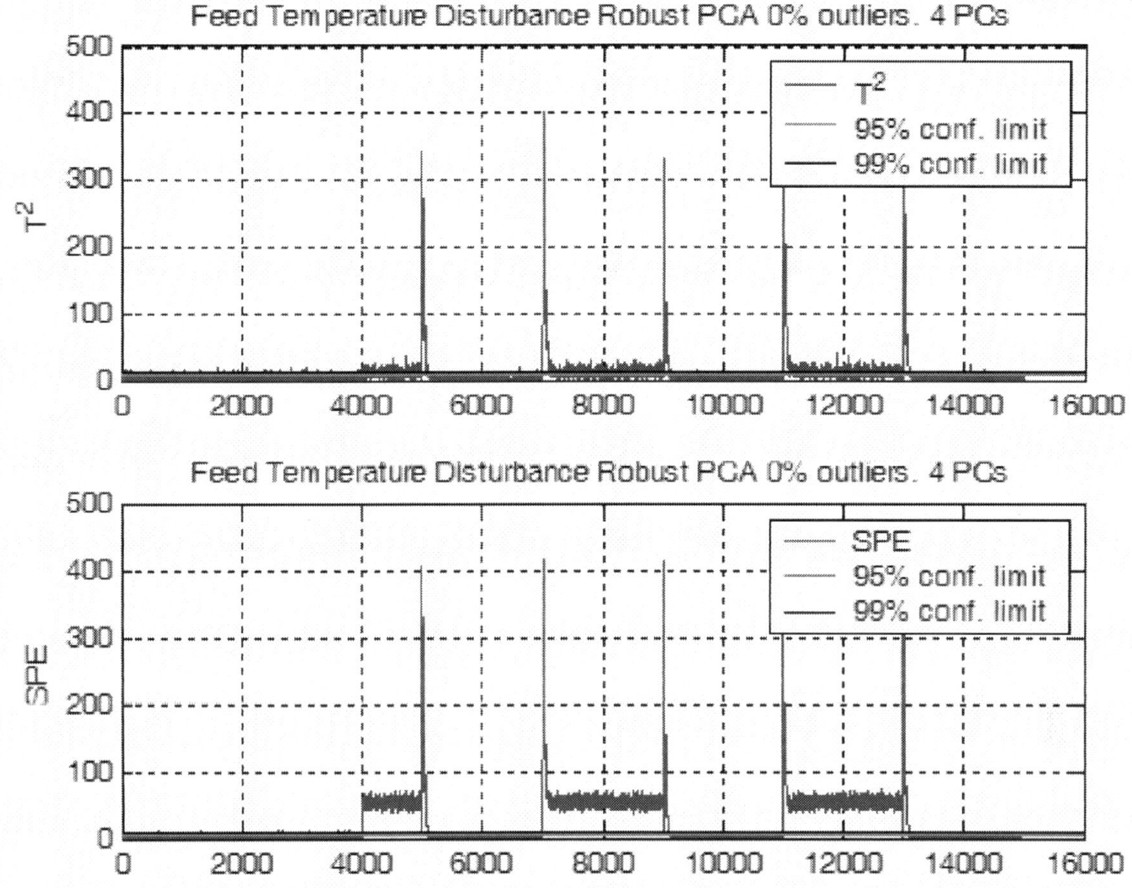

Figure 5. SPE and T^2 for Robust PCA applied to fault / disturbance data

Figure 5 shows the effect of the transducer fault and the temperature disturbances in the T^2 and SPE indices obtained from Robust PCA. It can be seen that when there is no fault present (0-4000 minutes), both indices predominantly lie within the confidence limits, therefore confirming normal operation. When the transducer fault is introduced at 4000 minutes both limits rise above the confidence limits. Robust PCA has therefore been more effective at detecting this fault than PCA and Dynamic PCA. When the feed temperature disturbances commence from 5000 minutes onwards, the manipulated variable is modified by the controller in order to bring the controlled variable back to the desired value. The correlation structure of the data changes and depending upon the level of feed temperature this can have the effect of masking the fault (for instance between 5000-7000 minutes) or enabling its detection (7000-9000 minutes).

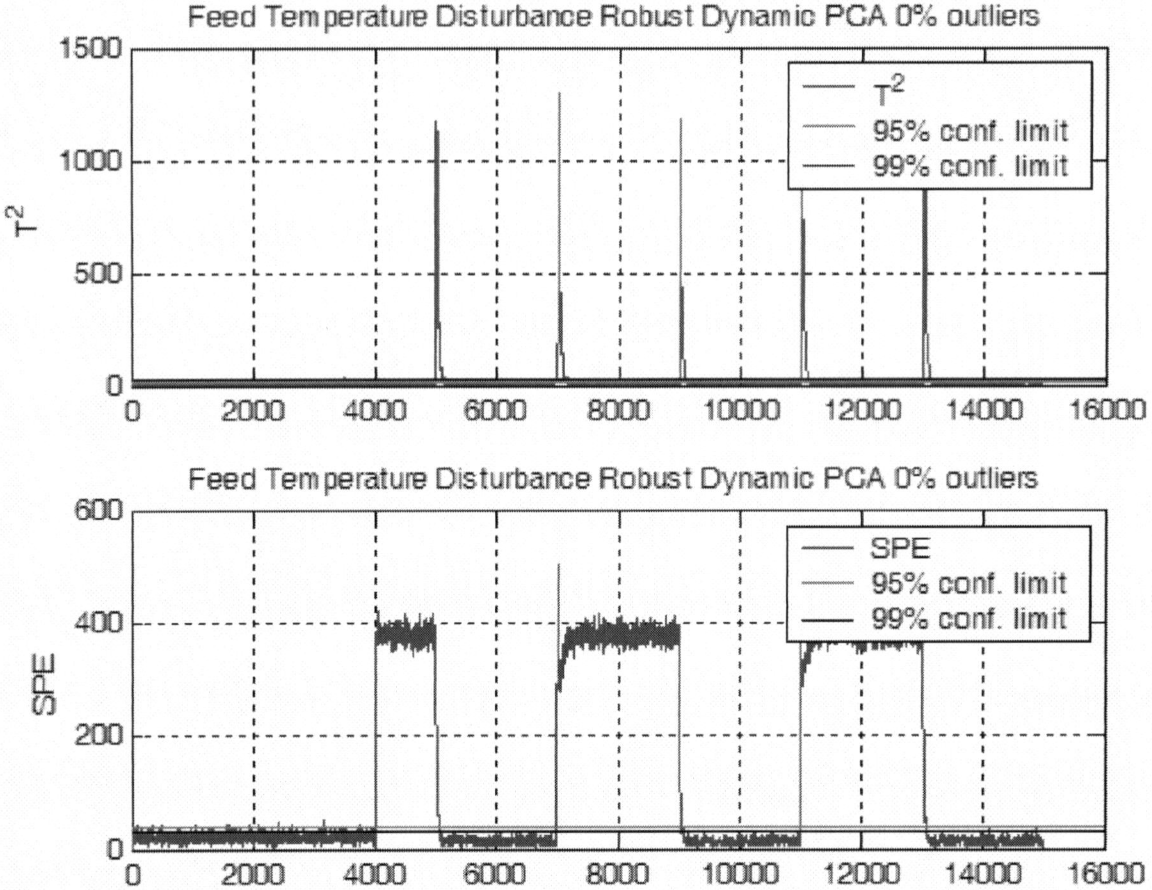

Figure 6. SPE and T^2 for Robust Dynamic PCA applied to fault / disturbance data

Figure 6 shows the effect that the transducer fault and temperature disturbance have on the indices obtained from Robust Dynamic PCA. The indices give qualitatively comparable performance to those obtained using robust PCA (Figure 5). The difference between Robust PCA and Robust Dynamic PCA lies in the fact that a more consistent fault signal is obtained by talking account of the dynamic effects through the dynamic data matrix containing delayed versions of the variables. However, the problem related to feed temperature disturbance indication is still present, with periods of the response where the fault signal is masked.

Figures 3-6 demonstrate the behavior of the fault detection methods when a transducer fault occurs and when feed disturbances impact on the process. Another common process occurrence is outlying data caused by instrument fault or data logging problems. To simulate these outlying data points, the level of noise used to simulate the transducer problem was magnified. The typical performance of the process in the presence of outliers can be seen in Figure 7 where feed temperature disturbances have been implemented. The same form of temperature disturbance pattern is used as in the case with no outliers.

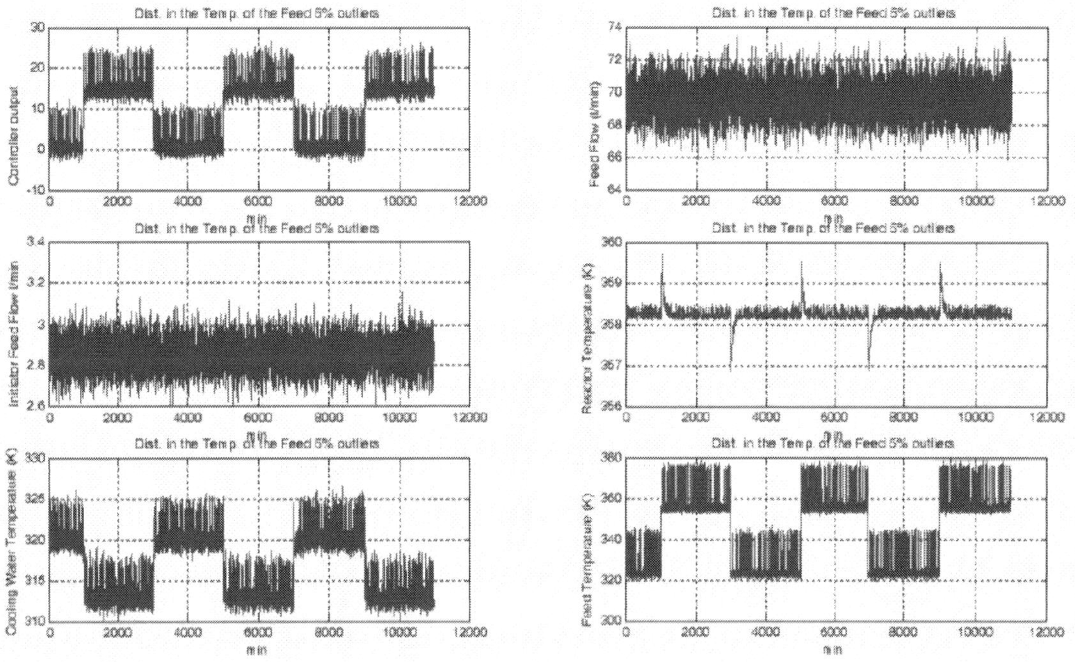

Figure 7. Process data with the measurement of the controller output corrupted by outliers.

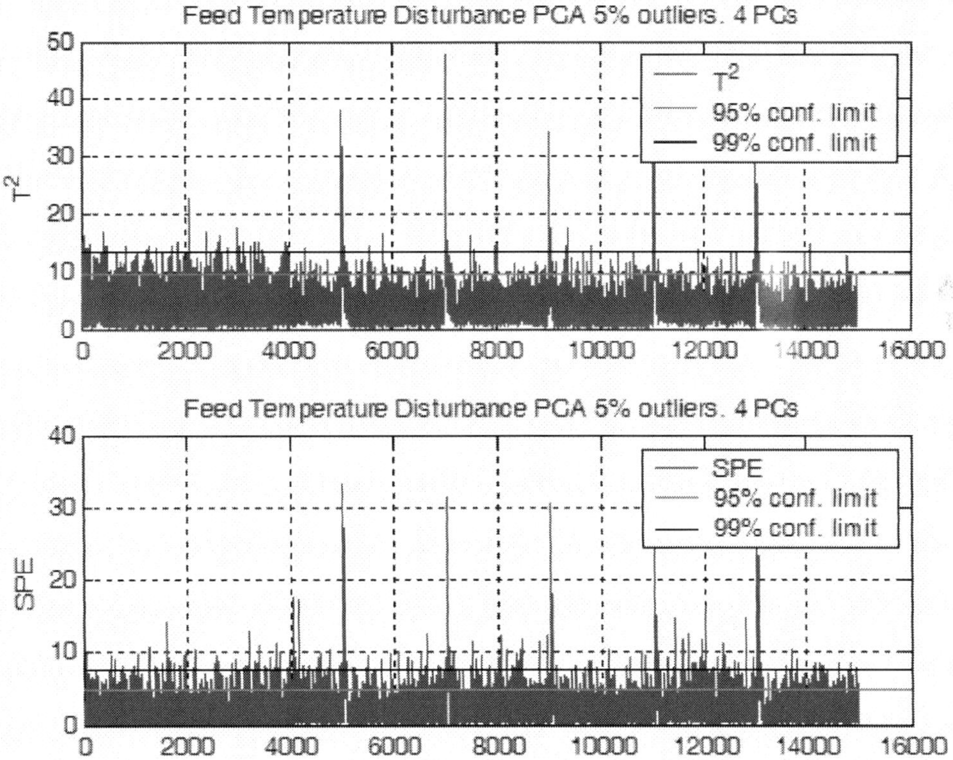

Figure 8. SPE and T^2 for PCA applied to fault / disturbance data in the presence of outliers

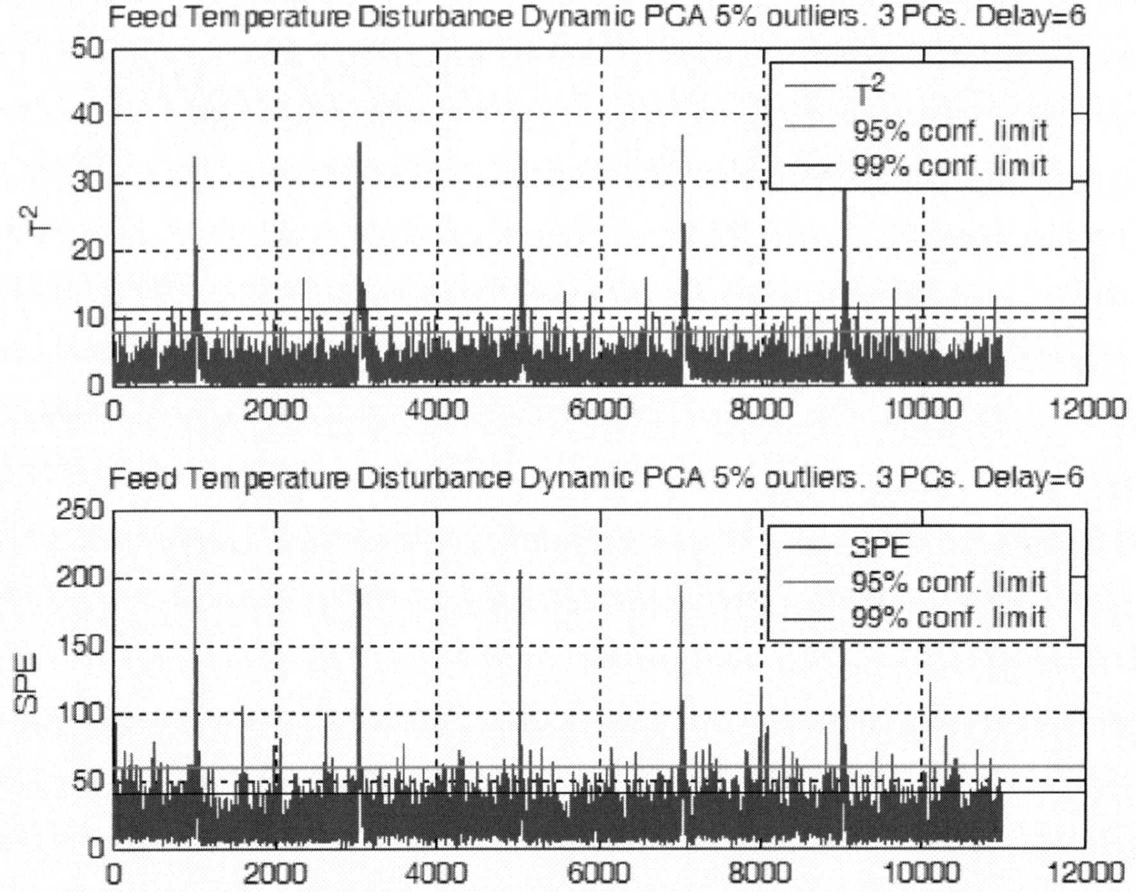

Figure 9 . SPE and T^2 for Dynamic PCA applied to fault / disturbance data in the presence of outliers

The results presented in Figures 8 and 9 for PCA and Dynamic PCA indicate that the fault is not detected in the presence of a 5% of outliers. As for the cases when no outliers where present (Figures 3 and 4), for both indices the violations of the confidence limits occur mainly when the step change disturbance occurs, causing the controller to take significant dynamic action. It is evident that PCA and Dynamic PCA fail to detect the faults present, with the exception of instantaneous but not prolonged indication of feed disturbance changes.

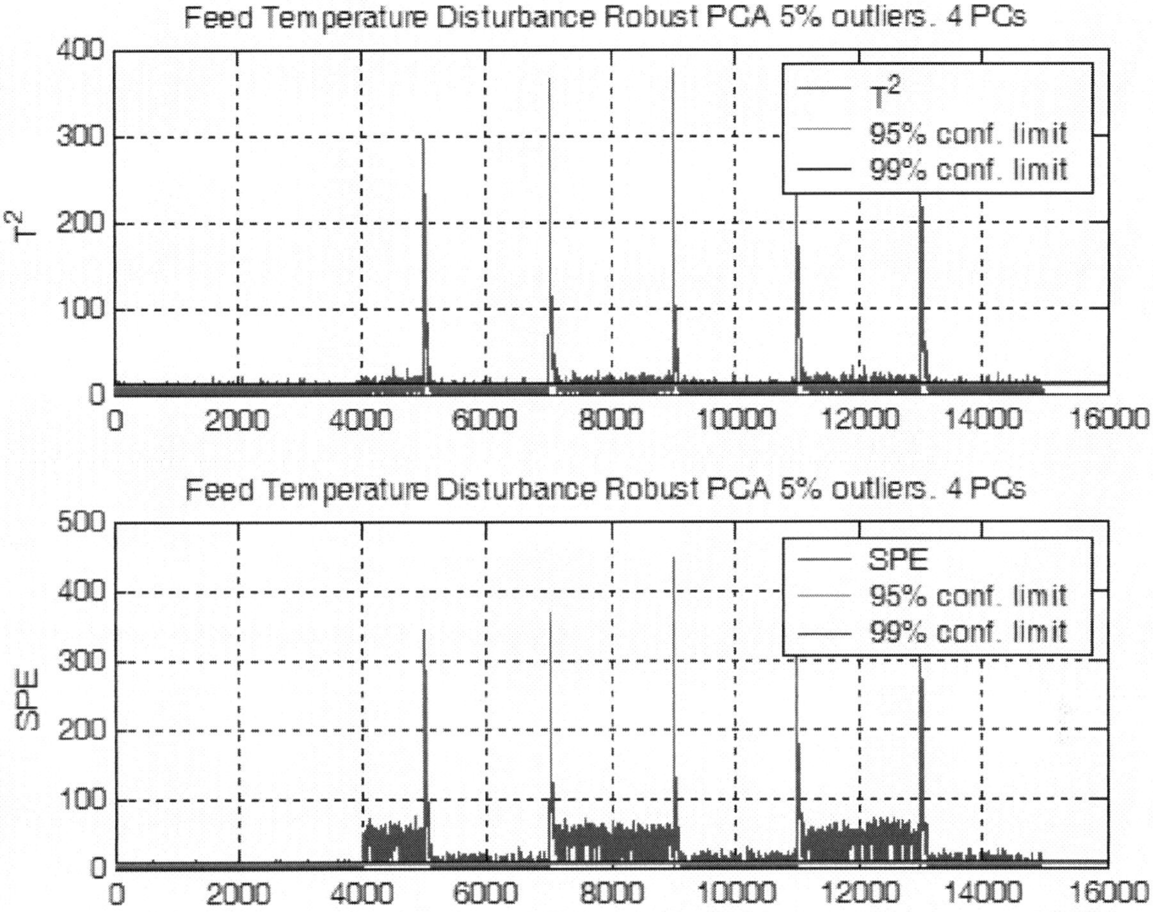

Figure 10. SPE and T^2 for Robust PCA applied to fault / disturbance data in the presence of outliers

Figure 10 shows that Robust PCA gives a more consistent fault indication in the presence of outliers as opposed to when no outliers were present. Importantly, compared with PCA and Dynamic PCA, Robust PCA gives improved results in terms of fault detection. It can be seen that the SPE index indicates the presence of a fault regardless of the magnitude of temperature disturbance. This more consistent indication of fault is due to the fact that the controller is not able to respond to the outlying data. The benefit of using the robust algorithm to obtain scores is clearly evident when comparing Figure 10 with the standard PCA approach shown in Figure 8. Faults are detected far more reliably using the robust algorithm.

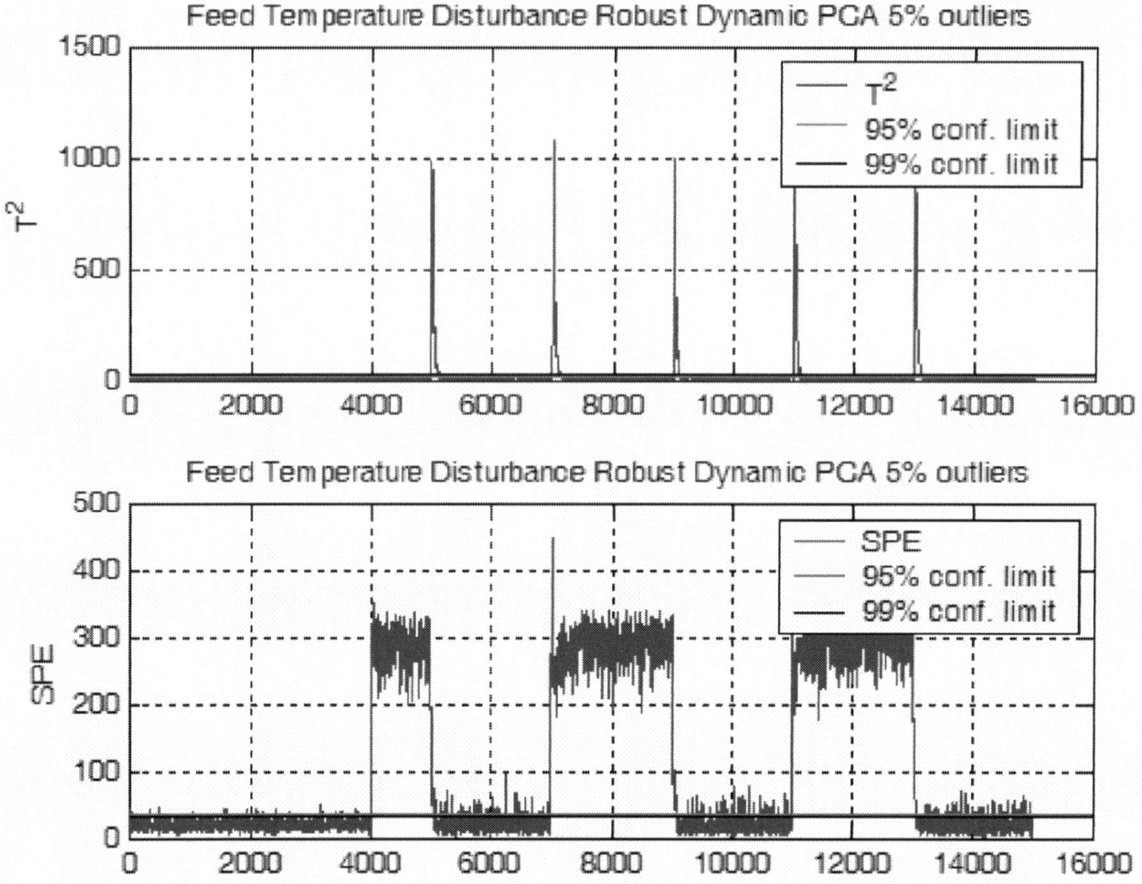

Figure 7 SPE and T^2 for Robust Dynamic PCA applied to fault / disturbance data in the presence of outliers

Figure 7 shows that Robust Dynamic PCA performs better than the other techniques under consideration. As in the case of Robust PCA, the SPE index indicates the presence of a fault even when the controller implements action to counteract the disturbance in the feed flow rate. The detected change indicates a variation in the correlation structure of the data caused by an unmodeled change in one of the variables. This can be used for the subsequent identification and diagnosis of process problems.

In the case of both robust PCA and robust dynamic PCA, the correlation matrix obtained is a better estimate of the true principal components, since the outliers are not used in the determination of the correlation matrix. Hence, the scores and residuals obtained provide more information about the process. As robust dynamic PCA uses a matrix with delayed versions of the variables, more information can be extracted from the data, making the fault detection superior to that obtained with robust PCA.

Conclusions

This paper has demonstrated that the robust dynamic PCA method is more sensitive than previously utilized methodologies for the detection of faults in dynamic process systems. The robust estimation of the correlation matrix employed helps to improve the accuracy of the calculated scores and residuals, resulting in improved fault detection when compared with standard PCA. The improved sensitivity of the SPE index can help to identify and diagnose the cause of the fault. The use of a data matrix which includes delayed variables allows the inclusion of more useful information regarding the condition of the process. However, simply accounting for dynamics using dynamic PCA is not sufficient. To obtain reliable fault detection, robust PCA should be coupled with a dynamic data matrix, as adopted in the robust dynamic PCA approach. The scores and residuals plots employed for the estimation of the T^2 and SPE indices from robust dynamic PCA provide reliable fault indicators when considered together.

In addition to the improved performance of robust dynamic PCA over other techniques, as a consequence of the improved fault indicators it should be possible to extract more valuable information from the process data. Future work will address this fault identification and diagnosis problem and the performance of the scheme when applied to closed loop systems. Contribution plots [2] and the variable reconstruction approach [12] which can be used with the proposed scheme are one possible solution. In addition, the effect of the closed loop controller action makes fault detection problematic. Studies are required to allow an unequivocal detection of a fault in closed loop systems. Fault masking by closed loop control systems is a common plant occurrence and it must be overcome if a practical and sensitive fault detection and identification scheme is to be realized.

Acknowledgements

The authors would like to acknowledge the financial support of the School of Chemical Engineering and Advanced Materials at the University of Newcastle upon Tyne, UK and the industrial members of the Centre for Process Analytics and Control Technology.

References

1. Castro, D., Deteccion y diagnostico de fallas mediante tecnicas estadisticas multivariables y clasificacion de patrones usando Support Vector Machines, in Facultad de Ingenieria. 2002, Universidad de Los Andes: Merida, Venezuela.
2. Chiang, L.H., E.L. Russell, and R.D. Braatz, Fault Detection and Diagnosis in Industrial Systems. Advanced Textbooks in Control and Signal Processing. 2001, London: Springer.
3. Hubert, M. and Engelen, S. Robust PCA and classification in biosciences. Bioinformatics. 2004. 30: p. 1728-1736.
4. Kourti, T., MacGregor, J., Process Analysis, monitoring and diagnosis using multivariate projection methods. Chemometrics and Intelligent Laboratory Systems, 1995. 28: p. 3-21.
5. Ku, W., R.H. Storer, and C. Georgakis, Disturbance detection and isolation by dynamic principal component analysis. Chemometrics and Intelligent Laboratory Systems, 1995. 30: p. 179-196.

6. Kano, M., Nagao, K., Hasebe, S., Hashimoto. I., Ohno, H., Strauss, R., Bakshi B., Comparison of multivariate statistical process monitoring methods with applications to the Eastman Challenge problem. Computers and Chemical Engineering, 2002. 26: p.161-174.
7. Jolliffe, I. T. (1986). Principal Component Analysis, Springer-Verlag.
8. Anderson, T. W. (1984). An Introduction to Multivariate Statistical Analysis, John Wiley and Sons.
9. Kano, M., Hasebe, S., Hashimoto, I., Ohno, H., A new nultivariate statistical process monitoring method using principal component analysis. Computers and Chemical Engineering, 2001. 25: p.1103-1113.
10. Stanimirova, I., Walczak B., Massart, D.L., Simeonov V., A comparison between two robust PCA algorithms. Chemometrics and Intelligent Laboratory Systems, 2004. 71:83-95.
11. Hubert, M., Rousseeuw, P.J., Verboven, S., A Fast Method for Robust Principal Components with Applications to Chemometrics. Chemometrics and Intelligent Laboratory Systems, 2002. 60:101-111.
12. Quin, S.J. Statistical process monitoring: basics and beyond. Journal of Chemometrics, 2003. 17:480-502.
13. Teymour, F. and H. W. Ray (1992). "The dynamic behaviour of continuous polymerisation reactors -V. Experimental investigation of limit-cycle behaviour for vynil acetate polimerization." Chemical Engineering Science **47**(15/16): 4121-4132.
14. Teymour, F. and H. W. Ray (1992). "The dynamic behaviour of continuous polymerisation reactors -VI. Complex dynamics in full-scale reactors." Chemical Engineering Science **47**(15/16): 4133-4140.
15. Smith, C. A. and A. B. Corripio (1997). Principles and Practice of Automatic Process Control. New York, Wiley.
16. Akaike, H., Canonical Correlation Analysis of Time Series and the use of an Information Criterion. *System identification: Advances and Case studies*, D. G. Lainiotis and R. K. Mehra eds., Academic Press, (1977) 27--96.

216c Wireless Instrumentation Enables New Best Practices in Monitoring and Automation

Clifford Lewis, Accutech, A Division Of Adaptive Instruments Corp., 577 Main Street, Hudson, MA 01749

Wireless Instrumentation provides a highly reliable and extremely economic implementation path to best practices in monitoring and automation. Wireless Instrumentation in enabling double digit improvements in efficiency and providing cost effective solutions for environmental compliance projects. Continuous monitoring of position, level, overfill and underfill, flow, shut-off emergency switches are some applications used to comply with ever increasing demands for tighter inventory control,safety and minimal environmental impact.

Wireless Instrumentation is particularly well suited to providing real-time, continuous data and is an ideal solution for retrofitting in existing plant structures that were built without conduit and wiring infrastructure. A cost saving of 90% in the design and installation of industrial grade measurement equipment can be realized. Wireless Instrumentation often can be deployed at dramatic savings and provide redundant monitoring when necessary.

This paper looks at the various wireless protocols and compares how well these protocols meet the needs for monitoring in the industrial process. In particular, new wireless sensors are surveyed that require no power wires out to the sensor. Common difficulties with wireless installations, such as line-of-sight, and Class I, Div I area classifications are discussed and solutions presented for these issues. Point-to-point radio solutions are contrasted to mesh network data communications solutions and hierarchical mesh networks.

216d Using Industrial Wireless Sensors to Monitor Safety Showers and Eye-Wash Stations

Doug Eberhart, Accutech, A Division Of Adaptive Instruments Corp., 577 Main Street, Hudson, MA 01749

Safety showers in industrial facilities provide first-line emergency treatment for employees that have come in contact hazardous materials. They are normally combined with eye-wash stations for emergency treatment for eyes. In virtually every instance where an employee has triggered the use of a safety-shower or eye-wash station, that employee will be in need of help and assistance for further treatment. Rapid assistance is often critical in minimizing or mitigating any potential injury and getting help to an injured employee can often make the difference between a serious injury and a minor incident.

Unfortunately, the cost to install wired monitoring switches to multiple safety-showers in a major production plant can be extremely expensive. Wiring monitoring switches entails the cost of running the wires to the switches in the field, often through areas classified as Class I, Div I or Class I, Div II. Wired switches also entail the expense of dedicating individual discrete inputs into plant DCS systems. These costs rapidly add up. One recent facility estimated the costs for wiring safety-showers to be over $4,000 per point. Per point monitoring costs are generally reduced by 75% using wireless sensors compared to hard wire.

A brief summary of this safety application will be reviewed along with other safety application using wireless instrumentation such as steam trap monitoring, on/off valve positioning, overfill protection, control valve status and pressure and temperature monitoring. This applications survey is intended to show that wireless solutions do have a place in the toolbox for those involved with plant safety and, when properly applied, this technology will solve difficult measurement problems not presently served with wired instrumentation or captured at all.

Advanced Control Strategy to a Fermentation Process to Obtain Ethanol

Duarte[1], E. R., Ender[2], L., Maciel Filho[1], R.

[1] LOPCA/FEQ/ State University of Campinas (UNICAMP) – Zip Code: 13.083-970 - Campinas – SP, Brazil.

[2] Chemical Engineering Department – Regional University of Blumenau – Zip Code: 89010-971 – Blumenau – SC, Brazil

Abstract

The aim of this work is to evaluate the performance of a non linear control strategy based on Artificial Neural Networks (ANN's) for an extractive alcoholic fermentation process (MIMO 3x3). The process is simulated through a validated deterministic model which takes into account the main phenomena taking place in the system. The control is build-up by a multilayered and multivariable ANN that represents the inverse dynamics of the system, which is on-line trained through an optimization routine. The optimization routine adjusts the weight of neural controller using the estimate global error of closed loop. The global error is obtained using a dynamic model of the process, to represent a prediction for the next sampling time. This process model is on-line trained with the process data. The obtained results have shown the efficiency of control strategy in multivariable system and the potential of proposed on-line learning.

1. Introduction

Nowadays days with the globalization, market competitivity and environmental concern, there is a real need of products with high quality, low cost and low emission of pollutants. In operational terms, the process control is the main form to reach such objectives. That is the reason for the great advance on researches in the investigation of new methodologies and conceptions about process development and process control in all areas.

Process control has been by far the most popular area of neural network application in chemical engineering. The neural network approach allows taking into account process non-linearities as well as variable interactions. Neural networks can learn sufficiently accurate models and give good non-linear control when model equations are not known or only partial state information is available.

Due to increase necessity of process control with better performance, it is indispensable to develop suitable strategies to deal whit non-linear process. A possible alternative is to design the control with the ability to adapt due to external perturbations that can to occur. on even changes in the metabolism of microorganisms.

In this context, control strategies based on real time optimization, where the artificial neural networks (ANN´s) are tools used to represent the process with complex dynamic appears to suitable.

Bearing, this is mind, in this work is proposed a control strategy based on artificial neural networks as dynamic model (indirect method) and as controller (direct model) (Psichogios and Ungar, 1991). The aim is to investigate a non-linear control strategy based on ANN applied in an alcoholic fermentation extractive process. The main advantage is the on line learning of networks according to modifications that occur in the process, so that the difficulties pointed out earlier may be overcome.

2. Process

The fermentation process is inhibited by ethanol and a way to improve the process is the selective extraction of the alcohol, during the fermentation, enhancing the productivity. In this context several schemes combining fermentation with a separation process have been developed (Minier and Goma, 1982; Maiorella, 1984; Christen et al., 1990; Taylor et al., 1998; Ishida and Shimizu, 1996; Silva et al. 1999).

As a study case its was used a 3x3 multivariable process, represented by an alcoholic fermentation plant, proposed by Silva et al. (1999), based on this scope and according to Costa et al. (2001) the process presented a high productive level compared to the convectional process.

For the extractive fermentations process, four interlinked units compose the process: fermentor (production unit of ethanol), centrifuge (phase separation unit), cube (microorganism treatment unit) and flash (separation unit). The flash is used to continually extract part of ethanol produced, improving the productivity, since it is expected a decrease in the inhibition. Figure 1 show schematically with process.

The process starts with the substrate (sugar-cane molasses) feed into the fermentor. The output stream of fermentor composed by substrate, cell and ethanol is sent to centrifuge and two phases are generated: cell suspension and liquid phase without cell. The cell suspension is sent to cell treatment where is diluted with water, and sulphuric acid is added in order to avoid bacterial contamination. The flash vessel is operated in a temperature range form 28 to 30 ºC, which is chosen in order to eliminate the necessity of a heat exchanger (Andrietta and Maugeri, 1994).

The associated pressure is in the range of 4-5.33 KPa. The proposed scheme was shown to be able to maintain suitable conditions for the growth of Sacharomyces cerevisae by keeping a constant temperature, which is possible to be controlled without any heat exchanger.

The vaporized stream leaving the flash vessel is fed in a rectification column with part of the liquid stream, whereas the other fraction of the liquid returns to the fermentor. It is important to have this liquid fraction refluxed to maintain the ethanol concentration in the fermentor in such value that it acts as antiseptic agent. According to practical knowledge in industrial units, alcohol concentration in the fermentor must be around 40 kg/m3, which has low inhibitory effect for alcoholic yeast but is highly inhibitory for most contaminating microorganisms (Silva et al., 1999).

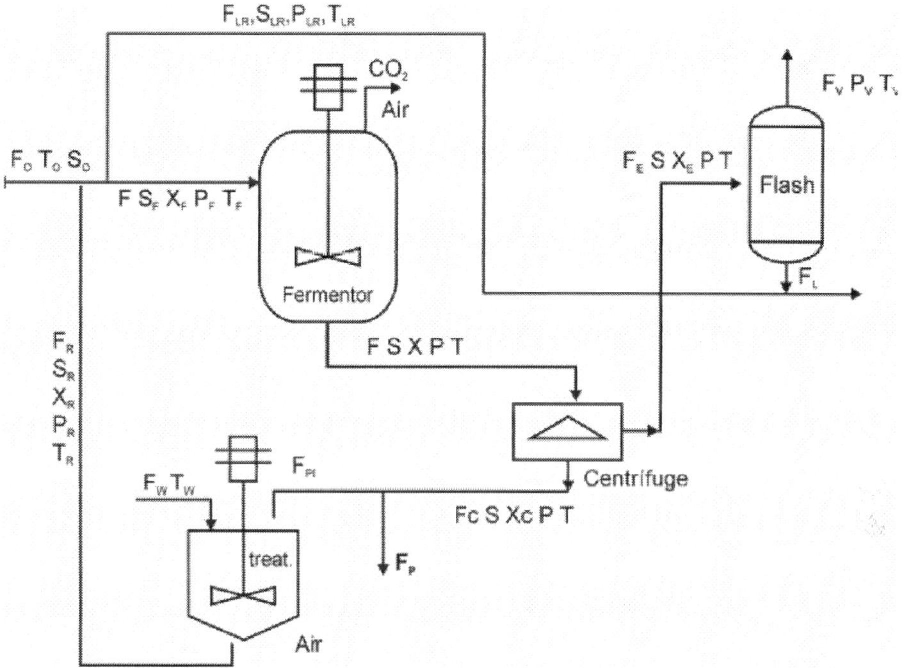

Figure 1. Extractive alcoholic fermentation process.

The mathematical modelling of the process of all equipment with the exception of the fermentor is modelled assuming the hypothesis of pseudo steady state. The fermentor model was based on the model developed by Costa et al. (2001). Such model simplifications approach is reasonable since the fermentation dynamics dominator the whole process dynamics.

To determine the feed rate and feed concentration of the fermentor, mass balances on the global process are necessary. The following considerations were made:

- The concentrations of substrate and product leaving the centrifuge are equal to the concentrations leaving the fermentor;
- The concentration of biomass in the cells recycle stream is fixed. To maintain the concentration in a fixed value, the flow rate of the water that dilutes the ferment (FW) is adjusted. The cell recycle flow rate (Fr) is maintained at a value fixed by the cell recycle rate (R) by adjusting the flow rate of the purge (Fp). The purge allows for cell renovation and the withdrawal of secondary products accumulate into the fermentor.

Assuming constant volume, the mass and energy balance equations for the fermentor with the intrinsic model approach are as follows:

- Viable Cells

$$\frac{dX_v}{dt} = r_x - r_d - \frac{F}{V}(X_v - X_{vF})$$

- Dead Cells

$$\frac{dX_d}{dt} = r_d - \frac{F}{V}(X_d - X_{dF})$$

- Substrate

$$\frac{d\left[\left(1 - \frac{X_t}{\rho}\right)SV\right]}{dt} = F(S_F - S) - Vr_s$$

- Product

$$\frac{d\left[\left(1 - \frac{X_t}{\rho}\right)PV + \frac{X_t}{\rho}\gamma PV\right]}{dt} = Vr_p + F(P_F - P)$$

- Energy

$$\frac{dT}{dt} = \frac{F}{V}(T_F - T) + \frac{\Delta H r_s}{\rho_m Cp}$$

In such equations ρ and γ are the ratio of dry cell weight per wet cell volume and the ratio of concentration of intracellular to extra cellular ethanol, respectively (Costa et al. 2001)

Table 1 showed the kinetic parameters as functions of the temperature. Experiments were performed at temperatures between 28 and 40 °C in a system with total cell recycling by tangential microfiltration. The substrate used was sugar-cane molasses (Atala *et al.* 2001). Table 2 depicts the process parameters at the steady state.

Table 1. Kinetic Parameters as functions of the Temperature

Parameter	*Expression or Value*
K_i	$1.393.10^{-4}\exp(0,1004\ T)$
m_p	0.1
m_x	0.2
m	1.0
n	1.5
K_s	4.1
ρ	390
Γ	0.78
μ_{max}	$1.57\exp\left(\frac{-41.47}{T}\right)-1.29.10^{4}\exp\left(\frac{431.4}{T}\right)$
	$-0.3279T^2+18.484T-191.06$
P_{max}	$-0.4421T^2+26.41T-279.75$
B	$2.704\ \exp(-0.1225T)$
Y_{PX}	$0.2556\exp(0.1086T)$
K_{dP}	
K_{dT}	$4.10^{13}\exp\left(\frac{41947}{1.987\ (T+273.15)}\right)$

Table 2. Stationary Parameters

Variável	*Stationary Parameters*
To	30 ° C
So	180 kg/m^3
Fo	100 kg/m^3
R	0.35
r	0.4
S	5.49 kg/m^3
P	41.12 kg/m^3
X	34.81 kg/m^3

3. Control Strategy

In process control applications neural networks can be incorporated in the control strategy in either direct or indirect methods (Psichogios & Ungar, 1991).

In the indirect method the neural network is trained with input-output data from the dynamic system to represent the forward dynamics; given the current state and the current control action, the network learns to predict the next state of the system. A process can be written in a discrete form as follows (Bulsari, 1995):

$$Y(k+1) = f(Y(k),...,Y(k-n+1),U(k),...,U(k-m+1)) \tag{1}$$

with

$$\begin{aligned} Y(k) &= [y(1,k),\ldots y(i,k)] \\ U(k) &= [u(1,k),\ldots u(j,k)] \end{aligned} \tag{2}$$

where $U(k)$ and $Y(k)(k=0,1,...)$ are input and output vectors at a time instant k, respectively; *i, j* are the number of controlled and manipulated variables and f is an unknown function. Consider a construction of an approximate function of f by using a neural network. Let $Y(k+1)$ and $Y(k),...,Y(k-n+1),U(k),...,U(k-m+1)$ be outputs and inputs of the neural network, and W be a set of the connection weights and the thresholds of the neural network. This network can construct an approximate dynamic model by adjusting W.

$$\hat{Y}(k+1) = \hat{f}(Y(k),...,Y(k-n+1),U(k),...,U(k-m+1);W) \tag{3}$$

This is a forward dynamic modelling previously denominated of indirect method.

In the direct method, a neural network is trained to represent the inverse dynamics of the system. In this case, given the current state of the dynamic system and the target state for the next sampling instant, the network is trained to produce the control action that drives the system to this target state. Solving Equation (1) with respect to $U(k)$, one can get

$$U(k) = \phi\begin{pmatrix} Y(k+1), Y(k),...,Y(k-n+1), U(k-1),... \\ ...,U(k-m+1) \end{pmatrix} \tag{4}$$

where ϕ is an unknown function. This function presents an inverse dynamics of the process. As the function ϕ is unknown as well as the function f, it is not possible to evaluate Equation (4). A neural network is used to generate an approximate function of ϕ. This is given by:

$$\hat{U}(k)=\hat{\phi}\begin{pmatrix}Y(k+1),Y(k),...,Y(k-n+1),U(k-1),...\\ ...,U(k-m+1);W\end{pmatrix} \tag{5}$$

Consider a control problem in which the output $Y(k)$ tracks a series of setpoints $R(k)$. Then:

$$Y(k+1)=R(k+1) \tag{6}$$

is desired at a next time interval. Substituting Eq. (5) into Eq. 6), the desirable input to the process is given by:

$$\hat{U}(k)=\hat{\phi}\begin{pmatrix}R(k+1),Y(k),...,Y(k-n+1),U(k-1),...\\ ...,U(k-m+1);W\end{pmatrix} \tag{7}$$

For the development of this work the feedforward architecture with backpropagation learning were used (Bhat & McAvoy, 1990 and Bhat *et al.*, 1990). The feedforward architecture has shown to be very robust when compared to other possible arrangements and its ability to approach complex functions as well as to deal with noise data, making it to be very interesting tool (Bulsari, 1995).

The dynamic model is constructed by using input and output data of the process. It is well known that the forward dynamic model has the sufficient accuracy if a large number of input and output data are available (Bulsari, 1995).

The inverse model network receives the inputs of the current and past system outputs, the reference signal $(R(k))$ and the past values of the system input. The nonlinear input-output mapping given by the network inverse model is given by the direct method, described in Equation (7).

The controller is based on a neural network that represents the inverse dynamics of the system, which is on-line trained through an optimization routine (Figure 2). The controller optimization routine adjusts the weight of the neural controller using the estimated global error of the closed loop at each sampling time, based on a dynamic model of the process, represented by a neural network with on-line learning. The dynamic model of the process is on-line trained with the last ones inputs/outputs data of the process stored in a vector. The controller design uses the same inputs of the neural controller at each sampling time. This optimization routine adjusts the weights of the neural network of the controller in such way to minimize the estimated global error. Considering that the estimated error is based on a neural model, it is necessary to have a model that represents with fidelity the dynamic behaviour of the process. When the quadratic error of the neural model outputs is smaller than the desired tolerance, this model is used in the optimization routine. If the quadratic error becomes larger in relation to a determined error, the controller makes use of the standard weight (weight of the off-line learning) to generate the control action for this sampling instant (Duarte, 2004).

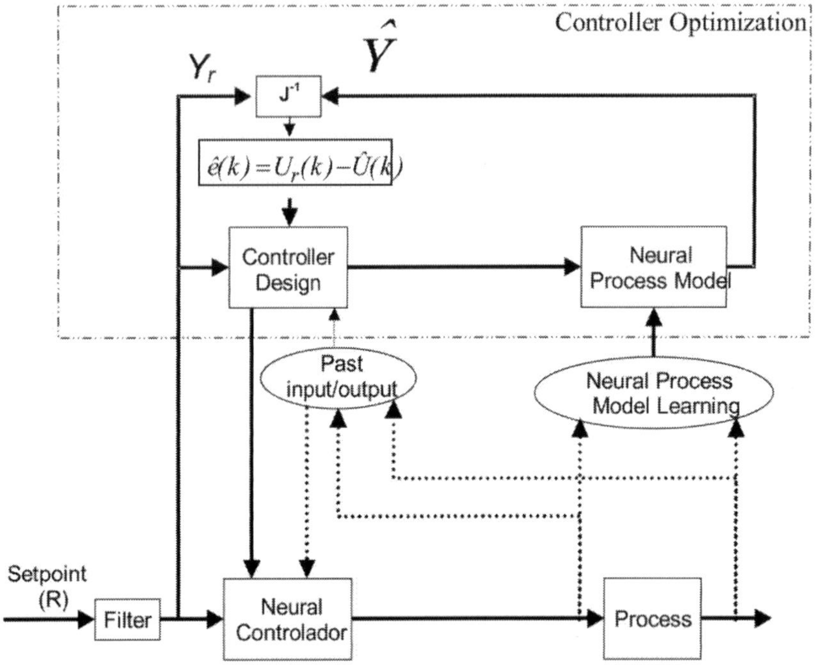

Figure 2 – Non Linear Controller

The global error cannot back propagate directly because of the location of the process information (it is available process exit data). The neural network of the controller is on-line optimized at each sampling time using the estimated variables, obtained by retro propagation the variables Y_r and $\hat{Y}$ through the jacobian matrix. Changing each input of the process model slightly at the operation point and measuring the change at the output it is possible to determine the Jacobian matrix.

$$J = \begin{pmatrix} \frac{\partial y_1}{\partial u_1} & \cdots & \frac{\partial y_1}{\partial u_i} \\ \vdots & \ddots & \vdots \\ \frac{\partial y_i}{\partial u_1} & \cdots & \frac{\partial y_i}{\partial u_i} \end{pmatrix} \tag{8}$$

were y_j and u_j are the estimated response of the closed loop and the control action which generated this estimated response of the closed loop. Then:

$$\begin{pmatrix} y_1 \\ \vdots \\ y_i \end{pmatrix} = \begin{pmatrix} \frac{\partial y_1}{\partial u_1} & \cdots & \frac{\partial y_1}{\partial u_i} \\ \vdots & \ddots & \vdots \\ \frac{\partial y_i}{\partial u_1} & \cdots & \frac{\partial y_i}{\partial u_i} \end{pmatrix} * \begin{pmatrix} u_1 \\ \vdots \\ u_i \end{pmatrix} \tag{9}$$

$$U = J^{-1} * Y \tag{10}$$

$$e(k) = U_r(k) - \hat{U}(k) \tag{11}$$

were U_r and $\hat{U}$ are the control actions obtained from $\mathbf{Y_r}$ and $\hat{Y}$, respectively, in agreement with equation (10).

These control actions deviations are used to on-line learning of the neural network of the controller in the controller optimization routine.

To guarantee the good dynamic representation of the process through neural networks, a strategy formed by three nets acting in parallel is proposed (Figure 3). The first (ANN 1) is formed by weights of the off-line learning, here denominated as standard weights; the second (ANN 2), is initialized with the standard weights and it is submitted at the on-line learning. Whenever the standard weights present better performance, the latter net has its weights substituted by the standard weights. The third (ANN 3) is initialized with the standard weights and continually is submitted to the on-line learning at each sampling time.

The neural network that presents the smallest quadratic error (eq. 12) in the representation of the vector that contains the last inputs/outputs of the process is used in the control strategy in this sampling time.

$$\varepsilon = \sum_{l=1}^{n_p} \left| d - y^p \right|^2 \tag{12}$$

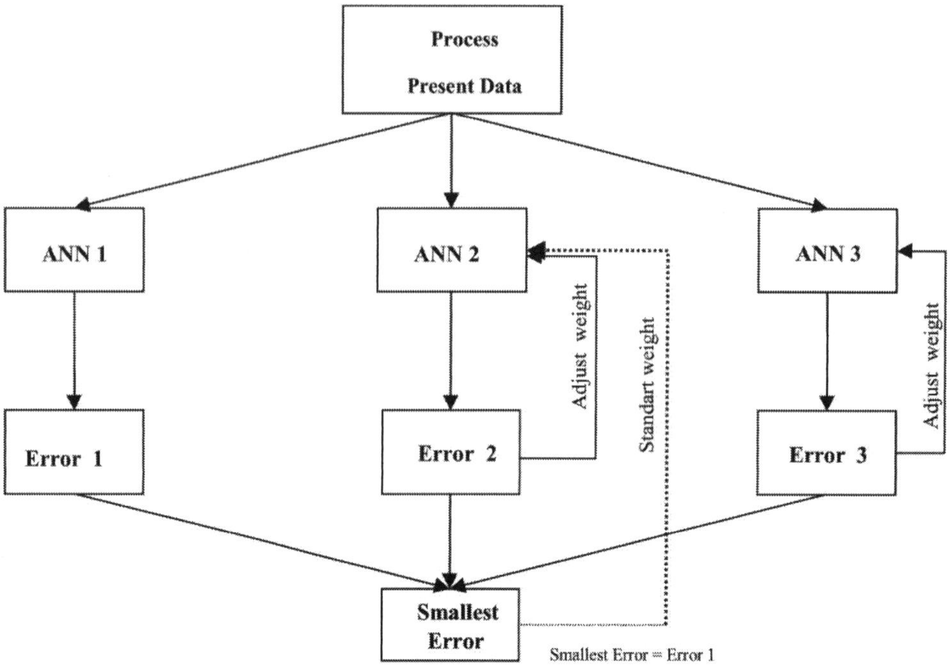

Figure 3.Three Nets Acting in Parallel.

The configuration for on-line learning is showed in Figure 4. Two vectors formed by the dynamic information of the system are used as patterns for the on-line learning and testing of the ANN's. The first vector, used in the learning phase, is formed by Pattern (t-*b*) until Pattern (t-*a*) where *a* and *b* are design parameters. The second vector, used in the testing phase, is formed by all the information on inputs/outputs, Pattern (t-*b*) until Pattern (t). In present configuration all the patterns are used for testing the ANN's and only part of the patterns are used for learning.

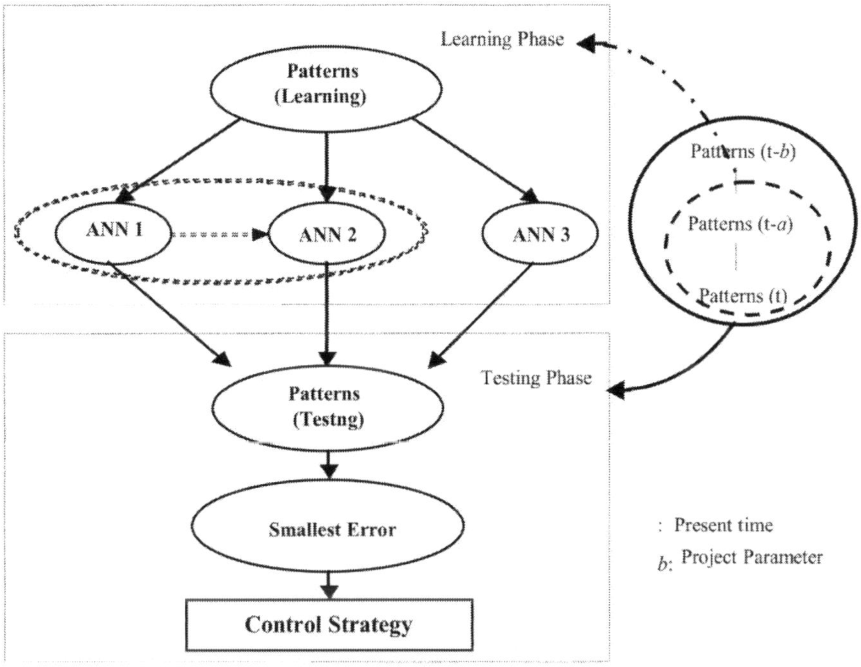

Figure 4. Configuration for On-line Learning.

4. Results

The variables (manipulated and control) were choose according to Costa et al. (2001), where the manipulated variables were the feed stream flow rate (F), cells recycle rate (R) and flash recycle rate (r). The control variables were: product concentration (P), substrate concentration (S) and cell concentration (X)..

The load changes were performed in feed substrate concentration (So) and feed temperature (To). These variables were considered for changes in load due to their importance to the process performance and the cells lifetime. Indeed care has to be taken with the temperature which leads the system to have higher performance in terms of conversion for relatively high temperatures (around 30ºC), but it may cause a reduction in the cell productivity with time (Atala et al., 2001).

The servo problem is presented in Table 3.

Tabela 3. Servo Problem.

Tempo (h)	*S*	*P*	*X*
0-50	10	45	37
50-100	3	40	32
100-150	8	43	36
150-200	5.5	42	35

Figures 5, 6 and 7, show the obtained results for set point and the controlled variables (substrate, cell and product concentration). It can be seen that control strategy presented good performance to lead the system to the new set points and to eliminate de influence of disturbances.

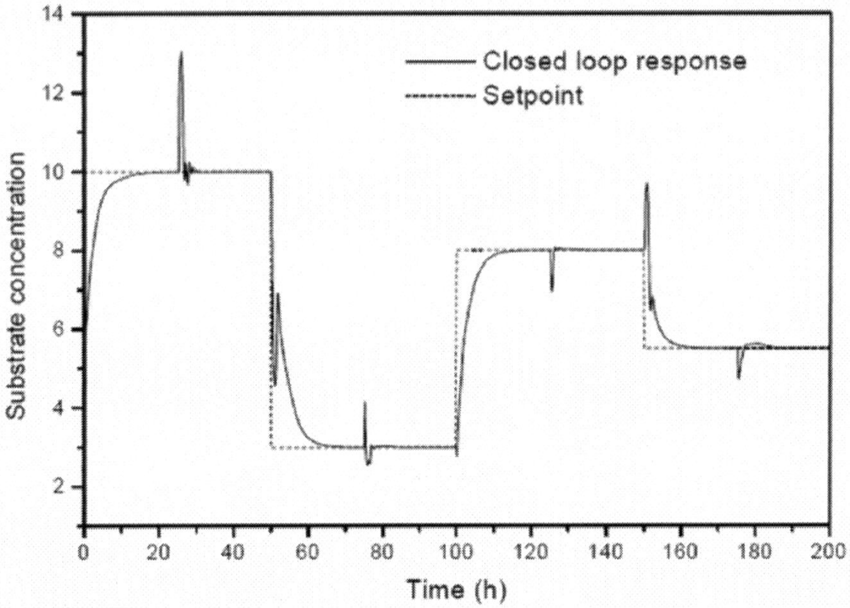

Figure 5 Closed loop response for substrate concentration

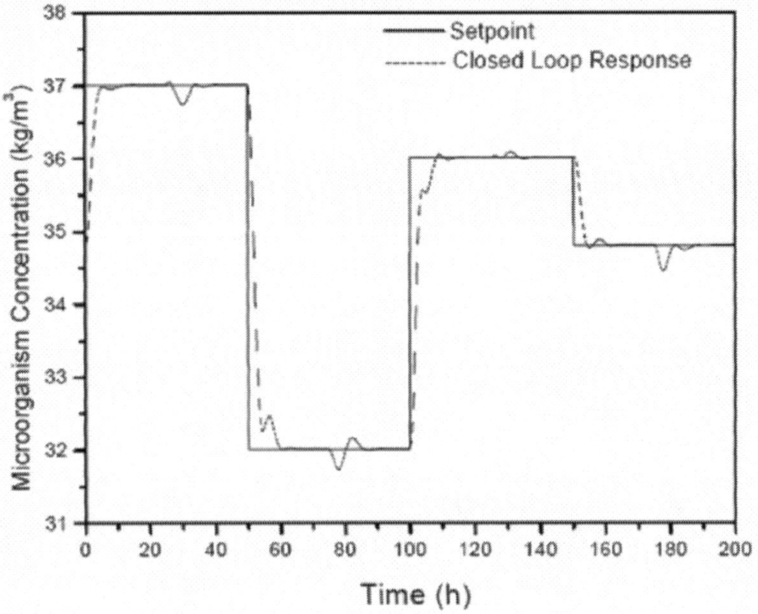

Figure 6 Closed loop response for microorganism concentration

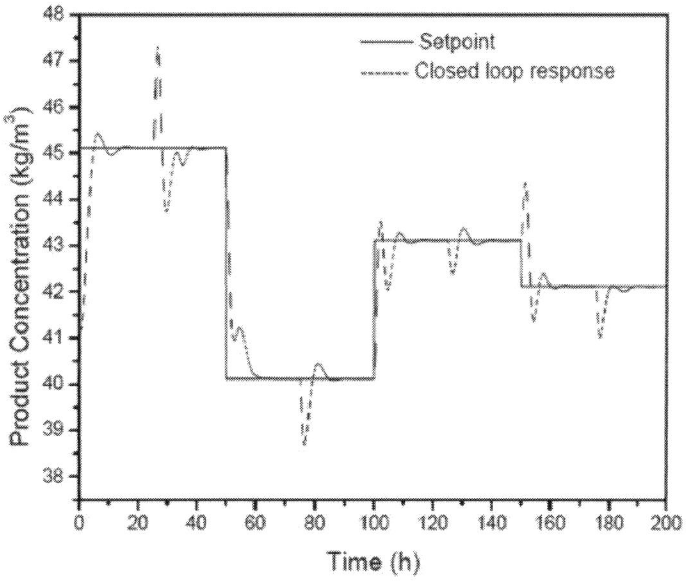

Figure 7 Closed loop response for product concentration

Figure 8 shows the closed-loop results for a servo problem of product concentration considering on-line learning of neural process model and without on-line learning. An off-set is observed when the control strategy uses a neural model of the process without on-line learning. In fact the control strategy with on-line learning is more robust since it allows tracking the system changes in real time.

Figure 9 depicted the results for product concentration using stochastic perturbations, in which it is observed the potential of control strategy.

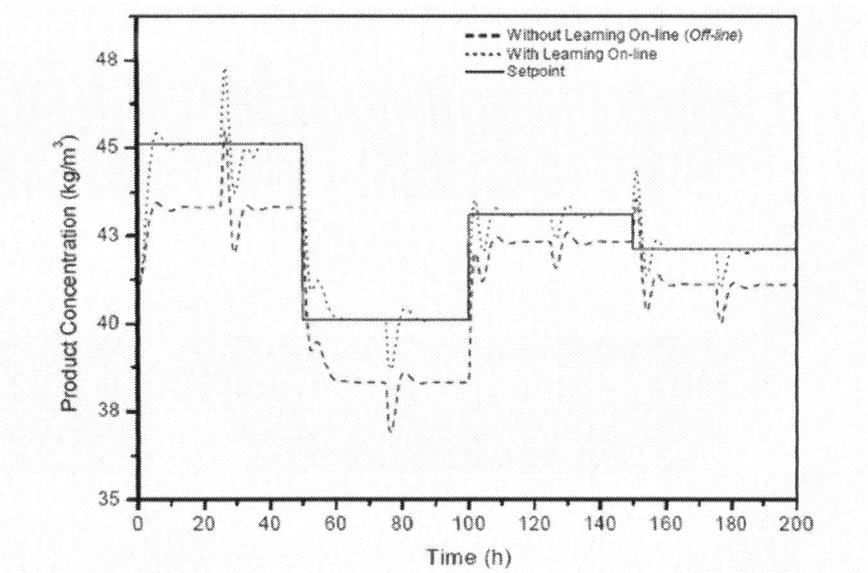

Figure 8.On-line, Off-line learning and set point for product concentration

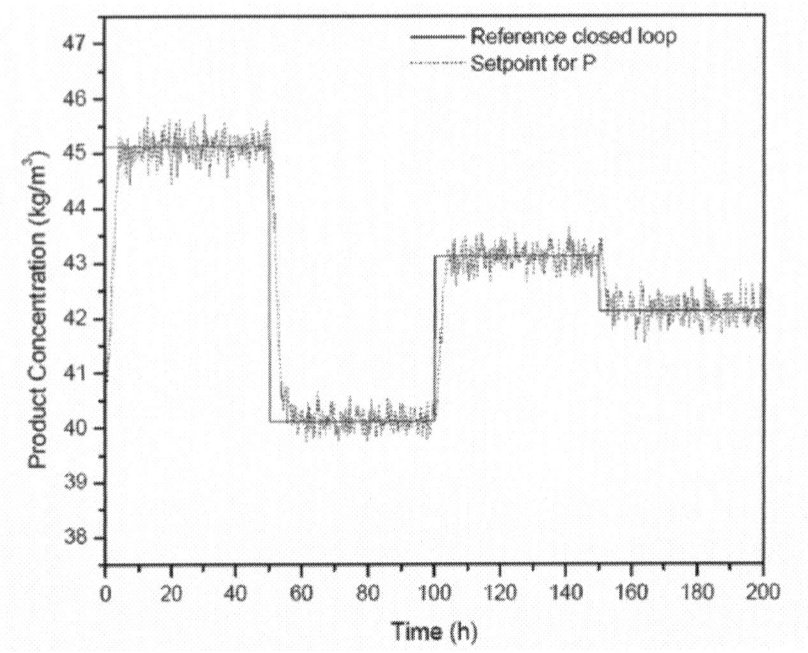

Figure 9. Results for product concentration with stochastic perturbations

5. Conclusions

In this work was proposed a non linear advanced control strategy based on on-line learning of ANN that are use is both way, direct in indirect form. As case study a plant of extractive fermentation to obtain ethyl alcohol was considered. A detailed dynamic model was solved and validated so that the virtual plant could represent well the pilot plant.

Next step will be to improve the monitoring facilities to on-line implementations. The virtual plant is an attractive alternative to carry out extensive tests so that a broader range of operational situations may be explored in such environment, to verify the potential of the proposed control strategy.

The results showed the control strategy efficiency in multivariable systems and the good results for on-line learning of networks. The proposed control strategy showed to be robust for all perturbations considered in this work. The obtained results showed the potential of on-line learning and the ability of the controller to follow the changes is process.

6. References

1. Bhat, N., & McAvoy (1990). "Use of neural nets for dynamic modeling and control of chemical process systems", *Comput. Chem. Eng.*, 14, 573-582.

2. Andrietta SR, Maugeri F.(1994). Optimum Design of a Continuous Fermentation Unit of an Industrial Plant for Alcohol Production. Adv. Bioprocess Eng., 47-52.

3. Atala, D. I. P. Costa, A. C., Maciel Filho, R. Maugeri, F. (2001). Kinetics of Ethanol Fermentation with High Biomass Concentration Considering the Effect of Temperature. Applied Biochemistry and Biotechnology, 91, 353-365.

4. Bhat N. and McAvoy T. (1990). Use of neural nets for dynamic modeling and control of chemical systems. Computers & Chemical Engineering. 14, 573-582.

5. Bulsari A. B. (1995). Neural Networks for Chemical Engineers. Amsterdam: Elsevier.

6. Christen M., Minier M., Renon H. (1990). Ethanol extraction by supported liquid membrane during fermentation. Biotechnol. Biong. 3, 116-123.

7. Cybeko, G. (1989). Approximation by superposition of a sigmoidal function. Mathematics of Control Signals and Systems. 2, 303-314.

8. Costa, A. C., Atala, D. I. P., Maciel Filho, R. Maugeri Filho. F.(2001). Factorial Design and simulation for the Optimization and Determination of Control Structures for an Extractive Alcoholic Fermentation. Process Biochemistry. 37(2) 125-137.

9. Duarte, E. R., (2004). Estratégia de Controle Não Linear Baseada em Redes Neurais Artificiais com Aprendizagem "On-line". Msc. Thesis. Brasil.

10. Ender, L., Maciel Filho, R.(2000). Design of multivariable controller based on neural networks. Computers & Chemical Engineering. 24, 937-943.

11. Fissore, D., Barresi A. A., Manca D. (2004). Modelling of methanol synthesis in a network of forced unsteady-state ring reactors by artificial neural networks for control pusposes. Chemical Engineering Science. 59, 4033-4041.

12. Hecht-Nielsen R. (1989). Neurocomputing, Addisin-Wesley Publishing Company, Reading, MA.

13. Hérnandez, E., & Arkun, Y. (1992). A study of control relevant properties of back-propagation neural netwok models of nonlinear dynamical system. Computers &Chemical Engineering, 16, 227-240.

14. Henson M. A. (1998). Non-linear model predictive control: current status and future directions. Computers & Chemical engineering. 23, 87-202.

15. Hunt, K. J. & Sbarbaro, D. (1992). Neural networks for nonlinear internal model control. IEE Proceedings (Part D), 138, 341-348

16. Ishida K. and Shimizu K. (1996). Novel repeated batch operation for flash fermentation system: experimental data and mathematical modeling. Journal of Chem. Technol. Biotechnol. 66, 340-346.

17. Jutan A. and Krishnapura V. G. (2000). A neural adaptive controller. Chemical Engineering Sciense. 55 3803-3812.

18. Jimenez-Marquez S. A., J. Thibault, C. Lacroix. (2004). Prediction of moisture in cheese of commercial production using neural networks. International Dairy Journal.

19. Maiorella B. L., Blanch H. W., Wilke C. R. (1984). Biotechonology report – Economic evaluation of alternative ethanol fermentation processes. Biotechnol. Bioeng. 26, 1003-1025.

20. Mjali F. S. (2005). Neural network model-based predictive control of liquid-liquid extraction contactors. Chemical Engineering Sciense. 60, 239-253.

21. Meleiro, L. A. C., Maciel Filho, R, Campello, R. J. G. B. e Amaral, W. C. (2001). Application of Neural Networks and Other Learning Technologies in Process Engineering, Imperial College Press, Capítulo: Hierarchical Neural Fuzzy Models as a Tool for Process Identification: A Bioprocess Aplication.

22. Minier M. and Goma G. (1982). Ethanol production by extractive fermentation. Biotechnol. Bioeng. 24, 1565-1579.

23. Psichogios, D. C., Ungar, L. H. (1991). Direct and Indirect Model based on Control Using Artificial Networks, Industrial Engineering & Chemical Research. 30, 2564-2573.

24. Sarkar D. Modak, J. M. (2003). ANNSA: a hybrid artificial neural network / simulated annealing algorithm for optimal control problems. Chemical Engineering Science, 58, 3131-3142.

25. Silva, F. L. H., Rodrigues M. I., Maugeri F. (1999). Dynamic Modeling, Simulation and Optimization of an Extractive Continuous Alcoholic Fermentation, Journal Chem. Tech. Biotechnol. 74 (2) 176- 182.

26. Taylor F. Kurantz M. J., Goldberg N., Craig J. C. (1998). Kinetics of continuous fermentation and stripping of ethanol. Bioechnol. Lett, 20, 67-72.

27. Ungar L. H., Powell, B. A. and Kamens S. N. (1990). Adaptive networks for fault diagnosis and process control. Computers & Chemical Engineering. 14, 561-572.

7 Nomenclature

Cp	*Heat Capacity, Kcal/(Kg°C)*
e(k)	Estimated globar error
F	*Feed stream flow rate, m³/h*
Fc	Cell suspension flow from centrifuge, m^3/h
F_L	Liquid outflow rate to flash tank, m^3/h
F_E	Liquid outflow rate to flash tank, m^3/h
F_{LR}	Light phase flow rate to flash tank, m^3/h
Fo	Fresh medium flow rate, m^3/h
Fp	Purge flow rate, m^3/h
F_R	Cell recycling flow rate, m^3/h
Fw	Water flow rate, m^3/h
Fv	Vapor outflow from the vacuum flash tank, m^3/h
i	The number of controlled variables
$Y(k)$	Output vectors at a time instant k
j	The number of manipulated variables
P	Product concentration fermentor, kg/m^3
P_F	Feed product concentration, kg/m^3
P_{LR}	Product concentration in the light phase from centrifuge, kg/m^3

P_V	Product concentration in the vapor phase from the flash tank, kg/m^3
Pr	Product concentration in the cells recycle, kg/m^3
R (F_R/F)	Cells recycle rate
r (F_{LR}/F_L)	Flash recycle rate
rd	Kinetic rate of death, $kg/(m^3/h)$
rp	Kinetic rate of production formation, $kg/(m^3/h)$
rs	Kinetic rate of substrate consumption, $kg/(m^3/h)$
rx	Kinetic rate of growth, $kg/(m^3/h)$
S	Substrate concentration into the fermentor, kg/m^3
S_F	Feed substrate concentration, kg/m^3
S_{LR}	Substrate concentration in the light phase from centrifuge, kg/m^3
So	Inlet substrate concentration, kg/m^3
S_R	Substrate concentration in the cells recycle, kg/m^3
T	Temperature into the fermentor, °C
T_F	Feed temperature, °C
T_{LR}	Light phase temperature, °C
To	Inlet temperature of the fresh medium, °C
T_r	Cells recycle temperature, °C
T_v	Temperature of vapor from the flash tank, °C
T_w	Water temperature, °C
$U(k)$	Input vectors at a time instant k

$U_R(k)$	Referenced control actions vectors at a time instant k
$U_R(k)$	Estimated control actions vectors at a time instant k
X_d	Dead biomass concentration into the fermentor, kg/m^3
X_{dF}	Dead biomass concentration in the feed stream, kg/m^3
X_E	Biomass concentration in the light phase flow rate to flash tank, kg/m^3
Xr	Cell recycling concentration, kg/m^3
X_F	Feed biomass concentration, kg/m^3
X_{LR}	Biomass concentration in the light phase flow centrifuge, kg/m^3
X_t	Total biomass concentration into the fermentor, kg/m^3
X_v	Viable biomass concentration into the fermentor, kg/m^3
X_{vF}	Viable biomass concentration in the feed stream, kg/m^3
w	Weights of ANN
ΔH	Reaction rate, Kcal(/kg TRS)
γ	Ratio of concentration of intracellular to extracelular ethanol, kg/m^3
ρ	Ratio of dry cell weight per wet cell volume, kg/m^3
ρ_m	Density, kg/m^3
ϕ	Unknown function.

PETROLEUM FOULING: CAUSES, TOOLS, and MITIGATION METHODS

Irwin A. Wiehe, Soluble Solutions, Gladstone, NJ

Abstract

With the high price of light crude oils most refineries are driven to purchase greater quantities of lower priced opportunity crudes that are heavier and contain higher concentrations of sulfur and of naphthenic acids. This has led to higher frequency of refinery fouling, just when incentives for refinery utilization and for energy conservation are at their peak. Fortunately, the understanding of the causes and mitigation methods of petroleum fouling has greatly improved recently through the development of tools for prediction and for identification.

Surprisingly, 90% of petroleum fouling in refining has only a few common causes. The analysis of the foulant will usually enable identifying the root cause that is confirmed by finding the precursor in the oil flowing through the fouled unit. By tracing the precursor to the source, one arrives at a number of potential mitigating methods from which a refinery can select the best to implement for the refinery. Case studies will be presented where this strategy was used to mitigate petroleum fouling for different causes. Since the most common cause of organic fouling is by insoluble asphaltenes, a tool, the Oil Compatibility Model, was developed to predict the insolubility of asphaltenes on blending of oils and from thermal conversion. The presence of the carbonaceous mesophase in a foulant shows that the asphaltenes became insoluble at thermal reaction temperatures to form coke. These and other tools now make asphaltene fouling the easiest form of fouling to predict, to detect, and to mitigate. As a result, fouling by inorganics, caused either by corrosion or by ineffective desalter operation, is often a greater challenge to mitigate.

Introduction

Fouling is defined as the formation of an unexpected phase that interferes with processing. While the fouling phase is often a solid, it could be a liquid when a gas is expected or it could be an emulsion when separate liquid phases are expected. Fouling during petroleum processing, from the well through the refinery, has been so common that petroleum processors expect that units need to be shut down periodically for cleaning. However, such fouling produces an unneeded expense because usually it can be greatly reduced or eliminated by a systematic procedure that determines the cause and suggests alternatives for its elimination.

There are large incentives to mitigate fouling in refining. Most only consider the maintenance cost of cleaning. However, in these times of high crude and energy costs with significant refinery margins, maintenance costs may only rank fourth among the costs of fouling. The lost production when units are shut down for cleaning is a significant cost because of refinery margins that have greatly increased as the demand for refinery products has begun to exceed refinery capacity. The insulating effect of layers of foulant on heat exchange surfaces can cost refineries large amounts of energy without it being realized. In addition, foulants reduce the efficiency of fractionators and reduce the reactivity in catalytic reactors. Finally, with the high price of light crude oils most refineries are driven to purchase greater quantities of lower priced opportunity crudes that usually are heavier and contain higher concentrations of reactive sulfur and of naphthenic acids. However, these opportunity crudes are often not purchased because of the fear of fouling or they are purchased and greater fouling results. In either case, fouling mitigation would reduce the cost of crude oil. Therefore, each refinery should carefully estimate these costs in determining their incentives for fouling mitigation. Most will conclude that they need not wait for a large fouling incident to justify a significant program on fouling mitigation.

Fouling Mitigation Strategy

The best strategy[1] to mitigate fouling is to elucidate the foulant chemistry and to use this basic knowledge to determine how and where to eliminate its formation. While there are methods to reduce the rate of fouling without knowing the cause, many of these merely pass the fouling problem on to the next unit. However, by stopping the foulant precursors from forming, the unexpected foulant phase can be eliminated from the refinery completely. Therefore, the recommended strategy is to use key indicators and knowledge of the most common causes of refinery fouling to determine the cause of fouling. This is usually determined by chemical analysis of the foulant but sometimes from analysis of the oil flowing through the fouled unit. Once the cause is determined, the foulant precursors should be traced to the source. The source may be in the fouled unit, in an upstream unit, or enter with the crude oil. While the most effective cure usually is to eliminate the formation of the foulant precursor, understanding the path from precursor source to foulant usually suggests a number of mitigation options that can be selected by the refinery.

Most Common Causes

While the general fouling mitigation strategy will solve any refinery fouling problem, the most common causes can be identified by using tests or indicators for the diagnosis and solved quickly. Surprising to most, 90% of refinery fouling is caused by only a few basic causes. Although some of these tests and indicators are proprietary, nonproprietary examples will be discussed. The most common causes of refinery fouling are:

A. Organic
    1. Insoluble Asphaltenes
        i. Incompatible Oils on Mixing
        ii. Self-Incompatible Oils

iii. Adsorption from Compatible, but Nearly Incompatible, Oils
iv. Caused by Cooling after a Conversion Unit
v. Coke: Insoluble Asphaltenes at Thermal Cracking Temperatures
2. Polymerization of Conjugated Olefins
B. Inorganic
1. Sea Salts: Sodium, Calcium, and Magnesium Chloride
2. Iron Sulfide and Iron Oxide: Corrosion Products
3. Ammonium Chloride after Conversion Unit
4. Aluminum Silicate: Clay or Catalyst Fines

Diagnosis

The diagnosis is arrived from three sources of evidence:

1. Process conditions/history
2. Analysis of Foulant
3. Analysis of oil flowing through fouling unit

Process Conditions/History

The process conditions/history should include the range and average conditions (temperature, pressure, flow rate, and feeds) and should include upstream units as well as the fouling unit. What is very revealing is determining the difference in conditions when the unit was not fouling and when the unit was fouling. Can the initiation of fouling be correlated with a particular incident, such as a process upset, a contaminant (slop or sludge) addition, or a new feed? The process and location of the foulant can suggest the most likely cause based upon past history and experience.

Analysis of the Foulant

The analysis of the foulant usually reveals the most information about the cause. As a result, the foulant sample should be taken with care while recording information as to its location, amount, and exposure to cleaning liquids prior to sampling. Pictures of the foulant in the unit prior to sampling are very revealing. If the foulant is a solid, it should be ground and mixed to homogenize to obtain an average sample. The sample should be washed with a solvent, like methylene chloride or toluene, to remove the oil and then dried. If the foulant is soluble in the solvent, like an asphaltene sediment, a different sample should be washed with heptane.

The first foulant analysis should be to determine how much is inorganic. An ash test, thermal gravimetric analysis (TGA), or elemental analysis may be used. If it is determined to be over 10 wt% inorganic, the author includes this as a possible cause. Often the inorganic foulant adsorbs organics out of the oil and elimination of the inorganic deposit also eliminates the organic portion of the foulant. However, if the foulant is over 50 wt% organic, one should also search for a cause of fouling by organics.

Organic Fouling during Crude Processing

In 1999 Lemke[2] determined that 2/3 of the cost of refinery fouling was in the crude unit and 76% of this was due to increased energy costs. Today, with higher refinery margins, with much higher crude costs, and with greater use of opportunity crudes, the cost of fouling in the crude unit probably is an even higher fraction of the total cost of refinery fouling. While in 1999 organic fouling was a mystery, today with the oil compatibility model and tests[3] refineries have the tools available to avoid organic fouling in the crude unit. All organic fouling in the crude unit is due to insoluble asphaltenes. There are three modes of insoluble asphaltenes in crude oils: asphaltenes may be insoluble in the crude oil as purchased (self-incompatible), the asphaltenes may precipitate when crude oils are mixed (incompatible), and the asphaltenes may adsorb out of the crude oil onto metal surfaces (nearly incompatible). Soluble Solutions has identified over 10 crude oils that are self-incompatible but the name of only one, now no longer in production, Yme, has been released by its customers. Likewise, Soluble Solutions has identified over 200 pairs of incompatible crude oils but the names of only a few have been published. Nevertheless, when insoluble asphaltenes are deposited on a hot heat exchanger surface they may form coke, depending on the time and temperature. Above 370ºC this reaction is almost instantaneous and below 200ºC it does not happen in many years. The significance is that once coke forms, it cannot be redissolved in the oil or any solvent. This is the reason that the fouling of the highest temperature heat exchanger in the crude unit is usually the greatest even though asphaltene solubility is more at higher temperatures. Once asphaltenes are insoluble at lower temperatures, the time scale of redissolution is much longer than the time it takes the crude to flow through the crude unit as the crude is heated to higher temperatures.

The Oil Compatibility Model and Crude Incompatibility

The oil compatibility model[3] is a solubility parameter based model that enables one to predict the compatibility or incompatibility of any mixture of any number of oils. This is based upon testing the compatibility of the individual oils with different proportions of a model solvent, such as toluene, and a model nonsolvent, such as n-heptane. These tests enable measuring the solubility parameter of the mixture at which asphaltenes just begin to precipitate. This solubility parameter on a reduced n-heptane-toluene scale is called the insolubility number, I_N. In addition, the tests measure the solubility parameter of the oil that on a reduced n-heptane-toluene scale is called the solubility blending number, S_{BN}. An example is shown in Figure 1 where the compatibility numbers are measured for the two crudes, Souedie and Forties, with the minimum two tests each. One test, the heptane dilution test, involves determining the maximum volume of n-heptane that can be added to a given volume of oil without precipitating asphaltenes. Insoluble asphaltenes are most accurately detected by observing a drop of the mixture between a glass slide and a cover slip under an optical microscope at 100 to 200X. Insoluble asphaltenes are observed throughout the entire view as yellow to brown curved chain agglomerates. In the other test, the toluene equivalence test, the minimum percent toluene in mixture with n-heptane to dissolve asphaltenes is determined at a concentration of two grams of oil and 10 ml. of toluene-n-heptane mixture (test liquid). As shown in Figure 1, the volume percent toluene in the test liquid is plotted versus 100 times the volume ratio of oil to test liquid. The insolubility number is the y-axis intercept of a line drawn through the two points. The solubility blending number is calculated by the equation given in Figure 1. There are other tests to determine the solubility blending number for oils that contain no asphaltenes[4], the case for about 25% of crude oils.

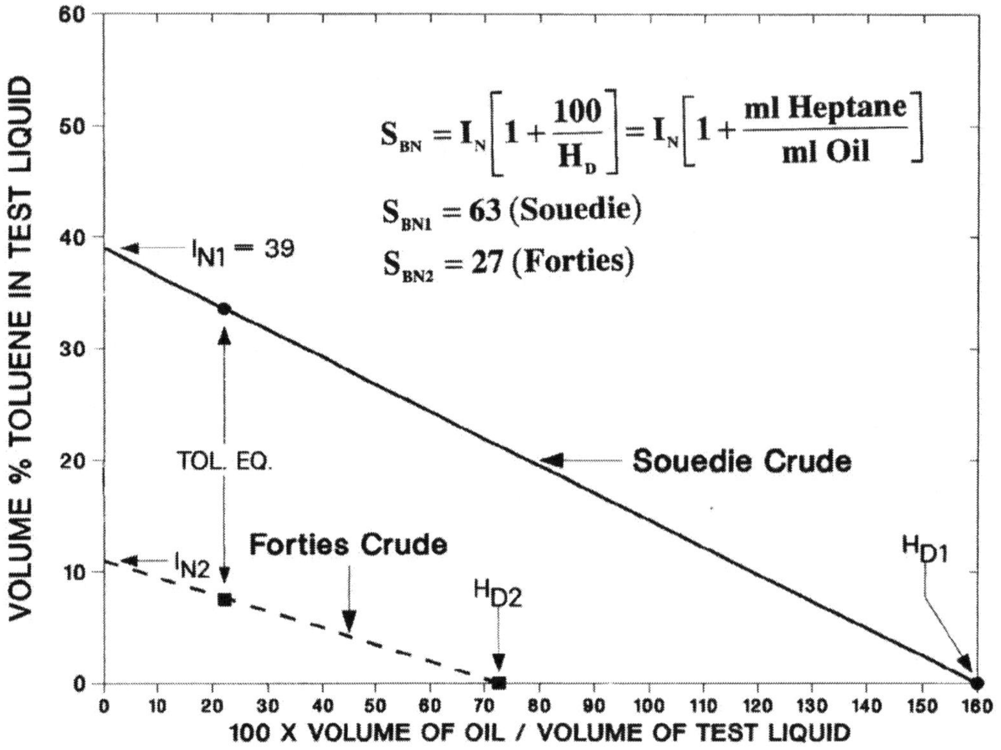

Figure 1. Determination of I_N, insolubility number, and S_{BN}, solubility blending number, for Forties and Souedie crude oils

Once these compatibility numbers are measured for the oil components, the criterion for compatibility of any blend is that the volume average solubility blending number is greater than the insolubility number of any component in the blend. Since the solubility blending number of Forties, 27, is less than the insolubility number of Souedie, 39, some proportions of these two crude oils are incompatible. The calculation of the solubility blending number of blends of the two crude oils is shown in Figure 2. If one started with a tank containing Souedie crude oil and added Forties, the solubility blending number would drop until reaching 67% Forties. Thereafter, any additional Forties would precipitate asphaltenes and cause catastrophic fouling. However, if one started with a tank containing Forties and added Souedie, the first drop of Souedie would precipitate asphaltenes and this would continue on adding more Souedie. When greater than 33% Souedie is added, no additional asphaltenes would precipitate. However, since asphaltenes in crude oil are a hybrid of a colloidal dispersion and a solution, the rate of dissolution is very slow, of the order of days. Meanwhile, insoluble asphaltenes can cause fouling and coking in a refinery that processes oil in the order of minutes. These very two crude oils were discovered to coke up a refinery vacuum furnace in hours when they were blended in the right proportions but the wrong order.

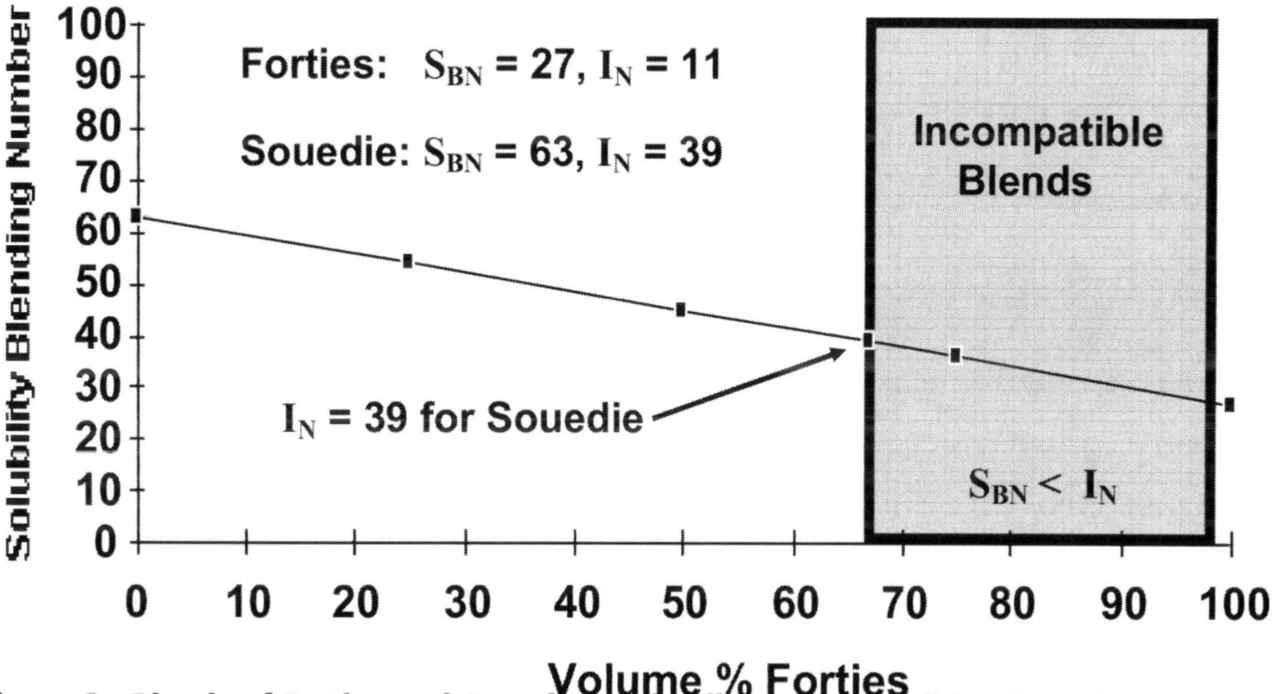

Figure 2. Blends of Forties and Souedie crude oils are compatible when the volumetric average solubility blending number is greater than the insolubility number of Souedie.

Nearly Incompatible Oils[5]

The ratio of the solubility blending number of a mixture of oils to the maximum insolubility number in the mixture is a measure of the distance from incompatibility. When this ratio is less than or equal to one, incompatibility is predicted. A value slightly greater than one predicts compatibile but nearly incompatible and a value much greater than one is very compatible. When one considers the adsorption of asphaltenes on a metal surface of process equipment, like a heat exchanger, the more soluble the asphaltenes are in the oil, the less tendency the asphaltenes will have to adsorb at a surface. To test this concept, the fouling by Souedie and Forties crude oils were measured as well as several blends[5]. These tests were conducted by F.A.C.T. using a Thermal Fouling/Coking Test Unit. Oil, at room temperature and under 700 psig. nitrogen pressure to prevent boiling, was pumped at 3 ml./min. through an annulus in which a carbon steel rod in the center was heated at a constant temperature of 760°F. As foulant built up on the rod surface, the insulating effect of the foulant reduced the ability to heat the flowing oil and caused the temperature at the outlet of the annulus to decrease. Therefore, the decrease in temperature of the flowing oil at the annulus outlet over a 3 hour period is a measure of the fouling rate of the oil. The variation in this fouling rate with volume percent Forties crude in blends with Souedie crude is shown in Figure 3.

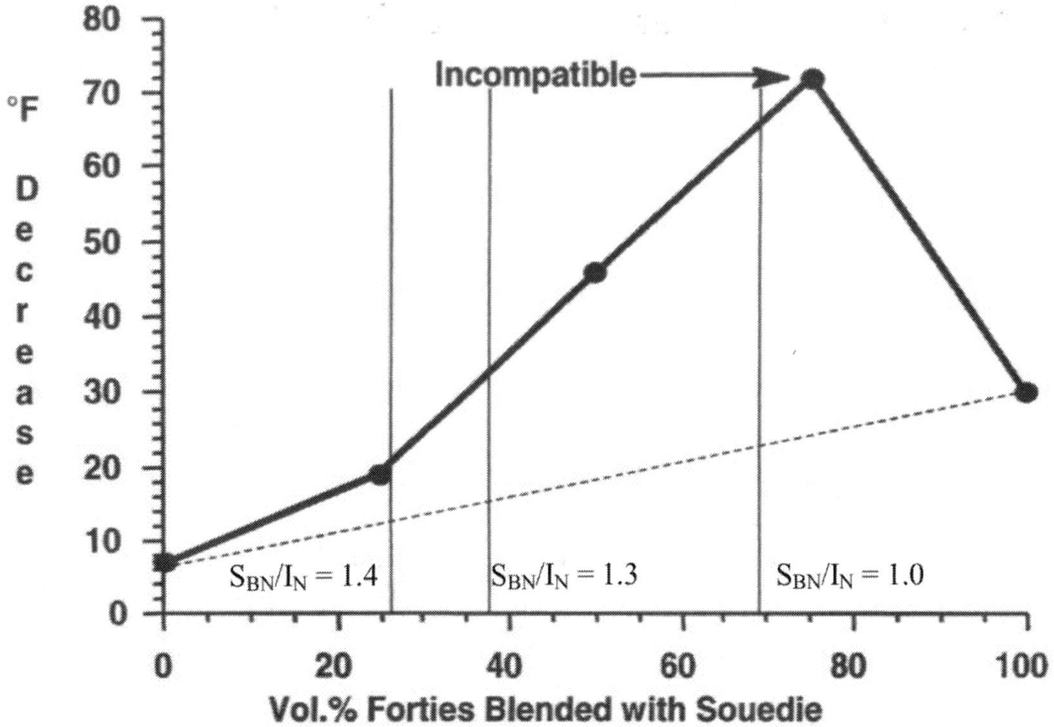

Figure 3. Thermal fouling data on Forties – Souedie blends show incompatible blends foul more but nearly incompatible blends also are high fouling.

The fouling rate at 25% Forties is only slightly higher than expected from a line drawn through the points at 0% Forties (only Souedie) and 100% Forties. Since at 75% Forties the blend is incompatible, it is not surprising that the fouling rate is the highest measured for this set. The surprise is that the rate of fouling at 50% Forties is higher than the rate of fouling of either pure component of this blend even though this blend is compatible. Based upon this limited data, blending oils such that the solubility blending number of the mixture is greater than 1.3 times the maximum insolubility number assures that the fouling rate is not greater than the fouling rate of either of the pure components. On the other hand, by maintaining the solubility blending number of the mixture to be greater than 1.4 times the maximum insolubility number keeps the fouling rate within the experimental error of the expected fouling rate. Therefore, nearly incompatible oils are in the range of S_{BN}/I_N between 1.0 and 1.4 but especially between 1.0 and 1.3.

Self-Incompatible Crude Oils[6,7]

Crude oils often contain salt, iron sulfide, rust, clays and other inorganic solids as well as wax that can be observed with an optical microscope. While these inorganic solids can be sources of fouling, waxes invariably dissolve on mild heating without fouling. However, we have discovered that commercial crude oils can contain insoluble asphaltenes that can cause catastrophic fouling and coking of crude heat exchangers and distillation furnace tubes. These insoluble asphaltenes can be observed with an optical microscope in a drop of the oil between a microscope slide and a cover slip and identified by a trained eye. However, they can be verified by diluting the oil in an equal volume of toluene and seeing the insoluble asphaltenes disappear but remain when diluting the oil in an equal volume of n-heptane. Another distinguishing feature of self-incompatible oils is that the volume percent toluene in the test liquid at the point of incipient asphaltene insolubility is independent of oil to test liquid volume ratio as shown in Figure 4 for Yme crude oil. This is because the chemical potential of a solute remains constant when the soluble and insoluble solute are in phase equilibrium. As a result, for a self-incompatible oil the insolubility number, the solubility blending number, and the toluene equivalence are all equal.

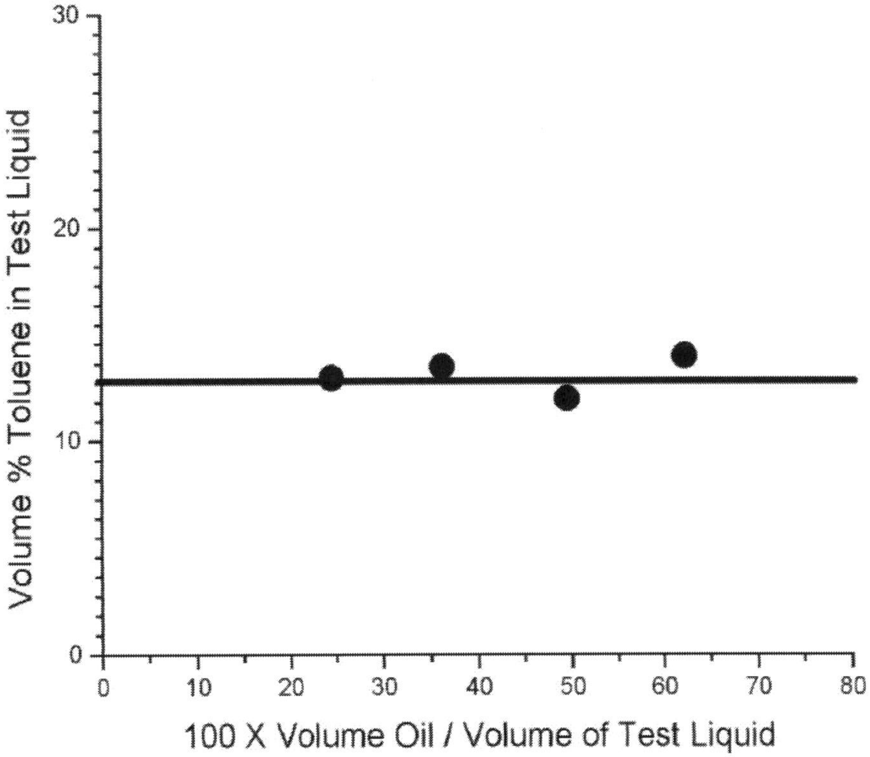

Figure 4. For self-incompatible oils like Yme crude oil the % toluene in test liquid is independent of oil concentration.

Mitigation of Organic Fouling in Crude Processing

As one can see from the above, all three modes of asphaltene fouling can be predicted once the insolubility number and solubility blending number of each potential crude oil are measured for any given refinery. Soluble Solutions provides this measurement service for refineries not wishing or equipped to do it for themselves. These measurements allow one immediately to identify self-incompatible crude oils. Soluble Solutions recommends that self-incompatible crude oils not be purchased. Any pair of crude oils in which the insolubility number of one crude oil is higher than the solubility blending number of the other crude oil is predicted to be incompatible in some proportion. Soluble Solutions recommends that such pairs of crude oils not be included in the same crude oil blend. Not only does this practice avoid incompatible crude blends but it almost always avoids nearly incompatible blends. Making these measurements and obeying these simple rules make avoiding organic fouling of crude preheat exchangers and distillation furnace tubes rather easy and straight forward. As a result, many refineries have adopted this as best practice and have a strong competitive edge over most refineries who continue to think fouling of their crude units is unavoidable. Of course, one still needs to address inorganic fouling that will be discussed later.

Organic Fouling Downstream of Crude Unit

Fouling by Coke

Coke is formed by the thermal cracking of insoluble asphaltenes[8]. If the asphaltenes are insoluble before being thermally cracked, such as on a hot heat exchanger surface or in a furnace tube, coke forms immediately and the coke is unordered (amporphous). However, if asphaltenes are soluble before reaching thermal cracking temperatures, such as in a visbreaker or a coker heater, and become insoluble during thermal cracking, the mechanism[8] is as is shown in Figure 5. Asphaltenes in the oil contain thermally stable, polynuclear aromatic cores with pendant groups connected to the core by thermally unstable bonds. When exposed to thermal cracking temperatures, these bonds break to form free radicals. As long as the asphaltenes are dispersed in the rest of the oil, these free radicals abstract hydrogen from hydroaromatics and terminate the free radicals and no coke forms (induction period). However, the loss of pendant groups make the asphaltenes less soluble. Eventually, the converted asphaltenes become insoluble and undergo a liquid-liquid phase separation. This asphaltene rich phase has little or no abstractable hydrogen. As a result, the asphaltene free radicals combine to form high molecular weight and insoluble coke. However, before coke is formed, the polynuclear aromatics in the converted asphaltenes tend to orient parallel to each other. This orientation can be detected by observing the coke with an optical microscope under crossed polarized light where ordered structures show up as bright while unordered (amorphous) structures are dark. Figure 6 shows coke formed from thermally cracking Cold Lake vacuum resid and dispersed in quinoline, an excellent solvent. The particles under normal light are spheres or agglomerate of spheres because of surface tension when it was a second liquid phase. Part of it is bright under cross polarized light because of the aromatic orientation. This is a liquid crystalline coke or the carbonaceous mesophase. Therefore, we use the presence of the carbonaceous mesophase in a foulant as an indicator to show that the foulant is coke that was formed by the asphaltene phase separation mechanism during thermal cracking. If the coke was formed by oil incompatibility on mixing and the insoluble asphaltenes

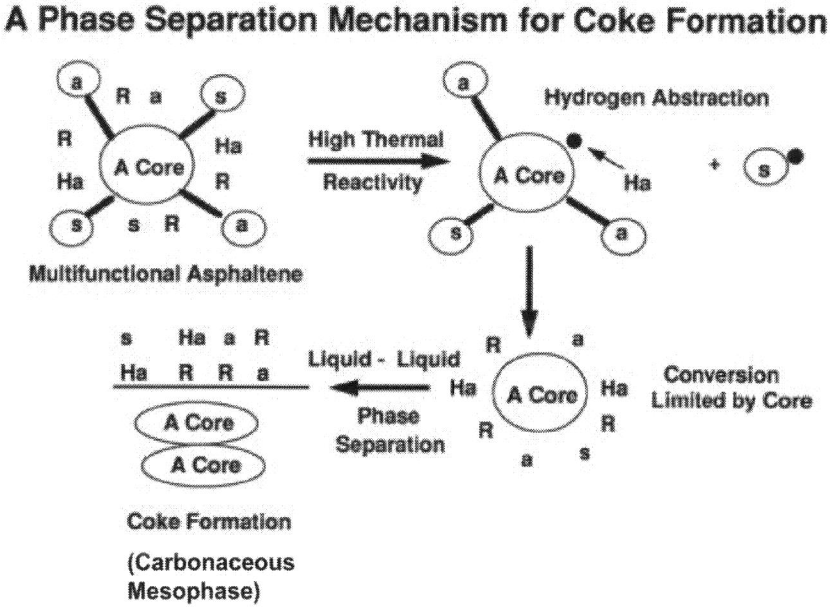

Figure 5. Mechanism of how coke forms from thermally cracking resid or oil.

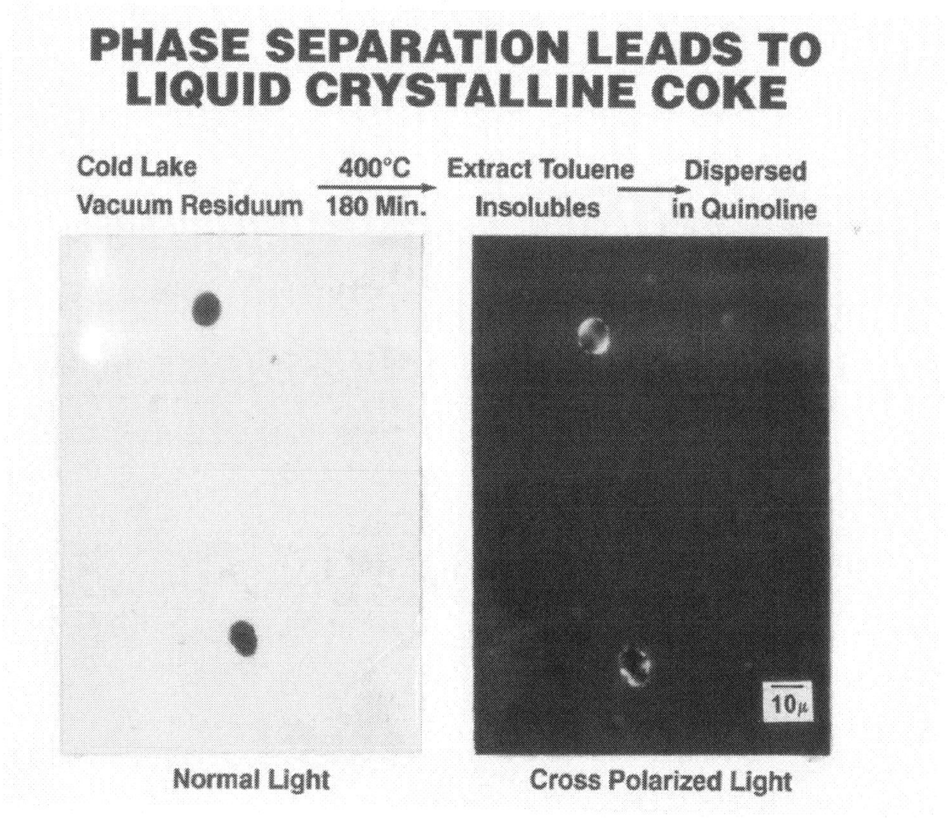

Figure 6. Micrograph showing spherical mesophase coke with bright areas under cross polarized light.

were later thermally cracked, no carbonaceous mesophase would be observed. We[9] have used the presence of carbonaceous mesophase in the heavy product of resid hydroconversion to detect that coke was forming in the reactor so that process conditions could be modified to eliminate further coke formation and avoided shutting the reactor down due to coke plugging. In addition, we have determined other indicators to detect the other common causes that enable quick diagnosis based upon foulant analysis. One of the simplest is to determine the hydrogen to carbon atomic ratio of the foulant (divide the wt.% hydrogen by wt.% carbon and multiply by 11.91):

H/C Atomic	Foulant
1.4 – 1.9	Wax
1.0 – 1.2	Unconverted Asphaltenes
0.7 – 1.0	Thermally Converted Asphaltenes
0.3 – 0.7	Coke

In visbreaking of resids it is critical to know the coke induction period for each resid feed. If the coke induction period is exceeded, the visbreaker quickly fills up with coke. However, if the conditions are far within the coke induction period, the visbreaker conversion is too low. Here is another application of the oil compatibility model. By measuring the insolubility number and the solubility blending number of the total liquid product, the coke induction period of a visbreaker can be predicted without coking up the visbreaker. Typically, the insolubility number is found to increase linearly with reaction time as is shown in Figure 7 for the visbreaking of Athabasca bitumen[10]. Although the data in Figure 7 were for various reaction temperatures, T° K, the actual reaction times, t, were transposed to an equivalent reaction time at 427°C, θ, using the Arrhenius relationship and an activation energy, E_A, of 53.1 kcal/g-mole:

$$\theta = t \ \exp[(E_A / R)(1/T - 1/700)]$$

Typically, the solubility blending number of a thermally converted resid slightly decreases linearly with reaction time. However, in this case with Athabasca bitumen the solubility blending number remained constant at 100. Nevertheless, since the oil compatibility model predicts incipient asphaltene precipitation when the insolubility number equals the solubility blending number, this is predicted to occur at a reaction time of 27.2 minutes at 427°C for Athabasca bitumen. Since coke will not form until asphaltenes become insoluble at reaction temperatures, coke will not form until slightly longer reaction times than when insoluble asphaltene sediments begin to form at room temperature. Thus, this same method of calculating when the insolubility number equals the solubility blending number provides a conservative prediction of the coke induction period.

Fouling at Inlet and Outlet of Hydrotreater

The oil compatibility model and tests also proved to be useful for determining the cause and mitigation solution for the fouling of the inlet of one hydrotreater and of the heat exchanger

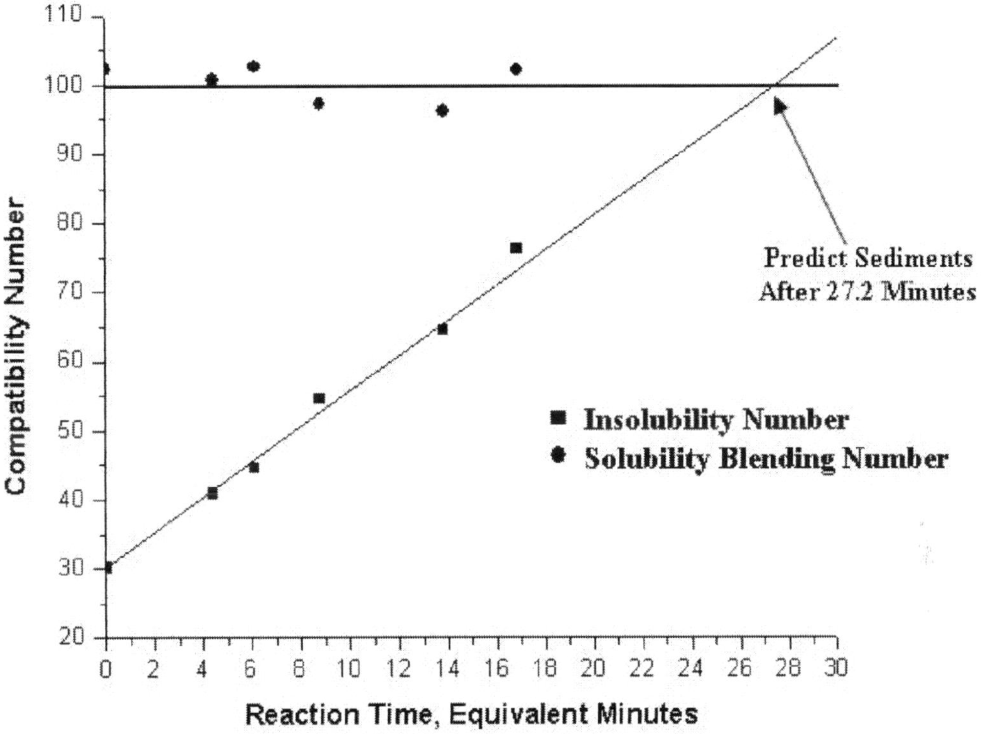

Figure 7. The Oil Compatibility Model predicts a coke induction period of 27.2 minutes for visbreaking of Athabasca bitumen at a temperature of 427°C when the insolubility number and the solubility blending number become equal.

after a different hydrotreater. The inlet of a fixed bed hydrotreater was plugging in only two weeks of operation after cleaning. While this hydrotreater had a wide mix of refinery feeds, by measuring the insolubility number and solubility blending number of the feed components, it was determined that the feed was incompatible because of the high insolubility number of slurry oil in the feed and the low solubility blending number of virgin gas oils in the feed. Actually, the insoluble asphaltenes in the feed coked the heat exchanger prior to the hydrotreater reactor. This coke flaked off the heat exchanger and plugged the inlet of the catalyst bed. Details can be found elsewhere[4]. However, by controlling the composition of the feed to maintain compatibility, the plugging problem was solved.

For an atmospheric resid hydrotreater the heat exchanger after the outlet severely fouled only near the end of one year cycles[7]. It was determined that the hydrotreater product at the beginning of the cycle was compatible but contained insoluble asphaltenes near the end of the cycle. Since the insoluble sediments were toluene soluble, it was clear that the sediments formed on cooling and were not coke (toluene insoluble) that would form at reaction temperatures. By separating the asphaltenes and resins from the hydrotreater feed and products and measuring their properties, it was determined that at the end of the cycle asphaltenes were bypassing the catalyst pores that became partially filled with vanadium and nickel. When the severity of the hydrotreater was increased to keep the product on sulfur spec, the natural asphaltene dispersant, resins, were converted. Thus, the asphaltenes became insoluble because of the loss of natural dispersant. This fouling mechanism gave the refinery a number of mitigation options in terms of changing catalysts, modifying the reactor, putting a limit on severity, monitoring the product with toluene equivalence, and improving asphaltene solubility in the product before cooling.

Coking of Coker Fractionator[11]

Coker gas oils typically contain high concentrations of hydrocarbons with a double bond one carbon away from an aromatic ring, called conjugated oleifins or dienes. These conjugated olefins can polymerize to form coke in the temperature range of 232–324°C (450-615°F). Unlike hard thermal coke, coke formed from the polymerization of conjugated olefins is brittle and is puffed, called "popcorn" coke. The best measure of the concentration of conjugated olefins in an oil is by the diene test (UOP method 326-82). Oils with a diene value of 4 or greater typically form popcorn coke in the temperature range of 232-324°C but particularly in 260–304°C (500–580°F). As a result, coker fractionators and heat exchangers before coker-gas-oil hydrotreaters are prone to form popcorn coke. Amazingly, this is one case that coke can be mitigated by raising the temperature (above the polymerization temperature window) and lowering the temperature can increase the rate of coke formation. This strongly points out the benefit of determining the cause of fouling before jumping to a mitigation solution.

Inorganic Fouling

Common inorganic foulants are iron sulfide, rust, sea salts, catalyst fines, clays or dirt, and ammonium chloride. Iron sulfide and rust are corrosion products. Therefore, they should be traced to the corrosion source or determined if they arrived in the crude oil. Iron sulfide is a black, granulated, insoluble solid that is often misidentified as coke. Iron sulfide can form directly from hydrogen sulfide reacting with iron or steel surfaces. Alternatively, iron naphthenate can form as an oil soluble salt by naphthenic acids in the oil reacting with steel surfaces and then decompose[12] and/or react with hydrogen sulfide, that is released when oil is thermally cracked. Usually, iron sulfide deposits can be found much before the reduction in the thickness of pipe walls by corrosion can be measured.

If sea salts, sodium, calcium, and magnesium chlorides, are not removed in the desalter, they can deposit wherever the water is evaporated in the preheat train. Since calcium and magnesium chloride are not thermally stable, they can decompose in the presence of water in resid conversion units, hydroconversion or coking, to form hydrogen chloride and react with ammonia released by the conversion to form the solid, ammonium chloride. This salt can be washed out of the unit, such as a coker fractionator, with water or caustic can be injected after the desalter to convert calcium and magnesium chloride to sodium chloride. However, the preferred solution is to correct the desalter operation to remove more effectively the sea salts from the crude oil. A typical desalter problem is the formation of stable oil-water emulsions (rag layer). These stable emulsions can be caused by incompatible or nearly incompatible oils, naphthenic acids or salts, clay, or iron sulfide. While inorganics are easily traced to their source, understanding the mechanism and solution of corrosion and desalting problems is now lagging behind that of organic fouling mitigation.

Analysis of the Oil: Tracing Precursors to the Source

The analysis of the oil flowing through the fouled unit should be based upon what was learned from the foulant analysis. If the foulant analysis indicated the cause of the foulant, the precursors should be sought in the oil to confirm the diagnosis. For example if the foulant is over 10 wt.% iron sulfide, the incoming oil should be analyzed for soluble and insoluble iron and for naphthenic acids[13] (Total Acid Number) and for reactive sulfur[14]. This should determine if the corrosion was in the unit or upstream of the unit and the type of corrosion if in the unit. On the other hand, if the foulant analysis is not able to distinguish among several causes, the oil should be analyzed for the complete range of possible precursors. Examples are shown below:

Common Cause	Precursor Analysis
1. Insol. converted asphaltenes	Asphaltenes with H/C atomic < 1.0, Oil Compat. Tests
2. Oil incompatibility on mixing	Oil Compatibility Tests
3. Coke	Conradson carbon, asphaltenes, carbon. mesophase
4. Polymerization of conj. olefins	Dienes
5. Inorganics	Metals found in foulant (IC Plasma Emission Spect.)

Once the precursor has been identified and measured, it can be traced from the fouled unit upstream until locating the source. This should provide enough information so that a number of ways of interrupting the path from precursor source to foulant will become obvious. The refinery then can select the mitigating action that best fits their situation.

Conclusions

A systematic procedure has been given for mitigating fouling by determining the cause of refinery fouling and tracing the precursors of fouling to the source, These are based upon understanding the chemical and physical mechanisms of the most common causes of refinery fouling and the use of key indicators that enable the quick diagnosis of the fouling and focusing on the correct precursors. The understanding of organic fouling mitigation is in good shape because of the oil compatibility model, the phase-separation mechanism of coke formation, and the mechanism of the polymerization of conjugated olefins to form popcorn coke. Inorganic fouling mitigation needs further emphasis to improve the removal of water soluble and oil insoluble inorganics in desalting and to prevent corrosion by naphthenic acids and hydrogen sulfide.

References

1. Wiehe, Irwin A. (2001), "The Chemistry of Petroleum Fouling", Proc. 4th Int'l Conference on Refinery Processing, AIChE, New York, pp. 204-210.

2. Lemke, Harald K. (1999), "Fouling in Refinery Equipment – An Overview", Proc. 1st Int'l Conference on Petroleum Phase Behavior and Fouling, AIChE, New York, pp. 375-382.

3. Wiehe, Irwin A. and Raymond J. Kennedy (2000), "The Oil Compatibility Model and Crude Oil Incompatibility", Energy & Fuels, 14, pp. 56-59.

4. Wiehe, Irwin A. and Raymond J. Kennedy (2000), "Application of the Oil Compatibility Model to Refinery Streams", Energy & Fuels, 14, pp. 60-63.

5. Wiehe, Irwin A., Raymond J. Kennedy, and G. Dickakian (2001), "Fouling of Nearly Incompatible Oils", Energy and Fuels, 15, pp. 1057-1058.

6. Wiehe, Irwin A. (2001), "Are You Processing Incompatible Crude Oils?", World Refining, October pp. 24-28.

7. Wiehe, Irwin A. (2004), "Self-Incompatible Crude Oils and Converted Petroleum Resids", J. Dispersion Sci. and Tech., 25, pp. 333-340.

8. Wiehe, Irwin A. (1993), "A Phase-Separation Kinetic Model for Coke Formation", Ind. Eng. Chem. Res., 32, pp. 2447-2454.

9. Wiehe, Irwin A. (1997), "Mechanisms of Coke and Sediment Resulting from Resid Conversion", Spring ACS Meeting, Petroleum Division, San Francisco.

10. Rahimi, Parviz M., Alan Teclemariam, E. Taylor, Theo deBruijn, and Irwin A. Wiehe (2003), "Determination of Coking Onset of Petroleum Feedstocks using Solubility Parameters", ACS Fuel Chem. Preprints, 48, pp. 103-105.

11. Wiehe, Irwin A. (2003), "Mitigation of the Fouling by Popcorn Coke", Pet. Sci. and Tech., 21, pp. 673-680.

12. Panchal, C.B. (1999), "Fouling Induced by Dissolved Metals", Proc. 1st Int'l Conference on Petroleum Phase Behavior and Fouling, AIChE, New York, pp. 367-372.

13. Robbins, Winston K. (1998), "Challenges in the Characterization of Naphthenic Acids in Petroleum", PREPRINTS, ACS Div. Petroleum Chem., 43, pp. 137-140.

14. Robbins, Winston K. (2000), "Challenges in the Analytical Characterization of Non-Thiophenic and Thiophenic Sulfur Compounds in Petroleum", PREPRINTS, ACS Div. Petroleum Chem., 45, pp. 68-71.

Newer Techniques to Control Fouling in Crude Pre-Heat Exchangers

Mahesh Subramaniyam PhD, Ramaswamy Perumangode

Fouling in the crude unit pre-heat exchanger train has always resulted in in-efficient operations often resulting in higher energy costs, lower throughputs or profits, increased cleaning cycles resulting in higher maintenance costs. In recent years, these problems have only increased and some of the causes are:

1. Increased energy costs due to rise in fuel oil/fuel gas costs
2. Increase in heavier crude oils being processed in blends or in blocked out operations
3. Increase in quantity of High acid crude processed in the refinery
4. Increased chemicals used in upstream oil production

There have been several methodologies used over the years to reduce the problems caused due to these problems. There have been hardware changes that have been used like using special inserts into the heat exchangers to reduce the rate of fouling. Having parallel heat exchangers that could allow the by-pass of fouled exchangers to increase furnace inlet temperatures have also been used to reduce the impact of fouling on normal operations.

With the use of various blends and possible compatible issues caused while blending the crude oil, there have been several established procedures to reduce incompatibility or near incompatibility in crude blends. Also, several newer crude oils also have an increased inorganic content like Iron Sulfide & Silica that contribute towards fouling.

Conventionally, success of anti-foulants has not been significantly recorded due to the higher cost of treatment or lower return on the investment. Typical anti-foulant efficiency rates have also been 50% +/- 10%. Furthermore, such anti-foulants have been promoted primarily to tackle phase separation due to asphaltene present in the crude oil. Typical deposit analyses in pre-heat exchangers would reveal:

Inorganics: 10% - 30%
Organics : 90% - 70%

The inorganic content could vary from salts to iron sulfide or silica depending on the location of the heat exchanger in the heat exchanger train. The inorganic content often can accelerate the rate of fouling in the pre-heat exchangers. The organic content would primarily show up as dehydrogenated coke particles. These deposits could also contain a considerable portion of aromatic or aliphatic molecules.

While it is an easy practice to suggest that anti-foulants can be used, it is very important that the return on investment for the use of such anti-foulants is maximized. Some of the focus areas that would result in such an improvement are:

1. Anti-foulant efficiency of minimum 75%
2. Improving performance of the heat exchanger train by effectively reducing fouling of heat exchangers with higher heat duties.
3. No adverse affects in downstream units
4. Quick laboratory screening process to determine product affectivity

Increased chemical efficiency for this application is extremely difficult to achieve due to a variety of reasons. With the onset of fouling, the heat exchanger train does not behave in the same way that it was originally designed to be. This results in shift of temperatures depending on how the heat configuration is designed. More often than not, fouling in the hottest exchangers results in increased temperatures in some of the warm exchangers thereby increasing the rate of fouling in such exchangers. It is hence very important to understand the heat configuration of each heat exchanger train to design a program that will provide maximum gain for each of the heat exchanger train.

Anti-foulants formulated to be an asphaltene dispersant had limited success, higher anti-foulant performance efficiencies has been achieved by:

- Formulating acidity modifiers in the treatment package
- Disperse iron sulfide, iron sulfide deposition on the heat exchanger surface is more damaging to the final pre-heat temperature.
- Identify heat exchangers that are doing maximum work in raising the pre-heat and reduce the rate of fouling effectively of such heat exchangers.

While using a thermal fouling test rig to screen potential fouling rates and response to anti-foulant treatment is very common, additional lab studies help to better understand chemical action. Some of the tests used are:

- Asphaltene dispersant test
- Iron Sulfide dispersant test

Fouling of the pre-heat train is a very dynamic problem and is very difficult to control. The techniques that shall be discussed during our session shall cover these techniques in detail with commercial examples.

220c Improved Crude Unit Design through Experimentation

Christopher A. Bennett[1], R. Stanley Kistler[1], Shakir Khambaty[2], Krish Nangia[2], Ahmad Al-Jemaz[3], and Walid Al-Ghawas[3]. (1) HTRI, 150 Ventura Drive, College Station,, TX 77845, (2) Fluor Daniel, One Fluor Daniel Drive, Aliso Viejo, CA 92698, (3) KNPC, P.O. Box 69, Safat, 13001, Kuwait

Raw (neither desalted nor flashed) crude oils were tested at similar conditions in the HTRI High Temperature Fouling Unit to obtain fouling rates and resistances. Although absolute fouling resistance can be difficult to apply, relative fouling rates were compared to known operating experience, allowing improved designs of heat exchangers and fired heaters in the crude preheat unit, thereby optimizing unit run length.

Fouling resistance as a function of time was measured for Eocene, Ratawi, Lower FARS, and Kuwait Export (KEC) crude oils at surface temperatures ranging between 360°C and 445°C, a nominal bulk temperature of 315°C, and a wall shear stress of 13 Pa. Coking was the dominant fouling mechanism for all tests, which were designed to model conditions in the fired heater. The fouling rates of all four crude oils were found to be exponentially dependent upon temperature with significantly different activation energies; as a result, the fouling propensity was divided into three regimes. At low solid-liquid interface temperatures (~370°C), the relative fouling rates were: Eocene ~ Lower FARS > KEC > Ratawi. At intermediate interface temperatures (~410°C), the fouling rates of all four crude oils were similar. At high surface temperatures (~455°C), the relative fouling rates were: Ratawi > KEC ~ Eocene > Lower FARS.

HEAT EXCHANGER TUBE INSERTS -

AN UPDATE WITH NEW APPLICATIONS IN CRUDE DISTILLATION UNIT

VACCUM APPLICATIO AND REBOILERS

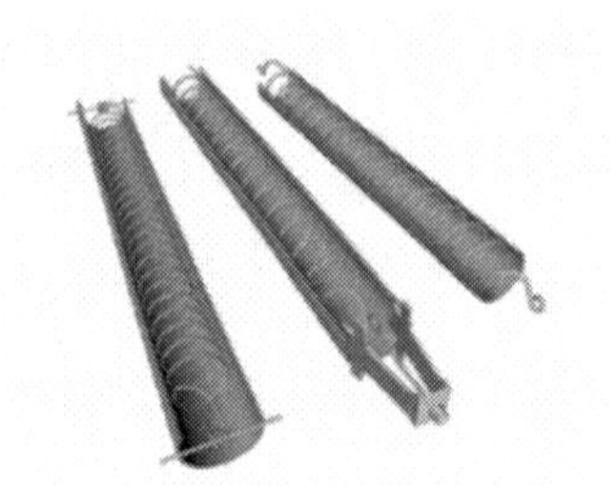

By:

Francois Pouponnot, Petroval, France

e-mail: f.pouponnot@petroval.com

Artur W. Krueger, Senior Consultant, Petroval, Houston

e-mail: art.krueger@att.net

Presented at:

AIChE Spring Meeting

April 23 - 27, 2006

Disney's Dolphin Hotel, Orlando, FL

ABSTRACT

Heat exchanger tube inserts have been used for many years as reliable means for heat transfer enhancement and fouling mitigation in petroleum refineries and chemical plants.

In this paper, we will present several new application examples for optimization of heat exchangers and related equipment, resulting in enhancement of the function of such equipment and improvement of on-line availability of the plant.

Usual insert applications in refining pre-heat trains continue to give good performances for fouling mitigation and heat transfer conservation. For the refiner, this improvement brings interesting energy savings but also longer run times with better overall performance of the pre-heat train. Tube inserts are largely used in CDU and VDU applications.

In some new applications such as reboilers (liquid and mixed phase flows), the observed efficiency was valuable because the shorter initial service times have been expanded substantially (in some cases 4 to 6 times longer than the initial service time), and with higher efficiency levels achieved during this extended service time.

Some of these applications and results are described in this paper.

I. TUBE INSERT DESCRIPTIONS

a. SPIRELF

The Spirelf system falls within the category of "on-line mechanical cleaning devices". The principle is based on the insertion of flexible metal devices, in spiral form, into the tubes of "shell and tube type" heat exchangers. The devices are stretched over the length of each heat exchanger tube and are held in place by straight wires at each end (inlet and outlet).

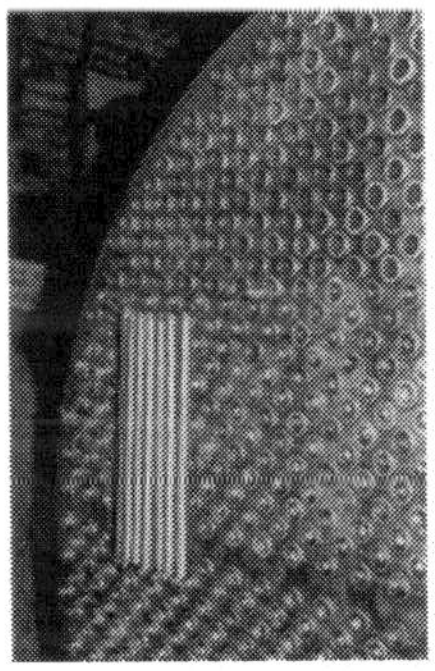

Fig. 1a

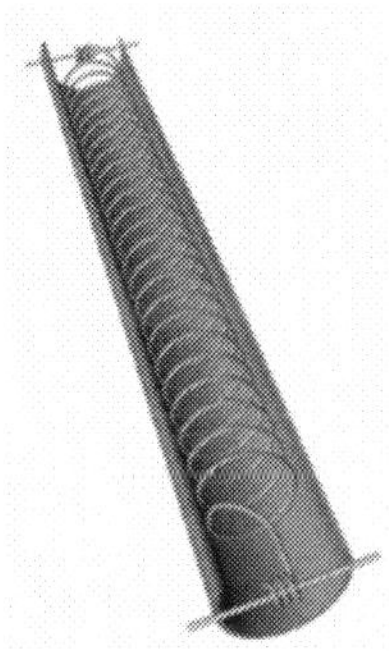

Fig. 1b

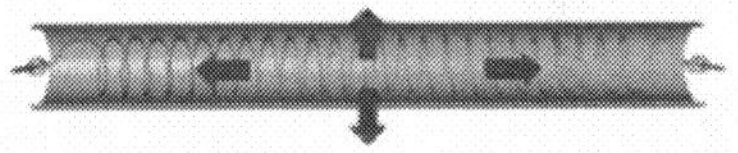

Fig 1c

OPERATING PRINCIPLE

After installation the insert stays under tension. Under the effect of the circulating flow, it enters into vibrations, radially and axially. The repeated contact between the loops of the devices and the inner tube wall has two effects:

- prevent the formation of deposits,
- break the boundary layer in the tube side flow.

As a consequence, the mechanical effect of the SPIRELF system reduces fouling build-up inside heat exchanger tubes and the turbulent effect achieves an improvement of the heat transfer rate (the internal heat transfer coefficient is multiplied by 1.8). Overall the sum of these two effects results in a reduction of the apparent fouling factor by as much as three-quarters.

b. TURBOTAL

The Turbotal system falls within the category of "on-line mechanical cleaning devices". The main aim of Turbotal is the control of fouling in heat exchangers. The principle is based on the insertion of rotating metal devices, in rigid helicoidal form, into the tubes of "shell and tube type" heat exchangers.

Fig. 2a

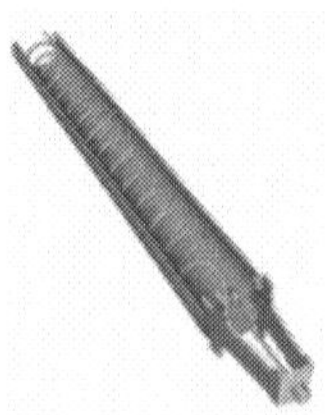

Fig. 2b

HELICOIDAL DEVICE ROTATING BY THE FLUID FLOW

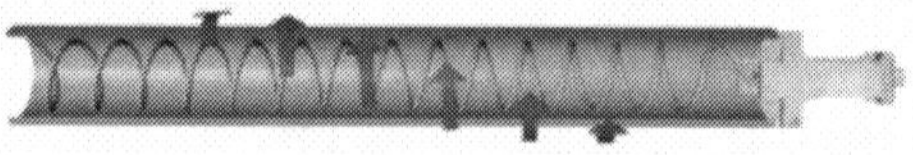

Fig.2 c

OPERATING PRINCIPLE

Turbotal is a system, which is held at the inlet of tubes by a fixing device allowing the mobile to rotate around its axis by means of the fluid flow.

This rotation causes a high turbulence in the flow and improves thus the internal heat transfer coefficient. As the boundary layer is continuously renewed, the wall temperature is lowered and fouling is slower. Combined with the mechanical effect of the rotation, TURBOTAL has a preventive effect against fouling with this two effects:

- prevent the formation of deposits,
- break the boundary layer in the tube side flow.

The TURBOTAL is extremely versatile. The rotation speed can adapt itself according to the operating conditions. The TURBOTAL is able to cope with :

- flow rate variation
- fluid properties variations (viscosity, density...)
- accidental vaporization (caused for instance by massive water occurrence)

The fixing device of the coil is placed at the outside of the tube. The design of the coil itself is adapted for each case to the geometry of the tubes and the process conditions. As a consequence, the mechanical effect of the Turbotal system reduces fouling build-up inside heat exchanger tubes and the turbulent effect achieves an improvement of the heat transfer rate (the internal heat transfer coefficient is multiplied by 1.8). Overall the sum of these two effects results in a reduction of the apparent fouling factor by as much as three-quarters.

c. AREAS OF APPLICATION OF TURBOTAL AND SPIRELF

Due to the operation principle, Turbotal and Spirelf devices can be used in all multi-tubular heat exchangers with straight tubes, where deposits within the tubes are found.

The Turbotal and Spirelf system applies:

- to all fluid up to 360°C-680°F,
- in a liquid phase,
- where flow velocities are between 2 and 6 ft/s (0.6 to 3.0 m/s) inclusive,
- in pipes of an outside diameter between 1/2 and 2 inches inclusive
- in tubes of length between 10 and 33 ft inclusive.

The first applications were in the refinery industry, particularly in atmospheric distillation preheat trains in which the fight against fouling is crucial. Nowadays Turbotal and Spirelf systems can also be applied to other industries, such as chemical and petrochemical applications.

d. FIXOTAL

Fig. 3a

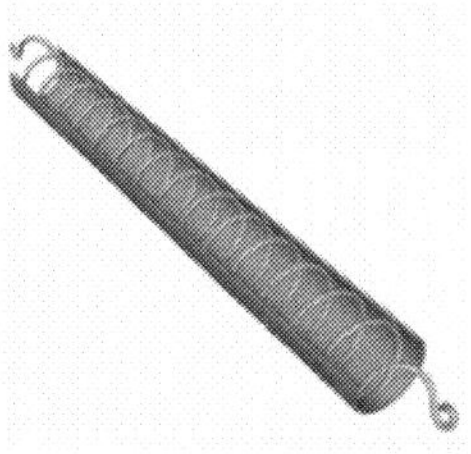

Fig. 3b

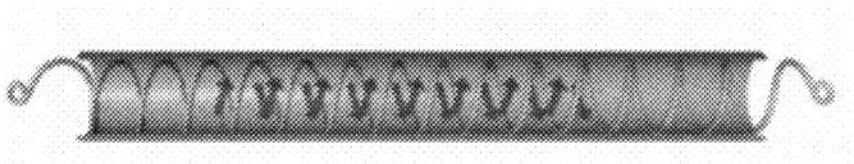

Fig. 3c

FIXOTAL is a fixed device, which has been developed by TOTAL. Its function is primarily to increase the thermal efficiency of tubular heat exchangers and also to reduce the tubes fouling rate.

DESCRIPTION OF THE DEVICE

FIXOTAL consists of a wire coil, which is inserted inside every tube, with the wire in firm contact with the inside tube wall. Once in place, the device has no possibility of the slightest displacement (no vibration, no rotation, and no translation), the device can be easily removed if necessary.

The main effect of the wire is to create a turbulence effect, thus decreasing the boundary layer at the wall. The result is a significant increase of the tube side heat transfer coefficient, resulting in of a moderate increase of pressure drop through the tube. This has two major consequences :

- Increase of the thermal efficiency of the exchanger, due to an increase of the overall heat transfer coefficient. The extent of the gain depends on the split of the overall heat transfer resistance between the inside and the outside tube surface.
- Decrease temperature gradient between tube wall and bulk. This leads to a decrease of the tube side surface fouling rate, especially with fouling which is wall temperature dependant.The parameters influencing the heat transfer enhancement and the pressure drop have been investigated at pilot scale for one & two-phase flows.

To summarize, FIXOTAL is easy to install technology, which increases the thermal efficiency of tubular heat exchangers and also reduces their tube side fouling rate.

Its key advantages are :

- sustain of heat duty
- reduction of fouling
- debottlenecking aspect for existing undersized equipment

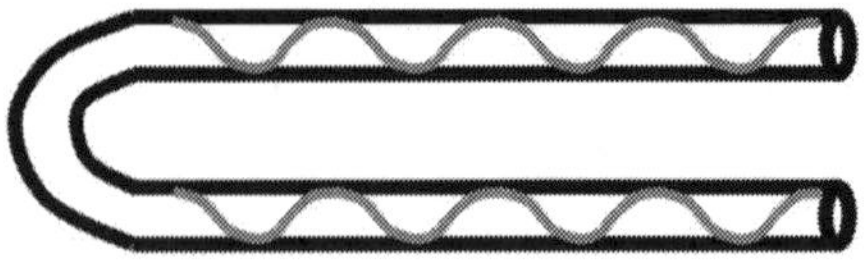

Fig. 4: U-tube equipped with Fixotal

II. APPLICATIONs OF TUBE INSERTS TO OIL REFINERIES

Spirelf, Turbotal and Fixotal are especially well adapted to the use in crude oil units, where the main fouling layer is composed by asphaltenes. In this application, the aim is primarily energy efficiency but also to improve maintenance management through extension of exchanger run times which at the same time results in improved throughput capacity.

Other units that can be equipped by Spirelf, Turbotal and Fixotal are desulphurisation units, hydrotreaters, with the condition of liquid phase tube side of the exchangers. Application on Naphtha, Kerosene, Gas Oil Light Crude Oil have been developed for several years successfully.

a. Main advantages of Spirelf, Turbotal and Fixotal

Spirelf, Turbotal and Fixotal installation and use do not require any modification of the heat exchanger.

Increased heat transfer rate :

- through higher flow turbulence (inner heat transfer coefficient -X 1.8)
- reduction of fouling and coking in heat exchanger tubes
- result : energy savings of up to US $ 500,000 per year, example : EEC installation -23,800 barrels or 3 400 tons of oil equivalent/year.
- Average cost/benefit ratio: 1 : 5

Reduction of maintenance work/ cost :

- decrease of exchanger cleaning through reduction of tube side fouling
- extended run time through increase of cleaning intervals example:
  for unequipped exchangers = 6 month cleaning cycles,
  for equipped exchangers = 12-20 months cleaning cycles
- extended tube life/reduced leakage due to reduction of under-layer corrosion
- designed to avoid mechanical tube erosion.

De-bottlenecking improved throughput capacity :

- improved operating conditions at main column through equipment of pump around/reflux line exchangers
- in case of limited furnace capacity, increase of crude temperature possible by equipping exchangers directly
- before furnace (29 degr. F temperature increase documented compared to unequipped train)
- improved throughput capacity due to reduced need for exchanger cleaning (runtimes multiplied for equipped exchangers)

- better cooling of outgoing products to storage

b. Pressure drop

- Delta-P caused by Spirelf, Turbotal and Fixotal devices = between 1 and 3 psi or 0.1 and 0.2 bar per pass (depending on tube dimensions, crude type, flow velocity) Delta-P for two-pass exchanger = 3 psi or 0.2 bar (average value at 1 m/s flow velocity)
- Delta-P caused by fouling to be considered for comparison
- Higher Delta-P resulting in increased flow turbulence (inner heat ~ transfer coefficient X 1.8)

c. New cleaning frequencies

By experience : the cycle duration is generally multiplied by 2 sometimes by 3; which translates in avoidance of frequent cleaning operation. (depending on the tube diameter, the fouling degree...)

Cleaning frequency before a tube inserts equipment	Service time of tube inserts or new cleaning frequency if chemical cleanings are not used
6 months (Turbotal, Spirelf and Fixotal)	1 year to 1 ,5 year
1 year (Turbotal, Spirelf and Fixotal)	2 years to 3 years
18 months (Spirelf and Fixotal)	3 years and more

d. Typical installations

- crude preheat train (with crude flow tube side) last exchangers before furnace, pumparound/reflux lines, product lines to storage
- Retrofit installation
- New exchangers to be equipped to improve exchanger performance/extend anticipated run time.

e. Possible alternative solution

In order to be able to run for longer time, there is possibility to perform some chemical cleaning on line with Spirelf and Fixotal in place. Tube inserts are used in order to keep high heat transfer efficiency and the chemical cleaning at regular period is a solution to recover the initial efficiency during an extended service time.

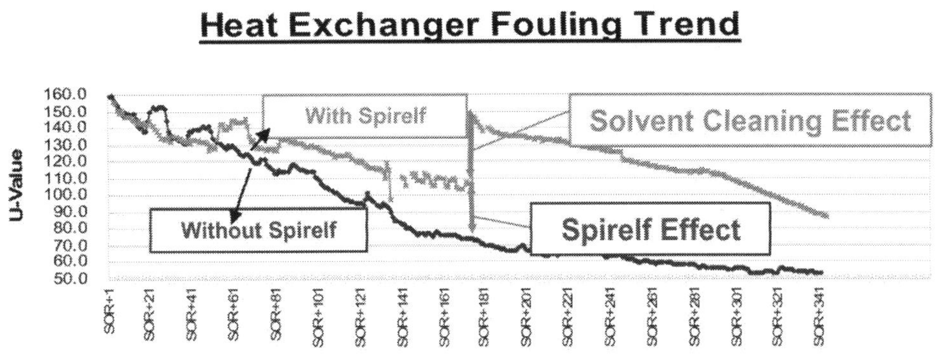

Fig. 6

III. INDUSTRIAL RESULTS : THE TURBOTAL SYSTEM - EFFICIENCY COMPARISON IN A CRUDE UNIT

a. Unit description

In November 2003, Turbotal has been installed in CDU 1 heat exchangers 8ABEF. In order to evaluate the efficiency of the Turbotal in these heat exchangers, a comparison study has been performed between the 2 branches of the pre heat train (one equipped 8AB, one unequipped 8EF, see fig below).

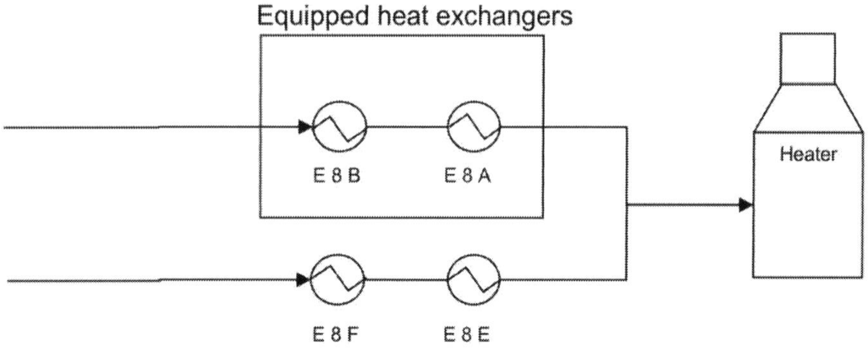

Fig. 7

After 1 year operation, an evaluation based on refinery results is prepared. Working performance of equipped heat exchangers is still around 40% higher than the unequipped ones. The observed extra DP (average value around 0.6 bars) due to Turbotal is in line with the normal extra DP evolution. Based on these experimental results, the energy savings generated by a better preheat train efficiency are already paid back.

b. Heat exchanger characteristics (tube side) (Used for expected for study/real)

- Unit: Crude Distillation Unit (Pre heat train)
- Tube number / bundle: 1 424
- Tube length: 6,100 mm
- OD / BWG: ¾" / 14
- Product tube / shell side: Crude / Atmospheric residue
- Flow velocity (tube side): 1.1 m/s
- Tube insert: Turbotal
- Replacement frequency: 2 years

The working conditions since the installation have been stabilized and generally match with the expected working conditions in tube side.

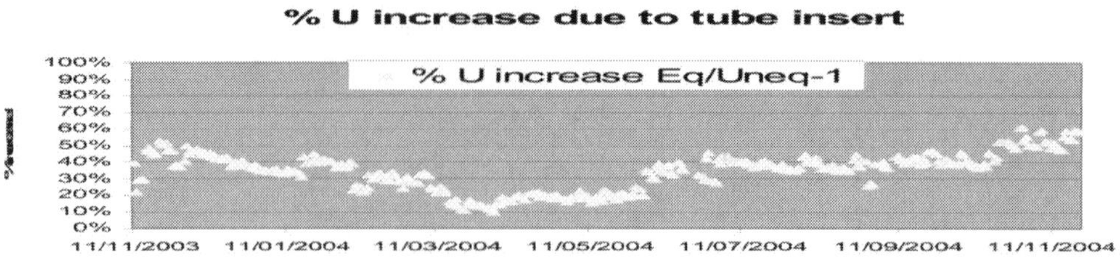

Fig. 8

Explanation of the evolution:

- Fouling part of the new tubes under Turbotal control,
- Fouling stabilization in equipped tubes and turbulent effect for heat transfer compared with fouling accumulation situation in unequipped ones.

According to the flow sheet, the shell side product is atmospheric residue which is usually a fouling product which strongly influences the global duty of the heat exchanger and against tube insert cannot work. The equipped heat exchangers are now running with 50% improvement for the duty compared than without. The fouling in tube side would be a very substancial limitation of the global heat transfer in these heat exchangers.

The Delta T (Outlet – Inlet) evolution between both branches has been quite significativelly improved due to Turbotal:

Average value for the first year	inlet	outlet	Delta T	DT after 1 year
Flow rate value in 8AB (°C, tube side)	203	242	39	37
Flow rate value in 8EF (°C, tube side)	212	242	30	27

Tube inserts installation in heat exchanger 8AB increased the average DT during the first year by 9°C. A conservation of 9°C is possibly observed at the furnace inlet by using tube insert during this period. The final monitored Delta T is still +10°C

c. Energy savings evaluation

The interest of Turbotal installation is both cleaning effect and turbulent effect in tube side by mechanical action. As it has been observed in this heat exchanger, it brings some better heat transfer performance which help to keep higher outlet temperature of the heat exchanger and when it is installed into the most strategic heat exchanger, this thermal improvement can bring some money savings due to energy savings among others.

In this particular case, energy savings can be calculated as follow:

Energy saving calculation:

Data: normal crude flow-rate : 800 metric t/hr
- specific heat of crude : 0.65 kcal/kg.°C at 210°C
- furnace yield : 0.85
- heating power of fuel : 9,700 kcal/kg

1°C higher at the furnace inlet corresponds to a daily savings of:

$$\frac{800{,}000 \times 24 \times 0.65}{0.85 \times 9{,}700} =$$ 1,500 kg of fuel saved per day per °C (9 bbl of fuel oil saved per day per °C)

Energy savings during one year of run : 3,285 bbl of fuel oil per year per °C
Based on 20 Euro / bbl for fuel oil: Euro 65,700 savings per year per °C

With Turbotal, it has been calculated during the 6 first months of last run that an improvement of 9°C is a realistic improvement in these heat exchangers.

Euro 591,300 savings per year per 9°C

Global energy savings over 1 year : Euro 591 300 during the first year

This estimate is based on energy savings during one year. Then the 18 additional months would generate only purely energy savings. Usually Turbotal installation generates also production loss savings and maintenance savings (if in between shutdown can be avoided).

d. Conclusion in this CDU

Results of efficiency improvement and technical characteristics of this evaluation are quite positive because it is in line with the expected results:

- The heat transfer improvement ("U" value) was at start of run around + 50% with an average value during the 6 first months around + 30 % and a final stabilization at + 20%. During the second semester, the duty increased from 20 % up to 50%. The increased was especially visible during a stabilized period between end of August and November 2004.
- This improvement results into higher outlet temperature of +10°C for equipped branch. And the average improvement is around +9°C.
- Energy savings are then better then initially expected
- Pay back time in less than 6 months.

IV. INDUSTRIAL RESULTS : SPIRELF IN CDU

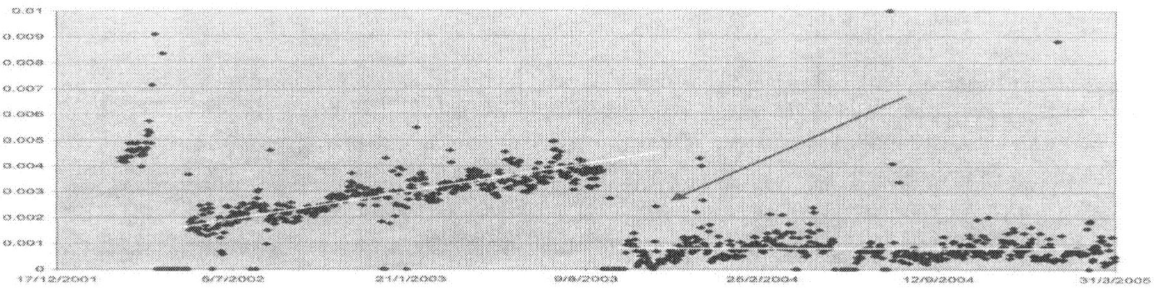

Fig 9.

General working conditions for this heat exchangers are:

- 800 tubes and 2 passes on tube side (for each bundle),
- **Flow rate: 249 t/h** (Fluid velocity:1.2 m/s),
- Desalted crude is flowing tube side and Top pump around is in shell side,
- Straight tubes 6,096 mm / OD 19,05 mm / BWG 14.
- Cleaning frequency of 1.5 years, when the hydraulic limit is reached.
- 0.3 bar SOR and 1.8 bar EOR
- Heat exchanger situated just after the desalter
- Strong fouling deposit has been observed at the cleaning time.

After Spirelf installation, the fouling factor had been reduced to the minimal value and kept under optimal value and optimized working conditions. Since it is installed, the fouling factor is stabilized to a low value and no excessive extra DP had been observed.

V. INDUSTRIAL RESULTS : THE TURBOTAL SYSTEM – INBETWEEN REMOVAL

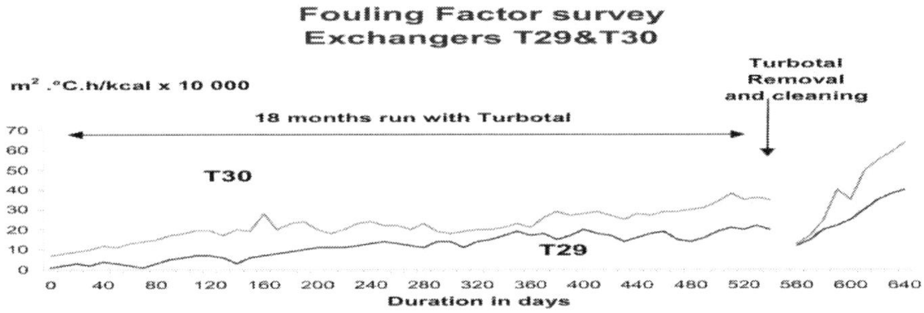

Fig. 10

After 18 months operation Turbotal had been removed because of production unexpected turnaround for a specific problem. The normal shutdown planned for 6 months later, the refinery decided not to install Turbotal for only 6 months but to go on without it during the final part of the run. The fouling factor increased dramatically without Turbotal where as it was under control during the period with Turbotal. At the shutdown, they reinstall the Turbotal in order to keep control with fouling factor. In this case the fouling factor evolution (slope) had been reduced by more than 3 time the one without tube inserts.

Since November 2002, this 2 heat exchangers are equipped with Spirelf with the idea to run and to perform some chemicals cleaning every years.

VI. INDUSTRIAL RESULTS : VDU APPLICATION - RELIABILITY ENHANCEMENTS WITH THE "FIXOTAL SYSTEM"

Heat exchanger characteristics

- Unit: Vacuum Distillation Unit (Pre heat train)
- Position in the pre heat train: Before the furnace
- Tube number / bundle: 1200
- Tube length: 6,100 mm
- OD: 1"
- Replacement frequency: 3 years (since 1998)
- Product tube / shell side: Atmospheric residue / bottom reflux
- Flow rate (tube side): 400 t/h

Installation problematics

Before installation, the heat exchanger was 6 passes with a strong problem of efficiency due to:

- Strong fouling even with high fluid velocity (around 2.2 m/s)
- Short residence time (compared with the effective heat transfer coefficient with fouling)

Observed results and production constraints:

- Poor heat transfer efficiency
- High extra DP due to fouling
- Cleaning frequency of 6 months
- Furnace bottleneck because of lower inlet temperature

Decided modifications:

- Pass configuration modification: from 6 to 2 passes (Flow velocity from 2.2 to 0.8 m/s)
- Fixotal installation

Expected results:

- Reduction of pass number: increase of residence time and global heat transfer. Strong increase of the fouling is foreseen (X3).
- Fixotal installation: fouling mitigation and heat transfer increase by efficient turbulent effect at the tube wall.

Observed results :

- Efficient heat transfer coefficient during the run
- Comparable service time
- Chemical cleaning every 6 months to recover periodically the initial performance.

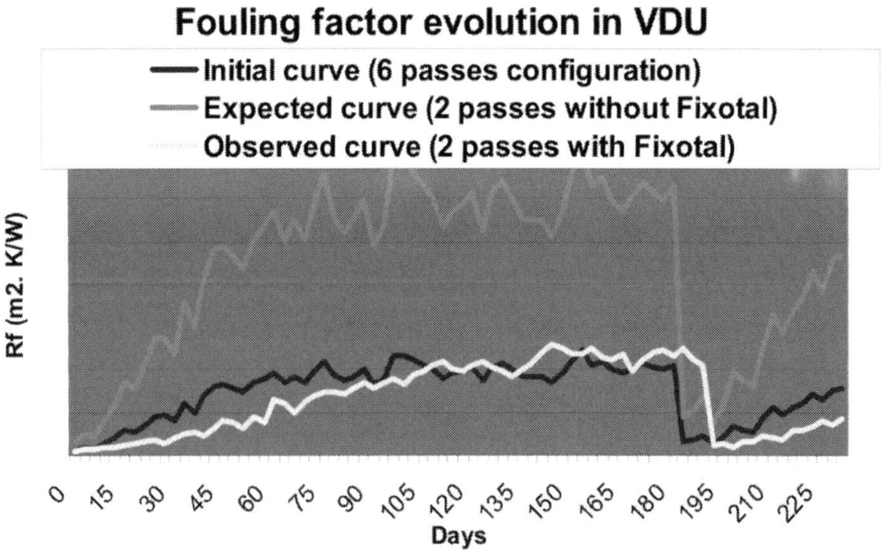

Fig. 11

Conclusion:

- Good thermal efficiency due to pass number reduction and Fixotal installation
- Good fouling mitigation due to Fixotal. Comparable fouling evolution whereas it should have been 3 times more severe and bring quite strong problem of the service time and production enhancement.

Chemical cleaning every 6 months in order to recover the initial efficiency but without removing Fixotal. A replacement is expected every 4years.

VII. INDUSTRIAL RESULTS : REBOILER APPLICATIONS

Generally reboilers are heated on shell side by steam, and the organic product flews tube side.

In terms of heat transfer, there is a big difference between the outside and the inner heat transfer coefficients, and this last one constitutes the main resistance for the heat transfer. This characteristic is even more pronounced in case of fouling coming from the organic product. Consequently, any increase of the inner coefficient (IHTC) translates to about the same improvement on the overall heat transfer coefficient (OHTC).

a. Action of the Spirelf System on vertical reboilers' performance at start-up.

- Improvements to be expected from Spirelf System.
- Spirelf devices have two actions, first a turbulent effect from the start-up of the reboiler, then a combinated effect, turbulent and mechanical, giving a reduction of the fouling layer during the service time of the reboiler.
- From the start-up, through the increase of turbulence coming from the shape of the devices, there are different types of action on almost each region of the boiling tube.
- Apart from the annular flow region where there is not enough liquid phase, Spirelf increases the performance of the convective boiling:
- In the slug flow zone, due to its presence (as a packing), Spirelf reduces the formation of large size bubbles (reduction of the coalescence of small bubbles). It maintains a longer time the bubbly flow, more efficient than the slug (or plug) flow.
- In the bubbly flow region, without change of the number of nucleation sites (mainly depending on the metal surface), the increase of turbulence associated to the motion of the spring going into contact with the wall, reduces the residence time of the bubbles on the metal surface, giving a better efficiency for each site of nucleation.
- In the submerged area, at the lower part of the reboiler when this one is vertical, Spirelf reduces the "laminar layer" thickness on the surface of the tubes, increasing the convection coefficient (see description in fig. 12)

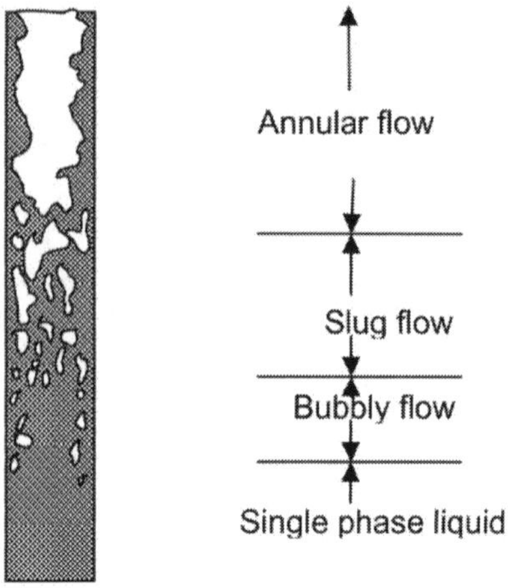

Fig. 12

b. Special case of thermosiphon reboilers.

In the case of thermosiphon reboilers, the increase of vaporization due to Spirelf leads normally to increase also the flow rate of product (when flowing in the tubes), but at the same time, the inherent extra pressure drop introduced by the springs has as effect to reduce it.

At the start-up, the global effect of Spirelf on a thermosiphon reboiler can be translated by:

Less liquid but more % vaporization = same duty

In fact, we have always noticed that at the start-up, the performance of a thermosiphon reboiler was not affected by Spirelf. This result has been observed every time when such type of reboiler has been equipped with Spirelf devices.

c. Spirelf effect in service

Generally fouling grows on the tube surface, and a source of fouling on the organic product side, comes from the high temperature of the wall, heated by medium pressure steam (around 200° C – 400° F). After forming of the first layer of deposit on the wall, the tube surface becomes rough which makes for easier formation of new deposits on it. As at the same time, the wall temperature becomes higher due to the resistance of the fouling layer, organic compounds are transformed into coke. It is important, in order to prevent later severe fouling, to reduce the start-up the fouling formation from the beginning.

Without to take into account the mechanical effect of the Spirelf springs through their contact with the wall, the increase of turbulence itself reduces the "laminar layer" thickness, reducing the wall temperature and consequently fouling phenomena from the beginning.

This reduction of fouling is illustrated hereafter in one industrial case, where a direct comparison of performance has been made between two identical parallel reboilers.

d. Unit description: Spirelf Effect for increased efficiency in duty and service time for thermosiphon reboiler

This is an application of Spirelf in a vertical reboiler, where strong fouling situation was a bottleneck issue for the chemical plant. This unit dealing with fluoric compound required a monthly water injection in order to clean them and to come back to normal situation.

The initial working conditions were:

- Natural thermosiphon
- 1 heat exchanger of 80 tubes and 1 pass on tube side,
- Flow rate: between 15 and 30 t/h,
- Boiling organic is flowing tube side,
- Straight tubes 1,500 mm / OD 27.45 mm / internal diameter 21.4 mm.

<u>Solution description</u>

Due to the low velocity in single liquid phase, the implementation of Spirelf in this reboiler was proposed in order to generate a turbulent effect in this part of the tubes. This effect was expected to increase the convection coefficient of the liquid and accelerate the gas formation.

Due to higher turbulence and reduction of boundary layer thickness because of Spirelf, the fouling reduction was estimated around 50% reduction.

Service time with Spirelf is usually doubled (theoretical expectation). As normal Spirelf life time is about few years, if fouling mitigation is strongly effective it can last around 5 times or more the initial service time.

<u>Observed results</u>

Initial evaluation after the first installation in February 2004 after 3 months installation the Spirelf was still in service and the 6 months service time without water cleaning was expected at the time of the evaluation. Observed results are a strong mitigation of the fouling deposition. Observed U value in equipped situation is now stabilised at higher U value than the expected stabilisation (around 200). As additional result, it is more than twice higher than the stabilisation without Spirelf.

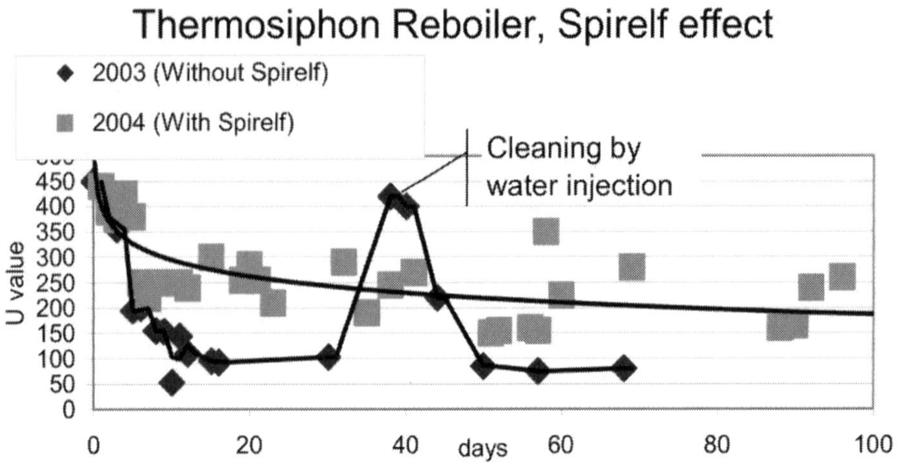

Fig. 13

There are 4 reboilers in the unit, and 2 of them were in bottleneck situation because of fouling. After this test, the chemical plant decided to equip both critical reboilers in order to be able to run 1 year with water injection every month.

List of Publications

The Petroval Tube Inserts Systems: "SPIRELF®, TURBOTAL, and FIXOTAL

have been described in the following publications:

Petroles et Techniques n°295, Janvier fevrier 1983

Chem. Eng. Progress, September 1991,"Heat Transfer Heads into 21st Century"

ASME 1995, National Heat transfer Conference, 7 August 1995, "Field of application of double enhanced tubes in shell and tube and air cooled heat exchangers",

Hydrocarbon Processing, February 1998, "Revamping Crude Units"

AIChE, March 8-12, 1998, "New Heat Integration Techniques for Boosting Profitability",

AIChE, March 14-18, 1999, "A Practical Approach to Fouling Mitigation in Refineries: Spirelf System".

ESDU Doc. 000 16,2000, "Heat exchanger fouling in crude oil distillation units",

Proceedings, August 2000, 2nd Intl. Conference on Petroleum and Gas Phase/Fouling, Copenhagen,

UEF Heat Exchanger Fouling Conference, 2001, Davos (Switzerland)

Publico, 2001, Heat Exchanger Fouling Mitigation and Cleaning Technologies, Ed. H.Mueller-Steinhagen.

AIChE Spring Meeting, March 2002

ARTC 5th Annual Meeting, April 2002, Session on Refining and Petrochemical Technology, Bangkok

ARTC Reliability Conference, November 2002 Singapore (SK Corp. Korea)

HDT Haus der Technik Heat Exchanger and Cleaning Seminar, Sept. 2003 Bad Duerkheim (Germ.)

ARTC 7th Annual Meeting/Reliability Conference, April 2004 Singapore (Thai Caprolactam/Petroval)

AIChE National Meeting, April 2004, New Orleans (Shell/SGS, Argonne, Petroval)

HOD Intl. Conference on Heavy Organic Depositions, Nov. 2004, Los Cabos (Mexico)

NPRA National Petrochemical and Refiners Assoc. Technical Q&A sessions: 1996, 1997, 1998, 1999, 2000, 2001, 2002, 2003, 2004, 2005

ECI Conference on Heat Exchanger Fouling and Cleaning, Bad Irsee, Germany, 2005

SGS, Shell Amsterdam Conference on Heat Exhanger Fouling and Cleaning, 2005

Heat Exchanger Fouling Monitoring and Cleaning Optimization Using Opticlean

Joseph L. Davis
Diego Polanco

KBC Advanced Technologies, Inc
14701 St. Mary's Lane, Suite 300
Houston, Texas 77079

Prepared for presentation at the AICHE 2006 Spring Meeting, Orlando, Florida
April 26, 2006
Fouling Mitigation Session

Heat Exchanger Fouling Monitoring and Cleaning Optimization Using Opticlean

ABSTRACT

Fouling petroleum deposits often found in crude unit preheat trains gradually reduce overall heat transfer coefficients, and thus heat recovery into the preheat train. This means that the furnace inlet temperature will drop over time, resulting in an increase in the furnace duty. This directly establishes the heating cost of fouling.

In addition, the pressure drop will increase due to a reduction in flow diameter. This is compounded by an increase in velocity (for the same flowrate), resulting in a further increase in pressure drop.

It has been common industry practice to clean crude preheat exchangers only when throughputs become constrained due to the increase in pressure drop. At today's fuel gas prices, however, waiting on crude unit rates to be impacted before cleaning is performed results in significant energy lost opportunity. The fact is, proper cleaning performed months before rates are impacted will result in real energy savings, enough to offset the exchanger cleaning costs and even the lost margin incurred due to having to slowdown the units during the several days that a particular exchanger bank is out of service for cleaning.

Proper fouling monitoring tools and practices enable the impact of fouling on exchanger performance to be assessed. This assessment must include the determination of not only the proper timing for cleaning, but must be able to specify which exchanger(s) should be cleaned for maximum benefit.

Without being too much of a "sales pitch", this presentation will discuss a new methodology for fouling monitoring and cleaning frequency optimization using a new product by Linnhoff March, a division of KBC Advanced Technologies Inc. known as OptiClean. OptiClean is an Excel-based application that retrieves and reconciles data from the plant data historian, and using a model of the exchanger network, quantifies the effect of each individual exchanger performance on the coil inlet temperature to identify the critical exchangers, seen in the context of the overall network.

Heat Exchanger Fouling Monitoring and Cleaning Optimization Using Opticlean

AICHE Spring Meeting
April 26, 2006

PROFIT. ABILITY.

KBC ADVANCED TECHNOLOGIES

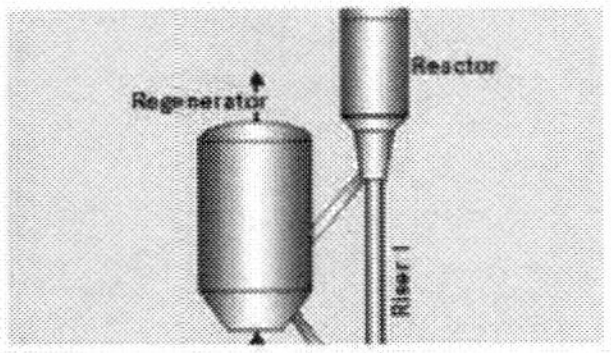

Background – Oil Refineries

- Crude streams leave deposits on exchangers
  - This increases fouling factor and reduces UA , which reduces heat transferred
  - Shortfall must be made up by utilities
  - Furnace may become bottlenecked leading to losses in throughput
  - Deposits reduce tube diameter and increase pressure drop
  - This may lead to bottlenecks in pumps and losses in throughput

KBC ADVANCED TECHNOLOGIES

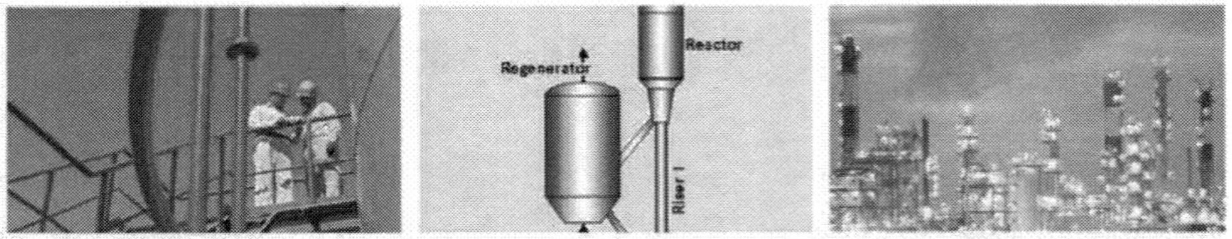

Effect of Fouling on Heat Recovery

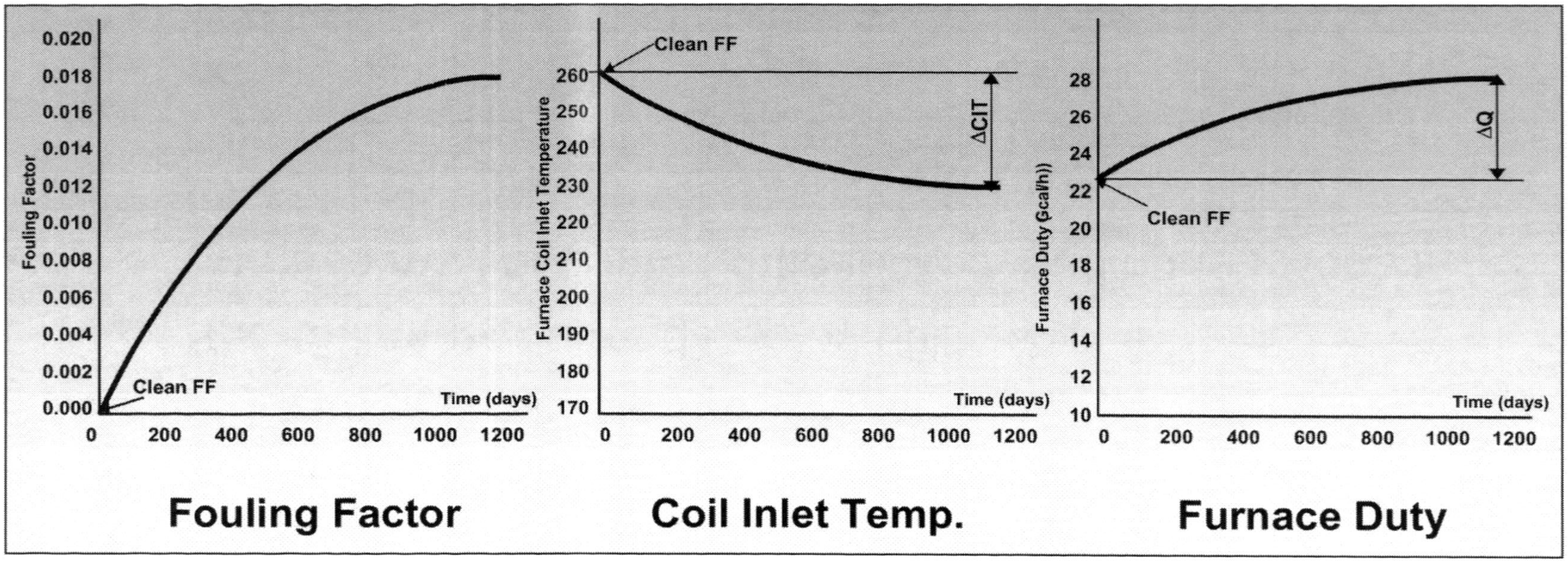

KBC ADVANCED TECHNOLOGIES

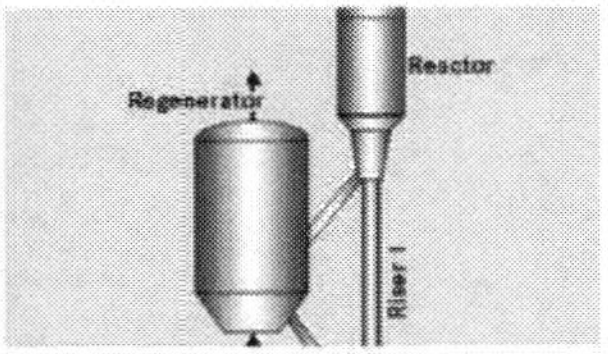

Typical Effect of Fouling on Energy Cost

250,000bpd Refinery

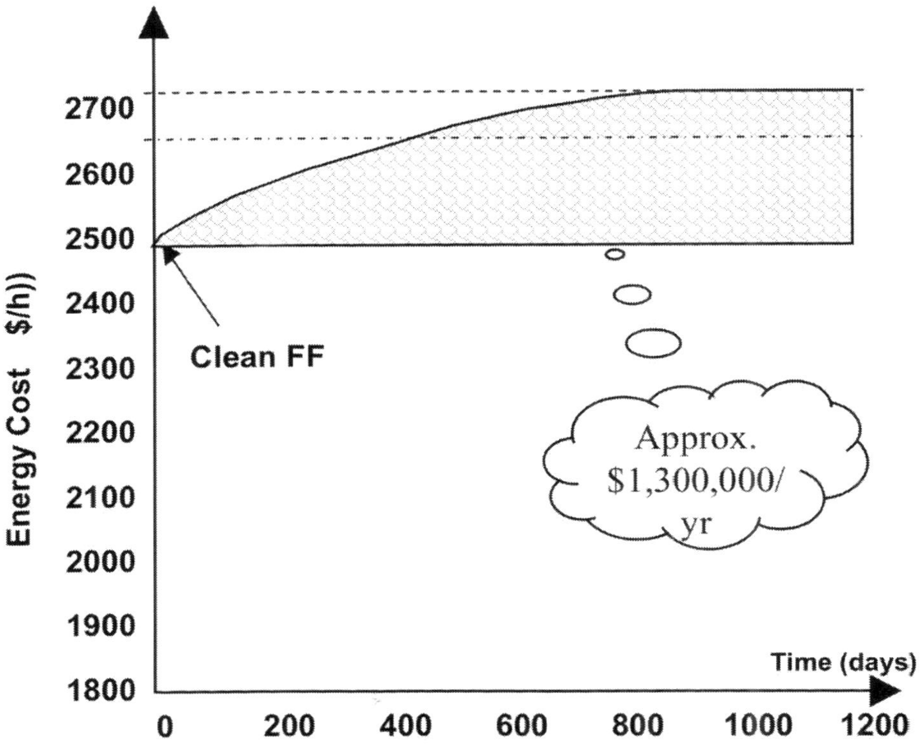

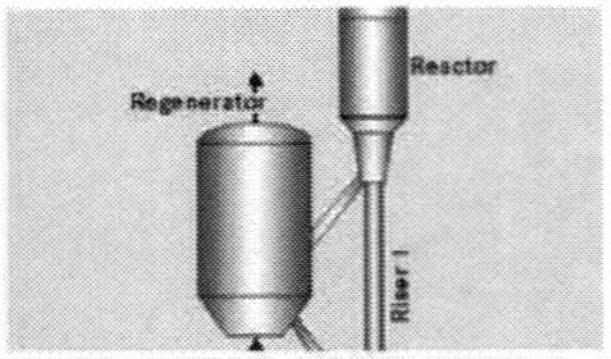

Questions

- Exchanger Cleaning
  - Which?
    - Individual exchanger ?
    - Group of exchangers ?
  - When ?
    - Individual times or all at once ?
    - In-between cycle ?
    - Only at end of cycle?
- Antifoulants
  - How much ?
  - For what benefit?

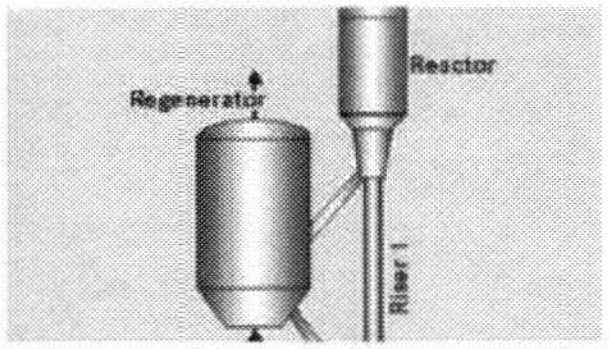

Need

- Fouling Monitoring
  - Establish relationship of fouling over time for each exchanger
  - Establish monetary cost of fouling over time for each exchanger and for whole network
- Cleaning Cycle Optimisation
  - Identify which exchangers to clean and when

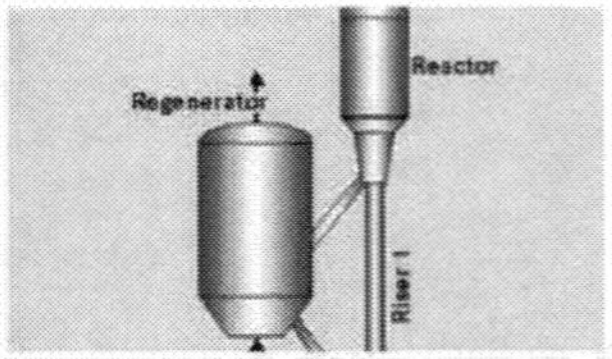

Benefits - Fouling Monitoring

- Gives insight to exchanger fouling characteristics
- Cost of fouling fully understood
- Enables fouling inhibitor assessment
  - Displays impact of inhibitor on fouling rates
  - Allows inhibitor quantities to be optimised

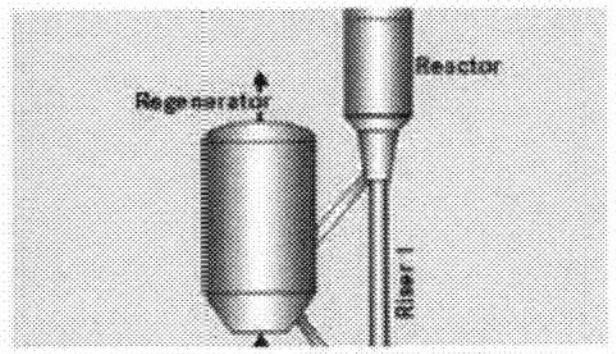

Benefits - Cleaning Cycle Optimisation

- Overcome bottlenecks in:
  - Furnace (winter operation)
  - Condenser and pumparounds (summer operation)
  - Pumps

- Energy savings
  - Reduce fuel consumption
  - Reduce power for pumps

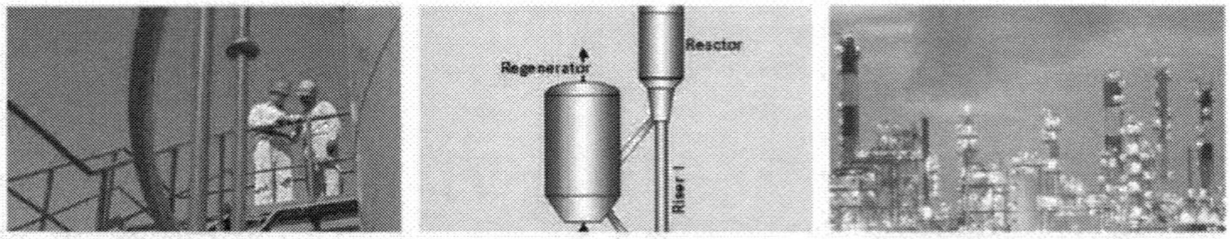

Fouling Monitoring - Approach

- Reconcile input data (flows and temperatures, hence duty) to ensure accuracy, for an extended time period

- Simulate network to determine fouling factors for given exchanger configurations

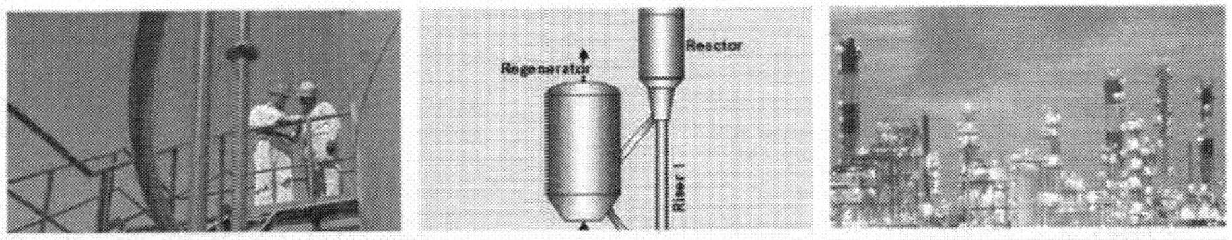

Fouling Monitoring System – "OptiClean"

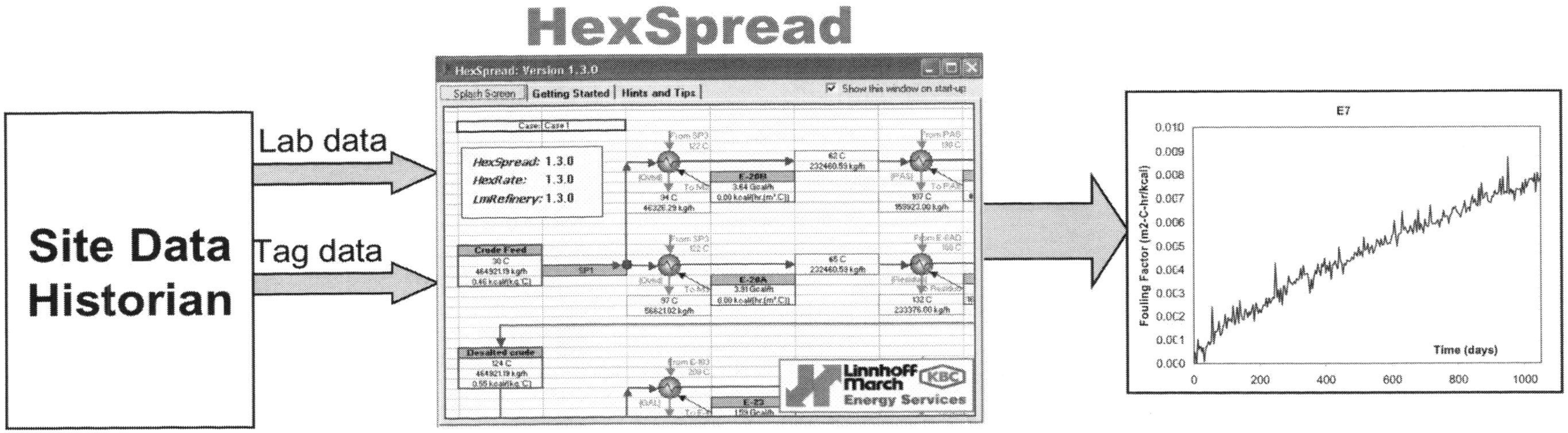

HexSpread

Heat exchanger network rating and simulation program

Data Reconciliation with gross error detection

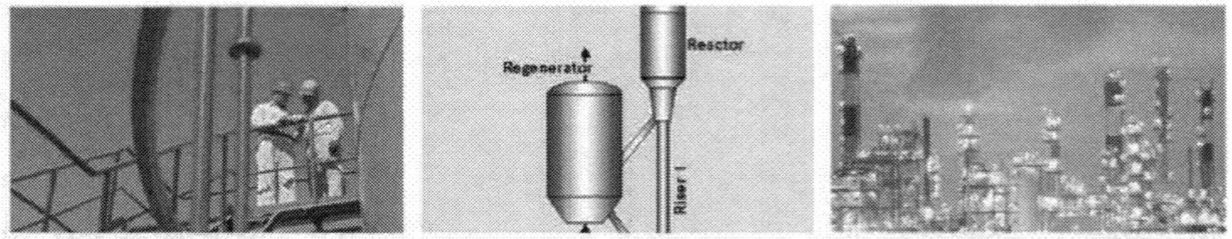

Data Reconciliation: Data Required

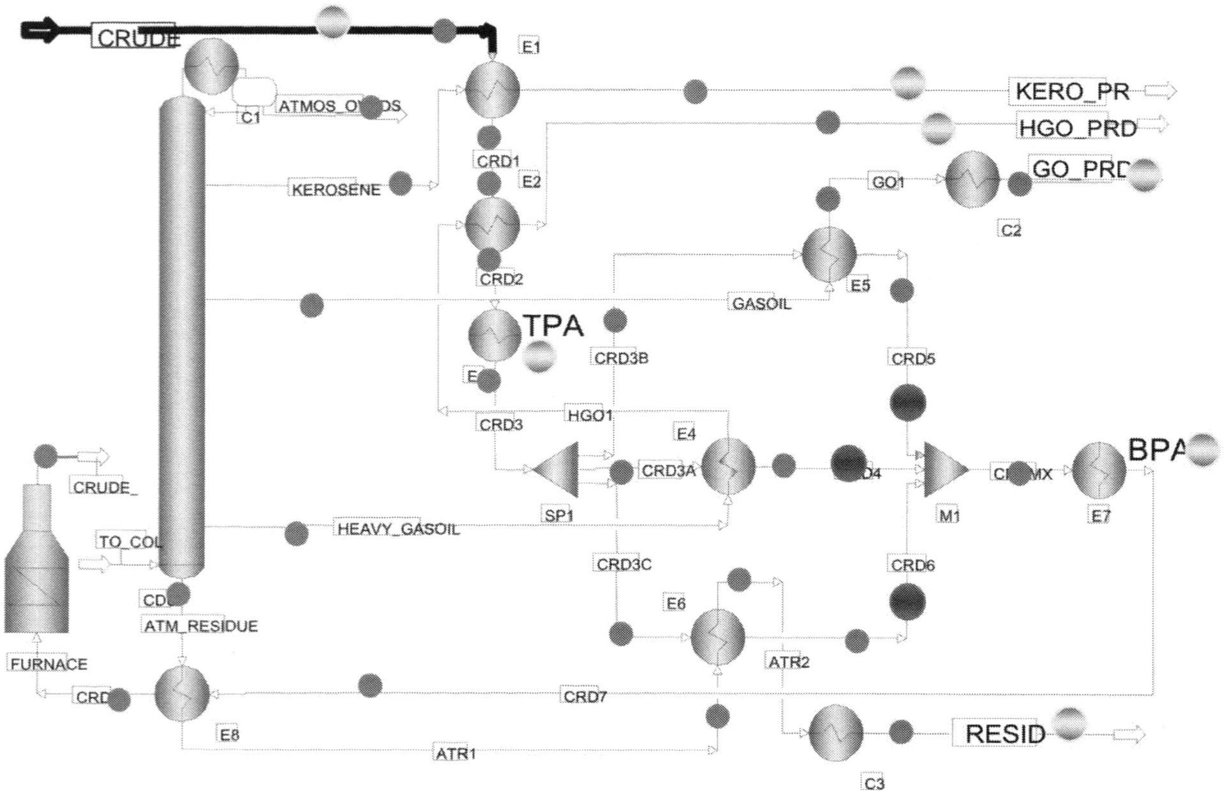

- Measured flow and composition for feed and product streams
- Measured flowrate for split branches
- Measured temperature for all streams wherever available

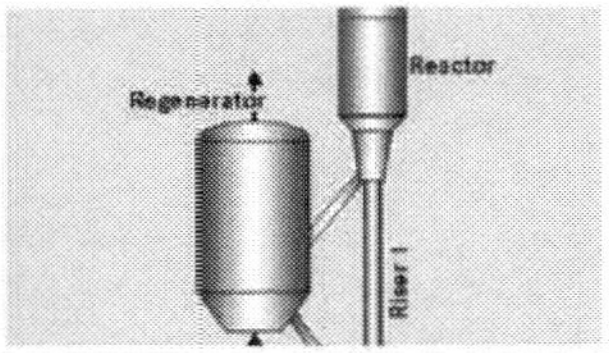

Data Reconciliation: Benefits

- Benefits
  - Produces reconciled heat and material balances
  - Data can be used for fouling calculations
  - Reconciled data can be used for other purposes, such as Pinch Analysis
  - Identifies critical areas where additional instrumentation are required
  - Identifies where existing tags are faulty and need re-calibration and/or repositioning

KBC ADVANCED TECHNOLOGIES

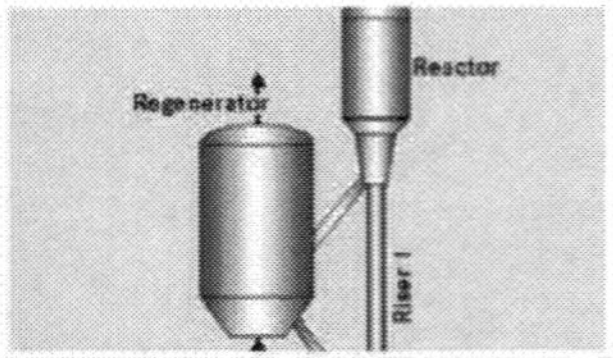

Cleaning Cycle Optimisation

Basic Approach

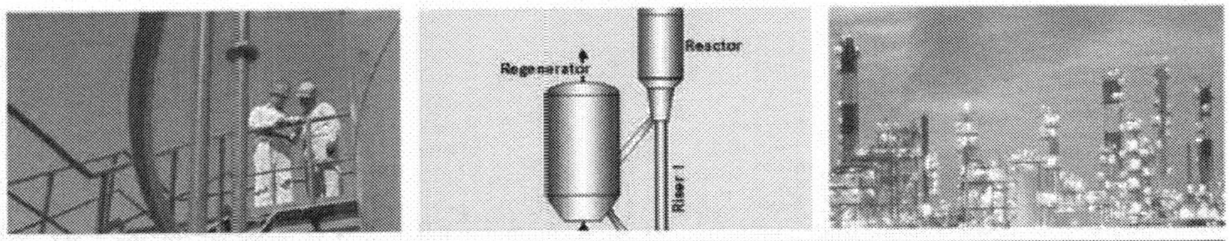

Cleaning Cycle Optimisation

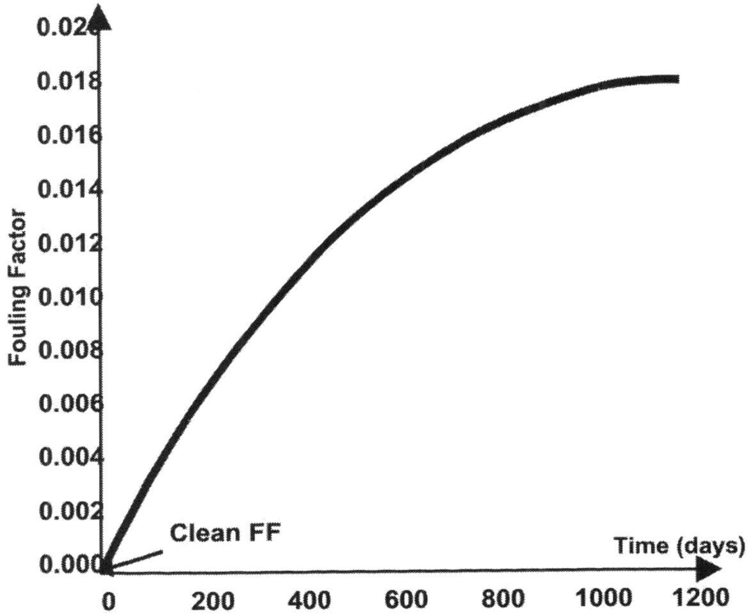

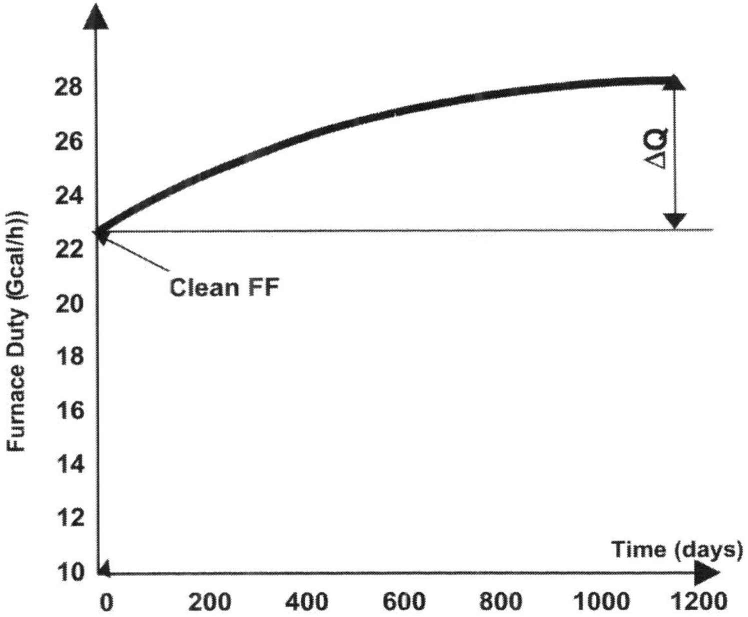

Increase in furnace duty establishes the **cost of fouling**

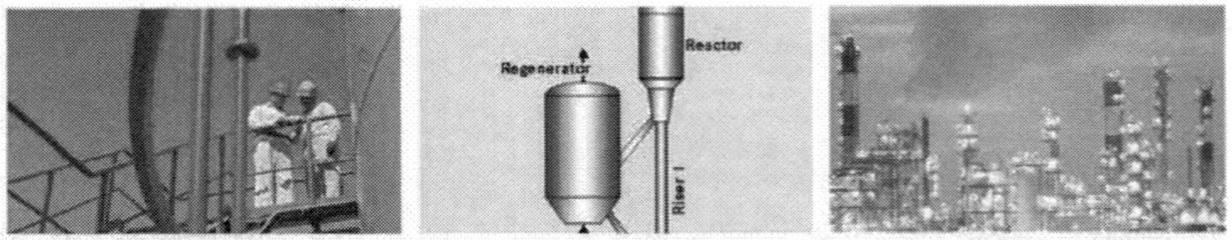

Cleaning Cycle Optimisation: Pumparound Cooling

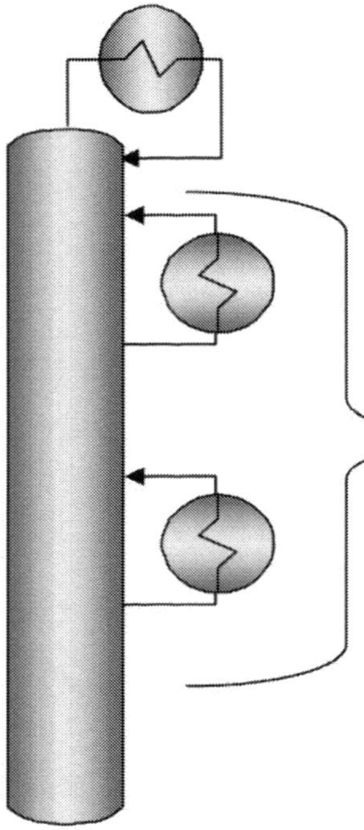

Maintain Heat Balance in the summer months

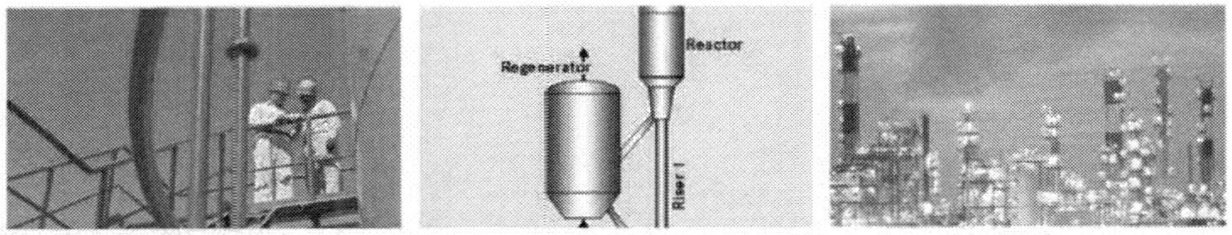

Cleaning Cycle Optimisation

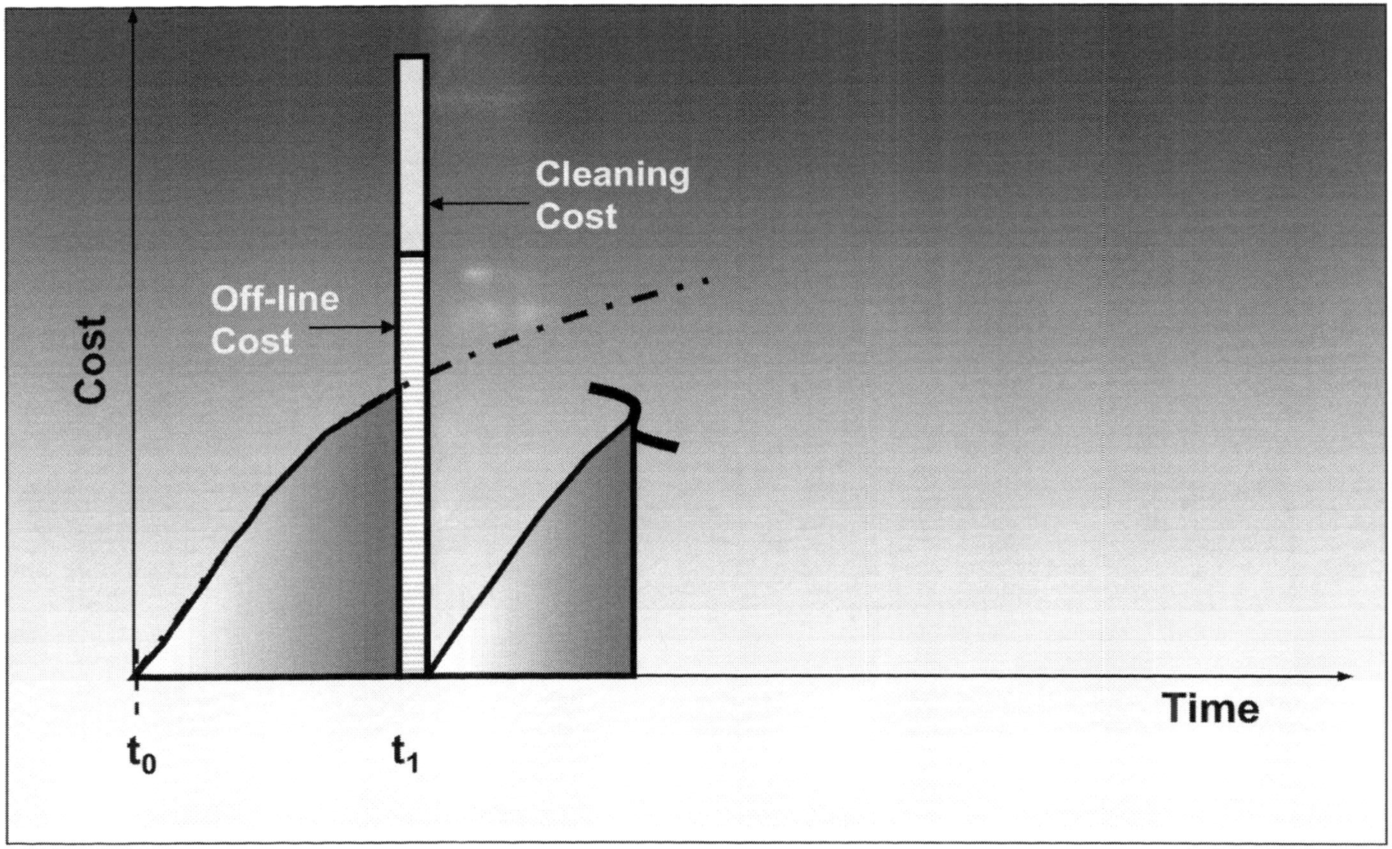

KBC ADVANCED TECHNOLOGIES

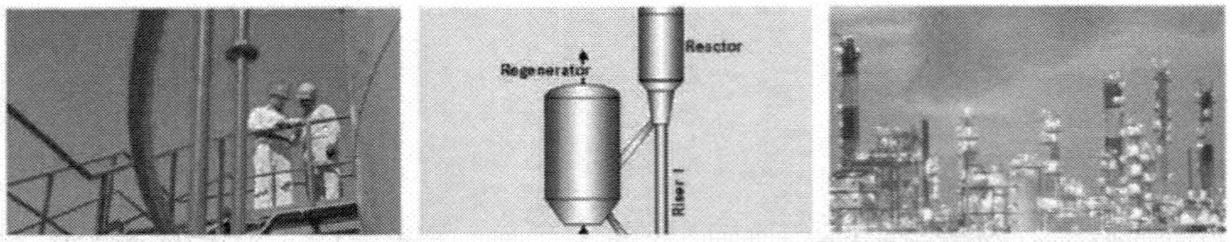

Cleaning Cycle Optimisation

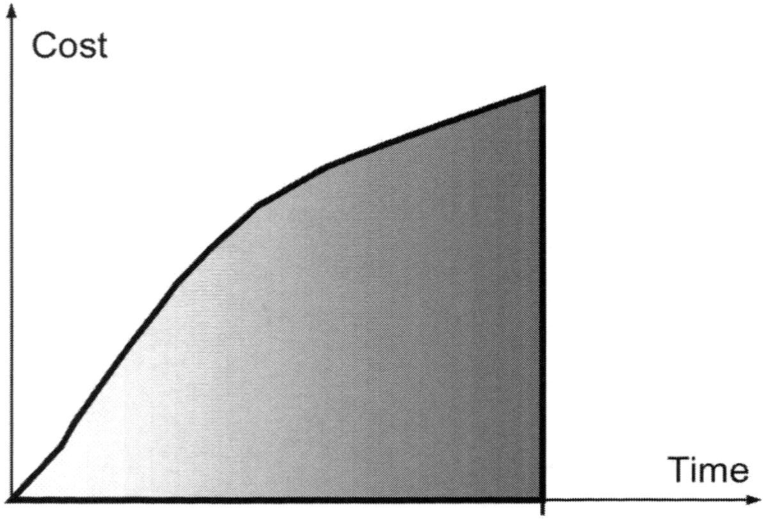

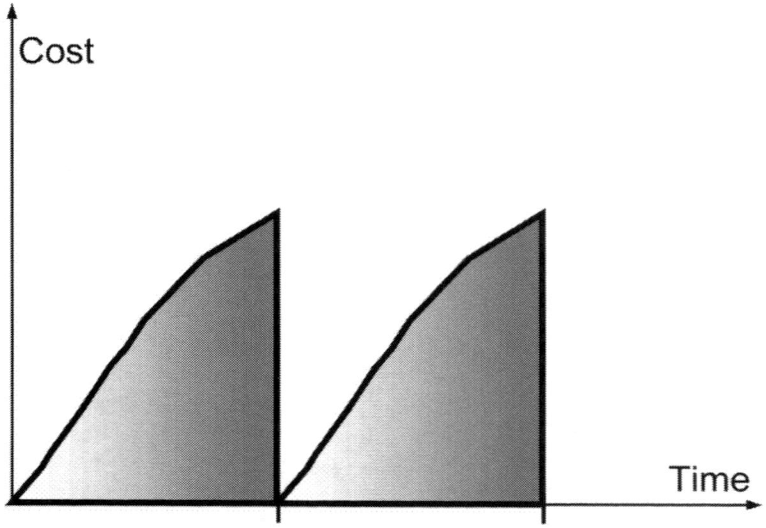

Operating Cost is reduced
if we clean more frequently

KBC ADVANCED TECHNOLOGIES

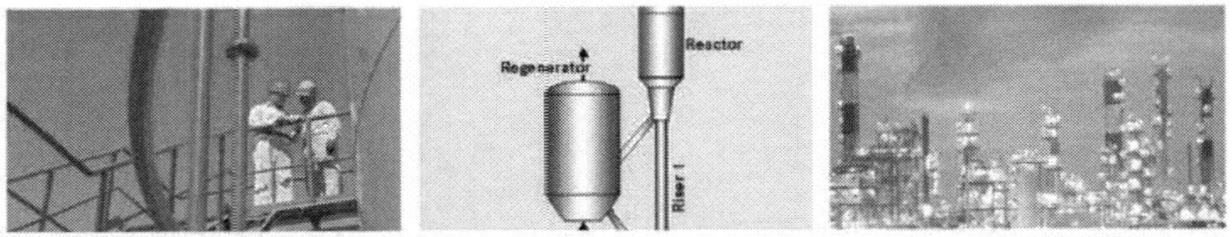

Cost of Fouling with intermediate cleaning

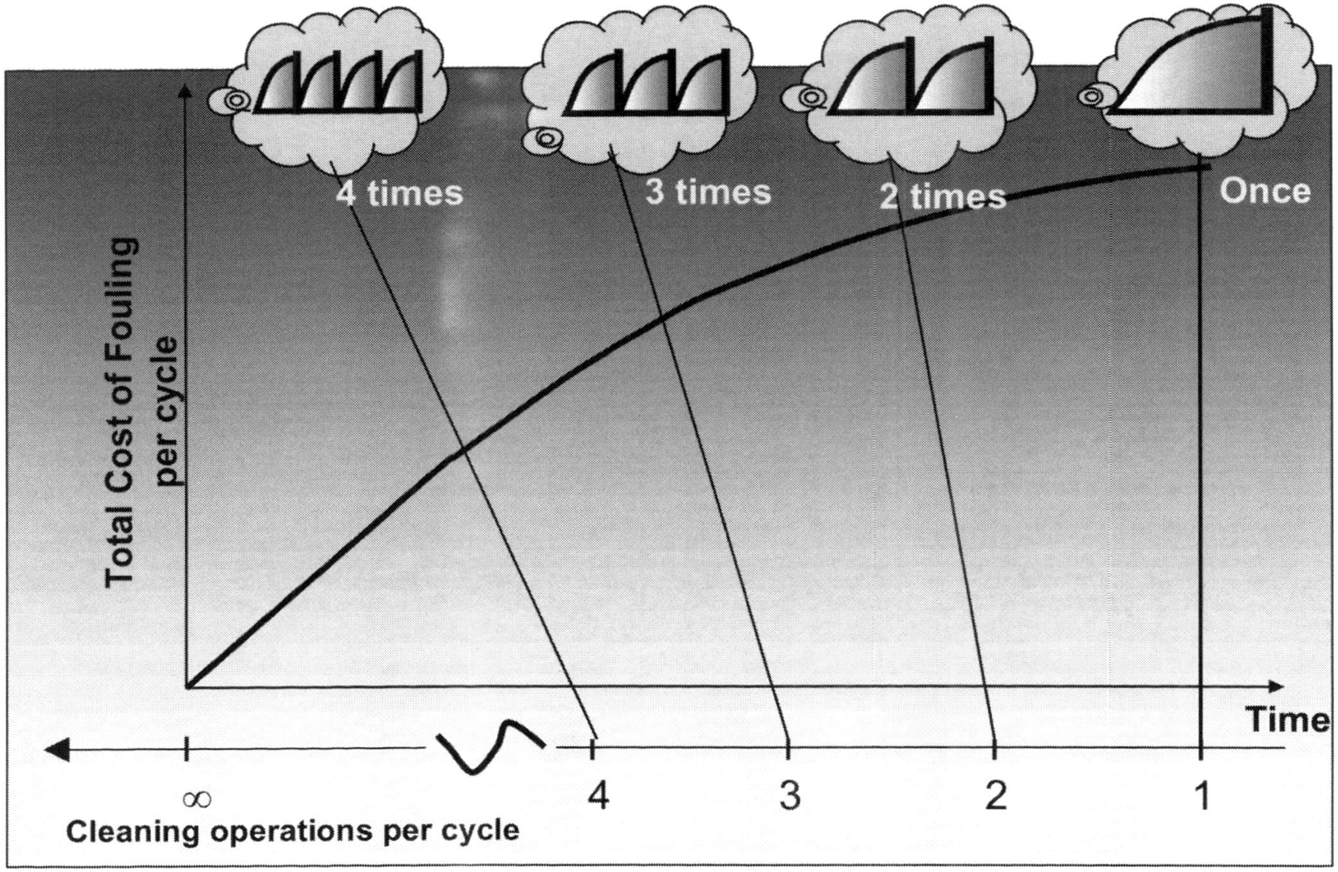

KBC ADVANCED TECHNOLOGIES

Optimum Cleaning Date

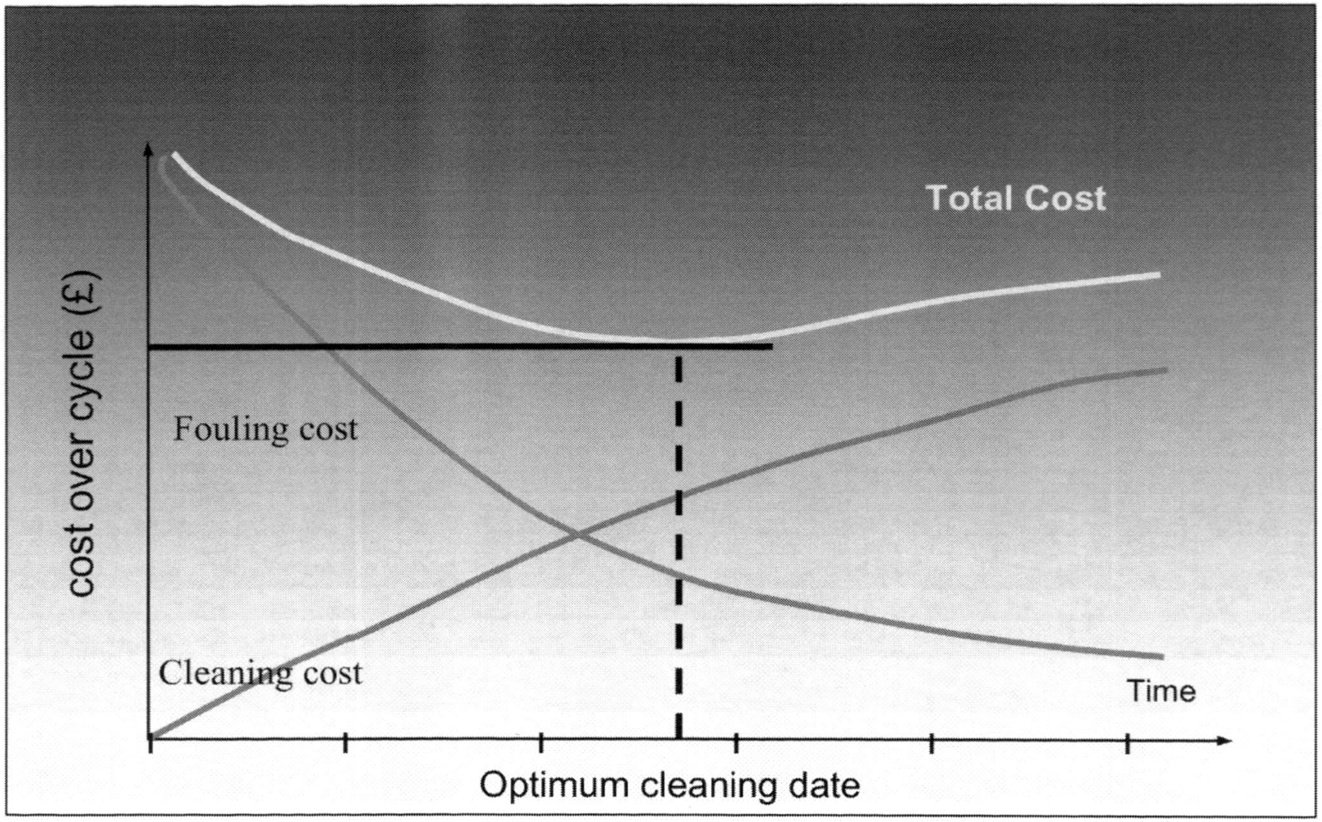

KBC ADVANCED TECHNOLOGIES

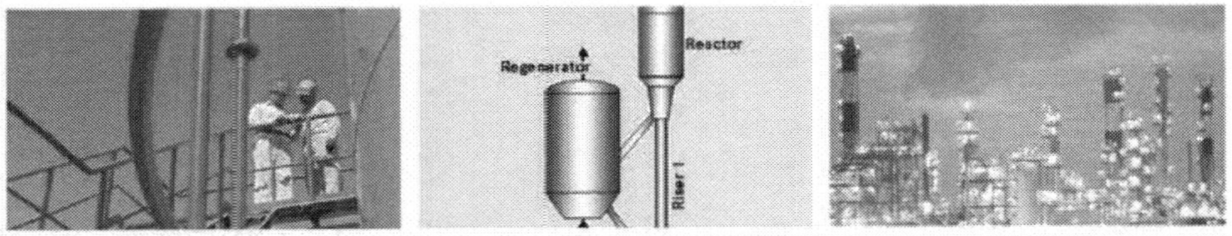

Optimum Cleaning Frequency

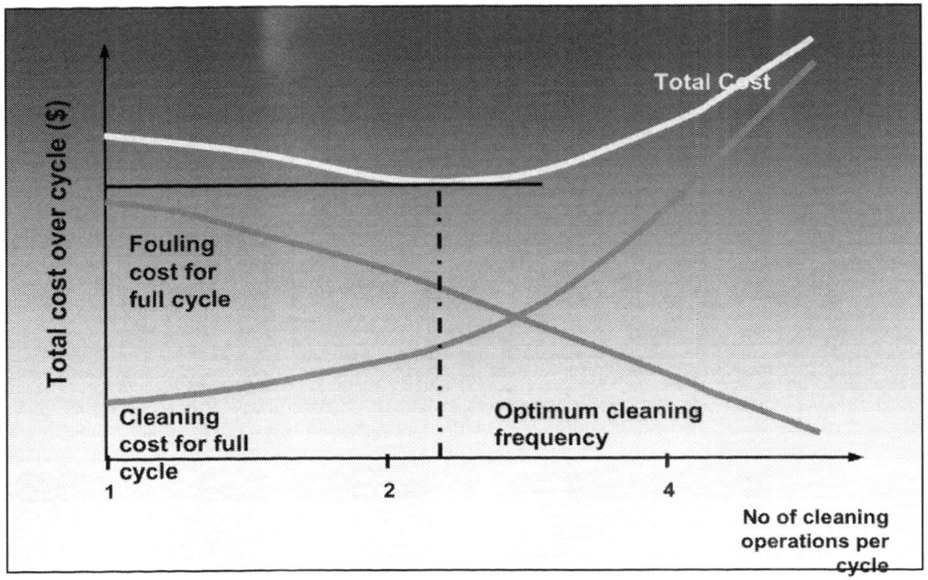

The basic principle is applied for the network as a whole (overall cycle time) and for individual exchangers (to be cleaned off cycle).

KBC ADVANCED TECHNOLOGIES

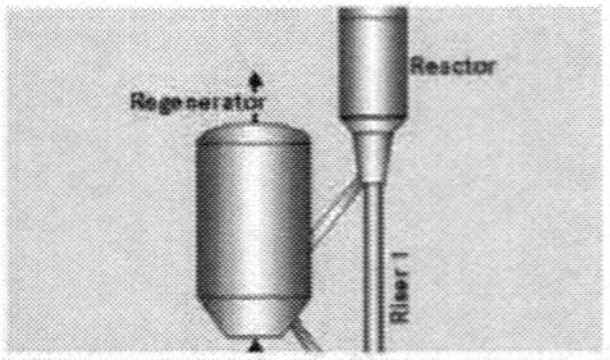

Fouling Monitoring and Cleaning Cycle Optimisation

A CASE STUDY

KBC ADVANCED TECHNOLOGIES

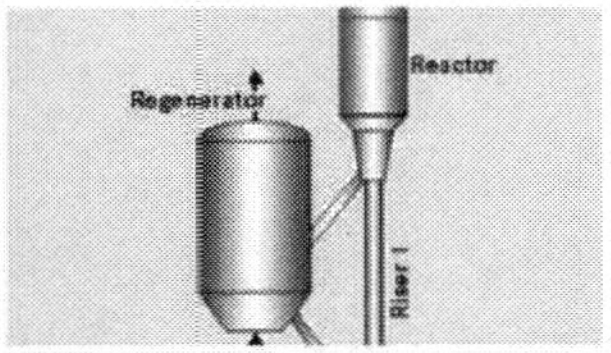

Highly Integrated Crude Preheat Train

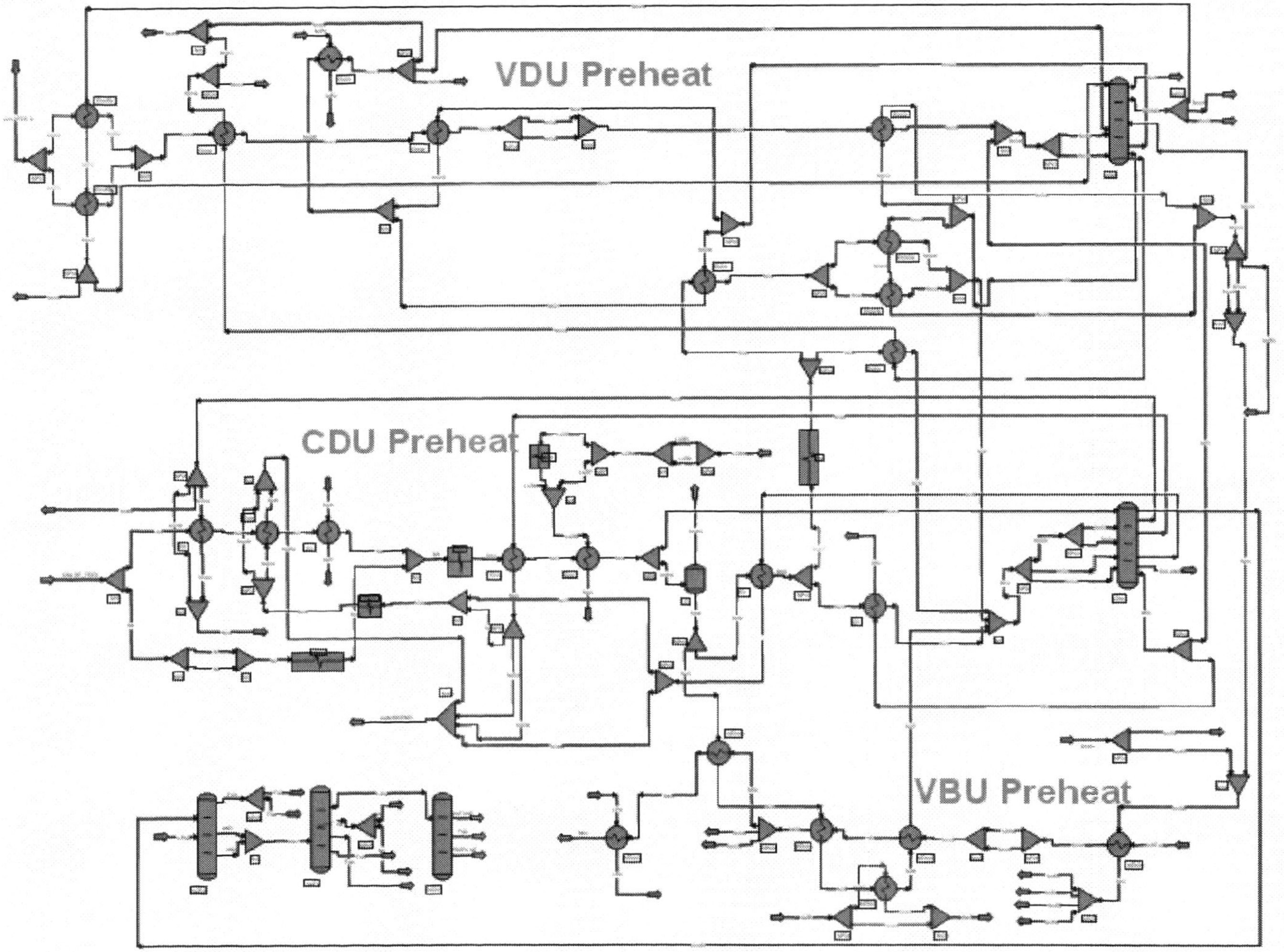

KBC ADVANCED TECHNOLOGIES

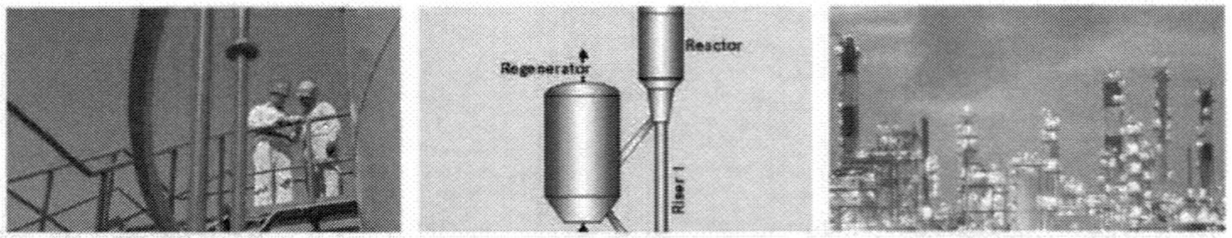

Fouling Monitoring: Typical Result

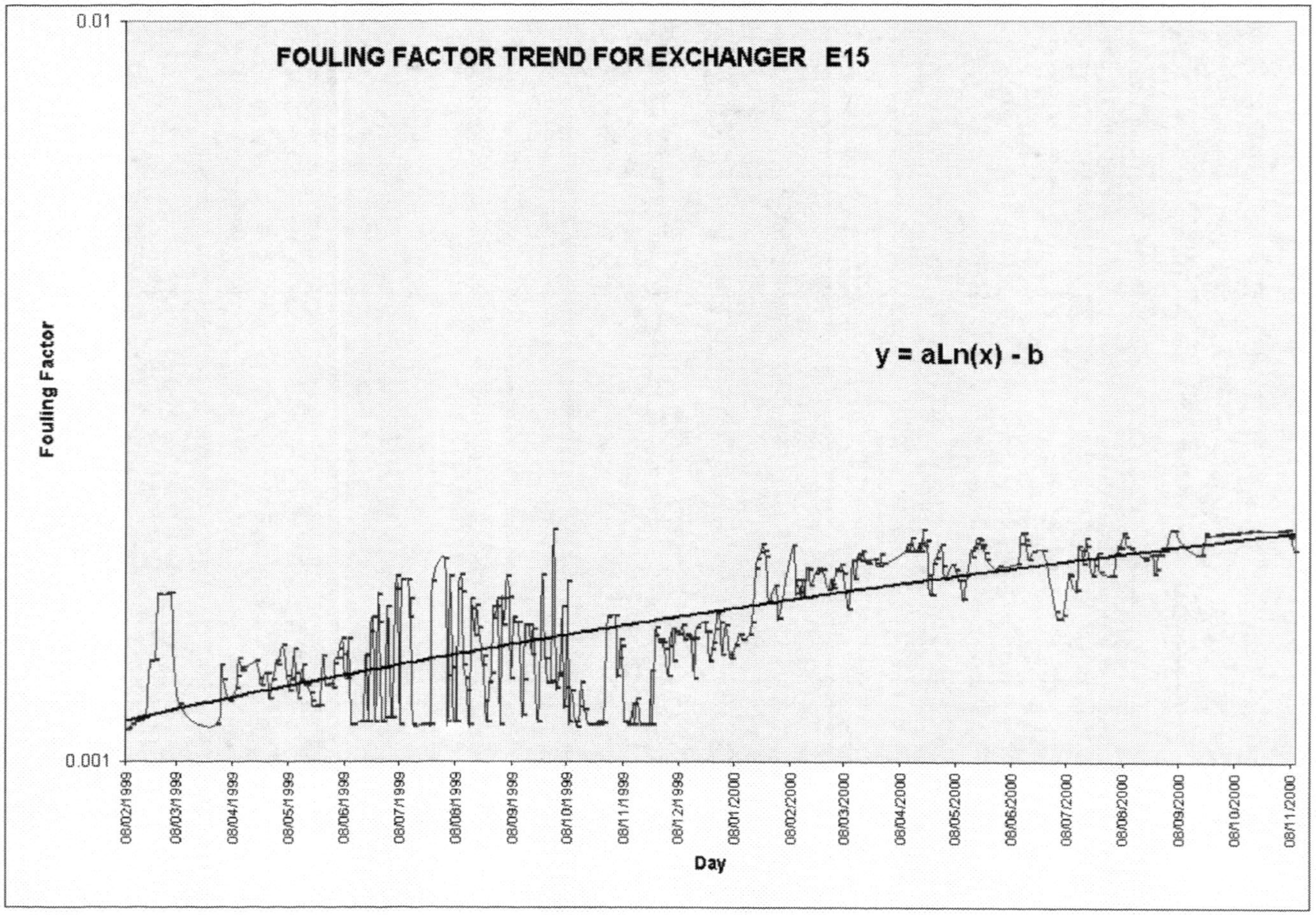

KBC ADVANCED TECHNOLOGIES

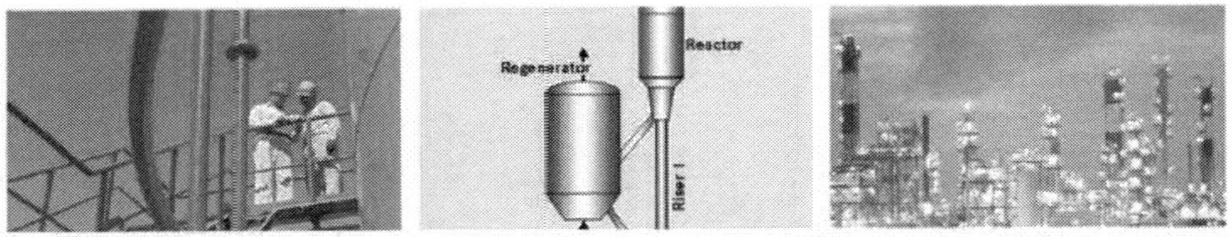

Increase in Fuel consumption for complete train Gcal/h

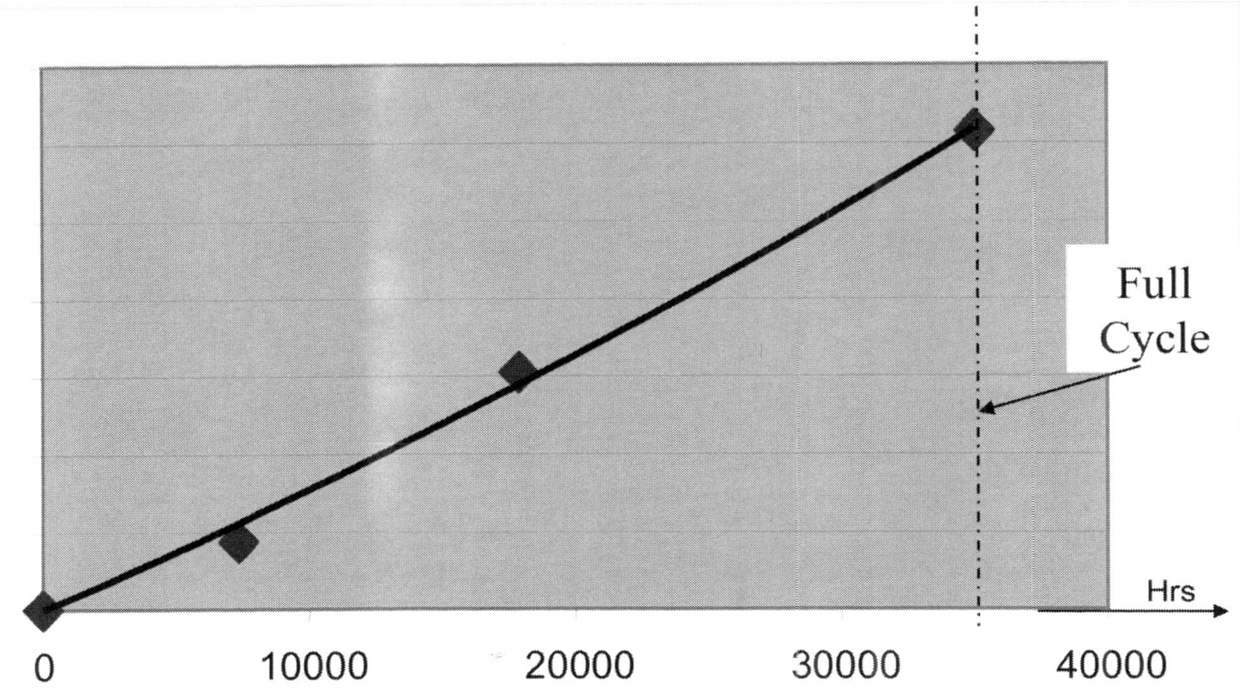

KBC ADVANCED TECHNOLOGIES

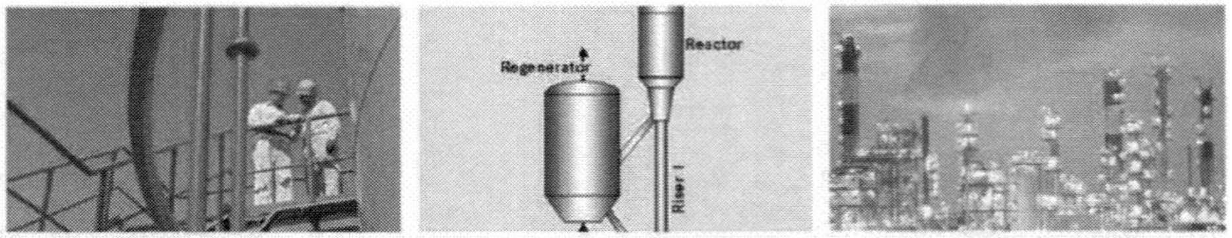

Cost of Fouling for complete train ($/h)

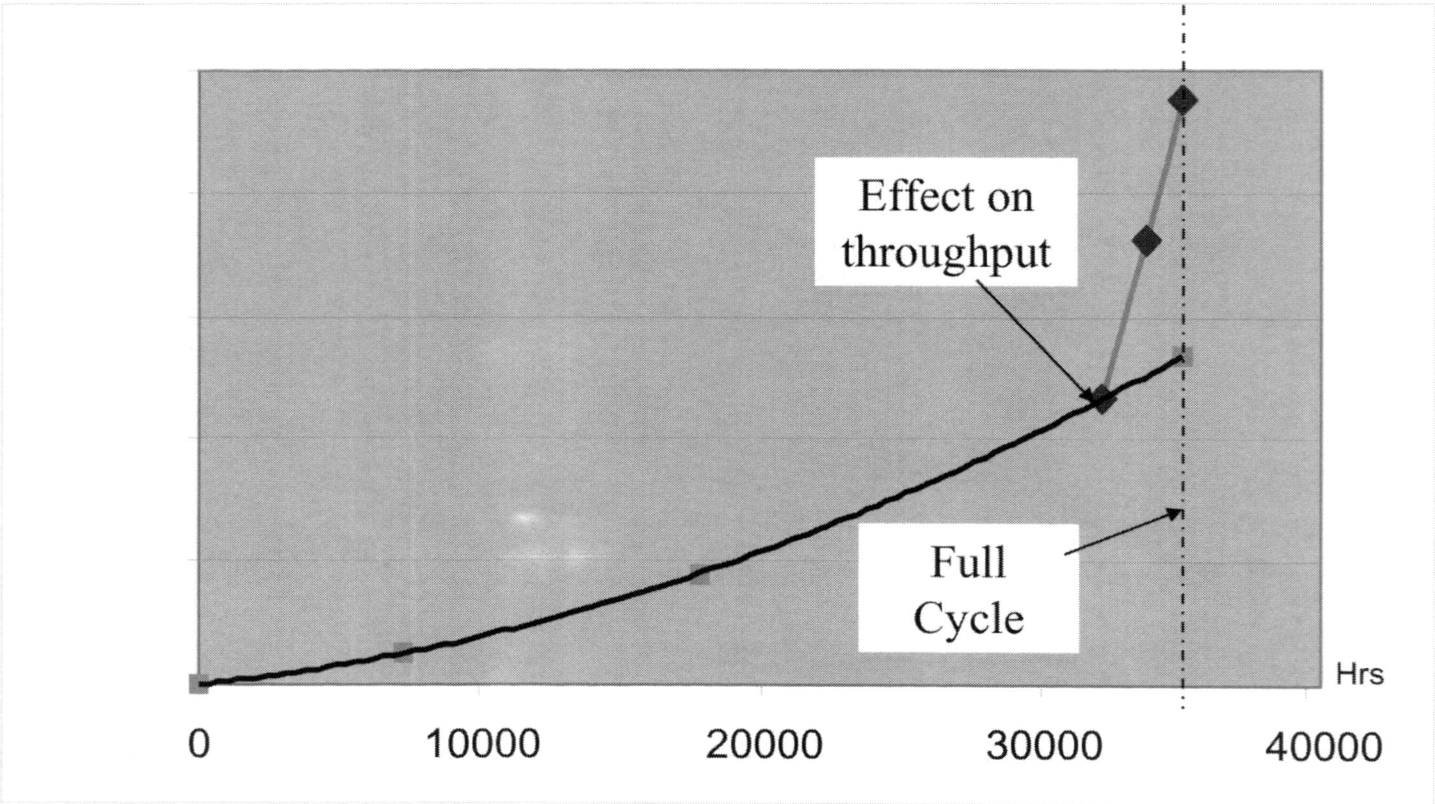

KBC ADVANCED TECHNOLOGIES

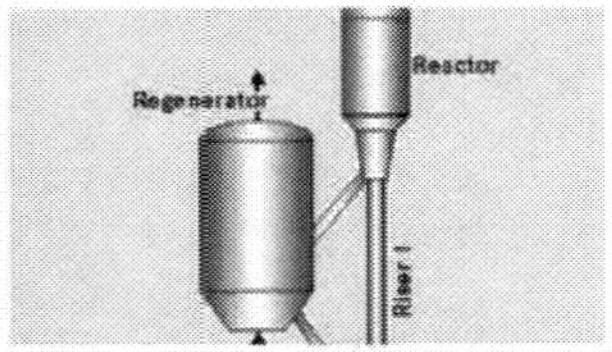

Cleaning Cycle Optimisation Graph for complete train

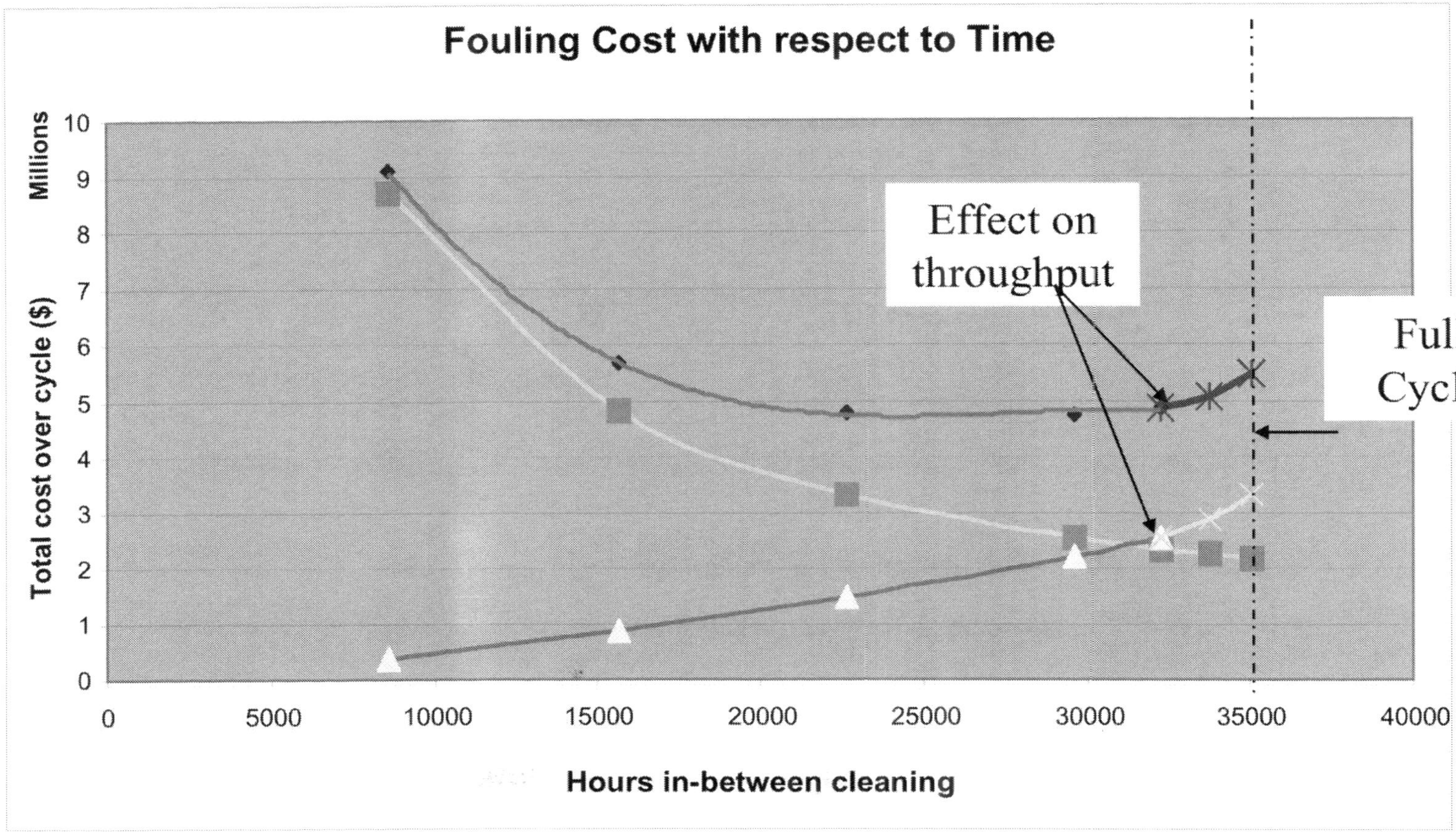

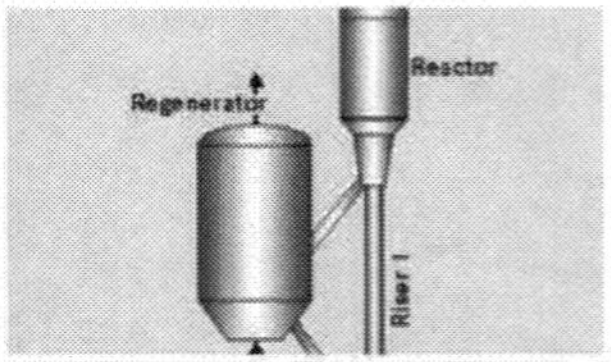

Cleaning Cycle Results – Individual Exchangers

Exchanger	Optimum Cycle Calculated yrs	Optimum Cycle Corrected Yrs	Total Cost Optimum $/TA	Total Cost Base Case $/TA	Total Saving £/TA
E1	5.2	4.00	567,780	567,780	0
E2	3.8	4	136,989	136,989	0
E3	7.0	4	104,159	104,159	0
E4	4.1	4	122,890	122,890	0
E5	1.8	2	1,149,118	1,728,997	579,878
E6	3.4	4	103,973	103,973	0
E7	10.3	4	153,484	153,484	0
E8	3.4	4	187,642	187,642	0
E9	8.7	4	179,616	179,616	0
E10	3.0	4	231,136	231,136	0
E11	6.4	4	88,694	88,694	0
E12	9.2	4	66,844	66,844	0
E13	9.8	4	65,069	65,069	0
E14	2.3	2	222,705	487,939	**265,234**
E15	1.5	2	279,962	1,249,000	**969,039**
Total Savings					**1,234,273**

Dependent

KBC ADVANCED TECHNOLOGIES

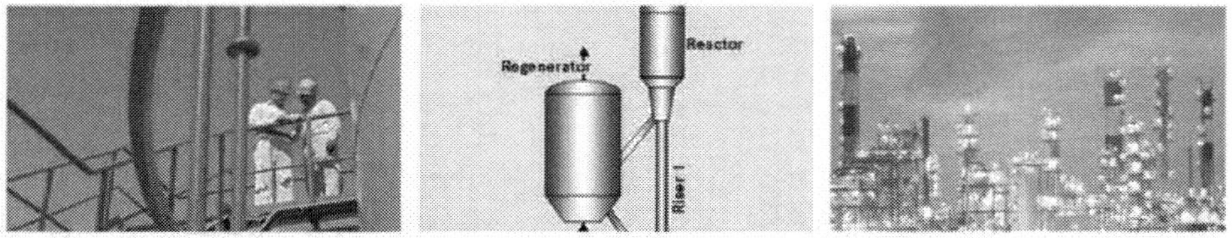

Cleaning Cycle Optimisation: Exchanger E15

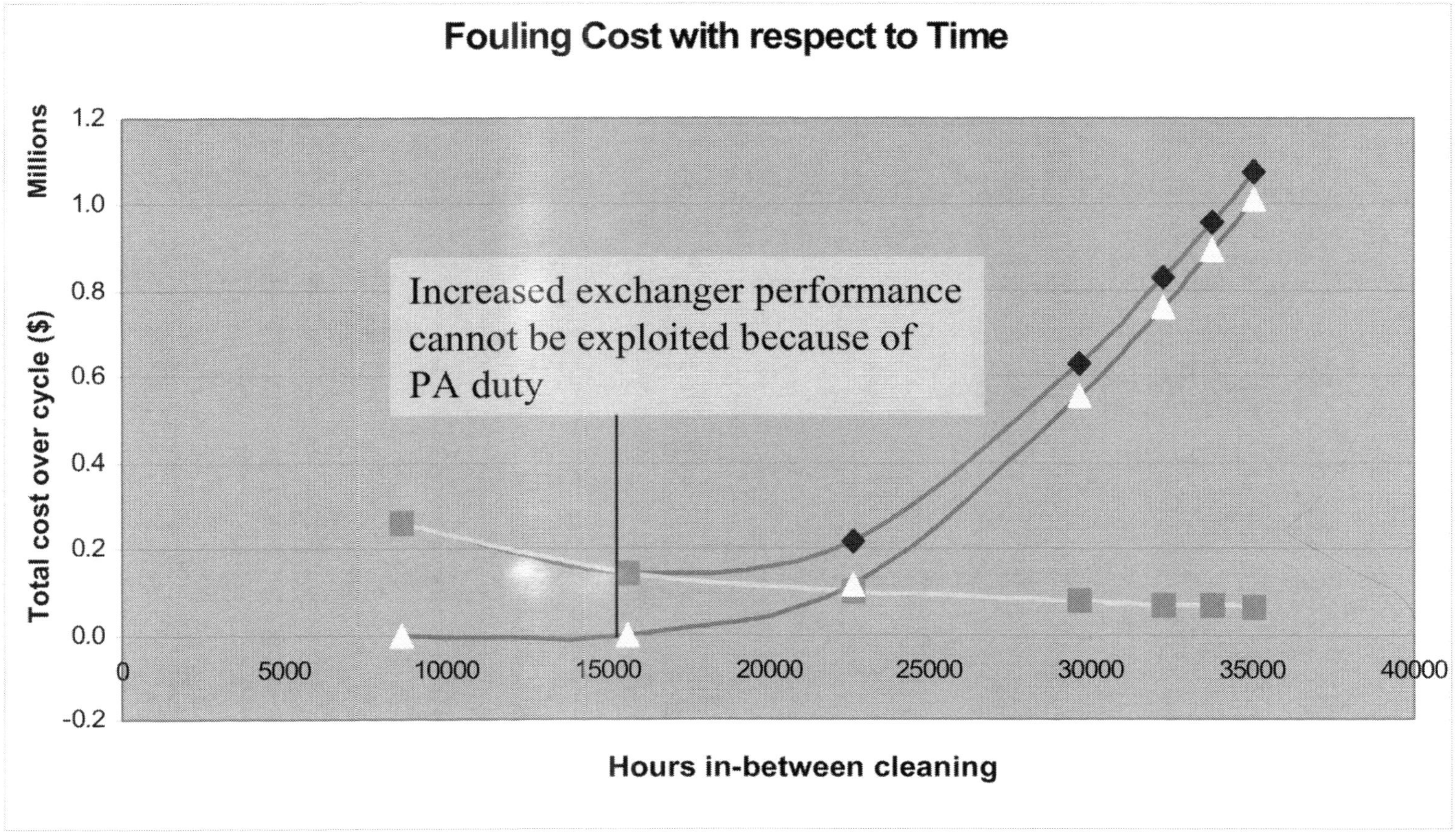

KBC ADVANCED TECHNOLOGIES

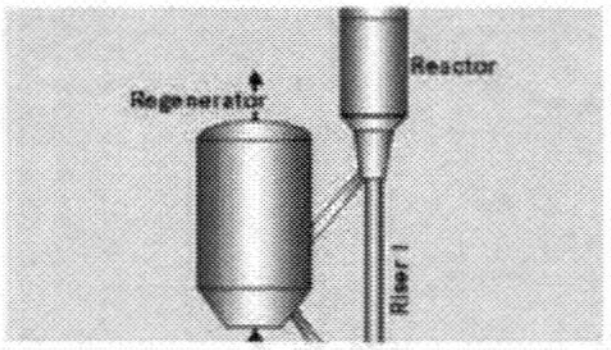

Cleaning Cycle: Results

Base Case: 4 year cycle for all exchangers

All exchangers cleaned only at end of cycle

Cost of fouling and cleaning = \$6.0 million per cycle

Optimised Case

Also clean 2 specific exchangers in-between cycle

Cost of fouling and cleaning = \$4.5 million per cycle

Energy Cost of fouling reduced by 25%

Optimum schedule eliminates throughput losses

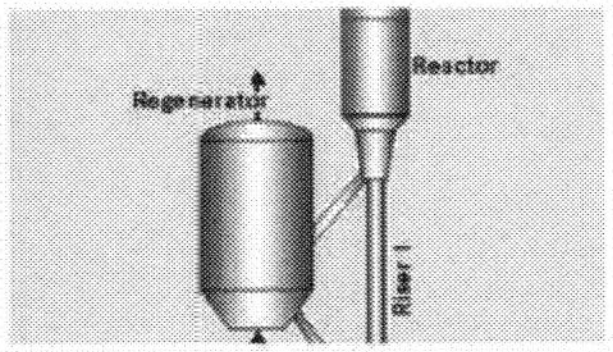

Summary

- Effect of throughput was not clear without "Monitoring"
- Not all exchangers have same cycle
- Throughput cost of fouling could be large
- Throughput can also be limited by bottleneck on condensers and pumparounds

American Institute of Chemical Engineers
3 Park Avenue, New York, NY
10016-5991

ISBN: 978-1-60423-528-9